Student's Solutions Manual to accompany Stein/Barcellos

CALCULUS

AND ANALYTIC GEOMETRY

FIFTH EDITION

ANTHONY BARCELLOS
American River College
Sacramento, California

Volume 1

McGRAW-HILL, INC.

New York St. Louis San Francisco Auckland Bogotá
Caracas Lisbon London Madrid Mexico Milan Montreal
New Delhi Paris San Juan Singapore Sydney Tokyo Toronto

Student's Solutions Manual
to accompany Stein/Barcellos:
CALCULUS AND ANALYTIC GEOMETRY, Fifth Edition,
Volume 1

1234567890 MAL MAL 9098765432

ISBN 0-07-061206-4

The editor was Maggie Lanzillo;
the production supervisor was Annette Mayeski.
Malloy Lithographing, Inc., was printer and binder.

Table of Contents

A Note on Notation

This manual uses the same notation and techniques as the text by Stein and Barcellos. However, for the sake of convenience and to save space, we also take two shortcuts. The symbol $\underset{H}{=}$ represents an equality that follows from an application of l'Hôpital's rule (Sec. 6.8 and thereafter). Also, the text writes out units of measurement in full, such as "meters per second per second," whereas we usually employ the standard abbreviations, such as "m/sec^2."

Preface

Der Irrnis und der
Leiden Pfade kam ich.
Act III, Scene 1, *Parsifal*

This manual is supposed to help you learn calculus. It contains worked-out solutions to approximately half of the exercises in the fifth edition of *Calculus and Analytic Geometry* by Sherman K. Stein and Anthony Barcellos. Volume 1 covers Chapters 1 through 11 and Volume 2 deals with Chapters 12 through 17.

Calculus is learned only through practice. No one would expect to learn to ride a bicycle by listening to lectures and reading a book. Calculus is probably somewhat harder than learning to ride a bike, so be prepared to work over the examples in the text and to do the homework exercises. *Always attempt to solve the exercises yourself before turning to this manual.* (Looking at a problem for 30 seconds is *not* the same as trying to solve it.) Refer also to the section *To the Student* in your textbook (p. *xxvii*), which gives detailed advice on studying and learning mathematics.

We provide solutions for practically all of the odd-numbered exercises and all of the Guide Quizzes. Some exercises, such as exploration problems (marked with a ✵ in the text), are included in the *Instructor's Manual* instead because giving their solutions would make them pointless. Don't worry; there aren't too many of these.

The solutions are complete, providing the steps necessary to permit you to follow the line of reasoning. Quite naturally, the solutions tend to be more detailed in the earlier parts of solution sets, where you are presumably just learning the material. Near the end of a set we may take somewhat bigger steps. In all cases, however, you should be able to take pencil and paper and fill in as many intermediate steps as you may need.

Be aware that numerical answers may vary slightly, depending on the accuracy of measurements, differing assumptions, and round-off errors in calculations. In such cases, do not take the solutions in the manual as the only "right" answers. Your results are just as good. In particular, a one-unit difference in the last decimal place is nothing to get exercised about.

All of the solutions have been checked and double-checked in hopes of eliminating all errors. With over three thousand solutions, however, it is quite possible that a few errors have managed to slip through anyway. Please bring any errors that you might find to the attention of Anthony Barcellos at P.O. Box 2249, Davis, CA 95617.

About This Book. Camera-ready copy was produced on a Hewlett-Packard LaserJet III from documents created with WordPerfect 5.1. The illustrations were created by CoPlot and CoDraw from CoHort Software (P.O. Box 1149, Berkeley, CA 94701, 415/524-9878), which generated .WPG files for direct import into WordPerfect.

Acknowledgments. We wish to thank the editorial staff at McGraw-Hill for their assistance with this student manual, including former math editors Robert Weinstein and Richard Wallis. Maggie Lanzillo was a constant source of encouragement even while gently reminding us about deadlines.

Many of the people who participated in preparation of the Stein/Barcellos textbook also contributed in various ways to this manual. We acknowledge in particular the efforts of **Keith Sollers** and **Mallory Austin** who drafted solutions to thousands of exercises over the past two years and were key contributors to this manual. Other significant contributions of solutions came from Dean Hickerson and Duane Kouba. Keith, Mallory, and Dean also provided the bulk of the proofreading, which they performed painstakingly, but are not to be held responsible for any errors left by the author. In addition to those who worked on the text, Travis Andrews and Kelly Riddle contributed to the answer-checking. Heroic efforts were made by **Richard Kinter**, **Judith Kinter**, and **Michael Kinter**, who typed the reams of solutions and became among the world's greatest experts in WordPerfect 5.1's equation editor in the process.

—Anthony Barcellos

1 An Overview of Calculus

1.1 The Derivative

1 (a)

x	$\sqrt{x^2+2x}$	$\sqrt{x^2+2x}-x$
1	1.7320508	0.7320508
5	5.9160798	0.9160798
10	10.9544512	0.9544512
100	100.9950494	0.9950494
1000	1000.999501	0.9995001

(b) The difference appears to approach 1.

3 (a)

x	x^3-1	$x-1$	$(x^3-1)/(x-1)$
0.5	−0.8750000	−0.500	1.7500000
0.9	−0.2710000	−0.100	2.7100000
0.99	−0.0297010	−0.010	2.9701000
0.999	−0.0029970	−0.001	2.9970010

(b) The ratio appears to approach 3.

5 (a) $(1.01)^3 - 1^3 = 1.030301 - 1 \approx 0.0303$ ft

(b) $\dfrac{0.0303}{0.01} = 3.03$ ft/min

(c) $\dfrac{1.001^3 - 1^3}{1.001 - 1} \approx \dfrac{0.003003}{0.001} = 3.003$ ft/min

(d) $\dfrac{1^3 - 0.999^3}{1 - 0.999} \approx \dfrac{0.002997}{0.001} = 2.997$ ft/min

(e) The shorter the time interval near $t = 1$ min, the closer the speed is to 3 ft/min.

7 (a)

x	2^x	2^x-1	$(2^x-1)/x$
1	2.0000000	1.0000000	1.0000000
0.5	1.4142136	0.4142136	0.8284271
0.1	1.0717735	0.0717735	0.7177346
0.01	1.0069556	0.0069556	0.6955550
0.001	1.0006934	0.0006934	0.6933875
−0.001	0.9993071	−0.0006929	0.6929070

(b) The ratio appears to approach 0.693.

9 (a) $\dfrac{\sin 1}{1} \approx 0.84147$, $\dfrac{\sin 0.1}{0.1} \approx 0.99833$,

$\dfrac{\sin 0.01}{0.01} \approx 0.99998$

(b) The ratio approaches 1.

11 (a) $\dfrac{(3^{2.1}-1)-(3^2-1)}{2.1-2} \approx \dfrac{1.0451}{0.1}$

$= 10.451$ m/sec

$\dfrac{(3^{2.01}-1)-(3^2-1)}{2.01-2} \approx \dfrac{0.09942}{0.01}$

$= 9.942$ m/sec

1.2 The Integral

1 (a)

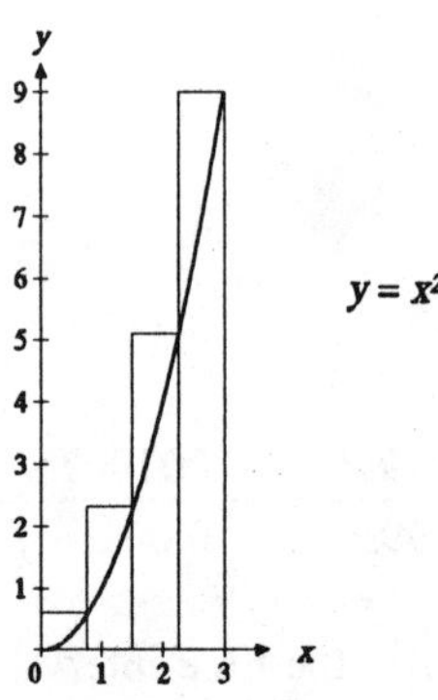

(b)

Rectangle	*Height*	*Width*	*Area*
First	9/16	3/4	27/64
Second	9/4	3/4	27/16
Third	81/16	3/4	243/64
Fourth	9	3/4	27/4

(c) $\frac{27}{64} + \frac{27}{16} + \frac{243}{64} + \frac{27}{4} = \frac{810}{64}$

3 (a)

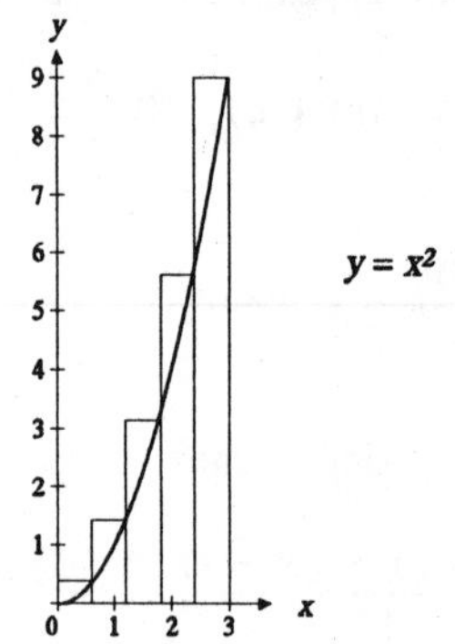

(b)

Rectangle	*Height*	*Width*	*Area*
First	9/25	3/5	27/125
Second	36/25	3/5	108/125
Third	81/25	3/5	243/125
Fourth	144/25	3/5	432/125
Fifth	9	3/5	27/5

(c) $\frac{27}{125} + \frac{108}{125} + \frac{243}{125} + \frac{432}{125} + \frac{27}{5} = \frac{1485}{125}$

$= 11.88$

5 See Figs. 8 and 9 in the text.

(a) The area of the 10 overestimating rectangles is $(0.3)^2(0.3) + (0.6)^2(0.3) + (0.9)^2(0.3) + (1.2)^2(0.3) + (1.5)^2(0.3) + (1.8)^2(0.3) + (2.1)^2(0.3) + (2.4)^2(0.3) + (2.7)^2(0.3) + (3.0)^2(0.3) = 10.395$, as claimed.

(b) The area of the 10 underestimating rectangles is $(0)^2(0.3) + (0.3)^2(0.3) + (0.6)^2(0.3) + (0.9)^2(0.3) + (1.2)^2(0.3) + (1.5)^2(0.3) + (1.8)^2(0.3) + (2.1)^2(0.3) + (2.4)^2(0.3) + (2.7)^2(0.3) = 7.695$, as claimed.

7 (a) The area of the four overestimating rectangles is $\left(\frac{1}{4}\right)^3 \cdot \frac{1}{4} + \left(\frac{1}{2}\right)^3 \cdot \frac{1}{4} + \left(\frac{3}{4}\right)^3 \cdot \frac{1}{4} + 1^3 \cdot \frac{1}{4} =$

$\frac{25}{64} = 0.390625.$

(b) The area of the eight underestimating rectangles is $0^3 \cdot \frac{1}{8} + \left(\frac{1}{8}\right)^3 \cdot \frac{1}{8} + \left(\frac{2}{8}\right)^3 \cdot \frac{1}{8} +$

$\left(\frac{3}{8}\right)^3 \cdot \frac{1}{8} + \left(\frac{4}{8}\right)^3 \cdot \frac{1}{8} + \left(\frac{5}{8}\right)^3 \cdot \frac{1}{8} + \left(\frac{6}{8}\right)^3 \cdot \frac{1}{8} +$

$\left(\frac{7}{8}\right)^3 \cdot \frac{1}{8} = \frac{49}{256} = 0.19140625$, as claimed.

9 (a) The area of the 10 underestimating rectangles

is $\frac{10}{33}\cdot\frac{3}{10} + \frac{10}{36}\cdot\frac{3}{10} + \frac{10}{39}\cdot\frac{3}{10} + \frac{10}{42}\cdot\frac{3}{10} +$

$\frac{10}{45}\cdot\frac{3}{10} + \frac{10}{48}\cdot\frac{3}{10} + \frac{10}{51}\cdot\frac{3}{10} + \frac{10}{54}\cdot\frac{3}{10} +$

$\frac{10}{57}\cdot\frac{3}{10} + \frac{10}{60}\cdot\frac{3}{10} \approx 0.6688.$

(b) The areas are the same.

11 If we let $f(x) = \frac{1}{3}\sqrt[3]{x^6 + 1}$, then the area of the 10

rectangles is $f(1.1)(0.2) + f(1.3)(0.2) + f(1.5)(0.2)$

$+ f(1.7)(0.2) + f(1.9)(0.2) + f(2.1)(0.2) +$

$f(2.3)(0.2) + f(2.5)(0.2) + f(2.7)(0.2) +$

$f(2.9)(0.2) \approx 2.9188.$

1.3 Survey of the Text

1 (a) $\sin 1 \approx 1 - \frac{1^3}{1\cdot 2\cdot 3} = 1 - \frac{1}{6} = \frac{5}{6}$

≈ 0.83333333

(b) $\sin 1 \approx 1 - \frac{1^3}{1\cdot 2\cdot 3} + \frac{1^5}{1\cdot 2\cdot 3\cdot 4\cdot 5}$

$= 1 - \frac{1}{6} + \frac{1}{120} = \frac{101}{120} \approx 0.84166667$

(c) $\sin 1 \approx$

$1 - \frac{1^3}{1\cdot 2\cdot 3} + \frac{1^5}{1\cdot 2\cdot 3\cdot 4\cdot 5} - \frac{1^7}{1\cdot 2\cdot 3\cdot 4\cdot 5\cdot 6\cdot 7}$

$= 1 - \frac{1}{6} + \frac{1}{120} - \frac{1}{5040} = \frac{4241}{5040}$

≈ 0.84146285

(d) $\sin 1 \approx 1 - \frac{1}{6} + \frac{1}{120} - \frac{1}{5040} + \frac{1}{362{,}880}$

≈ 0.84147101

(e) $\sin 1 \approx 0.84147098$

2 Functions, Limits, and Continuity

2.1 Functions

1

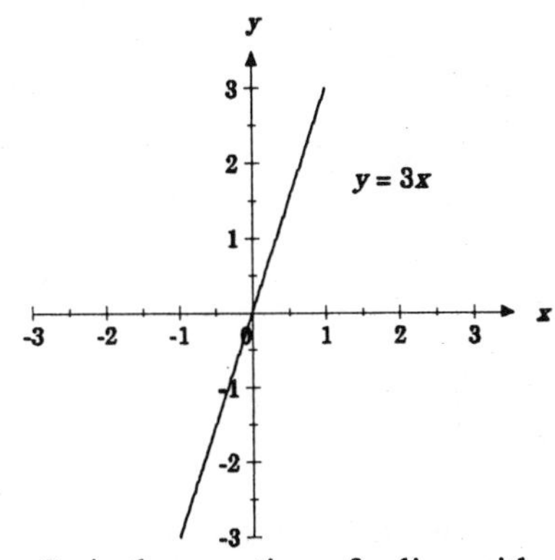

$f(x) = 3x$ is the equation of a line with slope 3 and y intercept 0.

3

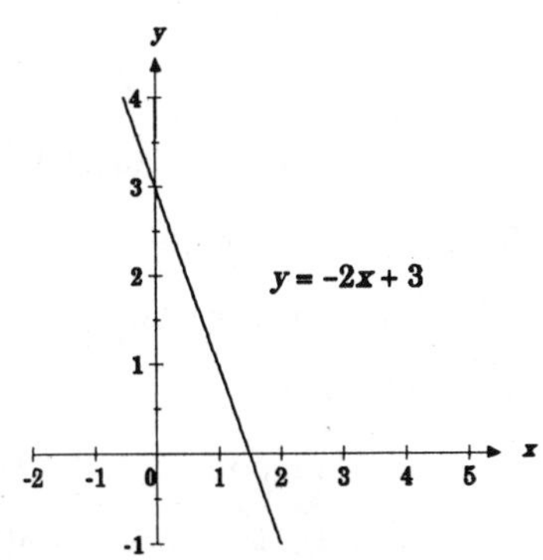

$f(x) = -2x + 3$ is the equation of a line with slope -2 and y intercept 3.

5

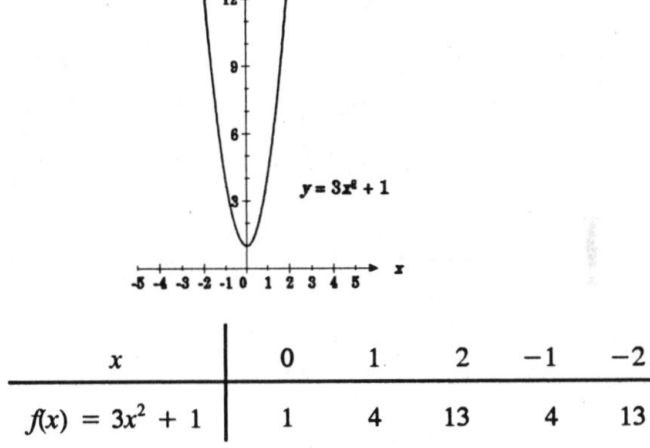

x	0	1	2	-1	-2
$f(x) = 3x^2 + 1$	1	4	13	4	13

$f(x) = 3x^2 + 1$ is the equation of a parabola with a minimum value at (0, 1).

7

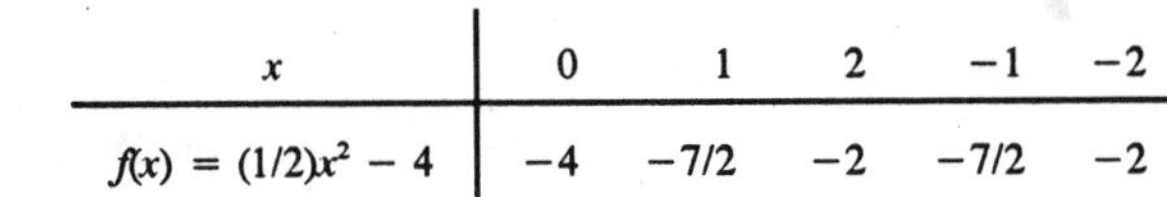

x	0	1	2	-1	-2
$f(x) = (1/2)x^2 - 4$	-4	$-7/2$	-2	$-7/2$	-2

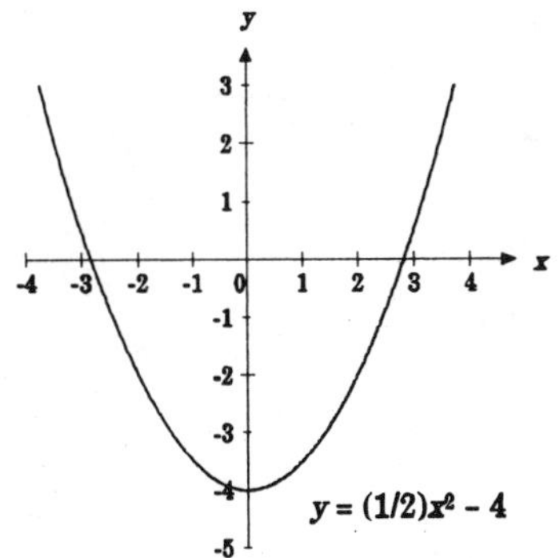

$f(x) = (1/2)x^2 - 4$ is the equation of a parabola with a minimum value at (0, -4).

9

x	0	1/2	1	2	$-1/2$	-1
$f(x) = x^2 - x$	0	$-1/4$	0	2	3/4	2

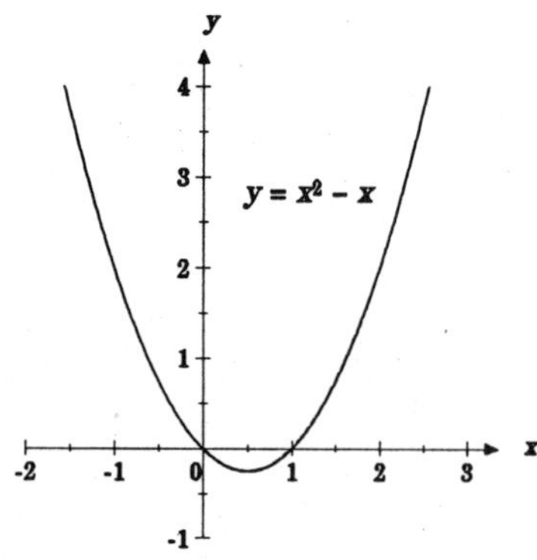

$f(x) = x^2 - x$ is the equation of a parabola with a minimum value at $(1/2, -1/4)$.

11

x	0	1/2	1	$-1/2$	-1	-2
$f(x) = x^2 + x + 1$	1	7/4	3	3/4	1	3

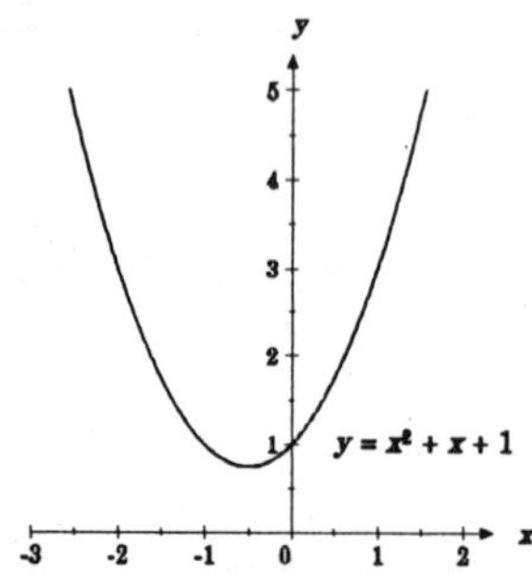

$f(x) = x^2 + x + 1$ is the equation of a parabola with a minimum value at $(-1/2, 3/4)$.

13

x	0	1	2	3	-1	-2	-3
$1/(2 + x^2)$	1/2	1/3	1/6	1/11	1/3	1/6	1/11

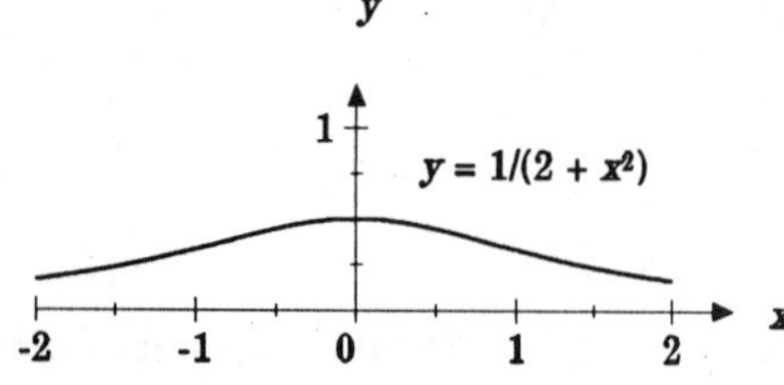

15

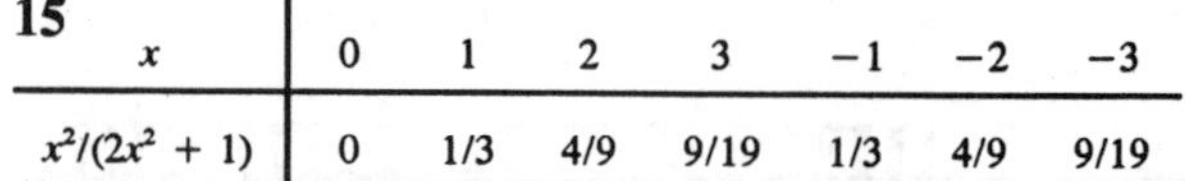
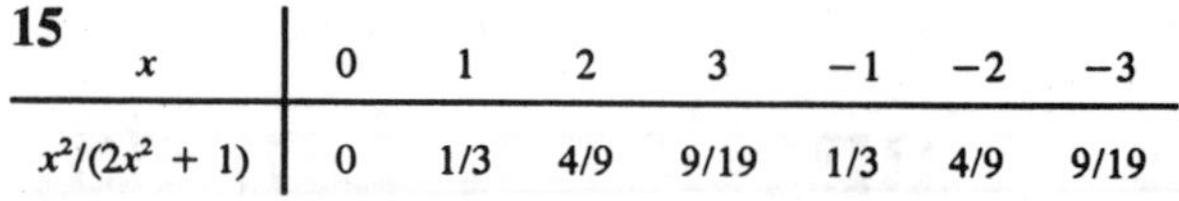

x	0	1	2	3	-1	-2	-3
$x^2/(2x^2 + 1)$	0	1/3	4/9	9/19	1/3	4/9	9/19

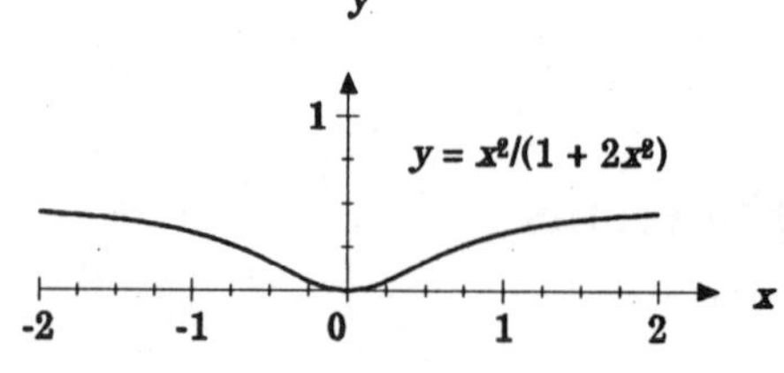

17 The domain consists of all non-negative real numbers; so does the range.

19 In order for $\sqrt{x + 1}$ to make sense, we must have $x + 1 \geq 0$, or $x \geq -1$. Therefore, the domain consists of all real numbers greater than or equal to -1; the interval is $[-1, \infty)$. The range consists of all non-negative real numbers since as x becomes arbitrarily large, the output becomes arbitrarily large; the interval is $[0, \infty)$.

21 We must have $4 - x^2 \geq 0$ for $f(x) = \sqrt{4 - x^2}$ to have a sensible output, so $x^2 \leq 4$ or $|x| \leq 2$. Therefore, the domain is the interval $[-2, 2]$. Clearly, the lower bound on the range must be 0 because $f(x)$ is always non-negative. Notice also that $f(x)$ is largest when $x = 0$, for which value we have $f(0) = 2$. Hence the range is $[0, 2]$.

23 The function $f(x) = 1/x$ is defined for all $x \neq 0$ (only division by zero is taboo). Thus the domain is all non-zero real numbers. Since all non-zero real numbers can be written as reciprocals of other numbers, the range is all non-zero real numbers.

25 We need $x + 1 \neq 0$ or $x \neq -1$; then the domain is all real numbers except -1. To get the range, we find all real b such that $b = 1/(x + 1)$ for some value of x in the domain. Solving for x, we have

$x + 1 = 1/b$, so $x = 1/b - 1$. This expression is valid so long as b is non-zero. Thus the range is all nonzero real numbers.

27 We want $1 - x^2 \neq 0$ or $x^2 \neq 1$. Hence $x \neq \pm 1$, and the domain is all real numbers whose magnitude is different from 1. To find the range, we plot several points and sketch the graph. As we can see, the range is $(-\infty, 0)$ and $[1, \infty)$.

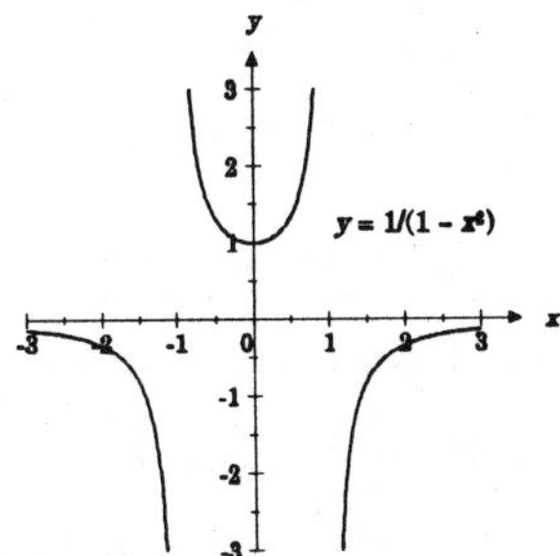

Alternatively, we can try find all real b such that $b = 1/(1 - x^2)$ for some value of x in the domain. This requires that $1 - x^2 = 1/b$, so $x^2 = 1 - 1/b$ and $x = \pm\sqrt{1 - \frac{1}{b}}$. Hence we must have $1 - \frac{1}{b} \geq 0$. Obviously $b < 0$ works. In the case of positive values of b, we need $\frac{1}{b} \leq 1$, so $b \geq 1$.

Putting these two results together, the domain consists of $(-\infty, 0)$ and $[1, \infty)$, as we saw earlier from the graph.

29 $f(x) = x + 1$

(a) $f(-1) = -1 + 1 = 0$

(b) $f(3) = 3 + 1 = 4$

(c) $f(1.25) = 1.25 + 1 = 2.25$

(d) $f(0) = 0 + 1 = 1$

31 $f(x) = x^3$

(a) $f(1 + 2) = f(3) = 3^3 = 27$

(b) $f(4 - 1) = f(3) = 3^3 = 27$

33 All of the following results may be obtained by plugging directly into a calculator. Alternatively, note that $\frac{f(3+h) - f(3)}{h} = \frac{(3+h)^2 - 3^2}{h}$

$$= \frac{9 + 6h + h^2 - 9}{h} = \frac{h^2 + 6h}{h} = h + 6.$$

(a) For $h = 1$, $h + 6 = 7$.

(b) For $h = 0.01$, $h + 6 = 6.01$.

(c) For $h = -0.01$, $h + 6 = 5.99$.

(d) For $h = 0.0001$, $h + 6 = 6.0001$.

(e) As h approaches 0, $h + 6 = \frac{f(3+h) - f(3)}{h}$ approaches 6.

35 $f(a + 1) - f(a) = (a + 1)^3 - a^3$

$= a^3 + 3a^2 + 3a + 1 - a^3 = 3a^2 + 3a + 1$

37 $$\frac{f(d) - f(c)}{d - c} = \frac{\frac{1}{d^2} - \frac{1}{c^2}}{d - c} = \frac{\frac{c^2 - d^2}{d^2c^2}}{d - c}$$

$$= \frac{c^2 - d^2}{(d - c)d^2c^2} = \frac{(c - d)(c + d)}{(d - c)d^2c^2}$$

$$= \frac{-(d - c)(c + d)}{(d - c)d^2c^2} = -\frac{c + d}{c^2d^2}$$

39 (a) is not the graph of a function, since some lines parallel to the y axis meet the graph in two places. (b) and (c) are the graphs of functions.

41 The circumference of a circle with radius x is $2\pi x$, so $f(x) = 2\pi x$. The domain is all nonnegative real numbers, since a circle with a negative radius makes no sense. (A circle with radius 0 is, of course, just a point. If you want to exclude this "degenerate" case, the stipulated domain would be all *positive* real numbers.)

43 The perimeter of a square with side x is $4x$, so $f(x) = 4x$. The domain is all nonnegative real numbers since a square with a negative side is meaningless. (If you want to exclude the "degenerate" case where the square reduces to a point, the domain is all *positive* real numbers.)

45 The total surface area of a cube of side x is the sum of the areas of the six sides, or $6x^2$. Thus $f(x) = 6x^2$. The domain is the interval $[0, \infty)$ because a negative-sided cube is ridiculous.

47

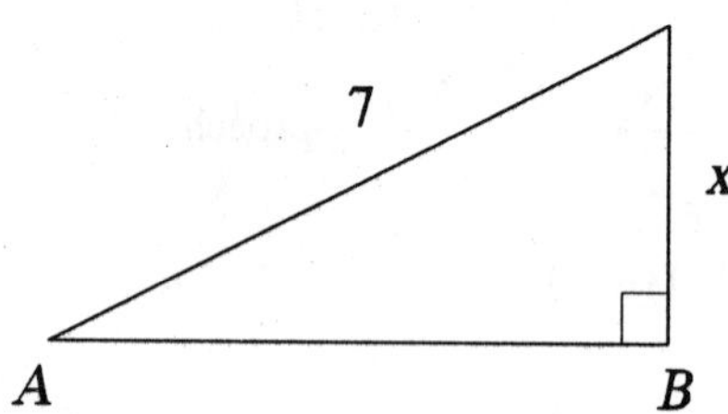

By the Pythagorean theorem, the length of AB is $\sqrt{7^2 - x^2}$, so $f(x) = \sqrt{49 - x^2}$. Now $f(x)$ can be computed whenever $49 - x^2 \geq 0$, or $|x| \leq 7$; but unless we want to allow triangles with sides of length 0, or negative length, we should limit the domain of $f(x)$ to $0 < x < 7$.

49

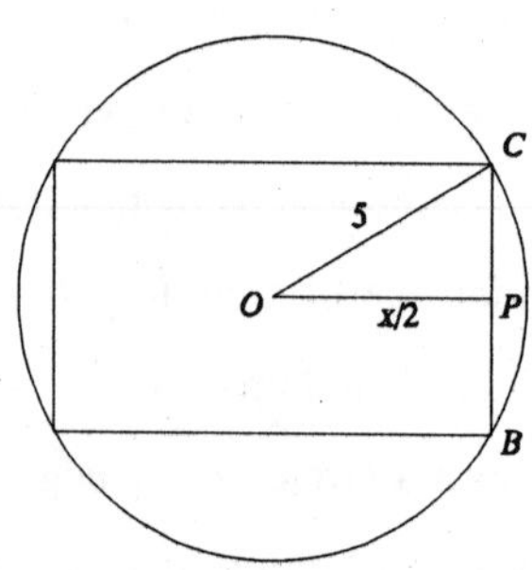

$\overline{BC} = 2\overline{PC} = 2\sqrt{5^2 - \overline{OP}^2} = 2\sqrt{25 - (x/2)^2} = \sqrt{100 - x^2}$, so $f(x) = 2\overline{AB} + 2\overline{BC} = 2x + 2\sqrt{100 - x^2}$.

51 $g(x) = \overline{AP} + \overline{PB} = \sqrt{2^2 + x^2} + \sqrt{(6 - x)^2 + 5^2}$ $= \sqrt{x^2 + 4} + \sqrt{x^2 - 12x + 61}$ miles.

53 (a) $f(0) = 0, f(h) = \pi a^2$

(b)

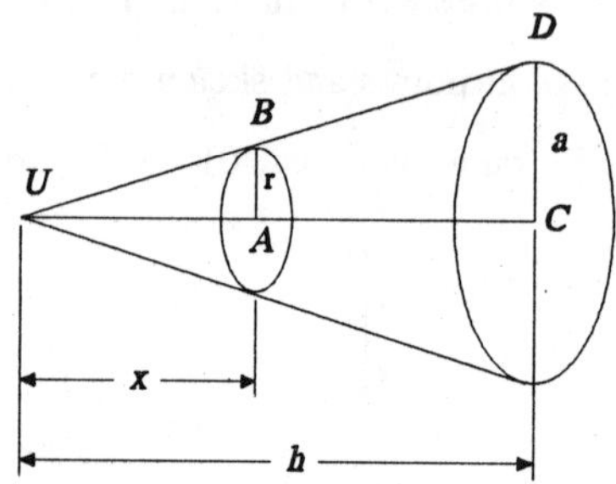

Draw in the altitude UC of the right circular cone. ΔUAB and ΔUCD are similar. Thus, $x/r = h/a$ or $r = (a/h)x$. But $f(x) = \pi r^2$, the cross-sectional area of the cone. By substitution, $f(x) = \pi\left(\frac{a}{h}x\right)^2 = \frac{\pi a^2 x^2}{h^2}$. This makes sense because $f(0) = 0$ and $f(h) = \pi a^2$, as in (a).

55 (a) Define the risk of driving function to be $r(t)$. Let the risk of driving be the ratio of accidents to traffic. Then $r(t) = \frac{g(t)}{f(t)}$.

(b) The most dangerous time to drive is when $r(t)$ is maximized and the safest time is when $r(t)$ is minimized. At 2AM, $r(t) = \frac{g(t)}{f(t)} = \frac{5}{1} = 5$. At 8AM, $r(t) = \frac{g(t)}{f(t)} = \frac{2.5}{5} = \frac{1}{2}$. Thus 2AM is the most dangerous time and 8AM is the safest. Driving at 2AM is about 10 times as risky.

57 (a) $f(0) = 0, f(1) = 2, f(2) = 5, f(3) = 9$

(b)

x	0	1	2	3
$f(x)$	0	2	5	9

(c)

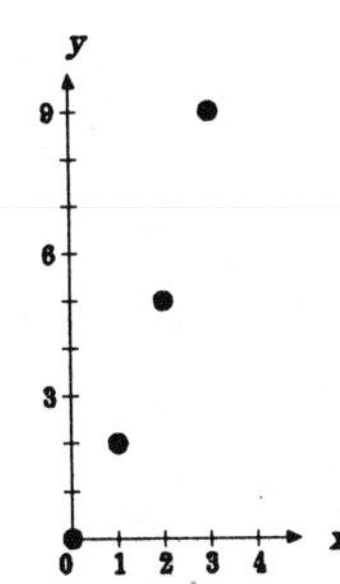

(d) $f(3) - f(2) = 9 - 5 = 4 > 3 = 5 - 2 = f(2) - f(1)$

59 (a) $f(1 \cdot 2) = f(2) = 6 \neq 18 = 3 \cdot 6 = f(1)f(2)$, so $f(ab)$ does not always equal $f(a)f(b)$.

(b) $f(ab) = (ab)^3 = a^3b^3 = f(a)f(b)$, so $f(ab) = f(a)f(b)$.

(c) $f(ab) = \dfrac{1}{ab} = \dfrac{1}{a} \cdot \dfrac{1}{b} = f(a)f(b)$, so $f(ab) = f(a)f(b)$.

(d) $f(ab) = \sqrt{ab} = \sqrt{a} \cdot \sqrt{b} = f(a)f(b)$, so $f(ab) = f(a)f(b)$.

(e) $f(1 \cdot 2) = f(2) = 3 \neq 2 \cdot 3 = (1 + 1)(2 + 1) = f(1)f(2)$, so $f(ab)$ is not, in general, equal to $f(a)f(b)$.

61 $f(x) = 137 \cdot 2^x$

$f(x) = 0$

$f(x) = 2^{\lfloor x \rfloor}$, where $\lfloor x \rfloor$ is the largest integer less than or equal to x.

63 (a) $f(x) = 0$ for $x = 1, 2, 3, \ldots$

$f(x) = -\pi x$ for $x = 1, 2, 3, \ldots$

$f(x) = \sqrt{3}x$ for $x = 1, 2, 3, \ldots$

2.2 Composite Functions

Note: The answers to Exercises 1 through 5 depend on the type of calculator used. In addition, there are often several different correct answers.

1 To evaluate $\sin x^2$ at $x = 3$ push [3] [x^2] [SIN].

3 To evaluate $\sqrt{1 + x^3}$ at $x = 4$ push [4] [y^x] [3] [+] [1] [=] [$\sqrt{x}$].

5 To evaluate $\sqrt[3]{\cos x^2}$ at $x = 2$ push [2] [x^2] [COS] [$\sqrt[x]{y}$] [3] [=].

7 $y = u^2 = (1 + x)^2$

9 $y = 1/u = 1/x^3$

11 $y = u^2 = (x^3)^2 = x^6$

13 $y = \sqrt{u} = \sqrt{\cos x}$

15 $y = u^3 = (1 + v)^3 = (1 + \sqrt{x})^3$

17 $y = \cos u = \cos(1 + \tan^2 x)$

19 Let $u = x^3 + x^2 - 2$. Then $y = u^{50}$.

21 Let $u = x + 3$. Then $y = \sqrt{u}$.

23 Let $u = 2x$. Then $y = \sin u$.

25 Let $u = 2x$ and $v = \cos u$. Then $y = v^3$.

27 (a) $f(g(7)) = f(3) = 9$

(b) $g(f(3)) = g(9) = 5$

29 $(f \circ g)(x) = f(g(x)) = f(4x^3 - 3x)$
$= 2(4x^3 - 3x)^2 - 1 = 32x^6 - 48x^4 + 18x^2 - 1$

$(g \circ f)(x) = g(f(x)) = g(2x^2 - 1)$
$= 4(2x^2 - 1)^3 - 3(2x^2 - 1)$
$= 32x^6 - 48x^4 + 18x^2 - 1$

31 We seek f such that $f(g(x)) = g(f(x))$, that is, $f(x^2) = g(ax + b)$. Thus $ax^2 + b = (ax + b)^2 = a^2x^2 + 2abx + b^2$. Since this is to be true for all x, we have $a = a^2$, $2ab = 0$, and $b = b^2$. Since $ab = 0$ either a or $b = 0$. But $a \neq 0$ so $b = 0$. Then a

$= 1$. Thus $f(x) = x$.

33 $(f\circ g)(x) = f(g(x)) = 2g(x) + 3 = 2(ax + b) + 3$ $= 2ax + (2b + 3)$. $(g\circ f)(x) = g(f(x)) = af(x) + b$ $= a(2x + 3) + b = 2ax + (3a + b)$. The two composite functions are equal if and only if $2b + 3$ $= 3a + b$, that is, $b = 3a - 3$. There are infinitely many such functions.

35 For any g, $(f\circ g)(x) = f(g(x)) = g(x)^5$, which equals x if and only if $g(x) = \sqrt[5]{x}$. So there is exactly one such function.

2.3 The Limit of a Function

1 $\lim_{x\to 5} (x + 7) = 12$, since $x + 7 \to 12$ as $x \to 5$.

3 $\lim_{x\to 2} \frac{x^2 - 4}{x - 2} = \lim_{x\to 2} \frac{(x + 2)(x - 2)}{x - 2} = \lim_{x\to 2} (x + 2)$; $x + 2 \to 4$ as $x \to 2$, so the limit is 4.

5 $\lim_{x\to 1} \frac{x^4 - 1}{x^3 - 1} = \lim_{x\to 1} \frac{(x - 1)(x^3 + x^2 + x + 1)}{(x - 1)(x^2 + x + 1)} =$

$\lim_{x\to 1} \frac{x^3 + x^2 + x + 1}{x^2 + x + 1}$; now $x^2 + x + 1 \to 3$ as $x \to 1$ and $x^3 + x^2 + x + 1 \to 4$ as $x \to 1$, so the limit is 4/3.

7 $\lim_{x\to 3} \frac{1}{x + 2} = \frac{1}{5}$, since $x + 2 \to 5$ as $x \to 3$.

9 $\lim_{x\to 3} 25 = 25$, since 25 is a constant.

11 $\lim_{x\to 0^+} \sqrt{x}$. Positive values of x near 0 yield values of $\sqrt{x}$ near 0, so the limit is 0.

13 $\lim_{x\to 1^+} \frac{x - 1}{|x - 1|}$. $x \to 1^+$ implies that $x > 1$, so $x - 1 > 0$ and $|x - 1| = x - 1$. Thus for $x > 1$, $\frac{x - 1}{|x - 1|} = \frac{x - 1}{x - 1} = 1$; therefore the limit is 1.

15 $\lim_{h\to 1} \frac{(1 + h)^2 - 1}{h} = \lim_{h\to 1} \frac{1 + 2h + h^2 - 1}{h} =$

$\lim_{h\to 1} \frac{2h + h^2}{h} = \lim_{h\to 1} (2 + h) = 3$, since $2 + h \to 3$ as $h \to 1$.

17 $\lim_{x\to 2} \frac{\frac{1}{x} - \frac{1}{2}}{x - 2} = \lim_{x\to 2} \frac{\frac{2 - x}{2x}}{x - 2} = \lim_{x\to 2} \frac{2 - x}{2x(x - 2)}$

$= \lim_{x\to 2} \frac{-1}{2x}$; $2x \to 4$ as $x \to 2$, so the limit is $-1/4$.

19 $\lim_{x\to 0} 64^x = 1$, since $64^x \to 64^0$ as $x \to 0$.

21 (a) $\lim_{x\to 0^+} f(x) = 2$, since as $x \to 0$ from the right, $f(x) \to 2$.

(b) $\lim_{x\to 1} f(x) = 1$, since as $x \to 1$ from either side, $f(x) \to 1$.

(c) $\lim_{x\to 2^-} f(x) = 1$, since as $x \to 2$ from the left, $f(x) = 1$.

(d) $\lim_{x\to 2^+} f(x) = 2$, since as $x \to 2$ from the right, $f(x) \to 2$.

23

x	1	0.1	0.01	−0.01	−0.1	−1
$3^x - 1$	2	0.1161	0.0110	−0.0109	−0.1040	−0.6667
$(3^x - 1)/x$	2	1.1612	1.1047	1.0926	1.0404	0.6667

$\frac{3^x - 1}{x}$ appears to approach a number near 1.10 as

$x \to 0$. (The limit exists.)

25 (a)

x	1	0.1	0.01	0.001	−1	−0.01	−0.001
$f(x)$	2	1.618	1.588	1.585	1.333	1.582	1.585

(b) It looks like $\lim_{x \to 0} \frac{3^x - 1}{2^x - 1}$ exists and is about 1.585, since $\frac{3^x - 1}{2^x - 1}$ is close to 1.585 for values of x near 0.

27 $\lim_{x \to 3} f(x) = 1$, since as $x \to 3$ from either side, $f(x) \to 1$.

29 (a)

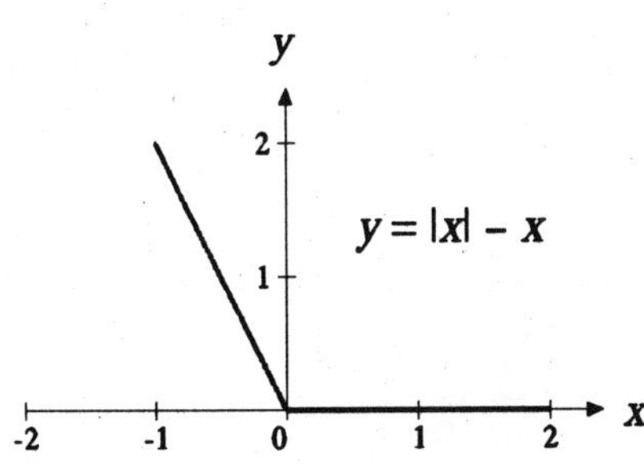

(b) $\lim_{x \to 0^+} f(x) = \lim_{x \to 0^-} f(x) = 0$, so $\lim_{x \to 0} f(x) = 0$.

For $x > 0$, $f(x) = |x| - x = 0$; hence if $a > 0$, $\lim_{x \to a} f(x) = 0$. For $x < 0$, $f(x) = -2x$, so if $a < 0$, $\lim_{x \to a} f(x) = \lim_{x \to a} (-2x) = -2a$. Thus $\lim_{x \to a} f(x)$ exists for *all a*.

31 (a)

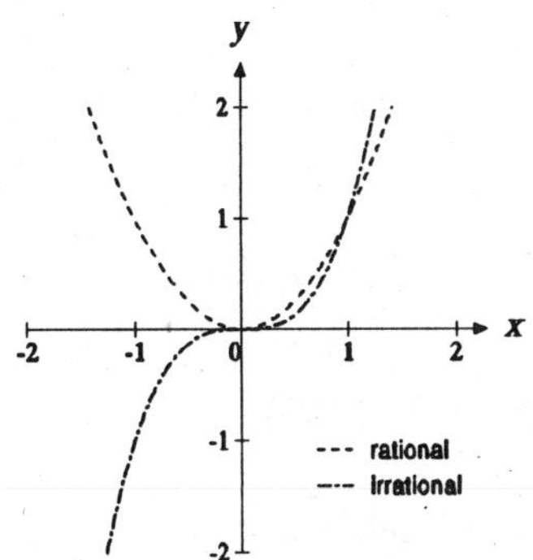

(b) For rational x, $f(x) \to 4$ as $x \to 2$. For irrational x, $f(x) \to 8$ as $x \to 2$. $f(x)$ does not approach one number as $x \to 2$, so $\lim_{x \to 2} f(x)$ does not exist.

(c) For rational x, $f(x) \to 1$ as $x \to 1$. For irrational x, $f(x) \to 1$ as $x \to 1$. Thus, $f(x) \to 1$ as $x \to 1$, so $\lim_{x \to 1} f(x)$ exists (and equals 1).

(d) For rational x, $f(x) \to 0$ as $x \to 0$. For irrational x, $f(x) \to 0$ as $x \to 0$. Thus $f(x) \to 0$ as $x \to 0$, so $\lim_{x \to 0} f(x)$ exists and equals 0.

(e) 0 and 1 are the only numbers a for which $f(x)$ approaches the same number as $x \to a$ for both rational and irrational x (since $x^2 = x^3$ only when $x = 0$ or 1), so $\lim_{x \to a} f(x)$ exists only for $a = 0, 1$.

33 (a)

x	1	0.1	0.01	−0.01
$4^x - 1$	3	0.1487	0.0140	−0.0138
$(4^x - 1)/x$	3	1.4870	1.3959	1.3767

x	-0.1	-1	0.001	-0.001
$4^x - 1$	-0.1294	-0.75	0.0014	-0.0014
$(4^x - 1)/x$	1.2945	0.75	1.3873	1.3853

(b) $\lim_{x \to 0} \frac{4^x - 1}{x} \approx 1.386$

(c) $1.386 = 2 \cdot (0.693)$

(d) It would seem that $\lim_{x \to 0} \frac{4^x - 1}{x} =$

$2 \lim_{x \to 0} \frac{2^x - 1}{x}$.

(e) First, observe that $4^x - 1 = 2^{2x} - 1 = (2^x + 1)(2^x - 1)$. Then we see that

$\lim_{x \to 0} \frac{4^x - 1}{x} = \lim_{x \to 0} \frac{(2^x + 1)(2^x - 1)}{x}$. But as

$x \to 0$, $2^x + 1 \to 2$, so we have

$\lim_{x \to 0} \frac{(2^x + 1)(2^x - 1)}{x} = 2 \lim_{x \to 0} \frac{2^x - 1}{x}$. This

implies that $\lim_{x \to 0} \frac{4^x - 1}{x} = 2 \lim_{x \to 0} \frac{2^x - 1}{x}$,

and we are done.

35 (a) $f(1) = \frac{1}{1} + \frac{1}{2} = \frac{3}{2}$

$f(2) = \frac{1}{2} + \frac{1}{3} + \frac{1}{4} = \frac{13}{12}$

$f(3) = \frac{1}{3} + \frac{1}{4} + \frac{1}{5} + \frac{1}{6} = \frac{19}{20}$

$f(4) = \frac{1}{4} + \frac{1}{5} + \cdots + \frac{1}{8} = \frac{743}{840}$

$f(5) = \frac{1}{5} + \frac{1}{6} + \cdots + \frac{1}{10} \approx 0.8456$

$f(6) \approx 0.8199$

$f(7) \approx 0.8016$

$f(8) \approx 0.7879$

$f(9) \approx 0.7773$

$f(10) \approx 0.7688$

(b) As n increases $f(n)$ decreases but by less and less each time.

(c) For large values of n, the sum is near 0.693. It appears that $\lim_{n \to \infty} f(n) \approx 0.693$.

2.4 Computations of Limits

1 $\lim_{x \to 2} (3x^2 + 2) = 3 \cdot 2^2 + 2 = 14$

3 $\lim_{x \to 2} \frac{3x^2 + 1}{x + 3} = \frac{3 \cdot 2^2 + 1}{2 + 3} = \frac{13}{5}$

5 $\lim_{x \to 1} [(4x^2 + x)(x + 3)] = (4 \cdot 1^2 + 1)(1 + 3) =$

$5 \cdot 4 = 20$

7 $\lim_{x \to \infty} (7x + 2) = \infty$, since $7x + 2$ grows without bound as $x \to \infty$.

9 $\lim_{x \to \infty} (4x^2 - x + 3) = \lim_{x \to \infty} x^2 \left(4 - \frac{1}{x} + \frac{3}{x^2}\right) = \infty$,

since $\lim_{x \to \infty} x^2 = \infty$ and $\lim_{x \to \infty} \left(4 - \frac{1}{x} + \frac{3}{x^2}\right) = 4$.

11 $\lim_{x \to \infty} (x^5 - 100x^4) = \lim_{x \to \infty} x^5 \left(1 - \frac{100}{x}\right)$ where

$\lim_{x \to \infty} \left(1 - \frac{100}{x}\right) = 1$ and $\lim_{x \to \infty} x^5 = \infty$, so

$\lim_{x \to \infty} (x^5 - 100x^4) = \infty$.

13 $\lim_{x \to -\infty} (6x^5 + 21x^3) = \lim_{x \to -\infty} x^5 \left(6 + \frac{21}{x^2}\right)$, where

$\lim_{x \to -\infty} \left(6 + \frac{21}{x^2}\right) = 6$ and $\lim_{x \to -\infty} x^5 = -\infty$, so

$\lim_{x \to -\infty} (6x^5 + 21x^3) = -\infty$.

15 $\lim_{x \to -\infty} (-x^3) = \lim_{x \to -\infty} (-1)x^3 = (-1) \lim_{x \to -\infty} x^3$, where

$\lim_{x \to -\infty} x^3 = -\infty$, so $\lim_{x \to -\infty} (-x^3) = \infty$.

17 $\lim_{x \to \infty} \frac{6x^3 - x}{2x^{10} + 5x + 8} = \lim_{x \to \infty} \frac{6x^3}{2x^{10}} = \lim_{x \to \infty} \frac{3}{x^7} = 0$

19 $\lim_{x \to \infty} \frac{x^4 + 1066x^2 - 1492}{2x^4 - 2001} = \lim_{x \to \infty} \frac{x^4}{2x^4} = \lim_{x \to \infty} \frac{1}{2}$

$= \frac{1}{2}$

21 $\lim_{x \to \infty} \frac{x^3 + 1}{x^4 + 2} = \lim_{x \to \infty} \frac{x^3}{x^4} = \lim_{x \to \infty} \frac{1}{x} = 0$

23 $\lim_{x \to 0^+} \frac{1}{x^3} = \infty$, since $x^3 \to 0$ as $x \to 0^+$ and $x^3 > 0$.

25 $\lim_{x \to 0^+} \frac{1}{x^4} = \infty$, since $x^4 \to 0$ as $x \to 0^+$ and $x^4 > 0$.

27 (a) $\lim_{x \to 1^+} \frac{1}{x - 1} = \infty$ since $x - 1 \to 0$ as $x \to 1^+$, and $x - 1 > 0$.

(b) $\lim_{x \to 1^-} \frac{1}{x - 1} = -\infty$ since $x - 1 \to 0$ as $x \to 1^-$, and $x - 1 < 0$.

(c) $\lim_{x \to 1} \frac{1}{x - 1}$ does not exist and is neither ∞ nor $-\infty$.

Note: A limit that equals ∞ or $-\infty$ (as in (a) or (b)) does not "exist" just because it is set equal to a symbol. "Infinity" is not a number, and we use the ∞ symbol as a shorthand notation. When, for example, a limit is said to equal ∞, that just means that the limit fails to exist for a specified reason: growth without bound.

29 $\lim_{x \to \infty} \frac{\sqrt{4x^2 + 2x + 1}}{3x} = \lim_{x \to \infty} \frac{\sqrt{x^2(4 + 2/x + 1/x^2)}}{3x}$

$= \lim_{x \to \infty} \frac{x\sqrt{4 + 2/x + 1/x^2}}{3x} = \lim_{x \to \infty} \frac{\sqrt{4 + 2/x + 1/x^2}}{3}$

$= \frac{\sqrt{4 + 0 + 0}}{3} = \frac{\sqrt{4}}{3} = \frac{2}{3}$

31 $\lim_{x \to \infty} \frac{\sqrt{4x^2 + x}}{\sqrt{9x^2 - 3x}} = \lim_{x \to \infty} \frac{\sqrt{x^2(4 + 1/x)}}{\sqrt{x^2(9 - 3/x)}} =$

$\lim_{x \to \infty} \frac{x\sqrt{4 + 1/x}}{x\sqrt{9 - 3/x}} = \lim_{x \to \infty} \frac{\sqrt{4 + 1/x}}{\sqrt{9 - 3/x}} = \frac{\sqrt{4}}{\sqrt{9}} = \frac{2}{3}$

33 Let a and b be the two numbers.

(a) If a and b are greater than 0, then the product ab is also a very small positive number. If $a > 0$ and $b < 0$ or $a < 0$ and $b > 0$, then the product ab is a negative number of very small absolute value. If a and b are less than 0, then the product ab is again a very small positive number.

(b) Nothing can be said about the quotient of the two numbers. For example, if $a = 0.001$ and $b = 0.000001$, then $\frac{a}{b} = \frac{0.001}{0.000001} = 1000$, a large number. Also, if $a = 0.000001$ and $b = 0.001$, then $\frac{a}{b} = \frac{0.000001}{0.001} = 0.001$, a small number. Finally, if $a = b = 0.001$, $\frac{a}{b} = \frac{0.001}{0.001} = 1$, a number that is neither large nor small. Since a can be positive when

b is negative, and vice-versa, we see that the quotient can be any number in $(-\infty, \infty)$.

35

x	1	5	10	100
2^x	2	32	1024	1.2677×10^{30}
$x/2^x$	0.5	0.1563	0.0098	7.8886×10^{-29}

It seems that $\lim_{x\to\infty} \dfrac{x}{2^x} = 0$.

37 (a) $\lim_{x\to 0} \dfrac{2x^3 + x^2 + x}{3x^3 - x^2 + 2x} = \lim_{x\to 0} \dfrac{x(2x^2 + x + 1)}{x(3x^2 - x + 2)}$

$= \lim_{x\to 0} \dfrac{2x^2 + x + 1}{3x^2 - x + 2} = \dfrac{0 + 0 + 1}{0 + 0 + 2} = \dfrac{1}{2}$

(b) Let $f(x) = \dfrac{2x^3 + x^2 + x}{3x^3 - x^2 + 2x}$. Then $f(0.01) =$

$\dfrac{2(0.01)^3 + (0.01)^2 + (0.01)}{3(0.01)^3 - (0.01)^2 + 2(0.01)} = \dfrac{0.010102}{0.019903}$

≈ 0.5076. This agrees with (a).

39 Since $\lim_{x\to\infty} f(x) = 0$ and $\lim_{x\to\infty} g(x) = \infty$, we need more information to determine $\lim_{x\to\infty} f(x)g(x)$. For example, if $f(x) = \dfrac{1}{x}$ and $g(x) = x$, then $\lim_{x\to\infty} f(x)g(x) = \lim_{x\to\infty} \dfrac{1}{x}\cdot x = 1$. But if $f(x) = \dfrac{1}{x^2}$ and $g(x) = x$, then $\lim_{x\to\infty} f(x)g(x) = \lim_{x\to\infty} \dfrac{1}{x^2}\cdot x = \lim_{x\to\infty} \dfrac{1}{x} = 0$. Also, if $f(x) = \dfrac{1}{x}$ and $g(x) = x^2$, then $\lim_{x\to\infty} f(x)g(x) = \lim_{x\to\infty} \dfrac{1}{x}\cdot x^2 = \lim_{x\to\infty} x = \infty$. So, depending on the choice of $f(x)$ and $g(x)$, any one of the three citizens could be correct. If we allow $f(x) = 2/x$ and $g(x) = x$, then $\lim_{x\to\infty} f(x)g(x) = \lim_{x\to\infty} \dfrac{2}{x}\cdot x = 2$, and all three would be incorrect.

41 (a) $\lim_{x\to\infty} (f(x) + g(x)) = \lim_{x\to\infty} f(x) + \lim_{x\to\infty} g(x)$

$= 0 + 1 = 1$

(b) $\lim_{x\to\infty} (f(x)/g(x)) = \dfrac{\lim_{x\to\infty} f(x)}{\lim_{x\to\infty} g(x)} = \dfrac{0}{1} = 0$

(c) $\lim_{x\to\infty} (f(x)g(x)) = \left(\lim_{x\to\infty} f(x)\right)\left(\lim_{x\to\infty} g(x)\right) = 0\cdot 1$

$= 0$

(d) The limit $\lim_{x\to\infty} (g(x)/f(x))$ does not exist. It may be ∞ or $-\infty$ or neither of these.

(e) The limit $\lim_{x\to\infty} g(x)/|f(x)|$ is ∞; it does not exist.

43 (a) Since $f(x) \to 1$ while $g(x) \to \infty$, the ratio $f(x)/g(x) \to 0$.

(b) Since $f(x) \to 1$ while $g(x) \to \infty$, the product $f(x)g(x) \to \infty$.

(c) Indeterminate. The limit $\lim_{x\to\infty} (f(x) - 1)g(x)$ may be any real number or may not exist. If it doesn't exist, it may be ∞, $-\infty$, or neither. The table below gives examples of each.

$f(x)$	$g(x)$	$\lim_{x\to\infty} (f(x) - 1)g(x)$
$1 + r/x$	x	r (any real number)
$1 + 1/x$	x^2	∞
$1 - 1/x$	x^2	$-\infty$
$1 + (\sin x)/x$	x	Does not exist

45 (a)

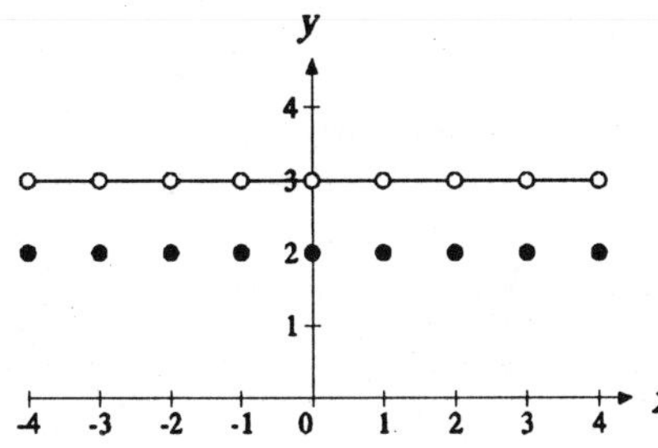

(b) The limit does not exist because $f(x)$ attains both of the values 2 and 3 for arbitrarily large values of x.

(c) $\lim_{x\to 2} f(x) = 3$ because the limit depends only on x values *near* 2, for which $f(x) = 3$. (The fact that $f(2) = 2$ is irrelevant.)

47 (a)

x	100	1000	10000	100000
$\sqrt{x^2 + 20x} - x$	9.5445	9.9505	9.9950	9.9995

(c) $\lim_{x\to\infty} (\sqrt{x^2 + 20x} - x)$

$$= \lim_{x\to\infty} \frac{(\sqrt{x^2 + 20x} - x)(\sqrt{x^2 + 20x} + x)}{(\sqrt{x^2 + 20x} + x)}$$

$$= \lim_{x\to\infty} \frac{x^2 + 20x - x^2}{\sqrt{x^2 + 20x} + x}$$

$$= \lim_{x\to\infty} \frac{20x}{\sqrt{x^2 + 20x} + x}$$

$$= \lim_{x\to\infty} \frac{20x}{\sqrt{x^2(1 + 20/x)} + x}$$

$$= \lim_{x\to\infty} \frac{20x}{x(\sqrt{1 + 20/x} + 1)}$$

$$= \lim_{x\to\infty} \frac{20}{\sqrt{1 + 20/x} + 1} = \frac{20}{\sqrt{1 + 0} + 1}$$

$$= \frac{20}{2} = 10$$

2.5 Some Tools for Graphing

1 (a) Since $f(x) = x^2 + 2 = (-x)^2 + 2 = f(-x)$, $f(x)$ is even.

(b) Since $f(x) = \sqrt{x^4 - 1} = \sqrt{(-x)^4 - 1} = f(-x)$, $f(x)$ is even.

(c) Since $f(x) = 1/x^2 = 1/(-x)^2 = f(-x)$, $f(x)$ is even.

3 (a) Since $f(x) = x^3 + x$, and $f(-x) = (-x)^3 - x = -x^3 - x = -(x^3 + x) = -f(x)$, $f(x)$ is odd.

(b) Since $f(x) = x + 1/x$, and $f(-x) = (-x) + 1/(-x) = -x - 1/x = -(x + 1/x) = -f(x)$, $f(x)$ is odd.

(c) Since $f(x) = \sqrt[3]{x}$, and $f(-x) = \sqrt[3]{-x} = \sqrt[3]{(-1)^3 x} = -\sqrt[3]{x} = -f(x)$, $f(x)$ is odd.

5 (a) We have $f(x) = 3 + x$ and $f(-x) = 3 - x$: Since $3 - x$ does not equal $3 + x$ or $-(3 + x) = -3 - x$, $f(x)$ is neither odd nor even.

(b) We have $f(x) = (x + 2)^2$ and $f(-x) = (-x + 2)^2$. Since $(-x + 2)^2 = x^2 - 4x + 4$ does not equal $(x + 2)^2 = x^2 + 4x + 4$ or

$-(x + 2)^2 = -x^2 - 4x - 4$, $f(x)$ is neither even nor odd.

(c) We have $f(x) = x/(x + 1)$ and $f(-x) = -x/(-x + 1)$. Since $-x/(-x + 1)$ does not equal $x/(x + 1)$ or $-[x/(x + 1)] = -x/(x + 1)$, $f(x)$ is neither even nor odd.

7 (a) $f(x) = x + x^3 + 5x^4$ and $f(-x) = -x + (-x)^3 + 5(-x)^4 = -x - x^3 + 5x^4$. Since $f(-x) \neq -f(x) = -x - x^3 - 5x^4$ and $f(-x) \neq f(x)$, $f(x)$ is neither odd nor even.

(b) Since all powers of x are even, $f(x) = 7x^4 - 5x^2$ is even.

(c) Since x^2 is even, $f(x) = \sqrt[3]{x^2 + 1}$ is even.

9 To find the x intercept(s), set $y = 0$ and solve for x: $0 = 2x + 3$, $-3 = 2x$, so $x = -3/2$. The x intercept is $-3/2$.

To find the y intercept, set $x = 0$ and find y: $y = 2x + 3 = 2\cdot 0 + 3 = 3$. The y intercept is 3.

11 To find the x intercept(s), solve $0 = x^2 + 3x + 2$ for x. Since $x^2 + 3x + 2 = (x + 2)(x + 1)$, we have $(x + 2)(x + 1) = 0$; thus $x = -1$ or -2. So the x intercepts are -1 and -2. To find the y intercept, let $x = 0$ and evaluate y: $y = 0^2 + 3\cdot 0 + 2 = 2$. The y intercept is 2.

13 Set $y = 0$ to find the x intercepts. We have $0 = 2x^2 + 1$, and it follows that $x^2 = -1/2$. But since the square or any real number is nonnegative, $y = 2x^2 + 1$ has no x intercept. Solve $y = 2x^2 + 1$ where $x = 0$ to get the y intercept. We have $y = 2\cdot 0 + 1 = 1$. So $y = 1$ is the intercept.

15 Examine $\lim_{x\to\infty} y$ and $\lim_{x\to-\infty} y$ to find any horizontal asymptotes. We see that $\lim_{x\to\infty} \frac{x + 2}{x - 2} = \lim_{x\to\infty} \frac{x}{x} = \lim_{x\to\infty} 1 = 1$ and $\lim_{x\to-\infty} \frac{x + 2}{x - 2} = \lim_{x\to-\infty} \frac{x}{x} = \lim_{x\to-\infty} 1 = 1$. So the line $y = 1$ is the only horizontal asymptote. To find any vertical asymptotes, find the values of x for which $y = \frac{x + 2}{x - 2}$ is undefined. This occurs when $x - 2 = 0$ or when $x = 2$. Thus, the line $x = 2$ is the only vertical asymptote.

17 To find all horizontal asymptotes, evaluate $\lim_{x\to\infty} y$ and $\lim_{x\to-\infty} y$. We have $\lim_{x\to\infty} \frac{x}{x^2 + 1} = \lim_{x\to\infty} \frac{x}{x^2} = \lim_{x\to\infty} \frac{1}{x} = 0$ and $\lim_{x\to-\infty} \frac{x}{x^2 + 1} = \lim_{x\to-\infty} \frac{x}{x^2} = \lim_{x\to-\infty} \frac{1}{x} = 0$. So $y = 0$ is the only horizontal asymptote. Since $\frac{x}{x^2 + 1}$ does not "blow up" for any real x, there are no vertical asymptotes.

19 Evaluate $\lim_{x\to\infty} y$ and $\lim_{x\to-\infty} y$; we have $\lim_{x\to\infty} \frac{x^2 + 1}{x^2 - 3} = \lim_{x\to\infty} \frac{x^2}{x^2} = 1$ and $\lim_{x\to-\infty} \frac{x^2 + 1}{x^2 - 3} = \lim_{x\to-\infty} \frac{x^2}{x^2} = 1$. So $y = 1$ is the solitary horizontal asymptote. Observe that the denominator $(x^2 - 3)$ equals zero when $x = \pm\sqrt{3}$. So $x = \pm\sqrt{3}$ are the vertical asymptotes.

21 Observe that y is neither even nor odd. However, $\lim_{x\to\infty} \frac{1}{x - 2} = 0$ and $\lim_{x\to-\infty} \frac{1}{x - 2} = 0$, so $y = 0$ is a horizontal asymptote. Furthermore, $\lim_{x\to 2^-} \frac{1}{x - 2}$

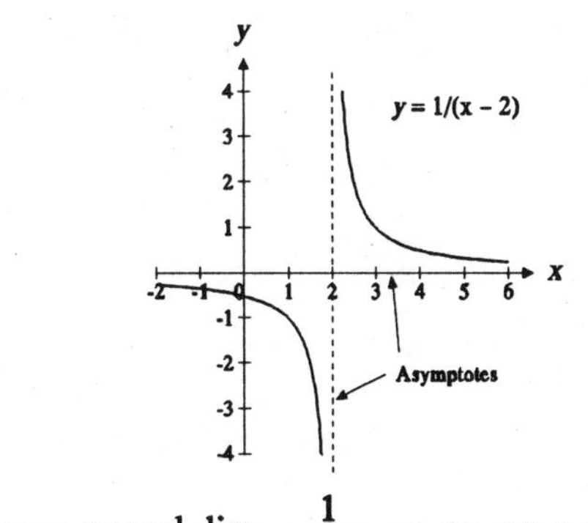

$= -\infty$ and $\lim_{x \to 2^+} \frac{1}{x - 2} = \infty$, so $x = 2$ is a vertical asymptote.

23 First, note that y is an even function, so we need

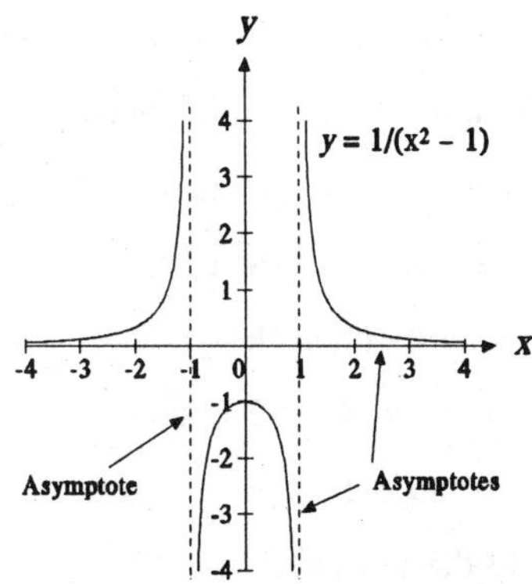

only concern ourselves with $x \geq 0$ and simply reflect that information across the y axis. Since $\lim_{x \to \infty} y = 0$, $y = 0$ is a horizontal asymptote. To see how y behaves near the $x = 1$ vertical asymptote, evaluate $\lim_{x \to 1} \frac{1}{x^2 - 1}$. The result is ∞. Noting that $y = -1$ when $x = 0$ and reflecting all information across the y axis, we are done.

25

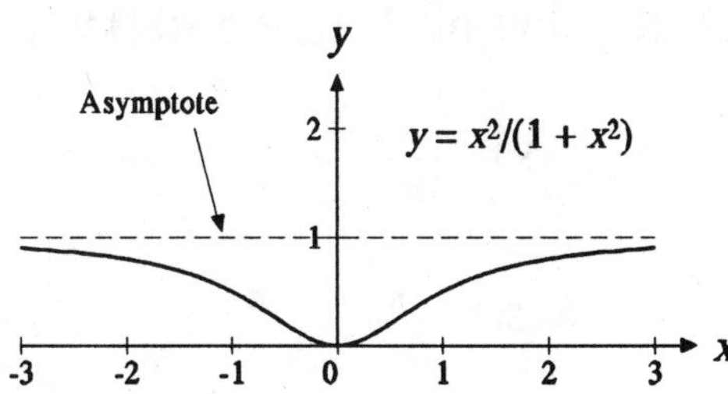

In this case, $y = \frac{x^2}{1 + x^2}$ is an even function. In addition, $\lim_{x \to \infty} \frac{x^2}{1 + x^2} = 1$, so $y = 1$ is a horizontal asymptote.

27

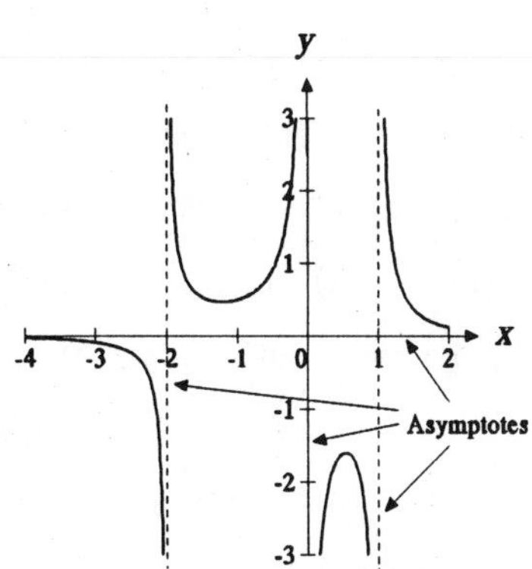

Now $y = \frac{1}{x(x - 1)(x + 2)}$ is neither odd nor even. However, $\lim_{x \to \infty} y = 0$ and $\lim_{x \to -\infty} y = 0$, so $y = 0$ is a horizontal asymptote. Note also that $x = 0$, $x = 1$, and $x = -2$ are vertical asymptotes.

29 Let the two odd functions be $f(x)$ and $g(x)$.

(a) Let $h(x) = f(x) + g(x)$. Therefore $h(-x) = f(-x) + g(-x) = -f(x) - g(x) = -[f(x) + g(x)] = -h(x)$. Since $-h(x) = h(-x)$, $h(x)$ is an odd function. Consequently, the sum of $f(x)$ and $g(x)$ is also odd.

(b) Let $h(x) = f(x) \cdot g(x)$. Then $h(-x) = f(-x)g(-x) = -f(x)[-g(x)] = f(x)g(x) = h(x)$. Since $h(-x) = h(x)$, $h(x)$ is an even function and so is the product of $f(x)$ and $g(x)$.

(c) Let $h(x) = f(x)/g(x)$. Then $h(-x) = f(-x)/g(-x) = -f(x)/(-g(x)) = f(x)/g(x) = h(x)$. Since $h(x) = h(-x)$, $h(x)$ is an even function and so is the quotient.

31 (a) Let $h(x) = f(x) + g(x)$. Then $h(-x) = f(-x) + g(-x) = -f(x) + g(x)$. Since $-f(x) + g(x) \neq -f(x) - g(x)$ and $-f(x) + g(x) \neq f(x) + g(x)$, their sum is neither even nor odd.

(b) Let $h(x) = f(x)g(x)$. Then $h(-x) = f(-x)g(-x) = -f(x)g(x) = -h(x)$. Since $h(-x) = -h(x)$, their product is odd.

(c) Let $h(x) = f(x)/g(x)$. Then $h(-x) = f(-x)/g(-x) = -f(x)/g(x) = -h(x)$. Since $h(-x) = -h(x)$, their quotient is odd.

33 Any odd polynomial can be expressed as $a_1x^1 + a_3x^3 + a_5x^5 + \cdots = p_0(x)$. Clearly $p_0(x) = -p_0(-x)$, and the addition of any x^n of even degree or a real number to $p_0(x)$ would destroy this property.

35 A function f that is both odd and even satisfies the following conditions: $f(x) = f(-x)$ and $f(x) = -f(-x)$. This implies that $f(-x) = -f(-x)$. The only f that is equal to the opposite of itself for *all* real x is $f(x) = 0$.

37

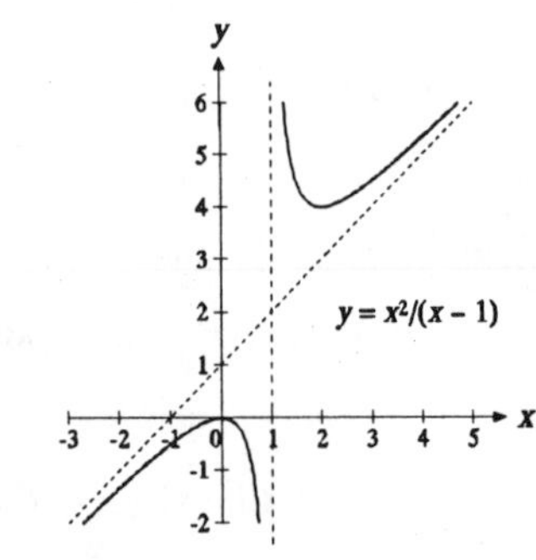

$$y = \frac{x^2}{x - 1} = x + 1 + \frac{1}{x - 1}$$

39

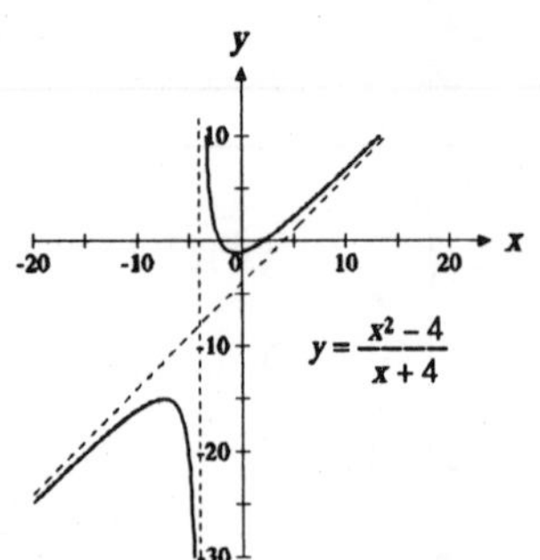

$$y = \frac{x^2 - 4}{x + 4} = x - 4 + \frac{12}{x + 4}$$

41 (a) One would add two to every y coordinate for which $f(x)$ yields an output in order to obtain the graph of $g(x)$, as the equation states.

(b) Similarly, one would subtract two from every y coordinate for which $f(x)$ is defined.

(c) One would obtain the graph of $g(x)$ by shifting $f(x)$ to the right by 2 units.

(d) One would obtain the graph of $g(x)$ by shifting $f(x)$ to the left by 2 units.

(e) One would multiply every y coordinate for which $f(x)$ is defined by 2 in order to obtain the graph of $g(x)$.

(f) From (c) and (e), we see that $g(x)$ can be obtained by shifting $f(x)$ to the right by 2, then multiplying the shifted version of $f(x)$ by 3.

43 Solutions to exploration problems are in the instructor's manual.

2.6 A Review of Trigonometry

1 (a) $90° = 90° \cdot \frac{\pi}{180°} = \frac{\pi}{2}$ (radians)

(b) $30° = 30° \cdot \frac{\pi}{180°} = \frac{\pi}{6}$

(c) $120° = 120° \cdot \dfrac{\pi}{180°} = \dfrac{2\pi}{3}$

(d) $360° \cdot \dfrac{\pi}{180°} = 2\pi$

3 (a) $\dfrac{3\pi}{4} \cdot \dfrac{180°}{\pi} = 135°$

(b) $\dfrac{\pi}{3} \cdot \dfrac{180°}{\pi} = 60°$

(c) $\dfrac{2\pi}{3} \cdot \dfrac{180°}{\pi} = 120°$

(d) $4\pi \cdot \dfrac{180°}{\pi} = 720°$

5 (a) $s = 5$ and $r = 3$, so $\theta = \dfrac{s}{r} = \dfrac{5}{3}$ (radians)

(b) By (a), the degree measure is $\dfrac{5}{3} \cdot \dfrac{180°}{\pi} =$

$\dfrac{300°}{\pi} \approx 95.49°.$

7 (a) $50° = 50° \cdot \dfrac{\pi}{180°} = \dfrac{5\pi}{18}$

(b) $2 = 2 \cdot \dfrac{180°}{\pi} = \dfrac{360°}{\pi} \approx 114.59°$

9 Given that $\theta = s/r$, $s = 2$ cm, and $r = 3$ cm, we find θ to be $\dfrac{2\text{ cm}}{3\text{ cm}} = \dfrac{2}{3}$.

11 (a)

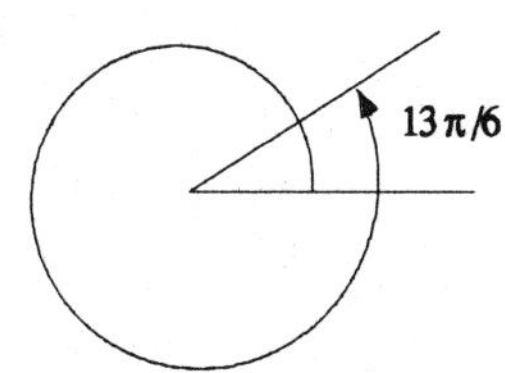

$13\pi/6$ (coterminal with $\pi/6$)

(b)

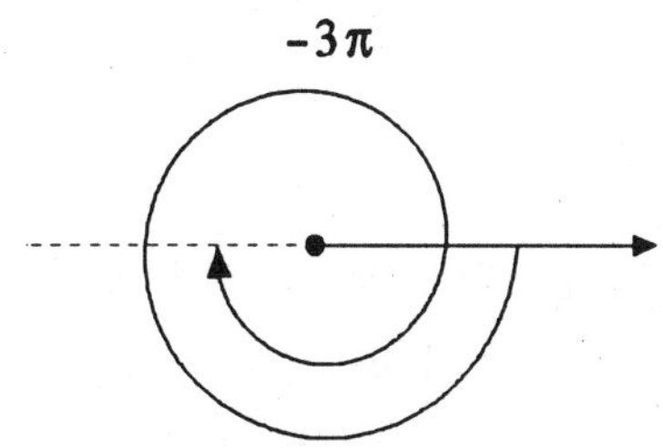

-3π (coterminal with $\pm\,\pi$)

13 $\sin\dfrac{\pi}{6} = \cos\left(\dfrac{\pi}{2} - \dfrac{\pi}{6}\right) = \cos\dfrac{\pi}{3} = \dfrac{1}{2}$

15

θ	0	$\pi/6$	$\pi/4$	$\pi/3$	$\pi/2$	π	$3\pi/2$	2π
$\sin\theta$	0	1/2	$1/\sqrt{2}$	$\sqrt{3}/2$	1	0	−1	0

17 $\cos\left(\dfrac{\pi}{6} + \dfrac{\pi}{3}\right) = \cos\dfrac{\pi}{6}\cos\dfrac{\pi}{3} - \sin\dfrac{\pi}{6}\sin\dfrac{\pi}{3} =$

$\dfrac{\sqrt{3}}{2} \cdot \dfrac{1}{2} - \dfrac{1}{2} \cdot \dfrac{\sqrt{3}}{2} = 0 = \cos\dfrac{\pi}{2}$

19 $\sin\dfrac{5\pi}{12} = \sin\left(\dfrac{\pi}{4} + \dfrac{\pi}{6}\right) =$

$\sin\dfrac{\pi}{4}\cos\dfrac{\pi}{6} + \cos\dfrac{\pi}{4}\sin\dfrac{\pi}{6} = \dfrac{1}{\sqrt{2}} \cdot \dfrac{\sqrt{3}}{2} + \dfrac{1}{\sqrt{2}} \cdot \dfrac{1}{2}$

$= \dfrac{\sqrt{3} + 1}{2\sqrt{2}}$

21 $\cos 2\theta = \cos(\theta + \theta) = \cos\theta\cos\theta - \sin\theta\sin\theta$ $= \cos^2\theta - \sin^2\theta$

23 $\cos 2\theta = \cos^2\theta - \sin^2\theta = \cos^2\theta - (1 - \cos^2\theta)$ $= \cos^2\theta - 1 + \cos^2\theta = 2\cos^2\theta - 1.$

25 As we saw in Exercise 21, $\cos 2\theta = 2\cos^2\theta - 1$, so $1 + \cos 2\theta = 2\cos^2\theta$ and therefore $\cos^2\theta = (1 + \cos 2\theta)/2$.

27 Examine Fig. 16 in the text.

(a) $-$

(b) $+$

(c) $+$

(d) $-$

29 (a) See Figure 21 in the text and the accompanying discussion.

(b) $\dfrac{\tan \pi/4}{1} = \dfrac{\sin \pi/4}{\cos \pi/4} = \dfrac{\sqrt{2}/2}{\sqrt{2}/2} = 1$

31 The slope of a line with angle of inclination θ is $\tan \theta$. Using a scientific calculator, we have, to four decimal places:

(a) $\tan 10° = 0.1763$

(b) $\tan 70° = 2.7475$

(c) $\tan 110° = -2.7475$

(d) $\tan 135° = -1.0000$

(e) $\tan 0° = 0.0000$

33 A line of slope m has angle of inclination $\theta = \tan^{-1} m$, so

(a) $\theta = \tan^{-1}(\sqrt{3}) = \dfrac{\pi}{3}$ (radians) or 60°;

(b) $\theta = \tan^{-1} \dfrac{1}{\sqrt{3}} = \dfrac{\pi}{6}$ (radians) or 30°.

35

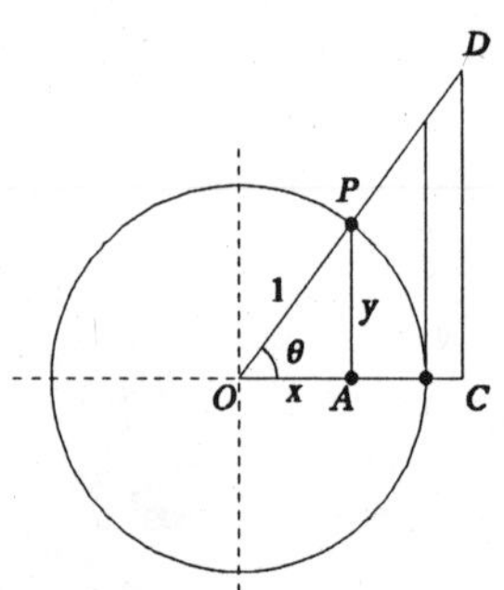

By the definition of sine and cosine, observe that the coordinates of P are $(x, y) = (\cos \theta, \sin \theta)$. Furthermore, $x = \overline{OA}$, $y = \overline{PA}$, and $\overline{OP} = 1$, so $\dfrac{\overline{OA}}{\overline{OP}} = \dfrac{x}{1} = \cos \theta$, and $\dfrac{\overline{PA}}{\overline{OP}} = \dfrac{y}{1} = \sin \theta$. By similar triangles, we now have $\dfrac{a}{c} = \dfrac{\overline{OC}}{\overline{OD}} = \dfrac{\overline{OA}}{\overline{OP}} = \cos \theta$, and $\dfrac{b}{c} = \dfrac{\overline{DC}}{\overline{OD}} = \dfrac{\overline{PA}}{\overline{OP}} = \sin \theta$. Since $\tan \theta = \dfrac{\sin \theta}{\cos \theta} = \dfrac{b/c}{a/c} = \dfrac{b}{a}$, we now have shown that

(a) $\cos \theta = a/c$,

(b) $\sin \theta = b/c$, and

(c) $\tan \theta = b/a$.

37 By the results of Exercise 35, we have

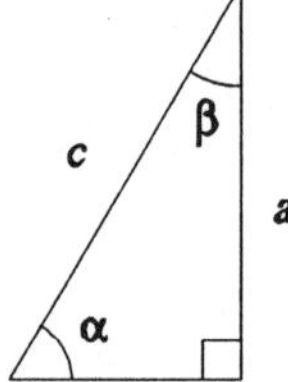

(a) $\cos \alpha = \dfrac{b}{c}$,

(b) $\sin \beta = \dfrac{b}{c}$, and

(c) $\tan \alpha = \dfrac{a}{b}$.

39 Observe that angle CDO is $\pi/6$. By the results of Exercise 35, $\cos \dfrac{\pi}{6} = \dfrac{\overline{CD}}{\overline{OD}} = \dfrac{\sqrt{3}}{2}$, $\sin \dfrac{\pi}{6} = \dfrac{\overline{OC}}{\overline{OD}} = \dfrac{1}{2}$, and $\tan \dfrac{\pi}{6} = \dfrac{\overline{OC}}{\overline{CD}} = \dfrac{1}{\sqrt{3}}$.

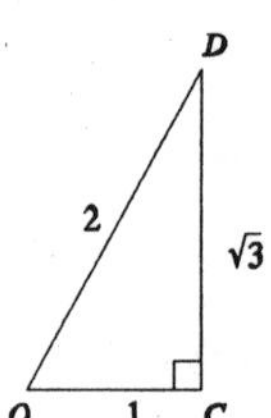

41 (a) Recall that $\sec \theta = \dfrac{1}{\cos \theta}$, so $\sec 0 = 1$, $\sec \dfrac{\pi}{6} = \dfrac{2}{\sqrt{3}}$, $\sec \dfrac{\pi}{4} = \sqrt{2}$, and $\sec \dfrac{\pi}{3} = 2$.

(b)

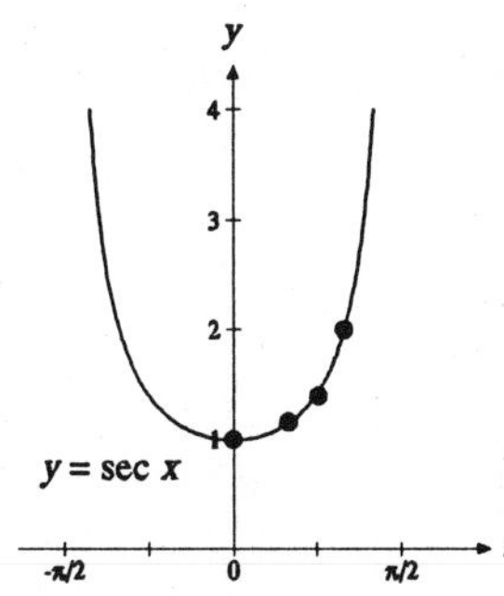

43 Refer to Figure 32 in the text.

(a) Note that $\overline{OP} = \overline{OQ} = \overline{OR} = \overline{OS} = 1$ and angle POQ is equal to angle ROS. Thus the triangles POQ and ROS are congruent, from which it follows that the corresponding sides PQ and RS have the same length.

(b) By definition of sine and cosine, $Q =$ $(\cos(A + B), \sin(A + B))$, $R =$ $(\cos(-A), \sin(-A)) = (\cos A, -\sin A)$, and $S = (\cos B, \sin B)$.

(c) $P = (1, 0)$ and $Q = (\cos(A + B),$ $\sin(A + B))$; by the distance formula $\overline{PQ}^2$ $= (1 - \cos(A + B))^2 + (\sin(A + B))^2 =$ $1 - 2\cos(A + B) + 1 = 2 - 2\cos(A + B)$

(d) $\overline{RS}^2 = (\cos A - \cos B)^2 + (-\sin A - \sin B)^2$ $= \cos^2 A - 2\cos A\cos B + \cos^2 B + \sin^2 A$ $+ 2\sin A\sin B + \sin^2 B$ $= 2 - 2\cos A\cos B + 2\sin A\sin B$.

(e) Using the results of (c) and (d) and the fact that $\overline{PQ}^2 = \overline{RS}^2$, $2 - 2\cos(A + B) =$ $2 - 2\cos A\cos B + 2\sin A\sin B$,

$-2\cos(A + B)$

$= -2\cos A\cos B + 2\sin A\sin B$,

$\cos(A + B) = \cos A\cos B - \sin A\sin B$.

45 $\cos\left(\frac{\pi}{2} - \theta\right) = \cos\frac{\pi}{2}\cos\theta + \sin\frac{\pi}{2}\sin\theta =$ $0 + 1\cdot\sin\theta = \sin\theta$

47 (a) Replacing θ by $A + B$ in the result of Exercise 45 yields $\sin(A + B) =$

$$\cos\left(\frac{\pi}{2} - (A + B)\right) = \cos\left(\left(\frac{\pi}{2} - A\right) - B\right).$$

(b) $\sin(A + B) = \cos\left(\left(\frac{\pi}{2} - A\right) - B\right) =$

$$\cos\left(\frac{\pi}{2} - A\right)\cos B + \sin\left(\frac{\pi}{2} - A\right)\sin B$$

(c) $\cos\left(\frac{\pi}{2} - A\right) = \sin A$ and

$\sin\left(\frac{\pi}{2} - A\right) = \cos A$, so $\sin(A + B) =$

$\sin A\cos B + \cos A\sin B$, as claimed.

49 (a) $\cos^2\theta = \frac{1}{2}(1 + \cos 2\theta)$, so $\cos\theta =$

$$\pm\sqrt{\frac{1}{2}(1 + \cos 2\theta)}.$$

(b) By (a), $\cos\frac{\pi}{4} = \pm\sqrt{\frac{1}{2}\left(1 + \cos\frac{\pi}{2}\right)} = \pm\sqrt{\frac{1}{2}}$

$= \pm\frac{1}{\sqrt{2}}$; since $\pi/4$ is in the first quadrant,

cosine is positive and $\cos\frac{\pi}{4} = \frac{1}{\sqrt{2}}$.

(c) $\cos\frac{3\pi}{4} = \pm\sqrt{\frac{1}{2}\left(1 + \cos\frac{3\pi}{2}\right)} = \pm\frac{1}{\sqrt{2}}$; since $\frac{3\pi}{4}$

is in the second quadrant, cosine is negative

and $\cos\frac{3\pi}{4} = -\frac{1}{\sqrt{2}}$.

51 $\tan(A - B) = \dfrac{\sin(A - B)}{\cos(A - B)}$

$= \dfrac{\sin A \cos B - \cos A \sin B}{\cos A \cos B + \sin A \sin B}$

$= \dfrac{\cos A \cos B\left(\dfrac{\sin A}{\cos A} - \dfrac{\sin B}{\cos B}\right)}{\cos A \cos B\left(1 + \dfrac{\sin A}{\cos A}\cdot\dfrac{\sin B}{\cos B}\right)}$

$= \dfrac{\tan A - \tan B}{1 + \tan A \tan B}$

53

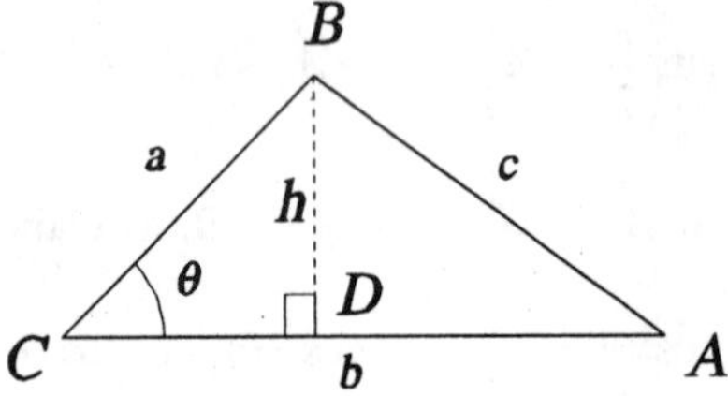

(a) By the result of Exercise 35, $\cos\theta = \dfrac{\overline{CD}}{\overline{CB}} = \dfrac{\overline{CD}}{a}$, so $\overline{CD} = a\cos\theta$. Since $\overline{AD} = \overline{CA} - \overline{CD} = b - \overline{CD}$, we have $\overline{AD} = b - a\cos\theta$.

(b) By the Pythagorean theorem applied to the triangles BCD and ADB, $h^2 = a^2 - \overline{CD}^2 = a^2 - a^2\cos^2\theta$ and $h^2 = c^2 - \overline{AD}^2 = c^2 - (b - a\cos\theta)^2$.

(c) By equating the two expressions for h^2 obtained in (b), we have $a^2 - a^2\cos^2\theta$ $= c^2 - (b - a\cos\theta)^2$ $= c^2 - b^2 + 2ab\cos\theta - a^2\cos^2\theta$, so a^2 $= c^2 - b^2 + 2ab\cos\theta$, or $c^2 = a^2 + b^2 - 2ab\cos\theta$, which is the law of cosines.

55 (a)

θ	1	0.1	−0.1	−1
$(\sin\theta)/\theta$	0.8415	0.9983	0.9983	0.8415

(b) Yes, $\lim_{\theta\to 0}\left(\dfrac{\sin\theta}{\theta}\right)$ seems to exist and appears to be approximately 1.

2.7 The Limit of $(\sin\theta)/\theta$ as θ Approaches 0

1 (a) $\dfrac{\theta r^2}{2} = \dfrac{(\pi/2)3^2}{2} = \dfrac{9\pi}{4}$

(b) $\dfrac{\theta r^2}{2} = \dfrac{\theta\cdot 1^2}{2} = \dfrac{\theta}{2}$

(c) $\dfrac{\theta r^2}{2} = \dfrac{\theta\cdot 2^2}{2} = 2\theta$

3 $\lim_{x\to 0}\dfrac{\sin x}{2x} = \dfrac{1}{2}\lim_{x\to 0}\dfrac{\sin x}{x} = \dfrac{1}{2}\cdot 1 = \dfrac{1}{2}$

5 $\lim_{x\to 0}\dfrac{\sin 3x}{5x} = \lim_{x\to 0}\dfrac{\sin 3x}{3x}\cdot\dfrac{3x}{5x} = \lim_{3x\to 0}\dfrac{\sin 3x}{3x}\cdot\dfrac{3}{5}$

$= 1\cdot\dfrac{3}{5} = \dfrac{3}{5}$

7 $\lim_{\theta\to 0}\dfrac{\sin^2\theta}{\theta} = \lim_{\theta\to 0}\dfrac{\sin\theta}{\theta}\cdot\sin\theta =$

$\lim_{\theta\to 0}\dfrac{\sin\theta}{\theta}\lim_{\theta\to 0}\sin\theta = 1\cdot 0 = 0$

9 $\lim_{\theta\to 0}\dfrac{\tan^2\theta}{\theta} = \lim_{\theta\to 0}\dfrac{1}{\theta}\dfrac{\sin^2\theta}{\cos^2\theta} = \lim_{\theta\to 0}\dfrac{\sin\theta}{\theta}\cdot\dfrac{\sin\theta}{\cos^2\theta}$

$= \lim_{\theta \to 0} \frac{\sin \theta}{\theta} \lim_{\theta \to 0} \frac{\sin \theta}{\cos^2 \theta} = 1 \cdot 0 = 0$

11 Since $\sin \theta \approx \theta$ for small θ (in radians), $\sin 0.05 \approx 0.05$. The calculator reveals all: $\sin 0.05 \approx 0.049979$.

13 First convert 7° to radians: $\frac{7°}{360°} = \frac{\theta}{2\pi}$, so $\theta = \frac{7}{360} \cdot 2\pi = \frac{7}{180}\pi$. Hence $\sin 7° = \sin \frac{7}{180}\pi \approx \frac{7}{180}\pi \approx \frac{7(3.14)}{180} \approx 0.12211$. The calculator says $\sin 7° \approx 0.12187$.

15 (a) $\frac{1 - \cos \theta}{\theta} \cdot \frac{1 + \cos \theta}{1 + \cos \theta} = \frac{1 - \cos^2 \theta}{\theta(1 + \cos \theta)} = \frac{\sin^2 \theta}{\theta(1 + \cos \theta)}$

(b) $\lim_{\theta \to 0} \frac{1 - \cos \theta}{\theta} = \lim_{\theta \to 0} \frac{\sin^2 \theta}{\theta(1 + \cos \theta)} = \lim_{\theta \to 0} \frac{\sin \theta}{\theta} \cdot \frac{\sin \theta}{1 + \cos \theta} = \left[\lim_{\theta \to 0} \frac{\sin \theta}{\theta}\right] \cdot \left[\lim_{\theta \to 0} \frac{\sin \theta}{1 + \cos \theta}\right] = 1 \cdot \frac{0}{2} = 0$

17 We know that $\frac{1 - \cos \theta}{\theta^2} \approx \frac{1}{2}$ for small θ (in radians). We can treat the approximation as an equality and solve for $\cos \theta$, as follows:

$\frac{1 - \cos \theta}{\theta^2} \approx \frac{1}{2}$, $1 - \cos \theta \approx \frac{\theta^2}{2}$, $1 - \frac{\theta^2}{2} \approx \cos \theta$. So $\frac{1 - \cos \theta}{\theta^2} \approx \frac{1}{2}$ is equivalent to $\cos \theta \approx 1 - \frac{\theta^2}{2}$ for θ small, and $1 - \frac{\theta^2}{2}$ is therefore a good estimate of $\cos \theta$ for θ small. The following table gives the required estimates: (to five decimal places)

θ	0.1	0.01	$\pi/60$
$\cos \theta$	0.99500	0.99995	0.99863
$1 - \theta^2/2$	0.99500	0.99995	0.99863

19 This exercise shows the need for caution when using a calculator. The quotient actually approaches $1/24 = 0.041666...$, but because of round-off errors, a calculator will give erroneous values for θ near 0, as the following table shows:

θ	1	0.5	0.05	0.0005	0.0001
$(\cos \theta - 1 + \theta^2/2)/\theta^4$	0.04030	0.04132	0.04166	0	0

Your own results may vary, depending on your calculator.

21 (a) The domain is all nonzero values of x, since we cannot divide by 0.

(b) $f(-x) = \frac{\sin(-x)}{-x} = \frac{-\sin x}{-x} = \frac{\sin x}{x} = f(x)$, so f is an even function.

(c) For all x, $|\sin x| \le 1$; thus $\left|\frac{\sin x}{x}\right| = \frac{|\sin x|}{|x|} \le \frac{1}{|x|}$. Since $\lim_{x \to \infty} \frac{1}{|x|} = \lim_{x \to \infty} \frac{1}{x} = 0$ and $\left|\frac{\sin x}{x}\right| \le \frac{1}{|x|}$, it follows that $\lim_{x \to \infty} \frac{\sin x}{x} = 0$.

(d) $f(x) = 0$ when f is defined and $\sin x = 0$. We have $\sin x = 0$ when $x = \pi n$, where n is any integer. However, $f(x)$ is not defined for $x = 0$, so $f(x) = 0$ when $x = \pi n$, n a nonzero integer.

(e)

x	0.1	$\pi/2$	$3\pi/2$	2π	$5\pi/2$	3π	$7\pi/2$
$\sin x$	0.10	1.00	-1.00	0	1.00	0	-1.00
$(\sin x)/x$	1.00	0.64	-0.21	0	0.13	0	-0.09

(f),(g) Since $f(x)$ is an even function, the graph of $f(x)$ is symmetric about the y axis:

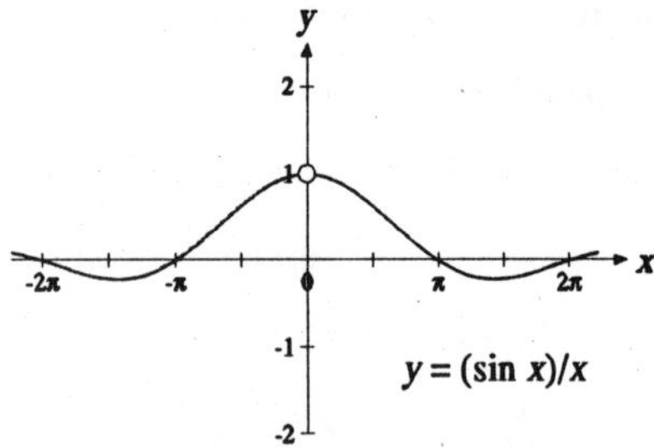

(h) $\lim_{x\to 0} f(x) = \lim_{x\to 0} \frac{\sin x}{x} = 1$

23 (a) If n is any nonzero integer, $f\left(\frac{1}{n\pi}\right) =$

$$\sin\left(\frac{1}{(1/n\pi)}\right) = \sin(n\pi) = 0.$$

25 (a) If n is any integer, $f\left(\frac{1}{(2n + 1/2)\pi}\right) =$

$$\sin\left(\left(2n + \frac{1}{2}\right)\pi\right) = \sin\left(2n\pi + \frac{\pi}{2}\right) = \sin\frac{\pi}{2}$$

$= 1.$

27 (b) As $x \to 0^+$, $1/x \to \infty$ and $\sin(1/x)$ oscillates infinitely often between -1 and 1. $\lim_{x\to 0^+} f(x)$ does not exist.

2.8 Continuous Functions

1 First, $a = 1/2$ is in the domain of the given function (call it $f(x)$). So condition 1 is satisfied. Second, $\lim_{x\to 1/2^-} f(x) = 1$ and $\lim_{x\to 1/2^+} f(x) = 1$, so $\lim_{x\to 1/2} f(x)$ exists and is 1, and condition 2 holds. Since $f(a) = f(1/2) = 1 = \lim_{x\to 1/2} f(x) = \lim_{x\to a} f(x)$, condition 3 is also satisfied. As a consequence of the validity of these three conditions, $f(x)$ is continuous at $a = 1/2$.

3 Condition 1 holds, because $f(a) = f(1/2) = 1/2$ implies that 1/2 is in the domain of $f(x)$. Condition 2 fails because $\lim_{x\to 1/2^-} f(x) = \frac{1}{2} \neq 1 = \lim_{x\to 1/2^+} f(x)$, and $\lim_{x\to 1/2} f(x)$ does not exist. Since $\lim_{x\to 1/2} f(x)$ does not exist, condition 3 automatically fails, and $f(x)$ is therefore not continuous at $a = 1/2$.

5 Note that $f(0) = 0$, and condition 1 holds. Condition 2 fails, however, because $\lim_{x\to 0^+} \sin(1/x)$ cannot be determined. (As $x \to 0^+$, $1/x \to \infty$, and $\sin(1/x)$ does not take on a determinable value.) If condition 2 fails, condition 3 also fails. So $f(x) = \sin(1/x)$ is not continuous at $a = 0$.

7 Clearly, condition 1 is satisfied since $f(1/4) = 1/2$. Also, condition 2 holds since $\lim_{x\to 1/4^-} f(x) = \frac{1}{4} = \lim_{x\to 1/4^+} f(x)$, and $\lim_{x\to 1/4} f(x)$ therefore exists. But $f(1/4) = \frac{1}{2} \neq \frac{1}{4} = \lim_{x\to 1/4} f(x)$, and condition 3 fails. Since condition 3 does not hold, $f(x)$ is not continuous at $a = 1/4$.

9 Since $f(a) = f(0)$ is not in the domain of f, to speak of continuity of f at $a = 0$ would be foolhardy.

11 (a)

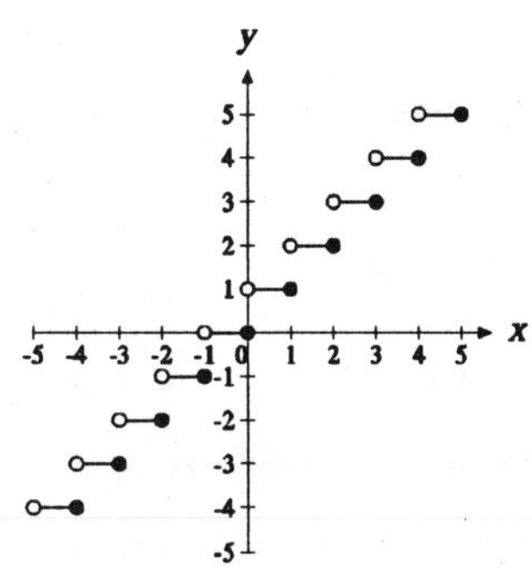

(b) $\lim_{x \to 4^-} f(x) = 4$

(c) $\lim_{x \to 4^+} f(x) = 5$

(d) $\lim_{x \to 4^-} f(x) \neq \lim_{x \to 4^+} f(x)$, so $\lim_{x \to 4} f(x)$ does not exist.

(e) f is not continuous at 4, since $\lim_{x \to 4} f(x)$ does not exist.

(f) f is continuous at noninteger x, since if a is not an integer, $\lim_{x \to a} f(x) = f(a)$.

(g) f is not continuous at integer x, since if n is an integer, $\lim_{x \to n^-} f(x) = n \neq n + 1 = \lim_{x \to n^+} f(x)$, so $\lim_{x \to n} f(x)$ does not exist.

13 $\lim_{x \to 0} f(x) = \lim_{x \to 0} \dfrac{1 - \cos x}{x} = 0$, so if we let $f(0) = 0$ then f will be continuous at 0.

15 We must pick $f(0)$ such that $f(0) = \lim_{x \to 0} \dfrac{x}{|x|}$. But

$$\lim_{x \to 0^-} \frac{x}{|x|} = \lim_{x \to 0^-} \frac{x}{-x} = -1 \neq 1 = \lim_{x \to 0^+} \frac{x}{x} = \lim_{x \to 0^+} \frac{x}{|x|},$$

so $\lim_{x \to 0} \dfrac{x}{|x|}$ does not exist. Therefore, no $f(0)$ can be chosen that will make $f(0) = \lim_{x \to 0} \dfrac{x}{|x|}$ a true statement.

17

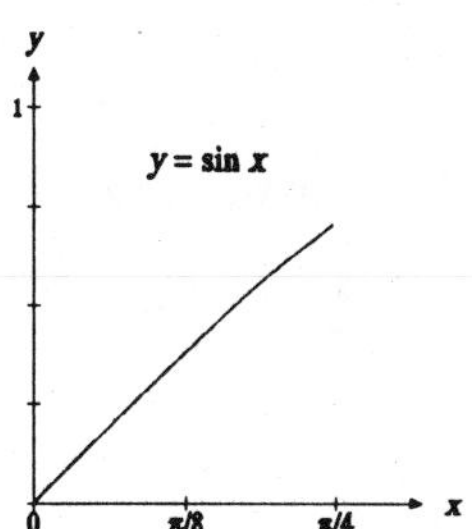

(a) From the graph, the maximum value of $\sin x$ over $[0, \pi/4]$ is $\dfrac{\sqrt{2}}{2}$, and it occurs at $x = \pi/4$.

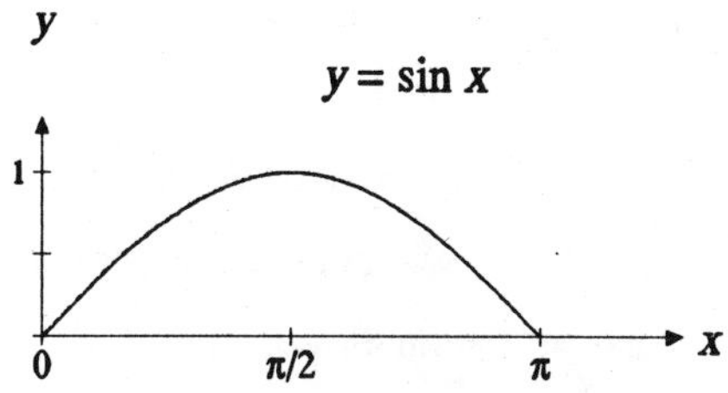

(b) From the graph, the maximum value of $\sin x$ over $[0, \pi]$ is 1, and it occurs at $x = \pi/2$.

19 For $x \neq 0$ we have $5x^2 > 0$ and $x^6 > 0$, hence $1 + 5x^2 + x^6 > 1 > 0$. Thus, $1 + 5x^2 + x^6 \neq 0$ for all x in $[1, 4]$, so $(x^3 + x^4)/(1 + 5x^2 + x^6)$ is continuous throughout at $[1, 4]$.

(a) Yes, by the maximum-value theorem the function attains a maximum in $[1, 4]$.

(b) Yes, by the minimum-value theorem the function attains a minimum in $[1, 4]$.

21 Note that x^3 is continuous for all x.

(a) Since $[2, 4]$ is a closed interval, $f(x) = x^3$ has a maximum value on $[2, 4]$; it occurs at

$x = 4$.

(b) Since $[-3, 5]$ is a closed interval, $f(x) = x^3$ has a maximum value on $[-3, 5]$; it occurs at $x = 5$.

(c) Observe that $x^3 < 6^3 = 216$ for all x in $(1, 6)$, but since x^3 takes on values arbitrarily close to 216 for inputs close to 6, the function does not have a maximum in the open interval.

23

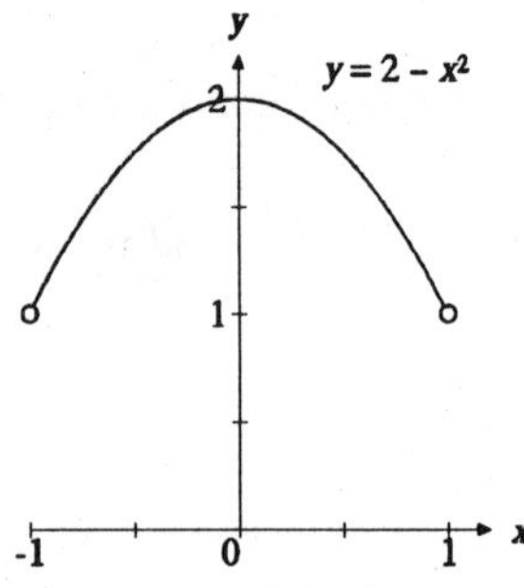

(a) From the graph, y has a maximum at $x = 0$.

(b) As $|x|$ nears 1, y takes on values arbitrarily close to 1 without ever attaining it. Hence, y has no minimum value on $(-1, 1)$.

25 At the endpoints of the interval $[0, 1]$, we have $0^5 + 3 \cdot 0^4 + 0 - 2 = -2$ and $1^5 + 3 \cdot 1^4 + 1 - 2 = 3$. The function $x^5 + 3x^4 + x - 2$ is continuous on $[0, 1]$, so by the intermediate-value theorem it takes on all values between -2 and 3 in the interval. But $-2 < 0 < 3$, so there is a number c such that $c^5 + 3c^4 + c - 2 = 0$ and $0 < c < 1$.

27 $3c + 5 = 10$ when $3c = 5$, so $c = 5/3$. Note that $1 < 5/3 < 2$.

29 Since $\sin x$ is continuous throughout $[\pi/2, 11\pi/2]$ and $f\left(\frac{\pi}{2}\right) = 1 \geq -1 \geq -1 = f\left(\frac{11\pi}{2}\right)$, the initial requirements for the intermediate-value theorem to be valid are satisfied. Note that $\sin x = -1$ when $x = \frac{3\pi}{2} + 2n\pi$, where n is an integer.

Thus, the acceptable values of x in $[\pi/2, 11\pi/2]$ are $\frac{3\pi}{2}, \frac{7\pi}{2}, \frac{11\pi}{2}$.

31 $c^3 - c = 0$ when $c(c^2 - 1) = 0$, so $c = 0, 1, -1$.

33 Yes. The function $x + \sin x$ is continuous; furthermore, $0 + \sin 0 = 0 + 0 = 0$ and $\frac{\pi}{2} + \sin\frac{\pi}{2} = \frac{\pi}{2} + 1$. Now $0 < 1 < \pi/2 + 1$, so by the intermediate-value theorem there is a number c such that $0 < c < \pi/2$ and $c + \sin c = 1$.

35 The function $3x^3 + 11x^2 - 5x$ is continuous; furthermore, $3 \cdot 0^3 + 11 \cdot 0^2 - 5 \cdot 0 = 0$ and $3 \cdot 1^3 + 11 \cdot 1^2 - 5 \cdot 1 = 9$. Now $0 < 2 < 9$, so by the intermediate-value theorem there is a number c such that $0 < c < 1$ and $3c^3 + 11c^2 - 5c = 2$.

37 (a)

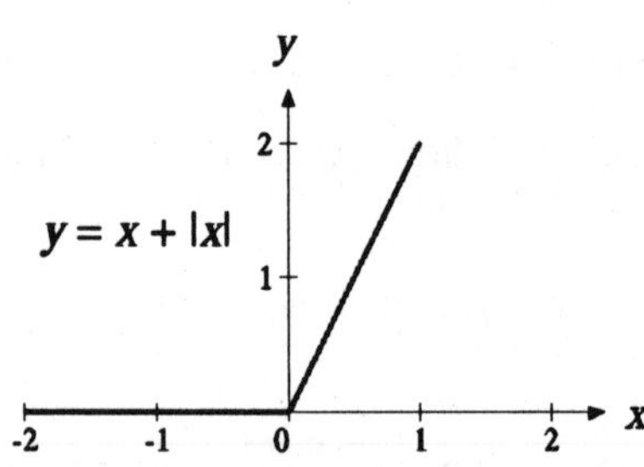

(b) Yes

39 Since $\lim\limits_{x \to -\infty} P(x) = -\infty$ and $\lim\limits_{x \to \infty} P(x) = \infty$, there are real numbers $a < b$ such that $P(a) < 0$ and $P(b) > 0$. By the intermediate-value theorem, $P(r) = 0$ for some r in $[a, b]$.

41 (a) Since L_x originates at $x = a$ and traverses through the convex set K to $x = b$, $A(a)$ has 0 area and $A(b)$ has an area K.

(b) We know that $A(x)$ is continuous on $[a, b]$ because K is a convex set. Since $A(a) = 0 \leq A/2 \leq A = A(b)$, there must be some c in $[a, b]$ such that $A(c) = A/2$.

(c) At the c identified in (b), there is a corresponding cross-sectional length L_c. We know that $A(c) + B(c) = A$, so $B(c) = A/2$ since $A(c) = A/2$. Clearly, L_c is parallel to L and divides K into two pieces of equal area.

43

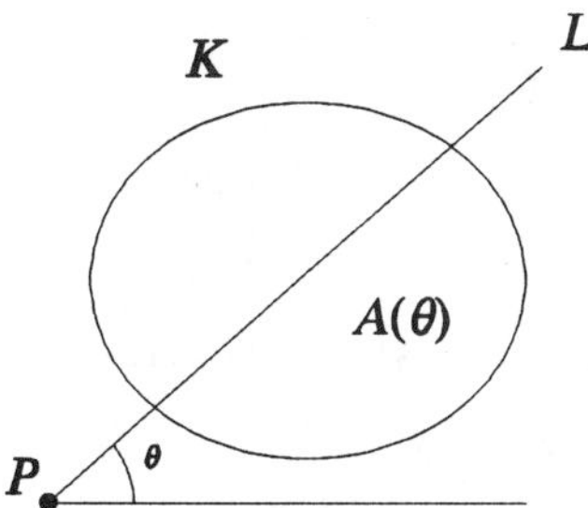

Yes. For $0 \leq \theta \leq \pi$ let L be the line through P with angle θ from the horizontal, and let $A(\theta)$ be the area of K to the right of L. (For $\theta = 0$, use the area below L. For $\theta = \pi$, use the area above L.) Then $A(\theta)$ is a continuous function of θ and $A(0) + A(\pi)$ equals the area of K. So either $A(0) \leq (1/2)(\text{Area of } K) \leq A(\pi)$ or $A(0) \geq (1/2)(\text{Area of } K) \geq A(\pi)$. In either case, the intermediate-value theorem implies that $A(\theta) = (1/2)(\text{Area of } K)$ for some θ in $[0, \pi]$.

45

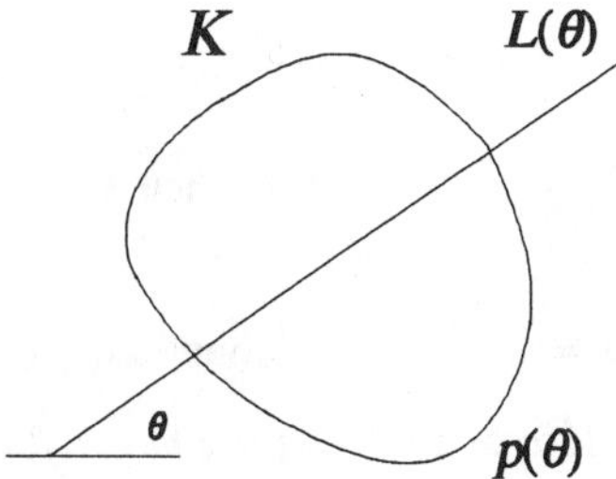

For $0 \leq \theta \leq \pi$, let $L(\theta)$ be the line which makes an angle θ with the horizontal and which divides K into two pieces of equal area. (By Exercise 41, $L(\theta)$ exists for each θ.) Let $p(\theta)$ be the length of that part of the boundary of K that lies to the right of $L(\theta)$ (below for $\theta = 0$, above for $\theta = \pi$). Then $p(\theta)$ is continuous and $p(0) + p(\pi)$ equals the perimeter of K. Reasoning as in Exercise 43 shows that $p(\theta) = (1/2)(\text{Perimeter of } K)$ for some θ in $[0, \pi]$. The desired line is $L(\theta)$ for that θ.

47

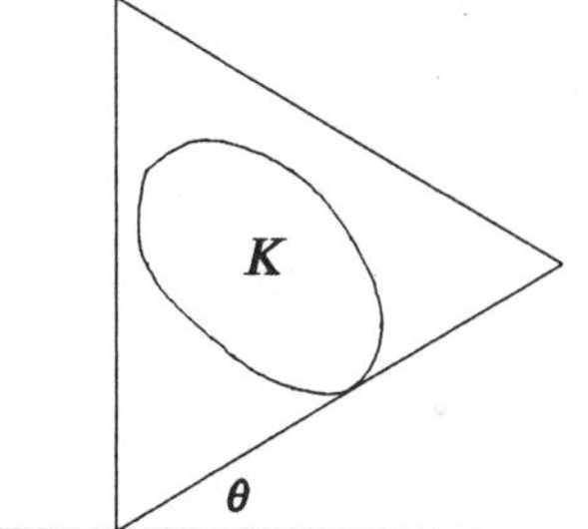

(a) Yes, there is always a circumscribing equilateral triangle. Fix an orientation for the surrounding triangle and choose the smallest such which contains K. You can always do this because each side of an equilateral triangle can be moved parallel to itself so that the resulting object is still an equilateral triangle. Eventually, all three sides of the triangle must touch K. There are infinitely many such triangles since one can be found for each possible orientation (or "tilt") of the triangle.

(b) There are an infinite number of circumscribing rectangles. The reasons are the same as in (a).

(c) Yes. By (b), for any angle θ there is a circumscribing rectangle one of whose sides has angle of inclination θ. Let $f(\theta)$ be the length of such a side and let $g(\theta)$ be the length of the other side. Let $h(\theta) = f(\theta) - g(\theta)$.

Since $f\left(\frac{\pi}{2}\right) = g(0)$ and $g\left(\frac{\pi}{2}\right) = f(0)$, $h\left(\frac{\pi}{2}\right) = -h(0)$. Hence $h(\theta) = 0$ for some θ in $[0, \pi/2]$. The corresponding rectangle is a square.

2.9 Precise Definitions of "$\lim_{x\to\infty} f(x) = \infty$" and "$\lim_{x\to\infty} f(x) = L$"

1 (a) $f(x) > 600$ when $3x > 600$, so $x > 200$. Let $D = 200$.

(b) Let D be any number greater than 200, for example, $D = 201$. Then $x > D = 201$ means that $f(x) = 3x > 3\cdot 201 = 603$, so $f(x) > 600$.

(c) $f(x) > 600$ *only* when $x > 200$, so 200 is the smallest number D such that, for $x > D$, it follows that $f(x) > 600$.

3 (a) $f(x) > 2000$ when $5x > 2000$, so $x > 400$. Let $D = 400$.

(b) $f(x) > 10{,}000$ when $5x > 10{,}000$ so $x > 2000$. Let $D = 2000$.

5 $3x > E$ when $x > E/3$, so let $D = E/3$.

7 $x + 5 > E$ when $x > E - 5$, so let $D = E - 5$.

9 $2x + 4 > E$ when $2x > E - 4$, $x > (1/2)(E - 4)$, so let $D = (1/2)(E - 4)$.

11 $\cos x \geq -1$, $100 \cos x \geq -100$, so we have $4x + 100 \cos x \geq 4x - 100$; thus $4x - 100 > E$ when $4x > E + 100$, $x > (1/4)(E + 100)$. Thus, let $D = (1/4)(E + 100)$.

13 (a) $f(x) > 100$ when $x^2 > 100$, so $\sqrt{x^2} > \sqrt{100} = 10$. Let $D = 10$.

(b) $f(x) > E$ when $x^2 > E$, so $\sqrt{x^2} > \sqrt{E}$. Let $D = \sqrt{E}$.

(c) $f(x) > E$ when $x^2 > E$; but $x^2 \geq 0$ and $E < 0$, so $x^2 > E$ for all x. Thus D can be any number.

(d) From (b) and (c), if E is any number, then if $x > \sqrt{|E|}$, it follows that $f(x) = x^2 > (\sqrt{|E|})^2 = |E| \geq E$.

15 (a) $|f(x) - 3| < 1/10$ when $|3 + 1/x - 3| < 1/10$, $|1/x| < 1/10$, $1 < \frac{|x|}{10}$, $|x| > 10$; let $D = 10$.

(b) Let D be any number greater than 10, for example, $D = 11$. Then $x > D = 11$ implies that $|f(x) - 3| = |3 + 1/x - 3| = |1/x| < 1/11 < 1/10$.

(c) $|f(x) - 3| < 1/10$ only when $|x| > 10$. We cannot use $D \leq -10$, since $1 > D \geq -10$, but $|f(1) - 3| = |3 + 1 - 3| = 1 > 1/10$. Thus 10 is the smallest number D such that, for $x > D$, it follows that $|f(x) - 3| < 1/10$.

(d) Let ϵ be a positive number. $|f(x) - 3| < \epsilon$ when $|3 - 1/x - 3| < \epsilon$, $|1/x| < \epsilon$, $1 < \epsilon|x|$, so let $D = 1/\epsilon$. Then $x > D = 1/\epsilon$ implies that $|f(x) - 3| = |1/x| < \left|\frac{1}{1/\epsilon}\right| = \epsilon$.

17 $|\sin x| \leq 1$ for all x, so $\left|\frac{\sin x}{x}\right| = \frac{|\sin x|}{|x|} \leq \frac{1}{|x|}$ for all $x > 0$. Let $\epsilon > 0$. $|f(x) - 0| < \epsilon$ when $\left|\frac{\sin x}{x} - 0\right| < \epsilon$. $\frac{1}{|x|} < \epsilon$ when $1 < \epsilon|x|$, $|x| > 1/\epsilon$. Let $D = 1/\epsilon$. Having found D, we verify that it works: If $x > D = 1/\epsilon$ we have $|f(x) - 0| =$

$$\left|\frac{\sin x}{x}\right| \le \frac{1}{|x|} < \epsilon.$$

19 Let $\epsilon > 0$. $|f(x) - 0| < \epsilon$ when $|4/x^2 - 0| =$ $4/x^2 < \epsilon$, $4 < \epsilon x^2$, $x^2 > 4/\epsilon$, $|x| > \sqrt{\frac{4}{\epsilon}} = \frac{2}{\sqrt{\epsilon}}$.

Let $D = \frac{2}{\sqrt{\epsilon}}$.

21 Let $\epsilon > 0$. $|f(x) - 0| < \epsilon$ when $\left|\frac{1}{x - 100} - 0\right|$ $= \frac{1}{|x - 100|} < \epsilon$, $1 < |x - 100|\epsilon$, $|x - 100|$ $> 1/\epsilon$. Let $D = 100 + 1/\epsilon$. Having found D, we verify that it works: If $x > D$, we then have $|x - 100| = x - 100 > D - 100 =$ $100 + 1/\epsilon - 100 = 1/\epsilon$, so $|f(x) - 0|$ $= \frac{1}{|x - 100|} < \frac{1}{|1/\epsilon|} = |\epsilon| = \epsilon$.

23 Suppose $\lim_{x\to\infty} \frac{x}{x+1} = \infty$. Let $E = 1$; there must be a D such that $\frac{x}{x+1} > 1$ for all $x > D$. But, for $x > 0$, $\frac{x}{x+1} < \frac{x+1}{x+1} = 1$; hence D does not exist and $\lim_{x\to\infty} \frac{x}{x+1} \ne \infty$.

25 Suppose $\lim_{x\to\infty} 3x = 6$. Let $\epsilon = 3$; there must be a D such that $|3x - 6| < 3$ for $x > D$. But this inequality holds only when $1 < x < 3$.

27

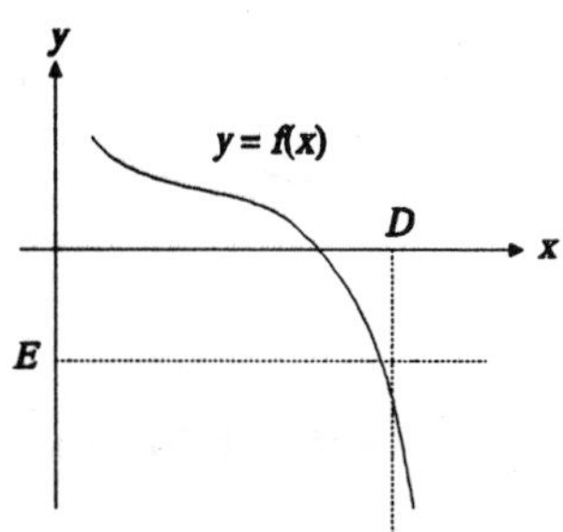

First, there is a number c such that $f(x)$ is defined for all $x > c$. Second, for each number E there is a number D such that for all $x > D$, it is true that $f(x) < E$.

29

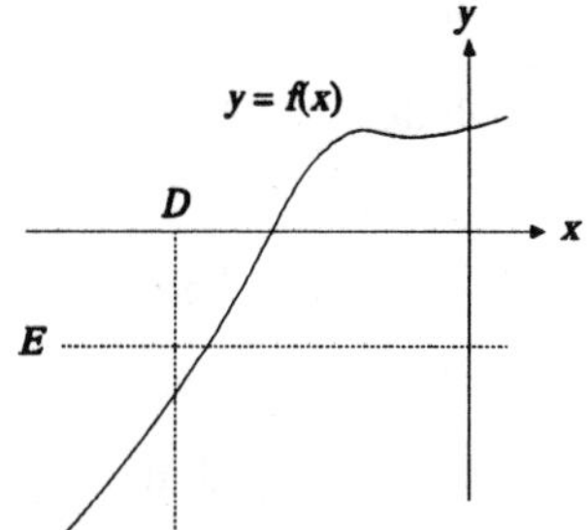

First, there is a number c such that $f(x)$ is defined for all $x < c$. Second, for each number E there is a number D such that $f(x) < E$ for all $x < D$.

30 First, there is a number c such that $f(x)$ is defined for all $x < c$. Second, for each positive number ϵ there is a number D such that, for all $x < D$, it is true that $|f(x) - L| < \epsilon$.

31 (a) First, $f(x)$ is defined for all x beyond some number c; in fact, any real number can be used as c. Given $\epsilon > 0$, let $D = -3$ (or any other real number). For all $x > D$, we have $f(x) = 5$, so $|f(x) - 5| = 0 < \epsilon$. Hence, $\lim_{x\to\infty} f(x) = 5$.

(b) (See Exercise 30, above, for the precise definition.) First, let $c = 0$ (or any other real number); then $f(x)$ is defined for all $x < c$. Next, given $\epsilon > 0$, let $D = 0$ (or any other real number). For all $x < D$, $f(x) = 5$, so $|f(x) - 5| = 0 < \epsilon$. Hence $\lim_{x \to -\infty} f(x) = 5$.

2.10 Precise Definition of "$\lim_{x \to a} f(x) = L$"

1 We have $a = 2$ and $L = 6$. We want a positive δ for which $0 < |x - 2| < \delta$ implies $|3x - 6| < \epsilon$. If we can modify the left-hand side of the second inequality so that it is in terms of $|x - 2|$, we would be done since the right-hand side would be in terms of ϵ and we could use it as the δ we desire. Accordingly, we rewrite $|3x - 6| < \epsilon$ as $3|x - 2| < \epsilon$, hence $|x - 2| < \epsilon/3$. Thus, we want $\delta = \epsilon/3$. To ensure that $0 < |x - 2| < \epsilon/3$ implies $|3x - 6| < \epsilon$, we verify the chain of reasoning by beginning with the assumption that $\delta = \epsilon/3$ and that $|x - 2| < \delta$. Then $|x - 2| < \epsilon/3$, so $3|x - 2| < \epsilon$ and $|3x - 6| < \epsilon$, as required. Our choice $\delta = \epsilon/3$ is correct.

3 Here $a = 1$ and $L = 3$. So we want $0 < |x - 1| < \delta$ to imply that $|(x + 2) - 3| < \epsilon$. Simplifying $|(x + 2) - 3| < \epsilon$, we have $|x - 1| < \epsilon$. Hence for $\delta = \epsilon$, we have $0 < |x - 1| < \epsilon$. Retracing our steps validates our choice for δ.

5 In this case, the challenge is $\epsilon = 0.01$ and $\lim_{x \to 2} f(x) = 2$. So we must find a δ for which $0 < |x - 2| < \delta$ implies $|(x/2 + 1) - 2| < 0.01$. Simplifying $|(x/2 + 1) - 2| < 0.01$, we have $|x/2 - 1| < 0.01$, so $|x - 2| < 0.02$. Thus $\delta = 0.02$ will work since, for $0 < |x - 2| < 0.02$, it follows that $|(x/2 + 1) - 2| < 0.01$.

7 Let $\epsilon > 0$. Then $\left|\frac{x^2}{4} - 0\right| = \frac{|x|^2}{4} < \epsilon$ if and only if $|x| < \sqrt{4\epsilon}$. Thus $\delta = \sqrt{4\epsilon} = 2\sqrt{\epsilon}$ suffices.

9 Let $\epsilon > 0$. Then $|(3x + 5) - 8| = 3|x - 1| < \epsilon$ if and only if $|x - 1| < \epsilon/3$. Thus $\delta = \epsilon/3$ suffices.

11 Let $\delta = 1/5$. If $|x - 2| < \delta$, then $|x^2 - 4| = |x - 2|\cdot|x + 2| < (1/5)|x + 2|$. But the largest $x + 2$ can be is $2 + 1/5 = 9/5$ (remember that x and 2 are within 1/5 of each other), so $(1/5)|x + 2| \leq (1/5)(9/5) = 9/25 < 1$. (In fact, any $\delta \leq \sqrt{5} - 2 \approx 0.236$ will work.)

13 Here, $\epsilon = 0.05$, $a = 1$, and $L = 2$. We must have $|2x^2 - 2| < 0.05 = \epsilon$ for some δ where $0 < |x - 1| < \delta$. Note that $|2x^2 - 2| < 0.05$ implies that $2|x^2 - 1| < 0.05$, so $2|x + 1||x - 1| < 0.05$. Observe that $|x - 1| < \delta$ (that is, x and 1 are within δ of each other), so $|x + 1| < 1 + \delta$. Therefore $|x + 1||x - 1| < (1 + \delta)\delta$ and $2|x + 1||x - 1| < 2\delta(1 + \delta)$. Now, the maximum δ we can choose will satisfy $2\delta(1 + \delta) = 0.05$, so $2\delta^2 + 2\delta - 1/20 = 0$ and $40\delta^2 + 40\delta - 1 = 0$. Let δ be the larger of the two roots, in this case $\delta = \frac{-40 + \sqrt{1600 + 160}}{80} \approx 0.0244$. We can use this value of δ or anything smaller.

15 (a) Suppose $0 < \delta < 1$ and $|x - 3| < \delta$. Then $-\delta < x - 3 < \delta$, so $3 - \delta < x < 3 + \delta$. Since $0 < \delta < 1$, $2 < x < 4$, so it follows that $5 < x + 3 < 7$ and $|x + 3| < 7$. Therefore $|x^2 - 9| = |(x - 3)(x + 3)| =$

$|x + 3|\cdot|x - 3| < 7\delta$.

(b) Let $\epsilon > 0$. If $\epsilon \geq 7$, then let $\delta = 1/2$. If $|x - 3| < \delta$, then by (a), $|x^2 - 9| < 7\delta = 7/2 < \epsilon$. If $\epsilon < 7$, then let $\delta = \epsilon/7$. From (a), if $0 < |x - 3| < \delta$, then $|x^2 - 9| < 7\delta = 7\cdot\epsilon/7 = \epsilon$. Thus, for any $\epsilon > 0$, there is a positive number δ such that if $0 < |x - 3| < \delta$, it follows that $|x^2 - 9| < \epsilon$. Thus $\lim_{x\to 3} x^2 = 9$.

17 (a) Suppose $0 < \delta < 1$ and $|x - 3| < \delta$. Then $-\delta < x - 3 < \delta$, so $3 - \delta < x < 3 + \delta$. Since $0 < \delta < 1$ we have $2 < x < 4$, so $10 < x + 8 < 12$ and $|x + 8| < 12$. Thus $|x^2 + 5x - 24| = |(x + 8)(x - 3)| = |x + 8|\cdot|x - 3| < 12\delta$.

(b) Let $\epsilon > 0$. If $\epsilon \geq 12$, then let $\delta = 1/2$. If $|x - 3| < \delta$, then, by (a), $|x^2 + 5x - 24| < 12\delta = 6 < \epsilon$. If $\epsilon < 12$, then let $\delta = \epsilon/12$. From (a), if $0 < |x - 3| < \delta$, then $|x^2 + 5x - 24| < 12\delta = 12\cdot\epsilon/12 = \epsilon$. Thus, for any $\epsilon > 0$, there is a positive number δ such that if $0 < |x - 3| < \delta$, it follows that $|x^2 + 5x - 24| < \epsilon$, so $\lim_{x\to 3} (x^2 + 5x) = 24$.

19 First, there is a number b, $a < b$, such that $f(x)$ is defined for all x in (a, b). Second, for each positive number ϵ there is a positive number δ such that for all x that satisfy $0 < x - a < \delta$, it is true that $|f(x) - L| < \epsilon$.

21 First, there exist numbers $c < a$ and $b > a$ such that $f(x)$ is defined in the intervals (c, a) and (a, b). Second, for any number E there is a number δ such that $f(x) > E$ for $0 < |x - a| < \delta$.

23 First, there is a number $b > a$ such that $f(x)$ is defined in the interval (a, b). Second, for each number E there is a number δ such that $f(x) > E$ for $a < x < a + \delta$.

25 (a) $|9x^2 - 0| = |9x^2| = 9x^2 < \dfrac{1}{100}$ when $x^2 < \dfrac{1}{900}$, $|x| = \sqrt{x^2} < \sqrt{\dfrac{1}{900}} = \dfrac{1}{30}$. Let $\delta = 1/30$.

(b) $|9x^2 - 0| = 9x^2 < \epsilon$ when $x^2 < \epsilon/9$, $|x| = \sqrt{x^2} < \sqrt{\epsilon/9} = \dfrac{\sqrt{\epsilon}}{3}$. Let $\delta = \sqrt{\epsilon}/3$.

(c) From (b), if $\epsilon > 0$, then for $0 < |x - 0| < \dfrac{\sqrt{\epsilon}}{3}$, it follows that $|9x^2 - 0| < \epsilon$, so $\lim_{x\to 0} 9x^2 = 0$.

27 Let $\epsilon = 1/2$. If $\lim_{x\to 2} 3x = 5$, then there is a positive number δ such that if $0 < |x - 2| < \delta$, then $|3x - 5| < 1/2$; thus $-1/2 < 3x - 5 < 1/2$ or, equivalently, $5 - 1/2 < 3x < 5 + 1/2$; that is, $9/2 < 3x < 11/2$. For each x satisfying $0 < |x - 2| < 1/6$, we have $x > 11/6$, $3x > 11/2$, and $|3x - 5| > 1/2$. If $0 < \delta < 1/6$, we therefore have, for each x satisfying $0 < |x - 2| < \delta < 1/6$, $|3x - 5| > 1/2$. Thus, there is no positive number δ such that if $0 < |x - 2| < \delta$, then $|3x - 5| < 1/2 = \epsilon$. Therefore $\lim_{x\to 2} 3x$ is *not* equal to 5.

2.S Guide Quiz

1 (a) Let $f(x) = 0$, $g(x) = x$, and $a = 0$.

(b) Let $f(x) = x$, $g(x) = x$, and $a = 0$.

(c) Let $f(x) = x$, $g(x) = x^3$, and $a = 0$.

2 $f(x) = \frac{x^2 - 4}{x^2 - 1} = \frac{(x + 2)(x - 2)}{(x + 1)(x - 1)}$, so $f(x) = 0$ at $x = \pm 2$ and blows up at $x = \pm 1$. Also $\lim_{x \to -1^-} f(x) = -\infty$, $\lim_{x \to -1^+} f(x) = \infty$, $\lim_{x \to 1^-} f(x) = \infty$, and $\lim_{x \to 1^+} f(x) = -\infty$. For large $|x|$, $\lim_{x \to -\infty} f(x) = 1$ and $\lim_{x \to \infty} f(x) = 1$. This all makes sense because $f(x) = f(-x)$. Note too that $f(0) = 4$.

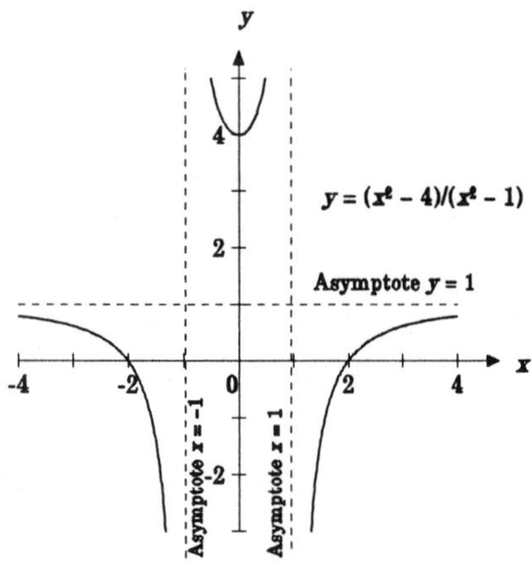

3 A polynomial of odd degree tends to $\pm\infty$ for negative x of large absolute value and $\mp\infty$ for large x. Since $f(-\infty) = \pm\infty$ and $f(\infty) = \mp\infty$, by the intermediate-value theorem there must be some m, $-\infty < m < \infty$, for which $f(m) = 0$, since $-\infty < 0 < \infty$.

4 (a) $\lim_{h \to 0} \frac{\sin h}{h} = 1$

(b) $5° = \pi/36 \approx 0.0872665$, so $\sin 5° = \sin \pi/36 \approx 0.0872665$ since $\sin h \approx h$ for h small.

5 (a) $\lim_{h \to 0} \frac{1 - \cos h}{h} = 0$

(b) $\lim_{h \to 0} \frac{1 - \cos h}{h^2} = \frac{1}{2}$ by Exercise 16 of Sec. 2.7.

(c) Since $\cos\theta \approx 1 - \frac{\theta^2}{2}$ for small θ (see Exercise 17 of Sec. 2.7), $\cos 0.2 \approx 1 - \frac{(0.2)^2}{2} = 1 - \frac{0.04}{2} = 1 - 0.02 = 0.98$.

(d) $20° = \pi/9 \approx 0.349066$, and $\cos 20° = \cos \pi/9 \approx 1 - \frac{(\pi/9)^2}{2} = 1 - \frac{\pi^2}{162} \approx 0.9390765$.

(e) The calculator says $\cos 20° \approx 0.9396926$.

6 An indeterminate limit is a limit that cannot be calculated unless it is manipulated in some way, for example, $\lim_{x \to 1} \frac{x^2 - 1}{x - 1}$, which yields 0/0 if one tries to evaluate directly. A determinate limit is one that can be determined simply by plugging in the appropriate value of x, for example, $\lim_{x \to 0} \frac{x^2 - 1}{x - 1} = \frac{0^2 - 1}{0 - 1} = \frac{-1}{-1} = 1$.

7 (a)

x	100	1000	10000	100000
$\sqrt{x^2 + 4x} - x$	1.9804	1.9980	1.9998	2.0000

(b) $\lim_{x \to \infty} \sqrt{x^2 + 4x} - x$

$$= \lim_{x \to \infty} \frac{(\sqrt{x^2 + 4x} - x)(\sqrt{x^2 + 4x} + x)}{\sqrt{x^2 + 4x} + x}$$

$$= \lim_{x \to \infty} \frac{x^2 + 4x - x^2}{x(\sqrt{1 + 4/x} + 1)} = \lim_{x \to \infty} \frac{4}{\sqrt{1 + 4/x} + 1}$$

$$= \frac{4}{1 + 1} = 2.$$

(c) The domain is all x such that $x^2 + 4x \geq 0$, or $x(x + 4) \geq 0$, so $x \geq 0$ or $x \leq -4$.

(d)

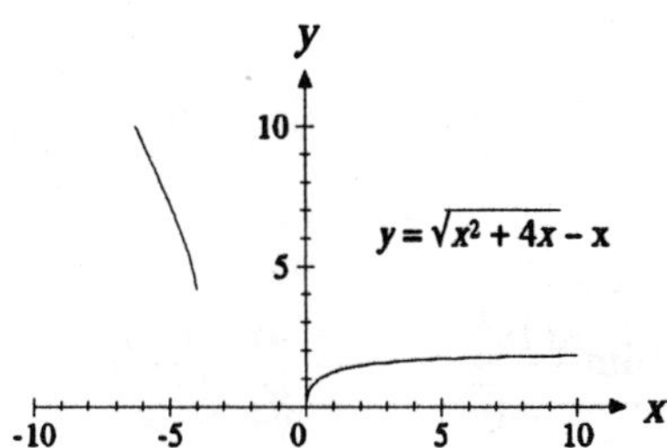

8 (a) $\dfrac{f(x + h) - f(x)}{h} = \dfrac{(x + h)^3 - x^3}{h}$

$$= \frac{x^3 + 3x^2h + 3xh^2 + h^3 - x^3}{h}$$

$$= \frac{h(3x^2 + 3xh + h^2)}{h} = 3x^2 + 3xh + h^2$$

(b) $\dfrac{f(x + h) - f(x)}{h} = \dfrac{\dfrac{1}{x + h} - \dfrac{1}{x}}{h} =$

$$\frac{\dfrac{x - (x + h)}{x(x + h)}}{h} = \frac{\dfrac{-h}{x(x + h)}}{h} = \frac{-1}{x(x + h)}$$

9 (a) 1. $f(1/2)$ is defined.

2. $\lim_{x \to 1/2} f(x)$ exists.

3. $\lim_{x \to 1/2} f(x) = f(1/2)$

(b) 1. $f(0)$ is defined.

2. $\lim_{x \to 0^+} f(x)$ exists.

3. $\lim_{x \to 0^+} f(x) = f(0)$

(c) 1. $f(2)$ is defined.

2. $\lim_{x \to 2^-} f(x)$ exists.

3. $\lim_{x \to 2^-} f(x) = f(2)$

10 (a) $\lim_{x \to 2^-} f(x) = 2 = \lim_{x \to 2^+} f(x)$, so $\lim_{x \to 2} f(x)$ exists and is equal to 2.

(b) $\lim_{x \to 2} f(x) = 2 \neq 1 = f(2)$, so $f(x)$ is not continuous at 2.

(c) $\lim_{x \to 3^-} f(x) = 1 \neq 2 = \lim_{x \to 3^+} f(x)$, so $\lim_{x \to 3} f(x)$ does not exist.

(d) $\lim_{x \to 5^-} f(x) = 1$, since $f(x) \to 1$ as $x \to 5^-$; however, $\lim_{x \to 5^+} f(x)$ does not exist because this involves points outside the domain of f. Hence $\lim_{x \to 5} f(x)$ does not exist.

(e) The function is not continuous on [1, 5] because it is not continuous at $x = 2$, 3, or 5.

(f) The function is continuous on (3, 5) because it is continuous at every point in (3, 5).

11 (a) $\lim_{x \to 1} (x^2 + 5x) = 1^2 + 5 \cdot 1 = 1 + 5 = 6$

(b) $\lim_{x \to \infty} \dfrac{3x^4 - 100x + 3}{5x^4 + 7x - 1} = \lim_{x \to \infty} \dfrac{3x^4}{5x^4} = \lim_{x \to \infty} \dfrac{3}{5}$

$$= \frac{3}{5}$$

(c) $\lim_{x\to 0} \frac{3x^4 - 100x + 3}{5x^4 + 7x - 1} = \frac{3}{-1} = -3$, since, as $x \to 0$, $3x^4 - 100x + 3 \to 3$ and $5x^4 + 7x - 1 \to -1$.

(d) $\lim_{x\to -\infty} \frac{500x^3 - x^2 - 5}{x^4 + x} = \lim_{x\to -\infty} \frac{500x^3}{x^4} =$

$\lim_{x\to -\infty} \frac{500}{x} = 0$

(e) $\lim_{x\to 0} \frac{\sin 3x}{6x} = \lim_{x\to 0} \frac{\sin 3x}{3x} . \frac{3x}{6x} =$

$\lim_{3x\to 0} \frac{\sin 3x}{3x} \cdot \frac{1}{2} = 1 \cdot \frac{1}{2} = \frac{1}{2}$

(f) $\lim_{x\to -\infty} \frac{-6x^5 + 4x}{x^2 + x + 5} = \lim_{x\to -\infty} \frac{-6x^5}{x^2} = \lim_{x\to -\infty} -6x^3$

$= -6 \lim_{x\to -\infty} x^3 = \infty$, since $\lim_{x\to -\infty} x^3 = -\infty$.

(g) $\lim_{x\to -\infty} 2^{-x} = 0$

(h) $\lim_{x\to 0} \frac{x^3 + 8}{x + 2} = \frac{8}{2} = 4$, since, as $x \to 0$, $x^3 + 8 \to 8$ and $x + 2 \to 2$.

(i) $\lim_{x\to -2} \frac{x^3 + 8}{x + 2} = \lim_{x\to -2} \frac{(x + 2)(x^2 - 2x + 4)}{x + 2}$

$= \lim_{x\to -2} (x^2 - 2x + 4) = (-2)^2 - 2(-2) + 4$

$= 12$

(j) $\lim_{x\to 0} \sin \frac{1}{x}$ does not exist, since $1/x \to \infty$ as $x \to 0^+$, $1/x \to -\infty$ as $x \to 0^-$, and $\sin x$ does not approach a limit in either case. It oscillates.

(k) $\lim_{x\to \infty} \sin x$ does not exist, since $\sin x$ continually runs through all values between -1 and 1 as $x \to \infty$. It oscillates.

(l) $\lim_{x\to \infty} \frac{1 + 3\cos x}{x^2} = 0$, since $|1 + 3\cos x| \le 1 + 3|\cos x| \le 4$ for all x, so $\left|\frac{1 + 3\cos x}{x^2}\right|$

$= \frac{|1 + 3\cos x|}{x^2} \le \frac{4}{x^2} \to 0$ as $x \to \infty$.

(m) $\lim_{x\to \infty} \left(\sqrt{4x^2 + 5x} - \sqrt{4x^2 + x}\right) =$

$\lim_{x\to \infty} \left(\sqrt{4x^2+5x} - \sqrt{4x^2+x}\right)\frac{\sqrt{4x^2+5x} + \sqrt{4x^2+x}}{\sqrt{4x^2+5x} + \sqrt{4x^2+x}}$

$= \lim_{x\to \infty} \frac{(4x^2 + 5x) - (4x^2 + x)}{\sqrt{4x^2 + 5x} + \sqrt{4x^2 + x}}$

$= \lim_{x\to \infty} \frac{4x}{\sqrt{x^2(4 + 5/x)} + \sqrt{x^2(4 + 1/x)}}$

$= \lim_{x\to \infty} \frac{4x}{x\sqrt{4 + 5/x} + x\sqrt{4 + 1/x}}$

$= \lim_{x\to \infty} \frac{4}{\sqrt{4 + 5/x} + \sqrt{4 + 1/x}} = \frac{4}{\sqrt{4} + \sqrt{4}}$

$= \frac{4}{4} = 1$

(n) $\lim_{x\to 16} \frac{\sqrt{x} - 4}{x - 16} = \lim_{x\to 16} \frac{\sqrt{x} - 4}{(\sqrt{x} - 4)(\sqrt{x} + 4)} =$

$\lim_{x\to 16} \frac{1}{\sqrt{x} + 4} = \frac{1}{\sqrt{16} + 4} = \frac{1}{4 + 4} = \frac{1}{8}$

12 (a) $\lim_{x\to a} f(x)g(x) = \lim_{x\to a} f(x) \lim_{x\to a} g(x) = 3\cdot 4 = 12$

(b) $\lim_{x\to a} f(x)/g(x) = \left(\lim_{x\to a} f(x)\right)/\left(\lim_{x\to a} g(x)\right) = 3/4$

$3 + 4 = 7$

(d) We cannot say anything about $\lim_{x \to a} \frac{f(x) - 3}{g(x) - 4}$.

It may have any finite value, or may not exist.

(e) $\lim_{x \to a} (f(x) - 3)^{g(x)}$ either equals 0 or does not exist. It is 0 if $f(x) \geq 3$ for all x near a. Otherwise it does not exist, since $(f(x) - 3)^{g(x)}$ is not defined when $f(x) < 3$.

13 (a) $\lim_{x \to a} (f(x) + g(x)) = \infty$, since $f(x) \to 0$ and $g(x) \to \infty$ as $x \to a$.

(b) $\lim_{x \to a} \frac{f(x)}{g(x)} = 0$, since $f(x) \to 0$ and $\frac{1}{g(x)} \to 0$ as $x \to a$.

(c) $\lim_{x \to a} f(x)^{g(x)}$ either equals 0 or does not exist.

(d) $\lim_{x \to a} (2 + f(x))^{g(x)} = \infty$, since $2 + f(x) \to 2$ and $g(x) \to \infty$ as $x \to a$.

(e) We cannot say anything about $\lim_{x \to a} f(x)g(x)$. It may have any finite value, or it may not exist.

16 (a) The box has height x, length $15 - 2x$, and width $10 - 2x$. Thus, the volume is $V(x) = x(15 - 2x)(10 - 2x)$.

(b) The domain of V is $0 \leq x \leq 5$, since $x > 5$ makes no sense (V would be negative).

17 Let $u = \sqrt{x}$. Let $v = \cos u$. Let $w = v^2$. Then $f(x) = w$.

18 The arc length of the sector is $r\theta = 3\theta$, so the perimeter is $3 + 3 + 3\theta = 3(2 + \theta) = P(\theta)$.

2.S Review Exercises

1 (a) Area of sector $= (1/2)\theta r^2 = (1/2)\pi/3 \cdot 5^2 = \frac{25\pi}{6}$

(b) Arc length $= r\theta = 5 \cdot \pi/3 = \frac{5\pi}{3}$

3 $x^{1/5}$. Any real number has a fifth root, so the domain is all real numbers. Any real number is obtainable as the fifth root of its fifth power, so the range is all real numbers.

5 $\cos x$. The domain is all real numbers. Since $-1 \leq \cos x \leq 1$ for all x, the range is $[-1, 1]$.

7 $1/\sqrt{x + 1}$. This may be computed only when $x + 1 > 0$, that is, the domain is $x > -1$. For large values of x the quantity $1/\sqrt{x + 1}$ becomes very small, though it never reaches zero. Near -1 the quantity grows arbitrarily large. Hence the range is all positive real numbers.

9 $f(x) = x^2$; $f(2 + 0.1) - f(2) = (2.1)^2 - (2)^2 = 4.41 - 4 = 0.41$

11 $f(x) = \frac{1}{x + 1}$; $\frac{f(a + h) - f(a)}{h} =$

$$\frac{\frac{1}{a + h + 1} - \frac{1}{a + 1}}{h} = \frac{1}{h} \cdot \frac{a + 1 - (a + h + 1)}{(a + h + 1)(a + 1)}$$

$$= \frac{-1}{(a + h + 1)(a + 1)}$$

13 $f(x) = x^3 - 3x - 2$; $\frac{f(u) - f(v)}{u - v} =$

$$\frac{(u^3 - 3u - 2) - (v^3 - 3v - 2)}{u - v}$$

$$= \frac{u^3 - v^3 - 3u + 3v}{u - v}$$

$$= \frac{(u - v)(u^2 + uv + v^2) - 3(u - v)}{u - v}$$

$$= u^2 + uv + v^2 - 3$$

15

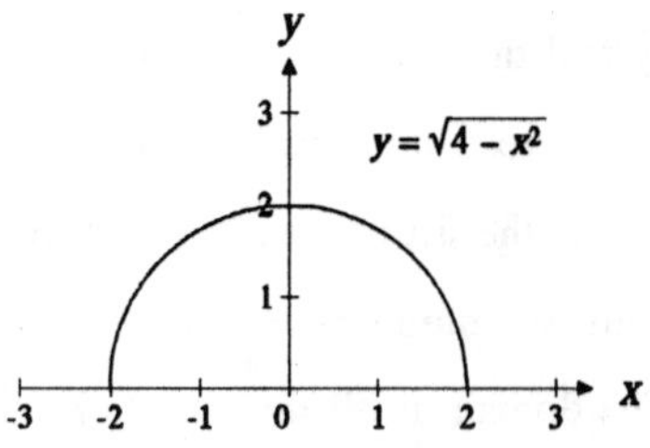

17

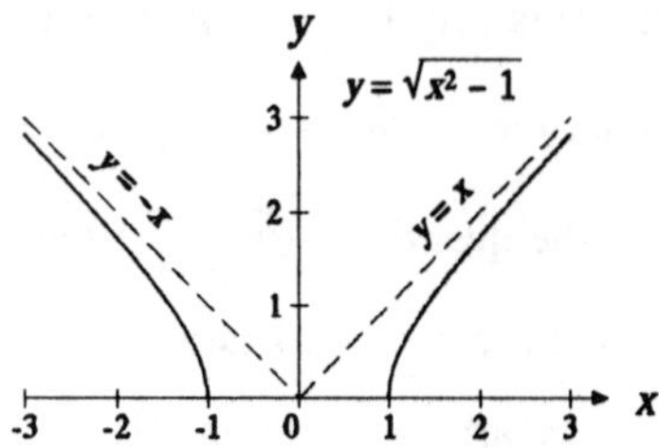

19

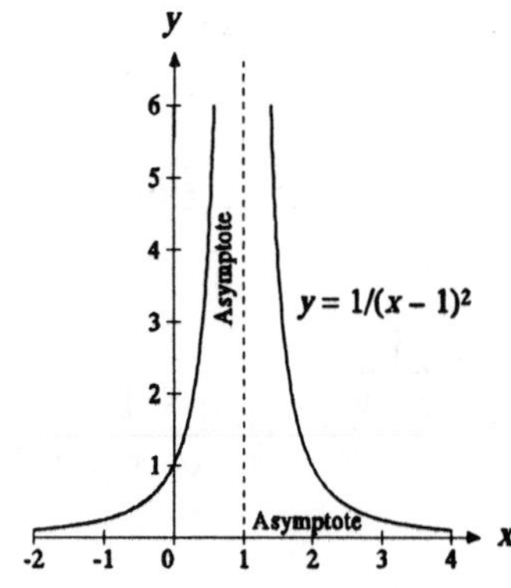

21

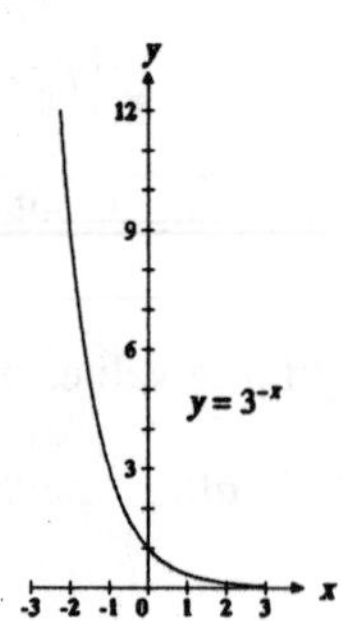

23

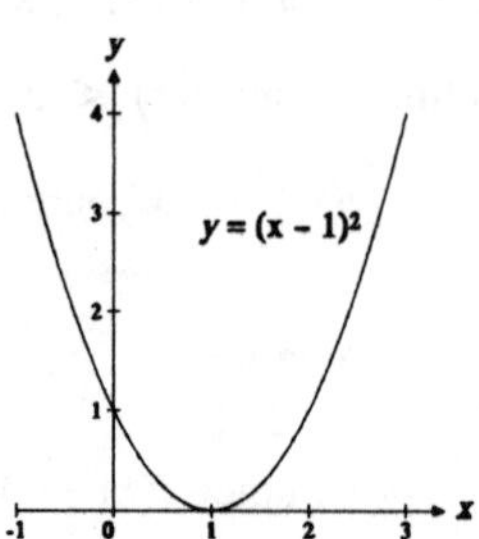

25 (a) $f(x) = 2^x$; $f(b + 1) = 2^{b+1} \neq 3 \cdot 2^b = 3f(b)$, so $f(b + 1) \neq 3f(b)$.

(b) $f(x) = 3^x$; $f(b + 1) = 3^{b+1} = 3 \cdot 3^b = 3f(b)$.

(c) $f(x) = \dfrac{3^x}{2}$; $f(b + 1) = \dfrac{3^{b+1}}{2} = 3 \cdot \dfrac{3^b}{2} = 3f(b)$.

(d) $f(x) = x/3$; $f(b + 1) = \dfrac{b + 1}{3}$ and $3f(b) = 3 \cdot b/3 = b$, so $f(b + 1) \neq 3f(b)$.

27 (a)

x	0.001	0.01	0.1	1	2	10	100
$(1 + x)^{1/x}$	2.717	2.705	2.594	2	1.732	1.271	1.047

(b)

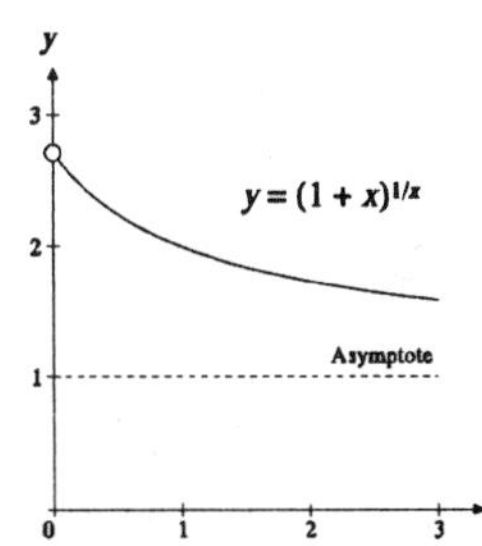

29 (a) Let the other side have length y inches. Then $2x + 2y = 100$, so $y = 50 - x$ and $f(x) = xy = x(50 - x)$.

(b) x and y must both be positive, so $0 < x < 50$. The domain is $(0, 50)$.

31 We must solve $\sin x - 3\cos x = 0$ to see if $\sin x = 3\cos x$ for some x. Let $f(x) = \sin x - 3\cos x$. Then $f(3\pi/4) > 0$ and $f(7\pi/4) < 0$. So by the intermediate-value theorem, there is some m in $[3\pi/4, 7\pi/4]$ such that $f(m) = 0$, and $\sin x$ does equal $3\cos x$ for at least one x.

33 $\lim_{x\to1} \dfrac{x^3+1}{x^2+1} = \dfrac{1^3+1}{1^2+1} = \dfrac{2}{2} = 1$

35 $\lim_{x\to2} \dfrac{x^4-16}{x^3-8} = \lim_{x\to2} \dfrac{(x-2)(x^3+2x^2+4x+8)}{(x-2)(x^2+2x+4)}$

$= \lim_{x\to2} \dfrac{x^3+2x^2+4x+8}{x^2+2x+4} = \dfrac{8+8+8+8}{4+4+4}$

$= \dfrac{32}{12} = \dfrac{8}{3}$

37 $\lim_{x\to\infty} \dfrac{x^7-x^2+1}{2x^7+x^3+300} = \lim_{x\to\infty} \dfrac{x^7}{2x^7} = \lim_{x\to\infty} \dfrac{1}{2} = \dfrac{1}{2}$

39 $\lim_{x\to-\infty} \dfrac{x^3+1}{x^2+1} = \lim_{x\to-\infty} \dfrac{x^3}{x^2} = \lim_{x\to-\infty} x = -\infty$

41 $\lim_{x\to4} \dfrac{\sqrt{x}-2}{x-4} = \lim_{x\to4} \dfrac{\sqrt{x}-2}{(\sqrt{x}-2)(\sqrt{x}+2)}$

$= \lim_{x\to4} \dfrac{1}{\sqrt{x}+2} = \dfrac{1}{\sqrt{4}+2} = \dfrac{1}{4}$

43 $\lim_{x\to\infty} \left(\sqrt{x^2+2x+3} - \sqrt{x^2-2x+3}\right)$

$= \lim_{x\to\infty} \dfrac{\left(\sqrt{x^2+2x+3}-\sqrt{x^2-2x+3}\right)\left(\sqrt{x^2+2x+3}+\sqrt{x^2-2x+3}\right)}{\sqrt{x^2+2x+3}+\sqrt{x^2-2x+3}}$

$= \lim_{x\to\infty} \dfrac{(x^2+2x+3)-(x^2-2x+3)}{\sqrt{x^2+2x+3}+\sqrt{x^2-2x+3}}$

$= \lim_{x\to\infty} \dfrac{4x}{\sqrt{x^2+2x+3}+\sqrt{x^2-2x+3}}$

$= \lim_{x\to\infty} \dfrac{4}{\sqrt{1+2/x+3/x^2}+\sqrt{1-2/x+3/x^2}}$

$= \dfrac{4}{\sqrt{1}+\sqrt{1}} = \dfrac{4}{2} = 2$

45 $\lim_{x\to4^+} \dfrac{1}{x-4} = \infty$

47 $\lim_{x\to3^-} [2x] = 5$, since as $x \to 3^-$, $2x \to 6$, but $2x < 6$.

49 $\lim_{x\to0^+} 2^{1/x} = \infty$, since $1/x \to \infty$ as $x \to 0^+$.

51 $\lim_{x\to\infty} 2^{1/x} = 1$, since $1/x \to 0$ as $x \to \infty$.

53 $\lim_{x\to\infty} \dfrac{(x+1)(x+2)}{(x+3)(x+4)} = \lim_{x\to\infty} \dfrac{x^2+3x+2}{x^2+7x+12} =$

$\lim_{x\to\infty} \dfrac{x^2}{x^2} = \lim_{x\to\infty} 1 = 1$

55 $\lim_{x\to\pi/2} \dfrac{\cos x}{1+\sin x} = \dfrac{0}{1+1} = 0$, since $\cos x \to 0$ as $\sin x \to 1$ as $x \to \pi/2$.

57 $\lim_{x\to0} \dfrac{\sin x}{3x} = \dfrac{1}{3} \lim_{x\to0} \dfrac{\sin x}{x} = \dfrac{1}{3}\cdot 1 = \dfrac{1}{3}$

59 $\lim_{x\to\pi/2^+} \cos x = \cos\frac{\pi}{2} = 0$

61 $\lim_{x\to 0^-} \sin x = \sin 0 = 0$

63 $\lim_{x\to\infty} \sin\frac{1}{x} = 0$, since $1/x \to 0$ as $x \to \infty$.

65 $\lim_{x\to\pi/4} x^2 \cos x = \left(\frac{\pi}{4}\right)^2 \cos\frac{\pi}{4} = \frac{\pi^2}{16}\cdot\frac{\sqrt{2}}{2} = \frac{\pi^2}{16\sqrt{2}}$

67 $\lim_{\theta\to\infty} (\cos^2\theta + \sin^2\theta) = \lim_{\theta\to\infty} 1 = 1$, since $\cos^2\theta + \sin^2\theta = 1$ for all θ.

69 Let $f(x) = 5x^2$, $g(x) = x^2$. Then $\lim_{x\to 0}\frac{f(x)}{g(x)} = \lim_{x\to 0}\frac{5x^2}{x^2} = \lim_{x\to 0} 5 = 5$.

71 Let $f(x) = 1/x$, $g(x) = x^2$. Then $\lim_{x\to\infty} f(x)g(x) = \lim_{x\to\infty} x = \infty$.

73 Let $f(x) = 2x$ and $g(x) = x$. Then $\lim_{x\to\infty} (f(x) - g(x)) = \lim_{x\to\infty} x = \infty$.

75 (a) Yes, $x + \sin x$ is continuous on $[0, 100]$, so by the maximum-value theorem it has a maximum value on the closed interval.

(b) No, $x + \sin x$ does not have a maximum on $[0, \infty)$ since when $x = 2n\pi$, $x + \sin x = 2n\pi$, which becomes arbitrarily large as n increases.

77 (a) $\frac{1}{1+x^2} \le \frac{1}{1+0^2} = 1$ for all x in $(-1, 1)$, so it has a maximum value when $x = 0$.

(b) $\frac{1}{1+x^2} > \frac{1}{1+1^2} = \frac{1}{2}$ for all x in $(-1, 1)$, but $\frac{1}{1+x^2}$ takes on values arbitrarily close to 1/2 for inputs close to 1 and -1. Thus $\frac{1}{1+x^2}$ has no minimum on $(-1, 1)$.

79 $x^5 = 2^x$ when $x^5 - 2^x = 0$. Let $f(x) = x^5 - 2^x$. Note that $f(x)$ is continuous for all x.

(a) $f(0) = 0 - 1 = -1$, $f(2) = 32 - 4 = 28$, and $-1 < 0 < 28$, so by the intermediate-value theorem there is a number c such that $-1 < c < 2$ and $c^5 - 2^c = 0$, so $c^5 = 2^c$.

(b) $f(2) = 28$, $f(23) = -1{,}952{,}265$, and $-1{,}952{,}265 < 0 < 28$, so by the intermediate-value theorem there is a number c such that $2 < c < 23$ and $c^5 - 2^c = 0$, so $c^5 = 2^c$.

81 (a) $(\sin x - \cos x)^2 = \sin^2 x + \cos^2 x - 2\sin x\cos x = 1 - \sin 2x$ is maximal when $\sin 2x$ is minimal. The minimum of $\sin 2x$ is -1 and occurs for $2x = -\pi/2 + 2\pi n$, where n is any integer. Hence the maximum of $(\sin x - \cos x)^2$ is $1 - (-1) = 2$.

(b) Similarly, the maximum of $\sin 2x$ is 1 (for $2x = \pi/2 + 2n\pi$), so the minimum value of $(\sin x - \cos x)^2$ is $1 - 1 = 0$.

83 $2^x = 4$ when $x = 2$; $1 < 2 < 8$.

85 $\tan x = 1$ when $x = \pi/4$; $0 < \pi/4 < \pi/3$.

87 No, as we saw in Exercise 27 of Sec. 2.7, there is no way to define $f(x)$ at $x = 0$ to make the function continuous.

89 Let g be defined by $g(x) = f(x) - x$. We wish to find a number c in $[0, 1]$ such that $f(c) = c$, so $g(c) = f(c) - c = 0$. $f(x)$ is a continuous function

on [0, 1], and so is x, so $g(x)$ is a continuous function on [0, 1]. The number 1 is in [0, 1], so $f(1)$ is in [0, 1]; thus $f(1) \leq 1$ and $g(1) = f(1) - 1 \leq 0$. The number 0 is in [0, 1], so $f(0)$ is in [0, 1]; thus $f(0) \geq 0$, and $g(0) - 0 \geq 0$. We have $g(1) \leq 0 \leq g(0)$, so by the intermediate-value theorem there is a number c such that $0 \leq c \leq 1$ and $g(c) = 0$.

91 Exploration exercises are in the *Instructor's Manual.*

3 The Derivative

3.1 Four Problems with One Theme

1 (a) We know $\theta = \tan^{-1} m$, where m is the slope of the line. Thus, $\theta = \tan^{-1} 1 = \pi/4$ or $\theta = 45°$.

(b) Since $\theta = \tan^{-1} m$, we have $\theta = \tan^{-1} 2 \approx 63.4°$ or $\theta \approx 1.11$ radians.

5 Consider the line through $P = (3, 3^2)$ and $Q = (3 + h, (3 + h)^2)$, where h is near 0. Its slope is

$$\frac{(3+h)^2 - 3^2}{(3+h) - 3} = \frac{9 + 6h + h^2 - 9}{h} = 6 + h.$$

As $h \to 0$, $6 + h \to 6$, which is the slope of the tangent line.

7 Consider the line through $P = (-2, (-2)^2)$ and $Q = (-2 + h, (-2 + h)^2)$, where h is near 0. Its slope is $\frac{(-2+h)^2 - (-2)^2}{(-2+h) - (-2)} = \frac{4 - 4h + h^2 - 4}{h}$ $= -4 + h$. As $h \to 0$, $-4 + h \to -4$, which is the slope of the tangent line.

9 Consider the line through $P = (2, 2^3)$ and $Q = (2 + h, (2 + h)^3)$, where h is near 0. Its slope is

$$\frac{(2+h)^3 - 2^3}{(2+h) - 2} = \frac{8 + 12h + 6h^2 + h^3 - 8}{h} =$$

$12 + 6h + h^2$. As $h \to 0$, $12 + 6h + h^2 \to 12$, which is the slope of the tangent line.

11 (a) Consider the line through $P = (0, 0)$ and $Q = (h, h^2)$, where h is near 0. The line has slope $\frac{h^2 - 0}{h - 0} = \frac{h^2}{h} = h$. As $h \to 0$, the slope approaches 0, which is the slope of the tangent line.

13 In the time interval from 3 to $3 + h$ seconds, where $h > 0$, the rock travels $16(3 + h)^2 - 16\cdot 3^2$ feet. The average velocity during this period is

$$\frac{16(3+h)^2 - 16\cdot 3^2}{(3+h) - 3} = 16\,\frac{9 + 6h + h^2 - 9}{h} =$$

$16(6 + h)$ ft/sec. As $h \to 0$, the average velocity approaches 96 ft/sec, which is the velocity after 3 seconds.

15 In the time interval from 1 to $1 + h$ seconds, where $h > 0$, the rock travels $16(1 + h)^2 - 16\cdot 1^2$ feet. The average velocity during this period is

$$\frac{16(1+h)^2 - 16\cdot 1^2}{(1+h) - 1} = 16\,\frac{1 + 2h + h^2 - 1}{h} =$$

$16(2 + h)$ ft/sec. As $h \to 0$, the average velocity approaches 32 ft/sec, which is the velocity after 1 second.

17 (a) $(2.1)^3 - 2^3 = 9.261 - 8 = 1.261$ feet

(b) $\frac{1.261}{2.1 - 2} = \frac{1.261}{0.1} = 12.61$ ft/sec

(c) $\frac{(2+h)^3 - 2^3}{(2+h) - 2} = \frac{8 + 12h + 6h^2 + h^3 - 8}{h}$

$= (12 + 6h + h^2)$ ft/sec

(d) As $h \to 0$, $12 + 6h + h^2 \to 12$ ft/sec.

19 (a) Slope $= \frac{2^2 - (1.99)^2}{2 - 1.99} = \frac{4 - 3.9601}{0.01}$

$= \frac{0.0399}{0.01} = 3.99$

(b) Slope $= \dfrac{2^2 - (2+h)^2}{2 - (2+h)} =$

$\dfrac{4 - (4 + 4h + h^2)}{-h} = 4 + h$

(c) As $h \to 0^-$, $4 + h \to 4$.

21 (a) Consider the pair of points $P(4, 16)$ and $Q(4 + h, (4 + h)^2)$ where h is a small nonzero quantity. Thus, the slope of the tangent line at $P(4, 16)$ is $\lim\limits_{h \to 0} \dfrac{(4+h)^2 - 4^2}{(4+h) - 4}$

$= \lim\limits_{h \to 0} \dfrac{8h + h^2}{h} = \lim\limits_{h \to 0} (8 + h) = 8.$

(b)

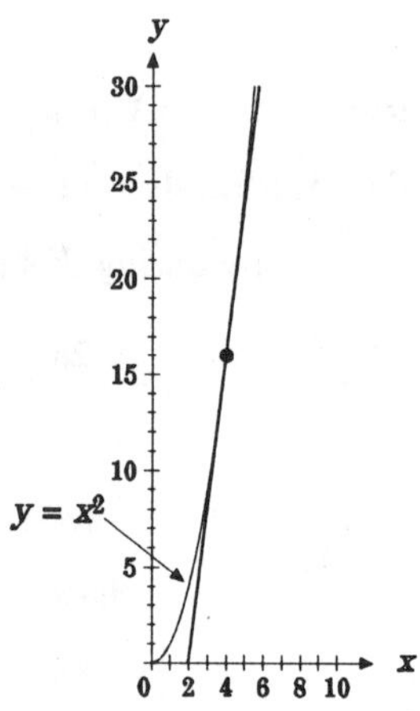

23 (a) The projection of [1, 1.1] is $[1, (1.1)^2]$.

Magnification is $\dfrac{(1.1)^2 - 1}{1.1 - 1} = \dfrac{1.21 - 1}{0.1}$

$= \dfrac{0.21}{0.1} = 2.1.$

(b) The projection of [1, 1.01] is $[1, (1.01)^2]$.

Magnification is $\dfrac{(1.01)^2 - 1}{1.01 - 1} = \dfrac{1.0201 - 1}{0.01}$

$= \dfrac{0.0201}{0.01} = 2.01.$

(c) The projection of [1, 1.001] is $[1, (1.001)^2]$.

Magnification is $\dfrac{(1.001)^2 - 1}{1.001 - 1} =$

$\dfrac{1.002001 - 1}{0.001} = \dfrac{0.002001}{0.001} = 2.001.$

(d) The projection of $[1, 1 + h]$ is $[1, (1 + h)^2]$.

Magnification is $\dfrac{(1+h)^2 - 1^2}{(1+h) - 1} =$

$\dfrac{1 + 2h + h^2 - 1}{h} = 2 + h$. At $x = 1$,

magnification is $\lim\limits_{h \to 0} (2 + h) = 2.$

25 (a) The projection of the interval [0.49, 0.5] is $[(0.49)^2, (0.5)^2]$. Magnification is

$\dfrac{(0.5)^2 - (0.49)^2}{0.5 - 0.49} = \dfrac{0.25 - 0.2401}{0.01} =$

$\dfrac{0.0099}{0.01} = 0.99.$

(b) The projection of [0.499, 0.5] is $[(0.499)^2, (0.5)^2]$. Magnification is

$\dfrac{(0.5)^2 - (0.499)^2}{0.5 - 0.499} = \dfrac{0.25 - 0.249001}{0.001}$

$= \dfrac{0.000999}{0.001} = 0.999.$

(c) The projection of $[0.5 + h, 0.5]$ is $[(0.5 + h)^2, (0.5)^2]$. Magnification is

$\dfrac{(0.5)^2 - (0.5+h)^2}{0.5 - (0.5+h)} =$

$\dfrac{0.25 - (0.25 + h + h^2)}{-h} = 1 + h.$

Magnification at $x = 0.5$ equals $\lim\limits_{h \to 0^-} (1 + h)$

$= 1.$

27 (a) Mass $= (3.01)^2 - 3^2 = 9.0601 - 9$

$= 0.0601$ g.

(b) Density in the interval is $\dfrac{0.0601}{3.01 - 3} =$

$\dfrac{0.0601}{0.01} = 6.01$ g/cm.

(c) Density in the interval is $\dfrac{3^2 - (2.99)^2}{3 - 2.99} =$

$\dfrac{9 - 8.9401}{0.01} = \dfrac{0.0599}{0.01} = 5.99$ g/cm.

(d) Density in the interval is $\dfrac{(3 + h)^2 - 3^2}{(3 + h) - 3} =$

$\dfrac{9 + 6h + h^2 - 9}{h} = (6 + h)$ g/cm. Density

at $x = 3$ is $\lim_{h \to 0^+} (6 + h) = 6$ g/cm.

(e) Density in the interval is $\dfrac{3^2 - (3 + h)^2}{3 - (3 + h)} =$

$\dfrac{9 - (9 + 6h + h^2)}{-h} = (6 + h)$ g/cm.

Density at $x = 3$ is $\lim_{h \to 0^-} (6 + h) = 6$ g/cm.

29 (a) Slope of the tangent line at (1, 1) is

$\lim_{h \to 0} \dfrac{(1 + h)^3 - 1^3}{(1 + h) - 1} = \lim_{h \to 0} \dfrac{3h + 3h^2 + h^3}{h}$

$= \lim_{h \to 0} (3 + 3h + h^2) = 3.$

(b)

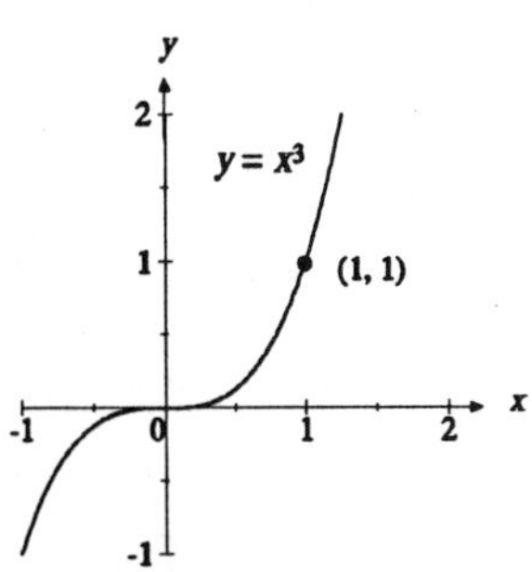

31 (a) Growth is $(2.01)^2 - 2^2 = 4.0401 - 4$

$= 0.0401$ grams.

(b) Rate of growth is $\dfrac{0.0401}{2.01 - 1} = \dfrac{0.0401}{0.01}$

$= 4.01$ g/min.

(c) Rate of growth in the interval $[2, 2 + h]$ is

$\dfrac{(2 + h)^2 - 2^2}{(2 + h) - 2} = \dfrac{4 + 4h + h^2 - 4}{h} =$

$4 + h$. At $t = 2$, the rate of growth is

$\lim_{h \to 0} (4 + h) = 4$ g/min.

33 (a) See the figure in the text's answer section.

(d) Slope $= \dfrac{[2(1 + h)^2 + (1 + h)] - 3}{(1 + h) - 1} =$

$\dfrac{(2 + 4h + 2h^2 + 1 + h) - 3}{h} = 5 + 2h$

(e) At $x = 1$, the slope is $\lim_{h \to 0} (5 + 2h) = 5$.

35 Slope of chord $= \dfrac{(x + h)^2 - x^2}{(x + h) - x} =$

$\dfrac{x^2 + 2xh + h^2 - x^2}{h} = 2x + h$. The slope at x is

$\lim_{h \to 0} (2x + h) = 2x$.

37 Magnification of interval $= \dfrac{(x + h)^2 - x^2}{(x + h) - x} =$

$\frac{2xh + h^2}{h} = 2x + h$. The magnification at x is

$\lim_{h \to 0} (2x + h) = 2x$.

39 (a)

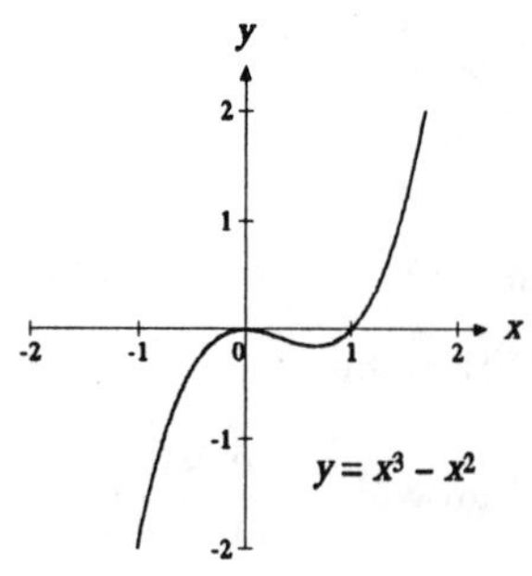

(b) The slope of the line through the points $(x, x^3 - x^2)$ and $(x + h, (x + h)^3 - (x + h)^2)$ is $\frac{[(x + h)^3 - (x + h)^2] - [x^3 - x^2]}{(x + h) - x}$

$= \frac{(x^3 + 3x^2h + 3xh^2 + h^3 - x^2 - 2xh - h^2) - (x^3 - x^2)}{h}$

$= 3x^2 + 3xh + h^2 - 2x - h$.

The slope of the tangent line at $(x, x^3 - x^2)$ is $\lim_{h \to 0} (3x^2 + 3xh + h^2 - 2x - h) = 3x^2 - 2x$.

(c) A tangent line is horizontal when its slope is 0. Note that $0 = 3x^2 - 2x = x(3x - 2)$ when $x = 0$ or $3x - 2 = 0$, which occurs when $x = 2/3$. Thus the tangent is horizontal at $(0, 0)$ and $\left(\frac{2}{3}, -\frac{4}{27}\right)$.

(d) A tangent line has slope 1 when $3x^2 - 2x = 1$; hence $3x^2 - 2x - 1 = 0$, $(3x + 1)(x - 1) = 0$, and $x = -1/3$ or 1. The slope is 1 at $\left(-\frac{1}{3}, -\frac{4}{27}\right)$ and $(1, 0)$.

41 Slope of the tangent line is $\lim_{h \to 0} \frac{(1 + h)^2 - 1^2}{(1 + h) - 1} =$ $\lim_{h \to 0} \frac{1 + 2h + h^2 - 1}{h} = \lim_{h \to 0} (2 + h) = 2$. An equation of the line is thus $y - 1 = 2(x - 1) = 2x - 2$, or $y = 2x - 1$. When $x = 6$, $y = 2 \cdot 6 - 1 = 11 \neq 12$. Thus, $(6, 12)$ is *not* on the line.

43 By Exercise 40, the slope of the tangent line at the point $(x_0, x_0^3 - x_0)$ is $3x_0^2 - 1$, so an equation of the tangent line is $y - (x_0^3 - x_0) = (3x_0^2 - 1)(x - x_0)$. Thus $y - x_0^3 + x_0 = (3x_0^2 - 1)x - 3x_0^3 + x_0$ and

$y = (3x_0^2 - 1)x - 2x_0^3$.

We wish to find x_0 such that the point $(2, 2)$ is on the line, so let $x = 2$ and $y = 2$; then $2 = (3x_0^2 - 1)2 - 2x_0^3 = 6x_0^2 - 2 - 2x_0^3$, so $0 = 2x_0^3 - 6x_0^2 + 4 = 2(x_0^3 - 3x_0^2 + 2)$. Observe that $x^0 = 1$ is a solution, which yield the factorization $2(x_0 - 1)(x_0^2 - 2x_0 - 2)$. Now $x_0^2 - 2x_0 - 2 = 0$ when $x_0 = \frac{2 \pm 2\sqrt{3}}{2} = 1 \pm \sqrt{3}$ and $x_0 - 1 = 0$ when $x_0 = 1$. But $1 + \sqrt{3} \approx 2.732 > 2$, which is too far along the trajectory, while $1 - \sqrt{3} \approx -0.732 < 2$, so she should stop the engine at either $(1 - \sqrt{3}, 9 - 5\sqrt{3})$ or $(1, 0)$.

45 Find the slope of the line through $\left(2, \frac{2 + 1}{2 + 2}\right) = \left(2, \frac{3}{4}\right)$ and $\left(2 + h, \frac{2 + h + 1}{2 + h + 2}\right) =$

$\left(2 + h, \frac{3+h}{4+h}\right)$. Hence the slope is $\dfrac{\frac{3+h}{4+h} - \frac{3}{4}}{(2+h) - 2}$

$= \frac{1}{h}\left(\frac{3+h}{4+h} - \frac{3}{4}\right) = m.$

(a) Here $h = 0.1$. Thus $m \approx 6.098 \times 10^{-2}$.

(b) Here $h = 0.01$. Thus $m \approx 6.234 \times 10^{-2}$.

(c) Here $h = 0.001$. Thus $m \approx 6.248 \times 10^{-2}$.

(d) Here $h = -0.001$. Thus $m \approx 6.252 \times 10^{-2}$.

3.2 The Derivative

1 $(2x)' = \lim_{h\to 0} \frac{2(x+h) - 2x}{h} = \lim_{h\to 0} 2 = 2$

3 $(4x + 4)' = \lim_{h\to 0} \frac{4(x+h) + 4 - (4x+4)}{h}$

$= \lim_{h\to 0} 4 = 4$

5 $(5x^2)' = \lim_{h\to 0} \frac{5(x+h)^2 - 5x^2}{h}$

$= \lim_{h\to 0} \frac{5x^2 + 10xh + 5h^2 - 5x^2}{h}$

$= \lim_{h\to 0} (10x + 5h) = 10x$

7 $(x^2 + 2x)'$

$= \lim_{h\to 0} \frac{(x+h)^2 + 2(x+h) - (x^2+2x)}{h}$

$= \lim_{h\to 0} \frac{x^2 + 2xh + h^2 + 2x + 2h - x^2 - 2x}{h}$

$= \lim_{h\to 0} (2x + 2 + h) = 2x + 2$

9 $(7\sqrt{x})' = \lim_{h\to 0} \frac{7\sqrt{x+h} - 7\sqrt{x}}{h}$

$= 7 \lim_{h\to 0} \left(\frac{\sqrt{x+h} - \sqrt{x}}{h} \cdot \frac{\sqrt{x+h} + \sqrt{x}}{\sqrt{x+h} + \sqrt{x}}\right)$

$= 7 \lim_{h\to 0} \frac{(x+h) - x}{h(\sqrt{x+h} + \sqrt{x})} = 7 \lim_{h\to 0} \frac{1}{\sqrt{x+h} + \sqrt{x}}$

$= 7 \cdot \frac{1}{2\sqrt{x}} = \frac{7}{2\sqrt{x}}$

11 $(x^2 + 3\sqrt{x})'$

$= \lim_{h\to 0} \frac{\left[(x+h)^2 + 3\sqrt{x+h}\right] - \left[x^2 + 3\sqrt{x}\right]}{h}$

$= \lim_{h\to 0} \left(2x + h + \frac{3(\sqrt{x+h} - \sqrt{x})}{h}\right)$

$= \lim_{h\to 0} \left(2x + h + \frac{3}{\sqrt{x+h} + \sqrt{x}}\right) = 2x + \frac{3}{2\sqrt{x}}$

13 $(x^3 + 3x)' = \lim_{h\to 0} \frac{(x+h)^3 + 3(x+h) - (x^3+3x)}{h}$

$= \lim_{h\to 0} \frac{x^3 + 3x^2h + 3xh^2 + h^3 + 3x + 3h - x^3 - 3x}{h}$

$= \lim_{h\to 0} (3x^2 + 3 + 3xh + h^2) = 3x^2 + 3$

15 $\left(x^2 + \frac{1}{x}\right)'$

$= \lim_{h\to 0} \left[\frac{(x+h)^2 + 1/(x+h) - (x^2 + 1/x)}{h}\right]$

$= \lim_{h\to 0} \left[\frac{(x+h)^2 - x^2}{h} + \frac{1/(x+h) - 1/x}{h}\right]$

$= \lim_{h\to 0} \left[2x + h + \frac{x - (x+h)}{h(x+h)x}\right]$

$= \lim_{h\to 0} \left(2x + h - \frac{1}{(x+h)x}\right) = 2x - \frac{1}{x^2}$

17 $(x^4)' = 4x^{4-1} = 4x^3$; when $x = -1$, this equals $4(-1)^3 = 4(-1) = -4$.

19 $(x^5)' = 5x^{5-1} = 5x^4$; when $x = a$, this equals $5a^4$.

21 $\left(\sqrt[3]{t}\right)' = (t^{1/3})' = \frac{1}{3}t^{1/3-1} = \frac{1}{3}t^{-2/3}$; when $t = -8$,

this equals $\frac{1}{3}(-8)^{-2/3} = \frac{1}{3(-8)^{2/3}} = \frac{1}{3\cdot 4} = \frac{1}{12}$.

23 $(x^\pi)'\big|_{x=2} = \pi x^{\pi-1}\big|_{x=2} = \pi\cdot 2^{\pi-1} \approx 13.862$

25 $(x^{2/3})'\big|_{x=8} = \frac{2}{3}x^{-1/3}\Big|_{x=8} = \frac{2}{3}\cdot 8^{-1/3} = \frac{1}{3}$

27 $(x^{5/4})'\big|_{x=16} = \frac{5}{4}x^{1/4}\Big|_{x=16} = \frac{5}{4}[16^{1/4}] = \frac{5}{2}$

29 Note that $f(x) = x^{-3}$. By Theorem 2, $f'(x) = -3x^{-4}$ and $f'(2) = -3(2^{-4}) = -3/16$.

31 $f'(x) = 2.7x^{1.7}$ and $f'(2) = 2.7(2^{1.7}) \approx 8.772$.

33 (a) Slope $= \frac{(1.1)^4 - 1^4}{1.1 - 1} = \frac{1.4641 - 1}{0.1} =$

$\frac{0.4641}{0.1} = 4.641$

(b) $(x^4)' = 4x^3$; when $x = 1$, this equals $4\cdot 1^3 = 4$, so the slope of the tangent line is 4.

35 (a) Magnification $= \frac{(1.01)^4 - 1^4}{1.01 - 1} =$

$\frac{1.04060401 - 1}{0.01} = \frac{0.04060401}{0.01} = 4.060401$

(b) $(x^4)' = 4x^3 = 4\cdot 1^3 = 4$ when $x = 1$, so the magnification is 4.

37 (a) $(x^3)' = 3x^2$; when $x = 1$, this equals $3\cdot 1^2 = 3$, so the slope of the tangent line is 3. An equation of the line is $y - 1 = 3(x - 1) = 3x - 3$, so $y = 3x - 2$. When $x = 2$, $y = 3\cdot 2 - 2 = 6 - 2 = 4$, so the line passes through (2, 4).

(b)

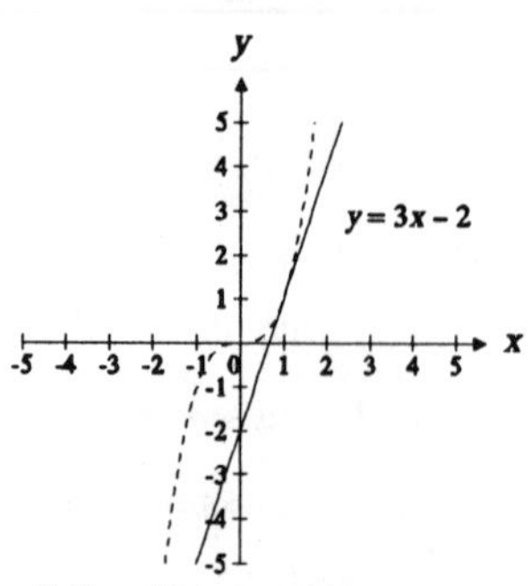

39 All of the following answers are approximations and may well differ from your own results.

(a) 0

(b) −2.5

(c) 0

(d) 0

(e) 1.75

(f) 0

(g)

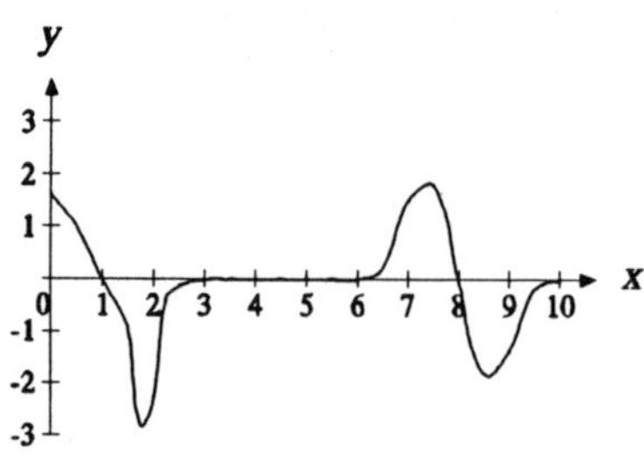

41 We must use the definition of the derivative. Let $f(x) = 3^x$. Then $f'(x) = \lim_{h\to 0}\frac{3^{x+h} - 3^x}{h} =$

$\lim_{h\to 0} 3^x\left(\frac{3^h - 1}{h}\right) = 3^x \lim_{h\to 0}\frac{3^h - 1}{h}$. But we are

given that $f'(0) \approx 1.0986$, so $1.0986 \approx f'(0) = 3^0 \lim_{h\to 0}\frac{3^h - 1}{h} = \lim_{h\to 0}\frac{3^h - 1}{h}$. Hence $f'(x) =$

$$3^x\left(\frac{3^h - 1}{h}\right) \approx 3^x(1.0986).$$

43 **Beware:** Your calculator may not give the appropriate answers here. Some refuse to take roots or powers of negative numbers, while others have a complex number mode.

(a) $(-8)^{7/5} = -(8^{7/5}) \approx -18.379$

(b) $(-8)^{10/7} = 8^{10/7} \approx 19.504$

(c) We note that $(-8)^{141/99} \approx -19.329$, while $(-8)^{141/100}$ is undefined. Also $8^{\sqrt{2}} \approx 18.931$. Let $x = m/n \approx \sqrt{2}$. For m even, n odd, $(-8)^x = 8^x \approx 8^{\sqrt{2}}$. For m odd, n odd, $(-8)^x = -(8^x) \approx -8^{\sqrt{2}}$. For m odd, n even, $(-8)^x$ is undefined. As we choose m and n such that $m/n \to \sqrt{2}$, we cannot know which of the above three cases will prevail. Thus $\lim_{x\to\sqrt{2}} (-8)^x$ is indeterminate.

3.3 The Derivative and Continuity

1 $$\frac{d}{dx}(x^3) = \lim_{\Delta x\to 0} \frac{(x + \Delta x)^3 - x^3}{\Delta x}$$

$$= \lim_{\Delta x\to 0} \frac{x^3 + 3x^2\Delta x + 3x\Delta x^2 + \Delta x^3 - x^3}{\Delta x}$$

$$= \lim_{\Delta x\to 0} \frac{1}{\Delta x}(3x^2\Delta x + 3x\Delta x^2 + \Delta x^3)$$

$$= \lim_{\Delta x\to 0} \frac{\Delta x}{\Delta x}(3x^2 + 3x\Delta x + \Delta x^2)$$

$$= \lim_{\Delta x\to 0} (3x^2 + 3x\Delta x + \Delta x^2) = 3x^2 + 3x\cdot 0 + 0^2$$

$$= 3x^2$$

3 $$\frac{d}{dx}(\sqrt{x}) = \lim_{\Delta x\to 0} \frac{\sqrt{x + \Delta x} - \sqrt{x}}{\Delta x}$$

$$= \lim_{\Delta x\to 0} \frac{(\sqrt{x + \Delta x} - \sqrt{x})(\sqrt{x + \Delta x} + \sqrt{x})}{\Delta x(\sqrt{x + \Delta x} + \sqrt{x})}$$

$$= \lim_{\Delta x\to 0} \frac{x + \Delta x - x}{\Delta x(\sqrt{x + \Delta x} + \sqrt{x})}$$

$$= \lim_{\Delta x\to 0} \frac{1}{\sqrt{x + \Delta x} + \sqrt{x}} = \frac{1}{\sqrt{x} + \sqrt{x}} = \frac{1}{2\sqrt{x}}$$

5 $$\frac{d}{dx}(5x^2) = \lim_{\Delta x\to 0} \frac{5(x + \Delta x)^2 - 5x^2}{\Delta x}$$

$$= \lim_{\Delta x\to 0} \frac{5(x^2 + 2x\Delta x + \Delta x^2) - 5x^2}{\Delta x}$$

$$= \lim_{\Delta x\to 0} \frac{5}{\Delta x}(x^2 + 2x\Delta x + \Delta x^2 - x^2)$$

$$= \lim_{\Delta x\to 0} \frac{5}{\Delta x}(2x\Delta x + \Delta x^2) = \lim_{\Delta x\to 0} (10x + 5\Delta x)$$

$$= 10x + 5\cdot 0 = 10x$$

7 $$D\left(\frac{3}{x}\right) = \lim_{\Delta x\to 0} \frac{\dfrac{3}{x + \Delta x} - \dfrac{3}{x}}{\Delta x}$$

$$= \lim_{\Delta x\to 0} \frac{1}{\Delta x}\left(\frac{3x - 3(x + \Delta x)}{(x + \Delta x)x}\right)$$

$$= \lim_{\Delta x\to 0} \frac{3x - 3x - 3\Delta x}{\Delta x(x + \Delta x)x} = \lim_{\Delta x\to 0} \frac{-3}{(x + \Delta x)x}$$

$$= \frac{-3}{(x + 0)x} = -\frac{3}{x^2}$$

9 $$\frac{d}{dx}\left(\frac{3}{x} - 4x + 2\right) =$$

$$\lim_{\Delta x\to 0} \frac{\dfrac{3}{x + \Delta x} - 4(x + \Delta x) + 2 - \left(\dfrac{3}{x} - 4x + 2\right)}{\Delta x}$$

$$= \lim_{\Delta x \to 0} \frac{1}{\Delta x}\left[\frac{3}{x + \Delta x} - \frac{3}{x} - 4(x + \Delta x) + 4x + 2 - 2\right]$$

$$= \lim_{\Delta x \to 0} \frac{1}{\Delta x}\left[\frac{-3\Delta x}{(x + \Delta x)x} - 4\Delta x\right]$$

$$= \lim_{\Delta x \to 0} \left[\frac{-3}{(x + \Delta x)x} - 4\right] = -\frac{3}{x^2} - 4$$

11 Since $f(x) = x^2$ we have $\Delta f = f(x + \Delta x) - f(x) = (x + \Delta x)^2 - x^2 = 2x\Delta x + \Delta x^2$.

(a) For $x = 1$ and $\Delta x = 0.1$, we have $\Delta f = 2 \cdot 1 \cdot 0.1 + (0.1)^2 = 0.21$.

(b) For $x = 3$ and $\Delta x = -0.1$, we have $\Delta f = 2 \cdot 3 \cdot (-0.1) + (-0.1)^2 = -0.59$.

13 (a) $\Delta y = \Delta f = f(x + \Delta x) - f(x) = f(2.1) - f(2) = 0.05$

(b)

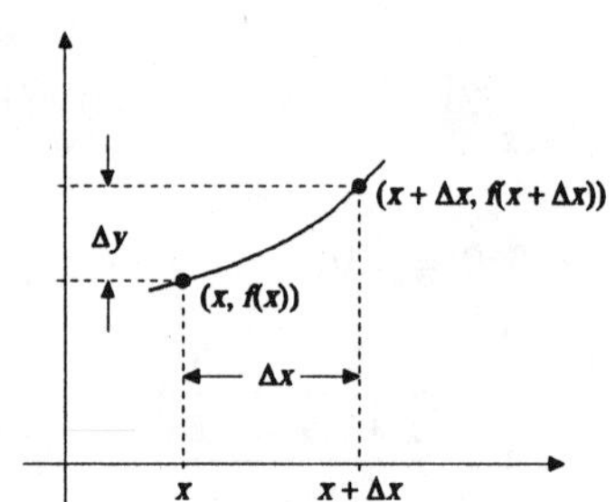

15 (a) For $a = -1$, $\lim_{x \to a} f(x) \neq f(a)$. For $a = 2$, $\lim_{x \to a} f(x)$ exists, but $f(a)$ does not exist.

(b) For $a = 1, 3$, $\lim_{x \to a} f(x) = f(a)$ but $\lim_{x \to a^-} f'(x) \neq \lim_{x \to a^+} f'(x)$.

17 (a) For $a = 5$, $\lim_{x \to a} f(x) = 1 \neq 2 = f(a)$.

(b) For $a = 2, 3$, $\lim_{x \to a} f(x) = f(a)$ but $\lim_{x \to a^-} f'(x) \neq \lim_{x \to a^+} f'(x)$.

19 $\lim_{h \to 0} \frac{[2(x+h)^3 - 3(x+h)^2 + 4(x+h) - 5] - [2x^3 - 3x^2 + 4x - 5]}{h}$

$$= \lim_{h \to 0} (6x^2 + 6xh + 2h^2 - 6x - 3h + 4)$$

$$= 6x^2 - 6x + 4$$

21 (a)

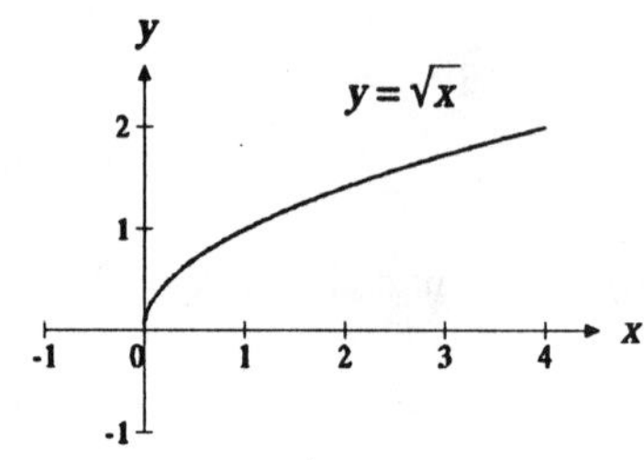

The domain consists of all nonnegative real numbers.

(b)

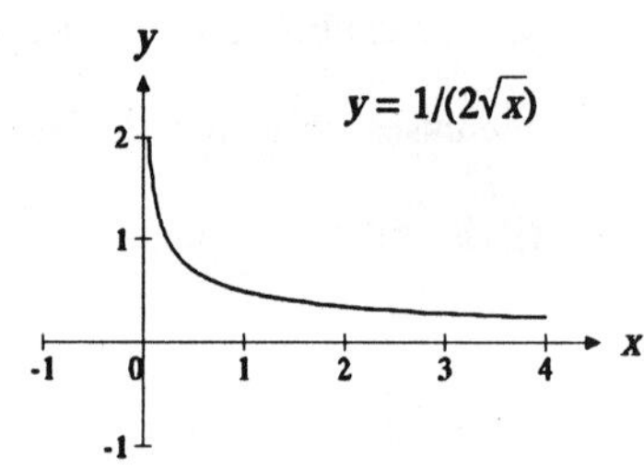

The domain consists of all positive real numbers.

23 (a) Any $x^5 + C$, where C is a constant.

(b) Any $x^7 + C$, where C is a constant.

25 (b)

x	1	2	3	4	5	6
Slope	1.8	0	−1.7	−1.7	0	1.8

(c)

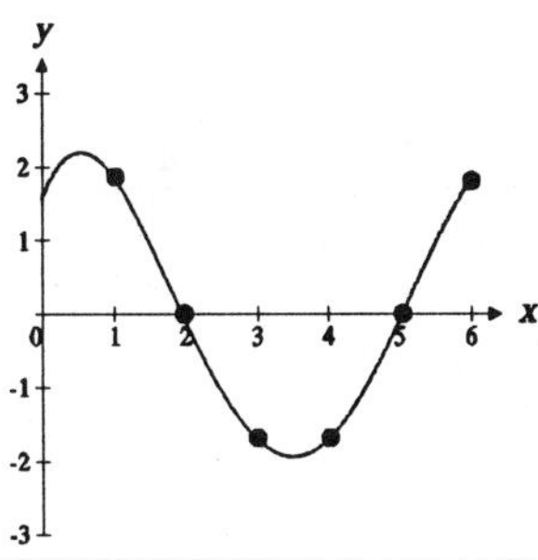

27

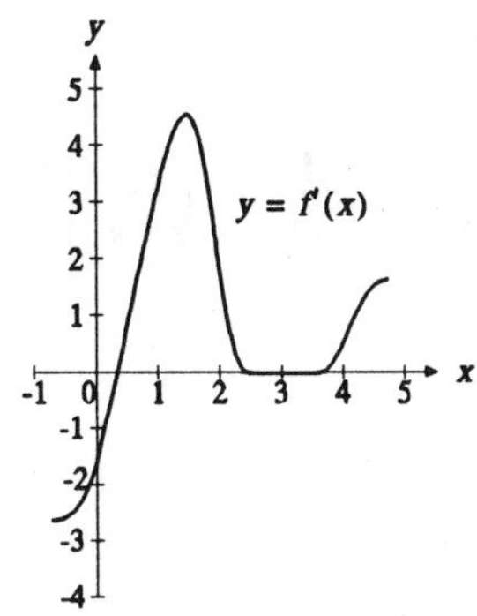

29 (a) $x^4 + C$; $D(x^4 + C) = 4x^3$

(b) $1/x + C$; $D(1/x + C) = D(x^{-1} + C)$

$= (-1)x^{-2} = -1/x^2$

31 (a) $x^3/3 + C$; $D(x^3/3 + C) = 3x^2/3 = x^2$

(b) $-\frac{1}{2x^2} + C$; $D\left(-\frac{1}{2x^2} + C\right) =$

$D\left(-\frac{1}{2}x^{-2} + C\right) = -\frac{1}{2}(-2)x^{-3} = \frac{1}{x^3}$

33 We know $f'(x) \approx \frac{f(x + \Delta x) - f(x)}{\Delta x}$ for small Δx.

Thus $f'(2.03) \approx \frac{f(2.05) - f(2.03)}{0.02} = 2.$

35 $D(\sin x) \approx \frac{\sin(x + \Delta x) - \sin x}{\Delta x}$

$D(\sin x)|_{x=2} \approx \frac{\sin(2.01) - \sin 2}{0.01} \approx -0.421$

3.4 The Derivatives of the Sum, Difference, Product, and Quotient

1 $D(x^2 + 5x) = D(x^2) + D(5x) = 2x + 5$

3 $D(x^3 + 5x^2 + 2) = D(x^3) + D(5x^2) + D(2)$

$= 3x^2 + 10x + 0 = 3x^2 + 10x$

5 $D(x^5 - 2x^2 + 3) = D(x^5) - D(2x^2) + D(3)$

$= 5x^4 - 4x + 0 = 5x^4 - 4x$

7 $D(3x^4 - 6\sqrt{x}) = D(3x^4) - D(6\sqrt{x})$

$= 12x^3 - 3x^{-1/2}$

9 $D(3/x + \sqrt[3]{x}) = D(3x^{-1}) + D(x^{1/3})$

$= -3x^{-2} + (1/3)x^{-2/3}$

11 $D[(2x + 1)(x - 4)]$

$= (2x + 1)\cdot D[x - 4] + (x - 4)\cdot D[(2x + 1)]$

$= (2x + 1)\cdot 1 + (x - 4)\cdot 2 = 2x + 1 + 2x - 8$

$= 4x - 7$

13 $D[(x^2 + x)(2x + 1)]$

$= (x^2 + x)\cdot D[2x + 1] + (2x + 1)\cdot D[x^2 + x]$

$= (x^2 + x)\cdot 2 + (2x + 1)(2x + 1)$

$= 2x^2 + 2x + 4x^2 + 4x + 1 = 6x^2 + 6x + 1$

15 $D[(x^3 - 2x)(2x^5 + 5)]$

$= (x^3 - 2x)D[2x^5 + 5] + (2x^5 + 5)D[x^3 - 2x]$

$= (x^3 - 2x)(10x^4) + (2x^5 + 5)(3x^2 - 2)$

$= 10x^7 - 20x^5 + 6x^7 - 4x^5 + 15x^2 - 10$

$= 16x^7 - 24x^5 + 15x^2 - 10$

17 $D[(4 + 5\sqrt{x})(3x - x^2)]$

$= (4 + 5\sqrt{x})D[3x - x^2] + (3x - x^2)D[4 + 5\sqrt{x}]$

$= (4 + 5x^{1/2})(3 - 2x) + (3x - x^2)\frac{5}{2}x^{-1/2}$

$= 12 - 8x + 15x^{1/2} - 10x^{3/2} + \frac{15}{2}x^{1/2} - \frac{5}{2}x^{3/2}$

$= -\frac{25}{2}x^{3/2} - 8x + \frac{45}{2}x^{1/2} + 12$

19 $$\frac{d}{dx}\left[\frac{2x + \sqrt[3]{x^2}}{4}\right] = \frac{1}{4}\frac{d}{dx}(2x + x^{2/3})$$

$$= \frac{1}{4}\left(2 + \frac{2}{3}x^{-1/3}\right) = \frac{1}{2} + \frac{1}{6}x^{-1/3}$$

21 $$\frac{d}{dx}\left[\frac{3 + x}{4 + x}\right] = \frac{(4 + x)(3 + x)' - (3 + x)(4 + x)'}{(4 + x)^2}$$

$$= \frac{(4 + x) - (3 + x)}{(4 + x)^2} = \frac{1}{(4 + x)^2}$$

23 $$\frac{d}{dx}\left[\frac{x^2 + x}{x + 3}\right]$$

$$= \frac{(x + 3)(x^2 + x)' - (x^2 + x)(x + 3)'}{(x + 3)^2}$$

$$= \frac{(x + 3)(2x + 1) - (x^2 + x)(1)}{(x + 3)^2}$$

$$= \frac{2x^2 + 7x + 3 - x^2 - x}{(x + 3)^2} = \frac{x^2 + 6x + 3}{(x + 3)^2}$$

25 $$\frac{d}{dt}\left[\frac{t^2 - 3t + 1}{t^2 + 1}\right] = \frac{d}{dt}\left[1 - \frac{3t}{t^2 + 1}\right]$$

$$= -\frac{(t^2 + 1)(3t)' - (3t)(t^2 + 1)'}{(t^2 + 1)^2}$$

$$= -\frac{(t^2 + 1)(3) - (3t)(2t)}{(t^2 + 1)^2} = -\frac{3t^2 + 3 - 6t^2}{(t^2 + 1)^2}$$

$$= \frac{3t^2 - 3}{(t^2 + 1)^2} = \frac{3(t^2 - 1)}{(t^2 + 1)^2}$$

27 $$\frac{d}{dx}\left[\frac{6 + \sqrt{x}}{2x + 3}\right]$$

$$= \frac{(2x + 3)(6 + x^{1/2})' - (6 + x^{1/2})(2x + 3)'}{(2x + 3)^2}$$

$$= \frac{(2x + 3)((1/2)x^{-1/2}) - (6 + x^{1/2})(2)}{(2x + 3)^2}$$

$$= \frac{x^{1/2} + (3/2)x^{-1/2} - 12 - 2x^{1/2}}{(2x + 3)^2}$$

$$= \frac{-x^{1/2} + (3/2)x^{-1/2} - 12}{(2x + 3)^2}$$

29 $$\frac{d}{dx}\left[\frac{1}{x} + \frac{1}{x^2}\right] = \frac{d}{dx}(x^{-1} + x^{-2}) = -x^{-2} - 2x^{-3}$$

31 $$\left(\frac{1}{x^3 + 2x + 1}\right) = -\frac{(x^3 + 2x + 1)'}{(x^3 + 2x + 1)^2}$$

$$= -\frac{3x^2 + 2}{(x^3 + 2x + 1)^2}$$

33 $$\frac{d}{dx}\left[\frac{(2x^2 - 3)(x^2 - 1)}{(5x + 7)}\right] =$$

$$\frac{(5x+7)[(2x^2-3)(x^2-1)]' - (2x^2-3)(x^2-1)(5x+7)'}{(5x + 7)^2} =$$

$$\frac{(5x+7)[(2x^2-3)(2x) + (x^2-1)(4x)] - (2x^2-3)(x^2-1)(5)}{(5x+7)^2}$$

$$= \frac{(5x + 7)(8x^3 - 10x) - 5(2x^4 - 5x^2 + 3)}{(5x + 7)^2}$$

$$= \frac{40x^4 + 56x^3 - 50x^2 - 70x - 10x^4 + 25x^2 - 15}{(5x + 7)^2}$$

$$= \frac{30x^4 + 56x^3 - 25x^2 - 70x - 15}{(5x + 7)^2}$$

35 $$\left(\frac{1 + \frac{1}{x}}{1 - \frac{1}{x}}\right)' = \left(\frac{x + 1}{x - 1}\right)' = \frac{(x - 1)(1) - (x + 1)(1)}{(x - 1)^2}$$

$= \dfrac{-2}{(x-1)^2}$

37 $(x^3 - x^2 + 2x)' = 3x^2 - 2x + 2$; when $x = 1$, this equals $3\cdot 1^2 - 2\cdot 1 + 2 = 3 - 2 + 2 = 3$, so the slope is 3. Thus the equation of the tangent is $y - 2 = 3(x - 1)$, or $y = 3x - 1$.

39 $D(\sqrt{x}(x^2 + 2)) = \sqrt{x}D(x^2 + 2) + (x^2 + 2)D(\sqrt{x})$

$= \sqrt{x}\cdot 2x + \dfrac{x^2+2}{2\sqrt{x}}$; when $x = 4$, this equals

$\sqrt{4}\cdot 2\cdot 4 + (4^2 + 2)\dfrac{1}{2\sqrt{4}} = 2\cdot 8 + 18\cdot\dfrac{1}{2\cdot 2} =$

$16 + \dfrac{18}{4} = \dfrac{41}{2}$, so the slope is $\dfrac{41}{2}$. Hence the equation of the tangent is $y - 36 = \dfrac{41}{2}(x - 4)$, or

$y = \dfrac{41}{2}x - 46$.

41 Velocity $= 8t^3 + 3t^2 + 2 = 13$

43 $D(\sqrt[3]{x}) = D(x^{1/3}) = \dfrac{1}{3}x^{-2/3} = \dfrac{1}{3x^{2/3}} = \dfrac{1}{3\cdot 8^{2/3}}$

$= \dfrac{1}{3\cdot 4} = \dfrac{1}{12}$ when $x = 8$, so magnification equals $1/12$.

45 $D(x\sqrt[3]{x}) = xD(x^{1/3}) + x^{1/3}D(x) =$

$x\cdot\dfrac{1}{3}x^{-2/3} + x^{1/3}\cdot 1 = \dfrac{4}{3}x^{1/3} = \dfrac{4}{3}\cdot 8^{1/3} = \dfrac{4}{3}\cdot 2$

$= \dfrac{8}{3}$ when $x = 8$, so the density is $8/3$.

47 (a) $\dfrac{x^3}{3} + C$ (Pick any two C's.)

(b) $x^4/4 + C$

(c) $-1/x + C$

(d) $x^3 + \dfrac{x^2}{2} + C$

49 (a) $(x)' = \lim\limits_{\Delta x\to 0} \dfrac{x + \Delta x - x}{\Delta x} = 1$

(b) $(x^2) = (x\cdot x)' = x\cdot(x)' + x\cdot(x') = 2x$

(c) $(x^3)' = (x\cdot x^2)' = x\cdot(x^2)' + x^2\cdot x' = 3x^2$

$(x^4)' = (x\cdot x^3)' = x\cdot(x^3)' + x^3\cdot x' = 4x^3$

(d) Let $P(n) = (x^n)' = nx^{n-1}$. We know $P(1) = (x)' = 1\cdot x^0 = 1$ is true. Assume $P(n)$ is true. Then $P(n + 1) = (x^{n+1})' = (x\cdot x^n)' = x\cdot P(n) + x^n\cdot P(1) = nx^n + x^n = (n + 1)x^{(n+1)-1}$. Since this fits the form of our P statement, we now know that $P(n)$ implies $P(n + 1)$. Because $P(n)$ is true for $n = 1$, it is therefore true for all $n > 1$ as well.

51 (a) $D(\sqrt[3]{x}\sqrt{x}) = \sqrt[3]{x}(\sqrt{x})' + \sqrt{x}(\sqrt[3]{x})$

$= \dfrac{\sqrt[3]{x}}{2\sqrt{x}} + \dfrac{\sqrt{x}}{3\sqrt[3]{x^2}} = \dfrac{1}{2}x^{1/3-1/2} + \dfrac{1}{3}x^{1/2-2/3}$

$= \dfrac{5}{6}x^{-1/6}$

(b) $D(\sqrt[3]{x}\sqrt{x}) = D(x^{1/3}\cdot x^{1/2}) = D(x^{5/6}) = \dfrac{5}{6}x^{-1/6}$

53 $(f + g + h)' = ((f + g) + h)' = (f + g)' + h'$

$= (f' + g') + h' = f' + g' + h'$

55 $(fgh)' = ((fg)h)' = (fg)h' + h(fg)' =$

$fgh' + h(fg' + gf') = fgh' + fg'h + f'gh$

57 The slope at $P = (a, 1/a)$ is $-\dfrac{1}{a^2}$, so the equation of the line is $y - \dfrac{1}{a} = -\dfrac{1}{a^2}(x - a)$; that is, $x = 2a - a^2y$. When $y = 0$, $x = 2a$, so $B = (2a, 0)$.

When $x = 0$, $y = 2/a$, so $A = (0, 2/a)$. The area of the triangle is $\frac{1}{2}\cdot\overline{OA}\cdot\overline{OB} = \frac{1}{2}\cdot\frac{2}{a}\cdot 2a = 2$.

59 (a) Slope $= 4x - 3 = 1$

Equation: $y - 0 = 1(x - 1)$; that is, $y = x - 1$.

(b) Slope $= -1/1 = -1$

Equation: $y - 0 = -1(x - 1)$; that is, $y = 1 - x$.

(c) If (x, y) is an intersection, $1 - x = y = 2x^2 - 3x + 1$. That is, $0 = 2x^2 - 2x = 2x(x - 1)$, so $x = 0$ or 1. The intersections are $(0, 1)$ and $(1, 0)$.

61 The tangent at $(x, x^4 - 8x^2)$ has slope $4x^3 - 16x$. Hence it passes through $(-11/3, 49)$ if and only if $\frac{(x^4 - 8x^2) - 49}{x - (-11/3)} = 4x^3 - 16x$. This simplifies to $9x^4 + 44x^3 - 24x^2 - 176x + 147 = 0$. Both $x = 1$ and $x = -3$ satisfy this equation. Factoring out $(x - 1)(x - (-3)) = x^2 + 2x - 3$, we obtain $(x - 1)(x + 3)(9x^2 + 26x - 49) = 0$, so the other roots are $\frac{-26 \pm \sqrt{26^2 - 4\cdot 9\cdot(-49)}}{2\cdot 9} = \frac{-13 \pm \sqrt{610}}{9}$.

3.5 The Derivatives of the Trigonometric Functions

1 $D(5 \sin x) = 5\, D(\sin x) = 5 \cos x$

3 $D(2 \tan x) = 2\, D(\tan x) = 2 \sec^2 x$

5 $D(3 \sec x) = 3\, D(\sec x) = 3 \sec x \tan x$

7 $(x^2 \sin x)' = (x^2)(\sin x)' + (\sin x)(x^2)'$
$= x^2 \cos x + 2x \sin x$

9 $\left(\frac{1 + \sin x}{\cos x}\right)'$

$$= \frac{\cos x\,(1 + \sin x)' - (1 + \sin x)(\cos x)'}{(\cos x)^2}$$

$$= \frac{\cos x\,(\cos x) - (1 + \sin x)(-\sin x)}{(\cos x)^2}$$

$$= \frac{\cos^2 x + \sin^2 x + \sin x}{\cos 2x} = \frac{1 + \sin x}{\cos^2 x}$$

11 $\left(\frac{1 + 3 \sec x}{\tan x}\right)'$

$$= \frac{\tan x\,(1 + 3 \sec x)' - (1 + 3 \sec x)(\tan x)'}{(\tan x)^2}$$

$$= \frac{\tan x\,(3 \sec x \tan x) - (1 + 3 \sec x)(\sec^2 x)}{\tan^2 x}$$

$$= \frac{3 \sec x \tan^2 x - \sec^2 x - 3 \sec^3 x}{\tan^2 x}$$

$$= \frac{3 \sec x\,(\tan^2 x - \sec^2 x) - \sec^2 x}{\tan^2 x}$$

$$= \frac{3 \sec x\,(-1) - \sec^2 x}{\tan^2 x} = \frac{-3 \sec x - \sec^2 x}{\tan^2 x}$$

$$= \frac{-3 \cos x - 1}{\sin^2 x}$$

13 $\left(\frac{\csc x}{\sqrt[3]{x}}\right)' = \frac{x^{1/3}(\csc x)' - \csc x\,(x^{1/3})'}{(x^{1/3})^2}$

$$= \frac{x^{1/3}(-\csc x \cot x) - \csc x\left(\frac{1}{3}x^{-2/3}\right)}{x^{2/3}}$$

$$= \frac{-x^{-2/3} \csc x\,(x \cot x + 1/3)}{x^{2/3}}$$

$= \dfrac{-\csc x\ (x \cot x + 1/3)}{x^{4/3}}$

15 $(\sin x \tan x)' = \sin x\ (\tan x)' + \tan x\ (\sin x)'$

$= \sin x\ (\sec^2 x) + \tan x \cos x$

$= \sin x \sec^2 x + \dfrac{\sin x}{\cos x} \cos x$

$= \sin x \sec^2 x + \sin x = \sin x\ (\sec^2 x + 1)$

17 $\left(\dfrac{\cot x}{1 + x^2}\right)' = \dfrac{(1 + x^2)(\cot x)' - \cot x\ (1 + x^2)'}{(1 + x^2)^2}$

$= \dfrac{(1 + x^2)(-\csc^2 x) - \cot x\ (2x)}{(1 + x^2)^2}$

$= \dfrac{-(1 + x^2) \csc^2 x - (2x \cot x)}{(1 + x^2)^2}$

19 $D(\sin x - x \cos x) = D(\sin x) - D(x \cos x)$

$= \cos x - [xD(\cos x) + \cos x\ D(x)]$

$= \cos x - [x(-\sin x) + \cos x\ (1)]$

$= \cos x + x \sin x - \cos x = x \sin x$

21 $D[2x \sin x + 2 \cos x - x^2 \cos x]$

$= D(2x \sin x) + D(2 \cos x) - D(x^2 \cos x)$

$= [2x\ D(\sin x) + \sin x\ D(2x)] + 2D(\cos x)$

$- [x^2\ D(\cos x) + \cos x\ D(x^2)]$

$= [2x \cos x + (\sin x)(2)] + 2(-\sin x)$

$- [x^2(-\sin x) + (\cos x)(2x)]$

$= x^2 \sin x$

23 $D(\tan x - x) = D(\tan x) - D(x) = \sec^2 x - 1$

$= \tan^2 x$

25 $D(2x \cos x - 2 \sin x + x^2 \sin x)$

$= D(2x \cos x) - D(2 \sin x) + D(x^2 \sin x)$

$= [2x(-\sin x) + (\cos x)(2)] - 2 \cos x$

$+ [x^2 \cos x + 2x \sin x]$

$= x^2 \cos x$

27 $D(\sin x) = \cos x$

(a) $D(\sin x)|_{x=\pi/6} = \cos \pi/6 = \dfrac{\sqrt{3}}{2}$

(b) $D(\sin x)|_{x=1.2} = \cos 1.2 \approx 0.362$

(c) $D(\sin x)|_{x=3\pi/4} = \cos 3\pi/4 = -\dfrac{\sqrt{2}}{2}$

29 $D(\tan x) = \sec^2 x$

(a) $D(\tan x)|_{x=\pi/4} = \sec^2(\pi/4) = 2$

(b) $D(\tan x)|_{x=\pi/6} = \sec^2(\pi/6) = 4/3$

(c) $D(\tan x)|_{x=3} = \sec^2 3 \approx 1.020$

31 $D(\csc \theta) = D\left(\dfrac{1}{\sin \theta}\right) = -\dfrac{\cos \theta}{\sin^2 \theta} = -\cot \theta \csc \theta$

33 (a) $\sin t \le 1$ for all t, so $3 \sin t \le 3$. Thus, 3 cm is the highest it goes.

(b) $\sin t \ge -1$ for all t, so $3 \sin t \ge -3$. Thus, -3 cm is the lowest it goes.

(c) $(3 \sin t)' = 3(\sin t)' = 3 \cos t$. Velocity when $t = 0$ is $3 \cos 0 = 3$ cm/sec. Velocity when $t = \pi$ is $3 \cos \pi = -3$ cm/sec.

(d) Speed when $t = 0$ is the absolute value of velocity when $t = 0$; speed therefore equals $|3| = 3$ cm/sec. Speed when $t = \pi$ is the absolute value of velocity when $t = \pi$; speed therefore equals $|-3| = 3$ cm/sec.

35 $(\tan x)' = \sec^2 x$. The graph of $y = \tan x$ crosses the x axis when $x = n\pi$. The tangent of the angle at which the graph crosses the x axis at $x = n\pi$ is the slope of the tangent line at $x = n\pi$; that is,

$\sec^2(n\pi) = \dfrac{1}{(\cos n\pi)^2} = \dfrac{1}{1} = 1$, since $\cos n\pi$ is ± 1. Thus, $\tan \theta = 1$, so $\theta = \pi/4$.

37 $\tan^{-1}((\sin x)') = \tan^{-1}(\cos x) = \tan^{-1}(\cos \pi)$

$= \tan^{-1}(-1) = -\pi/4$, so the angle of inclination is $3\pi/4$.

39 (a) $-3 \cos x$

(b) $4 \sin x$

41 (a) $2 \sec x$

(b) $-7 \csc x$

43 $(\cos x)' = \lim_{h\to 0} \frac{\cos(x+h) - \cos x}{h}$

$= \lim_{h\to 0} \frac{\cos x \cos h - \sin x \sin h - \cos x}{h}$

$= \lim_{h\to 0} \frac{(\cos x)(\cos h - 1) - \sin x \sin h}{h}$

$= \lim_{h\to 0} \left[\frac{\cos h - 1}{h}(\cos x) - \frac{\sin h}{h}(\sin x)\right]$

$= 0(\cos x) - 1(\sin x) = -\sin x$

45 $D(\cos 11x) = \lim_{\Delta x\to 0} \frac{\cos 11(x+\Delta x) - \cos 11x}{\Delta x} =$

$\lim_{\Delta x\to 0} \frac{\cos 11x \cos 11\Delta x - \sin 11x \sin 11\Delta x - \cos 11x}{\Delta x}$

$= \lim_{\Delta x\to 0} \cos 11x \frac{\cos 11\Delta x - 1}{\Delta x} - \lim_{\Delta x\to 0} \sin 11x \frac{\sin 11\Delta x}{\Delta x}$

$= (\cos 11x)(0) - (\sin 11x)(11) = -11 \sin 11x$

47 $(\text{Sin}\,\theta)' = \lim_{h\to 0} \frac{\text{Sin}(\theta + h) - \text{Sin}\,\theta}{h}$

$= \lim_{h\to 0} \frac{(\text{Sin}\,\theta)(\text{Cos}\,h - 1) + \text{Cos}\,\theta\ \text{Sin}\,h}{h}$

$= \left(-\frac{\pi}{180}\text{Sin}\,\theta\right) \lim_{h\to 0} \frac{1 - \cos(\pi h/180)}{\pi h/180} +$

$\left(\frac{\pi}{180}\text{Cos}\,\theta\right) \lim_{h\to 0} \frac{\sin(\pi h/180)}{\pi h/180}$

$= \left(-\frac{\pi}{180}\text{Sin}\,\theta\right)0 + \left(\frac{\pi}{180}\text{Cos}\,\theta\right)1 = \frac{\pi}{180}\text{Cos}\,\theta$

3.6 The Derivative of a Composite Function

1 Let $y = u^{100}$, where $u = 1 + 2x$. Then $\frac{dy}{dx} =$

$\frac{dy}{du}\cdot\frac{du}{dx} = 100u^{99}\cdot 2 = 200(1 + 2x)^{99}$.

3 Let $y = u^{40}$, where $u = 2x^3 - 1$. Then $\frac{dy}{dx} =$

$\frac{dy}{du}\cdot\frac{du}{dx} = 40u^{39}\cdot 6x^2 = 40(2x^3 - 1)^{39}\cdot 6x^2 =$

$240x^2(2x^3 - 1)^{39}$.

5 Let $y = 3 \sin u$, where $u = \sqrt{x}$. Then $\frac{dy}{dx} =$

$\frac{dy}{du}\cdot\frac{du}{dx} = 3 \cos u \left(\frac{1}{2\sqrt{x}}\right) = \frac{3 \cos \sqrt{x}}{2\sqrt{x}}$.

7 Let $y = u^3$, where $u = \sin x$. Then $\frac{dy}{dx} = \frac{dy}{du}\cdot\frac{du}{dx}$

$= 3u^2\cdot\cos x = 3 \sin^2 x \cos x$.

9 Let $y = u^4$, where $u = \cos v$ and $v = 3x$. Then

$\frac{dy}{dx} = \frac{dy}{du}\frac{du}{dv}\frac{dv}{dx} = 4u^3\cdot(-\sin v)\cdot 3 =$

$-12 \cos^3 v \sin v = -12 \cos^3 3x \sin 3x$.

11 Let $y = u^2$, where $u = \tan v$ and $v = 3x$. Then $dy/dx = 2u\cdot(\sec^2 v)\cdot(3) = 6 \tan v \sec^2 v$

$= 6 \tan 3x \sec^2 3x$.

13 Let $y = \sqrt{u}$, where $u = \cot x$. Then $\frac{dy}{dx} =$

$\frac{1}{2\sqrt{u}}(-\csc^2 x) = \frac{-\csc^2 x}{2\sqrt{\cot x}}$.

15 Let $y = \sqrt{u}$, where $u = x^3 + x + 2$; then

$\frac{d}{dx}\left(\sqrt{x^3 + x + 2}\right) = \frac{d}{du}(\sqrt{u}) \cdot \frac{d}{dx}(x^3 + x + 2) =$

$\frac{1}{2\sqrt{u}} \cdot (3x^2 + 1) = \frac{3x^2 + 1}{2\sqrt{x^3 + x + 2}}.$

17 Let $y = \sin u$, where $u = v^5$ and $v = 3x + 2$.

Then $\frac{dy}{dx} = (\cos u) \cdot 5v^4 \cdot 3$

$= 15(3x + 2)^4 \cos(3x + 2)^5.$

19 Let $y = \frac{u}{v}$, where $u = x^3 \tan x$ and $v = 1 + x^2$.

Then we have $u' = x^3(\tan x)' + (\tan x)(x^3)'$

$= x^3 \sec^2 x + 3x^2 \tan x$, $v' = 2x$, and it follows that

$\frac{dy}{dx} = \frac{vu' - uv'}{v^2}$

$= \frac{(1 + x^2)(x^3 \sec^2 x + 3x^2 \tan x) - x^3 \tan x\,(2x)}{(1 + x^2)^2}$

$= \frac{(1 + x^2)x^3 \sec^2 x + (3x^2(1 + x^2) - 2x^4) \tan x}{(1 + x^2)^2}$

$= \frac{x^2}{(1 + x^2)^2}[(x + x^3) \sec^2 x + (3 + x^2) \tan x].$

21 Let $y = u^5$, where $u = 2x + 1$; then $\frac{d}{dx}((2x + 1)^5)$

$= \frac{d}{du}(u^5) \cdot \frac{d}{dx}(2x + 1) = 5u^4 \cdot 2 = 10(2x + 1)^4.$

Now let $y = u^7$, where $u = 3x + 1$; then

$\frac{d}{dx}((3x + 1)^7) = \frac{d}{du}(u^7) \cdot \frac{d}{dx}(3x + 1) = 7u^6 \cdot 3 =$

$21(3x + 1)^6$. Thus, $\frac{d}{dx}((2x + 1)^5(3x + 1)^7)$

$= (2x + 1)^5 \frac{d}{dx}((3x + 1)^7) + (3x + 1)^7 \frac{d}{dx}((2x + 1)^5)$

$= (2x + 1)^5 \cdot 21(3x + 1)^6 + (3x + 1)^7 \cdot 10(2x + 1)^4$

$= (2x + 1)^4(3x + 1)^6[21(2x + 1) + 10(3x + 1)]$

$= (2x + 1)^4(3x + 1)^6(72x + 31).$

23 Let $y = u^5$, where $u = \sin v$ and $v = 3x$. Then

$\frac{dy}{dx} = \frac{d}{dx}(\sin^5 3x) = \frac{d}{du}(u^5) \cdot \frac{d}{dv}(\sin v) \cdot \frac{d}{dx}(3x)$

$= 5u^4(\cos v)(3) = 15u^4 \cos v =$

$15 \sin^4 3x \cos 3x$. Therefore, $\frac{d}{dx}(x^2 \sin^5 3x) =$

$x^2 \frac{d}{dx}(\sin^5 3x) + \frac{d}{dx}(x^2) \sin^5 3x$

$= 15x^2 \sin^4 3x \cos 3x + 2x \sin^5 3x.$

25 $\frac{1}{(2x + 3)^5} = (2x + 3)^{-5}$, so let $y = u^{-5}$, where

$u = 2x + 3$; then $\frac{d}{dx}((2x + 3)^{-5}) =$

$\frac{d}{du}(u^{-5}) \cdot \frac{d}{dx}(2x + 3) = (-5u^{-6}) \cdot 2 = \frac{-10}{(2x + 3)^6}.$

27 $\frac{(x^2 + 1)^3(x^2)' - x^2[(x^2 + 1)^3]'}{[(x^2 + 1)^3]^2}$

$= \frac{(x^2 + 1)^3 \cdot 2x - x^2 \cdot 3(x^2 + 1)^2 \cdot 2x}{(x^2 + 1)^6}$

$= \frac{2x}{(x^2 + 1)^4}((x^2 + 1) - 3x^2) = \frac{2x(1 - 2x^2)}{(x^2 + 1)^4}$

29 $\left[\frac{\sqrt{1 - x^2} \sec^2 x}{1 + x}\right]' =$

$\frac{(1 + x)\left[\sqrt{1 - x^2} \sec^2 x\right]' - \left(\sqrt{1 - x^2} \sec^2 x\right)[1 + x]'}{(1 + x)^2} =$

$\frac{(1 + x)\left[\sqrt{1 - x^2} \cdot 2 \sec^2 x \tan x + \sec^2 x \cdot \frac{-x}{\sqrt{1 - x^2}}\right] - \sqrt{1 - x^2} \sec^2 x}{(1 + x)^2}$

$$= \frac{\sec^2 x}{\sqrt{1-x^2}(1+x)^2}[2(1+x-x^2-x^3)\tan x - (1+x)]$$

$$= \frac{\sec^2 x}{\sqrt{1-x^2}(1+x)}\left[-1+2(1-x^2)\tan x\right].$$

31 $\left[\left(\frac{x^2+3x+5}{2x-1}\right)^5\right]'$

$$= 5\left(\frac{x^2+3x+5}{2x-1}\right)^4 \frac{(2x-1)(2x+3)-(x^2+3x+5)\cdot 2}{(2x-1)^2}$$

$$= \frac{5(2x^2-2x-13)(x^2+3x+5)^4}{(2x-1)^6}$$

33 $[(2x+1)^3 \cot^4 x^2]' =$

$3(2x+1)^2\cdot 2\cot^4 x^2 + (2x+1)^3\cdot 4\cot^3 x^2\cdot(-\csc^2 x^2)\cdot 2x$

$= (2x+1)^2 \cot^3 x^2\,[6\cot x^2 - 8x(2x+1)\csc^2 x^2]$

$= (2x+1)^2 \cot^3 x^2\,[6\cot x^2 - (16x^2+8x)\csc^2 x^2]$

35 $$\left(\frac{x}{\sqrt{1-x^2}}\right)' = \frac{\sqrt{1-x^2}\cdot 1 - x\cdot\frac{1}{2\sqrt{1-x^2}}(-2x)}{\left(\sqrt{1-x^2}\right)^2}$$

$$= \frac{1}{(1-x^2)^{3/2}}$$

37 Let $y = 5u^4 + 1$ and $u = 3x - 2$; then

$$\frac{d}{dx}(5(3x-2)^4+1) = \frac{d}{du}(5u^4+1)\cdot\frac{d}{dx}(3x-2)$$

$= 5\cdot 4u^3\cdot 3 = 60(3x-2)^3$. Now let $y = \sqrt{u}$ and $u = 5(3x-2)^4 + 1$. It follows that

$$\frac{d}{dx}\left(\sqrt{5(3x-2)^4+1}\right) = \frac{d}{du}\left(\sqrt{u}\right)\cdot\frac{d}{dx}(5(3x-2)^4+1)$$

$$= \frac{1}{2\sqrt{u}}\cdot 60(3x-2)^3 = \frac{30(3x-2)^3}{\sqrt{5(3x-2)^4+1}}.$$

39 Let $y = \cos u$, where $u = 3x$; then $\frac{d}{dx}(\cos 3x) =$

$$\frac{d}{du}(\cos u)\cdot\frac{d}{dx}(3x) = (-\sin u)(3) = -3\sin 3x.$$

Now let $y = \sin u$, where $u = 4x$; then $\frac{d}{dx}(\sin 4x)$

$$= \frac{d}{du}(\sin u)\cdot\frac{d}{dx}(4x) = (\cos u)(4) = 4\cos 4x.$$

Thus $\frac{d}{dx}(\cos 3x \sin 4x)$

$$= \cos 3x\,\frac{d}{dx}(\sin 4x) + \sin 4x\,\frac{d}{dx}(\cos 3x)$$

$$= 4\cos 3x\cos 4x - 3\sin 4x\sin 3x.$$

41 $$\left[\frac{2(9x-2)}{135}\sqrt{(3x+1)^3}\right]'$$

$$= \frac{2}{135}\left[(9x-2)\cdot\frac{3}{2}\sqrt{3x+1}\cdot 3 + 9\sqrt{3x+1)^3}\right]$$

$$= \frac{2}{135}\left[\frac{9}{2}(9x-2)\sqrt{3x+1} + 9(3x+1)\sqrt{3x+1}\right]$$

$$= \frac{1}{15}\sqrt{3x+1}\,[(9x-2)+2(3x+1)]$$

$$= x\sqrt{3x+1}$$

43 $$\left(\frac{3x}{8} - \frac{3\sin 5x\cos 5x}{40} - \frac{\sin^3 5x\cos 5x}{20}\right)'$$

$$= \frac{3}{8} - \frac{3}{40}(\sin 5x\,(-5\sin 5x) + \cos 5x\,(5\cos 5x)) -$$

$$\frac{1}{20}(\sin^3 5x\,(-5\sin 5x) + \cos 5x\,(3\sin^2 5x)(5\cos 5x))$$

$$= \frac{3}{8} + \frac{3}{8}\sin^2 5x - \frac{3}{8}\cos^2 5x + \frac{1}{4}\sin^4 5x -$$

$$\frac{3}{4}\cos^2 5x\sin^2 5x$$

$= \frac{3}{8}(1 + \sin^2 5x - \cos^2 5x) + \frac{1}{4}\sin^4 5x -$

$\frac{3}{4}(1 - \sin^2 5x)(\sin^2 5x)$

$= \frac{3}{8}(2\sin^2 5x) + \frac{1}{4}\sin^4 5x - \frac{3}{4}\sin^2 5x + \frac{3}{4}\sin^4 5x$

$= \sin^4 5x$

45 Note that $\frac{d}{dx}(1 + x^2) = 2x$. So $8x(1 + x^2)^3 =$ $2x \cdot 4(1 + x^2)^3$. Letting $u = 1 + x^2$, we have $2x \cdot 4(1 + x^2)^3 = \frac{du}{dx} \cdot 4u^3$. This suggests choosing $y = u^4$, since $\frac{dy}{dx} = \frac{d}{du}(u^4) \cdot \frac{du}{dx} = 4u^3 \cdot \frac{du}{dx}$. Thus, $y = (1 + x^2)^4$ is an antiderivative. (Any $y = (1 + x^2)^4 + C$ will do, where C is a constant.)

47 Note that $\frac{d}{dx}(1 + x^3) = 3x^2$. So $3x^2(1 + x^3)^4 =$ $(1/5)3x^2 \cdot 5(1 + x^3)^4$. Letting $u = 1 + x^3$, we have $\frac{1}{5} \cdot 3x^2 \cdot 5(1 + x^3)^4 = \frac{1}{5}\frac{du}{dx} \cdot 5u^4$. This suggests choosing $y = \frac{u^5}{5}$, since $\frac{dy}{dx} = \frac{1}{5}\frac{d}{du}(u^5) \cdot \frac{du}{dx} =$ $\frac{1}{5}5u^4 \cdot \frac{du}{dx}$. Thus, $y = \frac{1}{5}(1 + x^3)^5$ is an antiderivative. (Any $y = \frac{1}{5}(1 + x^3)^5 + C$ will do, where C is a constant.)

49 Since $f(3) = 2$ and $g(5) = 3$, we see that $f(g(5)) = 2$. Similarly, $f'(g(5)) = 4$. This suggests that it is possible to compute $(f \circ g)'(x) = [f(g(x))]'$ at $x = 5$, since $[f(g(x))]' = f'(g(x)) \cdot g'(x)$. Given $g'(5) = 7$ and $f'(g(5)) = 4$, we see $(f \circ g)'(5) = 4 \cdot 7 = 28$.

51 (a)

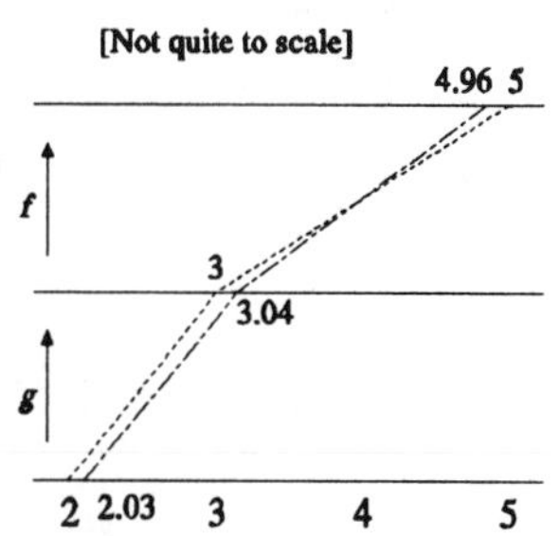

(b) $h'(2) = f'(g(2)) \cdot g'(2) = f'(3) \cdot g'(2)$. Since

$f'(3) \approx \frac{f(3.04) - f(3)}{0.04} = \frac{4.96 - 5}{0.04} = -1$

and $g'(2) \approx \frac{g(2.03) - g(2)}{0.03} = \frac{3.04 - 3}{0.03} =$ $\frac{4}{3}$, $h'(2) \approx -1 \cdot \frac{4}{3} = -\frac{4}{3}$.

53 Let $T(t)$ denote the temperature at any time t. Then $\frac{dT}{dt} = \frac{dT}{dx} \cdot \frac{dx}{dt}$. Now, $\frac{dx}{dt} = 2$ cm/min and $\left.\frac{dT}{dx}\right|_{x=3} = 2x|_{x=3} = 6°\ C/\text{cm}$. Therefore, $\frac{dT}{dt} =$ $(2\text{ cm/min})(6°\ C/\text{cm}) = 12°\ C/\text{min}$.

55 (a) The right-hand side of $h'(x) = f'(g(x)) \cdot g'(x)$ is 0, so we must prove $h'(x) = 0$; that is,

$\lim_{\Delta x \to 0} \frac{\Delta y}{\Delta x} = 0.$

(b) As $\Delta x \to 0$ through values of the first type, $\frac{\Delta u}{\Delta x} \to g'(x) = 0$, so $\frac{\Delta y}{\Delta x} = \frac{\Delta y}{\Delta u} \cdot \frac{\Delta u}{\Delta x} \to 0$.

(c) Observe that $\Delta y = h(x + \Delta x) - h(x)$

$= f(g(x + \Delta x)) - f(g(x))$

$= f(g(x) + \Delta u) - f(g(x)) = f(g(x)) - f(g(x))$

$= 0$.

3.S Guide Quiz

2 (a) $\frac{d}{dx}(5x^3 - 2x + 2)$

$= \lim_{h\to 0} \frac{[5(x+h)^3 - 2(x+h) + 2] - [5x^3 - 2x + 2]}{h}$

$= \lim_{h\to 0} \frac{[5x^3 + 15x^2h + 15xh^2 + 5h^3 - 2x - 2h + 2] - [5x^3 - 2x + 2]}{h}$

$= \lim_{h\to 0} (15x^2 + 15xh + 5h^2 - 2)$

$= 15x^2 - 2$

(b) $\frac{d}{dx}\left(\frac{5}{3x+2} + 6x\right) =$

$\lim_{h\to 0} \frac{\left[\frac{5}{3(x+h)+2} + 6(x+h)\right] - \left[\frac{5}{3x+2} + 6x\right]}{h}$

$= \lim_{h\to 0} \frac{1}{h}\left[\frac{5(3x+2) - 5(3x+3h+2)}{(3x+2)(3x+3h+2)} + 6x + 6h - 6x\right]$

$= \lim_{h\to 0} \left[\frac{-15}{(3x+2)(3x+3h+2)} + 6\right]$

$= -\frac{15}{(3x+2)^2} + 6$

(c) $\frac{d}{dx}(3 \sin 2x)$

$= \lim_{h\to 0} \frac{3 \sin(2(x+h)) - 3 \sin 2x}{h}$

$= 3 \lim_{h\to 0} \frac{\sin(2x+2h) - \sin 2x}{h}$

$= 3 \lim_{h\to 0} \frac{\sin 2x \cos 2h + \cos 2x \sin 2h - \sin 2x}{h}$

$= 3 \lim_{h\to 0} \frac{\sin 2x\ (\cos 2h - 1) + \cos 2x \sin 2h}{h}$

$= 3\left[(\sin 2x) \lim_{h\to 0} \frac{\cos 2h - 1}{h} + (\cos 2x) \lim_{h\to 0} \frac{\sin 2h}{h}\right]$

$= 3\left[(\sin 2x)(-2) \lim_{h\to 0} \frac{1 - \cos 2h}{2h} + (\cos 2x)\cdot 2 \lim_{h\to 0} \frac{\sin 2h}{2h}\right]$

$= 3[(\sin 2x)(-2)\cdot 0 + (\cos 2x)\cdot 2\cdot 1]$

$= 6 \cos 2x$

(d) $\frac{d}{dx}(x^{-2}) = \lim_{h\to 0} \frac{(x+h)^{-2} - x^{-2}}{h}$

$= \lim_{h\to 0} \frac{\frac{1}{(x+h)^2} - \frac{1}{x^2}}{h}$

$= \lim_{h\to 0} \frac{x^2 - (x+h)^2}{hx^2(x+h)^2}$

$= \lim_{h\to 0} \frac{x^2 - x^2 - 2xh - h^2}{hx^2(x+h)^2}$

$= \lim_{h\to 0} \frac{-2x - h}{x^2(x+h)^2} = \frac{-2x}{x^4} = -2x^{-3}$

3 (a) $D(5\sqrt{x}) = 5\,D(\sqrt{x}) = 5\,D(x^{1/2}) = 5\cdot\frac{1}{2}x^{(1/2)-1}$

$= \frac{5}{2}x^{-1/2}$

(b) Let $y = \sqrt{u}$ and $u = 3 - 2x^2$; then

$\frac{d}{dx}\left(\sqrt{3 - 2x^2}\right) = \frac{d}{du}(\sqrt{u})\cdot\frac{d}{dx}(3 - 2x^2)$

$= \frac{1}{2\sqrt{u}}\cdot(-4x) = \frac{-2x}{\sqrt{3 - 2x^2}}$. Thus

$\frac{d}{dx}\left(x^2\sqrt{3 - 2x^2}\right)$

$= x^2\frac{d}{dx}\left(\sqrt{3 - 2x^2}\right) + \sqrt{3 - 2x^2}\,\frac{d}{dx}(x^2)$

$$= x^2 \cdot \frac{-2x}{\sqrt{3 - 2x^2}} + \sqrt{3 - 2x^2} \cdot 2x$$

$$= \frac{x}{\sqrt{3 - 2x^2}}[-2x^2 + 2(3 - 2x^2)]$$

$$= \frac{x}{\sqrt{3 - 2x^2}}(-6x^2 + 6) = \frac{6x(1 - x^2)}{\sqrt{3 - 2x^2}}.$$

(c) Let $y = \cos u$, where $u = 5x$; then

$$\frac{d}{dx}(\cos 5x) = \frac{d}{du}(\cos u) \cdot \frac{d}{dx}(5x) =$$

$(-\sin u) \cdot 5 = -5 \sin 5x$.

(d) Let $y = u^{3/4}$ and $u = 1 + x^2$; then

$$\frac{d}{dx}((1 + x^2)^{3/4}) = \frac{d}{du}(u^{3/4}) \cdot \frac{d}{dx}(1 + x^2)$$

$$= \frac{3}{4}u^{(3/4)-1} \cdot 2x = \frac{3x}{2}u^{-1/4} = \frac{3x}{2(1 + x^2)^{1/4}}.$$

(e) Let $y = \sqrt[3]{u}$, $u = \tan v$, and $v = 6x$; then

$$\frac{d}{dx}\left(\sqrt[3]{\tan 6x}\right) = \frac{d}{du}\left(\sqrt[3]{u}\right) \cdot \frac{d}{dv}(\tan v) \cdot \frac{d}{dx}(6x)$$

$$= \frac{1}{3}u^{-2/3}(\sec^2 v) \cdot 6 = \frac{2 \sec^2 6x}{(\tan 6x)^{2/3}}.$$

(f) Let $y = \sin u$ and $u = 5x$; then $\frac{d}{dx}(\sin 5x) =$

$$\frac{d}{du}(\sin u) \cdot \frac{d}{dx}(5x) = (\cos u)(5) = 5 \cos 5x.$$

Therefore $\frac{d}{dx}(x^3 \sin 5x)$

$$= x^3 \cdot \frac{d}{dx}(\sin 5x) + (\sin 5x) \cdot \frac{d}{dx}(x^3)$$

$= x^3 \cdot 5 \cos 5x + (\sin 5x) \cdot 3x^2$

$= 5x^3 \cos 5x + 3x^2 \sin 5x$.

(g) $\frac{1}{\sqrt{2x + 1}} = (2x + 1)^{-1/2}$, so let $y = u^{-1/2}$ and

$u = 2x + 1$; then $\frac{d}{dx}((2x + 1)^{-1/2})$

$$= \frac{d}{du}(u^{-1/2}) \cdot \frac{d}{dx}(2x + 1) = -\frac{1}{2}u^{(-1/2)-1} \cdot 2$$

$$= -u^{-3/2} = \frac{-1}{(2x + 1)^{3/2}}.$$

(h) Let $y = u^{-4}$ and $u = 2x^5 - x^3$; then

$$\frac{d}{dx}((2x^5 - x^3)^{-4}) = \frac{d}{du}(u^{-4}) \cdot \frac{d}{dx}(2x^5 - x^3)$$

$$= -4u^{-5}(10x^4 - 3x^2) = \frac{-4(10x^4 - 3x^2)}{(2x^5 - x^3)^5}$$

$$= \frac{-4x^2(10x^2 - 3)}{x^{15}(2x^2 - 1)^5} = -\frac{4(10x^2 - 3)}{x^{13}(2x^2 - 1)^5}.$$

(i) Let $y = \sqrt[3]{u}$ and $u = x^3 - 3$; then

$$\frac{d}{dx}\left(\sqrt[3]{x^3 - 3}\right) = \frac{d}{du}\left(\sqrt[3]{u}\right) \cdot \frac{d}{dx}(x^3 - 3)$$

$$= \frac{1}{3}u^{-2/3} \cdot 3x^2 = \frac{x^2}{(x^3 - 3)^{2/3}}.$$

(j) $\left(\frac{2x^3 + 2}{3x + 1}\right)'$

$$= \frac{(3x + 1)(2x^3 + 2)' - (2x^3 + 2)(3x + 1)'}{(3x + 1)^2}$$

$$= \frac{(3x + 1)(6x^2) - (2x^3 + 2) \cdot 3}{(3x + 1)^2}$$

$$= \frac{18x^3 + 6x^2 - 6x^3 - 6}{(3x + 1)^2}$$

$$= \frac{12x^3 + 6x^2 - 6}{(3x + 1)^2} = \frac{6(2x^3 + x^2 - 1)}{(3x + 1)^2}$$

(k) $\left(\dfrac{1}{5x^2 + 1}\right)' = \dfrac{-(5x^2 + 1)'}{(5x^2 + 1)^2} = \dfrac{-10x}{(5x^2 + 1)^2}$

(l) $\dfrac{1}{(3x + 2)^{10}} = (3x + 2)^{-10}$, so let $y = u^{-10}$

and $u = 3x + 2$; then $\dfrac{d}{dx}((3x + 2)^{-10}) =$

$\dfrac{d}{du}(u^{-10}) \cdot \dfrac{d}{dx}(3x + 2) = -10u^{-11} \cdot 3 =$

$\dfrac{-30}{(3x + 2)^{11}}$.

(m) Let $y = u^5$ and $u = 1 + 2x$; then

$\dfrac{d}{dx}((1 + 2x)^5) = \dfrac{d}{du}(u^5) \cdot \dfrac{d}{dx}(1 + 2x) =$

$5u^4 \cdot 2 = 10(1 + 2x)^4$. Let $y = \sec u$ and $u =$

$3x$; then $\dfrac{d}{dx}(\sec 3x) = \dfrac{d}{du}(\sec u) \cdot \dfrac{d}{dx}(3x) =$

$(\sec u \tan u) \cdot 3 = 3 \sec 3x \tan 3x$. It follows

from Exercise 55 of Sec. 3.4 that

$\dfrac{d}{dx}((1 + 2x)^5 x^3 \sec 3x)$

$= \dfrac{d}{dx}((1 + 2x)^5) x^3 \sec 3x$

$+ (1 + 2x)^5 \cdot \dfrac{d}{dx}(x^3) \sec 3x + (1 + 2x)^5 x^3 \dfrac{d}{dx}(\sec 3x)$

$= 10(1 + 2x)^4 x^3 \sec 3x + (1 + 2x)^5 3x^2 \sec 3x$

$+ (1 + 2x)^5 x^3 \cdot 3 \sec 3x \tan 3x =$

$(1 + 2x)^4 x^2 \sec 3x\,[10x + (1 + 2x) \cdot 3 + (1 + 2x)\, 3x \tan 3x]$

$= x^2(1 + 2x)^4 \sec 3x\,[16x + 3 + 3x(1 + 2x) \tan 3x]$.

(n) Let $y = \csc u$ and $u = \sqrt{x}$; then $\dfrac{d}{dx}(\csc \sqrt{x})$

$= \dfrac{d}{du}(\csc u) \cdot \dfrac{d}{dx}(\sqrt{x}) = -\csc u \cot u \dfrac{1}{2\sqrt{x}}$

$= \dfrac{-\csc \sqrt{x} \cot \sqrt{x}}{2\sqrt{x}}$.

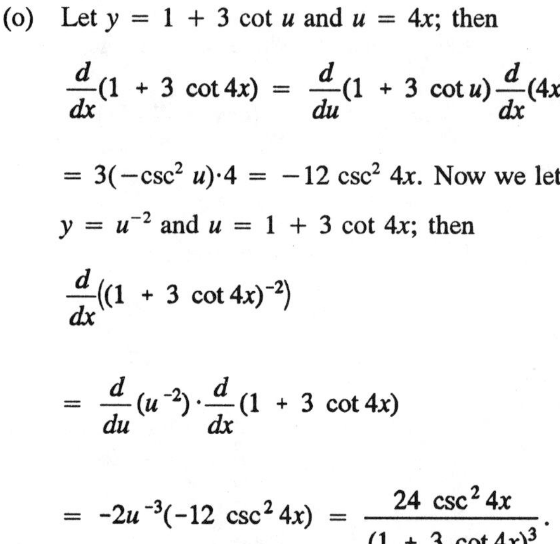

(o) Let $y = 1 + 3 \cot u$ and $u = 4x$; then

$\dfrac{d}{dx}(1 + 3 \cot 4x) = \dfrac{d}{du}(1 + 3 \cot u) \dfrac{d}{dx}(4x)$

$= 3(-\csc^2 u) \cdot 4 = -12 \csc^2 4x$. Now we let

$y = u^{-2}$ and $u = 1 + 3 \cot 4x$; then

$\dfrac{d}{dx}((1 + 3 \cot 4x)^{-2})$

$= \dfrac{d}{du}(u^{-2}) \cdot \dfrac{d}{dx}(1 + 3 \cot 4x)$

$= -2u^{-3}(-12 \csc^2 4x) = \dfrac{24 \csc^2 4x}{(1 + 3 \cot 4x)^3}$.

5 (a)

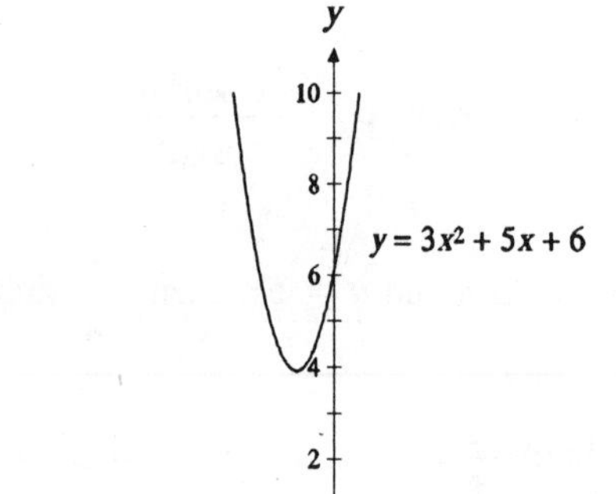

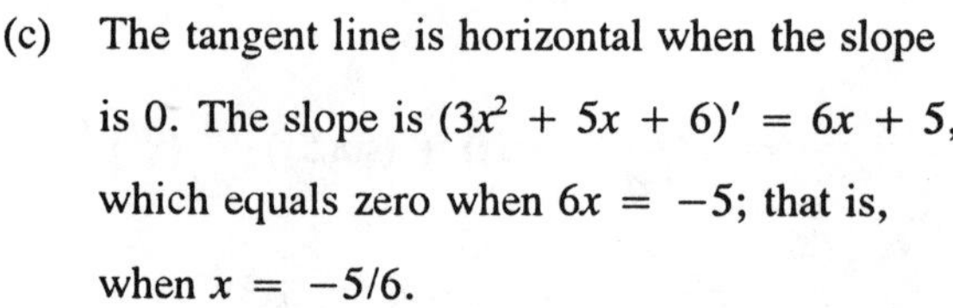

(c) The tangent line is horizontal when the slope is 0. The slope is $(3x^2 + 5x + 6)' = 6x + 5$, which equals zero when $6x = -5$; that is, when $x = -5/6$.

6 (a)

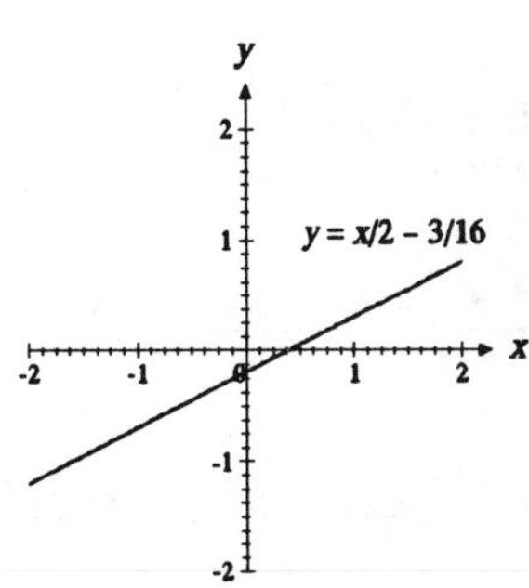

$(x^4)' = 4x^3 = 4\left(\frac{1}{2}\right)^3 = \frac{4}{8} = \frac{1}{2}$ when $x = \frac{1}{2}$, so the slope of the tangent line is $\frac{1}{2}$.

Knowing the slope and the fact that it passes through the point $\left(\frac{1}{2}, \left(\frac{1}{2}\right)^4\right) = \left(\frac{1}{2}, \frac{1}{16}\right)$ allows us to sketch the tangent line.

(b) Since the line has slope 1/2 and passes through $\left(\frac{1}{2}, \frac{1}{16}\right)$, its equation is $y - \frac{1}{16} = \frac{1}{2}\left(x - \frac{1}{2}\right)$, or $y = \frac{x}{2} - \frac{3}{16}$.

7 (a)

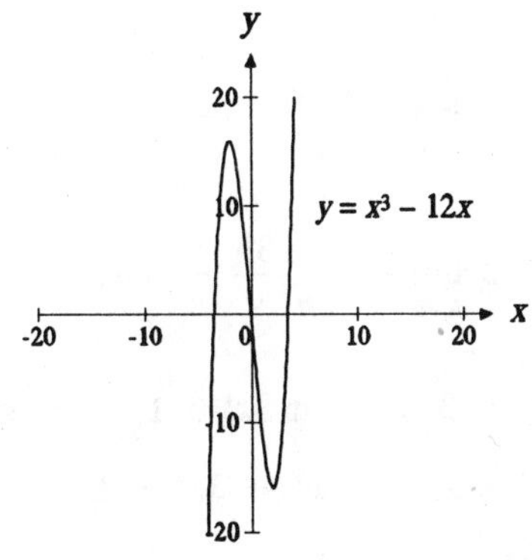

x	-2	-1	0	1	2	3
$x^3 - 12x$	16	11	0	-11	-16	-9

(b) $(x^3 - 12x)' = 3x^2 - 12$

(c) $3x^2 - 12 = 0$ when $3x^2 = 12$, $x^2 = 4$, $x = \pm 2$.

(d) The tangent line is horizontal when the derivative is 0; that is, at the points $(-2, 16)$ and $(2, -16)$.

8 (a) $f(x + \Delta x) - f(x)$ is the change in height from time x to time $x + \Delta x$ and Δx is the change in time from time x to time $x + \Delta x$. $\frac{f(x + \Delta x) - f(x)}{\Delta x}$ is the average rate of change of height from time x to time $x + \Delta x$.

(b) $f(x + \Delta x) - f(x)$ is the growth of the bacteria from time x to time $x + \Delta x$ and Δx is the change in time. $\frac{f(x + h) - f(x)}{h}$ is the average rate of growth from time x to time $x + \Delta x$.

(c) $f(x + \Delta x) - f(x)$ is the mass of the section of the rod between x and $x + \Delta x$ centimeters from the left end of the rod. Δx is the length of the section of the rod. $\frac{f(x + \Delta x) - f(x)}{\Delta x}$ is the average density of the section of the rod.

(d) $f(x + \Delta x) - f(x)$ is the distance between the images on the screen of the points x and $x + \Delta x$. Δx is the distance between the points x and $x + \Delta x$ on the slide. $\frac{f(x + \Delta x) - f(x)}{\Delta x}$ is the magnification of the interval $[x, x + \Delta x]$.

9 (a) $(t^2 - 2t)' = 2t - 2$, so the velocity at time t is $(2t - 2)$ m/sec.

(b) The velocity when $t = 1/4$ is $2\left(\frac{1}{4}\right) - 2$

$= -3/2$ m/sec.

(c) The speed when $t = 1/4$ is $|-3/2|$

$= 3/2$ m/sec.

(d) The bug is moving to the left when $t = 1/4$, since the velocity is negative.

10 (a) $\frac{x^6}{6} + C$

(b) $-\frac{1}{2}\cos 2x + C$

(c) $\sin x^2$

11 The answers to this problem are only approximate, and will depend on how you read the graph.

(a) We see that $f(0) = 1$, $f(1) = 2$, $f(2) = 2.5$, $f(3) = 1$, and $f(4) = 0$. The value of $f'(1)$ is approximately equal to the slope of the line segment connecting (0, 1) and (2, 2.5). Hence

$f'(1) \approx \frac{2.5 - 1}{2 - 0} = 0.75.$

(b) Since $g(x) = (f(x))^2$, we have $g'(x) = 2f(x)f'(x)$. Hence $g'(3) = 2f(3)f'(3)$. The tangent to $y = f(x)$ at $x = 3$ appears to pass through the point (3.5, 0), so $f'(3) \approx$

$\frac{0 - 1}{3.5 - 3} = -2$. Thus $g'(3) = 2\cdot1\cdot(-2)$

$= -4.$

(c) Since $h(x) = 1/f(x)$, we have $h'(x) = -f'(x)/((f(x))^2)$. For $x = 1$, we use the results of (a) to obtain $h'(1) \approx -0.75/(2^2) = -0.1875$. (It is unreasonable to suppose that all of these decimal places are significant, but it's fair to say that $h'(1) \approx -0.2$.)

3.S Review Exercises

1 $D(5x^3) = \lim_{h\to0} \frac{5(x+h)^3 - 5x^3}{h}$

$= 5\lim_{h\to0} \frac{x^3 + 3x^2h + 3xh^2 + h^3 - x^3}{h}$

$= 5\lim_{h\to0} (3x^2 + 3xh + h^2) = 5\cdot3x^2 = 15x^2$

3 $D\left(\frac{1}{x+3}\right) = \lim_{h\to0} \frac{\frac{1}{(x+h)+3} - \frac{1}{x+3}}{h}$

$= \lim_{h\to0} \frac{(x+3) - (x+h+3)}{h(x+3)(x+h+3)}$

$= \lim_{h\to0} \frac{-1}{(x+3)(x+h+3)} = -\frac{1}{(x+3)^2}$

5 $D(\cos 3x) = \lim_{h\to0} \frac{\cos 3(x+h) - \cos 3x}{h}$

$= \lim_{h\to0} \frac{\cos(3x+3h) - \cos 3x}{h}$

$= \lim_{h\to0} \frac{\cos 3x \cos 3h - \sin 3x \sin 3h - \cos 3x}{h}$

$= \lim_{h\to0} \frac{-\cos 3x\,(1 - \cos 3h) - \sin 3x \sin 3h}{h}$

$= -(\cos 3x)\lim_{h\to0} \frac{1-\cos 3h}{h} - (\sin 3x)\lim_{h\to0} \frac{\sin 3h}{h} =$

$-(\cos 3x)\cdot3\lim_{h\to0} \frac{1-\cos 3h}{3h} - (\sin 3x)\cdot3\lim_{h\to0} \frac{\sin 3h}{3h}$

$= -(\cos 3x)\cdot3\cdot0 - (\sin 3x)\cdot3\cdot1 = -3\sin 3x$

7 $(2x^5 + x^3 - x)' = 10x^4 + 3x^2 - 1$

9 $\left(\frac{x^2}{4x+1}\right)' = \frac{(4x+1)(x^2)' - x^2(4x+1)'}{(4x+1)^2}$

$= \dfrac{(4x+1)(2x) - x^2(4)}{(4x+1)^2} = \dfrac{8x^2 + 2x - 4x^2}{(4x+1)^2}$

$= \dfrac{4x^2 + 2x}{(4x+1)^2}$

11 $y = \sqrt{u}$, $u = 3x^2 + 2x + 4$: $\dfrac{d}{dx}\left(\sqrt{3x^2 + 2x + 4}\right)$

$= \dfrac{d}{du}(\sqrt{u}) \cdot \dfrac{d}{dx}(3x^2 + 2x + 4) = \dfrac{1}{2\sqrt{u}} \cdot (6x + 2)$

$= \dfrac{3x + 1}{\sqrt{u}} = \dfrac{3x + 1}{\sqrt{3x^2 + 2x + 4}}$.

13 $\sqrt[3]{(2t-1)^2} = (2t-1)^{2/3}$; let $y = u^{2/3}$ and $u =$ $2t - 1$; then $\dfrac{d}{dt}((2t-1)^{2/3}) =$

$\dfrac{d}{du}(u^{2/3}) \cdot \dfrac{d}{dt}(2t - 1) = \dfrac{2}{3}u^{(2/3)-1} \cdot 2 = \dfrac{4}{3}u^{-1/3}$

$= \dfrac{4}{3\sqrt[3]{2t-1}}$.

15 $y = u^2$, $u = \sin v$, and $v = 5x$: $\dfrac{d}{dx}(\sin^2 5x)$

$= \dfrac{d}{du}(u^2) \cdot \dfrac{d}{dv}(\sin v) \cdot \dfrac{d}{dx}(5x) = 2u(\cos v)(5)$

$= 10 \sin 5x \cos 5x$.

17 $y = \dfrac{u^4}{7}$, $u = 5x + 1$: $\dfrac{d}{dx}\left(\dfrac{(5x+1)^4}{7}\right)$

$= \dfrac{d}{du}\left(\dfrac{u^4}{7}\right) \cdot \dfrac{d}{dx}(5x + 1) = \dfrac{1}{7} \cdot 4u^3 \cdot 5$

$= \dfrac{20(5x+1)^3}{7}$.

19 Let $y = \sin u$, $u = 3x$; then $\dfrac{d}{dx}(\sin 3x) =$

$\dfrac{d}{du}(\sin u) \cdot \dfrac{d}{dx}(3x) = (\cos u)(3) = 3 \cos 3x$. Thus

$\dfrac{d}{dx}(x \sin 3x) = x \cdot \dfrac{d}{dx}(\sin 3x) + \sin 3x \dfrac{d}{dx}(x) =$

$x \cdot 3 \cos 3x + (\sin 3x) \cdot 1 = 3x \cos 3x + \sin 3x$.

21 Let $y = \sqrt[3]{u}$, $u = 1 + 2x$: $\dfrac{d}{dx}\left(\sqrt[3]{1 + 2x}\right) =$

$\dfrac{d}{du}(u^{1/3}) \cdot \dfrac{d}{dx}(1 + 2x) = \dfrac{1}{3}u^{-2/3} \cdot 2 =$

$\dfrac{2}{3}(1 + 2x)^{-2/3}$. Now instead let $y = u^2$, $u = \tan v$,

and $v = \sqrt[3]{1 + 2x}$; then $\dfrac{dy}{dx} = \dfrac{d}{dx}\left(\tan^2(\sqrt[3]{1 + 2x})\right)$

$= \dfrac{dy}{du} \cdot \dfrac{du}{dv} \cdot \dfrac{dv}{dx} = \dfrac{d}{du}(u^2) \cdot \dfrac{d}{dv}(\tan v) \cdot \dfrac{d}{dx}\left(\sqrt[3]{1 + 2x}\right)$

$= 2u(\sec^2 v) \cdot \dfrac{2}{3}(1 + 2x)^{-2/3}$

$= \dfrac{4 \tan\left(\sqrt[3]{1 + 2x}\right) \sec^2\left(\sqrt[3]{1 + 2x}\right)}{3(1 + 2x)^{2/3}}$.

23 Let $y = \cos u$, $u = 2x$; then $\dfrac{d}{dx}(\cos 2x) =$

$\dfrac{d}{du}(\cos u) \cdot \dfrac{d}{dx}(2x) = (-\sin u) \cdot 2 = -2 \sin 2x$.

Thus $\dfrac{d}{dx}\left(\dfrac{x^3 \cos 2x}{1 + x^2}\right)$

$= \dfrac{(1 + x^2)\dfrac{d}{dx}(x^3 \cos 2x) - x^3 \cos 2x \dfrac{d}{dx}(1 + x^2)}{(1 + x^2)^2}$

$= \dfrac{(1 + x^2)\left[x^3 \dfrac{d}{dx}(\cos 2x) + \cos 2x \dfrac{d}{dx}(x^3)\right] - x^3(\cos 2x)(2x)}{(1 + x^2)^2}$

$$= \frac{(1+x^2)[x^3(-2\sin 2x) + (\cos 2x)\cdot 3x^2] - 2x^4\cos 2x}{(1+x^2)^2}$$

$$= \frac{(1+x^2)(-2x^3\sin 2x + 3x^2\cos 2x) - 2x^4\cos 2x}{(1+x^2)^2}$$

$$= \frac{(x^4+3x^2)\cos 2x - (2x^5+2x^3)\sin 2x}{(1+x^2)^2}$$

25 $\frac{d}{dx}\left(\sqrt[3]{1+\sqrt{x^2+3}}\right) = \frac{d}{dx}\left(\left(1+\sqrt{x^2+3}\right)^{1/3}\right)$

$$= \frac{1}{3}\left(1+\sqrt{x^2+3}\right)^{-2/3}\cdot\frac{1}{2\sqrt{x^2+3}}\cdot 2x$$

$$= \frac{x}{3\sqrt{x^2+3}\left(1+\sqrt{x^2+3}\right)^{2/3}}$$

27 $\sqrt[3]{(\cot 5x)^7} = (\cot 5x)^{7/3}$; $y = u^{7/3}$, $u = \cot v$, $v = 5x$; $\frac{d}{dx}((\cot 5x)^{7/3}) = \frac{d}{du}(u^{7/3})\cdot\frac{d}{dv}(\cot v)\cdot\frac{d}{dx}(5x)$

$$= \frac{7}{3}u^{(7/3)-1}\cdot(-\csc^2 v)(5) = -\frac{35}{3}(\csc^2 5x)\cdot u^{4/3}$$

$$= -\frac{35}{3}(\csc^2 5x)(\cot 5x)^{4/3}$$

29 $y = \sqrt{u}$, $u = 8x+3$: $\frac{d}{dx}(\sqrt{8x+3}) =$

$$\frac{d}{du}(\sqrt{u})\cdot\frac{d}{dx}(8x+3) = \frac{1}{2\sqrt{u}}\cdot 8 = \frac{4}{\sqrt{8x+3}}.$$

31 $\left(\frac{x^2}{x^3+1}\right)' = \frac{(x^3+1)(x^2)' - x^2(x^3+1)'}{(x^3+1)^2}$

$$= \frac{(x^3+1)(2x) - x^2(3x^2)}{(x^3+1)^2} = \frac{2x^4+2x-3x^4}{(x^3+1)^2} =$$

$$\frac{2x-x^4}{(x^3+1)^2}$$

33 Let $y = u^4$, $u = x^2+3x$; then $\frac{d}{dx}((x^2+3x)^4)$

$$= \frac{d}{du}(u^4)\cdot\frac{d}{dx}(x^2+3x) = 4u^3\cdot(2x+3)$$

$= 4(2x+3)(x^2+3x)^3$. Thus $\frac{d}{dx}((x^2+3x)^4+x) =$

$$\frac{d}{dx}((x^2+3x)^4) + \frac{d}{dx}(x) = 4(2x+3)(x^2+3x)^3 + 1.$$

Now let $y = u^{-5/7}$ and $u = (x^2+3x)^4 + x$, so $\frac{dy}{dx}$

$$= \frac{d}{dx}(((x^2+3x)^4+x)^{-5/7})$$

$$= \frac{d}{du}(u^{-5/7})\cdot\frac{d}{dx}((x^2+3x)^4+x)$$

$$= -\frac{5}{7}u^{(-5/7)-1}\,[4(2x+3)(x^2+3x)^3+1]$$

$$= -\frac{5[4(2x+3)(x^2+3x)^3+1]}{7[(x^2+3x)^4+x]^{12/7}}.$$

35 Let $y = u^{1/2}$, where $u = 2x+1$. Then $\frac{dy}{dx} =$ $\frac{dy}{du}\frac{du}{dx} = \frac{1}{2}u^{-1/2}\cdot(2) = u^{-1/2} = (2x+1)^{-1/2}$. Let $y = \cos u$, where $u = 6x$. Then $\frac{dy}{dx} = \frac{dy}{du}\cdot\frac{du}{dx} =$ $-\sin u\cdot(6) = -6\sin 6x$. By Exercise 55 of Sec. 3.4, we have $\left[x\sqrt{2x+1}\cos 6x\right]' =$

$$(x)'\sqrt{2x+1}\cos 6x + x(\sqrt{2x+1})'\cos 6x + x\sqrt{2x+1}(\cos 6x)'$$

$$= \sqrt{2x+1}\cos 6x + x\cos 6x\,(2x+1)^{-1/2} + x\sqrt{2x+1}(-6\sin 6x)$$

$= \sqrt{2x+1}\cos 6x + x\cos 6x(2x+1)^{-1/2}$

$- 6x\sqrt{2x+1}\sin 6x.$

37 Let $y = \frac{1}{a^2}\sin u$, $u = ax$: $\frac{d}{dx}\left(\frac{1}{a^2}\sin ax\right) =$

$\frac{d}{du}\left(\frac{1}{a^2}\sin u\right)\cdot\frac{d}{dx}(ax) = \frac{1}{a^2}(\cos u)(a) = \frac{\cos ax}{a}.$

Thus $\frac{d}{dx}\left(\frac{1}{a^2}\sin ax - \frac{1}{a}x\cos ax\right)$

$= \frac{d}{dx}\left(\frac{1}{a^2}\sin ax\right) - \frac{d}{dx}\left(\frac{1}{a}x\cos ax\right)$

$= \frac{\cos ax}{a} - \left(\frac{1}{a}x\right)\frac{d}{dx}(\cos ax) - \cos ax\,\frac{d}{dx}\left(\frac{1}{a}x\right)$

$= \frac{\cos ax}{a} - \frac{x}{a}(-a\sin ax) - (\cos ax)\left(\frac{1}{a}\right)$

$= \frac{\cos ax}{a} + x\sin ax - \frac{\cos ax}{a} = x\sin ax.$

39 Let $y = \frac{\sin u}{4a}$, $u = 2ax$: $\frac{d}{dx}\left(\frac{\sin 2ax}{4a}\right) =$

$\frac{d}{du}\left(\frac{\sin u}{4a}\right)\cdot\frac{d}{dx}(2ax) = \frac{\cos u}{4a}\cdot 2a = \frac{\cos 2ax}{2}$; thus

$\frac{d}{dx}\left(\frac{x}{2} - \frac{\sin 2ax}{4a}\right) = \frac{d}{dx}\left(\frac{x}{2}\right) - \frac{d}{dx}\left(\frac{\sin 2ax}{4a}\right)$

$= \frac{1}{2} - \frac{\cos 2ax}{2} = \frac{1}{2}(1 - \cos 2ax)$

$= \frac{1}{2}(2\sin^2 ax) = \sin^2 ax.$

41 (a) Velocity $= (64t - 16t^2)' = 64 - 32t$ ft/sec

(b) The velocity when $t = 0$ is $64 - 32\cdot 0 =$ 64 ft/sec. When $t = 1$ the velocity is $64 - 32\cdot 1 = 32$ ft/sec. Velocity when $t = 2$ is $64 - 32\cdot 2 = 0$ ft/sec. Velocity when $t = 3$ is $64 - 32\cdot 3 = -32$ ft/sec.

(c) Speed when $t = 0$ is $|64| = 64$ ft/sec. Speed when $t = 1$ is $|32| = 32$ ft/sec. Speed when $t = 2$ is $|0| = 0$ ft/sec. Speed when $t = 3$ is $|-32| = 32$ ft/sec.

(d) The ball is rising when the velocity is positive, which occurs when $64 - 32t > 0$; so $64 > 32t$ and thus $0 < t < 2$. The ball is falling when the velocity is negative, which occurs when $64 - 32t < 0$, $32t > 64$, so $t > 2$.

43 The slope of the tangent line is $(x^3 - 2x^2)' = 3x^2 - 4x = 3\cdot 1^2 - 4\cdot 1 = 3 - 4 = -1$ when $x = 1$. The equation of the line is $y - (-1) = -1(x - 1)$, so $y = -x$.

45 (a) $(3x^4)' = 3\cdot 4x^3 = 12x^3 = 12\cdot 1^3 = 12$ when $x = 1$, so the density is 12 g/cm.

(b) A lens projects the point on a slide whose coordinate is x to the point on a screen with coordinate $3x^4$. What is the magnification at $x = 1$?

(c) An object travels $3t^4$ feet in t seconds. What is its velocity after 1 second?

47 Velocity $= (\sqrt{t})' = \frac{1}{2\sqrt{t}}$. Speed $= \left|\frac{1}{2\sqrt{t}}\right| = \frac{1}{2\sqrt{t}}$.

(a) Speed $= \frac{1}{2\sqrt{1/9}} = \frac{1}{2/3} = \frac{3}{2}.$

(b) Speed $= \frac{1}{2\sqrt{1}} = \frac{1}{2}.$

(c) Speed $= \frac{1}{2\sqrt{4}} = \frac{1}{4}.$

(d) Speed $= \dfrac{1}{2\sqrt{9}} = \dfrac{1}{6}$.

49 (a)

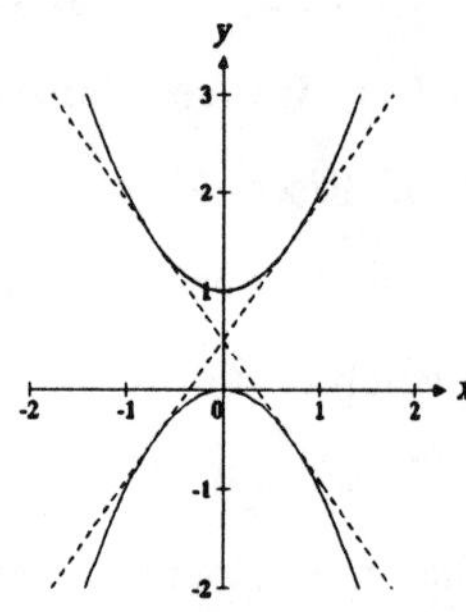

(b) The tangent line to $y = x^2 + 1$ at the point $(a, a^2 + 1)$ has slope $2a$, so its equation is $y - (a^2 + 1) = 2a(x - a)$, that is, $y = 2ax - a^2 + 1$. The tangent line at $(b, -b^2)$ has slope $-2b$; its equation is $y - (-b^2) = -2b(x - b)$, or $y = -2bx + b^2$. If these two lines coincide, then $2a = -2b$ and $-a^2 + 1 = b^2$. Hence $b = -a$ and $-a^2 + 1 = (-a)^2 = a^2$, so $2a^2 = 1$ and $a = \pm\dfrac{1}{\sqrt{2}}$. Thus the two simultaneous tangents are $y = \sqrt{2}x + \dfrac{1}{2}$ and $y = -\sqrt{2}x + \dfrac{1}{2}$.

51 (a) The image of [2, 2.1] is $[2^3, 2.1^3] = [8, 9.261]$. Magnification $= \dfrac{9.261 - 8}{2.1 - 2} = \dfrac{1.261}{0.1} = 12.61$.

(b) The image of [1.9, 2] is $[1.9^3, 2^3] = [6.859, 8]$. Magnification $= \dfrac{8 - 6.859}{2 - 1.9} = \dfrac{1.141}{0.1} = 11.41$.

(c) $(x^3)' = 3x^2$, when $x = 2$, this equals $3 \cdot 2^2 = 12$, so the magnification is 2.

53 (a) $C(0) = 1000 + 5 \cdot 0 + 0^2/200 = 1000$ dollars

(b) $C'(x) = (1000 + 5x + x^2/200)'$
$= 5 + 2x/200 = 5 + x/100$

(c) $C'(10) = 5 + 10/100 = 5.1$

(d) $C(11) - C(10) = (1000 + 5 \cdot 11 + 11^2/200) - (1000 + 5 \cdot 10 + 10^2/200)$

$$= 55 + \frac{121}{200} - \left(50 + \frac{100}{200}\right) = 5 + \frac{21}{200}$$

$= 5.105$

55 (a) Profit during the third year is $3^2 - 2^2 = 9 - 4 = 5$ million dollars.

(b) Profit from $t = 2$ to $t = 2.5$ is $(2.5)^2 - 2^2 = 6.25 - 4 = 2.25$ million dollars.

(c) Average rate of profit is $\dfrac{6.25 - 4}{2.5 - 2} = \dfrac{2.25}{0.5} = 4.5$ million dollars per year.

(d) Average rate of profit from time 2 to time t is

$$\frac{t^2 - 2^2}{t - 2} = \frac{(t + 2)(t - 2)}{t - 2} = t + 2.$$ Thus

the rate of profit at time 2 is $\lim_{t \to 2} (t + 2) = 2 + 2 = 4$ million dollars per year.

57 (a) Let $w = y^3$; then $\dfrac{dw}{dx} = \dfrac{dw}{dy} \cdot \dfrac{dy}{dx} = 3y^2 \cdot \dfrac{dy}{dx}$.

(b) Let $w = \cos y$; then $\dfrac{dw}{dx} = \dfrac{dw}{dy} \cdot \dfrac{dy}{dx}$

$$= (-\sin y) \cdot \frac{dy}{dx}.$$

(c) Let $w = 1/y$; then $\dfrac{dw}{dx} = \dfrac{dw}{dy} \cdot \dfrac{dy}{dx}$

$$= \frac{-1}{y^2} \cdot \frac{dy}{dx}.$$

59 (a) $D(x^4 + C) = 4x^3$, so $x^4 + C$ is an antiderivative of $4x^3$; choose two different values of C to obtain two distinct antiderivatives.

(b) We want $D(kx^4 + C) = 4kx^3 = x^3$, so choose $k = 1/4$. Thus $\frac{1}{4}x^4 + C$ provides antiderivatives of x^3.

(c) $D\left(\frac{x^5}{5}\right) = x^4$, $D\left(\frac{x^4}{4}\right) = x^3$, and $D(\sin x) = \cos x$, so $\frac{x^5}{5} + \frac{x^4}{4} + \sin x + C$ provides antiderivatives of $x^4 + x^3 + \cos x$.

(d) $D(x^4/4) = x^3$ and $D(-\cos x) = \sin x$ so $\frac{x^4}{4} - \cos x + C$ provides antiderivatives of $x^3 + \sin x$.

(e) $(x^2 + 1)^2 = x^4 + 2x^2 + 1$, while $D\left(\frac{x^5}{5}\right) = x^4$, $D\left(\frac{2x^3}{3}\right) = 2x^2$, and $D(x) = 1$, so $\frac{x^5}{5} + \frac{2x^3}{3} + x + C$ yields antiderivatives of $(x^2 + 1)^2$.

61 (a) The limit exists for all values of a.

(b) For $a = 1$, $\lim_{x \to a} f(x) = \lim_{x \to 1} f(x) = 1$, while $f(1) = 2$. At all other points $\lim_{x \to a} f(x) = f(a)$; that is, f is continuous at a for all values except 1.

(c) f is continuous but not differentiable at $a = 0$, 2, 3, and 4 because of the corners which occur at these points in the graph of f.

63 $x \cdot x = x + x + x + \cdots + x$ (x times) makes sense only when x is a positive integer. Since the right-hand side makes sense only for positive integers, it is meaningless to speak of differentiating it.

65 (a) $V'(r) = \left(\frac{4\pi r^3}{3}\right)' = \frac{4\pi}{3}(r^3)' = \frac{4\pi}{3} \cdot 3r^2$

$= 4\pi r^2 = S(r)$

(b)

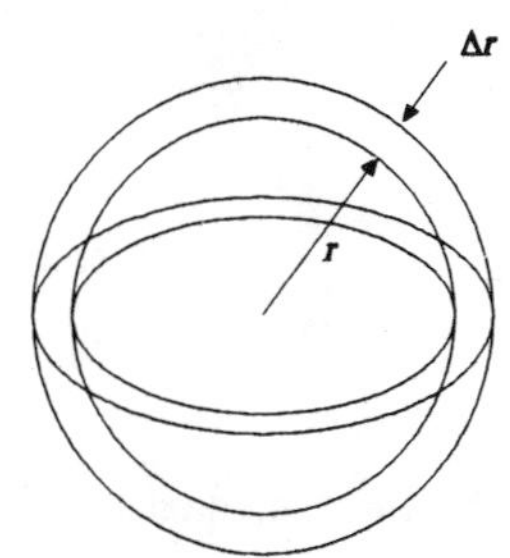

$V'(r) = \lim_{\Delta r \to 0} \frac{V(r + \Delta r) - V(r)}{\Delta r} = 4\pi r^2$, so for small positive values of Δr we have $V(r + \Delta r) - V(r) \approx 4\pi r^2 \Delta r$. This is plausible, as can be seen by examining the difference $V(r + \Delta r) - V(r)$, which is the volume of the region between concentric spheres with radii r and $r + \Delta r$. Partition the surface of the inner sphere into small, almost flat sections. Consider one of these sections, and let its area be denoted A. When Δr is small, the area of the section of the larger sphere above this section is only slightly larger than A. Since both sections are nearly flat, and since the distance between the sections is Δr, the volume of the region between the two sections is approximately $A \cdot \Delta r$. (The shape of the region resembles that of a cylinder, not necessarily circular, and its volume is the area of the base multiplied by its height.) The total surface area of the smaller sphere is $4\pi r^2$, so the volume of the region between the two spheres is approximately $4\pi r^2 \Delta r$.

67 $\lim_{\Delta x \to 0} \dfrac{\sin\sqrt{3 + \Delta x} - \sin\sqrt{3}}{\Delta x} = f'(3)$, where $f(x) =$ $\sin\sqrt{x}$. If we let $y = \sin u$, $u = \sqrt{x}$, we have

$$f'(x) = \frac{d}{dx}(\sin\sqrt{x}) = \frac{d}{du}(\sin u)\cdot\frac{d}{dx}(\sqrt{x}) =$$

$(\cos u)\cdot\dfrac{1}{2\sqrt{x}} = \dfrac{\cos\sqrt{x}}{2\sqrt{x}}$. When $x = 3$, $f'(3) =$

$\dfrac{\cos\sqrt{3}}{2\sqrt{3}}$, so the limit is equal to $\dfrac{\cos\sqrt{3}}{2\sqrt{3}}$.

69 (a),(b)

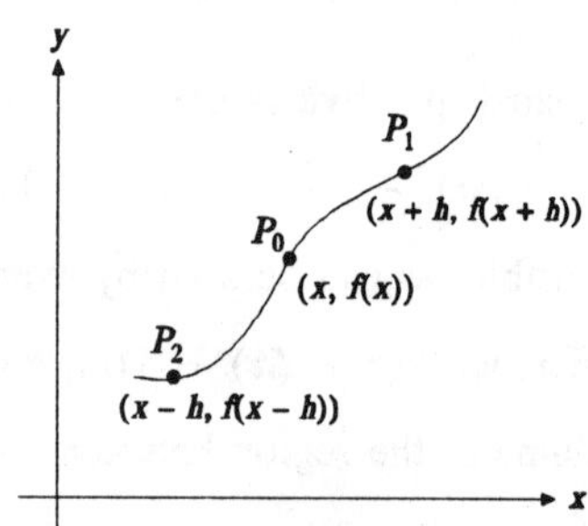

(c) The slope of the line through P_1 and P_2 is

$$\frac{f(x+h) - f(x-h)}{(x+h) - (x-h)} = \frac{f(x+h) - f(x-h)}{2h}.$$

(d) If f is differentiable, then

$$\lim_{h\to 0} \frac{f(x+h) - f(x-h)}{2h}$$

$$= \lim_{h\to 0} \frac{f(x+h) - f(x) + f(x) - f(x-h)}{2h}$$

$$= \lim_{h\to 0} \frac{f(x+h) - f(x)}{2h} + \lim_{h\to 0} \frac{f(x) - f(x-h)}{2h}$$

$$= \frac{1}{2}\lim_{h\to 0} \frac{f(x+h) - f(x)}{h} + \frac{1}{2}\lim_{h\to 0} \frac{f(x) - f(x-h)}{h}$$

$$= \frac{1}{2}f'(x) + \frac{1}{2}f'(x) = f'(x).$$

(e) $\lim_{h\to 0} \dfrac{(x+h)^3 - (x-h)^3}{2h} =$

$$\lim_{h\to 0} \frac{(x^3 + 3x^2h + 3xh^2 + h^3) - (x^3 - 3x^2h + 3xh^2 - h^3)}{2h}$$

$$= \lim_{h\to 0} \frac{6hx^2 + 2h^3}{2h} = \lim_{h\to 0}(3x^2 + h^2) = 3x^2$$

71 We wish to find an expression for the shaded area A in terms of r and Δr. This area is the difference between the area of a circle of radius r and a circle of radius $r + \Delta r$. Therefore $A = \pi(r + \Delta r)^2 - \pi r^2$. The approximate "length" of a circle of radius r can be thought of as the area A divided by the "width" Δr. Thus $l \approx \dfrac{A}{\Delta r} \approx$

$\dfrac{\pi(r + \Delta r)^2 - \pi r^2}{\Delta r}$. The exact value of l can be found by letting $\Delta r \to 0$. Therefore, $l =$

$$\lim_{\Delta r \to 0} \frac{\pi(r + \Delta r)^2 - \pi r^2}{\Delta r} =$$

$$\pi \lim_{\Delta r \to 0} \frac{(r + \Delta r)^2 - r^2}{\Delta r} = \pi\cdot 2r = 2\pi r.$$

This "length" l is referred to as the *circumference* C of a circle of radius r.

73 (a) $f'(1) \approx \dfrac{\log 1.001 - \log 1}{0.001} \approx 0.4341$

(b) $f'(2) \approx \dfrac{\log 2.001 - \log 2}{0.001} \approx 0.2171$

(c) $f'(1/2) \approx \dfrac{\log 0.5001 - \log 0.5}{0.0001} \approx 0.8685$

$f'(3) \approx \dfrac{\log 3.001 - \log 3}{0.001} \approx 0.1447$

$f'(4) \approx \dfrac{\log 4.001 - \log 4}{0.001} \approx 0.1086$

We fill in the table to obtain:

x	1/2	1	2	3	4
$f'(x)$	0.8685	0.4341	0.2171	0.1447	0.1086

(d) $f'(x) \approx \dfrac{0.434}{x}$

75 Figure 1 of Sec. 3.6 illustrates the case when Δu is nonzero. The approximate magnification of the lower projector at x is $\dfrac{\Delta u}{\Delta x}$. This projection is then magnified by $\dfrac{\Delta y}{\Delta u}$. The approximate overall magnification is $\dfrac{\Delta y}{\Delta u}\cdot\dfrac{\Delta u}{\Delta x} = \dfrac{\Delta y}{\Delta x}$. Taking the limit of both sides as Δx and Δu approach 0, we get $\dfrac{dy}{dx} = \dfrac{dy}{du}\cdot\dfrac{du}{dx}$, which is the chain rule.

When Δu is zero, the magnification of the first projector is 0, so $\dfrac{du}{dx} = 0$. Since there is no image to be magnified by the second projector, its projection is also of length 0; that is, Δy is 0. Hence $\dfrac{dy}{dx} = \lim\limits_{\Delta x \to 0} \dfrac{\Delta y}{\Delta x} = \lim\limits_{\Delta x \to 0} \dfrac{0}{\Delta x} = 0$, while $\dfrac{dy}{du}\cdot\dfrac{du}{dx} = \dfrac{dy}{du}\cdot 0 = 0$, so $\dfrac{dy}{dx} = \dfrac{dy}{du}\cdot\dfrac{du}{dx}$, as before.

4 Applications of the Derivative

4.1 Three Theorems About the Derivative

1 (a)

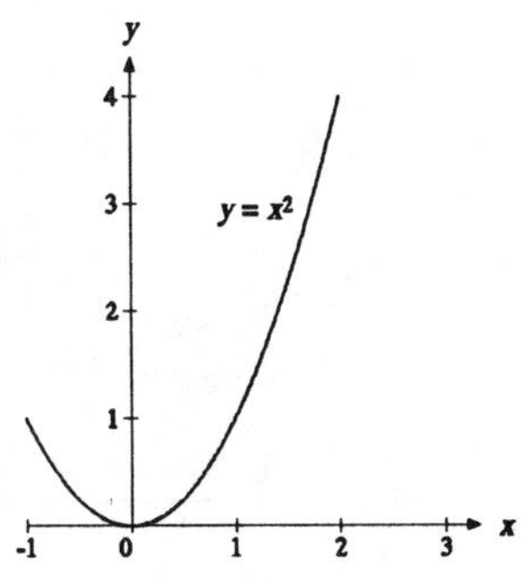

(b) From the graph we see that $f(x)$ attains the maximum value of 4 at the endpoint $x = 2$.

(c) $f'(2)$ exists, provided we allow for the fact that $f(x)$ is not defined for $x > 2$. If we compute $f'(2) = \lim_{x \to 2^-} \frac{x^2 - 2^2}{x - 2} =$

$\lim_{x \to 2^-} \frac{(x + 2)(x - 2)}{x - 2} = \lim_{x \to 2^-} (x + 2) = 4$, we get $f'(2) = 4$.

(d) From (c) we see that $f'(2) \neq 0$.

(e) The minimum occurs at $x = 0$, as is clear from the graph. Since $f'(x) = 2x$, we have $f'(0) = 2 \cdot 0 = 0$.

3 (a)

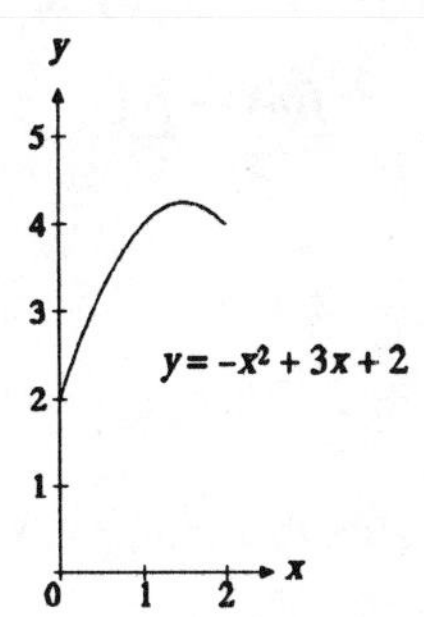

(c) $y = -x^2 + 3x + 2$, so $\frac{dy}{dx} = -2x + 3$, which is 0 when $x = 3/2$.

(d) The highest point on the graph occurs when $x = 3/2$ (where the slope is 0).

5 (a)

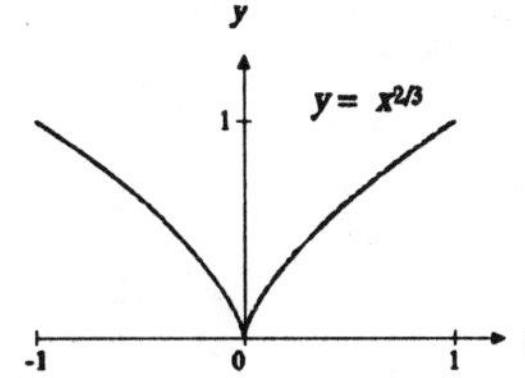

(b) $f(-1) = (-1)^{2/3} = [(-1)^{1/3}]^2 = [-1]^2 = 1$; $f(1) = 1^{2/3} = 1$.

(c) No, since $f'(x) = \frac{2}{3}x^{-1/3}$ is never equal to 0.

(d) $f'(x)$ is not defined at $x = 0$.

7 Since $f(x) = x^2 - 2x - 3$ is continuous on [0, 2], $f'(x) = 2x - 2$ is defined on (0, 2), and $f(0) = -3 = f(2)$, the conditions of Rolle's theorem are satisfied. Therefore, there exists at least one number c in [0, 2] such that $f'(c) = 0$. Hence $f'(c)$

$= 2c - 2 = 0$, so $c = 1$.

9 $f(x) = x^4 - 2x^2 + 1$ is continuous on $[-2, 2]$, $(x^4 - 2x^2 + 1)' = 4x^3 - 4x$ is defined on $(-2, 2)$, and $2^4 - 2\cdot2^2 + 1 = 9 = (-2)^4 - 2(-2)^2 + 1$; $f'(c) = 4c^3 - 4c = 4c(c^2 - 1) = 0$ when $c = 0$, $1, -1$.

11 Let $f(x) = x^2 - 3x$; then $f(1) = -2$ and $f(4) = 4$. Note that $f'(x) = 2x - 3$. By the mean-value theorem, $f'(c) = \dfrac{f(4) - f(1)}{4 - 1} = \dfrac{4 - (-2)}{3} = 2$.

Hence $2c - 3 = 2$ and $c = 5/2$.

13 Let $f(x) = 3x + 5$. Then $f(1) = 8$ and $f(3) = 14$. By the mean-value theorem, $f'(c) = \dfrac{f(3) - f(1)}{3 - 1} =$

$\dfrac{14 - 8}{2} = 3$. But $f'(x) = 3$, so we need to find c

such that $f'(c) = 3$, which is always true. Any value of c in $(1, 3)$ will serve.

15 (a),(b),(c)

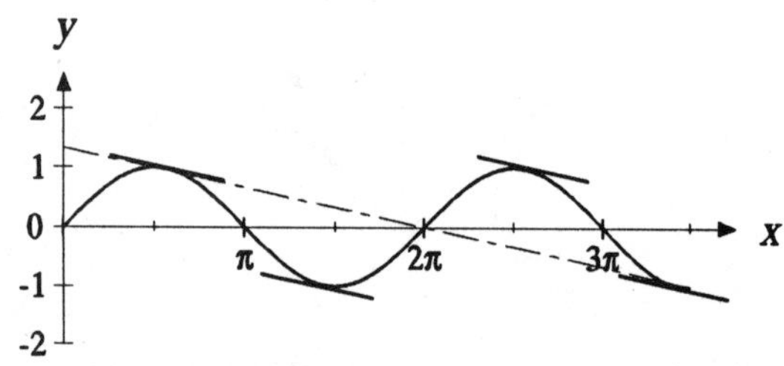

(d) Since $f(x) = \sin x$ and $f'(x) = \cos x$, the equation $f'(c) = \dfrac{f(7\pi/2) - f(\pi/2)}{7\pi/2 - \pi/2}$ can be

simplified to $\cos c = \dfrac{\sin 7\pi/2 - \sin \pi/2}{3\pi} =$

$\dfrac{-1 - 1}{3\pi} = -\dfrac{2}{3\pi} \approx -0.2122$. There are

four numbers in the interval $[\pi/2, 7\pi/2]$ for which the cosine is -0.2122.

(e) Estimating from the graph or with the aid of a calculator (recommended!), we find that the four possible values of c are 1.78, 4.50, 8.07, and 10.78.

17 (a) $(\sec^2 x)' = 2 \sec x (\sec x \tan x)$
$= 2 \sec^2 x \tan x$
$(\tan^2 x)' = 2 \tan x \sec^2 x$

(b) We know from trigonometry that $\tan^2 x + 1 = \sec^2 x$, so $C = 1$.

19

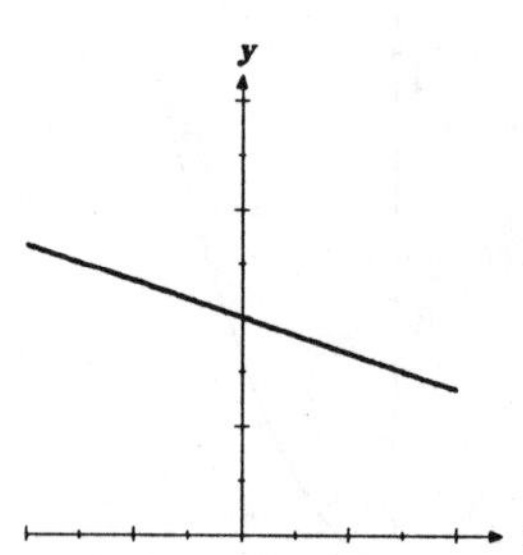

21

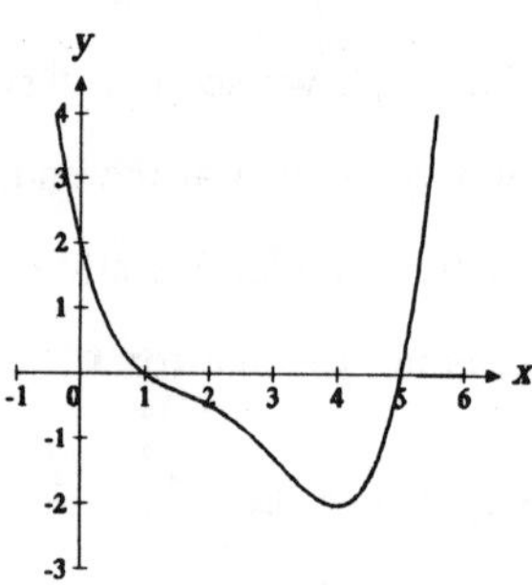

23

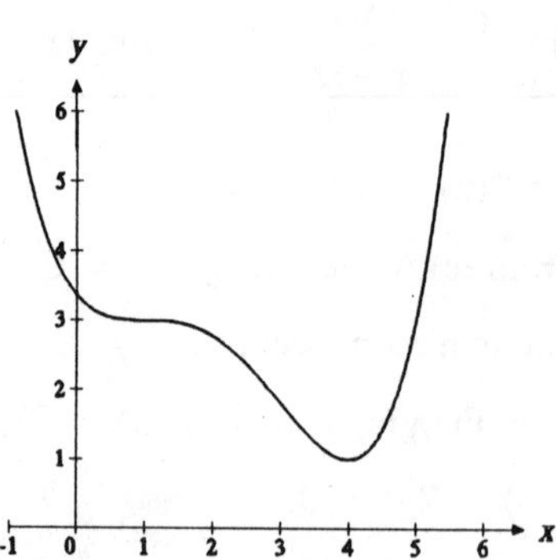

25 Note that $f(1) = 3 < 4 = f(2)$. If $f'(x)$ were always negative, then $f(x)$ would be a decreasing function; but this contradicts the requirement that

$f(x)$ somehow increase from 3 to 4 over the interval $[1, 2]$.

27 Since f is differentiable, it is continuous. Since $f(3) = -1 < 0 < 2 = f(4)$, the intermediate-value theorem says that there is a number c in $[3, 4]$ such that $f(c) = 0$. But then c would be an x intercept, and only 1 and 2 are supposed to be x intercepts.

29 The stated property is true of all smooth curves, by the theorem of the interior extremum. To quote Feynman further:

> They were all excited by this "discovery"—even though they had already gone through a certain amount of calculus and had already "learned" that the derivative (tangent) of the minimum (lowest point) of *any* curve is zero (horizontal). They didn't put two and two together. They didn't even know what they "knew."

31 (a) Note that $(2x^3 - 3x^2 + 12x - 5)' = 6x^2 - 6x + 12 = 6(x^2 - x + 2)$, which is always positive (since $x^2 - x + 2$ has no roots). Hence $2x^3 - 3x^2 + 12x - 5$ is an increasing function and therefore crosses the x axis at most once. In (b) we show that this solution actually exists.

(b) Plugging in $x = 0$ yields -5 and plugging in $x = 1$ yields 6. Hence there is a solution in the interval $[0, 1]$, and it is the only solution that exists.

(c) We evaluate f at the midpoint of the interval $(0, 1)$: $f\left(\frac{1}{2}\right) = 2\left(\frac{1}{2}\right)^3 - 3\left(\frac{1}{2}\right)^2 + 12\left(\frac{1}{2}\right) - 5$

$= \frac{1}{2}$. Since $f(1/2)$ is positive while $f(0)$ is negative, the solution lies in the interval $(0, 1/2)$. Now we check $x = 1/4$. We have $f\left(\frac{1}{4}\right)$

$= 2\left(\frac{1}{4}\right)^3 - 3\left(\frac{1}{4}\right)^2 + 12\left(\frac{1}{4}\right) - 5 = -\frac{69}{32}$, so

the solution is in the interval $(1/4, 1/2)$.

33

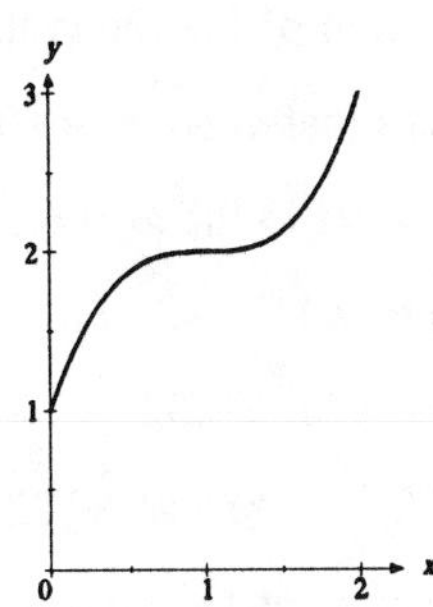

35

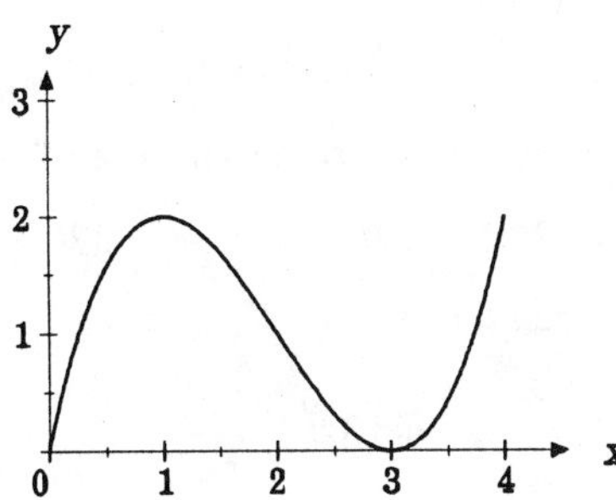

37 $(7x + k \sin 2x)' = 7 + 2k \cos 2x$, which must be nonnegative if $7x + k \sin 2x$ is always increasing. That is, $7 + 2k \cos 2x \geq 0$. The worst case occurs when $\cos 2x = \pm 1$, so we need $7 \pm 2k \geq 0$. Hence $|k| \leq 7/2$ will do the trick.

39 (a) Given a chord of a differentiable function f, the endpoints of the chord provide an interval within which there must be a number satisfying the mean-value theorem, that is, a point whose tangent is parallel to the chord.

(b) The tangent line to the graph of a differentiable function is not necessarily parallel to some chord of the function. Consider the function $y = x^3$ and the tangent line $y = 0$ at the origin; the tangent line is horizontal, but the function has no horizontal chords.

41 The interval $[a, b]$ is projected onto the interval $[f(a), f(b)]$ on the screen. Hence $\frac{f(b) - f(a)}{b - a}$ is the average magnification of the interval. The mean-value theorem says that at some point in $[a, b]$ the magnification is equal to the average magnification over the entire interval.

43 (a) Since $f(t) = -16t^2 + 32t + 40$, we have $f(0) = 40$ and $f(2) = -16\cdot 2^2 + 32\cdot 2 + 40 = 40$. Hence the ball is at the same height at times $t = 0$ and $t = 2$.

(b) Rolle's theorem says that the ball's velocity must be 0 for some t between 0 and 2.

(c) $(-16t^2 + 32t + 40)' = -32t + 32 = 0$ when $t = 1$.

45 (a) $\left(\sqrt{1 - x^2}\sin 3x\right)'$

$$= \sqrt{1 - x^2}(\sin 3x)' + \sin 3x\left(\sqrt{1 - x^2}\right)'$$

$$= \sqrt{1 - x^2}(3\cos 3x) + (\sin 3x)\frac{-2x}{2\sqrt{1 - x^2}}$$

$$= 3\sqrt{1 - x^2}\cos 3x - \frac{x}{\sqrt{1 - x^2}}\sin 3x$$

(b) $\left(\frac{\sqrt[3]{x}}{x^2 + 1}\right)' = \left(\frac{x^{1/3}}{x^2 + 1}\right)'$

$$= \frac{(x^2 + 1)(x^{1/3})' - x^{1/3}(x^2 + 1)'}{(x^2 + 1)^2}$$

$$= \frac{(x^2 + 1)\frac{1}{3}x^{-2/3} - x^{1/3}\cdot 2x}{(x^2 + 1)^2}$$

$$= \frac{\frac{1}{3}x^{-2/3}[(x^2 + 1) - 6x^2]}{(x^2 + 1)^2} = \frac{1 - 5x^2}{3x^{2/3}(x^2 + 1)^2}$$

(c) $\left(\tan\frac{1}{(2x + 1)^2}\right)' = \left(\tan(2x + 1)^{-2}\right)'$

$$= \left(\sec^2\frac{1}{(2x + 1)^2}\right)\cdot(-2)(2x + 1)^{-3}\cdot 2$$

$$= \frac{-4}{(2x + 1)^3}\left(\sec^2\frac{1}{(2x + 1)^2}\right)$$

47 (a) To obtain $\sqrt{x}$ by differentiation of $F(x)$, we need $F(x) = \frac{x^{3/2}}{3/2} + C = \frac{2}{3}x^{3/2} + C.$

(b) The most general antiderivative of $\sec^2 3x$ is $\frac{1}{3}\tan 3x + C.$

(c) The most general antiderivative of $\cos 3x$ is $\frac{1}{3}\sin 3x + C.$

(d) The most general antiderivative of $(2x + 1)^{10}$ is $\frac{1}{22}(2x + 1)^{11} + C.$

49 (a) $f'(x) = (x^3 - 3x)' = 3x^2 - 3 = 3(x^2 - 1) = 0$ when $x = \pm 1$.

(b) By the theorem of the interior extremum, the maximum value occurs at either an endpoint or at an interior point where the derivative is 0; but $f'(x) \neq 0$ for x in $(1, 5)$, so the maximum value occurs at either 1 or 5.

(c) $f(1) = 1 - 3 = -2$ and $f(5) = 125 - 15 = 110$, so by (b) the maximum value is 110.

51 The proposed proof is not correct. The tilted graph of a function is not necessarily still the graph of a function for the new axes.

4.2 The First Derivative and Graphing

1 $f'(x) = 5x^4 = 0$ when $x = 0$. But $f'(x) > 0$ for all nonzero values of x, so $x = 0$ is neither a local maximum nor a local minimum.

3 $f'(x) = [(x - 1)^3]' = 3(x - 1)^2 = 0$ when $x = 1$. But $f'(x) > 0$ for $x \neq 1$, so $x = 1$ is neither a local maximum nor a local minimum.

5 $f'(x) = (3x^4 + x^3)' = 12x^3 + 3x^2 = 3x^2(4x + 1) = 0$ for $x = 0$ or $x = -1/4$. Since $3x^2 \geq 0$ for all values of x, we see that the sign of the derivative is the same as that of $4x + 1$. Thus $f'(x) < 0$ for $x < -1/4$ and $f'(x) \geq 0$ for $x > -1/4$; it follows that there is a local minimum for $x = -1/4$. Since $f'(x) \geq 0$ for values of x near zero, we see that $x = 0$ is neither a local maximum nor minimum.

7 $(x \sin x + \cos x)' = x \cos x + (\sin x)\cdot 1 - \sin x = x \cos x = 0$ when $x = 0$ or $x = (n + 1/2)\pi$ for any integer n; $x \cos x < 0$ for $-\pi/2 < x < 0$ and $x \cos x > 0$ for $0 < x < \pi/2$, so $x = 0$ is a local minimum. If $n \geq 0$ is even, say $n = 2k$ with $k \geq 0$, then $x \cos x > 0$ for $2k\pi < x < (2k + 1/2)\pi$ and $x \cos x < 0$ for $(2k + 1/2)\pi < x < (2k + 1)\pi$, so $x = (n + 1/2)\pi$ is a local maximum. Similarly, for odd $n \geq 1$, $x = (n + 1/2)\pi$ is a local minimum; for even $n < 0$, $x = (n + 1/2)\pi$ is a local minimum; for odd $n < 0$, $x = (n + 1/2)\pi$ is a local maximum. More simply, there are local minima at $x = 0$ and at $x = \pm(4k + 3)\pi/2$ for $k \geq 0$; there are local maxima at $x = \pm(4k + 1)\pi/2$ for $k \geq 0$.

9 If (1, 2) is a critical point, $f'(1) = 0$. In addition, $f'(x) < 0$ for $x < 1$ and $f'(x) > 0$ for $x > 1$, implying that the critical point is a global minimum; indeed it is the only minimum.

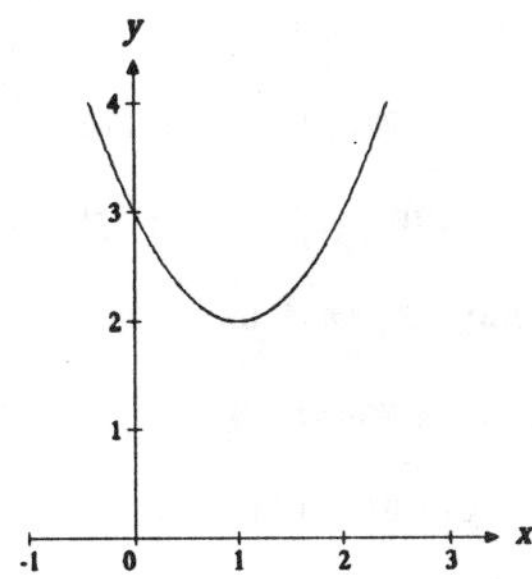

11 Since there are no critical points for $x < 1$, $f(-1) = 0$, and $f(1) = 3$, $f(x)$ must be increasing on $(-\infty, 1)$. Given $f(1) = 3$ and $f(2) = 1$, $f(x)$ must be decreasing over (1, 2) since there are no critical points in (1, 2). We know $y = 4$ is an asymptote for $x > 2$ and $y = -1$ is an asymptote for $x < -1$ from $\lim_{x\to\infty} f(x) = 4$ and $\lim_{x\to-\infty} f(x) = -1$.

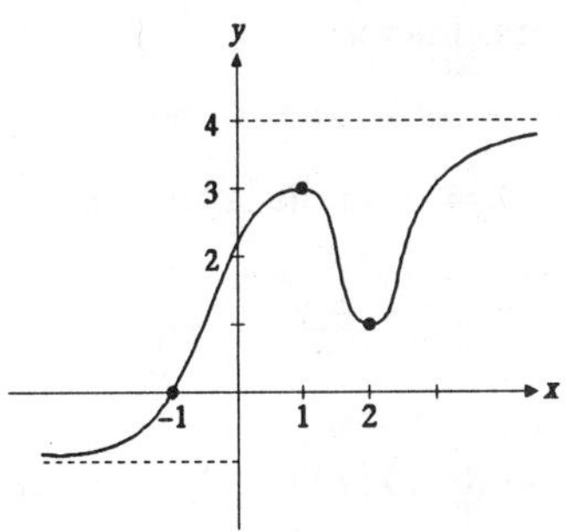

13 No critical points in $(-\infty, 1)$ implies $f(x)$ is increasing on $(-\infty, 1)$, given $\lim_{x\to\infty} f(x) = -\infty$.

Since $f(1) = 5$ and $f(2) = 4$, $f(x)$ is decreasing on (1, 2). At $x = 1$ there is a global maximum. There must be a minimum at $x = 2$ because $f(x) \to 5$ as $x \to \infty$. There is an x intercept at $x = -1$.

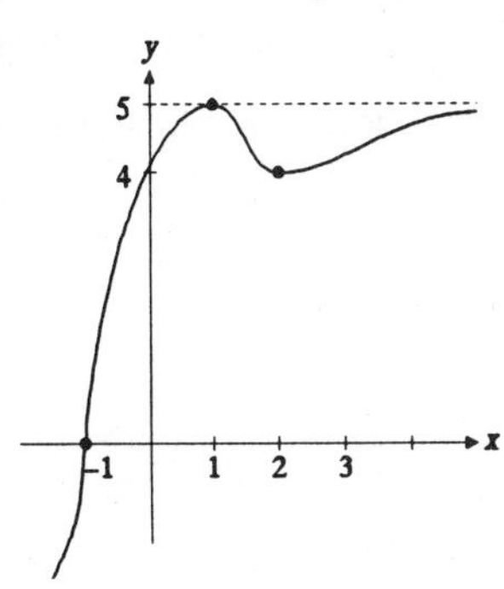

15 $f(x)$ must be increasing on $(3, \infty)$, given that $f(x) \to \infty$ as $x \to \infty$. $f(x)$ is decreasing on $(1, 3)$ since $f(1) = 3$ and $f(3) = -1$. $f(x)$ is increasing on $(-\infty, 1)$ since $f(x) \to 1$ as $x \to -\infty$ while $f(1) = 3$. Thus, $f(1)$ is a maximum, $f(3)$ is a minimum, and all intercepts are as stated.

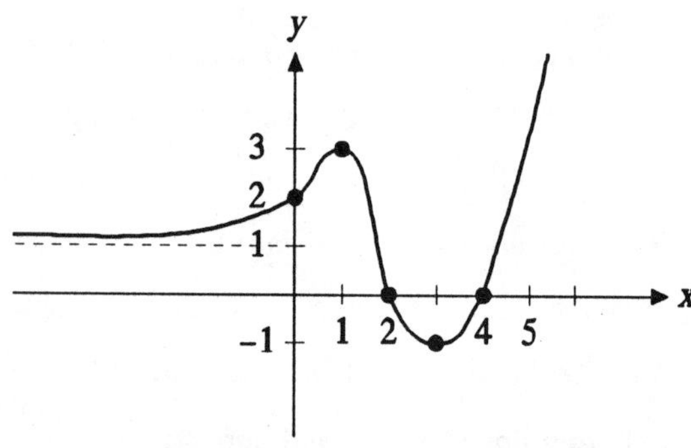

17 Let $f(x) = x^3 - 3x^2 + 3x$. The domain of f is all real numbers. Intercepts of the x axis occur when $f(x) = x(x^2 - 3x + 3) = 0$, hence for $x = 0$ or $x^2 - 3x + 3 = 0$; in the latter case, $x = \dfrac{3 \pm \sqrt{9 - 12}}{2} = \dfrac{3 \pm \sqrt{-3}}{2}$, yielding no real solutions, so $(0, 0)$ is the only x intercept. It is also the y intercept. Now $f'(x) = 3x^2 - 6x + 3 = 3(x^2 - 2x + 1) = 3(x - 1)^2 = 0$ when $x = 1$; thus $(1, 1)$ is a critical point. Observe that $(x - 1)^2 \geq 0$ for all x, so $f'(x) \geq 0$ for all x, and there is neither a local maximum nor a local minimum. $\lim_{x\to\infty} (x^3 - 3x^2 + 3x) = \infty$ and $\lim_{x\to-\infty} (x^3 - 3x^2 + 3x) = -\infty$, so there are no horizontal asymptotes; f is defined everywhere, so there are no vertical asymptotes.

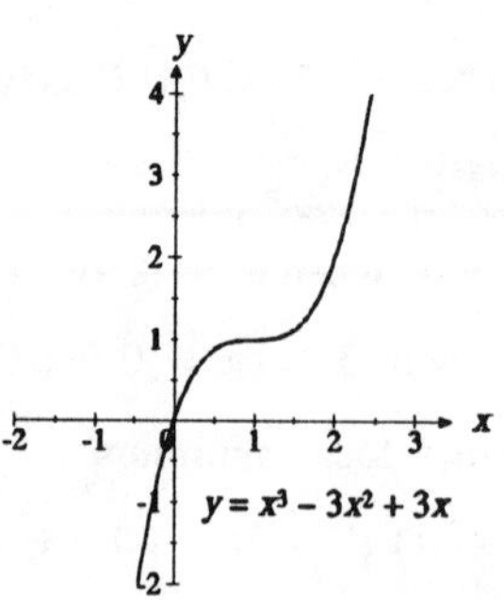

19 $f(x) = x^4 - 4x + 3$. Since f is defined for all x, there are no vertical asymptotes. Note that $f(1) = 0$, so $(1, 0)$ is an x intercept; $f(0) = 3$, so $(0, 3)$ is the y intercept. $f'(x) = 4x^3 - 4 = 4(x^3 - 1) = 0$ when $x = 1$, so $(1, 0)$ is the only critical point. $f'(x) < 0$ for $x < 1$ and $f'(x) > 0$ for $x > 1$, so $(1, 0)$ is a global minimum. (We now see that $(1, 0)$ is the *only* x intercept.) $\lim_{x\to\infty} (x^4 - 4x + 3) = \infty = \lim_{x\to-\infty} (x^4 - 4x + 3)$, so there are no horizontal asymptotes.

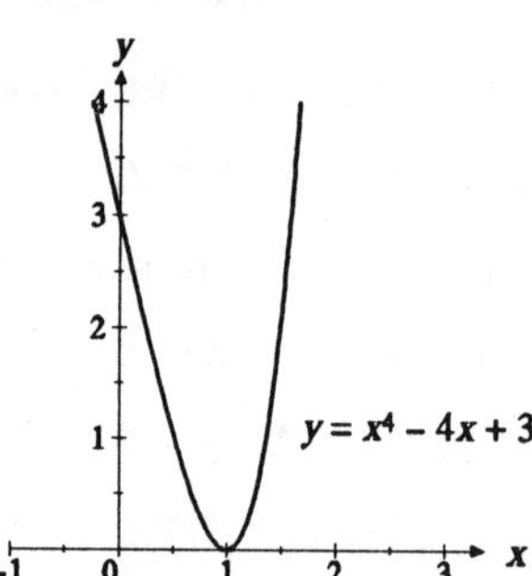

21 $f(x) = x^2 - 6x + 5$. Note that f is defined for all x, so there are no vertical asymptotes. $f(x) = (x - 5)(x - 1) = 0$ for $x = 1$ and 5, so $(1, 0)$ and $(5, 0)$ are the x intercepts; $f(0) = 5$, so $(0, 5)$ is the y intercept. $f'(x) = 2x - 6 = 2(x - 3) = 0$ when $x = 3$, and $f'(x) < 0$ for $x < 3$ and $f'(x) > 0$ for $x > 3$, so $(3, -4)$ is a global minimum. $\lim_{x\to\infty} (x^2 - 6x + 5) = \infty = \lim_{x\to-\infty} (x^2 - 6x + 5)$, so there are no horizontal asymptotes.

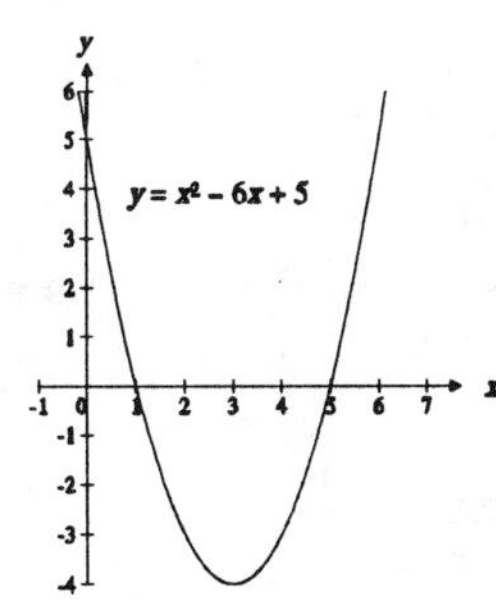

23 $f(x) = x^4 + 2x^3 - 3x^2$. Since f is defined for all x, there are no vertical asymptotes. Note that $f(x) = x^2(x + 3)(x - 1) = 0$ for $x = -3$, 0, and 1, so $(-3, 0)$, $(0, 0)$, and $(1, 0)$ are the x intercepts; $f(0) = 0$, so $(0, 0)$ is the y intercept. We have $f'(x) = 4x^3 + 6x^2 - 6x = 2x(2x^2 + 3x - 3) = 0$ when $x = 0$ or $2x^2 + 3x - 3 = 0$, so $x = \frac{-3 \pm \sqrt{9 + 24}}{4} = \frac{-3 \pm \sqrt{33}}{4}$, where $\frac{-3 + \sqrt{33}}{4} \approx 0.686$ and $\frac{-3 - \sqrt{33}}{4} \approx -2.186$. $f(0.686) \approx -0.545$ and $f(-2.186) \approx -12.393$, so $(0, 0)$, $(0.686, -0.545)$, and $(-2.186, -12.393)$ are critical points (the coordinates of the latter two being given only in approximation). For $x < \frac{-3 - \sqrt{33}}{4}$, $f'(x) < 0$, so $f(x)$ is decreasing. For $\frac{-3 - \sqrt{33}}{4} < x < 0$, $f'(x) > 0$, so $f(x)$ is increasing. For $0 < x < \frac{-3 + \sqrt{33}}{4}$, $f'(x) < 0$, so $f(x)$ is decreasing. For $x > \frac{-3 + \sqrt{33}}{4}$, $f'(x) > 0$, so $f(x)$ is increasing. Thus $(-2.186, -12.393)$ is a global minimum, $(0, 0)$ is a local maximum, and $(0.686, -0.545)$ is a local minimum.

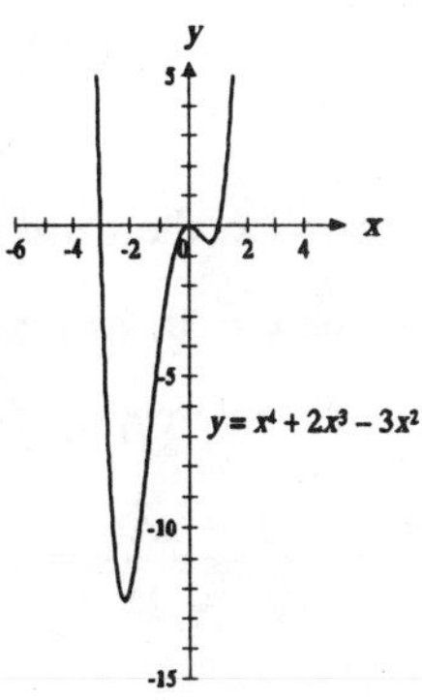

25 $f(x) = \frac{3x + 1}{3x - 1}$. Note that $f(x)$ is defined except where $3x - 1 = 0$, which occurs when $x = 1/3$. Now $f(x) = 0$ when $3x + 1 = 0$, which occurs when $x = -1/3$, so $(-1/3, 0)$ is an x intercept;

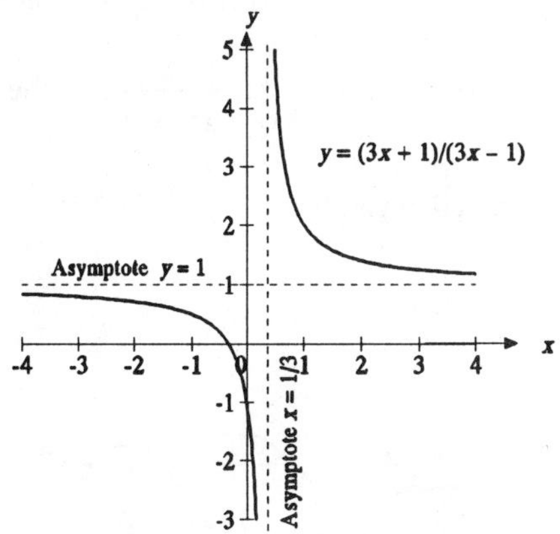

$f(0) = -1$, so $(0, -1)$ is the y intercept. $f'(x) = \frac{(3x - 1)\cdot 3 - (3x + 1)\cdot 3}{(3x - 1)^2} = \frac{-6}{(3x - 1)^2}$, which is never zero, so there are no critical points. Finally $\lim_{x\to\infty} \frac{3x + 1}{3x - 1} = 1 = \lim_{x\to-\infty} \frac{3x + 1}{3x - 1}$, so the line $y = 1$ is a horizontal asymptote; $\lim_{x\to 1/3^-} \frac{3x + 1}{3x - 1} = -\infty$ and $\lim_{x\to 1/3^+} \frac{3x + 1}{3x - 1} = \infty$, so the line $x = 1/3$ is a vertical asymptote.

27 $f(x) = \dfrac{x}{x^2 + 1}$. Since $x^2 + 1 > 0$ for all x, we see that f is defined for all x; there are no vertical asymptotes. $f(x) = 0$ for $x = 0$; hence $(0, 0)$ is both the x intercept and y intercept. $f'(x) = \dfrac{(x^2+1)\cdot 1 - x\cdot 2x}{(x^2+1)^2} = \dfrac{1 - x^2}{(x^2+1)^2} = 0$ when $1 - x^2 = 0$, so the critical points are $(-1, -1/2)$ and $(1, 1/2)$. $1 - x^2$ is positive when x is in the interval $(-1, 1)$ and is negative for $|x| > 1$. Thus $f'(x) \le 0$ for $|x| > 1$ (f decreases) and $f'(x) > 0$ for $|x| < 1$ (f increases). Thus $(1, 1/2)$ is a global maximum and $(-1, -1/2)$ is a global minimum.

$\lim_{x\to-\infty} \dfrac{x}{x^2+1} = 0 = \lim_{x\to\infty} \dfrac{x}{x^2+1}$, so the x axis is a horizontal asymptote.

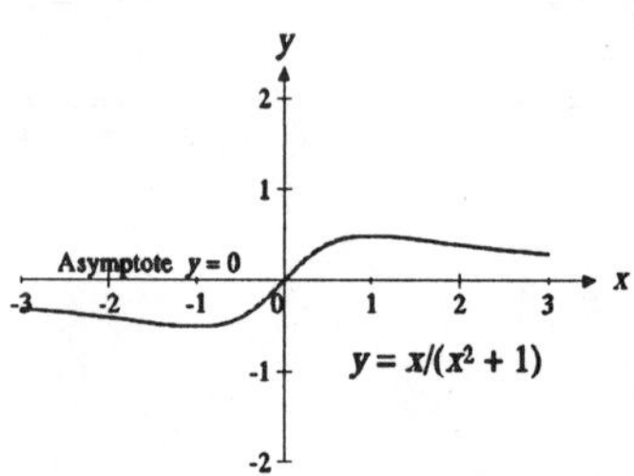

29 $f(x) = \dfrac{1}{2x^2 - x}$. $f(x)$ is defined except when $2x^2 - x = x(2x - 1) = 0$, which occurs when $x = 0$ or $1/2$. But $f(x)$ is never zero, so there are no x intercepts; f is not defined when $x = 0$, so there is no y intercept either. $f'(x) = \dfrac{-(4x-1)}{(2x^2-x)^2} = 0$ when $4x - 1 = 0$, which occurs when $x = 1/4$, so $(1/4, -8)$ is a critical point; $f'(x) > 0$ for $0 < x < 1/4$ and $f'(x) < 0$ for $1/4 < x < 1/2$, so $(1/4, -8)$ is a local maximum. $\lim_{x\to\infty} \dfrac{1}{2x^2 - x} = 0 = \lim_{x\to-\infty} \dfrac{1}{2x^2 - x}$, so the x axis is a horizontal asymptote. $\lim_{x\to0^-} \dfrac{1}{2x^2 - x} = \infty$, $\lim_{x\to0^+} \dfrac{1}{2x^2 - x} = -\infty$, so the y axis is a vertical asymptote. $\lim_{x\to1/2^-} \dfrac{1}{2x^2 - x} = -\infty$, $\lim_{x\to1/2^+} \dfrac{1}{2x^2 - x} = \infty$, so the line $x = 1/2$ is a vertical asymptote.

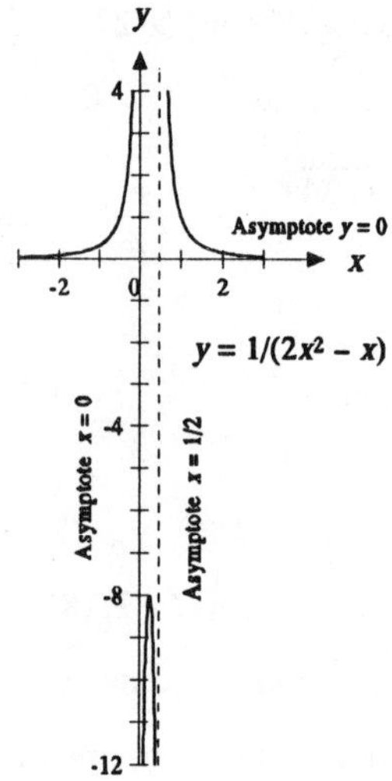

31 $f(x) = \dfrac{x^2 + 3}{x^2 - 4}$. $f(x)$ is defined except when $x^2 - 4 = (x - 2)(x + 2) = 0$, which occurs when $x = \pm 2$. But $x^2 + 3 \ge 3$, so there are no x intercepts; $f(0) = -3/4$, so $(0, -3/4)$ is the y intercept. $f'(x) = \dfrac{(x^2-4)2x - (x^2+3)\cdot 2x}{(x^2-4)^2} = \dfrac{-14x}{(x^2-4)^2} = 0$ when $x = 0$, so $(0, -3/4)$ is a critical point. $f'(x) > 0$ for $-2 < x < 0$ and $f'(x) < 0$ for $0 < x < 2$, so $(0, -3/4)$ is a local maximum. $\lim_{x\to\infty} \frac{x^2+3}{x^2-4} = 1 = \lim_{x\to-\infty} \frac{x^2+3}{x^2-4}$, so y $= 1$ is a horizontal asymptote. $\lim_{x\to-2^-} \frac{x^2+3}{x^2-4} =$ ∞, $\lim_{x\to-2^+} \frac{x^2+3}{x^2-4} = -\infty$, so $x = -2$ is a vertical asymptote. $\lim_{x\to2^-} \frac{x^2+3}{x^2-4} = -\infty$, $\lim_{x\to2^+} \frac{x^2+3}{x^2-4} =$ ∞, so $x = 2$ is a vertical asymptote.

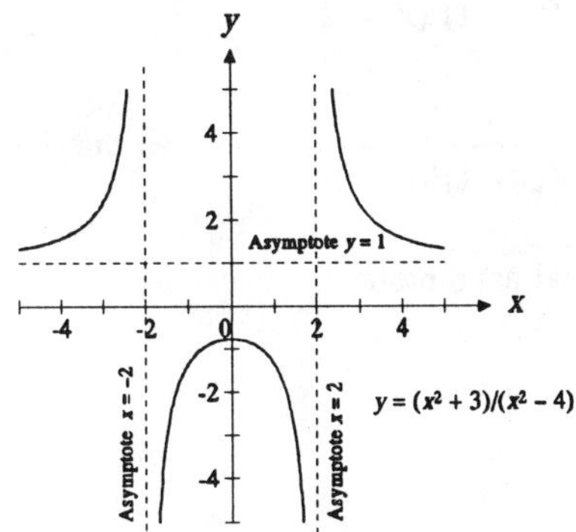

33 $f(x) = x^2 - x^4$. At the endpoints of [0, 1] we have $f(0) = 0$ and $f(1) = 0$. $f'(x) = 2x - 4x^3 = 2x(1 - 2x^2) = 0$ when $x = 0$ or $1 - 2x^2 = 0$, so $2x^2 = 1$, $x^2 = 1/2$, $x = \pm\frac{1}{\sqrt{2}}$; $\frac{1}{\sqrt{2}}$ is in [0, 1], and $f\left(\frac{1}{\sqrt{2}}\right) = \frac{1}{4}$. Thus, 1/4 is the maximum value and 0 is the minimum value.

35 $f(x) = 4x - x^2$. At the endpoints of [0, 1] we have $f(0) = 0$ and $f(1) = 3$. $f'(x) = 4 - 2x = 2(2 - x) = 0$ when $x = 2$, but 2 is not in [0, 1]. Thus 3 is the maximum value and 0 is the minimum value.

37 $f(x) = x^3 - 2x^2 + 5x$. At the endpoints of $[-1, 3]$ we have $f(-1) = -8$ and $f(3) = 24$. Now $f'(x) = 3x^2 - 4x + 5 = 0$ when $x = \frac{4 \pm \sqrt{16-60}}{6}$, so $f'(x) \neq 0$ for any real x. Thus 24 is the maximum value and -8 is the minimum value.

39 $f(x) = x^2 + x^4$. At the endpoints of [0, 1] we have $f(0) = 0$ and $f(1) = 2$. Note that $f'(x) = 2x + 4x^3 = 2x(1 + 2x^2) = 0$ only when $x = 0$, since $1 + 2x^2 > 0$ for all x. Thus 2 is the maximum value and 0 is the minimum value.

41 Let $f(x) = \sin x + \cos x$. Examine $f'(x) = \cos x - \sin x$. $f'(x) = 0$ when $\cos x = \sin x$. On $[0, \pi]$, this occurs at $x = \pi/4$. Since $f'(x) > 0$ for $x < \pi/4$ and $f'(x) < 0$ for $x > \pi/4$, $f(x)$ takes on a maximum value of $f(\pi/4) = \sin \pi/4 + \cos \pi/4 = \sqrt{2} \approx 1.414$. Checking the endpoints of the interval, we find $f(0) = 1$ and $f(\pi) = -1$. Thus, on $[0, \pi]$, $f(x)$ has a maximum value of $\sqrt{2}$ at $x = \pi/4$ and a minimum value of -1 at $x = \pi$.

43 $f(x) = \frac{\sin x}{1 + 2\cos x}$. $f(x)$ is defined except when $1 + 2\cos x = 0$, which occurs when $\cos x = -1/2$, so $x = (2n - 2/3)\pi$ or $(2n + 2/3)\pi$, for any integer n. $f(x) = 0$ when $\sin x = 0$, which occurs when $x = n\pi$, so the x intercepts are $(n\pi, 0)$, where n is any integer. $f(0) = 0$, so (0, 0) is the y intercept. $f'(x) =$

$$\frac{(1 + 2\cos x)\cos x - \sin x(-2\sin x)}{(1 + 2\cos x)^2} =$$

$$\frac{\cos x + 2(\cos^2 x + \sin^2 x)}{(1 + 2\cos x)^2} = \frac{\cos x + 2}{(1 + 2\cos x)^2} \neq 0,$$

since $\cos x + 2 \geq 1$ for all x. $f'(x) > 0$ for all x, so $f(x)$ is increasing whenever it is defined. Neither

$\lim_{x\to\infty} \frac{\sin x}{1 + 2\cos x}$ nor $\lim_{x\to-\infty} \frac{\sin x}{1 + 2\cos x}$ exists, so there are no horizontal asymptotes. Now $\frac{\sin x}{1 + 2\cos x}$ approaches ∞ as $x \to (2n - 2/3)\pi^-$ and $-\infty$ as $x \to (2n - 2/3)\pi^+$, so the lines $x = (2n - 2/3)\pi$ are vertical asymptotes. $\frac{\sin x}{1 + 2\cos x}$ approaches ∞ as $x \to (2n + 2/3)\pi^-$ and $-\infty$ as $x \to (2n + 2/3)\pi^+$, so the lines $x = (2n + 2/3)\pi$ are vertical asymptotes.

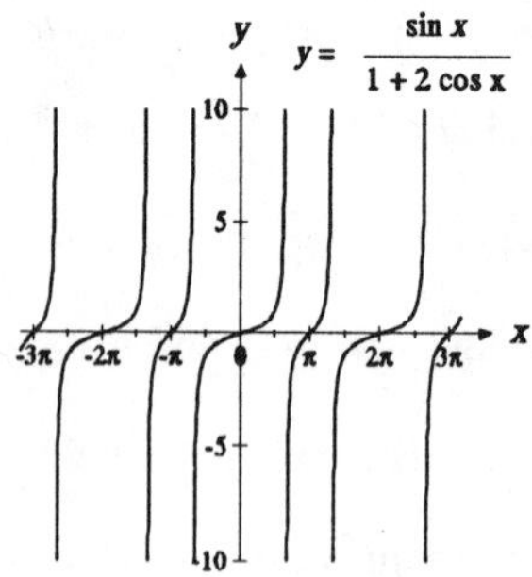

45 $f(x) = \frac{1}{(x - 1)^2(x - 2)}$. $f(x)$ is defined except where $(x - 1)^2(x - 2) = 0$, which occurs when $x = 1$ or 2. $f(x)$ is never zero, so there are no x intercepts; $f(0) = -1/2$, so $(0, -1/2)$ is the y intercept. $f'(x) = \frac{-[(x-1)^2\cdot 1 + (x-2)\cdot 2(x-1)]}{[(x - 1)^2(x - 2)]^2}$

$= \frac{-(3x - 5)}{(x-1)^3(x-2)^2} = 0$ when $3x - 5 = 0$, which occurs when $x = 5/3$, so $\left(\frac{5}{3}, -\frac{27}{4}\right)$ is the only critical point. $f'(x) < 0$ for $x < 1$, so $f(x)$ is decreasing; $f'(x) > 0$ for $1 < x < 5/3$, so $f(x)$ is increasing; $f'(x) < 0$ for $5/3 < x < 2$ and $x > 2$, so $f(x)$ is decreasing. Thus, $\left(\frac{5}{3}, -\frac{27}{4}\right)$ is a local maximum. $\lim_{x\to\infty} \frac{1}{(x - 1)^2(x - 2)} = 0 = \lim_{x\to-\infty} \frac{1}{(x - 1)^2(x - 2)}$, so the x axis is a horizontal asymptote. $\lim_{x\to1} \frac{1}{(x - 1)^2(x - 2)} = -\infty$ so the line $x = 1$ is a vertical asymptote.

$\lim_{x\to2^-} \frac{1}{(x - 1)^2(x - 2)} = -\infty$,

$\lim_{x\to2^+} \frac{1}{(x - 1)^2(x - 2)} = \infty$, so the line $x = 2$ is a vertical asymptote.

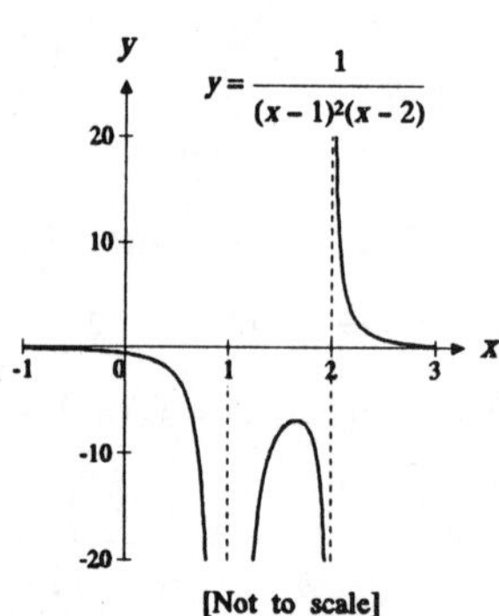

47 $f(x) = 2x^{1/3} + x^{4/3}$. $f(x)$ is defined for all x; there are no vertical asymptotes. $f(x) = x^{1/3}(2 + x) = 0$ when $x = 0$ or -2, so the x intercepts are $(0, 0)$ and $(-2, 0)$; $f(0) = 0$, so $(0, 0)$ is the y intercept. $f'(x) = (2/3)x^{-2/3} + (4/3)x^{1/3} = (2/3)x^{-2/3}(1 + 2x) = 0$ when $1 + 2x = 0$, which occurs when $x = -1/2$; thus $(-1/2, f(-1/2))$ is the only critical point. $f'(x) < 0$ for $x < -1/2$, so $f(x)$ is decreasing; $f'(x) > 0$ for $-1/2 < x < 0$ and $x > 0$, so $f(x)$ is increasing. Therefore the global minimum occurs for $x = -1/2$ and has a value of

$f(-1/2) \approx -1.191.$ $\lim_{x\to\infty} (2x^{1/3} + x^{4/3}) = \infty$

$\lim_{x\to-\infty} (2x^{1/3} + x^{4/3})$, so there are no horizontal

asymptotes.

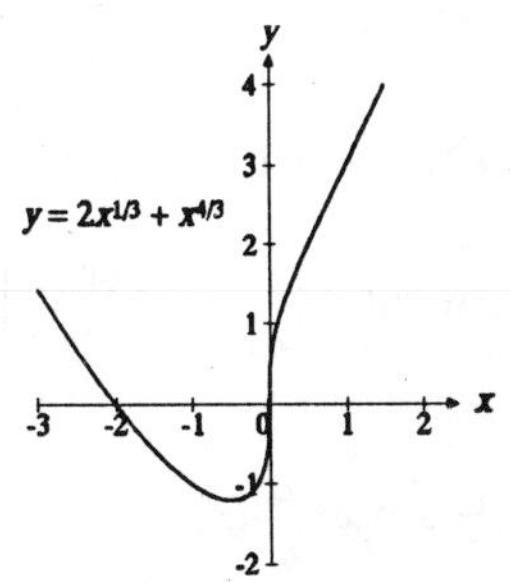

49 Observe that $f(x) = \sqrt{3}\sin x + \cos x =$

$$2\left(\frac{\sqrt{3}}{2}\sin x + \frac{1}{2}\cos x\right) =$$

$$2\left(\cos\frac{\pi}{6}\sin x + \sin\frac{\pi}{6}\cos x\right) = 2\sin(x + \pi/6).$$

Hence the graph of $y = f(x)$ is obtained by shifting the graph of $y = \sin x$ to the left by $\pi/6$ units and expanding it vertically by a factor of 2.

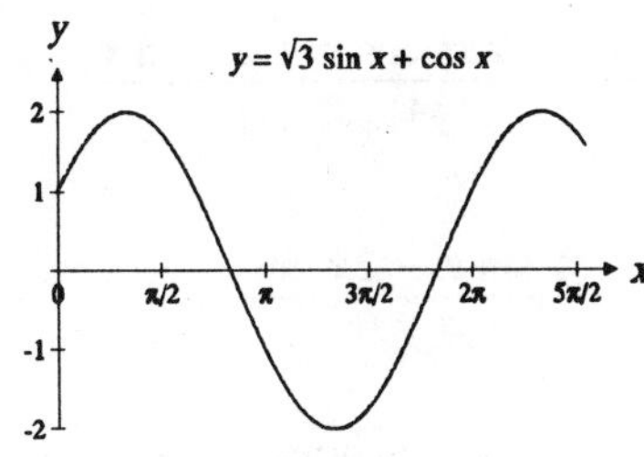

51 (a)

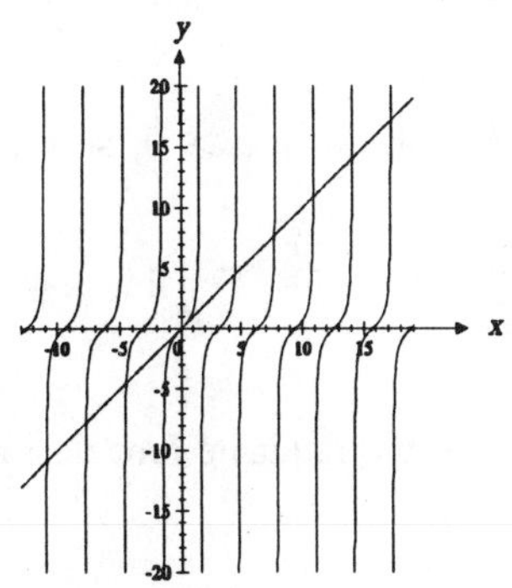

(b) There are infinitely many solutions.

(c)

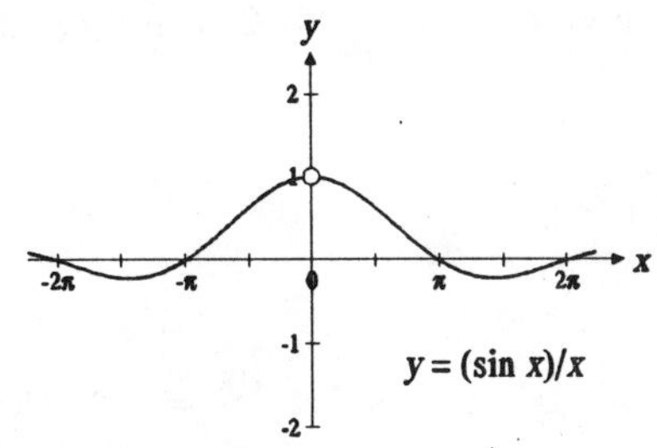

(d) Note that $\left(\frac{\sin x}{x}\right)' = \frac{x\cos x - \sin x(1)}{x^2} =$

$\frac{x\cos x - \sin x}{x^2} = 0$ when $x = \tan x$. From part (b) we know $x = \tan x$ has an infinite number of solutions, so $\frac{\sin x}{x}$ has an infinite number of critical numbers.

53 Let c denote the desired value of x. Since $y' =$

$$\left(\frac{x^3 - x}{x^2 - 4}\right)' = \frac{(x^2 - 4)(3x^2 - 1) - (x^3 - x)(2x)}{(x^2 - 4)^2}$$

$= \frac{x^4 - 11x^2 + 4}{(x^2 - 4)^2} = 0$ when $x^4 - 11x^2 + 4 = 0$,

we must have $x^2 = \frac{11 + \sqrt{121 - 16}}{2} =$

$\frac{11 + \sqrt{105}}{2}$, where the positive root was chosen because $c^2 > 4$. Also, $c > 0$, so we must have $x = c = \sqrt{\frac{11 + \sqrt{105}}{2}}$. For this c, $y = \frac{c(c^2 - 1)}{c^2 - 4} \approx 4.7356$, to the requisite four decimal places.

4.3 Motion and the Second Derivative

1 Since $y = 2x + 3$, $\frac{dy}{dx} = 2$. It follows that $\frac{d^2y}{dx^2} = \frac{d}{dx}\left(\frac{dy}{dx}\right) = \frac{d}{dx}(2) = 0$.

3 Since $y = x^5$, we see that $\frac{dy}{dx} = 5x^4$. Then $\frac{d^2y}{dx^2} = \frac{d}{dx}\left(\frac{dy}{dx}\right) = \frac{d}{dx}(5x^4) = 20x^3$.

5 $y = 2x^3 + x + 2$. Note that $\frac{dy}{dx} = 6x^2 + 1$. Thus $\frac{d^2y}{dx^2} = \frac{d}{dx}\left(\frac{dy}{dx}\right) = \frac{d}{dx}(6x^2 + 1) = 12x$.

7 By the quotient rule, $\frac{dy}{dx} = \left(\frac{x}{x+1}\right)' = \frac{(x+1)\cdot\frac{d}{dx}(x) - x\cdot\frac{d}{dx}(x+1)}{(x+1)^2} = \frac{(x+1)\cdot 1 - x\cdot 1}{(x+1)^2} = (x+1)^{-2}$. So $\frac{d^2y}{dx^2} = \frac{d}{dx}\left(\frac{dy}{dx}\right) = \frac{d}{dx}[(x+1)^{-2}] = -2(x+1)^{-3}$.

9 According to the product rule, $\frac{dy}{dx} = \frac{d}{dx}(x \cos x) = x\cdot(-\sin x) + (\cos x)\cdot 1 = -x \sin x + \cos x$. Therefore, $\frac{d^2y}{dx^2} = \frac{d}{dx}(-x \sin x + \cos x) = -x \cos x - \sin x - \sin x = -x \cos x - 2 \sin x$.

11 By the quotient rule, $\frac{dy}{dx} = \left(\frac{\sin x}{x}\right)' = \frac{x\cdot\frac{d}{dx}(\sin x) - \sin x\cdot\frac{d}{dx}(x)}{x^2} = \frac{x \cos x - \sin x}{x^2}$.

Now $\frac{d^2y}{dx^2} = \frac{d}{dx}\left(\frac{x \cos x - \sin x}{x^2}\right) = \frac{x^2\frac{d}{dx}(x \cos x - \sin x) - (x \cos x - \sin x)\frac{d}{dx}(x^2)}{x^4} = \frac{x^2(-x \sin x + \cos x - \cos x) - (x \cos x - \sin x)\cdot 2x}{x^4} = \frac{-x^3 \sin x - 2x^2 \cos x + 2x \sin x}{x^4} = \frac{(2 - x^2) \sin x - 2x \cos x}{x^3}$.

13 Using the chain rule with $u = x - 2$, we see that $\frac{dy}{dx} = \frac{dy}{du}\frac{du}{dx} = 4u^3\cdot 1 = 4(x-2)^3$. Similarly, $\frac{d^2y}{dx^2} = \frac{d}{dx}[4(x-2)^3] = 12(x-2)^2$.

15 Using the chain rule, with $u = 3x$, we see that $\frac{dy}{dx} = \frac{dy}{du}\cdot\frac{du}{dx} = \cos u \cdot 3 = 3 \cos 3x$. Similarly, $\frac{d^2y}{dx^2}$

$= \frac{d}{dx}[3 \cos 3x] = -9 \sin 3x.$

17 The angle $\theta = f(t)$ is increasing so $d\theta/dt = f'(t) > 0$. The rate of increase is itself increasing, so $\frac{d^2\theta}{dt^2} = f''(t) > 0$.

19 (a) The ball is at the top of the cliff when $y = 96$. Since $y = -16t^2 + 64t + 96$, this is equivalent to $-16t^2 + 64t = 0$, whose solutions are $t = 0$ and $t = 4$. The ball passes the top of the cliff after 4 seconds.

(b) The velocity is $v = -32t + 64$. When $t = 4$, this equals $-32 \cdot 4 + 64 = -64$ ft/sec. The speed is $|v| = 64$ ft/sec.

21 The constant term 96 gives the initial position. The term $64t$ provides the contribution to the motion of the initial velocity. The term $-16t^2$ is the acceleration term, indicating acceleration's effect on the motion.

23 Let y be the horizontal distance traveled by the plane. Initially, we have $y_0 = 0$, $v_0 = 500$ mi/hr, so $y = at^2/2 + 500t$. Now $v = y' = at + 500$, which equals 180 at touch-down time, so $at + 500 = 180$ and $at = -320$. At this time $y = 120$, so $120 = at^2/2 + 500t = (at)t/2 + 500t = -320t/2 + 500t = -160t + 500t = 340t$ and thus $t = 120/340 = 6/17$ hr ≈ 21.2 min.

25 At time $t = 0$, $y = 3$, $v = y' = -3$, and $a = v' = 6$ is constant. Since $dv/dt = 6$, we have $v = 6t + C$; plugging in $t = 0$ and $v = -3$ shows that $C = -3$: hence $v = 6t - 3$. Since $v = dy/dt$, we have $y = 3t^2 - 3t + K$; but $y = 3$ when $t = 0$, so $K = 3$: thus $y = 3t^2 - 3t + 3$.

27 At time $t = 0$, $v = y' = 0$ while $y'' = a = -32$ is constant. Since $a = dv/dt = -32$, it follows that $v = -32t + C$. Plugging in $t = 0$ and $v = y' = 0$ shows that $C = 0$, so $v = -32t$. Then $v = dy/dt = -32t$, so $y = -16t^2 + K$, where K is constant; since $y = 0$ when $t = 0$, we see that $y = -16t^2$.

29 Using feet and seconds as units, we have, in the notation of Exercise 24, $v_0 = 0$, $y_0 = 0$, and $f'(15) = 88$. Hence $f(t) = \frac{a}{2}t^2$, so $f'(t) = at$. Thus $88 = 15a$, $a = \frac{88}{15}$, and $f(t) = \frac{44}{15}t^2$. The distance traveled is $f(15) = \frac{44}{15} \cdot 15^2 = 660$ feet.

4.4 Related Rates

1 When $x = 1$ ft, $s = \sqrt{30^2 + 1^2} = \sqrt{901}$ ft.

Velocity $= \frac{dx}{dt} = -\frac{2s}{x} = -\frac{2\sqrt{901}}{1} = -2\sqrt{901}$ ft/sec, so it follows that speed equals $|-2\sqrt{901}| = 2\sqrt{901} \approx 60.03$ ft/sec.

3 (a)

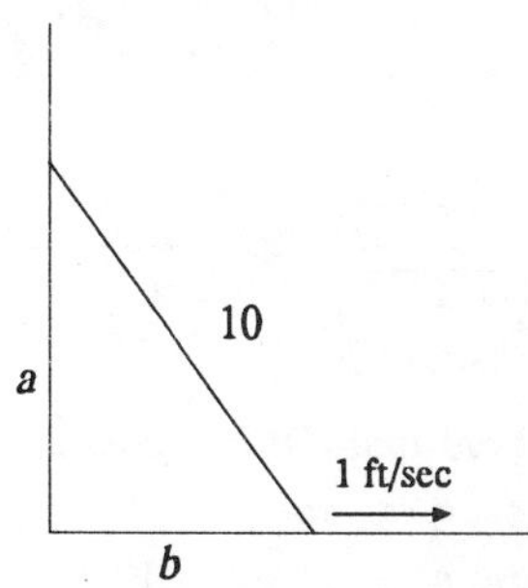

(b) $a^2 + b^2 = 100$

(c) $2a\frac{da}{dt} + 2b\frac{db}{dt} = 0$

(d) Solving the equation in (b) for da/dt, we see

$\frac{da}{dt} = -\frac{b}{a}\frac{db}{dt} = -\frac{b}{a}$ ft/sec. When $b = 6$, a

$= \sqrt{100 - b^2} = \sqrt{100 - 36} = \sqrt{64} = 8$,

and $\frac{da}{dt} = -\frac{3}{4}$ ft/sec. When $b = 8$, $a = 6$,

and $da/dt = -4/3$ ft/sec. When $b = 9$, $a =$

$\sqrt{100 - 9^2} = \sqrt{19}$ and $\frac{da}{dt} = -\frac{9}{\sqrt{19}}$ ft/sec

$\approx$ -2.06 ft/sec.

5 (b) From the graph we see that $\tan\theta = x/3$, or $\cos\theta = \frac{3}{\sqrt{9 + x^2}}$ or $\sin\theta = \frac{x}{\sqrt{9 + x^2}}$.

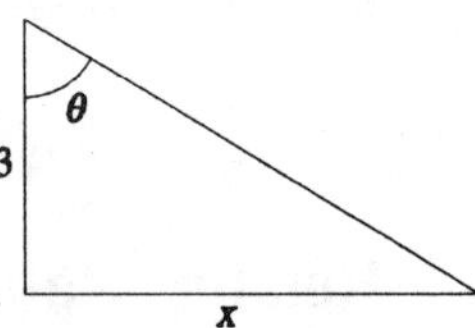

(c) $\frac{d}{dt}(\tan\theta) = \frac{d}{dt}\left(\frac{x}{3}\right)$, so $\sec^2\theta\,\frac{d\theta}{dt} = \frac{1}{3}\cdot\frac{dx}{dt}$,

and $\frac{d\theta}{dt} = \frac{1}{3\sec^2\theta}\cdot\frac{dx}{dt} = \frac{\cos^2\theta}{3}\cdot\frac{dx}{dt}$. Now $\cos\theta$

$= \frac{3}{\sqrt{9 + x^2}}$ and $\frac{dx}{dt} = 2$, so $\frac{d\theta}{dt} =$

$\frac{1}{3}\left(\frac{3}{\sqrt{9 + x^2}}\right)^2\cdot 2 = \frac{6}{9 + x^2}$, as claimed.

Observe that $d\theta/dt$ decreases as $x \to \infty$.

7 The volume of a sphere is $V = \frac{4}{3}\pi r^3$, so $\frac{dV}{dt} =$

$\frac{4}{3}\pi\cdot 3r^2\frac{dr}{dt} = 4\pi r^2\frac{dr}{dt}$. Hence $\frac{dr}{dt} = \frac{dV/dt}{4\pi r^2}$.

(a) When $r = 10$, $\frac{dr}{dt} = \frac{1}{4\pi\cdot 10^2}\cdot 100 =$

$\frac{1}{4\pi}$ ft/min.

(b) When $r = 20$, $\frac{dr}{dt} = \frac{1}{4\pi\cdot 20^2}\cdot 100 =$

$\frac{1}{16\pi}$ ft/min.

9 The radius r and height h of the conical hill are equal, so the hill's volume is $V = \frac{1}{3}\pi r^2 h =$

$\frac{1}{3}\pi r^3$, and $1000 = \frac{dV}{dt} = \pi r^2\cdot\frac{dr}{dt}$, so $\frac{dr}{dt}$

$= \frac{1000}{\pi r^2}$ yd/hr.

(a) When $r = 20$, $\frac{dr}{dt} = \frac{1000}{\pi\cdot 20^2} = \frac{1000}{400\pi}$

$= \frac{5}{2\pi}$ yd/hr.

(b) When $r = 100$, $\frac{dr}{dt} = \frac{1000}{\pi\cdot 100^2} = \frac{1000}{10000\pi}$

$= \frac{1}{10\pi}$ yd/hr.

11 The area is $A = \frac{1}{2}ab\sin\theta$, so $\frac{dA}{dt} =$

$\frac{1}{2}ab\cos\theta\frac{d\theta}{dt} + \frac{1}{2}a\cdot\frac{db}{dt}\sin\theta + \frac{1}{2}\cdot\frac{da}{dt}\cdot b\sin\theta$.

Since $a = 4$ ft, $b =$ 5 ft, $\theta = \frac{\pi}{4}$, $\frac{da}{dt} =$ 3 ft/sec, $\frac{db}{dt} =$

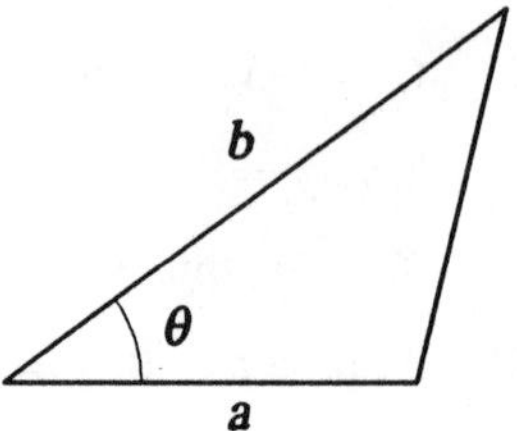

-2 ft/sec, and $\frac{d\theta}{dt} = 1$ radian/sec, we may

substitute to find $\frac{dA}{dt} =$

$$\left(\frac{1}{2}\cdot 4\cdot 5\frac{\sqrt{2}}{2}\cdot 1 + \frac{1}{2}\cdot 4(-2)\cdot\frac{\sqrt{2}}{2} + \frac{1}{2}\cdot 3\cdot 5\cdot\frac{\sqrt{2}}{2}\right) \text{ ft}^2/\text{sec}$$

$= \frac{27}{4}\sqrt{2}$ ft²/sec. $dA/dt > 0$, so the area is

increasing.

13 $x^2 + 30^2 = s^2$; differentiating we get $x\dot{x} = s\dot{s}$.

Differentiating again, $x\ddot{x} + (\dot{x})^2 = s\ddot{s} + (\dot{s})^2$. Since $\ddot{s}$

$= 0$ and $\dot{x} = \frac{s}{x}\dot{s} = -2\frac{s}{x}$, this simplifies to

$x\ddot{x} + 4\cdot\frac{s^2}{x^2} = 4$. Thus the acceleration is $\ddot{x} =$

$$\frac{4x^2 - 4s^2}{x^3} = \frac{-4(s^2 - x^2)}{x^3} = \frac{-4\cdot 30^2}{x^3}$$

$$= \frac{-3600}{x^3}.$$

(a) When $s = 300$, $x = \sqrt{s^2 - 30^2} =$

$\sqrt{90000 - 900} = \sqrt{89100}$, so $\ddot{x} =$

$\frac{-3600}{(89100)^{3/2}} \approx -0.00014$ ft/sec².

(b) When $s = 31$, $x = \sqrt{31^2 - 900} = \sqrt{61}$, so $\ddot{x}$

$= \frac{-3600}{(61)^{3/2}} \approx -7.56$ ft/sec².

15 (a) $y = x^2$, so $\dot{y} = \frac{d}{dt}(x^2) = 2x\dot{x} = 2x\cdot 3$

$= 6x$.

(b) $\ddot{y} = \frac{d}{dt}(\dot{y}) = \frac{d}{dt}(6x) = 6\dot{x} = 6\cdot 3 = 18.$

17 Label the lengths as shown. Then $L^2 =$

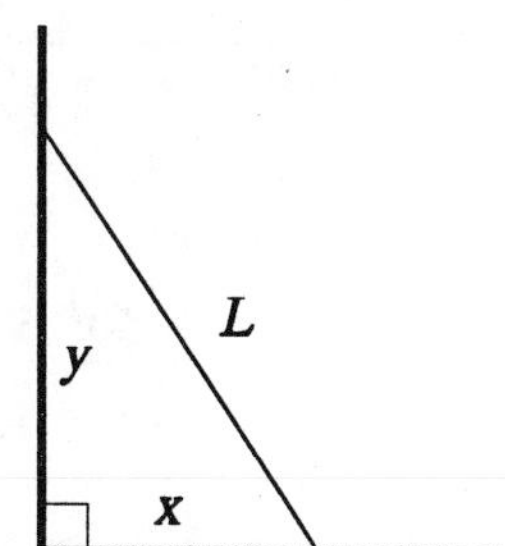

$x^2 + y^2$, so $2L\cdot\frac{dL}{dt} =$

$2x\cdot\frac{dx}{dt} + 2y\cdot\frac{dy}{dt}.$

Solving for $\frac{dy}{dt}$ gives

$\frac{dy}{dt} = \frac{1}{y}\left(L\cdot\frac{dL}{dt} - x\cdot\frac{dx}{dt}\right)$. We are told that $x = 6$,

$y = 8$, $\frac{dx}{dt} = 3$, and $\frac{dL}{dt} = -2$. Hence $L =$

$\sqrt{6^2 + 8^2} = 10$ and $\frac{dy}{dt} = \frac{1}{8}(10(-2) - 6\cdot 3) =$

$-\frac{19}{4}$. The top is moving down at 19/4 ft/sec.

19 (a) Let x be the horizontal distance from the boat to the bridge. Then the distance from the center

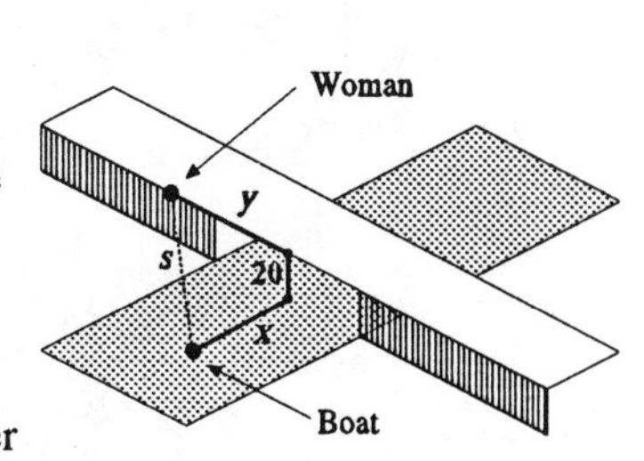

of the bridge to the boat is $\sqrt{20^2 + x^2} =$

$\sqrt{400 + x^2}$. When the boat is under the bridge, $x = 0$ ft and $dx/dt = 10$ ft/sec. Let y be the distance from the woman to the center of the bridge. When the boat is under the bridge, $y = 50$ ft and $dy/dt = -5$ ft/sec. Let s be the distance from the woman to the boat.

Then $s = \left[\left(\sqrt{400 + x^2}\right)^2 + y^2\right]^{1/2} =$

$\sqrt{400 + x^2 + y^2}$; $\dfrac{ds}{dt} = \dfrac{x\cdot\dfrac{dx}{dt} + y\cdot\dfrac{dy}{dt}}{\sqrt{400 + x^2 + y^2}} =$

$$\frac{0\cdot 10 + 50(-5)}{\sqrt{400 + 0^2 + 50^2}} = \frac{-250}{\sqrt{2900}} = \frac{-25}{\sqrt{29}} \text{ ft/sec.}$$

(b) Observe that $ds/dt < 0$, so the distance between the woman and the boat is decreasing. Their rate of *approach* is $\dfrac{25}{\sqrt{29}}$ ft/sec. Is this rate increasing or decreasing?

Now $\dfrac{ds}{dt} = \dfrac{x\dot{x} + y\dot{y}}{\sqrt{400 + x^2 + y^2}} = \dfrac{x\dot{x} + y\dot{y}}{s}$, so

$$\frac{d^2s}{dt^2} = \frac{s(x\ddot{x} + \dot{x}^2 + y\ddot{y} + \dot{y}^2) - (x\dot{x} + y\dot{y})\dot{s}}{s^2}.$$

The denominator is positive, so $\dfrac{d^2s}{dt^2}$ has the same sign as the numerator, which equals

$$\sqrt{2900}(0 + 100 + 0 + 25) - (0 - 250)\left(\frac{-25}{\sqrt{29}}\right) =$$

$125\sqrt{2900} - \dfrac{6250}{\sqrt{29}} > 0$. Thus $d^2s/dt^2 > 0$, so ds/dt is increasing with respect to t. But ds/dt is a negative quantity, so it will become a negative quantity of lesser magnitude. Thus the woman and the boat will continue to approach each other, but the rate of approach will *decrease*.

21 If θ is the angle shown in the figure, measured from the horizontal, and $y(\theta)$ the position of the shadow upon the ground relative to the shadow of the center of the Ferris wheel, then $y(\theta) = 25\cos\theta$. So $\dfrac{dy}{dt}$

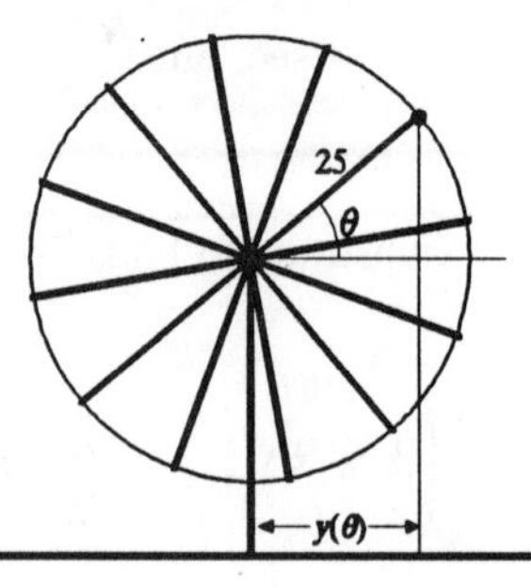

$= -25\sin\theta\,\dfrac{d\theta}{dt}$; since $\dfrac{d\theta}{dt} = (0.1)(2\pi) = \dfrac{\pi}{5}$,

$$\frac{dy}{dt} = -25(\sin\theta)\frac{\pi}{5} = -5\pi\sin\theta \text{ ft/sec.}$$

(a) At the two o'clock position, $\theta = 30° = \dfrac{\pi}{6}$ and $\dfrac{dy}{dt} = -5\pi\cdot\dfrac{1}{2} = \dfrac{-5\pi}{2}$ ft/sec, so the speed is $\dfrac{5\pi}{2}$ ft/sec.

(b) At the one o'clock position, $\theta = 60° = \dfrac{\pi}{3}$ and $\dfrac{dy}{dt} = -5\pi\cdot\dfrac{\sqrt{3}}{2} = -\dfrac{5\sqrt{3}\pi}{2}$ ft/sec, so the speed is $\dfrac{5\sqrt{3}\pi}{2}$ ft/sec.

(c) The shadow is moving fastest for $\sin\theta = \pm 1$, which corresponds to the top or bottom. The shadow stops moving—that is, has speed zero—for $\sin\theta = 0$, which gives the three o'clock and nine o'clock positions.

23 From the figure, $D^2 = (R + h)^2 - R^2 = h^2 + 2Rh$, so $2D \cdot \frac{dD}{dt} = 2h \cdot \frac{dh}{dt} + 2R \cdot \frac{dh}{dt}$; hence $\frac{dD}{dt} = \frac{h + R}{D} \cdot \frac{dh}{dt}$. We are given that $h = 1000$ feet, $R = 4000$ miles $= 4000 \cdot 5280$ feet, and $dh/dt = 10$ ft/sec. Hence $D = \sqrt{1000^2 + 2 \cdot 4000 \cdot 5280 \cdot 1000} = 1000\sqrt{42241}$ feet, and $\frac{dD}{dt} = \frac{1000 + 4000 \cdot 5280}{1000\sqrt{42241}} \cdot 10 = \frac{211210}{\sqrt{42241}} \approx 1027.7$ ft/sec.

(In practice, D increases discontinuously and may even decrease. It is usually almost constant, increasing suddenly when you rise high enough to see beyond one more mountain range.)

4.5 The Second Derivative and Graphing

1 First find the second derivative. Since $D(x^3 - 3x^2 + 2) = 3x^2 - 6x$, we have $D^2(x^3 - 3x^2 + 2) = D(3x^2 - 6x) = 6x - 6 = 6(x - 1)$. Now, $6(x - 1) > 0$ when $x > 1$, so $x^3 - 3x^2 + 2$ is concave up on $(1, \infty)$. Also, $6(x - 1) < 0$ when $x < 1$, so $x^3 - 3x^2 + 2$ is concave down on $(-\infty, 1)$. Since $x^3 - 3x^2 + 2$ changes concavity at $x = 1$, $(1, 0)$ is an inflection point.

3 Let $f(x) = x^2 + x + 1$. Then $f'(x) = 2x + 1$ and $f''(x) = 2$. Since $f''(x) > 0$ for all x, f is concave up for all x.

5 Let $f(x) = x^6$. Then $f'(x) = 6x^5$ and $f''(x) = 30x^4 > 0$ for all x. Thus, f is concave up for all x.

7 Let $f(x) = x^4 - 4x^3$. Then $f'(x) = 4x^3 - 12x^2$ and $f''(x) = 12x^2 - 24x = 12x(x - 2)$. Now, $f''(x) > 0$ either when $x > 0$ and $x - 2 > 0$ or when $x < 0$ and $x - 2 < 0$, or when $x > 2$ and $x < 0$. So f is concave up on $(-\infty, 0)$ and $(2, \infty)$. Consequently, f must be concave down on $(0, 2)$. Since f changes concavity at $x = 0$ and $x = 2$, $(0, 0)$ and $(2, -16)$ are inflection points.

9 Let $f(x) = \frac{1}{1 + x^2}$. Then $f'(x) = \frac{-2x}{(1 + x^2)^2}$ and

$$f''(x) = \frac{(1 + x^2)^2 \cdot (-2) - (-2x) \cdot 2(1 + x^2) \cdot 2x}{(1 + x^2)^4}.$$

The denominator is always positive and the numerator equals $-2(1 + x^2)^2 + 8x^2(1 + x^2)$, or $-2(1 + x^2)[1 - 3x^2]$. Now, $1 + x^2$ is always positive and $1 - 3x^2$ is negative whenever $|x| > \frac{1}{\sqrt{3}}$. So $f''(x) > 0$ for $|x| > 1/\sqrt{3}$, and f is concave up for $|x| > 1/\sqrt{3}$. Then f is concave down for $|x| < 1/\sqrt{3}$. So f changes concavity at $x = \pm 1/\sqrt{3}$. We conclude that f has inflection points at $(1/\sqrt{3}, 3/4)$ and $(-1/\sqrt{3}, 3/4)$. Note that this symmetry is a result of the fact that f is an even function.

11 Let $f(x) = x^3 - 6x^2 - 15x$. Then $f'(x) = 3x^2 - 12x - 15$ and $f''(x) = 6x - 12 = 6(x - 2)$. So $f''(x) > 0$ when $x > 2$, and f is concave up on $(2, \infty)$. Also, $f''(x) < 0$ when $x < 2$, so f is

concave down on $(-\infty, 2)$. Since f changes concavity at $x = 2$, f has an inflection point at $(2, -46)$.

13 Let $f(x) = \tan x$. Then $f'(x) = \sec^2 x$ and $f''(x) = 2\sec^2 x \tan x$. Now, we see that $f''(x) = 0$ whenever $\sec^2 x \tan x = \dfrac{\sin x}{\cos^3 x} = 0$ whenever $\sin x = 0$. This occurs at $x = n\pi$, where n is an integer. Furthermore, $\sin x < 0$ for values of x slightly less than $2n\pi$ and $\sin x > 0$ for values slightly greater than $2n\pi$, while $\cos^3 x$ remains positive. So $f''(x)$ goes from $-$ to $+$ at $x = 2n\pi$. Also, $\sin x > 0$ for values of x slightly less than $(2n + 1)\pi$ and $\sin x < 0$ for values slightly greater than $(2n + 1)\pi$, while $\cos x$ remains negative. So $f''(x)$ goes from $-$ to $+$ at $x = (2n + 1)\pi$. We finally note that $f''(x)$ goes from $+$ to $-$ at $x = (1/2)(2n + 1)\pi$ (though there is no inflection point there). We conclude that f has inflection points at $(n\pi, 0)$, f is concave up on $(n\pi, (n + 1/2)\pi)$ and concave down on $((n - 1/2)\pi, n\pi)$.

15 Let $f(x) = \cos x$. Then $f'(x) = -\sin x$ and $f''(x) = -\cos x = -f(x)$. So $f''(x) < 0$ whenever $\cos x < 0$, or when x is in $(-\pi/2 + 2k\pi, \pi/2 + 2k\pi)$, where k is an integer. Similarly, $f''(x) > 0$ when x is in $(\pi/2 + 2k\pi, 3\pi/2 + 2k\pi)$. So f has inflection points at $x = (1/2)(2k + 1)\pi$, is concave up on $(\pi/2 + 2k\pi, 3\pi/2 + 2k\pi)$, and is concave down on $(-\pi/2 + 2k\pi, \pi/2 + 2k\pi)$.

17 Let $f(x) = x^3 + 3x^2$. We find the critical points by setting $f'(x) = 0$ and solving for x. Here, $f'(x) = 3x^2 + 6x = x(3x + 6) = 0$ when $x = 0, -2$. We find the inflection points by setting $f''(x) = 0$ and solving for x. Now, $f''(x) = 6x + 6 = 6(x + 1) = 0$ when $x = -1$. We also see that $f''(x)$ goes from $-$ to $+$ at $x = -1$. So f is concave down on $(-\infty, -1)$ and concave up on $(-1, \infty)$. This implies that, because $x = 0$ and $x = -2$ are critical points, $f(-2)$ is a maximum and $f(0)$ is a minimum. We find the x intercepts by setting $f(x) = 0$ and solving for x. Then $f(x) = x^3 + 3x^2 = x^2(x + 3) = 0$ when $x = 0, -3$. We find the y intercept by setting $x = 0$ and finding $f(0)$: $f(0) = 0$.

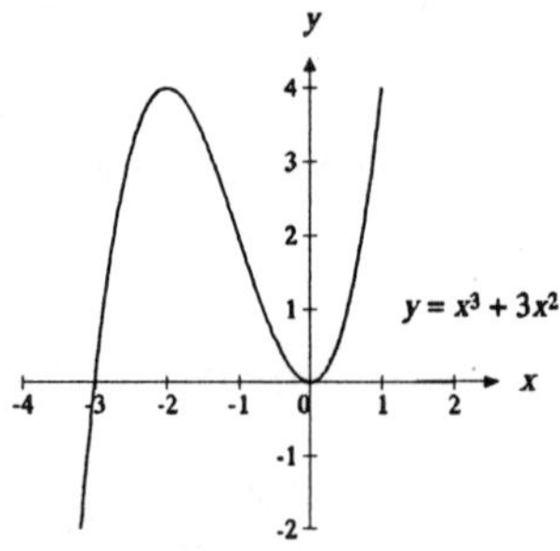

19 Let $f(x) = x^4 - 4x^3 + 6x^2$. Now, $f(x) = 0$ when $x^2(x^2 - 4x + 6) = 0$. This occurs when $x = 0$ or $x^2 - 4x + 6 = 0$. Now $x^2 - 4x + 6 = 0$ when $x = \dfrac{4 \pm \sqrt{-8}}{2}$. Thus, $x^2 - 4x + 6 \neq 0$ for any real

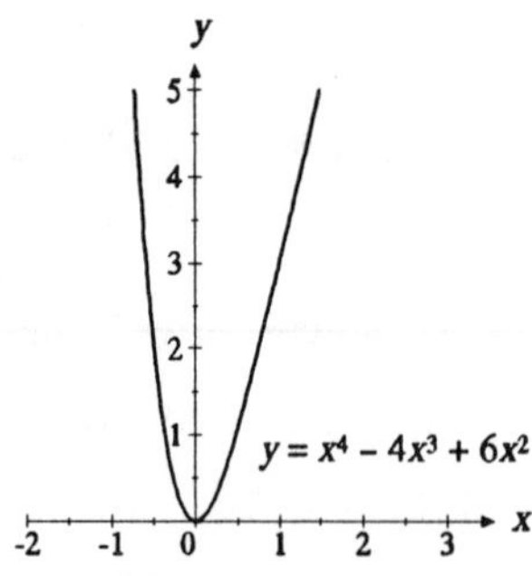

x. So $(0, 0)$ is the only x intercept. Since $f(0) = 0$, $(0, 0)$ is the y intercept. Also, observe that $f'(x) = 4x^3 - 12x^2 + 12x = 4x(x^2 - 3x + 3) = 0$ when $x = 0$ or $x^2 - 3x + 3 = 0$. Now $x^2 - 3x + 3 = 0$ when $x = \dfrac{3 \pm \sqrt{-3}}{2}$. So $x^2 - 3x + 3 \neq 0$ for

any real x. So (0, 0) is the only critical point. Finally, $f''(x) = 12x^2 - 24x + 12 = 12(x - 1)^2 = 0$ when $x = 1$. However, $f''(x)$ doesn't change sign at $x = 1$; there are no inflection points. In fact, $f''(x) > 0$ for all x, so f is concave up for all x.

21 $f'(1) = 0$, so there is a horizontal tangent at (1, 1); $f''(1) > 0$, so the curve is concave upward near (1, 1).

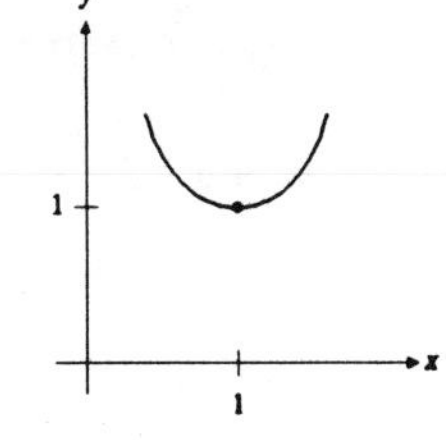

23 $f'(1) = 0$, so there is a horizontal tangent at (1, 1). $f''(x)$ can be positive, negative, or zero for $x < 1$ and it can be positive, negative, or zero for $x > 1$, so there are nine possibilities.

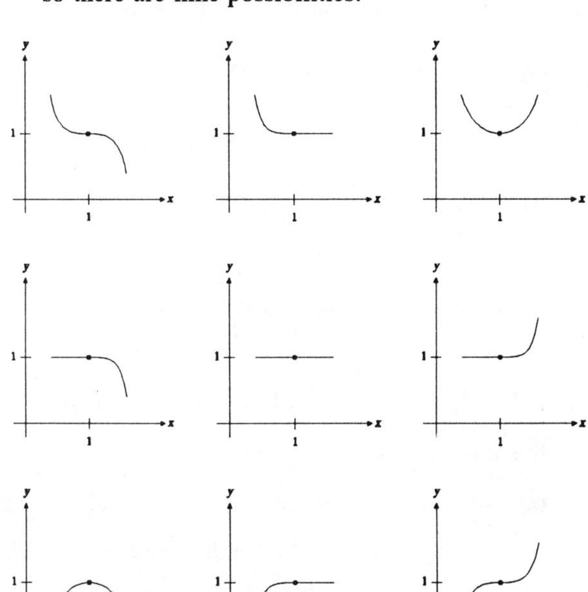

25 $f'(1) = 0$, so there is a horizontal tangent at (1, 1); $f''(x) < 0$ for x near 1, so the curve is concave downward near 1.

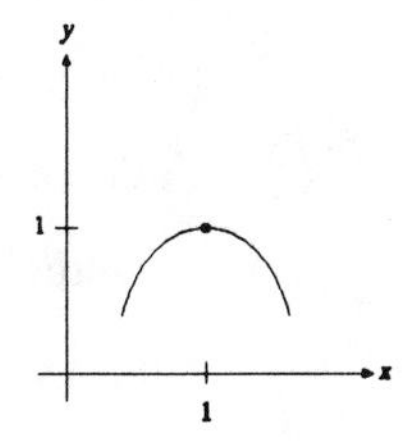

27 $f'(1) = 1$, so the function is increasing near $x = 1$. For $x < 1, f''(x) < 0$, so the curve is concave downward; for $x > 1, f''(x) > 0$, so the curve is concave upward.

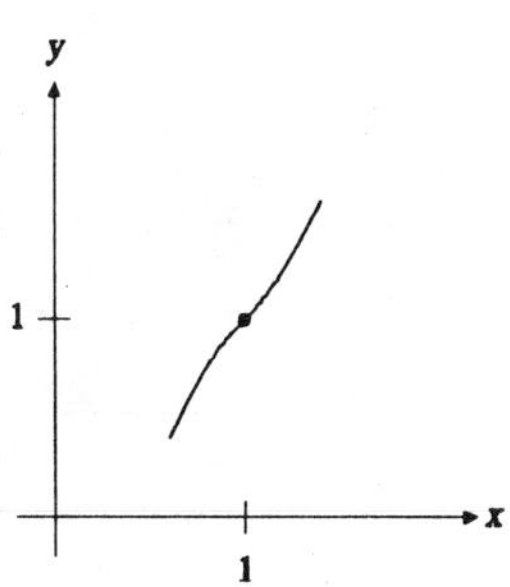

29 When $x = 0$, $y = 2(-1)^{5/3} + 5(-1)^{2/3} = 3$, so the y intercept is at 3. Letting $y = 0$, we have $2(x - 1)^{5/3} + 5(x - 1)^{2/3} = 0$ and therefore $(x - 1)^{2/3}[2(x - 1) + 5] = 0$. Thus $(x - 1)^{2/3} = 0$ or $2(x - 1) + 5 = 0$ and we see that $x = 1$ and $x = -3/2$ are the x intercepts. Now observe that $y = 2(x - 1)^{5/3} + 5(x - 1)^{2/3} = (x - 1)^{2/3}[2x + 3]$, and $y' = 2(x - 1)^{2/3} + (2x + 3)\cdot(2/3)(x - 1)^{-1/3}$ $= 2(x - 1)^{2/3} + \dfrac{2(2x + 3)}{3(x - 1)^{1/3}}.$ The derivative is 0 when $2(x - 1)^{2/3} + \dfrac{2(2x + 3)}{3(x - 1)^{1/3}} = 0$, so $\dfrac{2(2x + 3)}{3(x - 1)^{1/3}} = -2(x - 1)^{2/3}$, $2x + 3 = -3(x - 1) = -3x + 3$, and $5x = 0$, so $x = 0$. Hence (0, 3) is a critical point. Solving for y'' yields $y'' = (4/3)(x - 1)^{-1/3} - (2/9)(2x + 3)(x - 1)^{-4/3} + (4/3)(x - 1)^{-1/3} = (8/3)(x - 1)^{-1/3} - (2/9)(2x + 3)(x - 1)^{-4/3}$. Here, $y'' = 0$ when $\frac{8}{3}(x - 1)^{-1/3} = \frac{2}{9}(2x + 3)(x - 1)^{-4/3}$, $72(x - 1) = 12x + 18$, $60x = 90$, and $x = \frac{3}{2}$. Since y'' is undefined at $x = 1$, we see that these two points are of interest. The

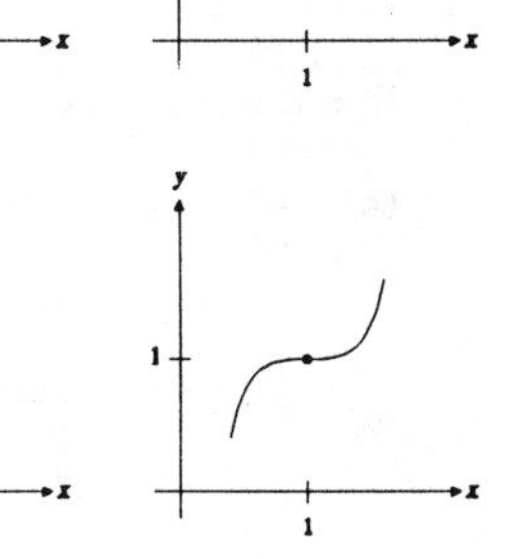

following table shows the sign of y'' at sample values of x in each of the intervals $(-\infty, 1)$, $(1, 3/2)$, and $(3/2, \infty)$.

	$(-\infty, 1)$	$(1, 3/2)$	$(3/2, \infty)$
x	0	5/4	2
y''	−	−	+
concavity	down	down	up

Since y'' is concave down in the neighborhood of $x = 0$, $(0, 3)$ is a maximum. Also note that y' is undefined at $x = 1$, $\lim_{x\to 1^-} y' = -\infty$, and $\lim_{x\to 1^+} y' = \infty$. Thus, a cusp (and minimum) exists at $(1, 0)$.

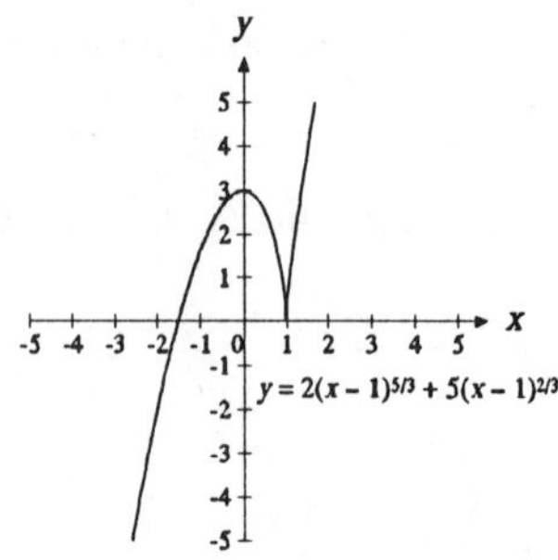

31 Since $\lim_{x\to\infty} f(x) = 0$ and $\lim_{x\to-\infty} f(x) = 0$, $y = 0$ is a horizontal asymptote. The tangent to the curve at $(2, 4)$ has slope 0 since $(2, 4)$ is a critical point. Also, the graph changes concavity at $(3, 1)$ and $(1, 1)$ because these are inflection points. See the figure for one possible graph.

33 (a) The positive derivative indicates that farm output is (and has been) increasing.

(b) The inflection point indicates that whereas before 1957 the rate of increase was rising, since 1957 the rate of increase has fallen.

35 To find the critical points, find where $f'(x) = -\sin^3 x + 2 \sin x \cos^2 x = 0$. Since $f'(x) = (\sin x)(2 \cos^2 x - \sin^2 x) = (\sin x)(2 - 3 \sin^2 x)$, $f'(x) = 0$ when $\sin x = 0$ or $\sin x = \pm\sqrt{2/3}$. In $[0, 2\pi]$, the critical numbers are 0, π, 2π, $\sin^{-1}\sqrt{2/3} \approx 0.955$, $\pi - \sin^{-1}\sqrt{2/3} \approx 2.186$, $\pi + \sin^{-1}\sqrt{2/3} \approx 4.097$, and $2\pi - \sin^{-1}\sqrt{2/3} \approx 5.328$. To find inflections points, find $f''(x) = -3 \sin^2 x \cos x + 2 \cos^3 x - 4 \sin^2 x \cos x = 2 \cos^3 x - 7 \sin^2 x \cos x = 0$. Writing $\cos^2 x$ as $1 - \sin^2 x$, this simplifies to $(\cos x)(2 - 9 \sin^2 x) = 0$, which is true when $\cos x = 0$ or $\sin x = \pm\sqrt{2}/3$. In the interval $[0, 2\pi]$, the inflection points are $x = \pi/2$, $3\pi/2$, $\sin^{-1}\sqrt{2}/3 \approx 0.491$, $\pi - \sin^{-1}\sqrt{2}/3 \approx 2.651$, $\pi + \sin^{-1}\sqrt{2}/3 \approx 2.632$, and $2\pi - \sin^{-1}\sqrt{2}/3 \approx 5.792$.

37 The second derivative of a polynomial of degree five is a polynomial of degree three. To find inflection points, we need to identify the roots of this third-degree polynomial.

(a) Since a third-degree polynomial must have an odd number of roots (by the intermediate-value theorem), it is impossible for a polynomial of degree three to have no real roots and thus impossible for a fifth-order polynomial to have no inflection points.

(b) Following the reasoning above, a fifth-order polynomial can have exactly one inflection point because a polynomial of degree three can have exactly one real root (for example, $f(x) = x^3$).

39 (a) $f''(x)$ changes from + to − at 1 and from − to + at 2, so f is concave upward for $x < 1$ and for $x > 2$.

(b) f is concave down for $1 < x < 2$.

(c) The two inflection numbers are 1 and 2.

(d) Since $f''(x) = (x - 1)(x - 2) = x^2 - 3x + 2$, $(x^3/3)' = x^2$, $(x^2/2)' = x$, and $(x)' = 1$ we have $f'(x) = x^3/3 - 3 \cdot x^2/2 + 2x + C$, where C is any constant. Also, $(x^4/4)' = x^3$, so $f(x)$

$$= \frac{1}{3}\frac{x^4}{4} - 3\frac{1}{2}\frac{x^3}{3} + 2\frac{x^2}{2} + Cx + D$$

$= x^4/12 - x^3/2 + x^2 + Cx + D$, where D is any constant. Choose any numbers you like for C and D to get a specific function for f.

41 (a) There must be a maximum at (1, 1) for if (1, 1) were a minimum, then another critical point would be necessary for $f(x) \to 0$ as $x \to \infty$, but (1, 1) is the only critical point. Since (1, 1) is thus a global maximum, f must be decreasing on $(1, \infty)$.

(b) From (a) we know (1, 1) is a maximum, so $f''(x) < 0$ in the neighborhood of $x = 1$ and $f'(x)$ is negative for values of $x > 1$ but near $x = 1$. If $f''(x)$ remains negative for all $x > 1$, then $f'(x)$ is negative and decreasing for all $x > 1$. However for $\lim_{x \to \infty} f(x) = 0$ to hold,

$\lim_{x \to \infty} f'(x) = 0$ as well. So an inflection point is necessary in $(1, \infty)$ so that $f''(x) > 0$ for all $x > a$, with a in $(1, \infty)$. This way, $f'(x)$ can increase through negative values and approach zero for large x. We conclude $f(x)$ must have an inflection point in $(1, \infty)$.

4.6 Newton's Method for Solving an Equation

1 In this case, $x_2 = x_1 - \dfrac{f(x_1)}{f'(x_1)} = 2 - \dfrac{0.3}{1.5} = 1.8.$

3 We want a function for which $x = \sqrt{a}$ is a root, so let $f(x) = x^2 - a$. Then $f'(x) = 2x$ and $x_{i+1} =$

$$x_i - \frac{f(x_i)}{f'(x_i)} = x_i - \frac{x_i^2 - a}{2x_i} = x_i - \frac{1}{2}x_i + \frac{a}{2x_i} = \frac{1}{2}\left(x_i + \frac{a}{x_i}\right).$$

5 $x_2 = \dfrac{x_1 + 15/x_1}{2} = \dfrac{4 + 15/4}{2} = \dfrac{31}{8} = 3.875$, x_3

$$= \frac{x_2 + 15/x_2}{2} = \frac{3.875 + 15/3.875}{2} \approx 3.873$$

7 Letting $a = 7$ and $x_1 = 2$ in the formula given in Exercise 4, we have $x_2 = \dfrac{2}{3} \cdot 2 + \dfrac{7}{3 \cdot 2^2} =$

$$\frac{4}{3} + \frac{7}{12} = \frac{23}{12} \approx 1.917 \text{ and } x_3 =$$

$$\frac{2}{3}\cdot\frac{23}{12} + \frac{7}{3 \cdot (23/12)^2} = \frac{23}{18} + \frac{7 \cdot 12^2}{3 \cdot 23^2}$$

$$= \frac{23}{18} + \frac{336}{529} \approx 1.913.$$

9 (a) In this case, $a = 5$ and $x_1 = 2$, we have $x_2 =$

$$\frac{1}{2}\left(2 + \frac{5}{2}\right) = \frac{9}{4} = 2.25,\ x_3 = \frac{1}{2}\left(\frac{9}{4} + 5 \cdot \frac{4}{9}\right)$$

$$= \frac{161}{72} \approx 2.2361111,\ x_4 =$$

$$\frac{1}{2}\left(\frac{161}{72} + 5 \cdot \frac{72}{161}\right) = \frac{51841}{23184} \approx 2.2360680,$$

and $x_5 = \frac{1}{2}\left(\frac{51841}{23184} + 5 \cdot \frac{23184}{51841}\right) \approx$ 2.2360680. Since x_4 and x_5 agree to four decimal places, we are done.

(b) From the calculator, $\sqrt{5} \approx 2.23606798$. Hence $|\sqrt{5} - x_5| \approx 2 \times 10^{-8}$, and x_5 and $\sqrt{5}$ agree to five decimal places.

11 (a) $f(0) = -1$ and $f(1) = 1$, so the intermediate-value theorem says that f has at least one root in [0, 1].

(b) $x_{i+1} = x_i - \dfrac{x_i^5 + x_i - 1}{5x_i^4 + 1} = \dfrac{5x_i^5 + x_i - x_i^5 - x_i + 1}{5x_i^4 + 1} = \dfrac{4x_i^5 + 1}{5x_i^4 + 1}$, so if $x_1 = 1/2$ then $x_2 = \dfrac{\frac{4}{32} + 1}{\frac{5}{16} + 1} = \dfrac{6}{7} \approx 0.857$.

(c) Since $f'(x) = 5x^4 + 1 > 0$ for x in [0, 1], f is increasing on [0, 1]. Thus the root in [0, 1] is unique.

13 Letting $a = 3$ and $x_1 = 10$ in the formula in Exercise 1, $x_2 = \frac{1}{2}\left(10 + \frac{3}{10}\right) = \frac{103}{20} = 5.15$, $x_3 = \frac{1}{2}\left(\frac{103}{20} + \frac{3 \cdot 20}{103}\right) \approx 2.8663$, and $x_4 = \frac{1}{2}\left(x_3 + \frac{3}{x_3}\right) \approx 1.9565$. Note that $\sqrt{3} \approx 1.7321$ and that x_4 isn't even accurate to one decimal place.

15 (a)

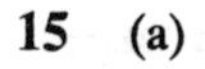

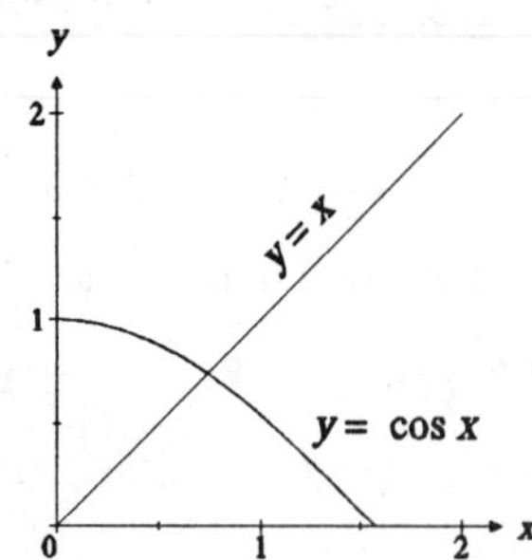

(b) The root of $x - \cos x = 0$ is approximately 3/4. There is no negative solution.

(c) Using $x_1 = 0.75$, we have $x_2 = x_1 - \dfrac{x_1 - \cos x_1}{1 + \sin x_1} = 0.75 - \dfrac{0.75 - 0.7316888}{1 + 0.6816387} \approx 0.7391111$, $x_3 \approx 0.7391111 - \dfrac{0.7391111 - 0.7390676}{1 + 0.6736312} \approx 0.7390851$, and $x_4 \approx 0.7390851 - \dfrac{0.7390851 - 0.7390851}{1 + 0.6736120} \approx$ 0.7390851. Here $x_3 \approx x_4$ and x_3 and x_4 agree to seven decimal places.

17 (a) $f(x) = 2x^3 - 4x + 1$, so $f(1) = -1$ and $f(0) = 1$; by the intermediate-value theorem, f must have a root between 0 and 1.

(b) Note that $f'(x) = 6x^2 - 4$, so we have $x_{i+1} = x_i - \dfrac{2x_i^3 - 4x_i + 1}{6x_i^2 - 4}$; if $x_1 = 1$, then $x_2 = 1 - \dfrac{-1}{2} = \dfrac{3}{2} = 1.5$ and $x_3 \approx 1.316$.

(c)

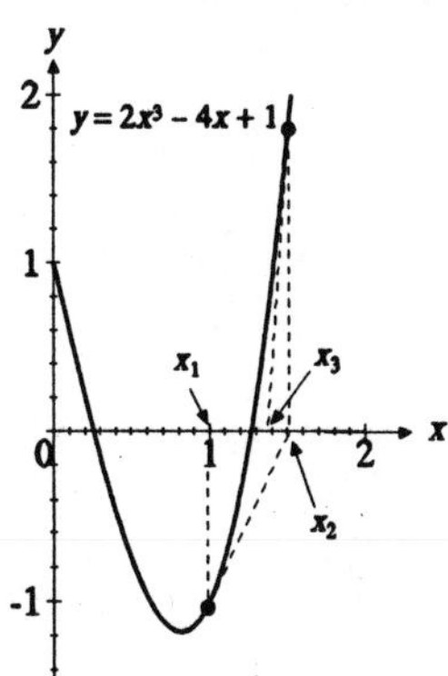

19 (a) In general, $x_{i+1} = x_i - \dfrac{x_i^3 - x_i}{3x_i^2 - 1} = \dfrac{2x_i^3}{3x_i^2 - 1}$. So $x_2 = \dfrac{2(1/\sqrt{5})^3}{3(1/\sqrt{5})^2 - 1} = \dfrac{2/(5\sqrt{5})}{3/5 - 1} = -\dfrac{1}{\sqrt{5}}$ and $x_3 = \dfrac{2(-1/\sqrt{5})^3}{3(-1/\sqrt{5})^2 - 1} = \dfrac{-2/(5\sqrt{5})}{3/5 - 1} = \dfrac{1}{\sqrt{5}}$.

(b) The x_i's oscillate between $\dfrac{1}{\sqrt{5}}$ and

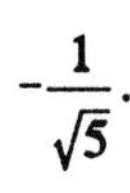

$-\dfrac{1}{\sqrt{5}}$.

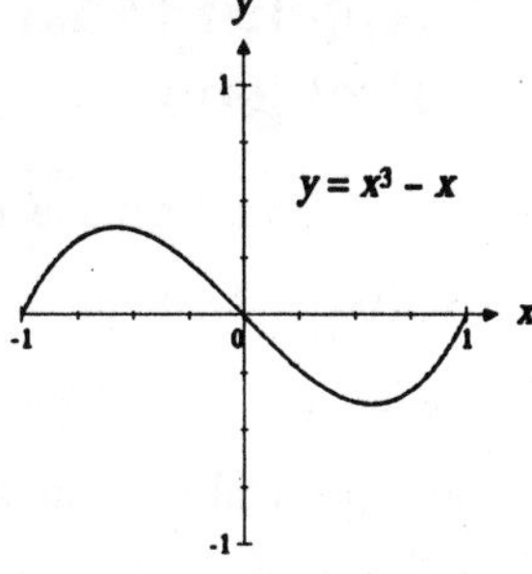

21 As we can see from the figure, the area of *triangle OCD* is $2 \cdot \dfrac{1}{2}(r \sin \theta)(r \cos \theta) = r^2 \sin \theta \cos \theta$, where r is the radius.

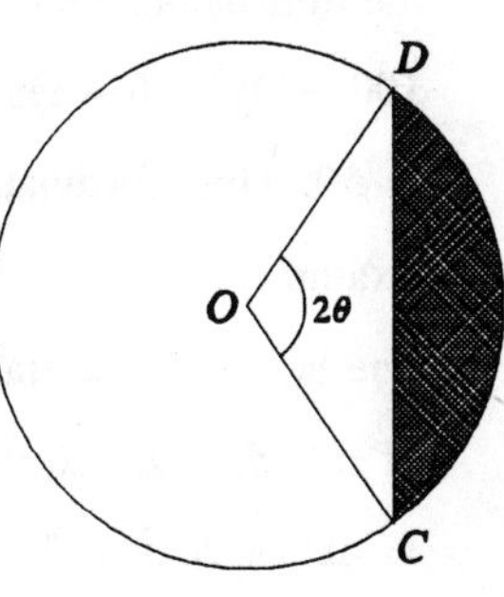

The area of the *sector OCD* is $\dfrac{1}{2}r^2(2\theta) = r^2\theta$. The triangle and the shaded area are supposed to be of equal area, which is the same as saying that the triangle has half the area of the sector. That is, $r^2 \sin \theta \cos \theta = \dfrac{1}{2}r^2\theta$. We can rewrite this as $2r^2 \sin \theta \cos \theta - r^2\theta = r^2(2 \sin \theta \cos \theta - \theta) = 0$; that is, $\sin 2\theta - \theta = 0$, for $0 < \theta < \pi/2$. Letting $f(\theta) = \sin 2\theta - \theta$, we have $f'(\theta) = 2 \cos 2\theta - 1$, which is positive for $0 \le \theta < \pi/6$ and negative for $\pi/6 < \theta \le \pi/2$. Hence $f(\theta) = 0$ has at most one root in $[0, \pi/6]$ and at most one in $[\pi/6, \pi/2]$. Since $f(0) = 0$, $f(\pi/6) = \dfrac{3\sqrt{3} - \pi}{6} > 0$, and $f(\pi/2) = -\pi/2 < 0$, we conclude that the equation has exactly one root for $0 < \theta < \pi/2$ (and, in fact, it is greater than $\pi/6$). To apply Newton's method, note that $\theta - \dfrac{f(\theta)}{f'(\theta)} = \theta - \dfrac{\sin 2\theta - \theta}{2 \cos 2\theta - 1} = \dfrac{2\theta \cos 2\theta - \sin 2\theta}{2 \cos 2\theta - 1}$. Letting $\theta_1 = 1$, we find that $\theta_2 \approx 0.9505$, $\theta_3 \approx 0.9478$, $\theta_4 \approx \theta_5 \approx 0.9477$. To two decimal places, $\theta \approx 0.95$.

23 (a) There is only one root. Since $0 \tan 0 = 0 < 1$ and $\lim_{x \to \pi/2^-} (x \tan x) = \infty > 1$, there is at least one root of $x \tan x = 1$ in $(0, \pi/2)$. For $0 < x < \pi/2$, $(x \tan x)' = x \sec^2 x + \tan x > 0$, so $x \tan x$ is an increasing function on $(0, \pi/2)$. Hence there is at most one root. Thus there is exactly one root.

(b) Letting $f(x) = x \tan x - 1$, we have $f'(x) =$

$x \sec^2 x + \tan x$, so $x - \dfrac{f(x)}{f'(x)} =$

$$x - \frac{x \tan x - 1}{x \sec^2 x + \tan x} = \frac{x^2 \sec^2 x + 1}{x \sec^2 x + \tan x} =$$

$\dfrac{x^2 + \cos^2 x}{x + \sin x \cos x}$. Letting $x_1 = 1$, Newton's method gives $x_2 \approx 0.888$, $x_3 \approx 0.861$, $x_4 \approx x_5 \approx 0.860$. To two decimal places, the root is 0.86.

25

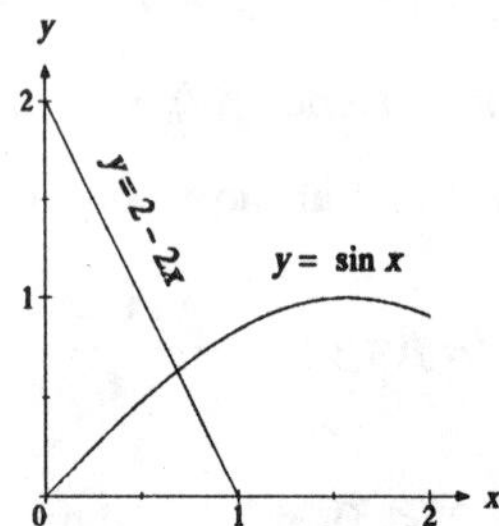

We can rewrite $2x + \sin x = 2$ as $\sin x = 2 - 2x$. Graphing, we see that $y = \sin x$ and $y = 2 - 2x$ coincide at only one point. So we conclude that $2x + \sin x - 2 = 0$ has only one solution. (Note also that $(2x + \sin x - 2)' = 2 + \cos x > 0$, so at most one root is possible.) The solution seems to be around 0.7, so we choose $x_1 = 0.7$. Now, $x_2 =$

$$x_1 - \frac{2x_1 + \sin x_1 - 2}{2 + \cos x_1} = \frac{x_1 \cos x_1 + 2 - \sin x_1}{2 + \cos x_1}$$

≈ 0.684. Hence $x_3 = \dfrac{x_2 \cos x_2 + 2 - \sin x_2}{2 + \cos x_2} \approx$

0.684. Since x_2 and x_3 agree to three places, we see the solution to $2x + \sin x - 2 = 0$ is 0.68 (to two decimal places).

27 We desire a function for which $x = \sqrt[5]{a}$ is a root, so let $f(x) = x^5 - a$. Then $x_{i+1} = x_i - \dfrac{x_i^5 - a}{5x_i^4}$

$$= x_i - \frac{x_i}{5} + \frac{a}{5x_i^4} = \frac{4}{5}x_i + \frac{a}{5x_i^4}.$$

29 (a)

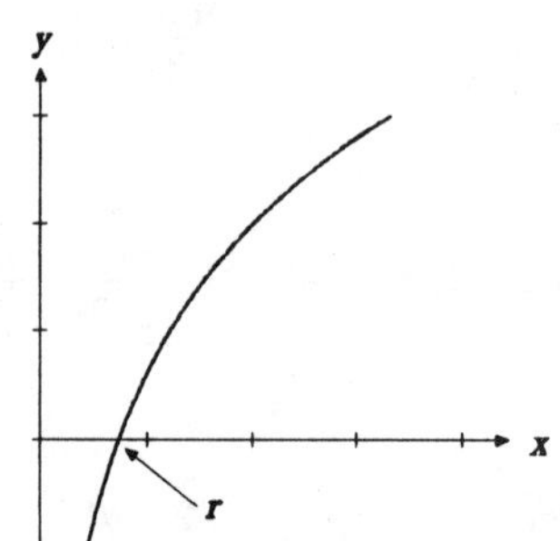

(b) The first choice, $x_1 > r$, produces an $x_2 < r$. All subsequent estimates are less than r but approach r.

(c) The first choice, $x_1 < r$, gives an $x_2 < r$. All subsequent estimates are less than r but approach r.

4.7 Applied Maximum and Minimum Problems

1 Recall from Example 1 that $2y + x = 100$; thus $x = 100 - 2y$. Expressing area in terms of y thus yields $A = xy = (100 - 2y)y = 100y - 2y^2$. Call this $g(y)$ and note that $0 \le y \le 50$, since none of the dimensions may be negative. Now $g'(y) = 100 - 4y = 0$ when $y = 25$ and $x = 100 - 2\cdot 25 = 50$. This, fortunately, agrees with the result of Example 1.

3 The height of the tray is x, and the base is a square of side $5 - 2x$, so the volume is $V(x) = x(5 - 2x)^2 = 4x^3 - 20x^2 + 25x$, where $0 < x < 5/2$. Since $V(0) = V(5/2) = 0$, the maximum must occur at

some critical number in (0, 5/2). Now $V'(x) = 12x^2 - 40x + 25 = (6x - 5)(2x - 5) = 0$ for $x = 5/6$ or $5/2$; $5/6$ lies in (0, 5/2), so it is the optimal length for the cut.

5 The height of the tray is x, and the rectangular base is $4 - 2x$ by $8 - 2x$, so the volume is $V(x) = x(4 - 2x)(8 - 2x) = 4x^3 - 24x^2 + 32x$, where $0 < x < 2$. Now $V(0) = V(2) = 0$, so the maximum must occur at some critical number in (0, 2). Observe that $V'(x) = 12x^2 - 48x + 32 = 4(3x^2 - 12x + 8)$, which equals 0 when $x = \frac{12 \pm \sqrt{144 - 4\cdot 3\cdot 8}}{6} = \frac{6 \pm 2\sqrt{3}}{3}$; since $(1/3)(6 + 2\sqrt{3}) \approx 3.15$ while $(1/3)(6 - 2\sqrt{3}) \approx 0.85$, $(1/3)(6 - 2\sqrt{3})$ is the desired length of the cut.

7 The equation for volume is $\pi r^2 h = 100$, so $r^2 = \frac{100}{\pi h}$ and $r = \frac{10}{\sqrt{\pi h}}$. Substitution into the surface area formula $S = 2\pi r^2 + 2\pi rh$ yields $S = 2\pi\cdot\frac{100}{\pi h} + 2\pi\cdot\frac{10}{\sqrt{\pi h}}\cdot h = \frac{200}{h} + 20\sqrt{\pi}\sqrt{h}$. Then $\frac{dS}{dh} = -\frac{200}{h^2} + \frac{20\sqrt{\pi}}{2\sqrt{h}}$. This equals 0 when $\frac{10\sqrt{\pi}}{\sqrt{h}} = \frac{200}{h^2}$, $h^{3/2} = \frac{20}{\sqrt{\pi}}$, $h = \left(\frac{20}{\sqrt{\pi}}\right)^{2/3} = \left(\frac{400}{\pi}\right)^{1/3}$.

Now $\frac{dS}{dh} > 0$ for $h > \left(\frac{400}{\pi}\right)^{1/3}$ and $\frac{dS}{dh} < 0$ for $h < \left(\frac{400}{\pi}\right)^{1/3}$, so $h = \left(\frac{400}{\pi}\right)^{1/3}$ does provide a minimum. Plugging h into the expression for r yields $r = \frac{10}{\sqrt{\pi h}} = \frac{10}{\sqrt{\pi}}h^{-1/2} = \frac{10}{\sqrt{\pi}}\left[\left(\frac{400}{\pi}\right)^{1/3}\right]^{-1/2}$

$$= \frac{10}{\sqrt{\pi}}\left[\left(\frac{400}{\pi}\right)^{-1/2}\right]^{1/3} = \frac{10}{\sqrt{\pi}}\left(\frac{\sqrt{\pi}}{20}\right)^{1/3} = \left(\frac{1000\sqrt{\pi}}{20\pi^{3/2}}\right)^{1/3}$$

$= \left(\frac{50}{\pi}\right)^{1/3}$, which agrees with the result of Example 3.

9 Let x be the length of a side of the square base and let y be the height of the box. The top and bottom of the box each have area x^2 and each of the four sides has area xy. The total area is thus $A = 2x^2 + 4xy$, which we must minimize in order to minimize the material used to make the box. Since the volume is $x^2y = 1000$, we have $y = 1000/x^2$ and $A = 2x^2 + 4x\left(\frac{1000}{x^2}\right) = 2x^2 + \frac{4000}{x}$. Now $\frac{dA}{dx} = 4x - \frac{4000}{x^2} = \frac{4(x^3 - 1000)}{x^2}$, which equals 0 for $x = 10$, and $\frac{dA}{dx} < 0$ when $x < 10$ and $\frac{dA}{dx} > 0$ when $x > 10$, so $x = 10$ provides a minimum. Thus, $y = \frac{1000}{x^2} = \frac{1000}{100} = 10$, so the optimal box is a cube of side 10.

11 Denote the length and width of the rectangle l and w, respectively. (See the figure.) From the figure, we see that $\cos\theta = \frac{l}{2a}$ and $\tan\theta = \frac{w/2}{l/2} = \frac{w}{l}$. So $l = 2a\cos\theta$ and $w = l\tan\theta$. Since $A = wl = (l\tan\theta)\cdot l = l^2\tan\theta = (2a\cos\theta)^2\tan\theta$

$= 4a^2 \cos^2\theta \tan\theta$

$= 4a^2 \cos\theta \sin\theta$, we now have A as a function of θ. Here $0 \le \theta \le \pi/2$ for sensible values and

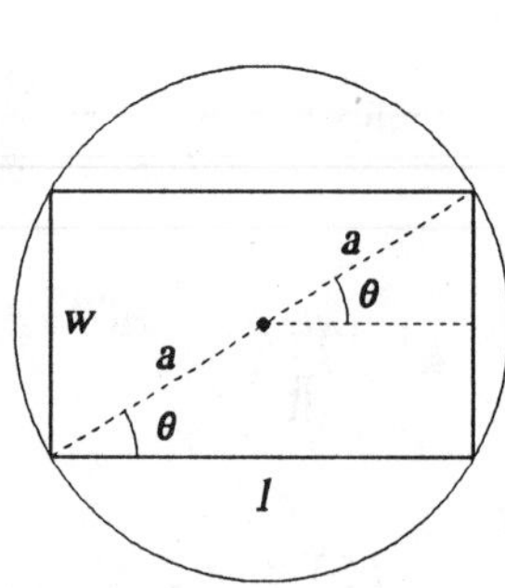

$\dfrac{dA}{d\theta} = 4a^2\cos^2\theta - 4a^2\sin^2\theta = 4a^2(\cos^2\theta - \sin^2\theta) = 4a^2\cos 2\theta = 0$ when $2\theta = \pi/2$ or when $\theta = \pi/4$. Also $\dfrac{d^2A}{d\theta^2} = -8a^2 \sin 2\theta < 0$ for $\theta = \pi/4$, so $\theta = \pi/4$ yields a maximum ($A(0) = A(\pi/2) = 0$). Now $l = 2a\cos\pi/4 = a\sqrt{2}$ and $w = a\sqrt{2} \tan\pi/4 = a\sqrt{2}$. So the dimensions of the inscribed rectangle of largest area are $a\sqrt{2} \times a\sqrt{2}$ (in other words, a square.)

13 Let x be the width, y the height, and p the perimeter of the rectangle. Then $p = 2x + 2y$, so $2x = p - 2y$, $x = (1/2)p - y$. Thus the area is $A = xy = ((1/2)p - y)y = (1/2)py - y^2$. Now $A = 0$ when $y = 0$ or $p/2$, so the maximal area must occur at a critical number in $(0, p/2)$. $\dfrac{dA}{dy} = \dfrac{1}{2}p - 2y = 0$ for $y = p/4$. It follows that $x = (1/2)p - (1/4)p = (1/4)p$, so the optimal rectangle is a square, as claimed.

15 Let x be the length of the sides to which the fence is parallel, and let y be the length of the other two sides. Then $3x + 2y = 240$, so $3x = 240 - 2y$, $x = 80 - (2/3)y$. The area enclosed is thus $A = xy = (80 - (2/3)y)y = 80y - (2/3)y^2$, where $0 < y < 120$. Now $A = 0$ for $y = 0$ or $y = 120$, so the maximum must occur at a critical number in $(0, 120)$.

$\dfrac{dA}{dy} = 80 - \dfrac{4}{3}y = 0$

when $(4/3)y = 80$, so $y = (3/4)\cdot 80 = 60$. Therefore $x = 80 - (2/3)\cdot 60 = 80 - 40 = 40$. The outer dimensions of the optimal corral are 40 by 60 feet.

17 (a) We wish to maximize $x^2 + y^2$ where $x + y = 1$ and x and y are nonnegative. Hence $y = 1 - x$, where $0 \le x \le 1$, and the function to be maximized is $f(x) = x^2 + (1 - x)^2 = x^2 + 1 - 2x + x^2 = 1 - 2x + 2x^2$. Note that $f(0) = f(1) = 1$. Checking for interior critical numbers, we have $f'(x) = -2 + 4x = 0$ for $x = 1/2$. But $f''(x) = 4 > 0$ for all x, so $x = 1/2$ is a minimum. The endpoints are therefore the maxima: $x = 1$ and $y = 0$, or $x = 0$ and $y = 1$.

(b) From part (a) we know that the minimum occurs for $x = y = 1/2$.

19 Let h be the height of the trapezoid. By the Pythagorean theorem,

$h^2 = 16 - \left(\dfrac{x-4}{2}\right)^2$

$= 16 - \dfrac{x^2 - 8x + 16}{4}$

$= \dfrac{64 - x^2 + 8x - 16}{4} = \dfrac{48 + 8x - x^2}{4}$. The area of the trapezoid is $A = \frac{1}{2}(x + 4)h$

$= \frac{1}{2}(x + 4)\frac{\sqrt{48 + 8x - x^2}}{2}$

$= \frac{1}{4}(x + 4)\sqrt{48 + 8x - x^2}$. We must maximize

this for $0 \le x \le 12$. Now $\frac{dA}{dx}$

$= \frac{1}{4}(x + 4)\frac{8 - 2x}{2\sqrt{48 + 8x - x^2}} + \frac{1}{4}\sqrt{48 + 8x - x^2}$

$= \frac{(x + 4)(4 - x) + 48 + 8x - x^2}{4\sqrt{48 + 8x - x^2}}$

$= \frac{16 - x^2 + 48 + 8x - x^2}{4\sqrt{48 + 8x - x^2}} = \frac{32 + 4x - x^2}{2\sqrt{48 + 8x - x^2}}$;

this equals 0 for $x^2 - 4x - 32 = (x - 8)(x + 4)$

$= 0$; that is, for $x = 8$ or -4. Since $A(0) = \sqrt{48}$,

$A(8) = 3\sqrt{48}$, and $A(12) = 0$, the maximum

occurs for $x = 8$.

21 Let the cylinder have radius r and height h. The girth is $2\pi r$, the circumference of the cylinder, so $2\pi r + h = 108$ and $h = 108 - 2\pi r$. The volume is $V = \pi r^2h = \pi r^2(108 - 2\pi r) = 108\pi r^2 - 2\pi^2r^3$. We must maximize this for $0 \le r \le 54/\pi$. Since $V = 0$ for $r = 0$ or $54/\pi$, the maximum must occur at a critical number in the interval $(0, 54/\pi)$.

$\frac{dV}{dr} = 216\pi r - 6\pi^2r^2 = 6\pi r(36 - \pi r)$, which

equals 0 when $r = 0$ or $r = 36/\pi$. Since $0 < 36/\pi < 54/\pi$, the maximum occurs for $r = 36/\pi$ inches and $h = 108 - 2\pi\cdot 36/\pi = 36$ inches.

23 Let the box have base b and height h. The girth is $4b$, so $4b + h = 108$ and $h = 108 - 4b$. The volume is $V = b^2h = b^2(108 - 4b) = 108b^2 - 4b^3$. We must maximize this for $0 \le b \le 27$. Since $V = 0$ for $b = 0$ or 27, the maximum must occur at a critical number in (0, 27). $\frac{dV}{db} =$

$216b - 12b^2 = 12b(18 - b)$, which equals 0 when $b = 0$ or 18. Since $0 < 18 < 27$, the maximum volume occurs for $b = 18$ inches and $h = 108 - 4\cdot 18 = 36$ inches.

25 (b) The top and bottom each have area πr^2, so the cost of the ends is $2k\cdot 2\pi r^2 = 4k\pi r^2$. The area of the side is $2\pi rh$, so the cost of the side is $k\cdot 2\pi rh = 2k\pi rh$. The total cost is therefore $C = 4k\pi r^2 + 2k\pi rh$.

(c) The volume is $V = \pi r^2h = 100$, so $h =$

$\frac{100}{\pi r^2}$. Thus $C = 4k\pi r^2 + 2k\pi r\cdot\frac{100}{\pi r^2} =$

$4k\pi r^2 + \frac{200k}{r}$ and $\frac{dC}{dr} = 8k\pi r - \frac{200k}{r^2} =$

$\frac{8k\pi r^3 - 200k}{r^2} = \frac{8k\pi}{r^2}\left(r^3 - \frac{25}{\pi}\right)$, which

equals 0 when $r^3 = 25/\pi$; that is, for $r =$

$(25/\pi)^{1/3}$. This is a minimum because $\frac{dC}{dr} <$

0 for smaller values of r and $\frac{dC}{dr} > 0$ for

larger values.

(d) $h = \frac{100}{\pi r^2} = \frac{100}{\pi}\cdot r^{-2} = \frac{100}{\pi}\cdot\left(\frac{25}{\pi}\right)^{-2/3}$

$= \frac{4\cdot 25^{1/3}}{\pi^{1/3}} = 4\left(\frac{25}{\pi}\right)^{1/3}$

27 Let x be the length of each side of the base and let

y be the height of the box. The top and bottom each have area x^2 and each of the four sides has area xy, so the total cost is $C = 2x^2 + 3 \cdot 4xy + 5x^2 = 7x^2 + 12xy$. The volume is $x^2y = 100$, so $y = 100/x^2$ and $C = 7x^2 + 12x \cdot 100/x^2 = 7x^2 + 1200/x$.

Then $\dfrac{dC}{dx} = 14x - \dfrac{1200}{x^2} = \dfrac{14x^3 - 1200}{x^2} =$

$\dfrac{14}{x^2}\left(x^3 - \dfrac{1200}{14}\right)$, which equals 0 when $x^3 = 1200/14 = 600/7$; that is, for $x = (600/7)^{1/3} = 2(75/7)^{1/3}$. This is a minimum since $\dfrac{dC}{dx} < 0$ for smaller values of x and $\dfrac{dC}{dx} > 0$ for larger values.

The corresponding value of y is $100x^{-2} =$

$100\left[2\left(\dfrac{75}{7}\right)^{1/3}\right]^{-2} = \dfrac{100}{4} \cdot \dfrac{7^{2/3}}{75^{2/3}} = 25 \cdot \dfrac{7}{75} \cdot \dfrac{75^{1/3}}{7^{1/3}} =$

$\dfrac{7}{3}\left(\dfrac{75}{7}\right)^{1/3}$.

29 There are 10,000 yd³ of earth to move, and each truck holds 10 yd³, so 1000 trucks are needed. Let x trucks be dispatched to the first road and y dispatched to the second; then $x + y = 1000$, or $y = 1000 - x$. The total cost is therefore $C = x(1 + 2x^2) + y(2 + y^2) = x + 2x^3 + 2y + y^3 = x + 2x^3 + 2(1000 - x) + (1000 - x)^3 = 2x^3 - x + 2000 + (1000 - x)^3$. Hence $\dfrac{dC}{dx} =$

$6x^2 - 1 + 3(1000 - x)^2(-1)$

$= 6x^2 - 1 - 3(1{,}000{,}000 - 2000x + x^2)$

$= 3x^2 + 6000x - 3{,}000{,}001$. $\dfrac{dC}{dx} = 0$ when $x =$

$\dfrac{-6000 \pm \sqrt{36{,}000{,}000 + 36{,}000{,}012}}{6} =$

$-1000 \pm \dfrac{1}{6}\sqrt{72{,}000{,}012}$. Since x must be positive, we take $x = -1000 + \dfrac{1}{6}\sqrt{72{,}000{,}012} \approx 414.2$.

Since the number of trucks must be an integer, we take $x = 414$ and $y = 1000 - 414 = 586$.

31 At an extremum, we must have $\dfrac{dI}{d\theta} = 0$; that is, $-2A\cos\theta\sin\theta + 2C\sin\theta\cos\theta - 2E(\cos^2\theta - \sin^2\theta) = 0$. But $\sin 2\theta = 2\sin\theta\cos\theta$ and $\cos 2\theta = \cos^2\theta - \sin^2\theta$, so this implies that $(C - A)\sin 2\theta = 2E\cos 2\theta$. If $\cos 2\theta = 0$, then $\sin 2\theta = \pm 1$, so $C - A = 0$, contradicting $A \neq C$. Hence we may divide by $(C - A)\cos 2\theta$, obtaining the desired result.

33 (a)

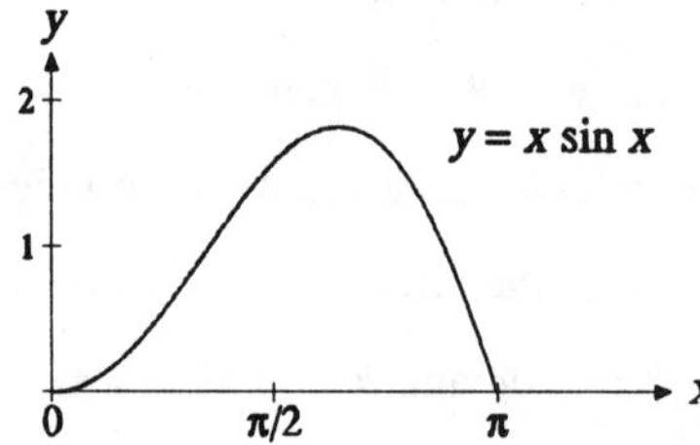

(b) For $0 < x < \pi/2$, $\sin x > 0$ and $\cos x > 0$, so $y' = \sin x + x\cos x > 0$; hence there is no local maximum in $[0, \pi/2]$. For $\pi/2 < x < \pi$, $\cos x < 0$ and $\sin x > 0$, so $y'' = 2\cos x - x\sin x < 0$. Hence y' is decreasing in $[\pi/2, \pi]$. Since $y'(\pi/2) = 1$ and $y'(\pi) = -\pi$, there is a unique x in $(\pi/2, \pi)$ such that $y'(x) = 0$. Since $y''(x) < 0$, this x provides the desired unique maximum.

(c) See the discussion in (b).

(d) $x_2 = x_1 - \dfrac{x_1 \cos x_1 + \sin x_1}{2\cos x_1 - x_1 \sin x_1} = \dfrac{\pi}{2} + \dfrac{2}{\pi}$

≈ 2.2074

(e) $x_3 \approx 2.0360$

35 Here $y'(x) = 2\cos x - 2x$. Since $y'(0) = 2 > 0 > -\pi = y'(\pi/2)$ and $y''(x) = -2\sin x - 2 < 0$ for all x in $[0, \pi/2]$, then $y'(c) = 0$ for some unique c in $(0, \pi/2)$. Using Newton's method to approximate this c with $x_1 = 0.75$, we have $x_2 = x_1 + \dfrac{2\cos x_1 - 2x_1}{2\sin x_1 + 2} = x_1 + \dfrac{\cos x_1 - x_1}{\sin x_1 + 1} =$

$\dfrac{x_1 \sin x_1 + \cos x_1}{\sin x_1 + 1} \approx 0.73911$, $x_3 \approx 0.73909$. So $c \approx 0.739$ to three decimal places. Now $y(c) \approx 0.801$, $y(0) = 0$, and $y(\pi/2) \approx -0.467$; hence $y(c)$ is the maximum on $[0, \pi/2]$.

37 Suppose that f feet are added, that the side opposite the existing wall is x feet long, and that the other side is y feet long. Then $f = 2x + 2y - 20$, so $y = f/2 + 10 - x$ and the area is given by $A = xy = (f/2 + 10)x - x^2$. We must maximize this for $20 \le x \le f/2 + 10$. (The second inequality ensures that $y \ge 0$.) Since $\dfrac{dA}{dx} = \dfrac{f}{2} + 10 - 2x$, the only critical value is $x = f/4 + 5$.

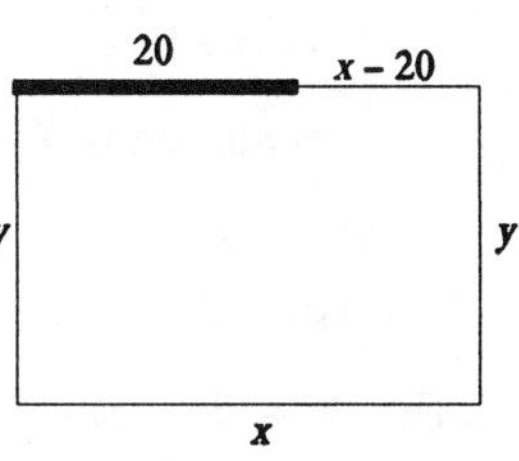

(a) $f = 40$. The critical value if $f/4 + 5 = 15$, which is not in the interval $(20, f/2 + 10) = (20, 30)$. Since $A(20) = 200$ and $A(30) = 0$, the maximum occurs at $x = 20$: the field should be 10 feet by 20 feet.

(b) $f = 80$. The critical value is 25, which is in the interval $(20, f/2 + 10) = (20, 50)$. Since $A(20) = 600$, $A(25) = 625$, and $A(50) = 0$, the maximum occurs at $x = 25$: the field should be 25 feet by 25 feet.

(c) $f = 60$. The critical value is 20, which is not in the interval $(20, f/2 + 10) = (20, 40)$. Since $A(20) = 400$ and $A(40) = 0$, the maximum occurs at $x = 20$: the field should be 20 feet by 20 feet.

[Intuitively these answers make sense. The field should be as close to square as possible, subject to one side's being at least 20 feet long.]

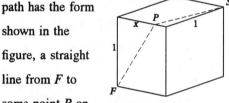

39 (a) Clearly such a path has the form shown in the figure, a straight line from F to some point P on an edge not containing F or S, followed by a straight line from P to S. If P is x inches from that end of its edge that is closest to F, then the length of the path, $l = \overline{FP} + \overline{PS}$, can be written as a function of x; that is, $l = f(x) = \sqrt{x^2+1} + \sqrt{(1-x)^2 + 1}$. This must be minimized for $0 \le x \le 1$. Solving $\dfrac{dl}{dx} = \dfrac{x}{\sqrt{x^2+1}} - \dfrac{1-x}{\sqrt{(1-x)^2+1}} = 0$ gives $x\sqrt{(1-x)^2+1} = (1-x)\sqrt{x^2+1}$; squaring

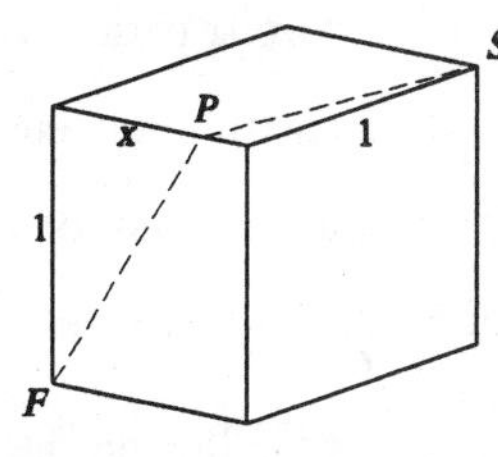

both sides, we obtain $x^2[(1-x)^2+1] = (1-x)^2(x^2+1)$, so $x^2 = (1-x)^2$, and $x = 1-x$, so $x = 1/2$. Hence the minimum occurs for $x = 0$, $1/2$, or 1. We find that $f(0) = f(1) = 1+\sqrt{2}$ and $f(1/2) = \sqrt{5}$. Note that $\sqrt{5} \approx 2.24$ while $1+\sqrt{2} \approx 2.41$. Hence a shortest path goes from F to the midpoint of an edge containing neither F nor S, and from there to S. (There are 6 such paths.)

(b) Assume that the cube is made of paper. If we unfold it along the edge containing P, the shortest path must be a straight line from F to S, so P must be the midpoint of its edge.

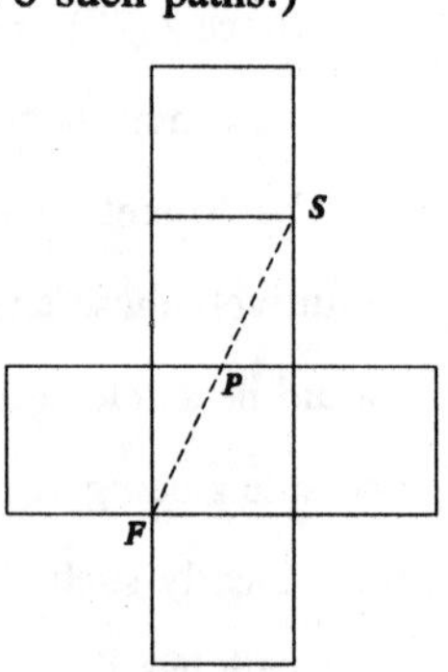

41 Let the woman walk $\sqrt{1+x^2}$ miles on the grass and $s-x$ miles on the sidewalk. (See the figure.) It takes her $\dfrac{\sqrt{1+x^2}}{3}$ hours to cross the grass and $\dfrac{s-x}{5}$ hours to travel down the sidewalk to B. The total travel time, which we wish to minimize, is thus given by $T(x) = \dfrac{\sqrt{1+x^2}}{3} + \dfrac{s-x}{5}$. Then $T'(x) = \dfrac{x}{3\sqrt{1+x^2}} - \dfrac{1}{5}$. (Note that the constant s dropped out!) Now $T'(x) = 0$ when $\dfrac{x}{3\sqrt{1+x^2}} - \dfrac{1}{5} = 0$, so $5x = 3\sqrt{1+x^2}$, $25x^2 = 9(1+x^2)$, $16x^2 = 9$, $x = \pm 3/4$. In our problem, only the critical number $x = 3/4$ is of interest. We can now examine the specific cases:

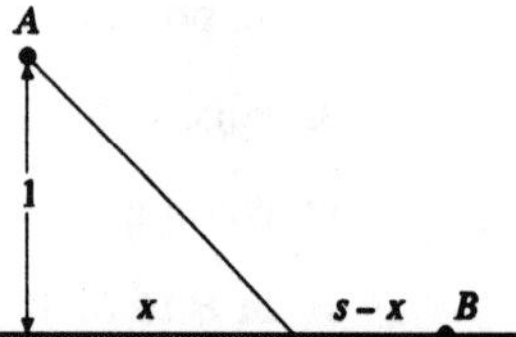

(a) $s = 1/2$. Since $3/4$ is not in the interval $(0, 1/2)$, we check the values of $T(x)$ at the endpoints: $T(0) = \dfrac{\sqrt{1+0}}{3} + \dfrac{\frac{1}{2}-0}{5} = \dfrac{1}{3} + \dfrac{1}{10} = \dfrac{13}{30}$ and $T\left(\dfrac{1}{2}\right) = \dfrac{\sqrt{1+1/4}}{3} + \dfrac{\frac{1}{2}-\frac{1}{2}}{5} = \dfrac{\sqrt{5}}{6} + 0 = \dfrac{\sqrt{5}}{6}$.

Now $\dfrac{\sqrt{5}}{6} \approx 0.37$ while $\dfrac{13}{30} \approx 0.43$, so the minimum is $T(1/2) = \dfrac{\sqrt{5}}{6}$. Thus, the woman should walk directly from A to B.

(b) $s = 3/4$. Since $3/4$ is not in the interval $(0, 3/4)$, we check the endpoints: $T(0) = \dfrac{\sqrt{1+0}}{3} + \dfrac{\frac{3}{4}-0}{5} = \dfrac{1}{3} + \dfrac{3}{20} = \dfrac{29}{60}$ and $T\left(\dfrac{3}{4}\right) = \dfrac{\sqrt{1+9/16}}{3} + \dfrac{\frac{3}{4}-\frac{3}{4}}{5} = \dfrac{5}{12}$.

Since $5/12 < 29/60$, the minimum is $T(3/4) = 5/12$. The woman should walk directly from A to B.

(c) $s = 1$. In this case, $3/4$ is in the interval $(0, 1)$, so we check the values of $T(x)$ at the endpoints 0 and 1 and at the critical number $3/4$. $T(0) = \frac{\sqrt{1+0}}{3} + \frac{1-0}{5} = \frac{1}{3} + \frac{1}{5} = \frac{8}{15}$, $T\left(\frac{3}{4}\right) = \frac{\sqrt{1+9/16}}{3} + \frac{1-\frac{3}{4}}{5} = \frac{5}{12} + \frac{1}{20} = \frac{28}{60} = \frac{7}{15}$, and $T(1) = \frac{\sqrt{1+1}}{3} + \frac{1-1}{5} = \frac{\sqrt{2}}{3}$. Now $\frac{\sqrt{2}}{3} \approx 0.4714 > 0.4667 \approx 7/15$, so the minimum is $T(3/4) = 7/15$. The woman should walk $\sqrt{1+(3/4)^2} = \frac{5}{4}$ miles on the grass and $1 - 3/4 = 1/4$ miles on the sidewalk.

43 Let r be the radius of the cylinder and h its height. By the Pythagorean theorem, $r^2 + (h/2)^2 = a^2$, so $h^2/4 = a^2 - r^2$, and $h = 2\sqrt{a^2 - r^2}$. The volume of the cylinder is $V = \pi r^2 h = 2\pi r^2\sqrt{a^2 - r^2}$, which we maximize for $0 \le r \le a$. Now $\frac{dV}{dr} = 2\pi r^2 \cdot \frac{-2r}{2\sqrt{a^2 - r^2}} + 4\pi r\sqrt{a^2 - r^2}$

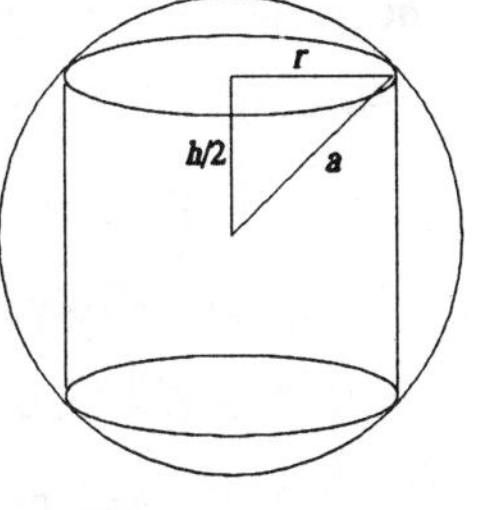

$= \frac{-2\pi r^3 + 4\pi r(a^2 - r^2)}{\sqrt{a^2 - r^2}} = \frac{4\pi a^2 r - 6\pi r^3}{\sqrt{a^2 - r^2}}$, which equals 0 whenever $4\pi a^2 r - 6\pi r^3 = 2\pi r(2a^2 - 3r^2) = 0$; that is, for $r = 0$ or $r = \sqrt{\frac{2a^2}{3}} = a\sqrt{\frac{2}{3}}$. Only $r = a\sqrt{\frac{2}{3}}$ is in the interval $(0, a)$. Note that for smaller values of r, $2a^2 - 3r^2 > 0$, so $\frac{dV}{dr} > 0$, and for larger values of r, $2a^2 - 3r^2 < 0$ and $\frac{dV}{dr} < 0$. Hence $r = a\sqrt{\frac{2}{3}}$ is a maximum. The corresponding value of h is $h = 2\sqrt{a^2 - \frac{2}{3}a^2} = 2\sqrt{\frac{1}{3}a^2} = \frac{2a}{\sqrt{3}}$.

45 Let k be the amount of heat per square foot that enters through the walls. Let x be the length of each side of the floor and let y be the height of the house. Each of the four walls has xy square feet and the roof has x^2 square feet, so the total amount of heat entering through the walls and roof is $H = k \cdot 4xy + 3k \cdot x^2 = 4kxy + 3kx^2$. Let V be the fixed volume; then $V = x^2y$, so $y = V/x^2$ and $H = 4kx \cdot V/x^2 + 3kx^2 = \frac{4kV}{x} + 3kx^2$. Then $\frac{dH}{dx} = -\frac{4kV}{x^2} + 6kx = \frac{6kx^3 - 4kV}{x^2} = \frac{6k\left(x^3 - \frac{2}{3}V\right)}{x^2}$, which equals 0 for $x = \sqrt[3]{\frac{2}{3}V}$. Note that $\frac{dH}{dx} < 0$ for smaller values of x and $\frac{dH}{dx} > 0$ for larger values, so $x = \sqrt[3]{\frac{2}{3}V}$ provides a minimum. The

corresponding value of y is $y = \frac{V}{x^2} = V\left(\frac{2}{3}V\right)^{-2/3}$ $= \frac{3^{2/3}}{2^{2/3}} \cdot V^{1/3} = \frac{3}{2}\sqrt[3]{\frac{2}{3}V}$. Since $V = 12{,}000$ ft³, we have $x = 20$ ft and $y = 30$ ft.

47 (a) The traffic density is D vehicles/mile and traffic speed is S miles/hour, so the number of vehicles that enter the tunnel in one hour is given by SD. To see this, note $S = \frac{ds}{dt}$, the change of distance with respect to time, and $D = \frac{dT}{ds}$, the density of traffic with respect to distance s. By the chain rule $\frac{dT}{dt} = \frac{dT}{ds}\cdot\frac{ds}{dt}$, where $\frac{dT}{dt}$ is the density of traffic T with respect to time t.

(b) Let N be the number of vehicles that enter the tunnel in an hour. Then $N = SD =$ $(42 - D/3)D = 42D - (1/3)D^2$ and $\frac{dN}{dD} =$ $42 - (2/3)D$, which equals 0 when $D =$ $(3/2)\cdot 42 = 63$. This is a maximum since $\frac{d^2N}{dD^2} = -\frac{2}{3} < 0$.

49 (a) The linear dimensions of a region grow in proportion to the square root of the region's area. For example, if a square's area is quadrupled, its sides are doubled. Transportation costs depend on linear distance traveled, so they too should be proportional to $\sqrt{A}$.

(b) The cost of warehousing an item is inversely proportional to A, so it equals w/A for some constant w. As noted in (a), transportation costs are proportional to $\sqrt{A}$, hence equal to $t\sqrt{A}$ for some constant t. The total cost is therefore $C = t\sqrt{A} + \frac{w}{A}$, as indicated.

(c) $\frac{dC}{dA} = \frac{t}{2\sqrt{A}} - \frac{w}{A^2} = \frac{tA^{3/2} - 2w}{2A^2}$, which equals 0 for $A^{3/2} = \frac{2w}{t}$; that is, when $A = \left(\frac{2w}{t}\right)^{2/3}$. For smaller values of A, $\frac{dC}{dA} < 0$, while for larger values, $\frac{dC}{dA} > 0$. Hence this is a minimum.

51 Observe that the longest pipe that may be carried horizontally around the corner is one whose maximum y value is 27, where y is the same as defined in Exercise 50. We have $y = 27$ and $a = 8$, so we solve $y = (b^{2/3} - a^{2/3})^{3/2}$ for b: $27 = (b^{2/3} - 8^{2/3})^{3/2}$, $27^{2/3} = b^{2/3} - 4$, $9 + 4 = b^{2/3}$, $13^{3/2} = b$. The length of the longest pipe is $13^{3/2} = 13\sqrt{13}$.

53 Let ϕ be as shown in the diagram. Then

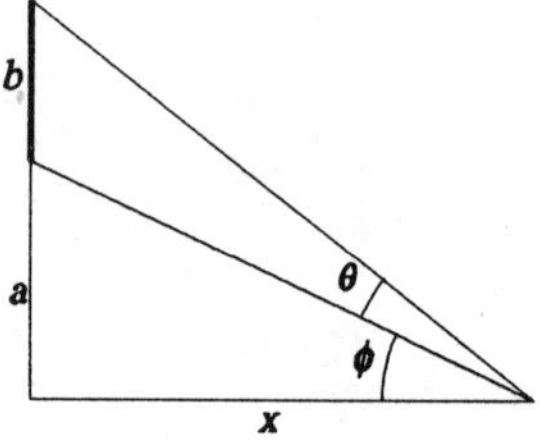

$\tan(\theta + \phi) = \frac{a + b}{x}$

and $\tan\phi = a/x$.

Hence $\tan\theta =$

$$\frac{\tan(\theta + \phi) - \tan\phi}{1 + \tan(\theta + \phi)\tan\phi}$$

$$= \frac{\frac{a+b}{x} - \frac{a}{x}}{1 + \frac{a+b}{x}\cdot\frac{a}{x}} = \frac{bx}{x^2 + a(a+b)}.$$ Recall that maximizing $\tan\theta$ is the same as maximizing θ. To find the value of x for which $\tan\theta$ is maximal, we compute $\frac{d}{dx}(\tan\theta) = \frac{(x^2 + a(a+b))b - bx(2x)}{(x^2 + a(a+b))^2}$, which equals 0 whenever the numerator is 0: $bx^2 + ab(a+b) - 2bx^2 = 0$, $bx^2 = ab(a+b)$, $x^2 = a(a+b)$, $x = \sqrt{a(a+b)}$.

55 The sale of y items at a price p dollars produces yp dollars in revenue. Since the cost of manufacture is $100 + 10y$, the profit is given by $P = yp - (100 + 10y) = (p - 10)y - 100$. Since $y = 250 - p$, we have $P = (p - 10)(250 - p) - 100$, so $P' = (p - 10)(-1) + (250 - p)(1) = -p + 10 + 250 - p = 260 - 2p$, which equals 0 when $2p = 260$ or $p = 130$. $P'' = -2 < 0$, so this provides a maximum. The optimal price is $130.

57 (a) $U = \frac{RLv}{A}n^{1/2} - nK - \frac{ALc}{v}n^{-1/2}$, so $\frac{dU}{dn} =$

$$\frac{RLv}{A}\cdot\frac{1}{2}\cdot n^{-1/2} - K - \frac{ALc}{v}\left(-\frac{1}{2}\right)n^{-3/2} =$$

$$\frac{RLv}{2A}n^{-1/2} - K + \frac{ALc}{2v}n^{-3/2},$$ as claimed.

(b) Simply replace $n^{-1/2}$ by x in the expression for dU/dn and set it equal to 0:

$$\frac{RLv}{2A}x - K + \frac{ALc}{2v}x^3 = 0.$$

4.8 Implicit Differentiation

1 The point (1, 4) lies on the graph of $xy = 4$, so we may seek the value of dy/dx at that point. Solving for y produces $y = \frac{4}{x}$, $\frac{dy}{dx} = -\frac{4}{x^2}$. At $x = 1$, $\frac{dy}{dx} = -4$. Differentiating implicitly produces

$$\frac{d}{dx}(xy) = \frac{d}{dx}(4),\ x\cdot\frac{dy}{dx} + y\cdot 1 = 0,\ x\cdot\frac{dy}{dx} = -y,$$

and $\frac{dy}{dx} = -\frac{y}{x}$ at the point (1, 4), $dy/dx = -4/1 = -4$.

3 Note that (3, 1) satisfies the equation $x^2y + xy^2 = 12$. Rearrange the equation to facilitate use of the quadratic formula: $xy^2 + x^2y - 12 = 0$. Thus $y =$

$$\frac{-x^2 \pm \sqrt{x^4 - 4x(-12)}}{2x} = -\frac{x}{2} \pm \frac{1}{2x}\sqrt{x^4 + 48x}.$$

To pick the correct expression for y, plug in $x = 3$ and choose the one that yields $y = 1$: $y =$

$$-\frac{3}{2} \pm \frac{1}{6}\sqrt{81 + 144} = -\frac{3}{2} \pm \frac{1}{6}\sqrt{225} =$$

$$-\frac{3}{2} \pm \frac{15}{6} = -\frac{3}{2} \pm \frac{5}{2},$$ which equals 1 or -4. We choose the plus sign since we want $y = 1$. Thus $y = -\frac{x}{2} + \frac{1}{2x}\sqrt{x^4 + 48x}$ and $\frac{dy}{dx} =$

$$-\frac{1}{2} + \frac{2x\cdot\frac{4x^3 + 48}{2\sqrt{x^4 + 48x}} - \sqrt{x^4 + 48x}\cdot 2}{(2x)^2}$$

$$= -\frac{1}{2} + \frac{4x^4 + 48x - 2(x^4 + 48x)}{4x^2\sqrt{x^4 + 48x}}$$

$= -\frac{1}{2} + \frac{x^3 - 24}{2x\sqrt{x^4 + 48x}}$, so for $x = 3$, $\frac{dy}{dx} =$

$-\frac{1}{2} + \frac{27 - 24}{6\sqrt{81 + 144}} = -\frac{1}{2} + \frac{1}{30} = -\frac{14}{30} = -\frac{7}{15}.$

If we use implicit differentiation instead, we have

$\frac{d}{dx}(x^2y + xy^2) = \frac{d}{dx}(12),$

$x^2 \cdot \frac{dy}{dx} + y \cdot 2x + x \cdot 2y\frac{dy}{dx} + y^2 = 0,$

$\frac{dy}{dx}(x^2 + 2xy) = -y^2 - 2xy$; so at (3, 1),

$\frac{dy}{dx}(9 + 6) = -1 - 6, \frac{dy}{dx} = -\frac{7}{15}.$

5 $\frac{d}{dx}\left(\frac{2xy}{\pi} + \sin y\right) = \frac{d}{dx}(2),$

$\frac{2x}{\pi} \cdot \frac{dy}{dx} + \frac{2y}{\pi} + (\cos y)\frac{dy}{dx} = 0, \frac{dy}{dx}\left(\frac{2x}{\pi} + \cos y\right)$

$= -\frac{2y}{\pi}$. To find $\frac{dy}{dx}$ at $(1, \pi/2)$, plug in $x = 1$

and $y = \pi/2$ and solve: $\frac{dy}{dx}\left(\frac{2 \cdot 1}{\pi} + 0\right) = -\frac{2}{\pi} \cdot \frac{\pi}{2},$

$\frac{dy}{dx} \cdot \frac{2}{\pi} = -1, \frac{dy}{dx} = -\frac{\pi}{2}.$

7 $\frac{d}{dx}(x^5 + y^3x + x + yx^2 + y^5) = \frac{d}{dx}(4),$

$5x^4 + y^3 + x \cdot 3y^2\frac{dy}{dx} + y \cdot 2x + x^2\frac{dy}{dx} + 5y^4\frac{dy}{dx} = 0,$

$(3xy^2 + x^2 + 5y^4)\frac{dy}{dx} = -5x^4 - y^3 - 2xy.$

Plug in (1, 1) and solve for dy/dx: $(3 + 1 + 5)\frac{dy}{dx}$

$= -5 - 1 - 2$, so $\frac{dy}{dx} = -8/9.$

9 $\frac{d}{dh}(\pi r^2h) = \frac{d}{dh}(100)$, so $\pi r^2 + 2\pi r \cdot \frac{dr}{dh} \cdot h = 0$

and $\frac{dr}{dh} = \frac{-\pi r^2}{2\pi rh} = -\frac{r}{2h}$. Now $\frac{dS}{dh} =$

$\frac{d}{dh}(2\pi r^2 + 2\pi rh) = 2\pi\left(2r \cdot \frac{dr}{dh}\right) + 2\pi r + 2\pi \cdot \frac{dr}{dh} \cdot h$

$= 4\pi r \cdot \frac{-r}{2h} + 2\pi r + 2\pi h \cdot \frac{-r}{2h} =$

$-\frac{2\pi r^2}{h} + 2\pi r - \pi r = \pi r - \frac{2\pi r^2}{h} =$

$\pi r\left(1 - \frac{2r}{h}\right)$, which equals 0 when $r = 0$ or $2r =$

h. Since $r = 0$ makes no sense, we take $2r = h$ as the answer. Recall that $\pi r^2h = 100$, so $\pi r^2h =$

$\pi r^2(2r) = 2\pi r^3 = 100$; hence $r = \left(\frac{100}{2\pi}\right)^{1/3} =$

$\left(\frac{50}{\pi}\right)^{1/3}$ and $h = 2r = 2(50/\pi)^{1/3}$. This agrees with

the result in Example 3.

11 In Example 1 of Sec. 4.7, we have the constraint

$2y + x = 100$. So $2\frac{dy}{dx} + 1 = 0$; hence $\frac{dy}{dx} =$

$-\frac{1}{2}$. Now $A = xy$, so $\frac{dA}{dx} = x \cdot \frac{dy}{dx} + y =$

$-\frac{1}{2}x + y$, which equals 0 when $x = 2y$. Therefore

$100 = 2y + x = x + x = 2x$, so $x = 50$ and $y =$

25.

13 In Exercise 13 of Sec. 4.7 we have the constraint

$2x + 2y = p$, so $2 + 2 \cdot \frac{dy}{dx} = 0$; hence $\frac{dy}{dx} =$

-1. Now $A = xy$, so $\dfrac{dA}{dx} = x\cdot\dfrac{dy}{dx} + y =$ $-x + y$, which equals 0 for $x = y$. Thus the rectangle is a square.

15 In Exercise 23 of Sec. 4.7 we have the constraint $4x + y = 108$, so $4 + \dfrac{dy}{dx} = 0$; hence $\dfrac{dy}{dx} = -4$. Since $V = x^2y$, $\dfrac{dV}{dx} = x^2\cdot\dfrac{dy}{dx} + 2xy = -4x^2 +$ $2xy = 2x(y - 2x)$, which equals 0 when $x = 0$ or $y = 2x$. Only $y = 2x$ makes sense, so $108 = 4x + y = 4x + 2x = 6x$; hence $x = 18$ and $y = 36$. The box therefore has dimensions $18 \times 18 \times 36$.

17 $0 = \dfrac{d}{dx}(1) = \dfrac{d}{dx}(xy^3 + \tan(x + y))$

$= x\cdot\dfrac{d}{dx}(y^3) + y^3 + \dfrac{d}{dx}(\tan(x + y)) =$

$x\cdot 3y^2\cdot\dfrac{dy}{dx} + y^3 + \sec^2(x + y)\cdot(1 + \dfrac{dy}{dx}) =$

$[3xy^2 + \sec^2(x + y)]\dfrac{dy}{dx} + [y^3 + \sec^2(x + y)]$.

Hence $\dfrac{dy}{dx} = -\dfrac{y^3 + \sec^2(x + y)}{3xy^2 + \sec^2(x + y)}$.

19 $0 = \dfrac{d}{dx}(25) = \dfrac{d}{dx}(-7x^2 + 48xy + 7y^2)$

$= -14x + 48(x\cdot\dfrac{dy}{dx} + y) + 14y\cdot\dfrac{dy}{dx}$

$= (48x + 14y)\dfrac{dy}{dx} + (48y - 14x)$. Hence $\dfrac{dy}{dx} =$

$\dfrac{14x - 48y}{48x + 14y} = \dfrac{7x - 24y}{24x + 7y}$.

21 (a) $\dfrac{d}{dx}(x^3y + y^4) = \dfrac{d}{dx}(2)$, so we have $x^3\cdot\dfrac{dy}{dx} + 3x^2y + 4y^3\cdot\dfrac{dy}{dx} = 0$, which can also be written as $x^3y' + 3x^2y + 4y^3y' = 0$.

(b) Since $y(1) = 1$, we can plug in $x = 1$ and $y = 1$ and solve for y': $y' + 3 + 4y' = 0$, $5y' = -3$, so $y' = -3/5$.

(c) $\dfrac{d}{dx}(x^3y' + 3x^2y + 4y^3y')$

$= x^3y'' + 3x^2y' + 3x^2y' + 6xy + 4y^3y'' + 12y^2(y')^2$

$= x^3y'' + 6x^2y' + 6xy + 4y^3y'' + 12y^2(y')^2$

$= 0$.

(d) We have $y(1) = 1$ and know from (b) that $y'(1) = -3/5$, hence

$y'' + 6(-3/5) + 6 + 4y'' + 12(-3/5)^2 = 0$,

$5y'' - 18/5 + 6 + \dfrac{108}{25} = 0$, $y'' = -\dfrac{168}{125}$.

23 $\dfrac{d}{dx}(\sin y) = \dfrac{d}{dx}(x - x^3)$, so $(\cos y)y' = 1 - 3x^2$. Since $y(1) = 0$, we have $(\cos 0)y' = 1 - 3$, so $y' = -2$. Differentiate implicitly again: $(\cos y)y'' - (\sin y)(y')^2 = -6x$. Now plug in $x = 1$, $y = 0$, and $y' = -2$: $(\cos 0)y'' - (\sin 0)(-2)^2 = -6$, $y'' = -6$.

25 $0 = \dfrac{d}{dx}(12) = \dfrac{d}{dx}(x^2 + xy + y^2)$

$= 2x + (x\cdot\dfrac{dy}{dx} + y) + 2y\cdot\dfrac{dy}{dx}$. At the highest and lowest points, $\dfrac{dy}{dx} = 0$, so $0 = 2x + y$. Hence $y = -2x$ and $12 = x^2 + xy + y^2 = x^2 - 2x^2 + 4x^2 = 3x^2$, $x = \pm 2$, and $y = \pm 4$. The highest point is $(-2, 4)$ and the lowest is $(2, -4)$.

27 $y^n = x$, so $ny^{n-1}y' = 1$; hence $y' = \dfrac{1}{ny^{n-1}} =$

$\frac{1}{n} \cdot y^{1-n}$. But $y = x^{1/n}$, so $y' = (1/n)(x^{1/n})^{1-n} =$ $(1/n)x^{(1/n)-1}$.

29 (a) Let $s = x^2 + y^2$. Differentiating $x^2 + 4y^2 =$ 16 with respect to x gives $2x + 8y\frac{dy}{dx} = 0$, so $2y\frac{dy}{dx} = -\frac{x}{2}$. Hence $\frac{ds}{dx} = 2x + 2y\frac{dy}{dx} =$ $2x - \frac{x}{2} = \frac{3}{2}x$. Setting $\frac{ds}{dx} = 0$ gives $x = 0$ so $y = \pm 2$ and $s = 4$. But this is not a maximum, since $s = 16$ when $x = \pm 4$ and $y = 0$.

(b) Note that, for any x and y satisfying the constraint, we have $s = x^2 + y^2 \le x^2 + 4y^2$ $= 16$, with equality for $y = 0$ and $x = \pm 4$. If $\frac{dy}{dx}$ existed at these points, we would have 0 $= \frac{d}{dx}(x^2 + 4y^2) = 2x + 8y \cdot \frac{dy}{dx} = 2x \ne 0$, a contradiction. Hence $\frac{dy}{dx}$ does not exist at either maximum.

(c) Since $x^2 + y^2$ is the square of the distance from the origin to (x, y), the problem can be stated as "Find the points on the ellipse $x^2 + 4y^2 = 16$ farthest from the origin." From (b), the desired points are $(\pm 4, 0)$. At these points, the ellipse has vertical tangents; that is, $\frac{dy}{dx}$ does not exist.

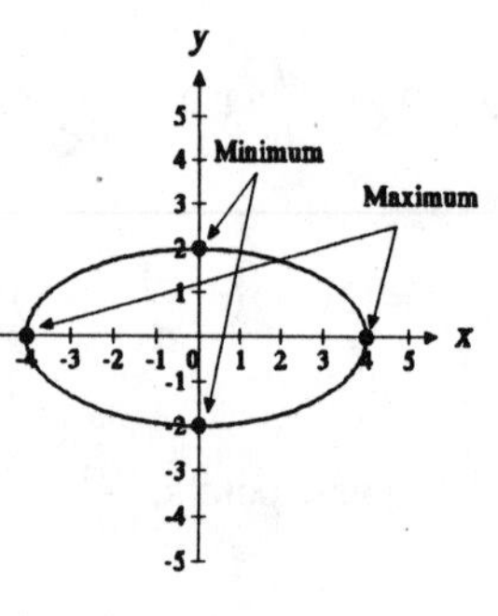

4.9 The Differential and Linearization

1 $y = x^2$ so $dy = 2x\,dx$. We are given that $x =$ 1 and $dx = \Delta x = 0.3$, so $dy = (2 \cdot 1)(0.3)$ $= 0.6$. The quantity Δy is given by $\Delta y =$ $(x + \Delta x)^2 - x^2 =$ $(1.3)^2 - 1^2 = 1.69 - 1 = 0.69$.

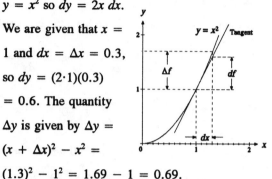

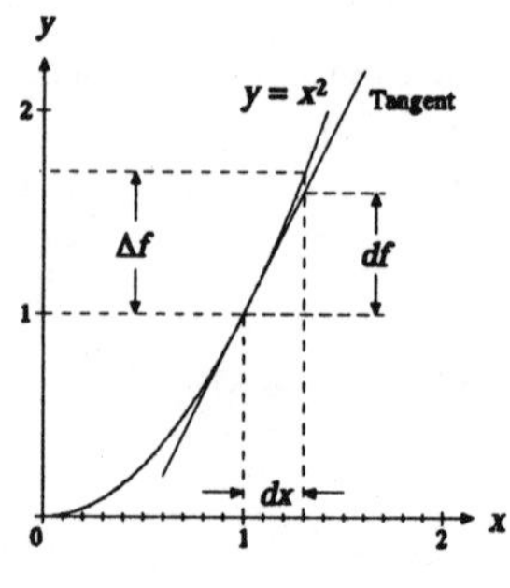

3 $f'(x) = \frac{1}{2\sqrt{x}} = \frac{1}{6}$ at x $= 9$, so $dy = f'(x)\,dx$ $= -1/3$. $\Delta y =$ $\sqrt{9-2} - \sqrt{9} = \sqrt{7}$ $- 3 \approx -0.354$.

5 Observe that we have $f'(x) = \sec^2 x = 4/3$ at $x = \pi/6$, so $dy =$ $(4/3) \cdot \pi/12 = \pi/9 \approx$ 0.349. By comparison, $\Delta y = \tan(\pi/6 + \pi/12)$ $- \tan(\pi/6) = \tan(\pi/4)$ $- \tan(\pi/6) = 1 - \frac{1}{\sqrt{3}} \approx 0.423$.

7 (a) $f(x) = \sqrt{x}$

(b) Since $f(100) = \sqrt{100} = 10$, $x = 100$ is a good choice.

(c) $f(98) \approx f(100) + f'(100)(98 - 100)$ $= \sqrt{100} + \frac{1}{2\sqrt{100}} \cdot (-2) = 10 - 1/10 = 9.9$

9 Let $f(x) = \sqrt{x}$, $x = 121$, and $dx = -2$, so $df = \frac{1}{2\sqrt{x}}\,dx = \frac{1}{2\sqrt{121}}(-2) = -\frac{1}{11} \approx -0.0909$. Then $\sqrt{119} = f(121 - 2) \approx f(121) + df = 11 - 0.0909 = 10.9091$.

11 Let $f(x) = \sqrt[3]{x}$, $x = 27$, and $dx = -2$, so $df = \frac{1}{3}x^{-2/3}\,dx = \frac{1}{3}(27)^{-2/3}(-2) = -\frac{2}{27}$. Then $\sqrt[3]{25} = f(27 - 2) \approx f(27) + df = 3 - \frac{2}{27} = \frac{79}{27} \approx 2.9259$.

13 Let $f(x) = \tan x$, $x = \pi/4$, and $dx = -0.01$, so $df = \sec^2 x\,dx = (\sec^2(\pi/4))(-0.01) = 2(-0.01) = -0.02$. Then $\tan(\pi/4 - 0.01) = f(\pi/4 - 0.01) \approx f(\pi/4) + df = \tan \pi/4 - 0.02 = 1 - 0.02 = 0.98$.

15 Let $f(x) = \sin x$, $x = \pi/3$, and $dx = -0.02$, so $df = (\cos x)\,dx = (\cos \pi/3)(-0.02) = (1/2)(-0.02) = -0.01$. Then $\sin(\pi/3 - 0.02) = f(\pi/3 - 0.02) \approx f(\pi/3) + df = \sin \pi/3 - 0.01 = \frac{\sqrt{3}}{2} - 0.01 \approx 0.8560$.

17 Let $f(x) = \sin x$, $x = 0$, and $dx = 0.13$, so $df = \cos x\,dx = (\cos 0)(0.13) = 0.13$. Then $\sin 0.13 = f(0 + 0.13) \approx f(0) + df = \sin 0 + 0.13 = 0.13$.

19 Let $f(x) = 1/x$, $x = 4$, and $dx = 0.03$, so $df = -\frac{1}{x^2}\,dx = -\frac{1}{16}(0.03) = -0.001875$. Then $\frac{1}{4.03} = f(4 + 0.03) \approx f(4) + df = 1/4 - 0.001875 = 0.248125$.

21 $\sin 32° = \sin \frac{32}{180}\pi = \sin\left(\frac{\pi}{6} + \frac{\pi}{90}\right)$. Since $(\sin x)' = \cos x$, $\sin 32° \approx \sin \frac{\pi}{6} + \left(\cos \frac{\pi}{6}\right)\frac{\pi}{90} = \frac{1}{2} + \frac{\sqrt{3}}{2}\cdot\frac{\pi}{90} \approx 0.5302$. (The actual value of $\sin 32°$ is approximately 0.5299.)

23 First, note that $\cos 28° = \cos \frac{28}{180}\pi = \cos\left(\frac{\pi}{6} - \frac{\pi}{90}\right)$. Since $(\cos x)' = -\sin x$, $\cos 28° \approx \cos \frac{\pi}{6} + \left(\sin \frac{\pi}{6}\right)\frac{\pi}{90} = \frac{\sqrt{3}}{2} + \frac{1}{2}\cdot\frac{\pi}{90} \approx 0.8835$. (The actual value is approximately 0.8829.)

25 Let $f(x) = 1/x$. Then $f'(x) = -\frac{1}{x^2}$, so $f'(1) = -1$ and $\frac{1}{1 + h} = f(1 + h) \approx f(1) + f'(1)\cdot h = 1 - h$.

27 Let $f(x) = \sqrt{x}$. Then $f'(x) = \frac{1}{2\sqrt{x}}$ so $f'(1) = 1/2$ and $\sqrt{1 + h} = f(1 + h) \approx f(1) + f'(1)\cdot h = 1 + h/2$.

29 $d(1/x^3) = -\frac{3}{x^4}\,dx$

31 $d(\sin 2x) = (\cos 2x)\cdot 2\,dx = 2 \cos 2x\,dx$

33 $d(\csc x) = -\csc x \cot x\,dx$

35 $d\left(\frac{\cot 5x}{x}\right) = \frac{x(-\csc^2 5x)(5) - (\cot 5x)(1)}{x^2}\,dx = -\frac{1}{x^2}(5x \csc^2 5x + \cot 5x)\,dx$

37 (a) Here $f(x) = x^{1/2}$ and $f'(x) = (1/2)x^{-1/2}$. Thus $f(a) = f(1) = 1$ and $f'(a) = f'(1) = 1/2$. Hence $p(x) = f(a) + f'(a)(x - a) =$

$$1 + \frac{1}{2}(x - 1) = \frac{1}{2}x + \frac{1}{2}.$$

(b) In this case, $f(a) = f(2) = \sqrt{2}$ and $f'(a) = f'(2) = \frac{1}{2\sqrt{2}}$. Hence $p(x) =$

$$\sqrt{2} + \frac{1}{2\sqrt{2}}(x - 2) = \frac{1}{2\sqrt{2}}x + \sqrt{2} - \frac{1}{\sqrt{2}} = \frac{1}{2\sqrt{2}}x + \frac{1}{\sqrt{2}}.$$

39

x	$f(x)$	$p(x)$	$f(x) - p(x)$
1.5	1.144714	1.166667	−0.021952
1.1	1.032280	1.033333	−0.001053
1.01	1.003322	1.003333	−0.000011
1.001	1.000333	1.000333	0
1.0001	1.000033	1.000033	0

Here $f(x) = x^{1/3}$, $f'(x) = (1/3)x^{-2/3}$, and $f(1) = 1$ while $f'(1) = 1/3$. Thus, $p(x) = 1 + (1/3)(x - 1) = (1/3)x + 2/3$.

41 Since $dT = \frac{k}{2\sqrt{l}}\,dl$, $\frac{dT}{T} = \frac{1}{2}\cdot\frac{dl}{l}$. So the relative error in the period is about half that in the length, hence at most about $p/2$ percent.

43 (a) $df = f'(x)\Delta x = 2x\Delta x$, $\Delta f = f(x + \Delta x) - f(x)$
$= (x + \Delta x)^2 - x^2$
$= x^2 + 2x\Delta x + (\Delta x)^2 - x^2 = 2x\Delta x + (\Delta x)^2$

(b)

(c)

45 (a) $d(f - g) = (f - g)'\,dx = (f' - g')dx$
$= f'\,dx - g'\,dx = df - dg$

(b) $d(fg) = (fg)'dx = (fg' + gf')dx$
$= fg'\,dx + gf'\,dx = f\,dg + g\,df$

(c) $d(f/g) = (f/g)'\,dx = \frac{gf' - fg'}{g^2}\,dx = \frac{gf'\,dx - fg'\,dx}{g^2} = \frac{g\,df - f\,dg}{g^2}$

47 Exploration exercises are in the instructor's manual.

49 (a) Since f is differentiable for all x, the mean-value theorem can be applied on the interval $[a, x]$. There is a number c_1 in this interval such that $f'(c_1) = \frac{f(x) - f(a)}{x - a}$ and $f(x) = f(a) + f'(c_1)(x - a)$.

(b) Since f' is differentiable for all x, the mean-value theorem can again be applied. We apply it on the interval (a, c_1). There is a number c_2 in the interval such that $f''(c_2) = \frac{f'(c_1) - f'(a)}{c_1 - a}$. It follows that $f'(c_1) = f'(a) + f''(c_2)(c_1 - a)$.

(c) Substituting the equation for $f'(c_1)$ in (b) into the equation for $f(x)$ in (a) gives $f(x) =$

$f(a) + [f'(a) + f''(c_2)(c_1 - a)](x - a) =$

$f(a) + f'(a)(x - a) + f''(c_2)(c_1 - a)(x - a)$.

(d) $|\Delta f - df| = |f(a + \Delta x) - f(a) - f'(a)\Delta x|$ $= |f(x) - f(a) - f'(a)(x - a)|$ since $\Delta x =$ $x - a$. By (c), $|f(x) - f(a) - f'(a)(x - a)|$ $= |f''(c_2)(c_1 - a)(x - a)| \le |f''(c_2)|(\Delta x)^2$, since c_1 is in $[a, x]$. It follows immediately that $|\Delta f - df| \le |f''(c_2)|(\Delta x)^2$, as claimed.

4.10 The Second Derivative and Growth of a Function

1 In this case, $y = (1/2)At^2 + v_0 t + y_0$ and $A = \frac{d^2y}{dt^2} \le 4$ ft/sec$^2 = M$. Since $v_0 = y_0 = 0$, $y(t) \le (1/2)Mt^2$ and $y(30) \le (1/2)\cdot 4\cdot(30)^2 = 1800$ feet.

3 Here $m = 5$ mi/hr^2 and $M = 12$ mi/hr^2 because $5 \le A \le 12$. From $v_0 = 0$ and $y_0 = 0$, we know that $(1/2)mt^2 \le y(t) \le (1/2)Mt^2$, so $(1/2)\cdot 5\cdot 4^2 \le y(4) \le (1/2)\cdot 12\cdot 4^2$, 40 mi $\le y(4) \le$ 96 mi.

5 Since $f(1) = 0$, $f'(1) = 0$, and $f''(x) \le 4$ for all x, the growth theorem can be applied with $t_0 = 1$, $t = 3$, and $M = 4$. No lower bound on $f''(x)$ is given, so we can conclude nothing about the lower bound of $f(x)$. We do know that $f(t) \le (1/2)M(t - t_0)^2$ and, in particular, $f(3) \le (1/2)\cdot 4(3 - 1)^2 = 8$. Therefore, we can say with certainty that $f(3) \le 8$.

7 Since $f(1) = 0$, $f'(1) = 0$, and $2.5 \le f''(x) \le 2.6$ for all x, the growth theorem can be applied with $t_0 = 1$, $t = 2$, $m = 2.5$, and $M = 2.6$. Thus

$\frac{1}{2}m(t - t_0)^2 \le f(t) \le \frac{1}{2}M(t - t_0)^2$,

$\frac{1}{2}(2.5)(2 - 1)^2 \le f(2) \le \frac{1}{2}(2.6)(2 - 1)^2$, so we can say that $1.25 \le f(2) \le 1.3$.

9 (a) Here $f'(x) = (1/2)x^{-1/2}$, so $f(4) = 2$ and $f'(4) = 1/4$. Thus $p(x) = f(4) + f'(4)(x - 4) = 2 + (1/4)(x - 4) = 1 + x/4$.

(b)

x	$E(x) = f(x) - p(x)$	$(x - 4)^2$	$E(x)/(x - 4)^2$
5	−0.013932	1.0000	−0.01393
4.1	−0.000154	0.0100	−0.01543
4.01	−0.000002	0.0001	−0.01561
3.99	−0.000002	0.0001	−0.01565

(c) $f''(x) = -\frac{1}{4}x^{-3/2}$, so $f''(4) = -\frac{1}{4}(4)^{-3/2} = -\frac{1}{32}$ and $\frac{f''(4)}{2} = -\frac{1}{64} = -0.015625$. Thus $\frac{f''(4)}{2}$ agrees with the fourth column in (b).

11 (a) Since $4 \le f''(x) \le 5$ for all x, we may apply Theorem 2 with $m = 4$, $M = 5$, and $a = 2$. Thus $\frac{m(x - a)^2}{2} \le E(x) \le \frac{M(x - a)^2}{2}$,

$\frac{4(x - 2)^2}{2} \le E(x) \le \frac{5(x - 2)^2}{2}$. So for any given x, $E(x)$ lies between $2(x - 2)^2$ and $(5/2)(x - 2)^2$.

(b) We know from (a) that $E(x) \ge 0$. Also, $E(x) \le (5/2)(x - 2)^2$ so $|E(x)| \le 5/2(x - 2)^2$ as well. We want $|E(x)| \le 0.01$, so let $(5/2)(x - 2)^2 = 0.01$. Then $(x - 2)^2 = 0.004$ and $x - 2 = \pm\sqrt{0.004} \approx \pm 0.0632$. We conclude that $|x - 2| \le 0.0632$ for $|E(x)|$

13 (a) Since f is differentiable for all x, we can apply the mean-value theorem with $a = 0$, $b = 3$, and $c = c_1$. Thus $f'(c_1) = \dfrac{f(3) - f(0)}{3 - 0} =$

$\dfrac{4 - 0}{3 - 0} = \dfrac{4}{3}$. So, for some c_1 in $(0, 3)$, $f'(c_1)$

$= 4/3$.

(b) Let m and M be minimum and maximum values, respectively, of f'' on the interval $[0, 3]$. Since $f''(x)$ is continuous for all x in $[0, 3]$, we may apply the growth theorem. Therefore we have $(1/2)m(x - x_0)^2 \le f(x) \le (1/2)M(x - x_0)^2$. With $x = 3$ and $x_0 = 0$, it follows that

$$(1/2)m(3 - 0)^2 \le f(3) \le (1/2)M(3 - 0)^2$$
$$(9/2)m \le 4 \le (9/2)M$$
$$m \le 8/9 \le M.$$

Now, since $f''(x)$ is continuous throughout $[0, 3]$, the intermediate-value theorem guarantees that it assumes all values between its minimum and maximum; in particular, for some c_2 in $[0, 3]$, $f''(c_2) = 8/9$.

15 (a) Under the assumption that $f(0) = 0$, $h(0) = f(0) - (1/2)M\cdot 0^2 = 0 - 0 = 0$. Now, $h'(t) = f'(t) - Mt$. From Exercise 14(c), $f'(t) \le Mt$, so $h'(t) \le 0$.

(b) Since $h'(t)$ is defined at $t = 0$, we know that

(from (a)) $h'(0) = \lim\limits_{t\to 0^+} \dfrac{h(t) - h(0)}{t - 0} =$

$\lim\limits_{t\to 0^+} \dfrac{h(t)}{t} \le 0$. Here, $t > 0$ so $h(t) \le 0$.

Because $f(t) = h(t) + (1/2)Mt^2$ and $h(t) \le 0$, we conclude that $f(t) \le (1/2)Mt^2$.

17 (a) With $b = r$ and $a = x_1$,

$$f(r) = f(x_1) + f'(x_1)(r - x_1) + \frac{f''(c)(r - x_1)^2}{2}$$

$$0 = \frac{f(x_1)}{f'(x_1)} + r - x_1 + \frac{f''(c)(r - x_1)^2}{2f'(x_1)}$$

$$x_1 - \frac{f(x_1)}{f'(x_1)} - r = \frac{f''(c)(r - x_1)^2}{2f'(x_1)}$$

$$x_2 - r = \frac{f''(c)(r - x_1)^2}{2f'(x_1)}.$$

(b) Since $f''(x) > 0$ for x in $[r, x_1]$, f is concave up on $[r, x_1]$. From (a) we see that the difference between x_2 and the actual root r varies as $(r - x_1)^2$. As n increases, the error decreases as the square of the difference between r and the last estimate. If x_1 is the first guess then the error $x_n - r$ is on the order of $(r - x_1)^2(r - x_2)^2(r - x_3)^2 \cdots (r - x_{n-1})^2$.

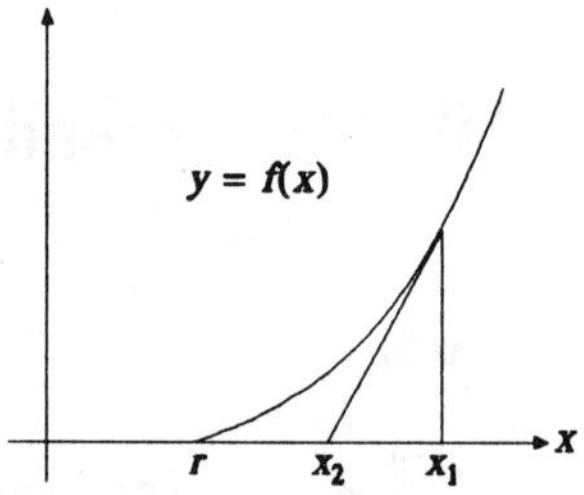

4.S Guide Quiz

2 (a) If $f(x)$ is continuous at $x = a$, then the graph of f crosses the x axis at $x = a$.

(b) If $f'(x)$ changes sign from positive to negative at $x = a$, then $f(a)$ is a maximum, provided that $f(x)$ is continuous at $x = a$.

(c) A sign change in f'' at $x = a$ indicates that a is an inflection point; since the change is from positive to negative, the curve changes from concave up to concave down.

3 There must exist a c in $[0, 6]$ such that $f'(c) = \frac{f(6) - f(0)}{6 - 0}$. Now, $2c - 2 = \frac{16 - (-8)}{6} = 4$, so $2c = 6$ and thus $c = 3$.

4 (a) $(\tan x - x)' = \sec^2 x - 1 = \tan^2 x \geq 0$ for $0 \leq x \leq \pi/2$, so $\tan x - x$ is an increasing function on that interval.

(b) Note that for $x = 0$, $\tan x - x = \tan 0 - 0 = 0$. By (a), $\tan x - x$ is increasing on $[0, \pi/2)$, so $\tan x - x > 0$ for x in $(0, \pi/2)$; that is $\tan x > x$.

(c) Since $\cos x > 0$ for x in $(0, \pi/2)$, we may multiply both sides of $\tan x > x$ by $\cos x$ and preserve the inequality; hence

$$\tan x \cos x > x \cos x$$
$$\sin x > x \cos x$$
$$0 > x \cos x - \sin x.$$

(d) $\left(\frac{\sin x}{x}\right)' = \frac{x \cos x - \sin x}{x^2}$, which is negative over $(0, \pi/2)$ by (c); hence $(\sin x)/x$ decreases over $(0, \pi/2)$.

5 (a) Since $(x^3)' = 3x^2$, all antiderivatives of $3x^2$ are of the form $x^3 + C$.

(b) Corollary 2 of Sec. 4.1 ensures that the result in (a) includes all antiderivatives.

6 Let the two numbers be x and $y = 1 - x$. We seek to minimize $f(x) = x^3 + y^2 = x^3 + (1 - x)^2 = x^3 + 1 - 2x + x^2$ for $0 \leq x \leq 1$. Then $f'(x) = 3x^2 - 2 + 2x$, which equals 0 when $x = \frac{-2 \pm \sqrt{4 - 4 \cdot 3 \cdot (-2)}}{6} = \frac{-2 \pm \sqrt{28}}{6} = \frac{1}{3}(-1 \pm \sqrt{7})$. We choose the + sign since x must be nonnegative. Since $f''(x) = 6x + 2 > 0$ for $0 \leq x \leq 1$, f' changes sign from $-$ to $+$ at $x = \frac{1}{3}(-1 + \sqrt{7})$. That is, f is decreasing for $0 \leq x \leq \frac{1}{3}(-1 + \sqrt{7})$ and increasing for $\frac{1}{3}(-1 + \sqrt{7}) \leq x \leq 1$. It follows that $x = \frac{1}{3}(-1 + \sqrt{7})$ should be cubed and $y = 1 - x = 1 + \frac{1}{3} - \frac{\sqrt{7}}{3} = \frac{1}{3}(4 - \sqrt{7})$ should be squared.

7 The perimeter of the track is $L = 2x + 2\pi r$, so $\frac{dL}{dx} = 2 + 2\pi \cdot \frac{dr}{dx}$; since L is constant, this equals 0 and $\frac{dr}{dx} = -\frac{2}{2\pi} = -\frac{1}{\pi}$. The area of the rectangle is $2xr$ and the area of each semicircle is $(1/2)\pi r^2$, so the total area is $A = 2xr + \pi r^2$. Thus $\frac{dA}{dx} = 2x \cdot \frac{dr}{dx} + 2r + 2\pi r \cdot \frac{dr}{dx} = \frac{-2x}{\pi} + 2r - 2r = \frac{-2x}{\pi} < 0$ for $x > 0$; A is a decreasing function of x for $x > 0$. Hence the area is maximal for $x = 0$, in which case the track is simply a circle. The minimum occurs when $r = 0$, in which case $A = 0$.

8 We are given that $|dr/r| \leq 0.02$. Now $V = (4/3)\pi r^3$ and $dV = 4\pi r^2\, dr$, so $\frac{dV}{V} = 3\frac{dr}{r}$ and $\left|\frac{dV}{V}\right| = 3\left|\frac{dr}{r}\right| \leq 0.06$. So the error in the volume calculation is at most 6%.

9 $f(x) = \dfrac{1}{x^2 - 3x + 2} = \dfrac{1}{(x - 1)(x - 2)}$, which is defined everywhere except at $x = 1$ or 2. $f(x)$ is never 0, so there are no x intercepts; $f(0) = 1/2$, so $(0, 1/2)$ is the y intercept. $f'(x) = \dfrac{-(2x - 3)}{(x^2 - 3x + 2)^2}$, which equals 0 when $2x - 3 = 0$; that is, for $x = 3/2$. Note that $f'(x) > 0$ (where defined) for $x < 3/2$ and $f'(x) < 0$ (where defined) for $x > 3/2$; hence $x = 3/2$ is a local maximum of value $f(3/2) = \dfrac{1}{(1/2)(-1/2)} = -4$. Since $\lim\limits_{x \to 1^-} f(x) = \infty$ and $\lim\limits_{x \to 1^+} f(x) = -\infty$, $x = 1$ is a vertical asymptote. Also, $\lim\limits_{x \to 2^-} f(x) = -\infty$ and $\lim\limits_{x \to 2^+} f(x) = \infty$, so $x = 2$ is another vertical asymptote. Finally, $\lim\limits_{x \to \infty} f(x) = \lim\limits_{x \to -\infty} f(x) = 0$, so the x axis is a horizontal asymptote.

10 $f(x) = 3x^4 - 16x^3 + 24x^2 = x^2(3x^2 - 16x + 24)$, which equals 0 when $x = 0$ or $3x^2 - 16x + 24 = 0$; in the latter case, $\sqrt{b^2 - 4ac} = \sqrt{(-16)^2 - 4 \cdot 3 \cdot 24} = \sqrt{-32}$, so $3x^2 - 16x + 24$ is never 0. The only x intercept is $(0, 0)$ (and also the y intercept). Now $f'(x) = 12x^3 - 48x^2 + 48x = 12x(x^2 - 4x + 4) = 12x(x - 2)^2$, which equals 0 when $x = 0$ or 2, so $(0, 0)$ and $(2, 16)$ are critical points. Next, $f''(x) = 36x^2 - 96x + 48 = 12(3x^2 - 8x + 4) = 12(x - 2)(3x - 2)$, which equals 0 when $x = 2$ or $2/3$. $f''(x) > 0$ for $x < 2/3$ and $x > 2$, and $f''(x) < 0$ for $2/3 < x < 2$, so $(2, 16)$ and $\left(\dfrac{2}{3}, \dfrac{176}{27}\right)$ are inflection points. Also $f''(0) = 48 > 0$, so $(0, 0)$ is a minimum. Since $f(x)$ is a polynomial of even degree $\lim\limits_{x \to \infty} f(x) = \lim\limits_{x \to -\infty} f(x) = \infty$.

11 Since $\dfrac{x^3 + 1}{x^2 + 1} = x - \dfrac{x - 1}{x^2 + 1}$, the line $y = x$ is an asymptote. For $x > 1$ the curve lies below the

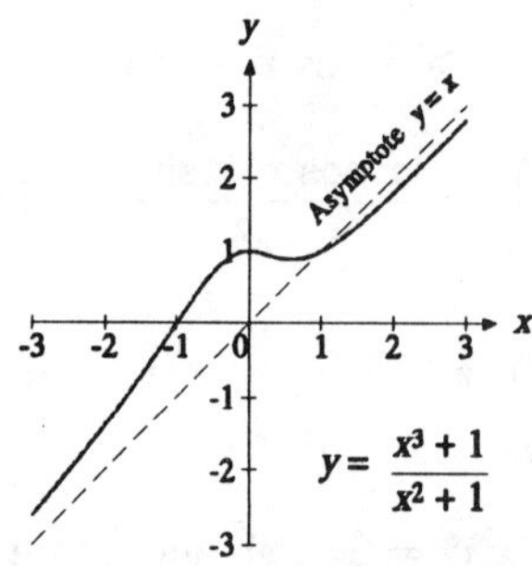

asymptote; for $x < 1$ it lies above it. The x intercepts are obtained by setting $y = 0$: If $\dfrac{x^3 + 1}{x^2 + 1} = 0$, then $x^3 + 1 = 0$, $x^3 = -1$, and $x = -1$. The y intercept is $y = \dfrac{0^3 + 1}{0^2 + 1} = 1$. Note that $\dfrac{dy}{dx} = \dfrac{(x^2 + 1) \cdot 3x^2 - (x^3 + 1) \cdot 2x}{(x^2 + 1)^2} = \dfrac{x^4 + 3x^2 - 2x}{(x^2 + 1)^2} = \dfrac{x(x^3 + 3x - 2)}{(x^2 + 1)^2}$. Hence there is a critical point at

$x = 0$ and at any root of $x^3 + 3x - 2 = 0$. Since $(x^3 + 3x - 2)' = 3x^2 + 3 > 0$ for all x, the equation $x^3 + 3x - 2 = 0$ has exactly one root. It is approximately 0.596, at which point $y \approx 0.894$. With this information, we can sketch the graph.

12 Since $V = x^3$, $\frac{dV}{dt} = 3x^2\frac{dx}{dt}$, so $\frac{dx}{dt} = \frac{1}{3x^2}\frac{dV}{dt}$.

When $x = 10$ and $\frac{dV}{dt} = 12$, $\frac{dx}{dt} = \frac{1}{3\cdot 10^2}\cdot 12 = \frac{1}{25}$ m/min.

13 Differentiating implicitly, $\frac{d}{dx}(y^2 + x^2y + x^3) = \frac{d}{dx}(8)$, so $2yy' + 2xy + x^2y' + 3x^2 = 0$ and $y' = -\frac{3x^2 + 2xy}{x^2 + 2y}$. Then $y'(0) = -\frac{3\cdot 0 + 2\cdot 0\cdot 2}{0 + 2\cdot 2} = 0$.

Differentiating implicitly again, we have $2(y')^2 + 2yy'' + 2xy' + 2y + 2xy' + x^2y'' + 6x = 0$, so $y'' = -\frac{6x + 4xy' + 2y + 2(y')^2}{2y + x^2}$.

Finally, $y''(0) = -\frac{6\cdot 0 + 4\cdot 0\cdot 0 + 2\cdot 2 + 2\cdot 0^2}{2\cdot 2 + 0^2} = -1$.

14 (a) $\frac{d^2}{dx^2}\left(2x^5 - \frac{1}{x}\right) = \frac{d}{dx}\left(10x^4 + \frac{1}{x^2}\right) = 40x^3 - \frac{2}{x^3}$

(b) $\frac{d^2}{dx^2}(\cos 2x) = \frac{d}{dx}(-2\sin 2x) = -4\cos 2x$

(c) $\frac{d^2}{dx^2}(17x^3 - 5x + 2) = \frac{d}{dx}(51x^2 - 5) = 102x$

(d) $\frac{d^2}{dx^2}(\sqrt{x}) = \frac{d}{dx}\left(\frac{1}{2\sqrt{x}}\right) = -\frac{1}{4\sqrt{x^3}}$

(e) $\frac{d^2}{dx^2}\left(\frac{\tan 3x}{1 + 2x}\right)$

$$= \frac{d}{dx}\left(\frac{(1 + 2x)(\sec^2 3x)(3) - (\tan 3x)(2)}{(1 + 2x)^2}\right)$$

$$= \frac{d}{dx}\left(\frac{3\sec^2 3x}{1 + 2x} - \frac{2\tan 3x}{(1 + 2x)^2}\right)$$

$$= \frac{3(1 + 2x)(2\sec 3x)(\sec 3x\tan 3x)(3) - 3(\sec^2 3x)(2)}{(1 + 2x)^2} - \frac{2(1 + 2x)^2(\sec^2 3x)(3) - 2(\tan 3x)\cdot 2(1 + 2x)\cdot 2}{(1 + 2x)^4}$$

$$= \frac{18(1 + 2x)\sec^2 3x\tan 3x - 6\sec^2 3x}{(1 + 2x)^2} - \frac{6(1 + 2x)^2\sec^2 3x - 8(1 + 2x)\tan 3x}{(1 + 2x)^4}$$

$$= \frac{18(1 + 2x)^2(\sec^2 3x\tan 3x) - 12(1 + 2x)\sec^2 3x + 8\tan 3x}{(1 + 2x)^3}.$$

(f) Let $y = \frac{\sin 2x}{1 + \sec 2x}$. Then $\frac{dy}{dx} = \frac{(1 + \sec 2x)(2\cos 2x) - (\sin 2x)(2\sec 2x\tan 2x)}{(1 + \sec 2x)^2} = \frac{2\cos 2x + 2 - 2\tan^2 2x}{(1 + \sec 2x)^2}$. Therefore

$$\frac{d^2y}{dx^2} = \frac{d}{dx}\left(\frac{2\cos 2x + 2 - 2\tan^2 2x}{(1 + \sec 2x)^2}\right)$$

$= \frac{vu' - uv'}{v^2}$, where $v^2 = (1 + \sec 2x)^4$ and

$vu' - uv' =$

$(1 + \sec 2x)^2(2\cos 2x + 2 - 2\tan^2 2x)'$

$- (2\cos 2x + 2 - 2\tan^2 2x)[(1 + \sec 2x)^2]'$

$= (1 + \sec 2x)^2[-4\sin 2x - 2\cdot 2(\tan 2x)(\sec^2 2x)(2)]$

$- (2\cos 2x + 2 - 2\tan^2 2x)2(1 + \sec 2x)(2\sec 2x\tan 2x)$

$= (1 + \sec 2x)[-4\sin 2x - 8\tan 2x\sec^2 2x - 4\tan 2x$

$- 8\tan 2x\sec^3 2x - 8\sec 2x\tan 2x - 8\sec 2x\tan^3 2x]$.

We can cancel a factor of $1 + \sec 2x$, leaving a denominator of $(1 + \sec 2x)^3$ and a numerator of

$-4\sin 2x - 8\tan 2x\sec^2 2x - 4\tan 2x - 8\tan 2x\sec^3 2x$

$- 8\tan 2x - 8\sec 2x\tan 2x + 8\sec 2x\tan^3 2x$

$= -4\sin 2x - 12\tan 2x - 8\tan 2x\sec^2 2x - 8\tan 2x\sec^3 2x$

$- 8\sec 2x\tan 2x + 8\sec 2x\tan^3 2x$

$= -4\sin 2x - 12\tan 2x - 8\tan 2x\sec^2 2x$

$- (8\tan 2x\sec 2x)(\sec^2 2x + 1 - \tan^2 2x)$

$= -4\sin 2x - 12\tan 2x - 8\tan 2x\sec^2 2x - 16\tan 2x\sec 2x$,

so $\frac{d^2y}{dx^2} =$

$$\frac{-4\sin 2x - 12\tan 2x - 8\tan 2x\sec^2 2x - 16\tan 2x\sec 2x}{(1 + \sec 2x)^3}.$$

15 (a) Find all critical points on the open interval, excluding the endpoints. Find which are relative maxima with the first or second derivative test, and compute those maximum values. Select the maximum number from the maxima on the open interval and the function values at the endpoints. This number is the global maximum on the closed interval.

(b) If $f'(c) = 0$ and f' changes from negative to positive at c, then $f(c)$ is a relative minimum.

(c) If $f'(c) = 0$ and $f''(c) > 0$, then $f(c)$ is a relative minimum.

16 (a) For a well-behaved function, Newton's method entails making a guess x_1 near the root r, finding the tangent line to the function at x_1 and extending this tangent line to where it intersects the x axis at, say, x_2. This x_2 is generally closer to r than x_1.

(b)

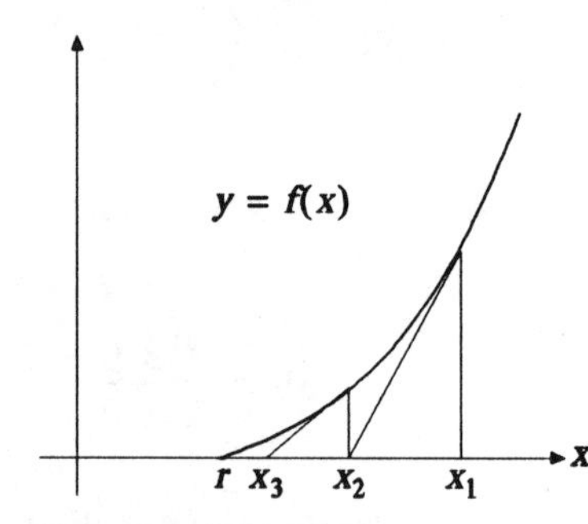

We see that $\lim_{n\to\infty} x_n = r$.

17 We must use $y = \frac{a}{2}t^2 + v_0t + y_0$. Here $y_0 = 0$, $a = -32$ ft/sec, and $v_0 = 90$ mi/hr $= 132$ ft/sec at a maximum height when $\frac{dy}{dt} = at + v_0 = 0$, or

when $t = \frac{-v_0}{a} = \frac{-132 \text{ ft/sec}}{-32 \text{ ft/sec}^2} = 4.125$ sec. At

this time, $y = -16(4.125)^2 + 132(4.125) + 0 =$ 272.25 ft.

18 (a) If $f(x) = \sqrt{x}$ then $f'(x) = \frac{1}{2\sqrt{x}}$, $f(100) = 10$,

$f'(100) = 1/20$, and $f(101.5) = f(100 + 1.5)$

$\approx f(100) + f'(100)\cdot 1.5 = 10.075$.

(b) Note that $f''(x) = -\frac{1}{4}x^{-3/2}$, which is a decreasing function. If we consider $f''(x)$ on the interval [81, 121] (chosen simply because it contains $a = 100$ and the values come out evenly), we see that the maximum value of $f''(x)$ is $M = f''(81) = -1/2916$, while the minimum value is $m = f''(121) = -1/5324$. Using Theorem 2 of Sec. 4.10, we have

$$\frac{(x - 100)^2}{2(-2916)} \le E(x) \le \frac{(x - 100)^2}{2(-5324)};$$ with $x =$ 101.5, these bounds become

$$\frac{(1.5)^2}{2(-2916)} \le E(101.5) \le \frac{(1.5)^2}{2(-5324)},$$ or

$-0.000386 \le E(101.5) \le -0.000211$. (We could, of course, have made these bounds even smaller by picking a shorter interval about $a = 100$.)

(c) Error $= \sqrt{101.5} - 10.075 = -0.000279$

19 (a) $p(x)$ is the formula for the tangent line at $f(a)$. From the graph, for x near a, $p(x)$ and $f(x)$ are very similar in appearance and their difference seems small.

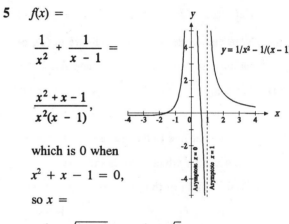

(b) Note that $p(a) = f(a)$ and $p'(a) = f'(a)$. So at a, there is no distinction between $p(a)$ and $f(a)$ when one looks only as far as their first derivatives. So for x near a, $p(x)$ should still remain a good approximation of $f(x)$.

4.S Review Exercises

1 Let $f(x) = x^5 + 2x^3 - 2$. Then $f(0) = -2$ and $f(1) = 1$; the intermediate-value theorem says that $f(x) = 0$ for some x in (0, 1), since f is continuous and 0 is between $f(0)$ and $f(1)$. But $f'(x) = 5x^4 + 6x^2 > 0$ for x in (0, 1). Hence $f(x)$ is an increasing function on [0, 1], so there is at most one root.

3 (a) $f(x) = 1/x$, so $f'(x) = -1/x^2$; $x^2 > 0$ for all x in the domain of f, so $f'(x) < 0$.

(b) Let $x_1 = 1$ and $x_2 = -1$, so $f(x_1) = 1$ and $f(x_2) = -1$; thus $x_1 > x_2$, but $f(x_1) > f(x_2)$. The answer to the question is therefore no.

5 $f(x) = \frac{1}{x^2} + \frac{1}{x-1} = \frac{x^2 + x - 1}{x^2(x-1)}$, which is 0 when $x^2 + x - 1 = 0$, so $x = \frac{-1 \pm \sqrt{1+4}}{2} = \frac{-1 \pm \sqrt{5}}{2}$; these are the x intercepts. $f(x)$ is not defined at $x = 0$ (or 1), so there is no y intercept; however, $x = 0$ and $x = 1$ are vertical asymptotes: $\lim_{x\to 0}\left(\frac{1}{x^2} + \frac{1}{x-1}\right) = \infty$,

$\lim_{x\to 1^-}\left(\frac{1}{x^2} + \frac{1}{x-1}\right) = -\infty$, $\lim_{x\to 1^+}\left(\frac{1}{x^2} + \frac{1}{x-1}\right) = \infty$. Also, $\lim_{x\to -\infty} f(x) = 0$ and $\lim_{x\to\infty} f(x) = 0$ so $y = 0$ is a horizontal asymptote.

7 (a)

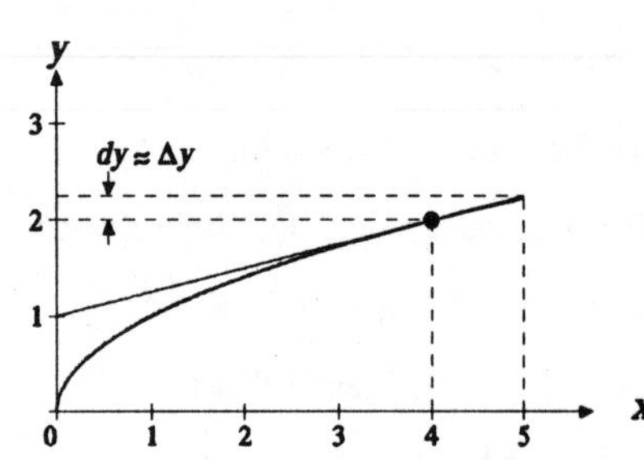

(b) $y = \sqrt{x}$, $x = 4$, and $dx = 1$, so $dy =$ $\frac{1}{2\sqrt{x}}\,dx = \frac{1}{2\sqrt{4}}\cdot 1 = \frac{1}{4}$.

(c) When $x = 4$ and $\Delta x = 1$, $\Delta y =$ $\sqrt{x + \Delta x} - \sqrt{x} = \sqrt{4 + 1} - \sqrt{4} = \sqrt{5} - 2$.

(d) Note that on the scale of the figure in (a), one cannot distinguish between dy and Δy.

9 (a) The tangent line to the curve $y = x^3$ at $x = 1$ also meets the curve at $(-2, -8)$.

(b) The tangent line to the curve $y = x^3$ at $x = 0$ is the x axis, which crosses the curve.

11 Let x be the length of the piece to be shaped into an equilateral triangle; then $y = L - x$ is the length of the piece to be shaped into a square. The formula for the area of an equilateral triangle of side a is $\frac{\sqrt{3}a^2}{4}$, so we have $\frac{\sqrt{3}}{4}\left(\frac{x}{3}\right)^2 = \frac{\sqrt{3}x^2}{36}$.

Each side of the square has length $\frac{L - x}{4}$, so the area is $\left(\frac{L - x}{4}\right)^2 = \frac{L^2 - 2Lx + x^2}{16}$. Thus the total area A is $\frac{\sqrt{3}x^2}{36} + \frac{L^2 - 2Lx + x^2}{16} =$ $\left(\frac{\sqrt{3}}{36} + \frac{1}{16}\right)x^2 - \frac{Lx}{8} + \frac{L^2}{16}$. Then $\frac{dA}{dx} =$ $2\left(\frac{\sqrt{3}}{36} + \frac{1}{16}\right)x - \frac{L}{8}$, which equals 0 when $x =$ $\frac{L}{8}\left(\frac{72}{4\sqrt{3} + 9}\right) = \frac{9L}{4\sqrt{3} + 9}$. Note that $\frac{d^2A}{dx^2} =$ $\frac{4\sqrt{3} + 9}{72} > 0$, so this provides a minimum.

(a) The area is minimized when $\frac{9L}{4\sqrt{3} + 9}$ is formed into the triangle and $L - \frac{9L}{4\sqrt{3} + 9} =$ $\frac{4\sqrt{3}L}{4\sqrt{3} + 9}$ is formed into the square.

(b) The only critical value for x in $(0, L)$ provided a minimum, so the maximum must occur at an endpoint. For $x = 0$, $A = \frac{L^2}{16}$, while for $x =$ L, $A = \frac{\sqrt{3}L^2}{36}$. Now $1/16 = 0.0625$ and $\frac{\sqrt{3}}{36}$ ≈ 0.0481, so the maximum $A = \frac{1}{16}L^2$ occurs for $x = 0$: all the wire is used for the square.

13 f' exists, so f is continuous. By the mean-value theorem, there is a number c in $(0, 3)$ such that $\frac{f(3) - f(0)}{3 - 0} = f'(c)$. Since $f(0) = 0$ and $f'(x) \geq 1$ for all x, we have $\frac{f(3)}{3} \geq 1$, so $f(3) \geq 3$.

15 (a) Let $f(x) = \sqrt{x}$, so $f'(x) = \frac{1}{2\sqrt{x}}$. Thus, for small x, $f(1 - x^2) \approx f(1) + f'(1)(-x^2) =$

$1 - (1/2)x^2$.

(b) Here $x = 0.2$, so $f(1 - (0.2)^2) \approx 1 - (1/2)(0.2)^2 = 0.98$. Exactly, $f(1 - (0.2)^2) = f(0.96) = \sqrt{.96} \approx 0.9798$. The estimate is quite good, only off by $|0.98 - 0.9798| = 0.0002$.

17 (a) Let $f(x) = \sqrt{x}$, so $f'(x) = \dfrac{1}{2\sqrt{x}}$. Thus, for small x, $f(2 + 7x) \approx f(2) + f'(2)(7x) = \sqrt{2} + \dfrac{1}{2\sqrt{2}}\cdot 7x$.

(b) Here $x = 0.1$, so $f(2 + 7\cdot 0.1) \approx \sqrt{2} + \dfrac{1}{2\sqrt{2}}\cdot 7\cdot 0.1 \approx 1.6617$. Exactly, $f(2 + 7\cdot 0.1) = f(2.7) = \sqrt{2.7} \approx 1.6432$. The estimate is not very close because though $x = 0.1$ is small relative to 2, $7x = 0.7$ is not, and the differential approximation fails.

19 We are given that $|\Delta\theta| = |d\theta| \le 1.6° \approx 0.02793$ radians. Now $d(\cos\theta) = -\sin\theta\, d\theta$ and $\dfrac{d(\cos\theta)}{\cos\theta} = -\tan\theta\, d\theta$, so $\left|\dfrac{d(\cos\theta)}{\cos\theta}\right| = |\tan\theta\, d\theta| \le 0.02793|\tan\theta|$, and this is the relative error in the calculation of $\cos\theta$.

(a) $\theta = 60° = \dfrac{\pi}{3}$: $\left|\dfrac{d(\cos\theta)}{\cos\theta}\right| \le 0.02793\left|\tan\dfrac{\pi}{3}\right| \approx 0.04838$ and the maximum error is 4.84%.

(b) $\theta = 45° = \dfrac{\pi}{4}$: $\left|\dfrac{d(\cos\theta)}{\cos\theta}\right| \le 0.02793\left|\tan\dfrac{\pi}{4}\right| \approx 0.02793$ and the maximum error is 2.79%.

(c) $\theta = 30° = \dfrac{\pi}{6}$: $\left|\dfrac{d(\cos\theta)}{\cos\theta}\right| \le 0.02793\left|\tan\dfrac{\pi}{6}\right| = 0.01613$ and the maximum error is 1.61%.

21 We are given $\left|\dfrac{dr}{r}\right| = 0.02$. Now $\dfrac{dV}{V} = \dfrac{4\pi r^2\, dr}{(4/3)\pi r^3} = 3\dfrac{dr}{r}$, so $\left|\dfrac{dV}{V}\right| = 3\cdot\left|\dfrac{dr}{r}\right|$. Thus, $\left|\dfrac{dV}{V}\right| = 3(0.02) = 0.06$, and a 6% error is induced in the measurement of the volume. Also $\dfrac{dS}{S} = \dfrac{8\pi r\, dr}{4\pi r^2} = 2\dfrac{dr}{r}$, so $\left|\dfrac{dS}{S}\right| = 2\left|\dfrac{dr}{r}\right|$. Hence, $\left|\dfrac{dS}{S}\right| = 2(0.02) = 0.04$, and a 4% error is induced in the surface area measurement.

23 If $y = x^a$, then $dy/dx = ax^{a-1}$. Given that $dy/dx = -y^2 = -(x^a)^2 = -x^{2a}$, we have $ax^{a-1} = -x^{2a}$, so $a = \dfrac{-x^{2a}}{x^{a-1}} = -x^{a+1}$. The function x^{a+1} can be constant only if $a + 1 = 0$, in which case $a = -1$ and $x^{a+1} = x^0 = 1$. Substituting $a = -1$ into $a = -x^{a+1}$ shows that it is indeed a solution.

25 First note that $p(x) = f(a) + f'(a)(x - a) = \tan\pi/4 - (\sec^2\pi/4)(x - \pi/4) = 1 - 2(x - \pi/4) = 1 + \pi/2 - 2x$, and $df = f'(a)\,dx = \sec^2(\pi/4)\,dx = 2\,dx = 2\Delta x$.

Δx	$f(a+\Delta x) - p(a+\Delta x)$	Δf	df	$\Delta f - df$
0.2	0.9085	0.5085	0.4	0.1085
0.1	0.4230	0.2230	0.2	0.0230
0.01	0.0402	0.0202	0.02	0.0002
−0.01	0.0398	−0.0198	−0.02	0.0002

27 $f(x) = \sqrt{x}/(1 + x)$ is defined for $x \ge 0$. $f(0) = 0$, so (0, 0) is both the x intercept and the y intercept.

$$f'(x) = \frac{(1+x)\frac{1}{2\sqrt{x}} - \sqrt{x}}{(1+x)^2} = \frac{1-x}{2\sqrt{x}(1+x)^2},$$ which

equals 0 when $x = 1$. Note that $f'(x) > 0$ for $x < 1$ and $f'(x) < 0$ for $x > 1$, so $x = 1$ is a maximum of value $f(1) = 1/2$, $\lim_{x\to\infty} f(x) = 0$, so the x axis is a horizontal asymptote.

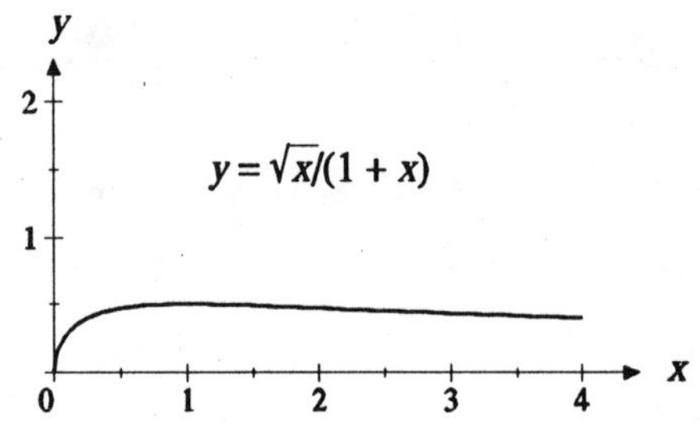

29 Let $f(x) = 2x^7 + 3x^5 + 6x + 10$. Then $f(0) = 10$, $f(-1) = -2 - 3 - 6 + 10 = -1$, and $-1 < 0 < 10$, so by the intermediate-value theorem, $f(c) = 0$ for some number c in $(-1, 0)$. But $f'(x) = 14x^6 + 15x^4 + 6 > 0$ for all x, so $f(x)$ is an increasing function; hence there is at most one root.

31 (a) $$\left(\frac{2x^3 - x}{x+2}\right)' = \frac{(x+2)(6x^2-1) - (2x^3-x)(1)}{(x+2)^2}$$

$$= \frac{4x^3 + 12x^2 - 2}{(x+2)^2}$$

(b) $\left(x^5\sqrt{1+3x}\right)'$

$$= x^5 \cdot \frac{3}{2\sqrt{1+3x}} + 5x^4\sqrt{1+3x}$$

(c) $$\left(\frac{(2x-1)^5}{7}\right)' = \frac{5}{7}(2x-1)^4(2)$$

$$= \frac{10}{7}(2x-1)^4$$

(d) $$\left(\sin^4\sqrt{x}\right)' = \left(4\sin^3\sqrt{x}\right)\left(\cos\sqrt{x}\right)\frac{1}{2\sqrt{x}}$$

$$= \frac{2}{\sqrt{x}}\sin^3\sqrt{x}\cos\sqrt{x}$$

(e) $(\cos(1/x^3))' = (-\sin(1/x)^3))(-3/x^4)$
$= (3/x^4)\sin(1/x^3)$

(f) $$\left(\tan\sqrt{1-x^2}\right)' = \left(\sec^2\sqrt{1-x^2}\right)\cdot\frac{-2x}{2\sqrt{1-x^2}}$$

$$= \frac{-x}{\sqrt{1-x^2}}\sec^2\sqrt{1-x^2}$$

33 The slope is given by $y' = 3x^2 - 18x + 15 = 3(x-3)^2 - 12$, so the minimum slope is -12 at $x = 3$.

35 (a) Let $f(x) = x^5$. Then $df = 5x^4\,dx$. For $x = 1$ and $dx = 0.002$ we have $(1.002)^5 = f(1 + 0.002) \approx f(1) + df = 1^5 + 5\cdot1^4(0.002) = 1 + 0.01 = 1.01$.

(b) Let $f(x) = x^3$. Then $df = 3x^2\,dx$. For $x = 1$ and $dx = -0.004$ we have $(0.996)^3 = f(1 - 0.004) \approx f(1) + df = 1^3 + 3\cdot1^2(-0.004) = 1 - 0.012 = 0.988$.

37 Since $x^2 + 2x = (x+1)^2 - 1$, it seems likely that $\sqrt{x^2+2x} \approx x + 1$ for large values of x. To prove this, note that $\lim_{x\to\infty}\left(\sqrt{x^2+2x} - (x+1)\right) =$

$$\lim_{x\to\infty}\frac{\left[\sqrt{x^2+2x} - (x+1)\right]\left[\sqrt{x^2+2x} + (x+1)\right]}{\sqrt{x^2+2x} + (x+1)}$$

$$= \lim_{x\to\infty}\frac{-1}{\sqrt{x^2+2x} + (x+1)} = 0.$$ Hence $y = x + 1$

is a tilted asymptote. Similarly, $y = -x$ is a tilted asymptote as $x \to -\infty$.

$$\lim_{x\to\infty} \frac{\left[\sqrt{x^2+2x}-(x+1)\right]\left[\sqrt{x^2+2x}+(x+1)\right]}{\sqrt{x^2+2x}+(x+1)}$$

$$= \lim_{x\to\infty} \frac{-1}{\sqrt{x^2+2x}+(x+1)} = 0.$$ Hence $y = x + 1$ is a tilted asymptote. Similarly, $y = -x$ is a tilted asymptote as $x \to -\infty$.

39 Since $\frac{dy}{dx} = 3y^2$, $\frac{d^2y}{dx^2} = \frac{d}{dx}\left(\frac{dy}{dx}\right) = \frac{d}{dx}(3y^2)$

$= 3\cdot 2y\cdot\frac{dy}{dx} = 3\cdot 2y\cdot 3y^2 = 18y^3$.

41 dy/dx is proportional to y^2; that is, $dy/dx = ky^2$ for some constant k. Thus $\frac{d^2y}{dx^2} = \frac{d}{dx}\left(\frac{dy}{dx}\right) = \frac{d}{dx}(ky^2)$

$= 2ky\cdot\frac{dy}{dx} = 2ky\cdot ky^2 = 2k^2y^3$; that is, $\frac{d^2y}{dx^2}$ is proportional to y^3.

43 $\frac{dy}{dx} = 3y^4$, so $\frac{d^2y}{dx^2} = \frac{d}{dx}\left(\frac{dy}{dx}\right) = \frac{d}{dx}(3y^4)$

$= 12y^3\cdot\frac{dy}{dx} = 12y^3\cdot 3y^4 = 36y^7$.

45 Let (x, y) be the first-quadrant corner of a rectangle inscribed in the ellipse $x^2/a^2 + y^2/b^2 = 1$. Then the rectangle has area $A = 4xy$, so $dA/dx = 4xy' + 4y = 0$, from which we obtain $y' = -y/x$. But $(x^2/a^2 + y^2/b^2)' = 2x/a^2 + 2yy'/b^2 = 0$, so $y' = -b^2x/(a^2y) = -y/x$ and $x^2/a^2 = y^2/b^2$. Hence $x^2/a^2 + y^2/b^2 = 2x^2/a^2 = 1$ or $x = a/\sqrt{2}$. Similarly, $y = b/\sqrt{2}$. The rectangle is therefore $\sqrt{2}a \times \sqrt{2}b$.

47 Let $f(x)$ be the square of the distance from (x, x^2) to $(3, 0)$; thus $f(x) = (x - 3)^2 + (x^2 - 0)^2 = x^4 + x^2 - 6x + 9$. Then $f'(x) = 4x^3 + 2x - 6 = (x - 1)(4x^2 + 4x + 6)$ changes sign from $-$ to $+$ at $x = 1$. The nearest point is $(1, 1)$.

49 (a) Let $f(x) = x^5 - 6x + 3$. Then $f'(x) = 5x^4 - 6$ is positive for $x < -\sqrt[4]{\frac{6}{5}}$, negative for $-\sqrt[4]{\frac{6}{5}} < x < \sqrt[4]{\frac{6}{5}}$, and positive for $x > \sqrt[4]{\frac{6}{5}}$. Since $\lim_{x\to-\infty} f(x) = -\infty$, $f\left(-\sqrt[4]{\frac{6}{5}}\right) = \frac{24}{5}\sqrt[4]{\frac{6}{5}} + 3 > 0$, $f\left(\sqrt[4]{\frac{6}{5}}\right) = 3 - \frac{24}{5}\sqrt[4]{\frac{6}{5}} < 0$, and $\lim_{x\to\infty} f(x) = \infty$, it follows that f has three roots, one in each of the intervals $\left(-\infty, -\sqrt[4]{\frac{6}{5}}\right)$, $\left(-\sqrt[4]{\frac{6}{5}}, \sqrt[4]{\frac{6}{5}}\right)$, and $\left(\sqrt[4]{\frac{6}{5}}, \infty\right)$.

51 Let $f(x) = x^3 - 3x - 3$. Then $f'(x) = 3x^2 - 3 = 3(x^2 - 1) = 3(x + 1)(x - 1)$ changes from $+$ to $-$ at $x = -1$ and from $-$ to $+$ at $x = 1$. So f is increasing on $(-\infty, -1)$, decreasing on $(-1, 1)$, and increasing on $(1, \infty)$. Since $f(-1) = -1$ and $\lim_{x\to-\infty} f(x) = -\infty$, there is no root in $(-\infty, -1)$. Also $f(1) = -5$, so while no root lies in $(-1, 1)$, exactly one root is in $(1, \infty)$ because $\lim_{x\to\infty} f(x) = \infty$. To estimate this root, apply Newton's method with $x_1 = 2$: $x_2 = x_1 - \frac{f(x_1)}{f'(x_1)} =$

53 $xy + e^y = e + 1$, where $x = 1$, $y = 1$, $\dot{x} = 2$, and $\ddot{x} = 3$ when $t = 0$. To find $\dot{y}$ when $t = 0$, we differentiate both sides with respect to t, obtaining $x\dot{y} + y\dot{x} + e^y\dot{y} = 0$. Hence $\dot{y} = \dfrac{-y\dot{x}}{x + e^y}$.

Substituting the given values, we have $\dot{y} = \dfrac{-1\cdot 2}{1 + e}$ $= -\dfrac{2}{1 + e}$. To obtain $\ddot{y}$, we differentiate again:

$\dfrac{d}{dt}(x\dot{y} + y\dot{x} + e^y\dot{y})$

$= x\ddot{y} + \dot{y}\dot{x} + y\ddot{x} + \dot{x}\dot{y} + e^y\ddot{y} + \dot{y}e^y\dot{y}$

$= (x + e^y)\ddot{y} + 2\dot{x}\dot{y} + y\ddot{x} + \dot{y}^2e^y$

$= (1 + e)\ddot{y} + 2\cdot 2\cdot\dfrac{-2}{1 + e} + 1\cdot 3 + \left(\dfrac{-2}{1 + e}\right)^2 e$

$= (1 + e)\ddot{y} - \dfrac{8}{1 + e} + 3 + \dfrac{4e}{(1 + e)^2} = 0$, so

$\ddot{y} = \dfrac{-(3e^2 + 2e - 5)}{(e + 1)^3}$.

55 If the cone has height h and radius r, then $a^2 = (h - a)^2 + r^2$. Hence $0 = 2(h - a) + 2r\dfrac{dr}{dh}$, so $\dfrac{dr}{dh} = \dfrac{a - h}{r}$. The volume is $V = \dfrac{1}{3}\pi r^2 h$, so $\dfrac{dV}{dh} = \dfrac{\pi}{3}\left(r^2 + 2rh\dfrac{dr}{dh}\right) = \dfrac{\pi}{3}(r^2 + 2h(a - h)) = \dfrac{\pi}{3}h(4a - 3h)$ changes sign from + to − when $h = (4/3)a$. Then $r = \dfrac{2\sqrt{2}}{3}a$ and $V = \dfrac{32\pi}{81}a^3$.

57 A cone of height h and radius r has slant height $l = \sqrt{r^2 + h^2}$. Recall that the area of the curved surface of a cone is πrl. Adding the area of the base πr^2, we obtain the total surface area of the cone: $A = \pi r^2 + \pi rl = \pi r^2 + \pi r\sqrt{r^2 + h^2}$. A is fixed, so $0 = \dfrac{dA}{dr} =$

$2\pi r + \pi r\cdot\dfrac{2r + 2h\dfrac{dh}{dr}}{2\sqrt{r^2 + h^2}} + \pi\sqrt{r^2 + h^2}$, so $0 =$

$2r + \dfrac{r^2 + rh\cdot\dfrac{dh}{dr}}{\sqrt{r^2 + h^2}} + \sqrt{r^2 + h^2}$. Multiply through by $\sqrt{r^2 + h^2}$ to obtain $0 =$

$2r\sqrt{r^2 + h^2} + r^2 + rh\cdot\dfrac{dh}{dr} + r^2 + h^2$, so $\dfrac{dh}{dr} =$

$-\dfrac{2}{h}\sqrt{r^2 + h^2} - \dfrac{2r}{h} - \dfrac{h}{r}$. We wish to maximize the volume, which is given by $V = \dfrac{1}{3}\pi r^2 h$, so $\dfrac{dV}{dr} =$

$\dfrac{1}{3}\pi r^2\cdot\dfrac{dh}{dr} + \dfrac{2}{3}\pi rh = \dfrac{1}{3}\pi\left(r^2\cdot\dfrac{dh}{dr} + 2rh\right)$, which equals 0 when $r^2\cdot\dfrac{dh}{dr} = -2rh$, so $\dfrac{dh}{dr} = -\dfrac{2h}{r}$.

Setting equal the two expressions for dh/dr produces $-\dfrac{2h}{r} = -\dfrac{2}{h}\sqrt{r^2 + h^2} - \dfrac{2r}{h} - \dfrac{h}{r}$, $-\dfrac{h}{r} =$

$-\dfrac{2}{h}\sqrt{r^2 + h^2} - \dfrac{2r}{h}$, $h^2 = 2r\sqrt{r^2 + h^2} + 2r^2$,

$h^2 - 2r^2 = 2r\sqrt{r^2 + h^2}$, $h^4 - 4h^2r^2 + 4r^4 =$

$4r^2(r^2 + h^2) = 4r^4 + 4r^2h^2$, $h^4 = 8r^2h^2$, $h^2 =$

$8r^2$, $h = \sqrt{8}r$.

59 (a) $y = 3 \sin t + 4 \cos t$, so $y' = 3 \cos t - 4 \sin t$, which equals 0 when $3 \cos t = 4 \sin t$, $3/4 = \tan t$, $\tan^{-1}(3/4) = t$. Note that t is the angle labeled in the accompanying diagram. Now we know that $y'' = -3 \sin t - 4 \cos t$, which, for $t = \tan^{-1}(3/4)$, is equal to $-3(3/5) - 4(4/5) = -5 < 0$, so $t = \tan^{-1}(3/4)$ provides a maximum. The maximum value of y is thus $y = 3(3/5) + 4(4/5) = 5$.

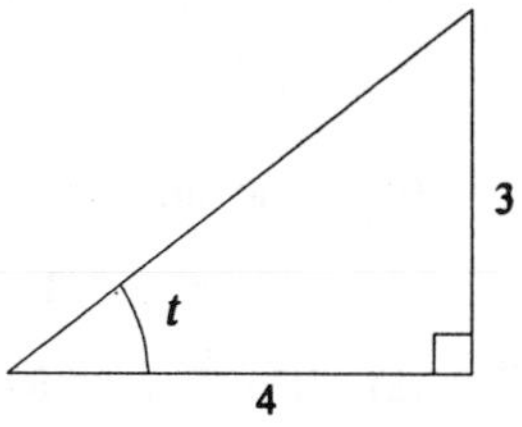

(b) Now we have $y = A \sin kt + B \cos kt$, and therefore $y' = Ak \cos kt - Bk \sin kt$, which equals 0 when $Ak \cos kt = Bk \sin kt$, so $A/B = \tan kt$ and $t = \dfrac{1}{k} \tan^{-1} \dfrac{A}{B}$; the angle kt is labeled in the diagram. As in (a), this provides a maximum. Its value is $y =$

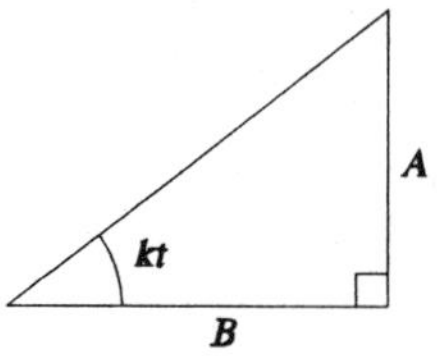

$$A \cdot \frac{A}{\sqrt{A^2 + B^2}} + B \cdot \frac{B}{\sqrt{A^2 + B^2}} = \frac{A^2 + B^2}{\sqrt{A^2 + B^2}}$$

$= \sqrt{A^2 + B^2}$. (Compare this to the result in (a).)

61 (a) Let $f(x) = x^3 + px + q$. $\lim_{x \to -\infty} f(x) = -\infty$, so there is a number $d_1 < 0$ for which $f(d_1) < 0$. Similarly, $\lim_{x \to \infty} f(x) = \infty$, so $f(d_2) > 0$ for some $d_2 > 0$. Now $f(d_1) < 0 < f(d_2)$, so by the intermediate-value theorem, there is a number c, $d_1 < c < d_2$, such that $f(c) = 0$. There is at least one real root. If there were two such roots, $f(c_1) = f(c_2) = 0$, $c_1 \neq c_2$, then Rolle's theorem would require the existence of a critical number between c_1 and c_2. However, $f'(x) = 3x^2 + p \geq p > 0$, so f' is never zero. Therefore the equation has precisely one real root.

(b) Suppose $4p^3 + 27q^2 < 0$. Then $4p^3 < -27q^2$ so $p < \left(\dfrac{-27q^2}{4}\right)^{1/3} = \dfrac{-3q^{2/3}}{2^{2/3}} < 0$, so $p < 0$. The critical points of $f(x) = x^3 + px + q$ occur when $f'(x) = 3x^2 + p = 0$; that is, when $x^2 = -p/3$, or $x = \pm\left(-\dfrac{p}{3}\right)^{1/2}$. Now $f''(x) = 6x$, so there is a maximum when $x = -(-p/3)^{1/2}$ (since $f''(x) < 0$) and a minimum when $x = (-p/3)^{1/2}$. Recall that $f(x) \to -\infty$ as $x \to -\infty$ and $f(x) \to \infty$ as $x \to \infty$. If we can show that $f\left(-\left(-\dfrac{p}{3}\right)^{1/2}\right) > 0$ and $f\left(\left(-\dfrac{p}{3}\right)^{1/2}\right) < 0$, then the intermediate-value theorem implies the existence of roots in $(-\infty, -(-p/3)^{1/2})$, $(-(-p/3)^{1/2}, (-p/3)^{1/2})$, and $((-p/3)^{1/2}, \infty)$. Since $f'(x)$ is always positive for x in the first or third interval, and always negative in the second, the root in each interval is unique. That would prove the existence of precisely three roots, as desired.

Now $f\left(\left(-\dfrac{p}{3}\right)^{1/2}\right) = \left(-\dfrac{p}{3}\right)^{3/2} + p\left(-\dfrac{p}{3}\right)^{1/2} + q$

$$= \left(-\frac{p}{3}\right)^{1/2}\left(-\frac{p}{3} + p\right) + q = \left(-\frac{p}{3}\right)^{1/2}\left(\frac{2p}{3}\right) + q$$

$$= \left(\frac{-4p^3}{27}\right)^{1/2} + q > \left(\frac{27q^2}{27}\right)^{1/2} + q =$$

$$|q| + q > 0 \text{ and } f\left(-\left(-\frac{p}{3}\right)^{1/2}\right) =$$

$$-\left(-\frac{p}{3}\right)^{3/2} - p\left(-\frac{p}{3}\right)^{1/2} + q =$$

$$\left(-\frac{p}{3}\right)^{1/2}\left(\frac{p}{3} - p\right) + q = \left(-\frac{p}{3}\right)^{1/2}\left(\frac{-2p}{3}\right) + q =$$

$$-\left(\frac{-4p^3}{27}\right)^{1/2} + q < -\left(\frac{27q^2}{27}\right)^{2} + q = q - |q|$$

≤ 0, as required. As explained above, it now follows that the equation $x^3 + px + q = 0$ has precisely three real roots.

63 $$\lim_{dx \to 0} \frac{\Delta y - dy}{dx} = \lim_{dx \to 0} \frac{f(x + dx) - f(x) - f'(x)\,dx}{dx}$$

$$= \lim_{dx \to 0} \left[\frac{f(x + dx) - f(x)}{dx} - f'(x)\right] = f'(x) - f'(x)$$

$= 0$

65 (a) Nothing, since $f(-1/2)$ is not defined.

(b) $$f'(x) = \frac{x^2 + x + 4}{(x + \frac{1}{2})^2} = 1 + \frac{15/4}{\left(x + \frac{1}{2}\right)^2} > 0$$

for all $x \neq -1/2$; $f'(x)$ is never 0.

67 (a) By inspection of the graph, relative maxima occur for $x = 2$ and $x = 11/2$.

(b) Relative minima occur for $x = 6$, $x = 4$, and $x \approx -3/4$.

(c) In addition to the points $-3/4$, 4, 11/2, and 6 from (a) and (b), $x = 1$ is also a critical point. (The point $x = 2$ from (a) does not belong in this list because it is a *cusp*—a sharp corner, rather than a point where the first derivative is 0.)

(d) The graph rises as $x \to \infty$ or $x \to -\infty$, so there is no global maximum.

(e) The smallest of the minima cited in (b) occurs at $x = -3/4$, so that is the global minimum.

69 (a) The density $\delta(x)$ is just the derivative of the mass function $18x^2 - x^3$, so $\delta(x) = (18x^2 - x^3)' = 36x - 3x^2$.

(b) $\delta'(x) = 36 - 6x$, which equals 0 when $x = 6$, so 6 is a critical number of the density function. $\delta''(x) = -6 < 0$, so $x = 6$ provides a maximum. The density is greatest 6 cm from the left-hand end.

5 The Definite Integral

5.1 Estimates in Four Problems

1 (a)

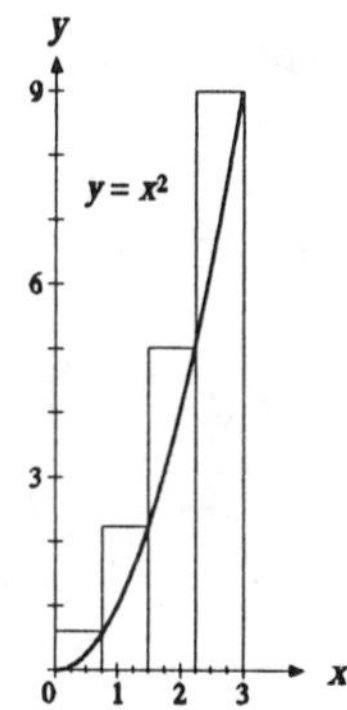

(c) The total area of the four rectangles is

$$\frac{3}{4}\left(\frac{3}{4}\right)^2 + \frac{3}{4}\left(\frac{6}{4}\right)^2 + \frac{3}{4}\left(\frac{9}{4}\right)^2 + \frac{3}{4}\left(\frac{12}{4}\right)^2$$

$$= \frac{3}{64}(3^2 + 6^2 + 9^2 + 12^2)$$

$$= \frac{27}{64}(1^2 + 2^2 + 3^2 + 4^2) = \frac{27}{64}\cdot 30 = \frac{810}{64}$$

$= 12.65625.$

3 (a)

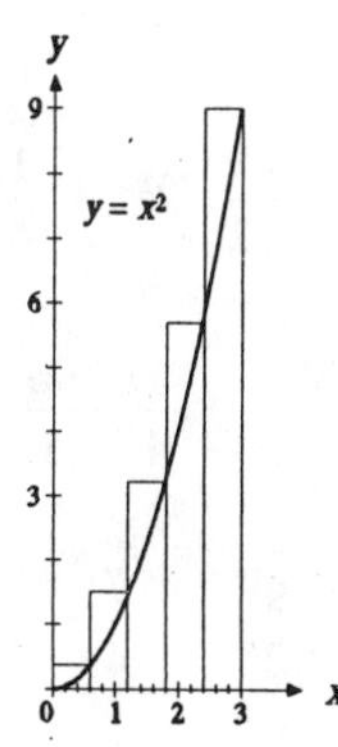

From the figure, the width of each rectangle is 3/5, and the height of each is determined by the value of the right endpoint squared, as displayed on the y axis. Thus the area of the five rectangles equals

$$\frac{3}{5}\left(\frac{3}{5}\right)^2 + \frac{3}{5}\left(\frac{6}{5}\right)^2 + \frac{3}{5}\left(\frac{9}{5}\right)^2 + \frac{3}{5}\left(\frac{12}{5}\right)^2 + \frac{3}{5}\left(\frac{15}{5}\right)^2$$

$$= \left(\frac{3}{5}\right)^3(1^2 + 2^2 + 3^2 + 4^2 + 5^2) = \frac{1485}{125}$$

$= 11.88.$

(b)

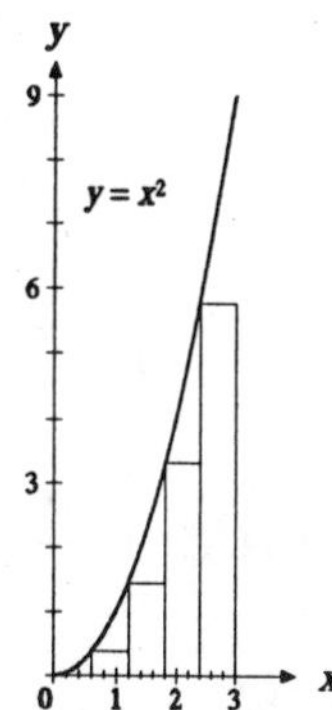

This time the height of each rectangle is determined by the value of the left endpoint squared, as indicated on the y axis, and the width of each rectangle remains 3/5. Hence the area of the five rectangles equals

$$\frac{3}{5}\left(\frac{0}{5}\right)^2 + \frac{3}{5}\left(\frac{3}{5}\right)^2 + \frac{3}{5}\left(\frac{6}{5}\right)^2 + \frac{3}{5}\left(\frac{9}{5}\right)^2 + \frac{3}{5}\left(\frac{12}{5}\right)^2$$

$$= \left(\frac{3}{5}\right)^3(0^2 + 1^2 + 2^2 + 3^2 + 4^2) = \frac{810}{125} =$$

6.48. We conclude that the area in Problem 1

is certainly less than 11.88 but larger than 6.48.

5 (a) The width of each section is $3/15 = 1/5$.

(b)

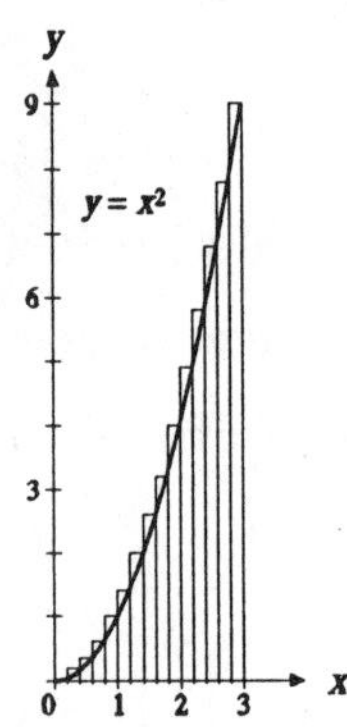

(c) The height of the smallest rectangle is $(3/15)^2$. The largest rectangle has height $(45/15)^2$.

(d) The total area of the estimate is

$$\frac{3}{15}\left(\frac{3}{15}\right)^2 + \frac{3}{15}\left(\frac{6}{15}\right)^2 + \frac{3}{15}\left(\frac{9}{15}\right)^2 + \dots + \frac{3}{15}\left(\frac{45}{15}\right)^2$$

$$= \left(\frac{3}{15}\right)^3(1^2 + 2^2 + 3^2 + \dots + 15^2)$$

$$= \frac{27}{3375}\cdot 1240 = \frac{33480}{3375} = 9.92.$$

7 The width of each rectangle is $\frac{2-1}{5} = \frac{1}{5}$, and the height of each rectangle equals the right endpoint squared. So the overestimate equals

$$\frac{1}{5}\left(\frac{6}{5}\right)^2 + \frac{1}{5}\left(\frac{7}{5}\right)^2 + \frac{1}{5}\left(\frac{8}{5}\right)^2 + \frac{1}{5}\left(\frac{9}{5}\right)^2 + \frac{1}{5}\left(\frac{10}{5}\right)^2 =$$

$$\frac{1}{5^3}(6^2 + 7^2 + 8^2 + 9^2 + 10^2) = \frac{330}{125} = 2.64.$$

9 (c),(d)

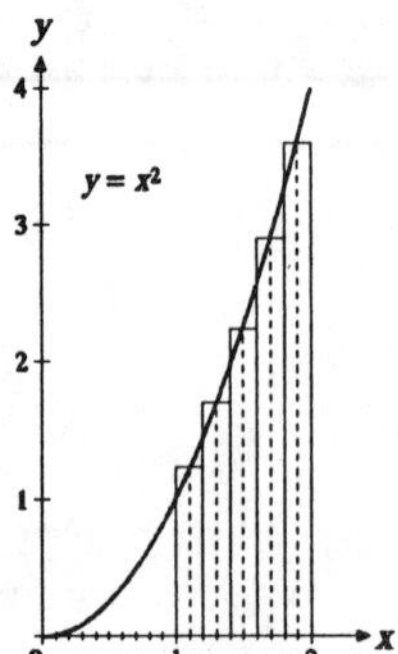

(e) The total area equals

$$\frac{1}{5}\left(\frac{11}{10}\right)^2 + \frac{1}{5}\left(\frac{13}{10}\right)^2 + \frac{1}{5}\left(\frac{15}{10}\right)^2 + \frac{1}{5}\left(\frac{17}{10}\right)^2 + \frac{1}{5}\left(\frac{19}{10}\right)^2$$

$$= \frac{1}{5\cdot 10^2}(11^2 + 13^2 + 15^2 + 17^2 + 19^2) = \frac{1165}{500}$$

$$= 2.33.$$

11

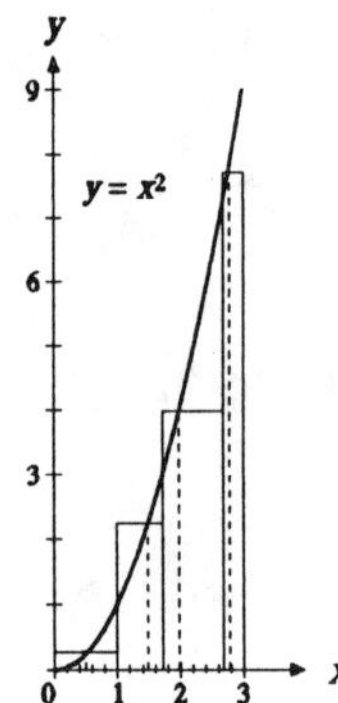

From the figure, we see that the area of the estimate is the sum of Width × Height for each rectangle:

$$1\cdot\left(\frac{1}{2}\right)^2 + \frac{2}{3}\cdot\left(\frac{3}{2}\right)^2 + \frac{13}{12}\cdot(2)^2 + \frac{1}{4}\cdot\left(\frac{14}{5}\right)^2$$

$$= \frac{1}{4} + \frac{2}{3}\cdot\frac{9}{4} + \frac{13}{12}\cdot 4 + \frac{1}{4}\cdot\frac{196}{25}$$

$$= \frac{1}{4} + \frac{6}{4} + \frac{13}{3} + \frac{49}{25} = \frac{7}{4} + \frac{13}{3} + \frac{49}{25}$$

$$= \frac{73}{12} + \frac{49}{25} = \frac{2413}{300} = 8.04\overline{3}.$$

13 (c) The estimate of the total area equals the sum of Width × Height for each rectangle:

$$\frac{1}{5}\left(\frac{5}{5}\right) + \frac{1}{5}\left(\frac{5}{6}\right) + \frac{1}{5}\left(\frac{5}{7}\right) + \frac{1}{5}\left(\frac{5}{8}\right) + \frac{1}{5}\left(\frac{5}{9}\right)$$

$$= \frac{1}{5} + \frac{1}{6} + \frac{1}{7} + \frac{1}{8} + \frac{1}{9} = \frac{1879}{2520}$$

$$\approx 0.7456.$$

15 (b)

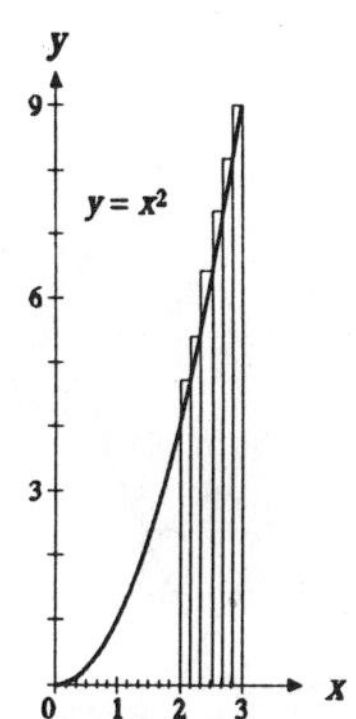

(d) The estimate of total area equals the sum of Width × Height for each rectangle:

$$\frac{1}{6}\left(\frac{13}{6}\right)^2 + \frac{1}{6}\left(\frac{14}{6}\right)^2 + \frac{1}{6}\left(\frac{15}{6}\right)^2 + \frac{1}{6}\left(\frac{16}{6}\right)^2 + \frac{1}{6}\left(\frac{17}{6}\right)^2 + \frac{1}{6}\left(\frac{18}{6}\right)^2$$

$$= \frac{1}{6^3}(13^2 + 14^2 + 15^2 + 16^2 + 17^2 + 18^2)$$

$$= \frac{1459}{216} \approx 6.7546.$$

17 (b)

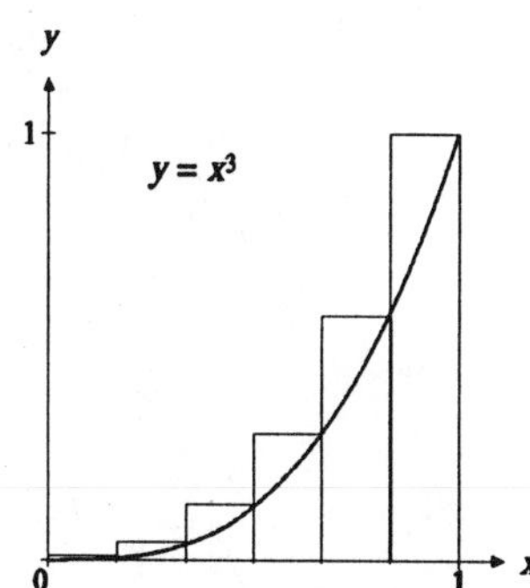

(d) The total area equals $\frac{1}{6}\left(\frac{1}{6}\right)^3 + \frac{1}{6}\left(\frac{2}{6}\right)^3 +$

$$\frac{1}{6}\left(\frac{3}{6}\right)^3 + \frac{1}{6}\left(\frac{4}{6}\right)^3 + \frac{1}{6}\left(\frac{5}{6}\right)^3 + \frac{1}{6}\left(\frac{6}{6}\right)^3$$

$$= \frac{1}{6^4}(1^3 + 2^3 + 3^3 + 4^3 + 5^3 + 6^3)$$

$$= \frac{441}{1296} \approx 0.34028.$$

19 (b)

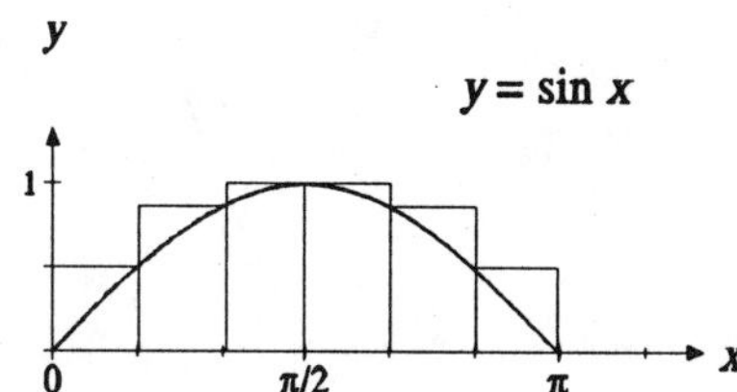

(d) The total area equals the sum of six terms:

$$\frac{\pi}{6}\sin\frac{\pi}{6} + \frac{\pi}{6}\sin\frac{\pi}{3} + \frac{\pi}{6}\sin\frac{\pi}{2} +$$

$$\frac{\pi}{6}\sin\frac{\pi}{2} + \frac{\pi}{6}\sin\frac{2\pi}{3} + \frac{\pi}{6}\sin\frac{5\pi}{6}$$

$$= \frac{2\pi}{6}\sin\frac{\pi}{6} + \frac{2\pi}{6}\sin\frac{\pi}{3} + \frac{2\pi}{6}\sin\frac{\pi}{2}$$

$$= \frac{\pi}{3}\left(\sin\frac{\pi}{6} + \sin\frac{\pi}{3} + \sin\frac{\pi}{2}\right)$$

$$= \frac{\pi}{3}\left(\frac{1}{2} + \frac{\sqrt{3}}{2} + 1\right) = \frac{(3 + \sqrt{3})\pi}{6} \approx 2.4777.$$

21 (a)

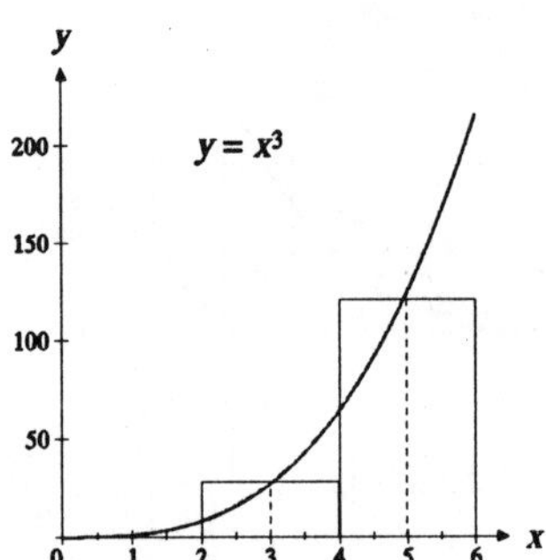

From the figure, we see that the width of each rectangle is 2, while the heights are 1^3, 3^3, and 5^3. Hence the total area equals $2\cdot1^3 + 2\cdot3^3 + 2\cdot5^3 = 2(1^3 + 3^3 + 5^3)$ $= 306$.

(b)

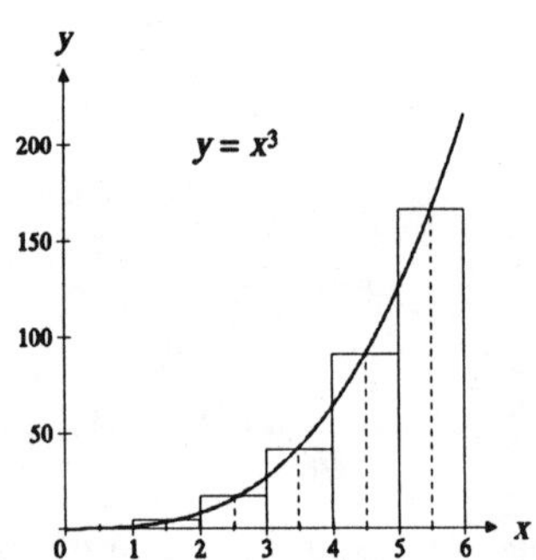

From the figure, the total area equals

$$1\cdot\left(\frac{1}{2}\right)^3 + 1\cdot\left(\frac{3}{2}\right)^3 + 1\cdot\left(\frac{5}{2}\right)^3 + 1\cdot\left(\frac{9}{2}\right)^3 + 1\cdot\left(\frac{11}{2}\right)^3$$

$$= \frac{1}{2^3}(1^3 + 3^3 + 5^3 + 7^3 + 9^3 + 11^3) = \frac{2556}{8}$$

$$= 319.5.$$

(c)

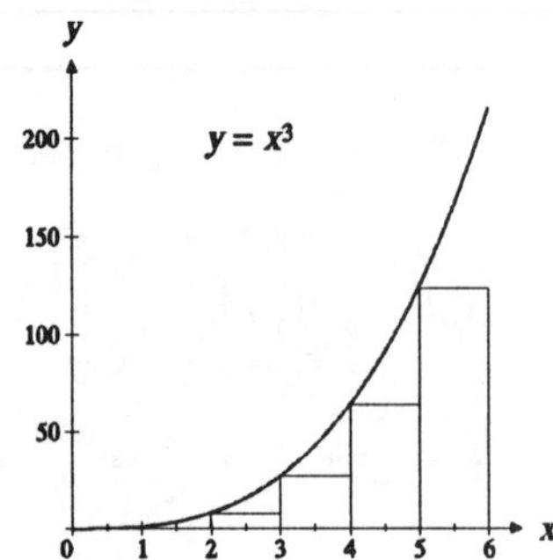

From the figure, the total area equals $1\cdot0^3 + 1\cdot1^3 + 1\cdot2^3 + 1\cdot3^3 + 1\cdot4^3 + 1\cdot5^3 = 225$.

(d)

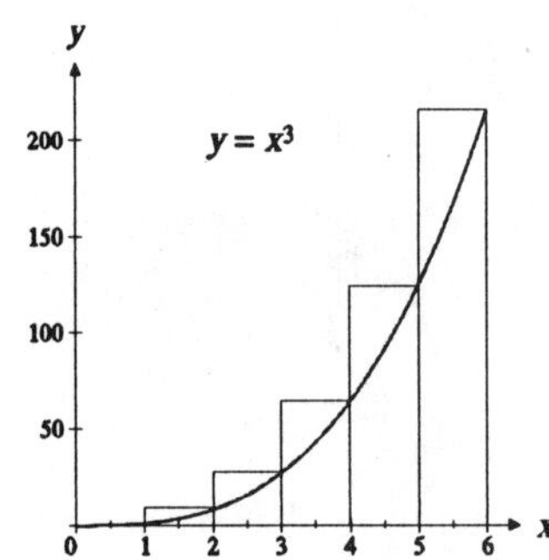

From the figure, the total area equals $1\cdot1^3 + 1\cdot2^3 + 1\cdot3^3 + 1\cdot4^3 + 1\cdot5^3 + 1\cdot6^3 = 441$.

23 (a) The width of each rectangle is again 3/50. This time, the height of each rectangle corresponds to the value of the right endpoint squared. The total area equals

$$\frac{3}{50}\left(\frac{3}{50}\right)^2 + \frac{3}{50}\left(\frac{6}{50}\right)^2 + \frac{3}{50}\left(\frac{9}{50}\right)^2 + \cdots + \frac{3}{50}\left(\frac{150}{150}\right)^2$$

$$= \left(\frac{3}{50}\right)^3(1^2 + 2^2 + 3^2 + \cdots + 50^2) = \frac{46359}{5000}$$

$$= 9.2718.$$

(b) The width of each rectangle is $\frac{3 - 0}{50} = \frac{3}{50}$.

The height of each rectangle is the value of

the left endpoint squared. The total area thus equals

$$\frac{3}{50}\left(\frac{0}{50}\right)^2 + \frac{3}{50}\left(\frac{3}{50}\right)^2 + \frac{3}{50}\left(\frac{6}{50}\right)^2 + \cdots + \frac{3}{50}\left(\frac{147}{150}\right)^2$$

$$= \left(\frac{3}{50}\right)^3 (0^2 + 1^2 + 2^2 + \cdots + 49^2) = \frac{43659}{5000}$$

$= 8.7318.$

25 The linear length of each section is $\frac{3 \text{ cm}}{5} =$

$\frac{3}{5}$ cm. The density at a distance x cm is x^2 g/cm.

The mass is length times density.

(a) Using the density at a midpoint of each section, the mass is approximately

$$\frac{3}{5}\left(\frac{3}{10}\right)^2 + \frac{3}{5}\left(\frac{9}{10}\right)^2 + \frac{3}{5}\left(\frac{15}{10}\right)^2 + \frac{3}{5}\left(\frac{21}{10}\right)^2 + \frac{3}{5}\left(\frac{27}{10}\right)^2$$

$$= \frac{3}{5}\cdot\left(\frac{3}{10}\right)^2 (1^2 + 3^2 + 5^2 + 7^2 + 9^2) = \frac{4455}{500} =$$

8.91 grams.

(b) Using the density at the right endpoint of each section, the mass is approximately

$$\frac{3}{5}\left(\frac{3}{5}\right)^2 + \frac{3}{5}\left(\frac{6}{5}\right)^2 + \frac{3}{5}\left(\frac{9}{5}\right)^2 + \frac{3}{5}\left(\frac{12}{5}\right)^2 + \frac{3}{5}\left(\frac{15}{5}\right)^2$$

$$= \left(\frac{3}{5}\right)^3 (1^2 + 2^2 + 3^2 + 4^2 + 5^2) = \frac{297}{25}$$

$= 11.88$ grams

(c) Using the density at the left endpoint of each section, the mass is approximately

$$\frac{3}{5}\cdot 0^2 + \frac{3}{5}\left(\frac{3}{5}\right)^2 + \frac{3}{5}\left(\frac{6}{5}\right)^2 + \frac{3}{5}\left(\frac{9}{5}\right)^2 + \frac{3}{5}\left(\frac{12}{5}\right)^2$$

$$= \left(\frac{3}{5}\right)^3 (0^2 + 1^2 + 2^2 + 3^2 + 4^2) = \frac{162}{25}$$

$= 6.48$ grams.

(d) We conclude that the mass in Problem 2 is less than 11.88 grams but larger than 6.48 grams.

27 (a)

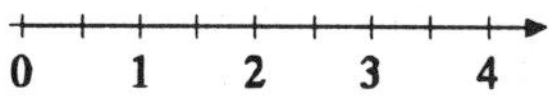

(b) To make an estimate that is too large, we assume that the electron is traveling throughout each section at the velocity given by the cube of the right endpoint of the section. Since distance is the product of velocity and time, we have the overestimate

$$\frac{4}{8}\left(\frac{4}{8}\right)^3 + \frac{4}{8}\left(\frac{8}{8}\right)^3 + \frac{4}{8}\left(\frac{12}{8}\right)^3 + \cdots + \frac{4}{8}\left(\frac{32}{8}\right)^3 =$$

$$\left(\frac{4}{8}\right)^4 (1^3 + 2^3 + 3^3 + \cdots + 8^3) = 81 \text{ km.}$$

(c) To make an estimate that is too small, we assume that the electron is traveling throughout each section at the velocity given by the cube of the left endpoint of the section. Hence distance is greater than

$$\frac{4}{8}\left(\frac{0}{8}\right)^3 + \frac{4}{8}\left(\frac{4}{8}\right)^3 + \frac{4}{8}\left(\frac{8}{8}\right)^3 + \cdots + \frac{4}{8}\left(\frac{28}{8}\right)^3 =$$

$$\left(\frac{4}{8}\right)^4 (0^3 + 1^3 + 2^3 + \cdots + 7^3) = 49 \text{ km.}$$

29 Dividing the 3-year period into six half-year intervals and approximating the rate of profit during each half-year by the rate of profit in the middle of that interval, total profit is approximately

$$\left(\frac{1}{4}\right)^2\left(\frac{1}{2}\right) + \left(\frac{3}{4}\right)^2\left(\frac{1}{2}\right) + \left(\frac{5}{4}\right)^2\left(\frac{1}{2}\right) + \left(\frac{7}{4}\right)^2\left(\frac{1}{2}\right) +$$

$$\left(\frac{9}{4}\right)^2\left(\frac{1}{2}\right) + \left(\frac{11}{4}\right)^2\left(\frac{1}{2}\right)$$

$$= \frac{1}{32}(1^2 + 3^2 + 5^2 + 7^2 + 9^2 + 11^2) = \frac{143}{16}$$

$= 8.9375$ million dollars.

31 First, partition the time interval [0, 10] minutes into x subintervals. Then find the midpoint of each subinterval, and evaluate 2^{-t} at each midpoint. This gives an estimate of the rate of leakage over each subinterval. Multiplying each estimate by its corresponding subinterval of time yields the approximate amount of oil that has leaked out during that subinterval. Summing up all these estimates produces an estimate of the total amount of oil that has leaked out of the tank in 10 minutes.

33 One possible approach: Divide [1, 2] into 5 sections and use midpoints. Then the area is approximately $\frac{1}{5}\left(\frac{2^{11/10}}{11/10}\right) + \frac{1}{5}\left(\frac{2^{13/10}}{13/10}\right) +$

$$\frac{1}{5}\left(\frac{2^{15/10}}{15/10}\right) + \frac{1}{5}\left(\frac{2^{17/10}}{17/10}\right) + \frac{1}{5}\left(\frac{2^{19/10}}{19/10}\right) =$$

$$4\left(\frac{1}{11}2^{1/10} + \frac{1}{13}2^{3/10} + \frac{1}{15}2^{5/10} + \frac{1}{17}2^{7/10} + \frac{1}{19}2^{9/10}\right)$$

≈ 1.9208. (A calculator gives 1.9224 for the area, so our result appears to be a good approximation.)

35 Many answers are possible, depending on your choices. In the following figure, left endpoints were used for the lower left corners of the rectangles on the curve $y = x^2$ and right endpoints were used for the upper right corners on the line $y = x$.

(a),(b)

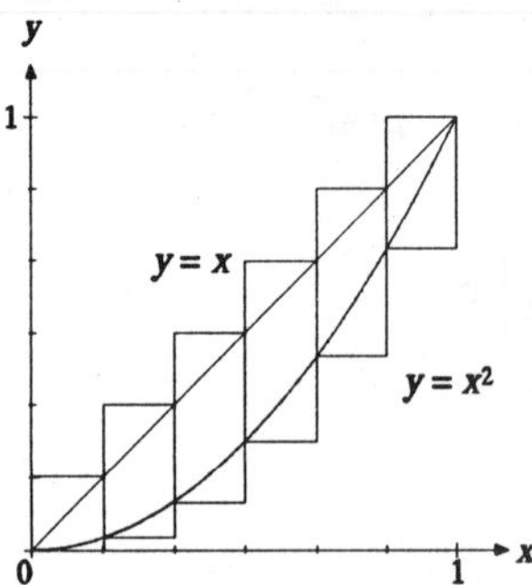

(d) The area of the six rectangles equals

$$\frac{1}{6}\left(\frac{1}{6} - 0^2\right) + \frac{1}{6}\left(\frac{2}{6} - \left(\frac{1}{6}\right)^2\right) + \frac{1}{6}\left(\frac{3}{6} - \left(\frac{2}{6}\right)^2\right) +$$

$$\frac{1}{6}\left(\frac{4}{6} - \left(\frac{3}{6}\right)^2\right) + \frac{1}{6}\left(\frac{5}{6} - \left(\frac{4}{6}\right)^2\right) + \frac{1}{6}\left(1 - \left(\frac{5}{6}\right)^2\right)$$

$$= \frac{1}{6}\left(\frac{6}{36} + \frac{11}{36} + \frac{14}{36} + \frac{15}{36} + \frac{14}{36} + \frac{11}{36}\right) = \frac{71}{216}$$

≈ 0.33, which overestimates the desired area.

37 (a) Our overestimate is $B(n) = \frac{3}{n}\left(\left(\frac{3}{n}\right)^2 + \left(\frac{6}{n}\right)^2 + \left(\frac{9}{n}\right)^2 + \cdots + \left(\frac{3n}{n}\right)^2\right)$. Our underestimate is $C(n) = \frac{3}{n}\left(\left(\frac{0}{n}\right)^2 + \left(\frac{3}{n}\right)^2 + \left(\frac{6}{n}\right)^2 + \cdots + \left(\frac{3(n-1)}{n}\right)^2\right)$. The difference between $B(n)$ and $C(n)$ involves much cancellation:

$$B(n) - C(n) = \frac{3}{n}\left(\frac{3n}{n}\right)^2 = \frac{27}{n}.$$

(b) We want $B(n) - C(n) \le 0.01$. Then $27/n \le 0.01$, and we should choose $n \ge 2700$.

39 (a) Since $(x^4)' = 4x^3$, a logical guess for an antiderivative of $(x + 2)^3$ is $\frac{(x+2)^4}{4}$; then

$$\left(\frac{(x+2)^4}{4}\right)' = (x+2)^3,$$ as desired. In fact,

any function of the form $\frac{(x+2)^4}{4} + C$,

where C is a constant, has $(x + 2)^3$ as its derivative.

(b) Note that $(x^2 + 1)^2 = x^4 + 2x^2 + 1$. It is easy to see that any function of the form

$\frac{x^5}{5} + \frac{2}{3}x^3 + x + C$, where C is a constant,

has $(x^2 + 1)^2$ as its derivative.

(c) Since $(\cos x^2)' = -2x \sin x^2$, division by -2 and addition of an arbitrary constant C yield

$$\left(-\frac{1}{2}\cos x^2 + C\right)' = x \sin x^2.$$

(d) Since $(x^4)' = 4x^3$ and $\left(\frac{1}{x^2}\right)' = -\frac{2}{x^3}$, it

follows immediately that $\left(\frac{x^4}{4} - \frac{1}{2x^2} + C\right)' =$

$x^3 + \frac{1}{x^3}$. Again, C is any constant.

(e) Since $\left(\sqrt{x}\right)' = \frac{1}{2\sqrt{x}}$, it is clear that

$(2\sqrt{x} + C)' = \frac{1}{\sqrt{x}}$, where C is any constant.

5.2 Summation Notation and Approximating Sums

1 (a) $\sum_{i=1}^{3} i = 1 + 2 + 3 = 6$

(b) $\sum_{i=1}^{4} 2i = 2 + 4 + 6 + 8 = 20$

(c) $\sum_{d=1}^{3} d^2 = 1^2 + 2^2 + 3^2 = 1 + 4 + 9 = 14$

3 (a) $\sum_{i=1}^{4} 1^i = 1^1 + 1^2 + 1^3 + 1^4 = 4$

(b) $\sum_{k=2}^{6} (-1)^k$

$= (-1)^2 + (-1)^3 + (-1)^4 + (-1)^5 + (-1)^6$

$= 1 - 1 + 1 - 1 + 1 = 1$

(c) $\sum_{j=1}^{150} 3 = (3)(150) = 450$, since $\sum_{i=1}^{n} c = cn$.

5 (a) $\sum_{k=1}^{100} 2^k$

(b) $\sum_{k=3}^{7} x^k$

(c) $\sum_{k=3}^{102} \frac{1}{k}$

7 (a) $\sum_{i=1}^{3} x_{i-1}^2(x_i - x_{i-1})$

(b) $\sum_{i=1}^{3} x_i^2(x_i - x_{i-1})$

9 (a) $\sum_{i=1}^{100} (2^i - 2^{i-1}) = (2^1 - 2^0) + (2^2 - 2^1) + \ldots$

$+ (2^{99} - 2^{98}) + (2^{100} - 2^{99})$. All but two terms cancel, leaving $-2^0 + 2^{100} = 2^{100} - 1$.

(b) $\sum_{i=2}^{100}\left(\frac{1}{i} - \frac{1}{i-1}\right) = \left(\frac{1}{2} - \frac{1}{1}\right) + \left(\frac{1}{3} - \frac{1}{2}\right) +$

$\cdots + \left(\frac{1}{99} - \frac{1}{98}\right) + \left(\frac{1}{100} - \frac{1}{99}\right) =$

$-1 + \frac{1}{100} = -\frac{99}{100}$

(c) $\sum_{i=1}^{50}\left(\frac{1}{2i+1} - \frac{1}{2(i-1)+1}\right) = \left(\frac{1}{3} - \frac{1}{1}\right) +$

$\left(\frac{1}{5} - \frac{1}{3}\right) + \cdots + \left(\frac{1}{99} - \frac{1}{97}\right) +$

$\left(\frac{1}{101} - \frac{1}{99}\right) = \frac{1}{101} - 1 = -\frac{100}{101}$

11 (a) Each summation expands to $a_1 + a_2 + a_3$. Hence the three are equal.

(b) $\sum_{i=1}^{3}(a_i + 4) = (a_1 + 4) + (a_2 + 4) + (a_3 + 4)$

$= (4 + 4 + 4) + (a_1 + a_2 + a_3)$

$= 12 + \sum_{j=2}^{4} a_{j-1}$

In Exercises 13 through 17, $x_0 = 1$, $x_1 = \frac{3}{2}$, $x_2 = 2$, $x_3 = \frac{5}{2}$, and $x_4 = 3$. Thus $x_i - x_{i-1} = \frac{1}{2}$ for $i = 1, 2, 3, 4$.

13 $\sum_{i=1}^{n} f(c_i)(x_i - x_{i-1}) = \sum_{i=1}^{4} (3x_i)\frac{1}{2} = \frac{3}{2}\sum_{i=1}^{4} x_i$

$= \frac{3}{2}\left(\frac{3}{2} + 2 + \frac{5}{2} + 3\right) = \frac{27}{2} = 13.5$

15 $\sum_{i=1}^{n} f(c_i)(x_i - x_{i-1}) = \sum_{i=1}^{4} (5)\frac{1}{2} = \sum_{i=1}^{4} \frac{5}{2} = 4\cdot\frac{5}{2}$

$= 10$

17 $\sum_{i=1}^{n} f(c_i)(x_i - x_{i-1}) = \sum_{i=1}^{4} \frac{1}{c_i}\cdot\frac{1}{2}$

$= \left(\frac{1}{1.25}\right)\frac{1}{2} + \left(\frac{1}{1.8}\right)\frac{1}{2} + \left(\frac{1}{2.2}\right)\frac{1}{2} + \left(\frac{1}{3}\right)\frac{1}{2}$

$= \frac{1}{2}\left(\frac{4}{5} + \frac{5}{9} + \frac{5}{11} + \frac{1}{3}\right) = \frac{1061}{990} \approx 1.07$

19 (a) $\sum_{i=1}^{100} i = \frac{100(100 + 1)}{2} = 5050$

(b) $\sum_{j=1}^{1000} 3j = 3\sum_{j=1}^{1000} j = 3\cdot\frac{1000(1000 + 1)}{2}$

$= 1{,}501{,}500$

21 We have $\sum_{i=1}^{n} [(i + 1)^4 - i^4] = (n + 1)^4 - 1$, so

$\sum_{i=1}^{n} (4i^3 + 6i^2 + 4i + 1) = n^4 + 4n^3 + 6n^2 + 4n.$

The left side of this equation equals

$4\sum_{i=1}^{n} i^3 + 6\sum_{i=1}^{n} i^2 + 4\sum_{i=1}^{n} i + \sum_{i=1}^{n} 1 =$

$4\sum_{i=1}^{n} i^3 + 6\left(\frac{n^3}{3} + \frac{n^2}{2} + \frac{n}{6}\right) + 4\left(\frac{n^2}{2} + \frac{n}{2}\right) + n$

$= 4\sum_{i=1}^{n} i^3 + 2n^3 + 5n^2 + 4n$. Hence $4\sum_{i=1}^{n} i^3 =$

$(n^4 + 4n^3 + 6n^2 + 4n) - (2n^3 + 5n^2 + 4n)$

$= n^4 + 2n^3 + n^2$. Dividing by 4 gives the desired result.

23 Since $f(x) = 11$ for all x, $f(c_i) = 11$ for any choice of c_i. Then $\sum_{i=1}^{n} f(c_i)(x_i - x_{i-1}) = \sum_{i=1}^{n} 11(x_i - x_{i-1})$

$= 11\sum_{i=1}^{n}(x_i - x_{i-1}) = 11(x_n - x_0) = 11(b - a).$

The last sum is evaluated as a telescoping sum.

25 $\sum_{i=1}^{n} f(c_i)(x_i - x_{i-1}) = \sqrt{1}(3 - 1) + \sqrt{4}(5 - 3)$

$= 1(2) + 2(2) = 6$

27 $\sum_{i=1}^{n} f(c_i)(x_i - x_{i-1}) =$

$\left(\frac{1}{1}\right)(0.25) + \left(\frac{1}{1.25}\right)(0.25) + \left(\frac{1}{1.6}\right)(0.25) + \left(\frac{1}{2}\right)(0.25)$

$= \frac{1}{4}\left(1 + \frac{4}{5} + \frac{5}{8} + \frac{1}{2}\right) = \frac{117}{160} = 0.73125$

29 Since $k \cdot 1000^2 = 3$, $k = 3 \times 10^{-6}$. The time required to evaluate $f(n)$ for all n from 1 to N is

$\sum_{i=1}^{N} kn^2 = k\left(\frac{N^3}{3} + \frac{N^2}{2} + \frac{N}{6}\right) \approx \frac{kN^3}{3}$ seconds for large N. In particular, for $N = 10^6$ the time needed is about $\frac{3 \cdot 10^{-6} \cdot (10^6)^3}{3} = 10^{12}$ seconds $\approx 32{,}000$ years.

31 (a) The appropriate Riemann sum is

$\sum_{i=1}^{100} f(c_i)(x_1 - x_{i-1}) = \sum_{i=1}^{100} x_i^2 \cdot \left(\frac{3}{100}\right)$. Now,

$\sum_{i=1}^{100} x_i^2\left(\frac{3}{100}\right) = \frac{3}{100}\sum_{i=1}^{100} x_i^2 = \frac{3}{100}\left(\left(\frac{3}{100}\right)^2 + \left(\frac{6}{100}\right)^2\right.$

$\left. + \left(\frac{9}{100}\right)^2 + \cdots + \left(\frac{300}{100}\right)^2\right)$

$= \left(\frac{3}{100}\right)^3(1^2 + 2^2 + \cdots + 100^2)$

$= \left(\frac{3}{100}\right)^3 \sum_{i=1}^{100} i^2 = \left(\frac{3}{100}\right)^3 \cdot \frac{100 \cdot 101 \cdot 201}{6}$

$= \frac{182709}{20000} = 9.13545.$

(b) The appropriate Riemann sum is

$\sum_{i=1}^{100} f(c_i)(x_i - x_{i-1}) = \sum_{i=1}^{100} x_{i-1}^2\left(\frac{3}{100}\right)$. Now,

$\sum_{i=1}^{100} x_{i-1}^2\left(\frac{3}{100}\right) = \frac{3}{100}\sum_{i=1}^{100} x_{i-1}^2 = \frac{3}{100}\left(\left(\frac{0}{100}\right)^2\right.$

$\left. + \left(\frac{3}{100}\right)^2 + \left(\frac{6}{100}\right)^2 + \cdots + \left(\frac{297}{100}\right)^2\right) =$

$\left(\frac{3}{100}\right)^3 \sum_{i=1}^{100} (i - 1)^2 = \left(\frac{3}{100}\right)^3 \cdot \frac{99 \cdot 100 \cdot 199}{6} =$

$\frac{177309}{20000} = 8.86545.$

33 (a) Here $n = 5$ and $f(x) = 1/x$,

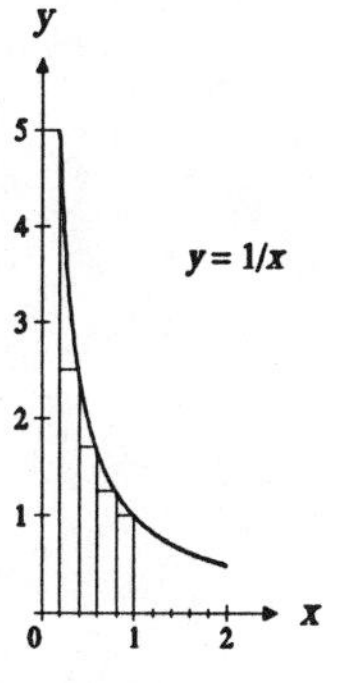

so $\sum_{i=1}^{n} f(i/n) \cdot 1/n$

$= \sum_{i=1}^{5} f(i/5) \cdot 1/5$

$= \frac{1}{5}\sum_{i=1}^{5} f(i/5) = \frac{1}{5}\sum_{i=1}^{5} \frac{5}{i}$

$= \frac{1}{5}\left(\frac{5}{1} + \frac{5}{2} + \frac{5}{3} + \frac{5}{4} + \frac{5}{5}\right) = \frac{137}{60} \approx 2.283.$

(b) Here $n = 5$ and $f(x) = \sin x$, so $\sum_{i=1}^{n} f(i/n) \cdot 1/n$

$$= \sum_{i=1}^{5} f(i/5)\cdot 1/5 = \frac{1}{5}\sum_{i=1}^{5} f(i/5)$$

$$= \frac{1}{5}\sum_{i=1}^{5} \sin\frac{i}{5} =$$

$$\frac{1}{5}\left(\sin\frac{1}{5} + \sin\frac{2}{5} + \sin\frac{3}{5} + \sin\frac{4}{5} + \sin\frac{5}{5}\right)$$

$\approx 0.542.$

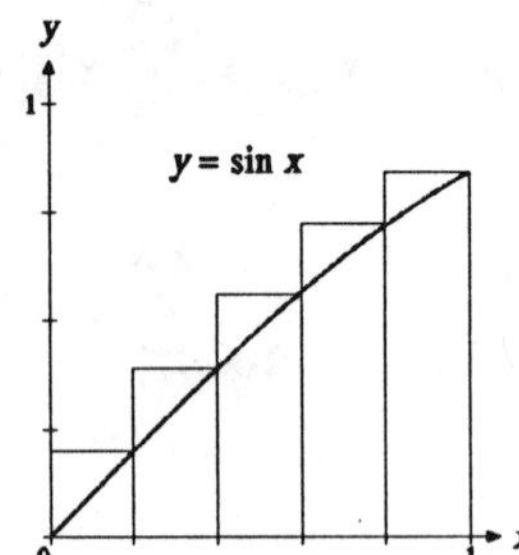

(c) Here $n = 5$ and $f(x) = 2^x$, so $\sum_{i=1}^{n} f(i/n)\cdot 1/n$

$$= \sum_{i=1}^{5} f(i/5)\cdot 1/5 = \frac{1}{5}\sum_{i=1}^{5} f(i/5) = \frac{1}{5}\sum_{i=1}^{5} 2^{i/5}$$

$$= \frac{1}{5}(2^{1/5} + 2^{2/5} + 2^{3/5} + 2^{4/5} + 2^{5/5}) \approx 1.5450.$$

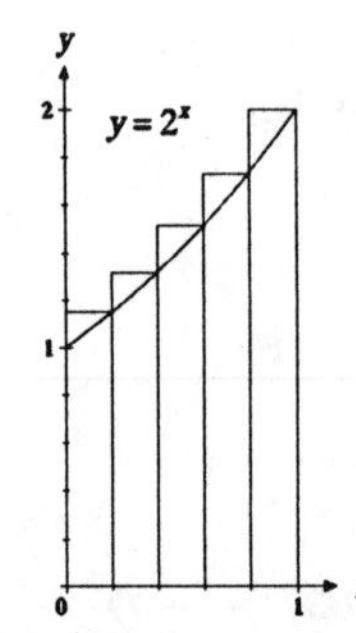

(d) The graphs are displayed above in the corresponding parts. Note how in (a) the region below $y = 1/x$ and above $[0, 1]$ is unbounded, so in this case the approximation can be computed easily, but the definite integral itself has major problems. (This kind of problem will be revisited in Sec. 8.8.)

5.3 The Definite Integral

1 (a) Note that $x_i - x_{i-1} = 1$ for each i. Hence the mesh of the partition is 1.

(b) Here $x_1 - x_0 = x_2 - x_1 = 2$, $x_3 - x_2 = 1$. Thus the mesh is 2.

(c) For this partition, $x_1 - x_0 = 3$, while $x_2 - x_1 = x_3 - x_2 = x_4 - x_3 = x_5 - x_4 = 0.5$. Hence the mesh is 3.

3 (a) $\int_0^5 x^2\,dx = \frac{5^3}{5} - \frac{0^3}{3} = \frac{125}{3}$

(b) $\int_0^4 x^2\,dx = \frac{4^3}{5} - \frac{0^3}{3} = \frac{64}{3}$

(c) $\int_4^5 x^2\,dx = \frac{5^3}{5} - \frac{4^3}{3} = \frac{61}{3}$

5 (a) $x_0 = a$

(b) $x_1 = a + \frac{b-a}{n}$

(c) $x_2 = a + 2\cdot\frac{b-a}{n}$, $x_i = a + i\cdot\frac{b-a}{n}$

(d) $x_i - x_{i-1} = \frac{b-a}{n}$, so $\sum_{i=1}^{n} x_i^2(x_i - x_{i-1}) =$

$$\sum_{i=1}^{n}\left(a + i\cdot\frac{b-a}{n}\right)^2\cdot\frac{b-a}{n}$$

$$= \frac{b-a}{n}\sum_{i=1}^{n}\left[a^2 + \frac{2a(b-a)}{n}i + \frac{(b-a)^2}{n^2}i^2\right]$$

$$= \frac{b-a}{n}\left[\sum_{i=1}^{n} a^2 + \frac{2a(b-a)}{n}\sum_{i=1}^{n} i + \frac{(b-a)^2}{n^2}\sum_{i=1}^{n} i^2\right]$$

$$= \frac{b-a}{n}\left[a^2 n + \frac{2a(b-a)}{n}\cdot\frac{n^2+n}{2} + \frac{(b-a)^2}{n^2}\cdot\frac{2n^3+3n^2+n}{6}\right] =$$

$$\frac{b-a}{6}\left[2(a^2+ab+b^2) + \frac{3(b^2-a^2)}{n} + \frac{(b-a)^2}{n^2}\right]$$

(e) $\int_a^b x^2\,dx =$

$$\lim_{n\to\infty}\frac{b-a}{6}\left[2(a^2+ab+b^2) + \frac{3(b^2-a^2)}{n} + \frac{(b-a)^2}{n^2}\right]$$

$$= \frac{b-a}{6}\cdot 2(a^2+ab+b^2) = \frac{b^3}{3} - \frac{a^3}{3}$$

7 (a) Duration of the ith time interval

(b) Speed at some time in the ith time interval

(c) Estimate of distance traveled during the ith time interval

(d) Estimate of total distance traveled

(e) Actual total distance traveled

9 Since $\int_a^b f(x)\,dx = \lim_{\text{mesh}\to 0}\sum_{i=1}^{n} f(c_i)(x_i - x_{i-1})$,

$\int_1^3 \frac{1}{x^2}\,dx = \lim_{\text{mesh}\to 0}\sum_{i=1}^{n}\frac{1}{c_i^2}\cdot(x_i - x_{i-1})$, where $x_0 = 1$ and $x_n = 3$. $(x_i - x_{i-1})$ represents the length of one of the n sections in the partition of [1, 3], and the sampling point c_i is contained in $[x_{i-1}, x_i]$. Now $\frac{1}{c_i^2}$ is the value of the function at c_i, and $\frac{1}{c_i^2}(x_i - x_{i-1})$ is an approximation of $\int_{x_{i-1}}^{x_i}\frac{1}{x^2}\,dx$. Hence $\sum_{i=1}^{n}\frac{1}{c_i^2}(x_i - x_{i-1})$ is an approximation of $\int_1^3\frac{1}{x^2}\,dx$, and taking the limit as mesh → 0 of this sum forces the sum to become exactly $\int_1^3\frac{1}{x^2}\,dx$.

11 The mass is $\int_1^3 x^2\,dx = \frac{3^3}{3} - \frac{1^3}{3} = \frac{26}{3} \approx 8.67$ grams.

13 Partitioning [1, 3] into four sections of equal length yields $x_0 = 1$, $x_1 = 3/2$, $x_2 = 2$, $x_3 = 5/2$, and $x_4 = 3$.

(a) Using left endpoints, $c_i = x_{i-1}$ and

$$\sum_{i=1}^{n} f(c_i)(x_i - x_{i-1}) = \sum_{i=1}^{4}\frac{1}{x_{i-1}}\cdot\frac{1}{2} =$$

$$\frac{1}{2}\left(1 + \frac{2}{3} + \frac{1}{2} + \frac{2}{5}\right) = \frac{77}{60} \approx 1.283.$$

(b) Using right endpoints, $c_i = x_i$ and

$$\sum_{i=1}^{n} f(c_i)(x_i - x_{i-1}) = \sum_{i=1}^{4}\frac{1}{x_i}\cdot\frac{1}{2} =$$

$$\frac{1}{2}\left(\frac{2}{3} + \frac{1}{2} + \frac{2}{5} + \frac{1}{3}\right) = \frac{19}{20} = 0.95.$$

15 Partitioning [0, 1] into 5 sections of equal length yields $x_0 = 0$, $x_1 = 1/5$, $x_2 = 2/5$, $x_3 = 3/5$, $x_4 = 4/5$, and $x_5 = 1$.

(a) With $c_i = x_{i-1}$, left endpoints, the approximating sum is $\sum_{i=1}^{5}\left(x_{i-1}^3\right)(x_i - x_{i-1}) =$

$$\frac{1}{5}\sum_{i=1}^{5} x_{i-1}^3 =$$

$$\frac{1}{5}\left(0^3 + \left(\frac{1}{5}\right)^3 + \left(\frac{2}{5}\right)^3 + \left(\frac{3}{5}\right)^3 + \left(\frac{4}{5}\right)^3\right)$$

$$= \frac{1}{5^4}(1^3 + 2^3 + 3^3 + 4^3) = \frac{4}{25} = 0.16.$$

(b) With $c_i = x_i$, right endpoints, the approximating sum is $\sum_{i=1}^{5} \left(x_i^3\right)(x_i - x_{i-1}) =$

$$\frac{1}{5}\sum_{i=1}^{5} x_i^3 = \frac{1}{5}\left(\left(\frac{1}{5}\right)^3 + \left(\frac{2}{5}\right)^3 + \left(\frac{3}{5}\right)^3 + \left(\frac{4}{5}\right)^3 + 1^3\right)$$

$$= \frac{1}{5^4}(1^3 + 2^3 + 3^3 + 4^3 + 5^3) = \frac{9}{25} = 0.36.$$

17 (a) $c_i = x_{i-1}$, so $\sum_{i=1}^{n} f(c_i)(x_i - x_{i-1}) =$

$$(\sin 0)\left(\frac{\pi}{6} - 0\right) + \left(\sin\frac{\pi}{6}\right)\left(\frac{\pi}{4} - \frac{\pi}{6}\right) +$$

$$\left(\sin\frac{\pi}{4}\right)\left(\frac{\pi}{3} - \frac{\pi}{4}\right) + \left(\sin\frac{\pi}{3}\right)\left(\frac{\pi}{2} - \frac{\pi}{3}\right)$$

$$= 0\left(\frac{\pi}{6}\right) + \frac{1}{2}\left(\frac{\pi}{12}\right) + \frac{\sqrt{2}}{2}\left(\frac{\pi}{12}\right) + \frac{\sqrt{3}}{2}\left(\frac{\pi}{6}\right)$$

$$= \frac{\pi}{24}(1 + \sqrt{2} + 2\sqrt{3}) \approx 0.77.$$

(b) $c_i = x_i$ and $\sum_{i=1}^{n} f(c_i)(x_i - x_{i-1})$ becomes

$$\left(\sin\frac{\pi}{6}\right)\left(\frac{\pi}{6} - 0\right) + \left(\sin\frac{\pi}{4}\right)\left(\frac{\pi}{4} - \frac{\pi}{6}\right) +$$

$$\left(\sin\frac{\pi}{3}\right)\left(\frac{\pi}{3} - \frac{\pi}{4}\right) + \left(\sin\frac{\pi}{2}\right)\left(\frac{\pi}{2} - \frac{\pi}{3}\right)$$

$$= \frac{1}{2}\left(\frac{\pi}{6}\right) + \frac{\sqrt{2}}{2}\left(\frac{\pi}{12}\right) + \frac{\sqrt{3}}{2}\left(\frac{\pi}{12}\right) + 1\left(\frac{\pi}{6}\right)$$

$$= \frac{\pi}{24}(6 + \sqrt{2} + \sqrt{3}) = 1.20.$$

19 (a) Using the same partition and sampling points, we find $\sum_{i=1}^{n} f(c_i)(x_i - x_{i-1}) = \sum_{i=1}^{n} \left(\frac{ib}{n}\right)^3 \cdot \frac{b}{n} =$

$$\frac{b^4}{n^4}\sum_{i=1}^{n} i^3 = \frac{b^4}{n^4}\left(\frac{n^4}{4} + \frac{n^3}{2} + \frac{n^2}{4}\right)$$

$= \frac{b^4}{4} + \frac{b^4}{2n} + \frac{b^4}{4n^2}$. As $n \to \infty$, this approaches $\frac{b^4}{4}$, so $\int_0^b x^3\,dx = \frac{b^4}{4}$.

(b) $\int_1^2 x^3\,dx = \int_0^2 x^3\,dx - \int_0^1 x^3\,dx =$

$$\frac{2^4}{4} - \frac{1^4}{4} = \frac{15}{4}$$

21 We have $x_0 = 1$, $x_1 = 3/2$, $x_2 = 2$, $x_3 = 5/2$, and $x_4 = 3$. Since the left endpoints are to be the sampling points, $c_i = x_{i-1}$ and the estimate is

$$\sum_{i=1}^{n} f(c_i)(x_i - x_{i-1}) = \sum_{i=1}^{4} (x_{i-1})^{-2}\frac{1}{2}$$

$$= \frac{1}{2}\left(\frac{1}{1} + \frac{4}{9} + \frac{1}{4} + \frac{4}{25}\right) = \frac{1669}{1800} \approx 0.9272.$$

23 The area of the trapezoid is $\int_a^b x\,dx$. The area of the trapezoid can also be written as $(b - a)\frac{a + b}{2}$

$$= \frac{1}{2}(b - a)(b + a) = \frac{1}{2}(b^2 - a^2) = \frac{b^2}{2} - \frac{a^2}{2}.$$

Hence $\int_a^b x\,dx = \frac{b^2}{2} - \frac{a^2}{2}$.

25 By interchanging the x and y axes, we see that $\text{Area}(ACD) = \int_0^a \sqrt{x}\,dx$. On the other hand, $\text{Area}(ACD) = \text{Area}(ABCD) - \text{Area}(ABC) =$

$$a\sqrt{a} - \int_0^{\sqrt{a}} x^2\,dx = a^{3/2} - \frac{(\sqrt{a})^3}{3} = \frac{2}{3}a^{3/2}.$$

27 (a)

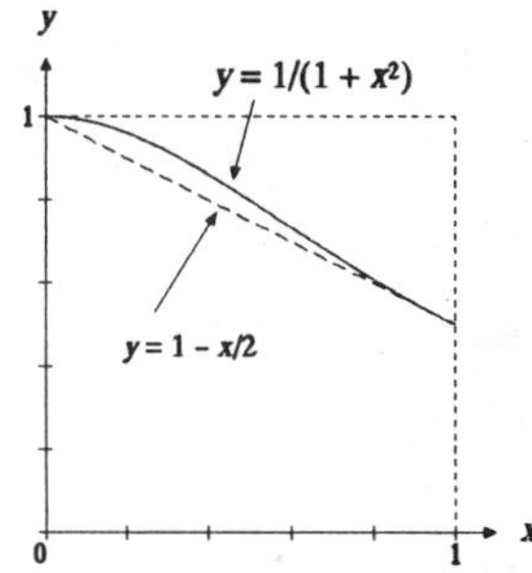

(b) The desired region is contained within the square with vertices (0, 0), (0, 1), (1, 0), and (1, 1). Hence $A < 1$.

(c) The line joining (0, 1) and (1, 1/2) has equation $y = 1 - x/2$; it crosses the graph of $y = \dfrac{1}{1+x^2}$ only at $x = 0$ and $x = 1$. This is seen by solving $\dfrac{1}{1+x^2} = 1 - \dfrac{x}{2}$ for x: $\dfrac{1}{1+x^2} = \dfrac{2-x}{2}$, $2 = (1 + x^2)(2 - x) = 2 + 2x^2 - x - x^3$. Therefore $x^3 - 2x^2 + x = x(x - 1)^2 = 0$ and $x = 0$ or $x = 1$. Since $\left(\dfrac{1}{2}, \dfrac{4}{5}\right)$ is on the graph of $y = \dfrac{1}{1+x^2}$ and above $\left(\dfrac{1}{2}, \dfrac{3}{4}\right)$ on the line, the curve lies above the line. Therefore A is larger than the trapezoid bounded by $x = 0$, $y = 0$, $x = 1$, and $y = 1 - x/2$. This area is $\dfrac{1 + 1/2}{2}\cdot 1 = \dfrac{3}{4}$. Hence $3/4 < A$.

(d) First, note that $\dfrac{1}{1+x^2}$ decreases as x increases. Hence the lower estimate is found using right endpoints and the upper estimate is found using left endpoints. A lower estimate is

$$\frac{1}{1+(1/5)^2}\cdot\frac{1}{5} + \frac{1}{1+(2/5)^2}\cdot\frac{1}{5} + \frac{1}{1+(3/5)^2}\cdot\frac{1}{5} + \frac{1}{1+(4/5)^2}\cdot\frac{1}{5} + \frac{1}{1+1^2}\cdot\frac{1}{5}$$

$$= \left(\frac{25}{26} + \frac{25}{29} + \frac{25}{34} + \frac{25}{41} + \frac{1}{2}\right)\frac{1}{5}$$

≈ 0.7337. An upper estimate is

$$\frac{1}{1+0^2}\cdot\frac{1}{5} + \frac{1}{1+(1/5)^2}\cdot\frac{1}{5} + \frac{1}{1+(2/5)^2}\cdot\frac{1}{5} + \frac{1}{1+(3/5)^2}\cdot\frac{1}{5} + \frac{1}{1+(4/5)^2}\cdot\frac{1}{5}$$

$$= \left(1 + \frac{25}{26} + \frac{25}{29} + \frac{25}{34} + \frac{25}{41}\right)\frac{1}{5} \approx 0.8337.$$

29 (a) As shown in 28(c), the area under the graph of $y = 1/x$ above $[1, 2]$ equals the area under the graph and above $[3, 6]$. Since the area above $[3, 6]$ is $A(6) - A(3)$, this says that $A(2) = A(6) - A(3)$.

(b) Let x and y be fixed. The area under the curve $y = 1/x$ and above $[y, xy]$ is $A(xy) - A(y)$, while the area below the graph and above $[1, x]$ is $A(x)$. Then an approximation for $A(x)$ is $\sum_{i=1}^{n} \dfrac{1}{c_i}(x_i - x_{i-1})$, where $1 = x_0 < x_1 < \ldots$

$< x_n = x$ and c_i is in $[x_{i-1}, x_i]$. But

$$\sum_{i=1}^{n} \frac{1}{c_i}(x_i - x_{i-1}) = \sum_{i=1}^{n} \frac{1}{c_i y}(x_i y - x_{i-1} y),$$

where $y = x_0 y < x_1 y < \ldots < x_n y = xy$ and $c_i y$ is in $[x_{i-1}y, x_i y]$. This last sum is an approximating sum for $A(xy) - A(y)$. As the mesh of the partition of $[1, x]$ goes to 0, so does the mesh of the corresponding partition of $[y, xy]$ (and vice versa). But

$$\lim_{\text{mesh}\to 0} \sum_{i=1}^{n} \frac{1}{c_i}(x_i - x_{i-1}) = \int_1^x \frac{1}{t}\, dt = A(x)$$

and $\lim_{\text{mesh}\to 0} \sum_{i=1}^{n} \frac{1}{c_i y}(x_i y - x_{i-1} y) = \int_y^{xy} \frac{1}{t}\, dt$

$= A(xy) - A(y)$. Since the sums are equal, $A(x) = A(xy) - A(y)$; equivalently, $A(xy) = A(x) + A(y)$.

(c) Logarithmic functions have this property.

31 To estimate the distance traveled, it is necessary to partition the interval [10, 20] into n sections. Choosing a point in each section, a sample of the velocity over that section of time can be determined. An estimate of the total distance traveled can then be made by multiplying the length of each section by the velocity at its corresponding sampling point, and summing up these n terms (remember velocity times time equals distance). So

$\sum_{i=1}^{n} v(T_i)(t_i - t_{i-1})$ is such an estimate, with $t_0 = 10$

$< t_1 < t_2 < \cdots < t_{n-1} < t_n = 20$ and $t_{i-1} \le T_i \le t_i$; this estimate becomes $\int_{10}^{20} v(t)\, dt$ as $n \to \infty$.

Applying the interpretation of the definite integral as an area under a curve, we see that $\int_{10}^{20} v(t)\, dt$ corresponds to the area in the figure.

33 n takes on its smallest value when every section is as long as possible, that is, when every section has length equal to the mesh, 0.02. If this is the case, then $n = \frac{5 - 1}{0.02} = 200$. We conclude that $n \ge 200$ since for a partition to have a mesh of 0.02 the only requirement is that there is at least one section of maximum length 0.02 (that is, one section of length 0.02 and an arbitrarily large number of sections with length less than 0.02).

5.4 Estimating the Definite Integral

1 Here $a = 0$, $b = 2$, $n = 2$, and $h = (b - a)/n = (2 - 0)/2 = 1$. So by the trapezoidal method,

$$\int_0^2 \frac{dx}{1 + x^2} \approx \frac{h}{2}[f(0) + 2f(1) + f(2)]$$

$$= \frac{1}{2}\left[\frac{1}{1 + 0^2} + 2\cdot\frac{1}{1 + 1^2} + \frac{1}{1 + 2^2}\right] = \frac{1}{2}\left[1 + 1 + \frac{1}{5}\right]$$

$$= \frac{1}{2}\cdot\frac{11}{5} = \frac{11}{10} = 1.1.$$

3 Here $a = 0$, $b = 2$, $n = 2$, and $h = (b - a)/n = (2 - 0)/2 = 1$. So, by the trapezoidal method,

$$\int_0^2 \sin\sqrt{x}\, dx \approx \frac{h}{2}[f(0) + 2f(1) + f(2)] =$$

$$\frac{1}{2}[\sin\sqrt{0} + 2\sin\sqrt{1} + \sin\sqrt{2}] = \sin 1 + \frac{1}{2}\sin\sqrt{2}$$

≈ 1.3354.

5 Here $a = 1$, $b = 3$, $n = 3$, and $h = (b - a)/n = (3 - 1)/3 = 2/3$. So, by the trapezoidal method

$$\int_1^3 \frac{2^x}{x}\,dx \approx \frac{h}{2}\left[f(1) + 2f\left(\frac{5}{3}\right) + 2f\left(\frac{7}{3}\right) + f(3)\right]$$

$$= \frac{2/3}{2}\left[\frac{2^1}{1} + 2\cdot\frac{2^{5/3}}{5/3} + 2\cdot\frac{2^{7/3}}{7/3} + \frac{2^3}{3}\right]$$

$$= \frac{1}{3}\left[2 + \frac{6}{5}\cdot 2^{5/3} + \frac{6}{7}\cdot 2^{7/3} + \frac{8}{3}\right] \approx 4.2654.$$

7 Here $a = 1$, $b = 3$, $n = 4$, and $h = (b - a)/n = (3 - 1)/4 = 1/2$. So, by the trapezoidal method,

$$\int_1^3 \cos x^2\,dx$$

$$\approx \frac{h}{2}[f(1) + 2f(\tfrac{3}{2}) + 2f(2) + 2f(\tfrac{5}{2}) + f(3)]$$

$$= \frac{1/2}{2}\left[\cos 1^2 + 2\cos\left(\frac{3}{2}\right)^2 + 2\cos 2^2 + 2\cos\left(\frac{5}{2}\right)^2 + \cos 3^2\right]$$

$$= \frac{1}{4}\left[\cos 1 + 2\cos\frac{9}{4} + 2\cos 4 + 2\cos\frac{25}{4} + \cos 9\right]$$

$$\approx -0.2339.$$

9 Here $a = 0$, $b = 1$, $n = 2$, and $h = (b - a)/2 = (1 - 0)/2 = 1/2$. So, by Simpson's method,

$$\int_0^1 \frac{dx}{x^3 + 1} \approx \frac{h}{3}[f(0) + 4f(\tfrac{1}{2}) + f(1)]$$

$$= \frac{1/2}{3}\left[\frac{1}{0^3 + 1} + 4\cdot\frac{1}{(1/2)^3 + 1} + \frac{1}{1^3 + 1}\right]$$

$$= \frac{1}{6}\left[1 + \frac{32}{9} + \frac{1}{2}\right] = \frac{1}{6}\cdot\frac{91}{18} = \frac{91}{108} \approx 0.8426.$$

11 Here $a = 0$, $b = 1$, $n = 4$, and $h = (b - a)/n = (1 - 0)/4 = 1/4$. So, by Simpson's method,

$$\int_0^1 \frac{dx}{x^4 + 1}$$

$$\approx \frac{h}{3}[f(0) + 4f(\tfrac{1}{4}) + 2f(\tfrac{2}{4}) + 4f(\tfrac{3}{4}) + f(1)]$$

$$= \frac{1/4}{3}\left[\frac{1}{0^4 + 1} + 4\cdot\frac{1}{(\frac{1}{4})^4 + 1} + 2\cdot\frac{1}{(\frac{2}{4})^4 + 1} + 4\cdot\frac{1}{(\frac{3}{4})^4 + 1} + \frac{1}{1^4 + 1}\right]$$

$$= \frac{1}{12}\left[1 + 4\cdot\frac{256}{257} + 2\cdot\frac{16}{17} + 4\cdot\frac{256}{337} + \frac{1}{2}\right]$$

$$= \frac{1}{12}\left[\frac{3}{2} + \frac{1024}{257} + \frac{32}{17} + \frac{1024}{337}\right] \approx 0.8671.$$

13 (a) By the trapezoidal method, the cross section's area is approximately (10/2)[0 + 2(22) + 2(27) + 2(30) + 2(27) + 2(22) + 0] = 5(256) = 1280 ft^2.

(b) By Simpson's method, the cross section's area is approximately (10/3)[0 + 4(22) + 2(27) + 4(30) + 2(27) + 4(22) + 0] = $\frac{10}{3}$(404) = $1346\frac{2}{3}$ ft^2.

15 The result will depend on the results of your measurements of Lake Tahoe's dimensions. In what follows, we take vertical cross sections with n = 12, which corresponds to the grid provided by the map. (We could also use horizontal cross sections with n = 22.) We obtain (1/3)[0 + 4(5) + 2(7.7) + 4(13) + 2(17.9) + 4(20.3) + 2(21.4) + 4(21.4) + 2(20.4) + 4(21.4) + 2(20.9) + 4(8.1) + 2.5] = (1/3)[535.9] = 178.6 mi^2.

17 With $f(a) = f(b)$, the left-point estimate is $h[f(a) + f(x_1) + \cdots + f(x_{n-1})]$, the right-point estimate is $h[f(x_1) + f(x_2) + \cdots + f(b)] = h[f(x_1) + f(x_2) + \cdots + f(a)]$, and the trapezoidal

estimate is $\frac{h}{2}[f(a) + 2f(x_1) + \cdots + 2f(x_{n-1}) + f(b)]$

$= \frac{h}{2}[2f(a) + 2f(x_1) + \cdots + 2f(x_{n-1})]$

$= h[f(a) + f(x_1) + \cdots + f(x_{n-1})]$.

Clearly these three estimates are the same for a given value of h.

19 We know $\int_0^1 x^2\,dx = \frac{1^3}{3} - \frac{0^3}{3} = \frac{1}{3}$. Now the trapezoidal estimate with $a = 1$, $b = 1$, and $h = 1$ gives $T = \frac{1}{2}[f(0) - f(1)] = \frac{1}{2}$. The error is

$\left|\int_0^1 x^2\,dx - T\right| = \left|\frac{1}{3} - \frac{1}{2}\right| = \frac{1}{6}$. Furthermore,

$\frac{d^2}{dx^2}(x^2) = \frac{d}{dx}(2x) = 2$, so M_2, the maximum possible value of $\left|\frac{d^2}{dx^2}(x^2)\right|$, is 2. Hence

$\frac{1}{12}(b - a)M_2h^2 = \frac{1}{12}(1 - 0)\cdot 2\cdot 1^2 = \frac{1}{6}$. The error does indeed equal $\frac{1}{12}(b - a)M_2h^2$.

21 $\int_{-h}^{h} f(x)\,dx = \int_{-h}^{h} (Ax^2 + Bx + C)\,dx =$

$\left(\frac{A}{3}x^3 + \frac{B}{2}x^2 + Cx\right)\Big|_{-h}^{h} = \frac{2A}{3}h^3 + 2Ch$. On the other hand, $\frac{h}{3}[f(-h) + 4f(0) + f(h)]$

$= \frac{h}{3}[(Ah^2 - Bh + C) + 4C + (Ah^2 + Bh + C)]$

$= \frac{2A}{3}h^3 + 2Ch$, which agrees with the value of $\int_{-h}^{h} f(x)\,dx$.

23 (a) From the figure, we can see that $r^2 = a^2 - x^2$ and that the volume of the sphere is equal to the definite integral of its circular cross-sectional area:

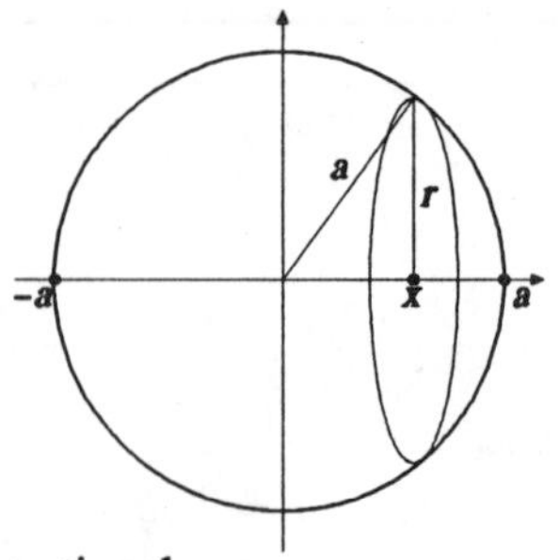

$\int_{-a}^{a} \pi r^2\,dx = \pi\int_{-a}^{a}(a^2 - x^2)\,dx$

$= \pi\cdot\frac{a}{3}[(a^2 - (-a)^2) + 4(a^2 - 0^2) + (a^2 - a^2)]$

$= \frac{4}{3}\pi a^3$.

(b) From the similar triangles in the figure, we see that $\frac{r}{x + h/2} = \frac{a}{h}$, so $r = \frac{a}{h}\left(x + \frac{h}{2}\right) = \frac{a}{h}x + \frac{a}{2}$. Then the volume is

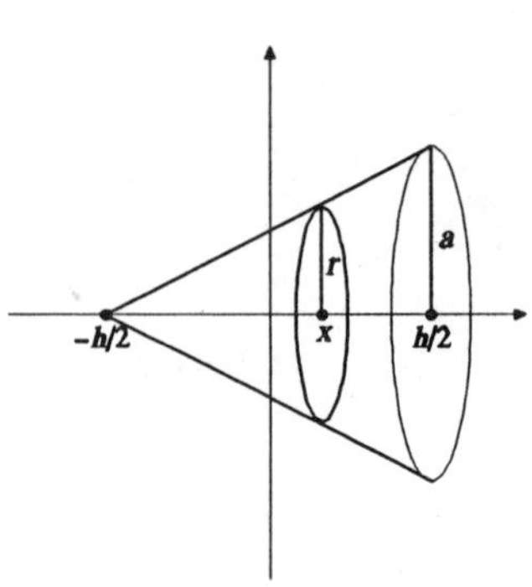

$\int_{-h/2}^{h/2} \pi r^2\,dx = \int_{-h/2}^{h/2} \pi\left(\frac{a}{h}x + \frac{a}{2}\right)^2 dx$

$= \pi\int_{-h/2}^{h/2}\left(\left(\frac{a}{h}x\right)^2 + \frac{a^2}{h}x + \frac{a^2}{4}\right) dx =$

$\pi\frac{h/2}{3}\left[\left(\left(\frac{a}{h}\left(-\frac{h}{2}\right)\right)^2 + \frac{a^2}{h}\left(-\frac{h}{2}\right) + \frac{a^2}{4}\right) + 4\cdot\frac{a^2}{4} + \left(\frac{a}{h}\cdot\frac{h}{2}\right)^2 + \frac{a^2}{h}\cdot\frac{h}{2} + \frac{a^2}{4}\right] = \frac{1}{3}\pi ha^2$

25 For $1 \le i \le \frac{n}{2}$, the ith pair of sections consists of the intervals $[x_{2i-2}, x_{2i-1}]$ and $[x_{2i-1}, x_{2i}]$. By the results of Exercises 22 and 24, the integral of the quadratic approximation on the two intervals is $\frac{h}{3}[f(x_{2i-2}) + 4f(x_{2i-1}) + f(x_{2i})]$. The sum of all these estimates is $\frac{h}{3}([f(x_0) + 4f(x_1) + f(x_2)] + [f(x_2) + 4f(x_3) + f(x_4)] + \cdots + [f(x_{n-2}) + 4f(x_{n-1}) + f(x_n)]) = \frac{h}{3}[f(x_0) + 4f(x_1) + 2f(x_2) + 4f(x_3) + \cdots + 2f(x_{n-2}) + 4f(x_{n-1}) + f(x_n)]$.

27 (a) For $f(x) = x$ or x^3, both sides equal 0. For $f(x) = 1$, both sides equal 2. For $f(x) = x^2$, both sides equal 2/3.

(b) The equation for x is $0 = a + b$, so $b = -a$. The equation for x^2 is $\frac{2}{3} = a^2 + b^2$. Hence $\frac{2}{3} = a^2 + b^2 = a^2 + (-a)^2 = 2a^2$, so $a = \pm\frac{1}{\sqrt{3}}$ and $b = \mp\frac{1}{\sqrt{3}}$.

(c) Since the equation $\int_{-1}^{1} f(x)\,dx = f\left(-\frac{1}{\sqrt{3}}\right) + f\left(\frac{1}{\sqrt{3}}\right)$ holds for each of the functions $1, x, x^2, x^3$, it also holds for any linear combination of them, that is, for any polynomial of degree at most 3.

(d) $\int_{-1}^{1} \frac{dx}{1 + x^2} \approx \frac{1}{1 + (-1/\sqrt{3})^2} + \frac{1}{1 + (1/\sqrt{3})^2} = \frac{3}{2}$

29 $\frac{2}{3}M + \frac{1}{3}T = \frac{2}{3}\sum_{i=1}^{n} f(c_i)\frac{b-a}{n} + \frac{1}{3}\cdot\frac{b-a}{2n}[f(x_0) + 2f(x_1) + \cdots + 2f(x_{n-1}) + f(x_n)] = \frac{b-a}{6n}\left[4\sum_{i=1}^{n} f(c_i) + f(x_0) + 2f(x_1) + \cdots + 2f(x_{n-1}) + f(x_n)\right]$

$= \frac{b-a}{6n}[f(x_0) + 4f(c_1) + 2f(x_1) + 4f(c_2) + \cdots + 4f(c_{n-1}) + 2f(x_{n-1}) + 4f(c_n) + f(x_n)] = S$

31 Your results will depend on the measurements you obtain from the figure. The following answers are therefore just approximate.

(a) Employing Simpson's method (on the x axis) with $h = 10$, $a = 0$, and $b = 60$, Area $\approx$ $(10/3)[4 + 4(31) + 2(37) + 4(38) + 2(33) + 4(25) + 2] = 1740$.

(b) Employing Simpson's method (on the y axis) with $h = 10$, $a = 0$, and $b = 40$, Area $\approx$ $(10/3)[0 + 4(43) + 2(56) + 4(58) + 8] = 1746\frac{2}{3}$.

The results in (a) and (b) agree well enough so that we can be reasonably confident that the true area is somewhere between 1740 and 1750 square units.

33 (a)

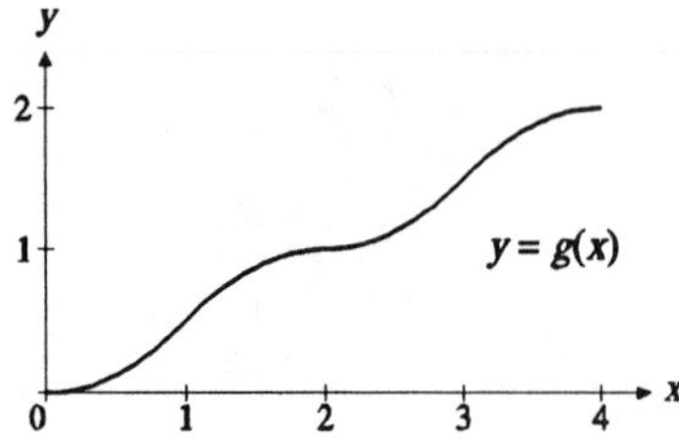

We see that $g(1) = \int_0^1 f(t)\,dt = \frac{1}{2}\cdot 1\cdot 1 = \frac{1}{2}$, $g(2) = \int_0^2 f(t)\,dt = \frac{1}{2}\cdot 1\cdot 2 = 1$, $g(3) = \int_0^3 f(t)\,dt = \frac{1}{2}\cdot 1\cdot 2 + \frac{1}{2}\cdot 1\cdot 1 = \frac{3}{2}$, $g(4) = \int_0^4 f(t)\,dt = \frac{1}{2}\cdot 1\cdot 2 + \frac{1}{2}\cdot 1\cdot 2 = 2$. Also note that $g(x)$ increases most rapidly at $x = 1$ and $x = 3$, while $g(x)$ increases least rapidly at $x = 0$, $x = 2$, and $x = 4$. Hence $g(x)$ has a high slope at $x = 1$ and $x = 3$, yet it is relatively flat at $x = 0$, $x = 2$, and $x = 4$. Since $g(x)$ is always increasing, its slope is never negative.

(b)

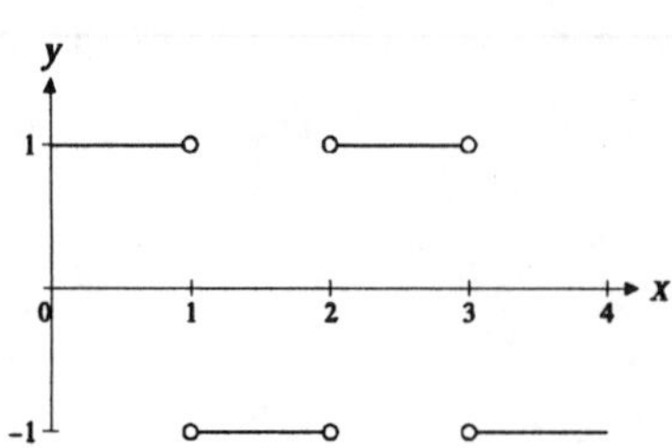

The slope of $f(x)$ alternates between 1 and -1 and is undefined for integral values of x.

5.5 Properties of the Antiderivative and the Definite Integral

1 $\int 5x^2\,dx = 5\int x^2\,dx = 5\frac{x^3}{3} + C = \frac{5}{3}x^3 + C.$

Checking by differentiating, $\left(\frac{5}{3}x^3 + C\right)' = \frac{5}{3}\cdot 3x^2 + 0 = 5x^2$, as expected.

3 $\int (2x - x^3 + x^5)\,dx$

$= \int 2x\,dx - \int x^3\,dx + \int x^5\,dx$

$= 2\int x\,dx - \int x^3\,dx + \int x^5\,dx$

$= 2\frac{x^2}{2} - \frac{x^4}{4} + \frac{x^6}{6} + C$

$= x^2 - \frac{1}{4}x^4 + \frac{1}{6}x^6 + C$. Checking by differentiating, $\left(x^2 - \frac{1}{4}x^4 + \frac{1}{6}x^6 + C\right)' = 2x - x^3 + x^5$, as expected.

5 (a) $\int \cos x\,dx = \sin x + C$. Checking by differentiating, we see $(\sin x + C)' = \cos x$.

(b) $\int \cos 2x\,dx = \frac{\sin 2x}{2} + C$. Checking by differentiating, we see $\left(\frac{1}{2}\sin 2x + C\right)' = \frac{1}{2}\cdot 2\cos 2x = \cos 2x$.

7 (a) $\int (2\sin x + 3\cos x)\,dx$

$= 2\int \sin x\,dx + 3\int \cos x\,dx$

$= 2(-\cos x) + 3(\sin x) + C$

$= 3 \sin x - 2 \cos x + C$. Checking by differentiating, $(3 \sin x - 2 \cos x + C)'$ $= 3 \cos x + 2 \sin x$.

(b) $\int (\sin 2x + \cos 3x)\, dx =$

$\int \sin 2x\, dx + \int \cos 3x\, dx =$

$-\frac{1}{2} \cos 2x + \frac{1}{3} \sin 3x + C$. Checking by differentiating, we see

$\left(-\frac{1}{2} \cos 2x + \frac{1}{3} \sin 3x + C\right)'$

$= -\frac{1}{2}(-2 \sin 2x) + \frac{1}{3}(3 \cos 3x)$

$= \sin 2x + \cos 3x$.

9 $\int \sec^2 x\, dx = \tan x + C$, since $(\tan x + C)'$ $= \sec^2 x$.

13 (a) $\int_2^5 x^2\, dx = \frac{5^3}{3} - \frac{2^3}{3} = \frac{125}{3} - \frac{8}{3} = \frac{117}{3}$ $= 39$

(b) $\int_5^2 x^2\, dx = -\int_2^5 x^2\, dx = -\frac{117}{3} = -39$

(c) $\int_5^5 x^2\, dx = 0$, by definition.

15 (a) $\int x\, dx = \frac{x^2}{2} + C$

(b) $\int_3^4 x\, dx = \frac{4^2}{2} - \frac{3^2}{2} = \frac{16}{2} - \frac{9}{2} = \frac{7}{2}$

17 Applying Property 7 with $m = 2$, $M = 3$, $a = 1$, and $b = 6$, we transform $m(b - a) \le \int_a^b f(x)\, dx$ $\le M(b - a)$ into $2(6 - 1) \le \int_1^6 f(x)\, dx \le$ $3(6 - 1)$, so $10 \le \int_1^6 f(x)\, dx \le 15$. Hence $\int_1^6 f(x)\, dx$ lies in $[10, 15]$.

19 First, $\int_1^5 2x\, dx = 5^2 - 1^2 = 24$. Hence $24 =$ $\int_1^5 2x\, dx = f(c)(5 - 1)$ for some c in $[a, b]$. Thus $24 = f(c)\cdot 4$ and $f(c) = 6$. Furthermore, since $f(c)$ $= 2c$, we see that $2c = 6$ so $c = 3$.

21 First, $\int_0^4 x^2\, dx = \frac{4^3}{3} - \frac{0^3}{3} = \frac{64}{3}$. Hence $\frac{64}{3} =$ $\int_0^4 x^2\, dx = f(c)(4 - 0)$ and $f(c) = \frac{16}{3}$. Since $f(c)$ $= c^2$, we see that $c^2 = \frac{16}{3}$ and $c = \pm\sqrt{\frac{16}{3}} =$ $\pm\frac{4}{\sqrt{3}}$. We choose $c = \frac{4}{\sqrt{3}}$ because it is the only value that lies in $[0, 4]$.

23 (a) $\int_2^1 f(x)\, dx = -\int_1^2 f(x)\, dx = -3$

(b) By Property 6, $\int_2^1 f(x)\, dx + \int_1^5 f(x)\, dx =$ $\int_2^5 f(x)\, dx$. Thus $\int_2^5 f(x)\, dx = -3 + 7$ $= 4$.

25 (a) Applying Property 7 with $m = 3$ and $M = 7$, we get $3(b - a) \le \int_a^b f(x)\, dx \le 7(b - a)$.

(b) Since $3 \le f(x) \le 7$ on $[a, b]$, the average value of $f(x)$ on $[a, b]$ must lie in $[3, 7]$ as well.

27 First we find the minimum and maximum values of $f(x) = x$ on $[1, 3]$. Since $f'(x) = 1 > 0$, $f(x)$ is increasing and has no critical points for all x. So

the minimum of $f(x)$ on $[1, 3]$ occurs at the left endpoint where $f(1) = 1$, and the maximum occurs at the right endpoint where $f(3) = 3$. Now the average value of $f(x)$ on $[1, 3]$ is given by

$$\frac{\int_1^3 f(x)\,dx}{3-1} = \frac{\int_1^3 x\,dx}{3-1} = \frac{3^2/2 - 1^2/2}{2} =$$

$$\frac{9}{4} - \frac{1}{4} = 2.$$

29 Since $f'(x) = 2x > 0$ for all x in $[2, 3]$, we conclude that $f(x)$ is increasing and has no critical points in $[2, 3]$. Thus the minimum of $f(x)$ on $[2, 3]$ is $f(2) = 2^2 = 4$, and the maximum on $[2, 3]$ occurs at $f(3) = 3^2 = 9$. The average value of $f(x)$ on $[2, 3]$ is given by $\frac{\int_2^3 f(x)\,dx}{3-2} = \frac{\int_2^3 x^2\,dx}{3-2} =$

$$\frac{3^3/3 - 2^3/3}{1} = \frac{27}{3} - \frac{8}{3} = \frac{19}{3}.$$

31 Employing Simpson's method with $a = 1$, $b = 7$, $n = 6$, and $h = \frac{b-a}{n} = 1$, we have $\int_1^7 f(x)\,dx$

$$\approx \frac{h}{3}[f(1) + 4f(2) + 2f(3) + 4f(4) + 2f(5) + 4f(6) + f(7)]$$

$$= \frac{1}{3}[3 + 4\cdot 1 + 2\cdot 4 + 4\cdot 5 + 2\cdot 2 + 4\cdot 2 + 6]$$

$= \frac{53}{3}$. So the average value of $f(x)$ on $[1, 7]$ is

$$\frac{\int_1^7 f(x)\,dx}{7-1} \approx \frac{53/3}{6} = \frac{53}{18}.$$

33 (a)

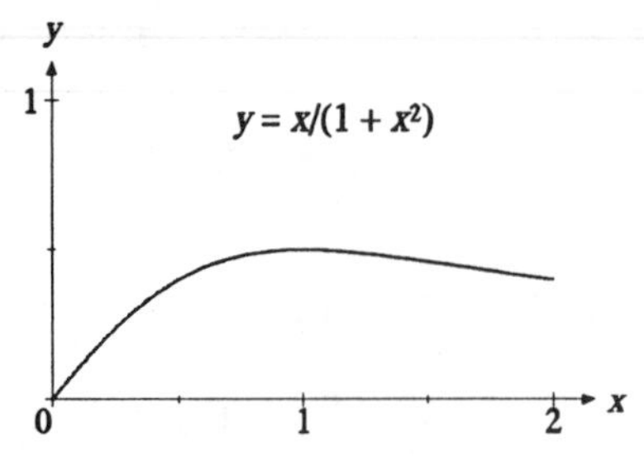

$$f'(x) = \left(\frac{x}{x^2+1}\right)' = \frac{x^2 + 1 - x\cdot 2x}{(x^2+1)^2} =$$

$\frac{1 - x^2}{(1 + x^2)^2}$, which equals 0 when $x = 1$; and $f'(x)$ changes from $+$ to $-$ at $x = 1$, so $f(1) = 1/2$ is a maximum. Also $f''(x) =$

$$\left(\frac{1 - x^2}{(1 + x^2)^2}\right)' =$$

$$\frac{(1 + x^2)^2(-2x) - (1 - x^2)\cdot 2(1 + x^2)2x}{(1 + x^2)^4} = 0$$

when $-2x(1 + x^2)^2 - 4x(1 - x^2)(1 + x^2) = 0$, or when $-2x(1 + x^2)[1 + x^2 + 2(1 - x^2)] = 0$. Hence $-2x(1 + x^2)[3 - x^2]$, which occurs when $x = 0$ or $x = \sqrt{3}$. Since $f''(x)$ changes sign from $+$ to $-$ at $x = 0$ and from $-$ to $+$ at $x = \sqrt{3}$, f is concave down on $[0, \sqrt{3}]$ and concave up on $[\sqrt{3}, 2]$.

(b) Referring to the figure in (a), we see that the minimum value of $f(x)$ on $[0, 2]$ is $f(0) = 0$ and the maximum is $f(1) = 1/2$.

(c) We know the average value of $f(x)$ on $[0, 2]$ is between 0 and 1/2 since $0 \le f(x) \le 1/2$ on

$[0, 2]$. Thus $0 \le \dfrac{\int_0^2 \frac{x}{x^2+1}\,dx}{2-0} \le \dfrac{1}{2}$ and

$$0 \le \int_0^2 \frac{x}{x^2+1}\,dx \le 1.$$

35 (a) $\left(\dfrac{\sin^3 x}{3}\right)' = \dfrac{1}{3}(\sin^3 x)' = \dfrac{1}{3}(3\sin^2 x\cos x)$

$= \sin^2 x\cos x \ne \sin^2 x$

(b) $\int \sin^2 x\,dx = \int \dfrac{1-\cos 2x}{2}\,dx$

$= \int \dfrac{1}{2}\,dx - \int \dfrac{\cos 2x}{2}\,dx$

$= \dfrac{1}{2}\int dx - \dfrac{1}{2}\int \cos 2x\,dx$

$= \dfrac{1}{2}x - \dfrac{1}{2}\dfrac{\sin 2x}{2} + C = \dfrac{1}{2}x - \dfrac{\sin 2x}{4} + C$

(c) $\left(\dfrac{1}{2}x - \dfrac{1}{4}\sin 2x + C\right)' = \dfrac{1}{2} - \dfrac{1}{4}\cdot 2\cos 2x$

$= \dfrac{1}{2} - \dfrac{1}{2}\cos 2x = \dfrac{1-\cos 2x}{2} = \sin^2 x$

37 $\left(\dfrac{x^2}{4} - \dfrac{x\sin 2ax}{4a} - \dfrac{\cos 2ax}{8a^2} + C\right)' =$

$\dfrac{x}{2} - \dfrac{1}{4a}[x(2a\cos 2ax) + \sin 2ax] - \dfrac{1}{8a^2}(-2a\sin 2ax)$

$= \dfrac{x}{2} - \dfrac{x}{2}\cos 2ax - \dfrac{\sin 2ax}{4a} + \dfrac{\sin 2ax}{4a} =$

$x\left(\dfrac{1}{2} - \dfrac{1}{2}\cos 2ax\right) = x\sin^2 ax$

39 (a) Average value of x^2 on $[0, 3]$ =

$\lim_{n\to\infty} \dfrac{1}{n+1}\sum_{i=0}^{n} f(x_i) = \lim_{n\to\infty}\sum_{i=0}^{n}\dfrac{1}{n+1}(x_i)^2$

$= \lim_{n\to\infty}\sum_{i=0}^{n}\dfrac{1}{n+1}\left(\dfrac{3i}{n}\right)^2 = \lim_{n\to\infty}\sum_{i=0}^{n}\dfrac{9i^2}{n^3+n^2}$

$= \lim_{n\to\infty}\dfrac{9}{n^3+n^2}\sum_{i=0}^{n} i^2$

$= \lim_{n\to\infty}\left(\dfrac{9}{n^3+n^2}\cdot\dfrac{2n^3+3n^2+n}{6}\right)$

$= \dfrac{9}{6}\lim_{n\to\infty}\dfrac{2n^3+3n^2+n}{n^3+n^2} = \dfrac{3}{2}\cdot 2 = 3$

(b) Yes. $\dfrac{1}{3}\int_0^3 x^2\,dx = \dfrac{1}{3}\cdot\dfrac{x^3}{3}\Big|_0^3 = \dfrac{1}{9}(27-0)$

$= 3.$

(c) Let $f(x)$ be a continuous function defined on $[a, b]$. Then the average value of $f(x)$ on $[a, b]$ is defined to be $\lim_{n\to\infty}\dfrac{1}{n+1}\sum_{i=0}^{n} f(x_i) =$

$\dfrac{1}{b-a}\lim_{n\to\infty}\dfrac{n}{n+1}\sum_{i=0}^{n} f(x_i)\dfrac{b-a}{n}$

$= \dfrac{1}{b-a}\lim_{n\to\infty}\sum_{i=0}^{n} f(x_i)\dfrac{b-a}{n}$

$= \dfrac{1}{b-a}\lim_{n\to\infty}\left(\sum_{i=0}^{n-1} f(x_i)\dfrac{b-a}{n} + f(b)\dfrac{b-a}{n}\right)$

$= \dfrac{1}{b-a}\lim_{n\to\infty}\sum_{i=0}^{n-1} f(x_i)\dfrac{b-a}{n} + 0$

$= \dfrac{1}{b-a}\lim_{n\to\infty}\sum_{i=0}^{n-1} f(x_i)\,\Delta x$

$= \dfrac{1}{b-a}\int_a^b f(x)\,dx.$

5.6 Background for the Fundamental Theorems of Calculus

1 (a)

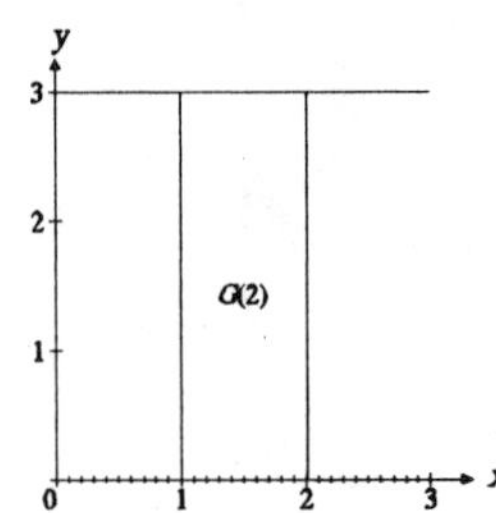

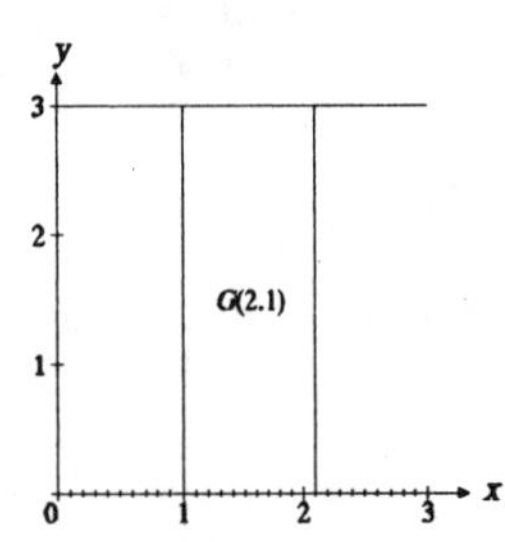

(b) Since $G(2) = 3(2 - 1) = 3$ and $G(2.1) = 3(2.1 - 1) = 3.3$, we see that

$$\frac{G(2.1) - G(2)}{0.1} = \frac{3.3 - 3}{0.1} = \frac{0.3}{0.1} = 3.$$

(c) $G(x) = \int_1^x f(t)\,dt = \int_1^x 3\,dt = 3x - 3$

(d) $G'(x) = (3x - 3)' = 3 = f(x)$

3 (a) $G(3) \approx 1.0986$, $G(3.1) \approx 1.1314$

(b)

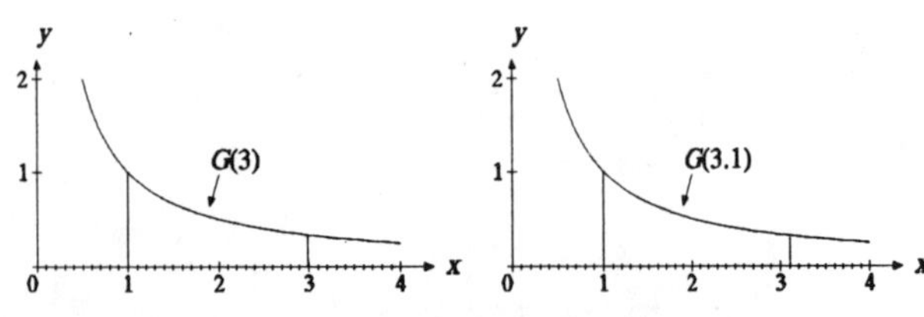

(c) $G'(3) \approx \dfrac{G(3.1) - G(3)}{0.1} \approx$

$$\frac{1.1314 - 1.0986}{0.1} = \frac{0.0328}{0.1} = 0.328$$

(d) $f(3) = \dfrac{1}{3} \approx 0.333$, and $|G'(3) - f(3)| \approx 0.005$.

5 (a) $G(1) \approx 0.8356$, $G(1.1) \approx 0.8820$

(b)

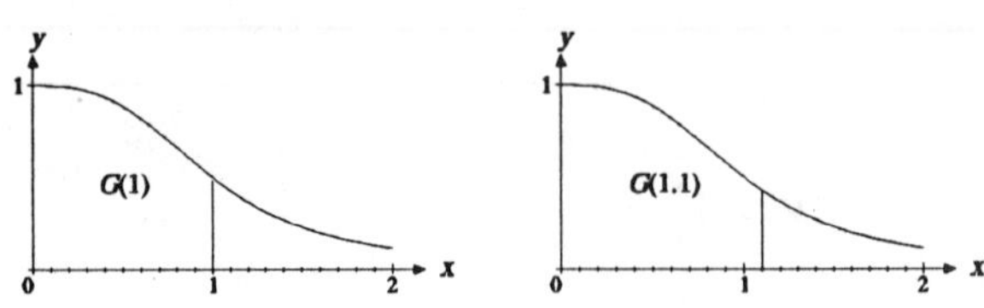

(c) $G'(1) \approx \dfrac{G(1.1) - G(1)}{0.1} \approx$

$$\frac{0.8820 - 0.8356}{0.1} = \frac{0.0464}{0.1} = 0.464$$

(d) $f(1) = \dfrac{1}{1 + 1^3} = \dfrac{1}{2}$

7 (a) $G(2) \approx 0.3607$, $G(1.9) \approx 0.3348$

(b)

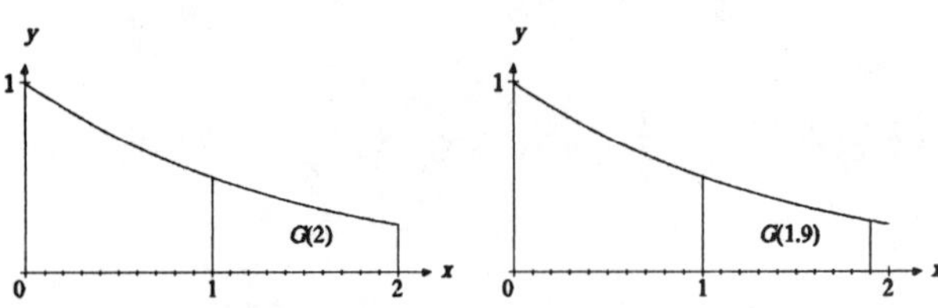

(c) $G'(2) \approx \dfrac{G(1.9) - G(2)}{-0.1} \approx$

$$\frac{0.3348 - 0.3607}{-0.1} = \frac{-0.0259}{-0.1} = 0.259$$

(d) $f(2) = 2^{-2} = 1/4$

9 (a) $G(x) = \int_1^x t\,dt = \dfrac{x^2}{2} - \dfrac{1^2}{2} = \dfrac{1}{2}x^2 - \dfrac{1}{2}$

(b) $G'(x) = x$, so $G'(2) = 2$.

(c) Yes, since $G'(2) = 2 = f(2)$.

(d) Yes, since $G'(3) = 3 = f(3)$.

11 (a),(b)

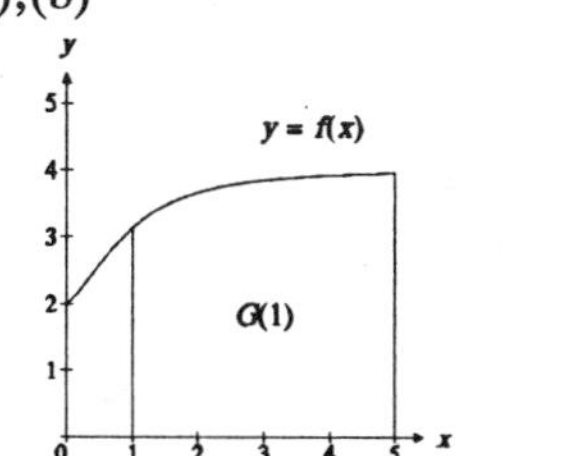

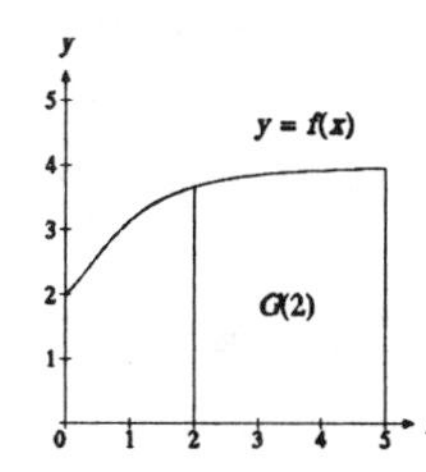

(c) Since $G(x) < G(y)$, when $y < x \leq 5$, G is a decreasing function.

(d) The derivative of a decreasing function is negative, so we expect G' to be negative.

13 (a)

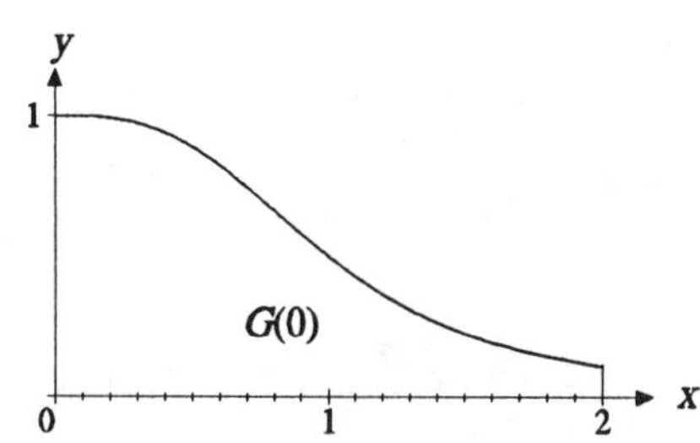

$G(0.01)$ is almost the same as $G(0)$, except that a very slender strip is missing from the left edge of the region.

(b) $G(0) > G(0.01)$ because the latter is the area of a smaller region.

(c) $G(x)$ is a decreasing function, since as x increases, the region shrinks.

(d) From (c), G' is negative.

(e) Here $G(0.01) \approx 1.0800$ while $G(0) \approx 1.0900$. Thus $G'(0) \approx \dfrac{G(0.01) - G(0)}{0.01}$

$$\approx \frac{1.0800 - 1.0900}{0.01} = \frac{-0.01}{0.01} = -1.$$

(f) Since $f(0) = 1$, $G'(0)$ seems to be the opposite of $f(0)$.

15 (a),(b),(c)

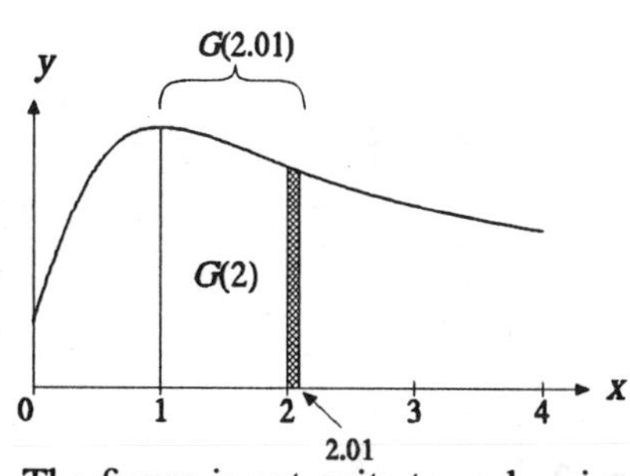

The figure is not quite to scale, since otherwise 2 and 2.01 would be impossible to distinguish.

(d) $\dfrac{G(2.01) - G(2)}{0.01}$ is the average value of $f(t)$ on [2, 2.01]. This should be roughly $f(2)$ since f varies only slightly on [2, 2.01].

(e) As the interval $[2, 2 + \Delta x]$ is chosen smaller and smaller, f varies on $[2, 2 + \Delta x]$ less and less. Thus we expect $G'(2) = f(2)$.

5.7 The Fundamental Theorems of Calculus

1 (a) $x^3|_1^2 = 2^3 - 1^3 = 8 - 1 = 7$

(b) $x^3|_{-1}^2 = 2^3 - (-1)^3 = 8 + 1 = 9$

(c) $\cos x\,|_0^\pi = \cos \pi - \cos 0 = -1 - 1 = -2$

3 The derivative of the area of the region under the graph of f and above $[0, x]$ with respect to x is the value of f at x.

5 $\displaystyle\int_1^2 5x^3\,dx = \frac{5x^4}{4}\bigg|_1^2 = \frac{5}{4}(16 - 1) = \frac{75}{4}$

7 $\displaystyle\int_1^4 (x + 5x^2)\,dx = \left(\frac{x^2}{2} + \frac{5x^3}{3}\right)\bigg|_1^4 =$

$$\frac{16}{2} + \frac{5\cdot 64}{3} - \left(\frac{1}{2} + \frac{5}{3}\right) = \frac{225}{2} = 112\tfrac{1}{2}$$

9 $\int_{\pi/6}^{\pi/3} 5\cos x\,dx = 5\sin x\Big|_{\pi/6}^{\pi/3} = 5\left(\frac{\sqrt{3}}{2} - \frac{1}{2}\right)$

$= \frac{5}{2}(\sqrt{3} - 1) \approx 1.83$

11 $\int_0^{\pi/2} \sin 2x\,dx = -\frac{1}{2}\cos 2x\Big|_0^{\pi/2}$

$= -\frac{1}{2}(\cos\pi - \cos 0) = -\frac{1}{2}(-1 - 1) = 1$

13 $\int_4^9 5\sqrt{x}\,dx = \int_4^9 5x^{1/2}\,dx = 5\frac{x^{3/2}}{3/2}\Big|_4^9$

$= \frac{10}{3}(9^{3/2} - 4^{3/2}) = \frac{10}{3}(27 - 8) = \frac{190}{3}$

15 $\int_1^8 \sqrt[3]{x^2}\,dx = \int_1^8 x^{2/3}\,dx = \frac{x^{5/3}}{5/3}\Big|_1^8$

$= \frac{3}{5}(8^{5/3} - 1^{5/3}) = \frac{3}{5}(31) = \frac{93}{5}$

17 $\frac{1}{5-3}\int_3^5 x^2\,dx = \frac{1}{2}\frac{x^3}{3}\Big|_3^5 = \frac{1}{6}(125 - 27)$

$= \frac{98}{6} = \frac{49}{3}$

19 $\frac{1}{\pi - 0}\int_0^{\pi} \sin x\,dx = \frac{1}{\pi}(-\cos x)\Big|_0^{\pi}$

$= \frac{-1}{\pi}(-1 - 1) = \frac{2}{\pi}$

21 $\frac{1}{\pi/4 - \pi/6}\int_{\pi/6}^{\pi/4} \sec^2 x\,dx = \frac{12}{\pi}\tan x\Big|_{\pi/6}^{\pi/4}$

$= \frac{12}{\pi}\left(1 - \frac{1}{\sqrt{3}}\right) = \frac{12(\sqrt{3} - 1)}{\sqrt{3}\,\pi}$

23 The area is given by $\int_1^4 3x^2\,dx = x^3\Big|_1^4$

$= 4^3 - 1^3 = 64 - 1 = 63.$

25 The area is given by $\int_{-1}^1 6x^4\,dx = 6\frac{x^5}{5}\Big|_{-1}^1$

$= \frac{6}{5}(1^5 - (-1)^5) = \frac{12}{5}.$

27 The distance is given by $\int_1^2 t^5\,dt = \frac{t^6}{6}\Big|_1^2$

$= \frac{1}{6}(2^6 - 1^6) = \frac{63}{6} = \frac{21}{2} = 10.5$ ft.

29 The mass is given by $\int_1^2 5x^3\,dx = 5\frac{x^4}{4}\Big|_1^2$

$= \frac{5}{4}(2^4 - 1^4) = \frac{75}{4} = 18.75$ grams.

31 The volume is given by $\int_1^5 6x^3\,dx = 6\frac{x^4}{4}\Big|_1^5$

$= \frac{3}{2}(5^4 - 1^4) = 936\text{ cm}^3.$

33 (a) $\int x^2\,dx$ is a function. (In particular, it is a function of the form $x^3/3 + C$.)

(b) $\int x^2\,dx\Big|_1^3$ is a number, since it is the difference of two particular values of the antiderivative $\int x^2\,dx$. (We even know what this number has to be in this case: $3^3/3 - 1^3/3 = 26/3$.)

(c) $\int_1^3 x^2\,dx$ is a definite integral; it is a number. (We can even evaluate this number by the second fundamental theorem of calculus, using the result in (b). It equals 26/3.)

35 (a) True. Every elementary function has an elementary derivative.

(b) False. Many elementary functions do not have elementary antiderivatives.

37 (a) $\frac{d}{dx}\left(\int \sin x^2\, dx\right) = \sin x^2$, by the definition of antiderivative.

(b) $\frac{d}{dx}\left(3x + \int_{-2}^{3} \sin x^2\, dx\right) = 3 + 0 = 3$, since the definite integral is a constant and the derivative of a constant is 0.

(c) $\frac{d}{dx}\left(\int_{-2}^{x} \sin t^2\, dt\right) = \sin x^2$, by the first fundamental theorem. (In this case the definite integral isn't a constant because it has a variable upper limit of integration.)

39 (a) $\frac{d}{dx}\left(\int_1^x \sqrt[3]{1 + \sin t}\, dt\right) = \sqrt[3]{1 + \sin x}$, by the first fundamental theorem.

(b) Let $y = \int_1^{x^2} \sqrt[3]{1 + \sin t}\, dt$ and $u = x^2$. Then

$$\frac{dy}{dx} = \frac{dy}{du}\frac{du}{dx} = \left(\sqrt[3]{1 + \sin u}\right)(2x)$$

$= 2x\sqrt[3]{1 + \sin x^2}$, by a combination of the chain rule and the first fundamental theorem.

41 $\frac{d}{dx}\left(\int_{2x}^{3x} t \tan t\, dt\right)$

$$= \frac{d}{dx}\left(\int_{2x}^{0} t \tan t\, dt + \int_0^{3x} t \tan t\, dt\right)$$

$$= \frac{d}{dx}\left(-\int_0^{2x} t \tan t\, dt + \int_0^{3x} t \tan t\, dt\right)$$

$$= -(2x \tan 2x)(2) + (3x \tan 3x)(3)$$

$= -4x \tan 2x + 9x \tan 3x$, by a combination of the chain rule and the first fundamental theorem.

43 From Figure 9 we see that the formula for the function is $f(x) = 2 - x$ for x in $[1, 3]$, so $G(x)$

$$= \int_1^x f(t)\, dt = \int_1^x (2 - t)\, dt = 2t - \left(\frac{1}{2}t^2\right)\Big|_1^x$$

$$= 2x - \frac{1}{2}x^2 - \left(2 - \frac{1}{2}\right) = -\frac{1}{2}x^2 + 2x - \frac{3}{2}.$$

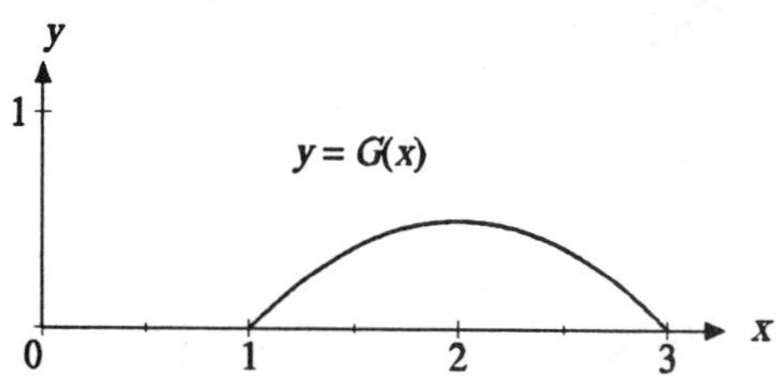

45 Partially fill the measuring glass with water and slowly dunk the stone into the water. You can use the measuring glass to find $V(x)$, the volume of the submerged portion of the rock as a function of x, the depth to which you have dunked the rock into the water in the measuring glass. Let $A(x)$ be the area of the cross section which the rock makes with the plane of the water's surface. Then $A(x) = V'(x)$, so use your recorded values of $V(x)$ to estimate the derivative of V at the depth x corresponding to the desired cross section. (If you are using a measuring glass calibrated in milliliters, recall that milliliters correspond to cubic centimeters, indicating that x should be measured in centimeters.)

47 If you let $f(x) = \int_4^x \sqrt[3]{1 + t^2}\, dt$, then $f(4) = \int_4^4 \sqrt[3]{1 + t^2}\, dt = 0$, as desired, and, by the first fundamental theorem, $f'(x) = \sqrt[3]{1 + x^2}$.

49 (a) $\int_0^1 \sqrt{x}\,dx = \frac{2}{3}x^{3/2}\Big|_0^1 = \frac{2}{3}(1 - 0) = \frac{2}{3}$

(b) Let $S = \int_0^1 \sqrt{x}\,dx$. When $h = 1/2$, $S \approx$

$$\frac{1}{6}\left(0 + \frac{4}{\sqrt{2}} + 1\right) \approx 0.6380712.$$

When $h = 1/4$, $S \approx$

$$\frac{1}{12}\left(0 + \frac{4}{2} + \frac{2}{\sqrt{2}} + \frac{4\sqrt{3}}{2} + 1\right) \approx 0.6565263.$$

When $h = 1/8$, $S \approx$

$$\frac{1}{24}\left(0 + \frac{4}{\sqrt{8}} + \frac{2}{2} + \frac{4\sqrt{3}}{\sqrt{8}} + \frac{2}{\sqrt{2}} + \frac{4\sqrt{5}}{\sqrt{8}} + \frac{2\sqrt{3}}{2} + \frac{4\sqrt{7}}{\sqrt{8}} + 1\right)$$

$$= \frac{1}{24}\left(\sqrt{2} + 1 + \sqrt{6} + \sqrt{2} + \sqrt{10} + \sqrt{3} + \sqrt{14} + 1\right)$$

$\approx 0.6630793.$

When $h = 1/16$, $S \approx$

$$\frac{1}{48}\left(0 + \frac{4}{\sqrt{16}} + \frac{2\sqrt{2}}{\sqrt{16}} + \cdots + \frac{4\sqrt{15}}{\sqrt{16}} + 1\right)$$

$\approx 0.6653982.$

When $h = 1/32$, $S \approx$

$$\frac{1}{96}\left(0 + \frac{4}{\sqrt{32}} + \cdots + \frac{4\sqrt{31}}{\sqrt{32}} + 1\right)$$

$\approx 0.6662182.$

When $h = 1/64$, $S \approx$

$$\frac{1}{192}\left(0 + \frac{4}{\sqrt{64}} + \cdots + \frac{4\sqrt{63}}{\sqrt{64}} + 1\right)$$

$\approx 0.6665081.$

When we subtract each of these results from 2/3, we obtain the following table.

h	*Simpson's estimate*	*Error*
1/2	0.6380712	0.0285955
1/4	0.6565263	0.0101404
1/8	0.6630793	0.0035874
1/16	0.6653982	0.0012685
1/32	0.6662182	0.0004485
1/64	0.6665081	0.0001586

(c) $\frac{E(1/4)}{E(1/2)} = 0.3546156$, $\frac{E(1/8)}{E(1/4)} = 0.3537716$,

$\frac{E(1/16)}{E(1/8)} = 0.3535939$, $\frac{E(1/32)}{E(1/16)} =$

0.3535607, $\frac{E(1/64)}{E(1/32)} = 0.3535547$.

(d) All the rates in part (c) are near $\frac{1}{2\sqrt{2}} \approx$ 0.3535534. Look at one, say, the last one:

$$\frac{E(1/64)}{E(1/32)} \approx \frac{A(1/64)^k}{A(1/32)^k} = \frac{(1/2)^k(1/32)^k}{(1/32)^k} =$$

$\left(\frac{1}{2}\right)^k$. So we have $\frac{1}{2\sqrt{2}} \approx \left(\frac{1}{2}\right)^k$ or, equivalently, $2\sqrt{2} \approx 2^k$. Therefore, $\ln 2\sqrt{2} \approx k \ln 2$ and thus $k \approx \frac{\ln 2\sqrt{2}}{\ln 2} = \frac{3}{2}$. To find A, solve $E(h) \approx Ah^{3/2}$ to obtain $A \approx \frac{E(h)}{h^{3/2}}$.

Now plug in any corresponding values of h and $E(h)$ from the table in (b). The solutions vary slightly, but they are all near 0.081.

51 $\int_1^3 R''(x)\,dx = R'(x)|_1^3 = R'(3) - R'(1) = 8 - 6$

$= 2$

5.S Guide Quiz

2 (b) The trapezoidal estimate is exact for constant and linear functions.

(c) Since the error using the trapezoidal method is less than $(1/12)(b - a)M_2h^2$, the error reduces as the square of h.

(d) Here $M_2 = 4$, $a = 1$, and $b = 8$. We desire $(1/12)(b - a)M_2h^2 = (7/12)\cdot 4\cdot h^2 < 0.001$, so $h < \sqrt{\frac{3}{7000}}$. Hence $n = \frac{7}{h} > \sqrt{\frac{7000}{3}} \approx 338.1$, so choose $n \geq 339$.

3 (b) Simpson's estimate is exact for constant, linear, quadratic, and cubic functions.

(c) Since the error using Simpson's method is less than $\frac{1}{180}(b - a)M_4h^4$, the error reduces as the fourth power of h.

(d) Here $M_4 = 7$, $a = 1$, and $b = 8$. We desire $\frac{1}{180}(b - a)M_4h^4 = \frac{7}{180}\cdot 7\cdot h^4 < 0.001$, so $h < \sqrt[4]{\frac{180}{49000}} = \sqrt[4]{\frac{9}{2450}}$. Hence $n = \frac{7}{h} > 7\cdot\sqrt[4]{\frac{2450}{9}} \approx 28.4$. So choose $n \geq 30$ (and even).

4 (a) $\int_0^3 \frac{dx}{1 + x^3}$

$$\approx \frac{1}{2}\left[\frac{1}{1 + 0^3} + \frac{1}{1 + (1/2)^3} + \frac{1}{1 + (2/2)^3} + \frac{1}{1 + (3/2)^3} + \frac{1}{1 + (4/2)^3} + \frac{1}{1 + (5/2)^3}\right]$$

$$= \frac{1}{2}\cdot\left[1 + \frac{8}{9} + \frac{1}{2} + \frac{8}{35} + \frac{1}{9} + \frac{8}{133}\right]$$

$$= \frac{1}{2}\cdot\frac{3709}{1330} = \frac{3709}{2660} \approx 1.3944$$

(b) $$\int_0^3 \frac{dx}{1 + x^3} \approx \frac{1/2}{2}\left[\frac{1}{1 + (0/2)^3} + \frac{2}{1 + (1/2)^3} + \frac{2}{1 + (2/2)^3} + \frac{2}{1 + (3/2)^3} + \frac{2}{1 + (4/2)^3} + \frac{2}{1 + (5/2)^3} + \frac{1}{1 + (6/2)^3}\right]$$

$$= \frac{1}{4}\left[1 + \frac{16}{9} + 1 + \frac{16}{35} + \frac{2}{9} + \frac{16}{133} + \frac{1}{28}\right]$$

≈ 1.1533

(c) $$\int_0^3 \frac{dx}{1 + x^3} \approx \frac{1/2}{3}\left[\frac{1}{1 + (0/2)^3} + \frac{4}{1 + (1/2)^3} + \frac{2}{1 + (2/2)^3} + \frac{4}{1 + (3/2)^3} + \frac{2}{1 + (4/2)^3} + \frac{4}{1 + (5/2)^3} + \frac{1}{1 + (6/2)^3}\right]$$

$$= \frac{1}{6}\left[1 + \frac{32}{9} + 1 + \frac{32}{35} + \frac{2}{9} + \frac{32}{133} + \frac{1}{28}\right]$$

≈ 1.1614

5 (a) $G(3.02) - G(3) \approx G'(3)\cdot(0.02) = f(3)\cdot(0.02)$

$= 5\cdot 0.02 = 0.1$

(b) $G'(3) = f(3) = 5$ and $G'(1) = f(1) = 4$.

(c) For x in $[1, 3]$, $4 \le f(x) \le 5$. Employing the mean-value theorem for definite integrals with $m = 4$, $M = 5$, $a = 1$, and $b = 3$ gives $4(3 - 1) < \int_1^3 f(t)\,dt < 5(3 - 1)$. (There can be no equality since f is an increasing function.) Hence, $8 < \int_1^3 f(t)\,dt < 10$. But $G(3) = \int_1^3 f(t)\,dt$, so we conclude that $8 < G(3) < 10$.

7 (a) $\left(\frac{1}{2x + 3}\right)' = ((2x + 3)^{-1})' = -1(2x + 3)^{-2}(2) = -\frac{2}{(2x + 3)^2}$

(b) $\int_0^1 \frac{dx}{(2x + 3)^2} = -\frac{1}{2}\frac{1}{2x + 3}\Big|_0^1 = -\frac{1}{2}\left[\frac{1}{2 + 3} - \frac{1}{0 + 3}\right] = -\frac{1}{2}\left[\frac{1}{5} - \frac{1}{3}\right] = \frac{1}{15}$

8 The average is $\frac{1}{\pi/8 - \pi/12}\int_{\pi/12}^{\pi/8} \sec^2 2x\,dx$

$= \frac{24}{\pi}\left(\frac{1}{2}\tan 2x\right)\Big|_{\pi/12}^{\pi/8} = \frac{12}{\pi}\left(\tan\frac{\pi}{4} - \tan\frac{\pi}{6}\right)$

$= \frac{12}{\pi}\left(1 - \frac{1}{\sqrt{3}}\right) = \frac{4}{\pi}(3 - \sqrt{3})$.

9 First note that $D\left(\int_{x^2}^{x^3} \cos 3t\,dt\right)$

$= D\left(\int_{x^2}^{0} \cos 3t\,dt + \int_0^{x^3} \cos 3t\,dt\right)$

$= D\left(-\int_0^{x^2} \cos 3t\,dt + \int_0^{x^3} \cos 3t\,dt\right)$

$= -D\left(\int_0^{x^2} \cos 3t\,dt\right) + D\left(\int_0^{x^3} \cos 3t\,dt\right)$. These last two derivatives are found by using the chain rule and the first fundamental theorem of calculus. For the first, let $u = x^2$; then $\frac{d}{dx}\left(\int_0^{x^2} \cos 3t\,dt\right) = \frac{d}{du}\left(\int_0^{u} \cos 3t\,dt\right)\cdot\frac{du}{dx} = (\cos 3u)\cdot 2x = 2x\cos 3x^2$.

Similarly, $\frac{d}{dx}\left(\int_0^{x^3} \cos 3t\,dt\right) = 3x^2\cos 3x^3$. Thus $D\left(\int_{x^2}^{x^3} \cos 3t\,dt\right) = -2x\cos 3x^2 + 3x^2\cos 3x^3$.

Differentiating again gives $D^2\left(\int_{x^2}^{x^3} \cos 3t\,dt\right) = -2[x\cdot D(\cos 3x^2) + D(x)\cos 3x^2] + 3[x^2\cdot D(\cos 3x^3) + D(x^2)\cos 3x^3] = -2[x(-\sin 3x^2)\cdot 6x + \cos 3x^2] + 3(x^2(-\sin 3x^3))\cdot 9x^2 + 2x\cos 3x^3] = 12x^2\sin 3x^2 - 2\cos 3x^2 - 27x^4\sin 3x^3 + 6x\cos 3x^3$.

10 $G(x) = \int_a^x f(t)\,dt$ is the distance traveled between time $t = a$ and $t = x$ by an object whose velocity at time t is $f(t)$. Hence $\Delta G = G(x - \Delta x) - G(x) = \int_a^{x+\Delta x} f(t)\,dt - \int_a^x f(t)\,dt = \int_x^{x+\Delta x} f(t)\,dt$ is the distance traveled during the time interval $[x, x + \Delta x]$, so $\Delta G/\Delta x$ is the average velocity over this interval. By the mean-value theorem for definite integrals, there is a time $t = c$ in the interval $[x, x + \Delta x]$ where $\Delta G = \int_x^{x+\Delta x} f(t)\,dt = f(c)\,\Delta x$; that is, average velocity $\Delta G/\Delta x$ equals instantaneous velocity $f(c)$. Thus $G'(x) = \lim_{\Delta x \to 0}\frac{\Delta G}{\Delta x} = \lim_{\Delta x \to 0} f(c) = f(x)$, since $x < c < x + \Delta x$ and therefore $c \to x$ as $\Delta x \to 0$. In this case the first fundamental theorem of calculus states that the

velocity $G'(x)$ is equal to the original velocity function, $f(x)$.

11 (a) Any function that can be expressed in terms of polynomials, trigonometric functions (such as $\sin x$), radicals, exponentials or logarithms, or any combination of the above. (See comments immediately following Example 1, Sec. 5.7.)

(b) $\dfrac{\sin x}{x}$ is just one example of an elementary function with no elementary antiderivative.

(c) $\int_0^x \dfrac{\sin u}{u}\,du$ is just one of many such examples.

5.S Review Exercises

1 The area is given by $\int_1^2 2x^3\,dx = \left.\dfrac{x^4}{2}\right|_1^2$

$= \dfrac{16}{2} - \dfrac{1}{2} = \dfrac{15}{2}.$

3 Area $= \int_0^{\pi/6} \sin 3x\,dx = \left.-\dfrac{1}{3}\cos 3x\right|_0^{\pi/6}$

$= -\dfrac{1}{3}\left[\cos\dfrac{\pi}{2} - \cos 0\right] = -\dfrac{1}{3}[0 - 1] = \dfrac{1}{3}$

5 Area $= \int_2^3 \dfrac{1}{x^3}\,dx = \left.\dfrac{x^{-2}}{-2}\right|_2^3 = \left.-\dfrac{1}{2}\cdot\dfrac{1}{x^2}\right|_2^3$

$= -\dfrac{1}{2}\left[\dfrac{1}{9} - \dfrac{1}{4}\right] = -\dfrac{1}{2}\cdot\dfrac{-5}{36} = \dfrac{5}{72}$

7 $\int \sec^2 x\,dx = \tan x + C$

9 $\int \sec x \tan x\,dx = \sec x + C$

11 $\int 4 \csc x \cot x\,dx = -4\int(-\csc x \cot x)\,dx$

$= -4\csc x + C$

13 $\int (x^3 + 1)^2\,dx = \int (x^6 + 2x^3 + 1)\,dx$

$= \dfrac{x^7}{7} + 2\dfrac{x^4}{4} + x + C = \dfrac{x^7}{7} + \dfrac{x^4}{2} + x + C$

15 $\int 100x^{19}\,dx = 100\cdot\dfrac{x^{20}}{20} + C = 5x^{20} + C$

17 (a) $[(x^3 + 1)^6]' = 6(x^3 + 1)^5(3x^2)$

$= 18x^2(x^3 + 1)^5$

(b) From (a) we know that $\dfrac{1}{18}(x^3 + 1)^6$ is an antiderivative of $x^2(x^3 + 1)^5$. Thus

$\int (x^3 + 1)^5 x^2\,dx = \dfrac{1}{18}(x^3 + 1)^6 + C.$

19 (a) $\sum_{j=1}^{3} d^j = d^1 + d^2 + d^3 = d + d^2 + d^3$

(b) $\sum_{k=1}^{4} x^k = x^1 + x^2 + x^3 + x^4$

$= x + x^2 + x^3 + x^4$

(c) $\sum_{i=0}^{3} i2^{-i} = 0\cdot 2^{-0} + 1\cdot 2^{-1} + 2\cdot 2^{-2} + 3\cdot 2^{-3}$

$= 0 + \dfrac{1}{2} + \dfrac{2}{4} + \dfrac{3}{8} = \dfrac{11}{8}$

(d) $\sum_{i=2}^{5} \dfrac{i+1}{i} = \dfrac{2+1}{2} + \dfrac{3+1}{3} + \dfrac{4+1}{4} + \dfrac{5+1}{5}$

$= \dfrac{3}{2} + \dfrac{4}{3} + \dfrac{5}{4} + \dfrac{6}{5} = \dfrac{317}{60}$

(e) $\sum_{i=2}^{4}\left(\frac{1}{i}-\frac{1}{i+1}\right)$

$= \left(\frac{1}{2}-\frac{1}{3}\right)+\left(\frac{1}{3}-\frac{1}{4}\right)+\left(\frac{1}{4}-\frac{1}{5}\right) = \frac{1}{2}-\frac{1}{5}$

$= \frac{3}{10}$

(f) $\sum_{i=1}^{4}\sin\frac{\pi i}{4}$

$= \sin\frac{\pi}{4}+\sin\frac{2\pi}{4}+\sin\frac{3\pi}{4}+\sin\frac{4\pi}{4}$

$= \frac{1}{\sqrt{2}}+1+\frac{1}{\sqrt{2}}+0 = 1+\frac{2}{\sqrt{2}}$

$= 1+\sqrt{2}$

21 We know that $\int_0^4 4x^2\,dx =$ $\lim_{\text{mesh}\to 0}\sum_{i=1}^{n} f(c_i)(x_i - x_{i-1})$, where $x_0 = 0$ and $x_n =$ 4. Now partition [0, 4] into n sections, all of length $x_i - x_{i-1} = \Delta x_i = \frac{x_n - x_0}{n} = \frac{4}{n}$ and select $c_i =$ $x_i = x_0 + \frac{4}{n}i = \frac{4}{n}i$ as the sampling point. We then have $\lim_{\text{mesh}\to 0}\sum_{i=1}^{n} f(c_i)(x_i - x_{i-1}) =$ $\lim_{\text{mesh}\to 0}\sum_{i=1}^{n} 4\left(\frac{4i}{n}\right)^2\cdot\frac{4}{n} = \lim_{n\to\infty}\frac{64}{n^3}\sum_{i=1}^{n} i^2$, since the mesh $4/n$ approaches zero as $n\to\infty$. Using the formula for $\sum_{i=1}^{n} i^2$ derived in Sec. 5.2, $\lim_{n\to\infty}\frac{64}{n^3}\sum_{i=1}^{n} i^2$ reduces to $\lim_{n\to\infty}\frac{64}{n^3}\left(\frac{n^3}{3}+\frac{n^2}{2}+\frac{n}{6}\right)$

$= \lim_{n\to\infty}\left(\frac{64}{3}+\frac{32}{n}+\frac{32}{3n^2}\right) = \frac{64}{3}$. Thus $\int_0^4 4x^2\,dx$

$= \frac{64}{3}.$

23 $\int_1^3 (4x^3 - x^2)\,dx = 4\int_1^3 x^3\,dx - \int_1^3 x^2\,dx$

$= 4\cdot\frac{x^4}{4}\bigg|_1^3 - \frac{x^3}{3}\bigg|_1^3 = 4\left(\frac{3^4}{4}-\frac{1^4}{4}\right)-\left(\frac{3^3}{3}-\frac{1^3}{3}\right)$

$= 80 - \frac{26}{3} = \frac{214}{3}$

25 $\int_{1/2}^{2}\frac{1}{x^2}\,dx = -\int_{1/2}^{2}(-x^{-2})\,dx = -(x^{-1})\big|_{1/2}^{2}$

$= -\left(2^{-1}-\left(\frac{1}{2}\right)^{-1}\right) = -\left(\frac{1}{2}-2\right) = \frac{3}{2}$

27 $\frac{d}{dx}\left(\int_3^x \sqrt{1+t^2}\,dt\right) = \sqrt{1+x^2}$

29 $\frac{d}{dx}\left(\int_{3x}^{4}\tan 3t\,dt\right) = \frac{d}{du}\left(\int_u^4 \tan 3t\,dt\right)\cdot\frac{du}{dx}$

$= -(\tan 3u)(3) = -3\tan 9x$

31 $\frac{d}{dx}\left(\int_4^{\tan x}\cos t\,dt\right) = \frac{d}{du}\left(\int_4^u \cos t\,dt\right)\cdot\frac{du}{dx}$

$= (\cos u)(\sec^2 x) = \cos(\tan x)\sec^2 x$

33 $\int_0^0 2^{x^2}\,dx = 0$

35 $\int_2^1 (12x^3 - 2x)\,dx = -\int_1^2 (12x^3 - 2x)\,dx$

$= -(3x^4 - x^2)\big|_1^2 = -[3\cdot 2^4 - 3\cdot 1^4 - (2^2 - 1^2)]$

$= -(45 - 3) = -42$

37 (a)

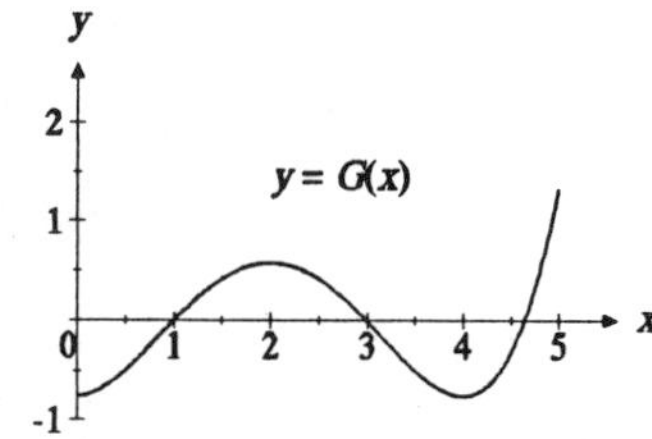

(b) Since the slope of $G(x)$ at $x = a$ corresponds to $G'(a) = f(a)$, we see that the slope when $x = 2$ is $G'(2) = f(2) = 0$ and when $x = 1$ is $G'(1) = f(1) = 1$.

(c) $G(0) = \int_1^0 f(t)\,dt = -\int_0^1 f(t)\,dt \approx -\frac{2}{3}$

(d) $G(3.01) - G(3) \approx G'(3)\cdot(0.01) = f(3)\cdot(0.01)$ $= -0.01$

39 (a) With $h = 1/2$: $\int_1^2 x^4\,dx \approx$

$$\frac{1/2}{3}\left[1^4 + 4\left(\frac{3}{2}\right)^4 + 2^4\right] = \frac{1}{6}\left[1 + \frac{81}{4} + 16\right]$$

$$= \frac{149}{24} \approx 6.208333.$$

With $h = 1/4$: $\int_1^2 x^4\,dx \approx$

$$\frac{1/4}{3}\left[1^4 + 4\cdot\left(\frac{5}{4}\right)^4 + 2\cdot\left(\frac{3}{2}\right)^4 + 4\cdot\left(\frac{7}{4}\right)^4 + 2^4\right]$$

$$= \frac{1}{12}\left[1 + \frac{625}{64} + \frac{81}{8} + \frac{2401}{64} + 16\right]$$

$$= \frac{2381}{384} \approx 6.200521.$$

(b) Since $\int_1^2 x^4\,dx = \left.\frac{x^5}{5}\right|_1^2 = \frac{2^5}{5} - \frac{1^5}{5} = \frac{31}{5}$ $= 6.2$, the errors in (a) are $E(1/2) \approx$ $6.208333 - 6.2 = 0.008333$ and $E(1/4) \approx$ $6.20051 - 6.2 = 0.000521$.

(c) Their ratio is $\frac{E(1/4)}{E(1/2)} \approx \frac{0.000521}{0.008333} =$ 0.0625, that is, 1 to 16.

(d) With $h = 1/20$, the Simpson's estimate is 6.20000083. The ratio of $E(1/20)$ to $E(1/2)$ is 0.0001.

41 (a) With $h = 1/2$, $\int_1^2 x^4\,dx \approx \frac{1}{2}\left[\left(\frac{2}{2}\right)^4 + \left(\frac{3}{2}\right)^4\right]$

$$= \frac{1}{2}\left[1 + \frac{81}{16}\right] = \frac{97}{32} = 3.031250.$$

With $h = 1/4$, $\int_1^2 x^4\,dx \approx$

$$\frac{1}{4}\left[\left(\frac{4}{4}\right)^4 + \left(\frac{5}{4}\right)^4 + \left(\frac{6}{4}\right)^4 + \left(\frac{7}{4}\right)^4\right] =$$

$$\frac{1}{4}\left[\frac{256}{256} + \frac{625}{256} + \frac{1296}{256} + \frac{2401}{256}\right] = \frac{2289}{512}$$

$\approx 4.470703.$

(b) As usual, let $E(h)$ denote the error when sections of length h are used. Then $E(1/2) =$ $3.031250 - 6.2 = -3.168750$ and $E(1/4) \approx$ $4.470703 - 6.2 = -1.729297$.

(c) $\frac{E(1/4)}{E(1/2)} = \frac{-1.729297}{-3.168750} \approx 0.54573$

(d) With $h = 1/20$, the left endpoint method yields the estimate 5.83083. Hence $E(1/20) \approx$ $5.83083 - 6.2 = -0.36917$ and $\frac{E(1/20)}{E(1/2)} \approx$ $\frac{-0.369170}{-3.168750} \approx 0.11650$. In other words, the error with $h = 1/20$ is only about one-tenth of the error with $h = 1/2$.

43 (a) $\frac{d}{dx}\left(\int_4^{x^2} \frac{\sqrt{1 + u^2}}{2\sqrt{u}}\, du\right) = \frac{\sqrt{1 + (x^2)^2}}{2\sqrt{x^2}}\cdot 2x$

$= \sqrt{1 + x^4}$

(b) $\frac{d}{dx}\left(\int_2^x \sqrt{1 + t^4}\, dt\right) = \sqrt{1 + x^4}$

(c) Both sides of the equation must hold for all values of x. Let $x = 2$. Then both integrals are 0, so $C = 0$.

45 (a) $F(b) - F(a)$

(b) $\int_a^b f(x)\, dx$

47 Since total distance is the definite integral of velocity, the distance traveled in the first 3 minutes is $\int_0^3 \frac{1}{(t + 1)^2}\, dt = \int_0^3 (t + 1)^{-2}\, dt$

$= -(t + 1)^{-1}\Big|_0^3 = -4^{-1} + 1^{-1} = \frac{3}{4}$ ft.

49 $y = 3x^2 - 2x^3 = x^2(3 - 2x)$ intercepts the x axis at $x = 0$ and $x = 3/2$. So the area is given by $\int_0^{3/2} c(x)\, dx$, where $c(x) = 3x^2 - 2x^3$ is the cross-sectional length. Thus, $\int_0^{3/2} (3x^2 - 2x^3)\, dx =$

$\left(x^3 - \frac{1}{2}x^4\right)\Big|_0^{3/2} = \left(\frac{3}{2}\right)^3 - \frac{1}{2}\left(\frac{3}{2}\right)^4 - (0 - 0) =$

$\frac{27}{8} - \frac{81}{32} = \frac{27}{32}$.

51 $y = -x^2 + 6x - 7$ intercepts the x axis when $-x^2 + 6x - 7 = 0$, or when $x =$

$\frac{-6 \pm \sqrt{36 - 4(-1)(-7)}}{2(-1)} = 3 \mp \sqrt{2}$. So the area is given by $\int_{3-\sqrt{2}}^{3+\sqrt{2}} c(x)\, dx$, where $c(x) = -x^2 + 6x - 7$ is the cross-sectional length. Thus

$\int_{3-\sqrt{2}}^{3+\sqrt{2}} (-x^2 + 6x - 7)\, dx$

$= \left(-\frac{x^3}{3} + 3x^2 - 7x\right)\Big|_{3-\sqrt{2}}^{3+\sqrt{2}}$

$= -\frac{(3 + \sqrt{2})^3}{3} + 3(3 + \sqrt{2})^2 - 7(3 + \sqrt{2})$

$- \left(-\frac{(3 - \sqrt{2})^3}{3} + 3(3 - \sqrt{2})^2 - 7(3 - \sqrt{2})\right)$

$= \frac{8}{3}\sqrt{2}$.

53 (a) $\int_0^1 \sqrt{1 - x}\, dx = \int_0^1 (1 - x)^{1/2}\, dx$

$= -\frac{2}{3}(1 - x)^{3/2}\Big|_0^1 = \frac{2}{3}$

(b) $\int_0^1 \sqrt[3]{1 - x^2}\, dx$ cannot be evaluated by the fundamental theorem. We therefore employ Simpson's method with $a = 0$, $b = 1$, $n = 4$, and $h = \frac{1}{4}$: $\int_0^1 (1 - x^2)^{1/3}\, dx$

$\approx \frac{1/4}{3}[f(0) + 4f(1/4) + 2f(1/2) + 4f(3/4) + f(1)]$

$= \frac{1}{12}\left[1 + 4\left(1 - \left(\frac{1}{4}\right)^2\right)^{1/3} + 2\left(1 - \left(\frac{1}{2}\right)^2\right)^{1/3}\right.$

$\left. + 4\left(1 - \left(\frac{3}{4}\right)^2\right)^{1/3} + 0\right]$

$= \frac{1}{12}\left[1 + 4\left(\frac{15}{16}\right)^{1/3} + 2\left(\frac{3}{4}\right)^{1/3} + 4\left(\frac{7}{16}\right)^{1/3}\right]$

≈ 0.8140.

(c) $\int_0^1 \sqrt[3]{1+x}\,dx = \int_0^1 (1+x)^{1/3}\,dx$

$= \frac{3}{4}(1+x)^{4/3}\Big|_0^1 = \frac{3}{4}(2)^{4/3} - \frac{3}{4}(1)^{4/3}$

$= \frac{3}{4}\left[2\sqrt[3]{2} - 1\right]$

55 (a) Partition [0, 1] by utilizing the points $x_i = i/n$; that is, $x_0 = \frac{0}{n} < x_1 = \frac{1}{n} < x_2 = \frac{2}{n} < \cdots < x_{n-1} = \frac{n-1}{n} < x_n = \frac{n}{n} = 1$. Each section of length $x_i - x_{i-1}$ equals $1/n$. With $c_i = x_i$, the right endpoint, the approximating sum for the definite integral $\int_0^1 f(x)\,dx$ is

$$\sum_{i=1}^{n} f(c_i)(x_i - x_{i-1}) = \sum_{i=1}^{n} f\left(\frac{i}{n}\right)\frac{1}{n}.$$

(b) The ith section $[x_i, x_{i-1}]$ has length $x_i - x_{i-1} = \frac{i}{n} - \frac{i-1}{n} = \frac{1}{n}$.

(c) The mesh, the length of the largest section, is $1/n$.

(d) The sampling number c_i is the right endpoint of $[x_{i-1}, x_i]$.

57 (a) Compare $\sum_{i=1}^{200}\left(\frac{i}{100}\right)^3\frac{1}{100}$ to the expression $\sum_{i=1}^{n} f\left(\frac{i}{n}\right)\frac{1}{n}$ from Exercise 55(a) and observe that it corresponds to an approximating sum using right endpoints over a partition of the interval [0, 2] into 200 equal sections. The function being integrated is x^3, so we have an approximating sum for $\int_0^2 x^3\,dx$.

(b) The expression $\sum_{i=1}^{100}\left(\frac{i-1}{100}\right)^4\frac{1}{100}$ is an approximating sum (using left endpoints) for the definite integral of x^4 over the interval [0, 1], that is, $\int_0^1 x^4\,dx$.

(c) The sum $\sum_{i=101}^{300}\left(\frac{i}{100}\right)^5\frac{1}{100}$ is a right-endpoint approximating sum for the definite integral of x^5 over the interval [1, 3], that is, $\int_1^3 x^5\,dx$.

59 Consider the volume that is swept out by the jeep's windshield as it travels on its one-mile journey. The raindrops are uniformly distributed and falling at a constant rate so that the total number of drops within the specified volume is essentially constant with respect to time. (As raindrops fall out of the volume the same number fall in.) Therefore the number of drops that spatter on the windshield during the trip will be the same as if the drops were not falling, but merely suspended in air. Since time does not matter, neither does the speed of the jeep.

61 The average velocity over the time interval $[a, b]$ is $\frac{1}{b-a}\int_a^b v(t)\,dt$. Since this equals $v\left(\frac{a+b}{2}\right)$, we have Eq. (1): $\int_a^b v(t)\,dt = (b-a)v\left(\frac{a+b}{2}\right)$.

Holding a fixed and differentiating this equation with respect to b yields $v(b) =$

$(b-a)\cdot\frac{d}{db}\left[v\left(\frac{a+b}{2}\right)\right] + v\left(\frac{a+b}{2}\right)\frac{d}{db}(b-a)$, so we obtain Eq. (2): $v(b) =$

$$(b - a)\frac{1}{2}v'\left(\frac{a + b}{2}\right) + v\left(\frac{a + b}{2}\right).$$ Also, since

$\int_a^b v(t)\,dt = -\int_b^a v(t)\,dt$, holding b fixed and differentiating Eq. (1) with respect to a yields

$-v(a) =$

$$(b - a)\frac{d}{da}\left[v\left(\frac{a + b}{2}\right)\right] + v\left(\frac{a + b}{2}\right)\frac{d}{da}(b - a),$$ so

$$-v(a) = (b - a)\frac{1}{2}v'\left(\frac{a + b}{2}\right) - v\left(\frac{a + b}{2}\right).$$

Subtracting this equation from Eq. (2) yields

$v(b) + v(a) = 2v\left(\frac{a + b}{2}\right)$. Now hold a fixed and differentiate this equation with respect to b: $v'(b)$

$= 2\cdot\frac{1}{2}v'\left(\frac{a + b}{2}\right) = v'\left(\frac{a + b}{2}\right)$. Since this is true for any a and b, $v'(x)$ is a constant function; say $v'(x) = c$ for all x. Then $v(t) =$

$v(0) + \int_0^t v'(x)\,dx = v(0) + \int_0^t c\,dx =$

$ct + v(0) = ct + d$, where $d = v(0)$.

63 We seek dy/dt, evaluated at $y = \sqrt{7}$ meters, since this is the rate of change in the height of the water. We know that $dV/dt = 2\text{ m}^3/\text{hr}$, because the tank is being filled at a rate of 2 cubic meters per hour. We also know the volume as a function of the height y, given the radius $r = (1 + y^2)^{1/3} - 1$. It is

$$V = \int_0^y A(u)\,du = \int_0^y \pi[(1 + u^2)^{1/3} - 1]^2\,du.$$

Using the fact that $\frac{dy}{dt} = \frac{dV/dt}{dV/dy}$, if we knew dV/dy, we could find dy/dt as a function of y (since dV/dt is constant). But we know dV/dy from the first fundamental theorem. Hence $\frac{dV}{dy} =$

$$\frac{d}{dy}\left(\int_0^y \pi[(1 + u^2)^{1/3} - 1]^2\,du\right) = \pi[(1 + y^2)^{1/3} - 1]^2.$$

Finally, $$\frac{dy}{dt} = \frac{dV/dt}{dV/dy} = \frac{2\text{ m}^3/\text{hr}}{\pi[(1 + y^2)^{1/3} - 1]^2}.$$

So dy/dt evaluated at $y = \sqrt{7}$ meters is

$$\frac{2}{\pi[(1 + (\sqrt{7})^2)^{1/3} - 1]^2} = \frac{2}{\pi[8^{1/3} - 1]^2}$$

$$= \frac{2}{\pi}\text{ m/hr}.$$

65 (a) The mean-value theorem applied to $F(x)$ over the interval $[x_{i-1}, x_i]$ implies the existence of a number c_i in (x_{i-1}, x_i) such that $F'(c_i) =$ $\frac{F(x_i) - F(x_{i-1})}{x_i - x_{i-1}}$. Therefore $F(x_i) - F(x_{i-1}) =$

$F'(c_i)(x_i - x_{i-1})$.

(b) Since $F' = f$, $F'(c_i) = f(c_i)$, so the result in (a) becomes $F(x_i) - F(x_{i-1}) = f(c_i)(x_i - x_{i-1})$ for some c_i in (x_{i-1}, x_i).

(c) With the sampling points c_i chosen as in (b),

$$\sum_{i=1}^{n} f(c_i)(x_i - x_{i-1}) = \sum_{i=1}^{n} (F(x_i) - F(x_{i-1})).$$

This telescoping sum equals $F(x_n) - F(x_0) =$ $F(b) - F(a)$.

(d) We know that $\int_a^b f(x)\,dx$ exists. Therefore

$\lim_{\text{mesh}\to 0} \sum_{i=1}^{n} f(c_i)(x_i - x_{i-1})$ exists, no matter how the c_i are chosen. This limit must then equal the limit where the c_i are chosen as in (b). Therefore $\lim_{\text{mesh}\to 0} \sum_{i=1}^{n} f(c_i)(x_i - x_{i-1}) =$

$\lim_{\text{mesh}\to 0} (F(b) - F(a)) = F(b) - F(a)$, which completes the proof of the second fundamental theorem of calculus.

67 (a) $\frac{1}{5}\left(\frac{1}{1} + \frac{1}{1.2} + \frac{1}{1.4} + \frac{1}{1.6} + \frac{1}{1.8}\right) = \frac{1879}{2520}$

(b) $\frac{3}{5}\left(\frac{1}{3} + \frac{1}{3.6} + \frac{1}{4.2} + \frac{1}{4.8} + \frac{1}{5.4}\right) = \frac{1879}{2520}$

(c) Yes. Areas of corresponding rectangles are equal.

(d) The second region is obtained from the first by expanding horizontally by a factor of 3 and compressing vertically by a factor of 3.

(e) The second region is obtained from the first by expanding horizontally by a factor of b and compressing vertically by a factor of b.

(f) $G(ab)$ = (Area above $[1, b]$) + (Area above $[b, ab]$) $= G(b) + G(a)$

(g) $\log_r x$ for any base r has the property that the log of any product is the sum of the logs of the factors.

69 (a) Let $g(t) = \left[\int_0^t f(x)\,dx\right]^2 - \int_0^t [f(x)]^3\,dx$. We will show that $g(t)$ is nonnegative. Since $g(0) = 0$, it suffices to show that $g'(t) \geq 0$. Note that $g'(t) = 2\left[\int_0^t f(x)\,dx\right]f(t) - [f(t)]^3 =$

$f(t)\left[2\int_0^t f(x)\,dx - [f(t)]^2\right]$. Since $f(t) \geq 0$, $g'(t)$ has the same sign as

$2\int_0^t f(x)\,dx - [(f(t)]^2$, which, for convenience, we call $h(t)$. Note that $h(0) = 0$. Then $h'(t) = 2f(t)[1 - f'(t)] \geq 0$ for $t > 0$, so $h(t)$ is increasing; that is, $h(t) \geq 0$. Therefore $g'(t) \geq 0$, from which it follows that $g(t) \geq 0$, as previously observed.

(b) Setting $f(x) = 0$ or $f(x) = x$ gives equality in (a).

71 (a) We use the given table of values to plot the function $y = a(t)$.

t	0	1	2	3	4	5	6	7	8	9	10
$a(t)$	10	11	12	8	6	4	1	−3	−4	−5	−6

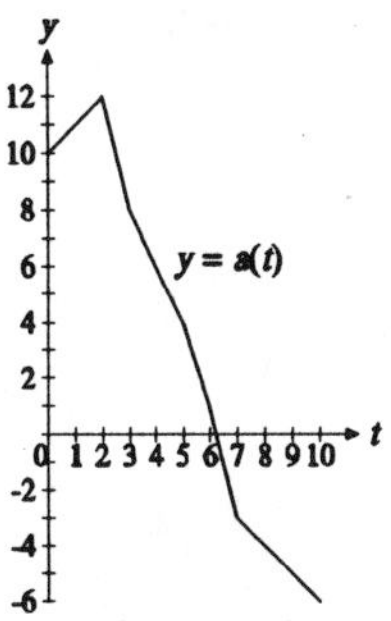

(b) Since the velocity function $v(t)$ is given by $v(t) = \int_0^t a(x)\,dx$, we can compute it as the area under the curve $y = a(t)$. Although somewhat tedious because of the number of computations, it is fairly straightforward to obtain the following graph of velocity. (All results are, of course, approximate.)

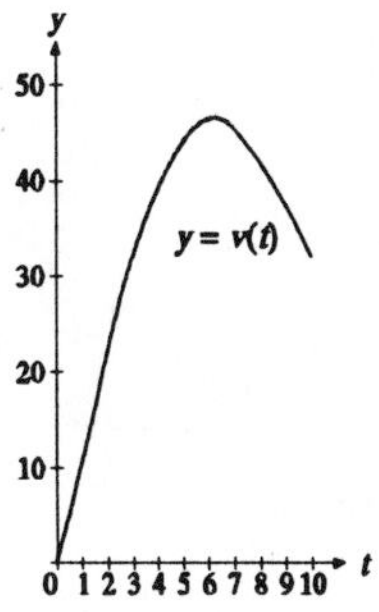

Note that $v(10) = 30$ mi/hr, the velocity of the car after 10 seconds.

(c) Since the distance function $s(t)$ is the definite integral of the velocity function, the area below the curve graphed in (b) yields values for the distance. We have $s(5) = \int_0^5 v(t)\, dt$, which is approximately 126. The units are miles per hour multiplied by seconds, so we need to divide by 3600 sec/hr to obtain 0.035 mi or 184.8 ft. Furthermore, $s(10) = \int_0^{10} v(t)\, dt \approx 336$, which converts to 0.093 mi or 492.8 ft.

6 Topics in Differential Calculus

6.1 Logarithms

1 The expression $b^x = y$ is equivalent to $x = \log_b y$.

(a) $5 = \log_2 32$ (b) $4 = \log_3 81$

(c) $-3 = \log_{10}(0.001)$ (d) $0 = \log_5 1$

(e) $1/3 = \log_{1000}(10)$ (f) $1/2 = \log_{49} 7$

3 (a)

x	1/9	1/3	1	3	9
$\log_3 x$	−2	−1	0	1	2

(b)

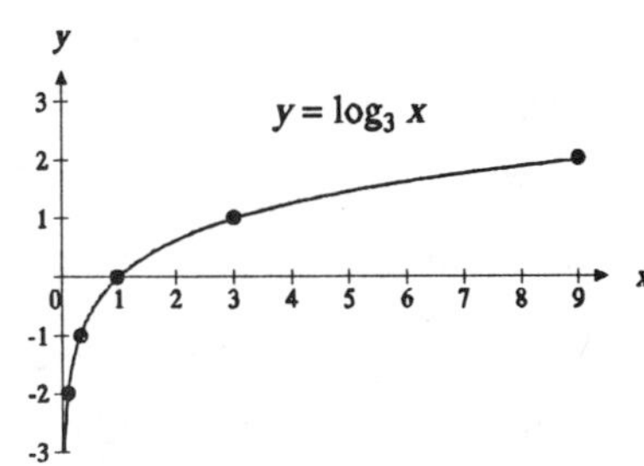

5 $\log_b y = x$ is equivalent to $b^x = y$.

(a) $2^x = 7$ (b) $5^s = 2$

(c) $3^{-1} = 1/3$ (d) $7^2 = 49$

7 Recall that $b^{\log_b c} = c$.

(a) $2^{\log_2 16} = 16$ (b) $2^{\log_2 (1/2)} = \frac{1}{2}$

(c) $2^{\log_2 7} = 7$ (d) $2^{\log_2 g} = g$

9 (a) Since $3^{1/2} = \sqrt{3}$, $\log_3 \sqrt{3} = \frac{1}{2}$.

(b) $\log_3 (3^5) = 5 \log_3 3 = 5 \cdot 1 = 5$. (In general, $\log_b b = 1$.)

(c) $\log_3 (1/27) = \log_3 (3^{-3}) = -3 \log_3 3 = -3 \cdot 1 = -3$

11 Division by 2 yields $3^x = 7/2$. Hence $x = \log_3 (7/2)$, which, as in Example 5, we may write as $x = \dfrac{\log_{10}(7/2)}{\log_{10} 3}$.

Note: We could just as well have used logs to other bases.

13 Taking the common logarithms of $3^{5x} = 2^{7x}$, we obtain

$\log_{10}(3^{5x}) = \log_{10}(2^{7x})$

$5x \log_{10} 3 = 7x \log_{10} 2$ (log of a power)

$5x \log_{10} 3 - 7x \log_{10} 2 = 0$

$x(5 \log_{10} 3 - 7 \log_{10} 2) = 0.$

Since $5 \log_{10} 3 - 7 \log_{10} 2 = \log_{10} 3^5 - \log_{10} 2^7 = \log_{10} \dfrac{3^5}{2^7}$ is nonzero, x must be 0.

15 (a) Since $2^3 = 8$, $\log_2 8 = 3$. On the other hand, $2 = \sqrt[3]{8} = 8^{1/3}$, so $\log_8 2 = 1/3$. Then $(\log_2 8)(\log_8 2) = 1$.

(b) Let a and b be any two nonzero bases. We know that $a^{\log_a b} = b$. Taking $\log_b$ of this equation, $\log_b\left[a^{\log_a b}\right] = \log_b b$. Recall that $\log_b b = 1$ and that $\log_b\left[a^{\log_a b}\right] = (\log_a b)(\log_b a)$ because $\log_b(c^m) = m \log_b c$. Combining these, $(\log_a b)(\log_b a) = 1$.

17 (a) $\log_{10} 4 = \log_{10} (2^2) \approx 2\cdot(0.30) = 0.60$

(b) $\log_{10} 5 = \log_{10} (10/2) = \log_{10} 10 - \log_{10} 2$

$= 1 - \log_{10} 2 \approx 1 - 0.30 = 0.70$

(c) $\log_{10} 6 = \log_{10} (2\cdot 3) = \log_{10} 2 + \log_{10} 3$

$\approx 0.30 + 0.48 = 0.78$

(d) $\log_{10} 8 = \log_{10} (2^3) = 3 \log_{10} 2 \approx 3\cdot 0.30$

$= 0.90$

(e) $\log_{10} 9 = \log_{10} (3^2) = 2 \log_{10} 3 \approx 2\cdot 0.48$

$= 0.96$

(f) $\log_{10} 1.5 = \log_{10} (3/2) = \log_{10} 3 - \log_{10} 2$

$\approx 0.48 - 0.30 = 0.18$

(g) $\log_{10} 1.2 = \log_{10} \left(\frac{2^2\cdot 3}{10}\right)$

$= 2 \log_{10} 2 + \log_{10} 3 - \log_{10} 10$

$\approx 2\cdot 0.30 + 0.48 - 1 = 0.08$

(h) $\log_{10} 1.33 \approx \log_{10} (4/3) = \log_{10} (2^2/3) =$

$2 \log_{10} 2 - \log_{10} 3 \approx 2\cdot 0.30 - 0.48 = 0.12$

(i) $\log_{10} 20 = \log_{10} (10\cdot 2) = \log_{10} 10 + \log_{10} 2$

$= 1 + \log_{10} 2 \approx 1 + 0.30 = 1.30$

(j) $\log_{10} 200 = \log_{10} (100\cdot 2)$

$= \log_{10} 100 + \log_{10} 2 = 2 + \log_{10} 2$

$\approx 2 + 0.30 = 2.30$

(k) $\log_{10} 0.006 = \log_{10} \left(\frac{2\cdot 3}{1000}\right)$

$= \log_{10} 2 + \log_{10} 3 - \log_{10} 1000$

$\approx 0.30 + 0.48 - 3 = -2.22$

19 (a) $\log_3 5 = \frac{\log_{10} 5}{\log_{10} 3} \approx \frac{0.69897}{0.47712} \approx 1.46$

(b) $\log_2 3 = \frac{\log_{10} 3}{\log_{10} 2} \approx \frac{0.47712}{0.30103} \approx 1.58$

21 (a) $\log_{1/2} x = \log_{1/2} 2 \cdot \log_2 x = -\log_2 x$

(b)

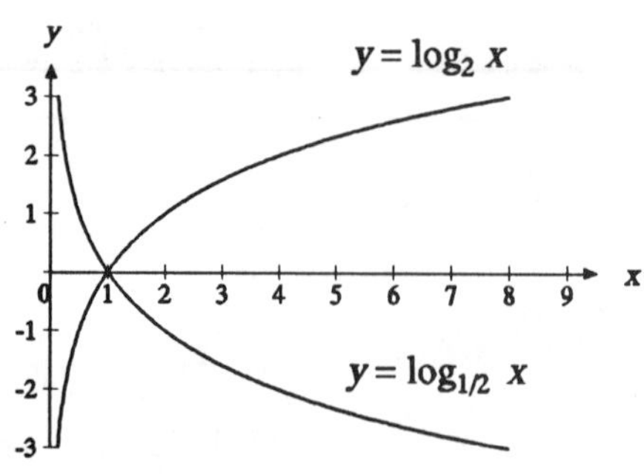

23 $c = b^{\log_b c}$, $c^m = \left[b^{\log_b c}\right]^m = b^{(\log_b c)m} = b^{m \log_b c}$

and $\log_b(c^m) = \log_b\left[b^{m \log_b c}\right] = m \log_b c$.

25 $\log_{10} f(x) = \log_{10} (\cos^7 x) + \log_{10}\sqrt{(x^2 + 5)^3} -$

$\log_{10}(4 + \tan^2 x) = 7 \log_{10}(\cos x) +$

$(3/2)\log_{10}(x^2 + 5) - \log_{10}(4 + \tan^2 x)$

27 $\log_{10} f(x) = x^2 \log_{10} (x\sqrt{2 + \cos x})$

$= x^2(\log_{10} x + \log_{10} \sqrt{2 + \cos x})$

$= x^2(\log_{10} x + (1/2)\log_{10} (2 + \cos x))$

29 $\log_2(\log_2(\log_2 2^{1024}))) = \log_2(\log_2 1024) = \log_2 10$

31 To find the product of a and b use the fact that $\log_{10} ab = \log_{10} a + \log_{10} b$ (since the calculator has a "log" key). We know $\log_{10} a$, $\log_{10} b$, and $\log_{10} a + \log_{10} b$ can be found on the calculator (since the "+" key still works). Hence $\log_{10} ab$ is known. Using the "10^x" key with $x = \log_{10} ab$ gives the product ab because $10^{\log_{10} ab} = ab$. To find a/b, recall that $\log_{10}(a/b) = \log_{10} a - \log_{10} b$. Again, $\log_{10} a - \log_{10} b$ can be easily found on the calculator (since the "−" key still works) so $\log_{10} (a/b)$ is known. Using the "10^x" key with $x = \log_{10} (a/b)$ gives a/b, because $10^{\log_{10} (a/b)} = a/b$. Note that b/a could be found through a similar process or through use of the "$1/x$" key with $x = a/b$ (if a/b is known).

33 (a) Substituting $a = 1.5$ and $E_0 = 2.5 \times 10^{11}$ ergs into the equation given and solving for $\log_{10} E$ yields

$$1.5M = \log_{10}\left(\frac{E}{2.5 \times 10^{11}}\right)$$

$1.5M = \log_{10} E - \log_{10} (2.5 \times 10^{11})$

$1.5M + \log_{10} (2.5 \times 10^{11}) = \log_{10} E$

$1.5M + \log_{10} 2.5 + \log_{10} 10^{11} = \log_{10} E$

Thus, $\log_{10} E = 11 + \log_{10} 2.5 + 1.5M$

$\approx 11 + 0.4 + 1.5M \approx 11.4 + 1.5M.$

(b) Let E_J denote the energy of the earthquake in Japan and E_S the energy of the earthquake in San Francisco. Let R equal their ratio: $R = E_J/E_S$. Therefore $\log_{10} R = \log_{10} (E_J/E_S) = \log_{10} E_J - \log_{10} E_S$. Now, from (a), with $M_J = 8.9$ and $M_S = 7.1$, we have

$\log_{10} R = \log_{10} E_J - \log_{10} E_S$

$\approx 11.4 + 1.5M_J - (11.4 + 1.5M_S)$

$= 1.5(M_J - M_S) = 1.5(8.9 - 7.1)$

$= 1.5(1.8) = 2.7.$

Hence, $R \approx 10^{2.7} \approx 500$. We conclude that the earthquake in Japan involved 500 times the energy of the earthquake in San Francisco.

(c) This is similar to (b) with $M_{1906} = 8.3$ corresponding to M_J and $M_{1989} = 7.1$ corresponding to M_S. Thus $\log_{10} R \approx 1.5(M_{1906} - M_{1989}) = 1.5(8.3 - 7.1) = 1.5(1.2) = 1.8$. Hence $R \approx 10^{1.8} \approx 63$. So the 1906 earthquake was 63 times as powerful as the 1989 earthquake.

(d) Since $\log_{10} E \approx 11.4 + 1.5M$, it immediately follows that $E \approx 10^{11.4 + 1.5M} = 10^{11.4} \cdot 10^{1.5M} \approx (2.5 \times 10^{11})\cdot(31.623)^{M}$.

(e) Let E_W and M_W correspond to the weaker earthquake and E_S and M_S correspond to the stronger. We are given $M_S = M_W + 1$. So the ratio of their energies, E_W/E_S, from (d),

becomes $\dfrac{E_S}{E_W} \approx \dfrac{(2.5 \times 10^{11})(31.623)^{M_s}}{(2.5 \times 10^{11})(31.623)^{M_W}} =$

$$\frac{(31.623)^{M_W + 1}}{(31.623)^{M_W}} = 31.623.$$

(f) The energy of a 10-megaton H-bomb is

10×10^6 tons times $\dfrac{4.2 \times 10^6 \text{ ergs}}{\text{ton}} =$

4.2×10^{13} ergs. Since, from (a), $\log_{10} E \approx 11.4 + 1.5M$, we know E and can solve for its corresponding M. Therefore

$\log_{10}(4.2 \times 10^{13}) \approx 11.4 + 1.5M,$

$\log_{10} 4.2 + \log_{10} 10^{13} \approx 11.4 + 1.5M,$

$0.623 + 13 \approx 11.5 + 1.5M$, $2.223 \approx 1.5M$, and $M \approx 2.223/1.5 \approx 1.482$. So the Richter rating of a 10-megaton H-bomb is about 1.5.

35 (a) $\lim_{x\to\infty} \log_b x = -\lim_{x\to\infty} \log_{1/b} x = -\infty$

(b) $\lim_{x\to 0^+} \log_b x = -\lim_{x\to 0^+} \log_{1/b} x = -(-\infty)$

$= \infty$

6.2 The Number *e*

1

x	0.1	0.01	0.001	0.0005
$(1 + x)^{1/x}$	2.59374	2.70481	2.71692	2.71760

3 $\lim_{t\to 0} (1 + t)^{1000} = (1 + 0)^{1000} = 1$

5 $\lim_{h\to 0} (1 + 3h)^{1/(4h)} = \lim_{h\to 0}[(1 + 3h)^{1/(3h)}]^{3/4} = e^{3/4}$

7 $\lim_{\Delta x \to 0}\left(1 + \frac{\Delta x}{2x}\right)^{\frac{x}{\Delta x}} = \lim_{\Delta x \to 0}\left[\left(1 + \frac{\Delta x}{2x}\right)^{\frac{2x}{\Delta x}}\right]^{\frac{1}{2}} = e^{1/2}$

9

x	-0.9	-0.99	-0.999
$(1 + x)^{1/x}$	12.915	104.76	1006.9

$(1 + x)^{1/x}$ seems to approach ∞.

11 In the notation of Example 3, we have $A = \$1000$ and $r = 0.5$, so the final amount is

$1000\left(1 + \frac{0.5}{n}\right)^n$ dollars.

(a) $n = 1$: $1000\left(1 + \frac{1}{2}\right)^1 = 1500$ dollars.

(b) $n = 2$: $1000\left(1 + \frac{1}{4}\right)^2 = 1562.50$ dollars.

(c) $n = 12$: $1000\left(1 + \frac{1}{24}\right)^{12} \approx 1632.09$ dollars.

(d) $n = 365$: $1000\left(1 + \frac{1}{730}\right)^{365} \approx 1648.16$ dollars.

(e) As $n \to \infty$: $1000\left(1 + \frac{1}{2n}\right)^n \to 1000e^{1/2} \approx 1648.72$ dollars.

13

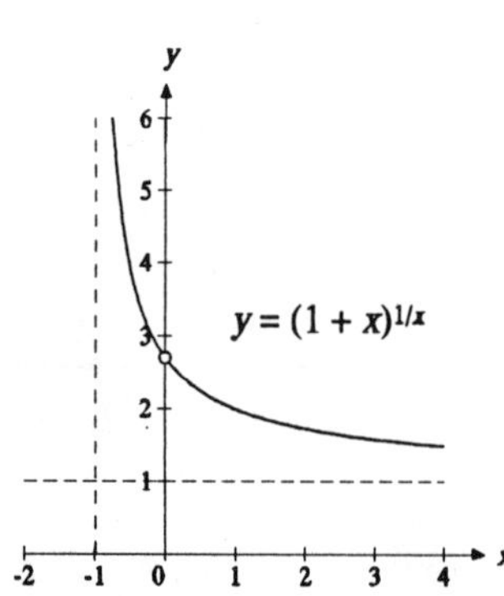

15 (a) $h = 0.1$: $\frac{2^{0.1} - 1}{0.1} \approx 0.7177$

$h = 0.001$: $\frac{2^{0.001} - 1}{0.001} \approx 0.6934$

$h = -0.001$: $\frac{2^{-0.001} - 1}{-0.001} \approx 0.6929$

(b) Let $f(x) = 2^x$. Then for small h, $f'(0) \approx \frac{f(0 + h) - f(0)}{h} = \frac{2^h - 1}{h}$, which is approximately 0.6934 when $h = 0.001$.

(c) Let $f(x) = 3^x$. Then, for small h, $f'(0) \approx \frac{f(0 + h) - f(0)}{h} = \frac{3^h - 1}{h}$, which is approximately 1.0992 when $h = 0.001$.

17 Each of the n periods is now t/n instead of $1/n$. So the total amount compounded continuously is

$$\lim_{n \to \infty} A\left(1 + \frac{rt}{n}\right)^n = A \lim_{n \to \infty}\left[\left(1 + \frac{rt}{n}\right)^{n/(rt)}\right]^{rt} = Ae^{rt}.$$

6.3 The Derivative of a Logarithmic Function

1 $(\ln(1 + x^2))' = \frac{1}{1 + x^2}(1 + x^2)' = \frac{2x}{1 + x^2}$

3 $(x^2 \ln x)' = x^2(\ln x)' + \ln x\,(x^2)'$

$= x^2 \cdot \frac{1}{x} + \ln x \cdot (2x) = x + 2x \ln x$

$= x(1 + 2 \ln x)$

5 $\left(\frac{\ln x}{x}\right)' = \frac{x \cdot \frac{1}{x} - (\ln x) \cdot 1}{x^2} = \frac{1 - \ln x}{x^2}$

7 $(\ln 5x \sin 2x)' = \ln 5x\,(\sin 2x)' + \sin 2x\,(\ln 5x)'$

$= \ln 5x\,(2\cos 2x) + \sin 2x \cdot \dfrac{1}{5x} \cdot 5$

$= 2\cos 2x \ln 5x + \dfrac{\sin 2x}{x}$

9 $[\ln(\sin x)]' = \dfrac{1}{\sin x} \cdot (\sin x)' = \dfrac{\cos x}{\sin x} = \cot x$

11 $\dfrac{d}{dx}(\ln(2x+3)) = \dfrac{1}{2x+3} \cdot (2x+3)' = \dfrac{2}{2x+3}$

13 First, note that $\dfrac{2}{25(5x+2)} = \dfrac{2}{25}(5x+2)^{-1}$. Its derivative is now easily computed as

$\dfrac{2}{25}(-1)(5x+2)^{-2}(5) = -\dfrac{2}{5}(5x+2)^{-2} = -\dfrac{2}{5(5x+2)^2}$. To compute the derivative of $\dfrac{1}{25}\ln(5x+2)$, let $u = 5x+2$ and use the chain rule: $\dfrac{d}{dx}\left(\dfrac{1}{25}\ln(5x+2)\right) = \dfrac{1}{25} \cdot \dfrac{d}{dx}(\ln(u)) =$

$\dfrac{1}{25} \cdot \dfrac{d}{du}(\ln(u)) \cdot \dfrac{du}{dx} = \dfrac{1}{25} \cdot \dfrac{1}{u} \cdot 5 = \dfrac{1}{5}\left(\dfrac{1}{5x+2}\right)$.

Combining these results:

$\left(\dfrac{2}{25(5x+2)} + \dfrac{1}{25}\ln(5x+2)\right)'$

$= \dfrac{-2}{5(5x+2)^2} + \dfrac{1}{5}\left(\dfrac{1}{5x+2}\right)$

$= \dfrac{1}{5(5x+2)^2}(-2 + (5x+2)) = \dfrac{5x}{5(5x+2)^2}$

$= \dfrac{x}{(5x+2)^2}$.

15 Let $u = x + \sqrt{x^2-5} = x + (x^2-5)^{1/2}$; then $\dfrac{du}{dx}$

$= 1 + \dfrac{1}{2}(x^2-5)^{-1/2}(2x) = 1 + x(x^2-5)^{-1/2}$. (We used the chain rule.) Now $\dfrac{d}{dx}\left[\ln\left(x + \sqrt{x^2-5}\right)\right] =$

$\dfrac{d}{dx}(\ln u) = \dfrac{d}{du}(\ln u)\dfrac{du}{dx} = \dfrac{1}{u}\left[1 + \dfrac{x}{\sqrt{x^2-5}}\right]$

$= \left[\dfrac{1}{x + \sqrt{x^2-5}}\right]\left[\dfrac{\sqrt{x^2-5} + x}{\sqrt{x^2-5}}\right] = \dfrac{1}{\sqrt{x^2-5}}$.

17 The logarithm of a quotient is a difference of logarithms, so

$\dfrac{1}{5}\ln\dfrac{x}{3x+5} = \dfrac{1}{5}(\ln x - \ln(3x+5))$

$= \dfrac{1}{5}\ln x - \dfrac{1}{5}\ln(3x+5)$.

Then $\dfrac{d}{dx}\left(\dfrac{1}{5}\ln\dfrac{x}{3x+5}\right) =$

$\dfrac{1}{5}\dfrac{d}{dx}(\ln x) - \dfrac{1}{5}\dfrac{d}{dx}(\ln(3x+5))$. Let $u = 3x+5$, so $\dfrac{du}{dx} = 3$. Then, by the chain rule $\dfrac{d}{dx}(\ln(3x+5))$

$= \dfrac{d}{dx}(\ln u) = \dfrac{d}{du}(\ln u) \cdot \dfrac{du}{dx} = \dfrac{1}{3x+5} \cdot 3 =$

$\dfrac{3}{3x+5}$. Hence $\dfrac{d}{dx}\left(\dfrac{1}{5}\ln\dfrac{x}{3x+5}\right) =$

$\dfrac{1}{5x} - \dfrac{1}{5} \cdot \dfrac{3}{3x+5} = \dfrac{1}{5} \cdot \dfrac{3x+5-3x}{x(3x+5)}$

$= \dfrac{1}{x(3x+5)}$.

19 By the properties of logarithms, $\dfrac{1}{10}\ln\left(\dfrac{5+x}{5-x}\right) =$

$\dfrac{1}{10}(\ln(5+x) - \ln(5-x))$, and so

$$\frac{d}{dx}\left[\frac{1}{10}\ln\left(\frac{5+x}{5-x}\right)\right]$$

$$= \frac{1}{10}\left[\frac{d}{dx}(\ln(5+x)) - \frac{d}{dx}(\ln(5-x))\right]$$

$$= \frac{1}{10}\left[\frac{1}{5+x}(5+x)' - \frac{1}{5-x}(5-x)'\right]$$

$$= \frac{1}{10}\left[\frac{1}{5+x}(1) - \frac{1}{5-x}(-1)\right]$$

$$= \frac{1}{10}\left[\frac{1}{5+x} + \frac{1}{5-x}\right] = \frac{1}{10}\cdot\frac{(5-x)+(5+x)}{(5+x)(5-x)}$$

$$= \frac{1}{10}\cdot\frac{10}{25-x^2} = \frac{1}{25-x^2}.$$

21 We have $\ln[(x^2 + 1)^3(x^5 + 1)^4] =$ $3 \ln (x^2 + 1) + 4 \ln(x^5 + 1)$. By the chain rule, $[\ln(f(x))]' = \dfrac{f'(x)}{f(x)}$, so $(\ln[(x^2 + 1)^3(x^5 + 1)^4])' =$ $(3 \ln(x^2 + 1) + 4 \ln (x^5 + 1))' =$ $3\cdot\dfrac{2x}{x^2+1} + 4\cdot\dfrac{5x^4}{x^5+1} = \dfrac{6x}{x^2+1} + \dfrac{20x^4}{x^5+1}$.

23 Recall that $(\log_b x)' = \dfrac{\log_b e}{x}$. Then $\dfrac{d}{dx}\left(\log_{10}\sqrt[3]{x}\right)$ $= \dfrac{\log_{10} e}{\sqrt[3]{x}}\cdot(x^{1/3})' = \dfrac{\log_{10} e}{\sqrt[3]{x}}\cdot\dfrac{1}{3}x^{-2/3} =$ $\dfrac{\log_{10} e}{x^{1/3}}\cdot\dfrac{1}{3x^{2/3}} = \dfrac{\log_{10} e}{3x}$. Since $\log_{10} e =$ $1/(\ln 10)$, our answer can also be written as $\dfrac{1}{3(\ln 10)x}$.

25 As in Example 3, we find that both axes are asymptotes. Also, (1, 0) is the intercept, and the graph crosses the axis from negative to positive there. Next, $D\left[\dfrac{\ln x}{x^2}\right] = \dfrac{1 - 2\ln x}{x^3}$ changes from positive to negative when $x = e^{1/2}$; thus $\left(\sqrt{e}, \dfrac{1}{2e}\right)$ is a maximum. Finally, $D^2\left[\dfrac{\ln x}{x^2}\right] = \dfrac{6\ln x - 5}{x^4}$ changes from negative to positive when $x = e^{5/6}$; thus $\left[e^{5/6}, \dfrac{5}{6e^{5/3}}\right]$ is an inflection point.

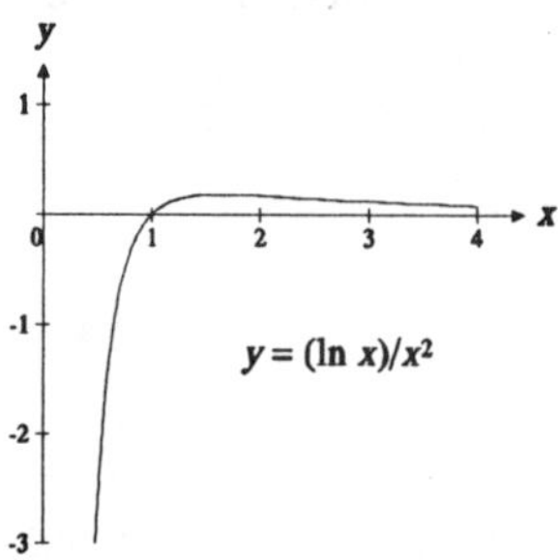

27 Let $y = (1 + 3x)^5(\sin 3x)^6$. Then $\ln y =$ $5 \ln(1 + 3x) + 6 \ln (\sin 3x)$. Realizing that $[\ln(f(x))]' = \dfrac{f'(x)}{f(x)}$, implicit differentiation yields $\dfrac{1}{y}\cdot y' = 5\left(\dfrac{3}{1+3x}\right) + 6\left(\dfrac{3\cos 3x}{\sin 3x}\right)$, so $y' =$ $15(1+3x)^4(\sin 3x)^6 + 18(1+3x)^5(\sin 3x)^5 \cos 3x$.

29 Let $y = \dfrac{(\sec 4x)^{5/3}\sin^3 2x}{\sqrt{x}}$; we have $\ln y =$ $\dfrac{5}{3}\ln(\sec 4x) + 3\ln(\sin 2x) - \dfrac{1}{2}\ln x$. Since $[\ln(f(x))]' = \dfrac{f'(x)}{f(x)}$, implicit differentiation yields

$$\frac{1}{y}\cdot y' = \frac{5}{3}\left(\frac{4\sec 4x\tan 4x}{\sec 4x}\right) + 3\left(\frac{2\cos 2x}{\sin 2x}\right) - \frac{1}{2x}$$

$= \frac{20}{3}\tan 4x + 6\cot 2x - \frac{1}{2x}$. Solving for y', y'

$= y\left(\frac{20}{3}\tan 4x + 6\cot 2x - \frac{1}{2x}\right) =$

$\frac{(\sec 4x)^{5/3}\sin^3 2x}{\sqrt{x}}\left(\frac{20}{3}\tan 4x + 6\cot 2x - \frac{1}{2x}\right)$.

31 (a) $\int_1^4 \frac{1}{x}\,dx \approx \frac{1/2}{3}\left[\frac{1}{2/2} + 4\cdot\frac{1}{3/2} + 2\cdot\frac{1}{4/2} +\right.$

$\left. 4\cdot\frac{1}{5/2} + 2\cdot\frac{1}{6/2} + 4\cdot\frac{1}{7/2} + \frac{1}{8/2}\right] =$

$\frac{1}{6}\cdot\frac{3497}{420} = \frac{3497}{2520} \approx 1.3877$

(b) Since the error in Simpson's method (with $a = 1$, $b = 4$, and $f(x) = 1/x$) is at most $\frac{1}{180}(4 - 1)h^4M_4 = \frac{h^4}{60}\cdot M_4$, we must first find M_4, the maximum value of $|f^{(4)}(x)|$ over $[1, 4]$, before solving for h. Now $f^{(1)}(x) = -\frac{1}{x^2}$, $f^{(2)}(x) = \frac{2}{x^3}$, $f^{(3)}(x) = -\frac{6}{x^4}$, and $f^{(4)}(x)$ $= \frac{24}{x^5}$. Hence $M_4 = |f^{(4)}(1)| = 24$. We want to bound the error, so $\frac{24}{60}\cdot h^4 = \frac{2}{5}h^4 <$ 0.0005; solving for h gives $h < \sqrt[4]{\frac{1}{800}} \approx$ 0.1880. Therefore, we should choose $h <$ 0.1880.

33 (a) First, let $f(x) = \ln(1 + x) - (1 - x) =$ $\ln(1 + x) + x - 1$. Then we have $f(0) =$ $\ln 1 + 0 - 1 = -1 < 0$ while $f(1) =$ $\ln 2 + 1 - 1 = \ln 2 > 0$. Since f is continuous, the intermediate-value theorem says that there exists a number r in $[0, 1]$ such that $f(r) = 0$, or $\ln(1 + r) = 1 - r$.

(b) $f'(x) = \frac{1}{1 + x} + 1 > 0$ for x in $[0, 1]$. Since f is increasing on $[0, 1]$, there can be at most one number r in $[0, 1]$ such that $f(r) = 0$. By (a), we know r exists.

(c) Applying Newton's method with $f(x) =$ $\ln(1 + x) + x - 1$ and $x_1 = 0.5$, we see that

$x_2 = x_1 - \frac{f(x_1)}{f'(x_1)}$

$= x_1 - \frac{\ln(1 + x_1) + x_1 - 1}{1/(1 + x_1) + 1}$

$= 0.5 - \frac{\ln 1.5 + 0.5 - 1}{1/(1.5) + 1} \approx 0.5567209$.

(d) Continuing the application of Newton's method yields $x_3 \approx 0.5571456$ and $x_4 \approx$ 0.5571456. Therefore, to four decimal places, $r \approx 0.5571$.

35 (a) Let $y = fg$. Then $\ln y = \ln f + \ln g$, so

$\frac{1}{y}\cdot\frac{dy}{dx} = \frac{1}{f}\cdot\frac{df}{dx} + \frac{1}{g}\cdot\frac{dg}{dx} = \frac{f'}{f} + \frac{g'}{g}$.

Multiplying by $y = fg$ gives $(fg)' = \frac{dy}{dx} =$

$fg\left(\frac{f'}{f} + \frac{g'}{g}\right) = gf' + fg' = fg' + gf'$.

(b) Let $y = f/g$. Then $\ln y = \ln f - \ln g$, so

$\frac{1}{y}\cdot\frac{dy}{dx} = \frac{1}{f}\cdot\frac{df}{dx} - \frac{1}{g}\cdot\frac{dg}{dx} = \frac{f'}{f} - \frac{g'}{g}$.

Multiplying by $y = f/g$ gives $(f/g)' = \dfrac{dy}{dx} =$

$$\frac{f}{g}\left(\frac{f'}{f} - \frac{g'}{g}\right) = \frac{f'}{g} - \frac{fg'}{g^2} = \frac{gf' - fg'}{g^2}.$$

37 (a) $\displaystyle\int \frac{5\,dx}{5x + 1} = \ln|5x + 1| + C$

(b) $\displaystyle\int \frac{x\,dx}{x^2 + 5} = \frac{1}{2}\int \frac{2x\,dx}{x^2 + 5}$

$$= \frac{1}{2}\ln(x^2 + 5) + C$$

(c) $\displaystyle\int \frac{\cos x\,dx}{\sin x} = \ln|\sin x| + C$

(d) $\displaystyle\int \frac{(1/x)\,dx}{\ln x} = \ln|\ln x| + C$

39 $[(\ln x)^2]' = 2(\ln x)(\ln x)' = \dfrac{2\ln x}{x}$, so

$$\int_e^{e^2} \frac{\ln x}{x}\,dx = \frac{1}{2}(\ln x)^2\Big|_e^{e^2} = \frac{1}{2}[(\ln e^2)^2 - (\ln e)^2]$$

$= \dfrac{1}{2}[2^2 - 1^2] = \dfrac{3}{2}$, which is the desired area.

41 (a) The area of the rectangles is

$\dfrac{1}{1} + \dfrac{1}{2} + \dfrac{1}{3} + \cdots + \dfrac{1}{n-1}$ and the area under the curve $y = 1/x$ and above $[1, n]$ is

$\displaystyle\int_1^n \frac{1}{x}\,dx = \ln n$, so the area we want is the difference of the two,

$\dfrac{1}{1} + \dfrac{1}{2} + \dfrac{1}{3} + \cdots + \dfrac{1}{n-1} - \ln n$, as claimed.

(b) Observe that in Figure 4 all of the blue regions can be translated to the left so as to fit without overlapping within the unit square with base $[1, 2]$; hence their total area is less than 1.

(c) If we "straighten" the curved edge of each blue region we obtain a right triangle that lies entirely within the region. Adding up the area of the triangles will give a lower bound on the total area of the blue regions. Hence

$$\frac{1}{2}(1)\left(1 - \frac{1}{2}\right) + \frac{1}{2}(1)\left(\frac{1}{2} - \frac{1}{3}\right) +$$

$$\frac{1}{2}(1)\left(\frac{1}{3} - \frac{1}{4}\right) + \cdots + \frac{1}{2}(1)\left(\frac{1}{n-1} - \frac{1}{n}\right) =$$

$$\frac{1}{2}\left(1 - \frac{1}{2} + \frac{1}{2} - \frac{1}{3} + \frac{1}{3} - \frac{1}{4} + \cdots + \frac{1}{n-1} - \frac{1}{n}\right)$$

$= \dfrac{1}{2}\left(1 - \dfrac{1}{n}\right)$ is less than the area of the blue regions for all choices of n. As $n \to \infty$, this shows that $1/2$ is a lower bound on γ. Combining this result with (b), we see that γ is between $1/2$ and 1.

43 (a) If we divide $[1, 2]$ into n sections of equal size and use left endpoints as sampling points, we can construct an overestimate for $\displaystyle\int_1^2 \frac{dx}{x}$.

The simplest partition uses one section of length $2 - 1 = 1$, where the height at the left endpoint is $1/1 = 1$, so the overestimate is $1 \cdot 1 = 1$. Therefore, $\displaystyle\int_1^2 \frac{dx}{x} < 1$.

(b) If we divide $[1, 3]$ into n sections of equal size and use right endpoints as sampling points, we can construct an underestimate for $\displaystyle\int_1^3 \frac{dx}{x}$. The partition points are $1 = n/n$,

$(n + 2)/n$, $(n + 4)/n$, ..., $(3n)/n = 3$ and the underestimating approximating sum is

$$\sum_{i=1}^{n} f(x_i)(x_i - x_{i-1}) = \sum_{i=1}^{n} \frac{1}{x_i}\left(\frac{2}{n}\right)$$

$$= \left(\frac{n}{n+2} + \frac{n}{n+4} + \frac{n}{n+6} + \cdots + \frac{n}{3n}\right)\frac{2}{n}$$

$$= \frac{2}{n+2} + \frac{2}{n+4} + \frac{2}{n+6} + \cdots + \frac{2}{3n}.$$ In the case when $n = 8$, we have

$$\frac{1}{5} + \frac{1}{6} + \cdots + \frac{1}{12} = \frac{28{,}271}{27{,}720} > 1,$$ so

$$\int_1^3 \frac{dx}{x} > 1.$$

45 (a) The area of the region under the curve $y = 1/x$ and above $[1, b]$ is $\int_1^b \frac{1}{x}\,dx = \ln x\big|_1^b = \ln b - \ln 1 = \ln b$.

(b) The area under $y = 1/x$ and above $[1, \infty)$ is infinite because $\ln b \to \infty$ as $b \to \infty$.

(c) The volume of the solid of revolution is

$$\int_1^b \pi\left(\frac{1}{x}\right)^2 dx = -\frac{\pi}{x}\bigg|_1^b = \pi\left(1 - \frac{1}{b}\right).$$

(d) As $b \to \infty$, the area approaches π, a finite value.

(e) The confusion lies in the idea of "painting" a surface. A coat of paint is supposed to have a certain uniform thickness. Although the infinite region of (b) fits inside the finite volume of (d), the solid of revolution becomes arbitrarily narrow as $b \to \infty$. So if that volume were filled with paint, the infinite region of (b) would be getting a thinner and thinner coating as b grows.

6.4 One-to-One Functions and Their Inverses

1 (a) No, since $(-1)^4 = 1^4$.

(b) Yes. The inverse is $x = \sqrt[4]{y}$ for y in $[0, 16]$.

3 (a) Yes. The inverse is $x = \sqrt[5]{y - 1}$ for y in $[1, 2]$.

(b) Yes. The inverse is $x = \sqrt[5]{y - 1}$ for y in $[1 - 100^5, 1 + 100^5]$.

5 (a) Yes. Its inverse is $x = \sqrt[3]{y^5 - 1}$ for y in $(-\infty, \infty)$.

(b) Yes. Its inverse is $x = \sqrt[3]{y^5 - 1}$ for y in $[-1, \infty)$.

7 (a) Yes. Its inverse is $x = y^{3/5}$ for y in $(-\infty, \infty)$.

(b) Yes. Its inverse is $x = y^{3/5}$ for y in $[0, \infty)$.

9 Yes.

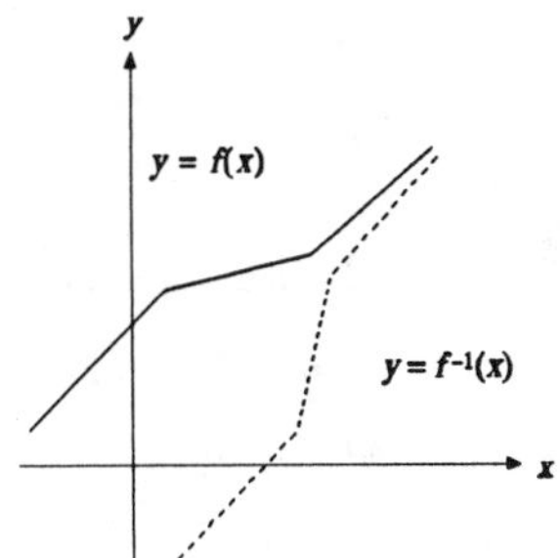

11

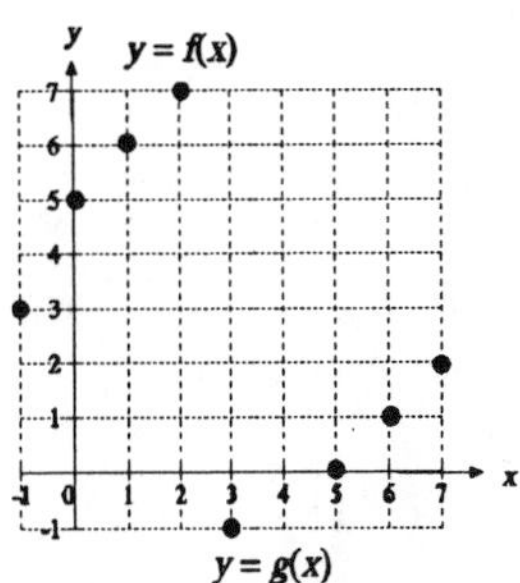

13 First, note that $y = 2^x$ is increasing for all x; hence it is a one-to-one function on $(-\infty, \infty)$. To find its

inverse, take the logarithm to the base 2 of both sides of $y = 2^x$. This gives $\log_2 y = x$, so the inverse of $y = 2^x$ is $x = \log_2 y$ for y in $(0, \infty)$.

15 (a) Since $\frac{dy}{dx} = \cos x$, we see that $\frac{dy}{dx} > 0$ for x in $[0, \pi/2)$ and $\frac{dy}{dx} < 0$ for x in $(\pi/2, \pi]$.

Thus $y = \sin x$ is increasing on $[0, \pi/2]$ and decreasing on $[\pi/2, \pi]$. So $y = \sin x$ is one-to-one on $[0, \pi/2]$, while it is not one-to-one on $[0, \pi]$.

(b)

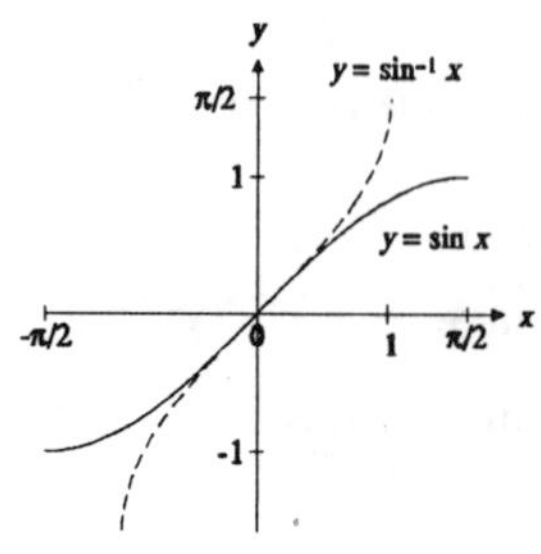

17 (a) Since $\frac{dy}{dx} = \sec^2 x > 0$ for all x not of the form $n\pi + \pi/2$ with n an integer, $y = \tan x$ is increasing on $(-\pi/2, \pi/2)$. Hence it is one-to-one on $(-\pi/2, \pi/2)$.

(b)

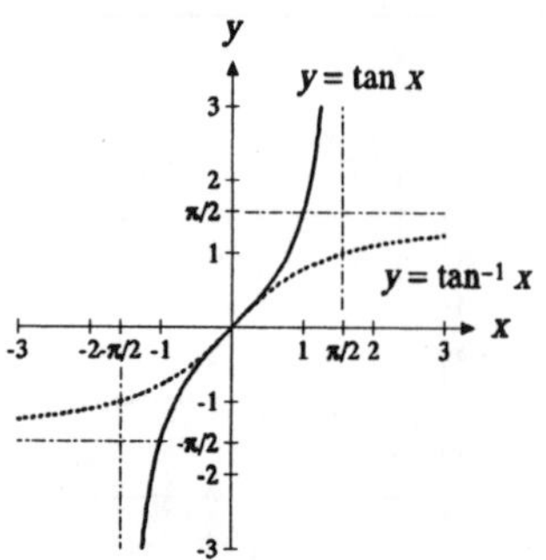

19 (a) Since $\frac{dy}{dx} = 3x^2 + 2 > 0$ for all x, the value of $y = x^3 + 2x$ is increasing for all x. We conclude that $y = x^3 + 2x$ is one-to-one.

(b) Since $\frac{dy}{dx} = 4x^3 + 3$ changes from $-$ to $+$ at

$x = -\sqrt[3]{\frac{3}{4}}$, $y = x^4 + 3x$ has a minimum at

$x = -\sqrt[3]{\frac{3}{4}}$. Since y is continuous, we may

conclude that there exists an $a < -\sqrt[3]{\frac{3}{4}}$ and a

$b > -\sqrt[3]{\frac{3}{4}}$ such that $a^4 + 3a = b^4 + 3b$.

Therefore, $y = x^4 + 3x$ cannot be one-to-one.

21 (a) Let $f(x) = x^3 + kx^2 + x$. Then $f'(x) = 3x^2 + 2kx + 1$. In order for $f(x)$ to be increasing on $(-\infty, \infty)$ and therefore be one-to-one, we must guarantee that $f'(x) \geq 0$ for all x. So let's minimize $f'(x)$. We find that $f''(x) = 6x + 2k$ changes sign from $-$ to $+$ at $x = -k/3$, so the minimum of $f'(x)$ is $f'(-k/3) = 1 - k^2/3$. We want this to be 0 or greater, which occurs if and only if $|k| \leq \sqrt{3}$. Therefore $f(x)$ is one-to-one if and only if

$|k| \leq \sqrt{3}$.

(b) Let $f(x) = x^3 + kx^2 - x$. If f is one-to-one then $f'(x)$ must never change sign. But $f'(x) = 3x^2 + 2kx - 1$ is positive for large values of x, while $f'(0) = -1$ is negative. Hence f is not one-to-one for any choice of k.

23 If $x_1 < x_2$ then $g(x_1) > g(x_2)$ and $f(g(x_1)) < f(g(x_2))$. Thus $f \circ g$ is increasing.

25 (a) Let $f(x) = x^n + 6x^3$. Then we have $f'(x) = nx^{n-1} + 18x^2$. If $n \geq 1$ is an odd integer then $f'(x) = n(x^{(n-1)/2})^2 + 18x^2 \geq 0$ and f is increasing, hence one-to-one. If $n \geq 4$ is even then $f'(1) = n + 18 > 0$ while $\lim_{x\to-\infty}(nx^{n-1} + 18x^2) =$

$$\lim_{x\to-\infty} x^{n-1} \lim_{x\to-\infty}\left(n + \frac{18}{x^{n-3}}\right) = -\infty.$$ Thus $f'(x)$ changes sign, so f is not one-to-one. Finally, if $n = 2$, then $f'(1) = 20$ while $f'(-1/18) = -1/18$. Again $f'(x)$ changes sign. Hence f is one-to-one if and only if n is odd.

(b) For $n = 3$, $y = 7x^3$ so $x = \sqrt[3]{y/7}$. For other values of n, solving for x involves solving a polynomial of degree 3 or greater (which is *not* easy).

6.5 The Derivative of b^x

1 $f(x) = e^{x^2}$, $f'(x) = e^{x^2}\cdot 2x = 2xe^{x^2}$

3 $f(x) = x^2e^{2x}$, $f'(x) = x^2\cdot 2e^{2x} + 2x\cdot e^{2x} = 2(x^2 + x)e^{2x}$

5 $f(x) = 2^{-x^2} = (e^{\ln 2})^{-x^2}$, $f'(x) =$

$$-(\ln 2)\cdot 2x\cdot e^{-(\ln 2)x^2} = -2(\ln 2)x\cdot 2^{-x^2}$$

7 Let $y = x^{(x^2)}$, so $\ln y = \ln x^{(x^2)} = x^2 \ln x$. Differentiating with respect to x yields

$$\frac{1}{y}\cdot y' = x^2\cdot\frac{1}{x} + 2x\ln x = x + 2x\ln x,$$ so $y' = y(x + 2x\ln x) = x^{(x^2)}(x + 2x\ln x)$.

9 Let $y = x^{\tan 3x}$, so $\ln y = \ln(x^{\tan 3x}) = \tan 3x \cdot \ln x$. Differentiating with respect to x yields $1/y\cdot y' = (\tan 3x)\cdot 1/x + (\ln x)(3\sec^2 3x)$.

Thus, $y' = y((1/x)\tan 3x + 3\sec^2 3x \ln x)$ $= x^{\tan 3x}((1/x)\tan 3x + 3\sec^2 3x\ln x)$

11 $$f(x) = \frac{e^{-4x}}{1 + e^x},\ f'(x) = \frac{(1 + e^x)(-4e^{-4x}) - e^{-4x}e^x}{(1 + e^x)^2}$$

$$= -\frac{4e^{-4x} + 5e^{-3x}}{(1 + e^x)^2}$$

13 $\left(e^{x^2}\right)' = 2x\cdot e^{x^2}$ and $\left(x^{\sqrt{3}}\right)' =$. Thus, if $f(x) = x^{\sqrt{3}}(\sin 3x)e^{x^2}$, then $f'(x) = x^{\sqrt{3}}(\sin 3x)2x\cdot e^{x^2} + x^{\sqrt{3}}(3\cos 3x)e^{x^2} + \sqrt{3}x^{\sqrt{3}-1}(\sin 3x)e^{x^2} = x^{\sqrt{3}-1}e^{x^2}(2x^2\sin 3x + 3x\cos 3x + \sqrt{3}\sin 3x)$.

15 $$\frac{d}{dx}\left(\ln\left(x + \sqrt{1 + e^{3x}}\right)\right)$$

$$= \frac{1}{x + \sqrt{1 + e^{3x}}}\left(1 + \frac{1}{2\sqrt{1 + e^{3x}}}\cdot 3e^{3x}\right)$$

$$= \frac{1}{x + \sqrt{1 + e^{3x}}}\left(1 + \frac{3e^{3x}}{2\sqrt{1 + e^{3x}}}\right)$$

17 $$f(x) = \frac{e^{ax}(ax - 1)}{a^2},$$

$$f'(x) = \frac{1}{a^2}\left[e^{ax}a + (ax - 1)ae^{ax}\right]$$

$$= \frac{1}{a^2}\left[ae^{ax} + a^2xe^{ax} - ae^{ax}\right] = \frac{1}{a^2}\left[a^2x\,e^{ax}\right]$$

$$= xe^{ax}$$

19 $$f(x) = \frac{e^{ax}}{a^2 + b^2}(a\sin bx - b\cos bx).$$

$$f'(x) = \frac{1}{a^2 + b^2}[e^{ax}(ab\cos bx + b^2\sin bx)$$

$$+ ae^{ax}(a\sin bx - b\cos bx)]$$

$$= \frac{e^{ax}}{a^2+b^2}[ab\cos bx + b^2\sin bx + a^2\sin bx - ab\cos bx]$$

$$= \frac{e^{ax}}{a^2+b^2}[(a^2+b^2)\sin bx] = e^{ax}\sin bx$$

21 Since $(e^x)' = e^x = 1$ for $x = 0$, we have $e^x = e^{0+x} \approx e^0 + 1\cdot x = 1 + x$ for small x.

23 Since $(10^x)' = (\ln 10)10^x = \ln 10$ for $x = 0$, we have $10^x = 10^{0+x} \approx 10^0 + (\ln 10)x \approx 1 + 2.30x$ for small x.

25 Since $(\ln x)' = 1/x = 1$ at $x = 1$, $\ln(1 + x) \approx \ln 1 + 1\cdot x = x$ for small x.

27 Since $(\log_{10} x)' = \frac{\log_{10} e}{x} = \log_{10} e$ at $x = 1$, we have $\log_{10}(1 + x) \approx \log_{10} 1 + (\log_{10} e)x \approx 0.43x$ for small x.

29 (a) $f(x) = (1 + x)e^{-x} = 0$ when $x = -1$, so $(-1, 0)$ is the x intercept. Since $f(0) = (1 + 0)e^{-0} = 1$, $(0, 1)$ is the y intercept.

(b) $f'(x) = (1 + x)(-e^{-x}) + e^{-x} = e^{-x}(1 - (1 + x)) = -xe^{-x} = 0$ when $x = 0$, so $(0, 1)$ is the only critical point.

(c) $f'(x) > 0$ for $x < 0$ and $f'(x) < 0$ for $x > 0$, so $(0, 1)$ is a local (and global) maximum.

(d) $f''(x) = -x(-e^{-x}) - e^{-x} = e^{-x}(x - 1) = 0$ when $x = 1$; $f''(x) < 0$ for $x < 1$ and $f''(x) > 0$ for $x > 1$, so $(1, 2/e)$ is an inflection point.

(e) $\lim_{x\to\infty} (1 + x)e^{-x} = 0$, so the x axis is a horizontal asymptote as $x \to \infty$.

$\lim_{x\to-\infty} (1 + x)e^{-x} = -\infty$, so there is no horizontal asymptote as $x \to -\infty$.

(f)

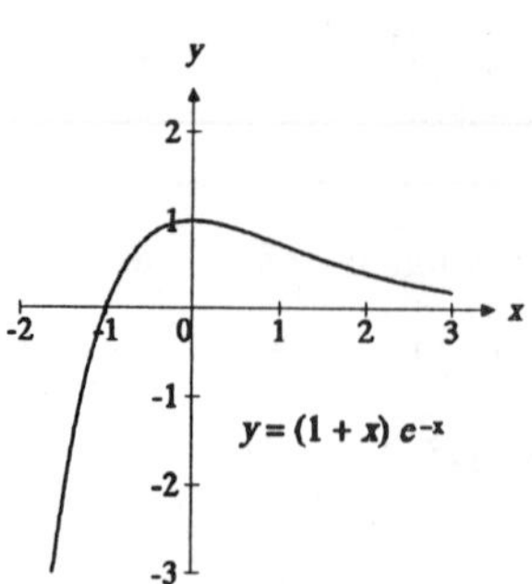

31 (a) $f(x) = x^3e^{-x} = 0$ when $x = 0$, so $(0, 0)$ is both the x intercept and the y intercept.

(b) $f'(x) = -x^3e^{-x} + 3x^2e^{-x} = (3x^2 - x^3)e^{-x} = x^2(3 - x)e^{-x} = 0$ when $x = 0$ or 3, so $(0, 0)$ and $(3, 27e^{-3})$ are the critical points.

(c) $f'(x) > 0$ for $x < 0$ and $0 < x < 3$ and $f'(x) < 0$ for $x > 3$, so $(3, 27e^{-3})$ is a local maximum.

(d) $f''(x) = -(3x^2 - x^3)e^{-x} + (6x - 3x^2)e^{-x} = (x^3 - 6x^2 + 6x)e^{-x} = x(x^2 - 6x + 6)e^{-x}$, which equals 0 when $x = 0$ or $x^2 - 6x + 6 = 0$; that is, when $x = \frac{6 \pm \sqrt{36 - 24}}{2} = 3 \pm \sqrt{3}$. $f''(x) < 0$ for $x < 0$ and $3 - \sqrt{3} < x < 3 + \sqrt{3}$ and $f''(x) > 0$ for $0 < x < 3 - \sqrt{3}$ and $x > 3 + \sqrt{3}$, so 0, $3 - \sqrt{3}$, and $3 + \sqrt{3}$ are inflection numbers.

(e) $\lim_{x\to\infty} x^3e^{-x} = 0$, so the x axis is a horizontal asymptote as $x \to \infty$. $\lim_{x\to-\infty} x^3e^{-x} = -\infty$, so there is no horizontal asymptote as $x \to -\infty$.

(f)

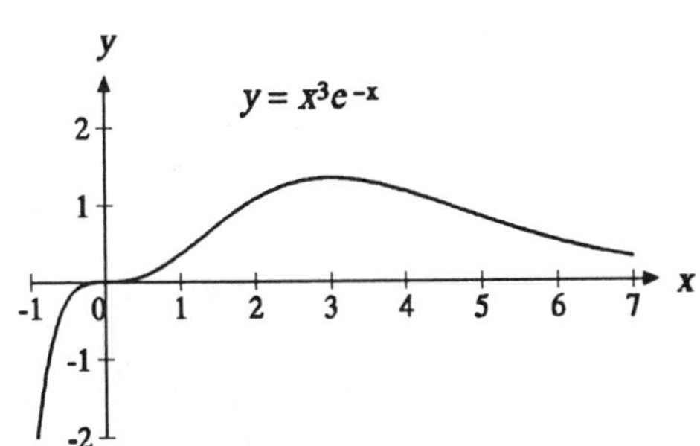

33 (a) $f(x) = (x - x^2)e^{-x} = x(1 - x)e^{-x} = 0$ when $x = 0$ or 1, so (0, 0) and (1, 0) are the x intercepts, and (0, 0) is also the y intercept.

(b) $f'(x) = -(x - x^2)e^{-x} + (1 - 2x)e^{-x} = (x^2 - 3x + 1)e^{-x} = 0$ when $x^2 - 3x + 1 = 0$, which occurs when $x = \frac{3 \pm \sqrt{9 - 4}}{2} = \frac{3 \pm \sqrt{5}}{2}$. Thus $x = \frac{3 - \sqrt{5}}{2}$ and $x = \frac{3 + \sqrt{5}}{2}$ are critical numbers.

(c) $f'(x) > 0$ for $x < \frac{3 - \sqrt{5}}{2}$ and $x > \frac{3 + \sqrt{5}}{2}$ and $f'(x) < 0$ for $\frac{3 - \sqrt{5}}{2} < x < \frac{3 + \sqrt{5}}{2}$, so $x = \frac{3 - \sqrt{5}}{2}$ provides a local maximum of approximately 0.161 and $x = \frac{3 + \sqrt{5}}{2}$ provides a local minimum of approximately -0.309.

(d) $f''(x) = -(x^2 - 3x + 1)e^{-x} + (2x - 3)e^{-x} = -(x^2 - 5x + 4)e^{-x} = -(x - 4)(x - 1)e^{-x} = 0$ when $x = 1$ or 4. $f''(x) < 0$ for $x < 1$ and $x > 4$ and $f''(x) > 0$ for $1 < x < 4$, so (1, 0) and $(4, -12e^{-4})$ are inflection points.

(e) $\lim_{x\to\infty} (x - x^2)e^{-x} = \lim_{x\to\infty} xe^{-x} - \lim_{x\to\infty} x^2e^{-x} = 0 - 0 = 0$, so the x axis is a horizontal asymptote as $x \to \infty$. $\lim_{x\to-\infty} (x - x^2)e^{-x} = -\infty$, so there is no horizontal asymptote as $x \to -\infty$.

(f)

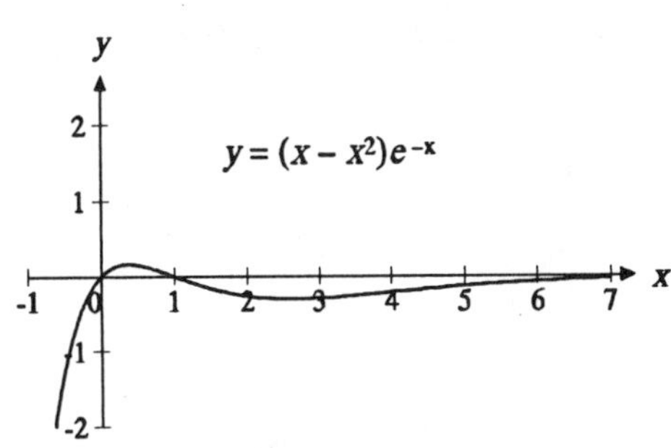

35 (a) Area $= \int_1^5 e^{3x}\,dx = \frac{1}{3}e^{3x}\Big|_1^5 = \frac{1}{3}(e^{15} - e^3)$

37 Since $(10^x)' = (\ln 10)10^x$, it follows that $\frac{10^x}{\ln 10}$ is an antiderivative of 10^x, so the area is $\int_0^3 10^x\,dx = \frac{10^x}{\ln 10}\Big|_0^3 = \frac{10^3}{\ln 10} - \frac{10^0}{\ln 10} = \frac{999}{\ln 10}$.

39 (a) Let $f(x) = e^x$. Then $f'(0) = \lim_{h\to0} \frac{f(0 + h) - f(0)}{h} = \lim_{h\to0} \frac{e^{0+h} - e^0}{h} = \lim_{h\to0} \frac{e^h - 1}{h}$. $f'(x) = e^x$, so the limit is equal to $f'(0) = e^0 = 1$.

(b) Let $f(x) = 2^x$. Then $f'(1) = \lim_{x\to1} \frac{f(x) - f(1)}{x - 1}$

$= \lim_{x \to 1} \dfrac{2^x - 2}{x - 1}.$

$f'(x) = (\ln 2)2^x$, so the limit is equal to $f'(1)$ $= (\ln 2)2^1 = 2 \ln 2$.

(c) Let $f(x) = 10^x$. Then $f'(0) =$

$$\lim_{h \to 0} \frac{f(0 + h) - f(0)}{h} = \lim_{h \to 0} \frac{10^{0+h} - 10^0}{h} =$$

$\lim_{h \to 0} \dfrac{10^h - 1}{h}$. $f'(x) = (\ln 10)10^x$, so the limit is equal to $f'(0) = (\ln 10)10^0 = \ln 10$.

41 (a)

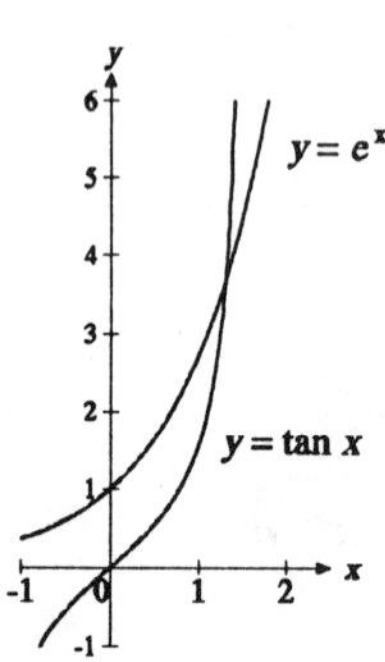

(b) Let $f(x) = e^x - \tan x$. Then $f(0) = 1$ and $f(x) \to -\infty$ as $x \to \pi/2^-$, so f must possess a root between 0 and $\pi/2$.

(c) $f'(x) = e^x - \sec^2 x$, so $x_{i+1} = x_i - \dfrac{f(x_i)}{f'(x_i)} =$

$x_i - \dfrac{e^{x_i} - \tan x_i}{e^{x_i} - \sec^2 x_i}$. If $x_1 = 1.3$, then

$x_2 \approx 1.307$.

43 (a)

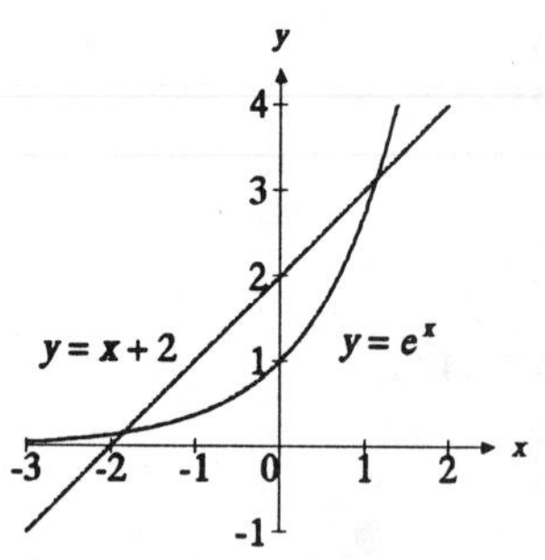

(b) From the graph we see that one root is slightly greater than 1 and the other is slightly greater than -2.

(c) With $f(x) = e^x - x - 2$ and $f'(x) = e^x - 1$, we have $x_{i+1} = x_i - \dfrac{e^{x_i} - x_i - 2}{e^{x_i} - 1}$ so with x_1 $= 1$, a calculator then gives, to three places, $x_2 = 1.164$, $x_3 = 1.146$, and $x_4 = 1.146$. To two decimal places, the root is 1.15. With x_1 $= -2$, we obtain, to three places, $x_2 =$ -1.843, $x_3 = -1.841$, and $x_4 = -1.841$. To two decimal places, the other root is -1.84.

45 (a)

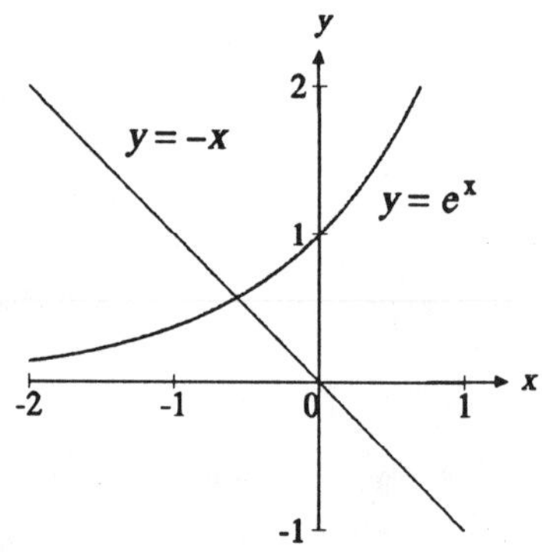

(b)

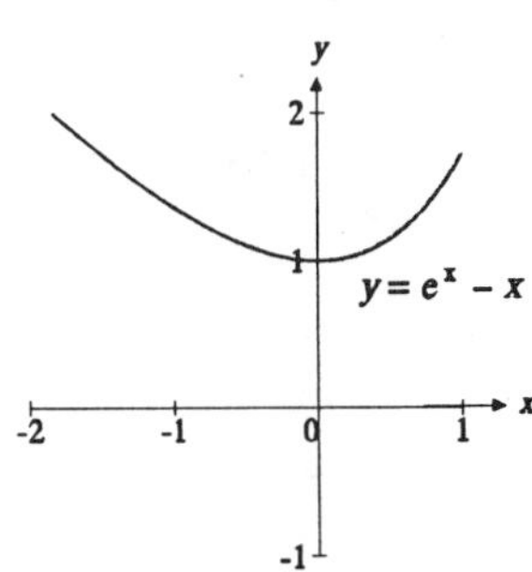

(c) $y' = e^x - 1 = 0$. The tangent line is horizontal at (0, 1).

47 (a) Let V be volume and C be cost. Then $\frac{V}{C} = \frac{bh^3}{ae^{h/20}} = \frac{b}{a}h^3e^{-h/20}$ so $\frac{d}{dh}\left(\frac{V}{C}\right) = \frac{b}{a}\cdot\frac{1}{20}h^2(60 - h)e^{-h/20}$ changes from positive to negative at $h = 60$. Hence a height of 60 meters maximizes the volume-to-cost ratio.

(b) $\frac{A}{C} = \frac{ch^2}{ae^{h/20}} = \frac{c}{a}\cdot h^2e^{-h/20}$ so $\frac{d}{dh}\left(\frac{A}{C}\right) = \frac{c}{a}\cdot\frac{1}{20}h(40 - h)e^{-h/20}$ changes from positive to negative at $h = 40$. The area-to-height ratio is maximal when the height is 40 meters.

49 We have $b^n = \sum_{i=0}^{n}\binom{n}{i}c^i > \sum_{i=0}^{3}\binom{n}{i}c^i = 1 + nc + \frac{n(n-1)}{2}c^2 + \frac{n(n-1)(n-2)}{3!}c^3$. Hence $b^n > \frac{n(n-1)(n-2)}{6}c^3$ and

$$0 < \frac{n^2}{b^n} < \frac{n^2}{\frac{n(n-1)(n-2)}{6}c^3} = \frac{6n}{(n-1)(n-2)c^3}.$$

As $n \to \infty$, $\frac{6n}{(n-1)(n-2)c^3} \to 0$, so $\frac{n^2}{b^n} \to 0$.

Now, for large x, let n be the smallest integer that is greater than or equal to x. Thus $0 < \frac{x^2}{b^x} < \frac{n^2}{b^{n-1}} = b\frac{n^2}{b^n}$. Since $\frac{n^2}{b^n} \to 0$ as $x \to \infty$, $\frac{x^2}{b^x} \to 0$ as $x \to \infty$. The case when $a \le 0$ is trivial because we would have $\lim_{x\to\infty}\frac{1}{b^x}$ or $\lim_{x\to\infty}\frac{1}{x^{-a}b^x}$, which are both clearly 0.

6.6 The Derivatives of the Inverse Trigonometric Functions

1 To find the inverse tangent of a number, we mark off a distance on the unit circle's tangent line at (1, 0), where the distance equals the number whose inverse tangent we want. We can then construct a triangle whose angle at the origin is the desired angle.

(a)

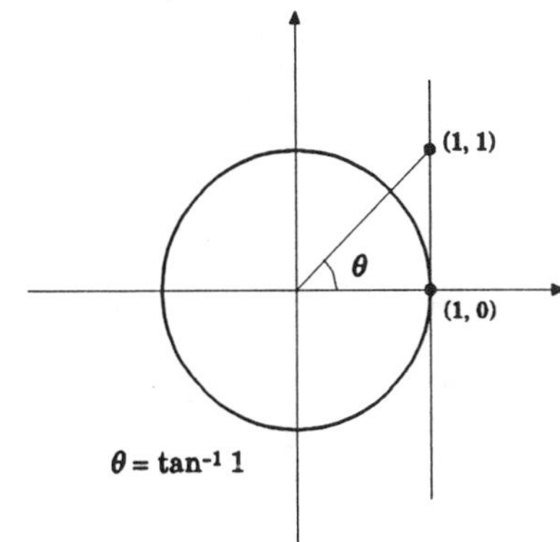

(b)

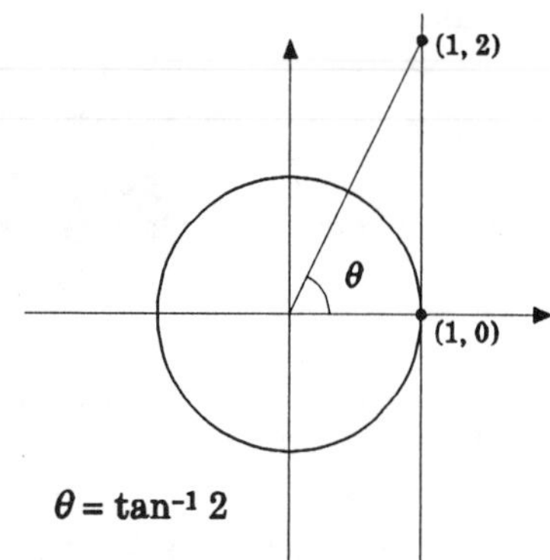

(c)

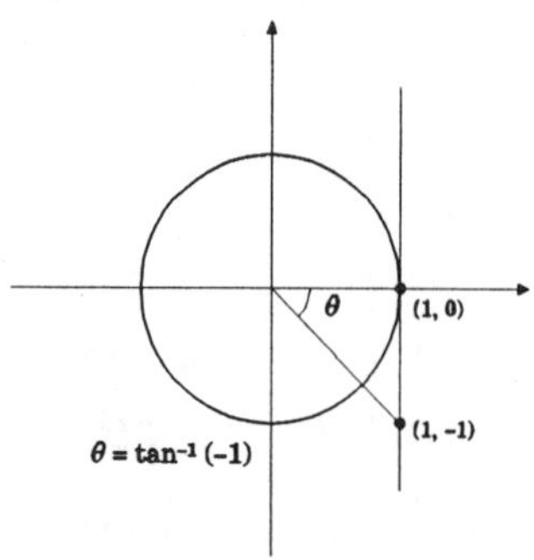

3

x	-4	-3	-2	-1
$\tan^{-1} x$	-1.326	-1.249	-1.107	-0.785

x	0	1	2	3	4
$\tan^{-1} x$	0	0.785	1.107	1.249	1.326

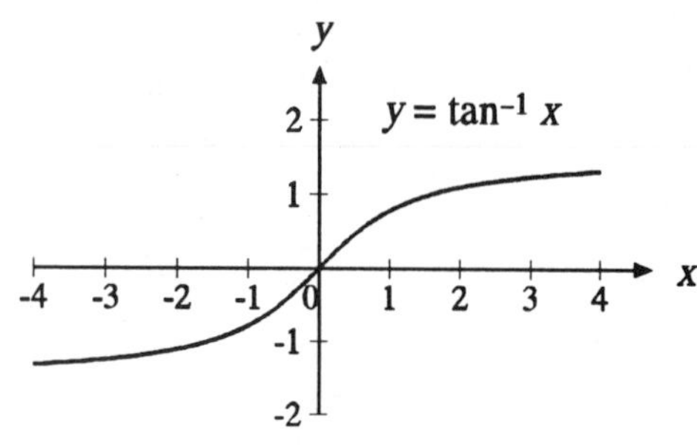

5

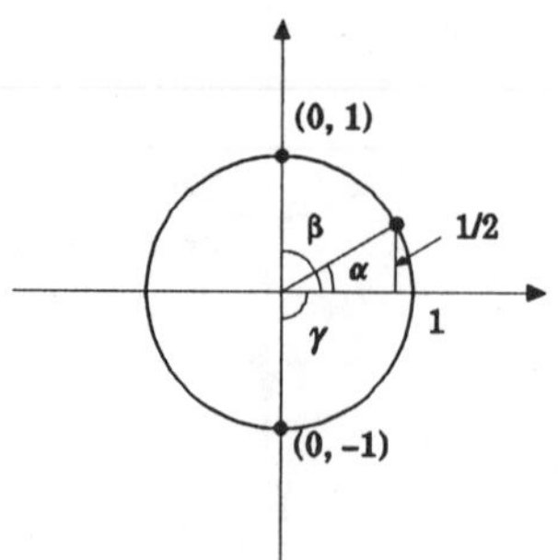

To find $\sin^{-1} y$, we find the angle corresponding to the point (x, y) on the right half of the unit circle. Hence the angle α is $\sin^{-1}(1/2)$, β is $\sin^{-1} 1$, and γ is $\sin^{-1}(-1)$.

7

x	-1	-0.8	-0.6	-0.4	-0.2
$\sin^{-1} x$	-1.571	-0.927	-0.644	-0.412	-0.201

x	0	0.2	0.4	0.6	0.8	1
$\sin^{-1} x$	0	0.201	0.412	0.644	0.927	1.571

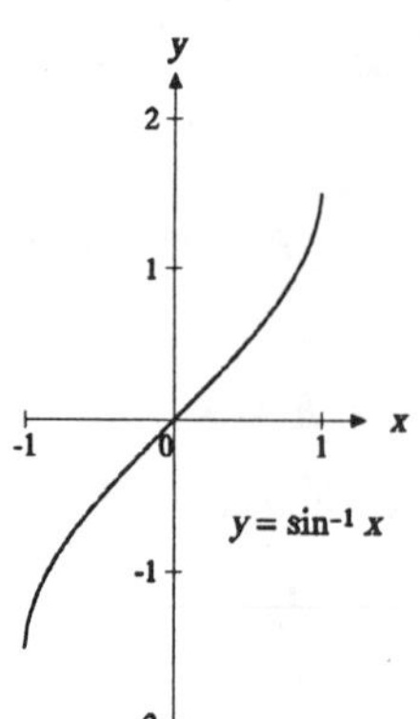

9

x	1	2	3	4
$\sec^{-1} x$	0	1.047	1.231	1.318

x	-1	-2	-3	-4
$\sec^{-1} x$	3.142	2.094	1.911	1.823

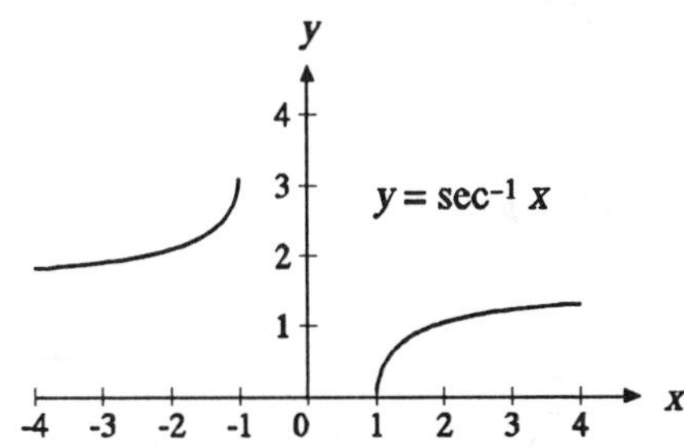

11 (a) $\sin^{-1}(1/2)$ is the angle whose sine is 1/2 (in the interval $[-\pi/2, \pi/2]$). Since $\sin \pi/6 = 1/2$, $\sin^{-1}(1/2) = \pi/6$.

(b) $\tan^{-1}\frac{1}{\sqrt{3}}$ is the angle whose tangent is $\frac{1}{\sqrt{3}}$ (in $(-\pi/2, \pi/2)$). Since $\tan \pi/6 = \frac{1}{\sqrt{3}}$, $\tan^{-1}\frac{1}{\sqrt{3}} = \pi/6$.

(c) $\sin^{-1}\left(-\frac{\sqrt{3}}{2}\right)$ is the angle whose sine is $-\frac{\sqrt{3}}{2}$ (in $[-\pi/2, \pi/2]$). Since $\sin(-\pi/3) = -\frac{\sqrt{3}}{2}$,

$\sin^{-1}\left(-\frac{\sqrt{3}}{2}\right) = \pi/3$.

(d) $\tan^{-1}(-\sqrt{3})$, is the angle whose tangent is $-\sqrt{3}$ (in $(-\pi/2, \pi/2)$). Since $\tan(-\pi/3) = -\sqrt{3}$, $\tan^{-1}(-\sqrt{3}) = -\pi/3$.

(e) $\sec^{-1}\sqrt{2}$ is the angle whose secant is $\sqrt{2}$ or the angle whose cosine is $1/\sqrt{2}$ (in $[0, \pi]$). Since $\cos \pi/4 = 1/\sqrt{2}$, $\sec^{-1}\sqrt{2} = \pi/4$.

13 $\sin(\tan^{-1} 1)$ is the sine of the angle whose tangent is 1 in $(-\pi/2, \pi/2)$, that is, the sine of $\pi/4$. Thus, $\sin(\tan^{-1} 1) = \sin \pi/4 = \sqrt{2}/2$.

15 $\tan[\sin^{-1}(-\sqrt{2}/2)]$ is the tangent of the angle whose sine is $-\sqrt{2}/2$ in $[-\pi/2, \pi/2]$, that is, the tangent of $-\pi/4$. Thus, $\tan[\sin^{-1}(-\sqrt{2}/2)] = \tan(-\pi/4) = -1$.

17 $\sin(\sin^{-1} 0.3)$ is the sine of the angle whose sine is 0.3, which is by definition 0.3. Thus $\sin(\sin^{-1} 0.3) = 0.3$.

19 $\sin^{-1}(\sin \pi) = \sin^{-1} 0$, so we seek the angle in $[-\pi/2, \pi/2]$ whose sine is 0. Since $\sin 0 = 0$, $\sin^{-1}(\sin \pi) = \sin^{-1} 0 = 0$.

21 $(\sin^{-1} 5x)' = \dfrac{5}{\sqrt{1 - (5x)^2}} = \dfrac{5}{\sqrt{1 - 25x^2}}$

23 $(\sec^{-1} 3x)' = \dfrac{3}{|3x|\sqrt{(3x)^2 - 1}} = \dfrac{1}{|x|\sqrt{9x^2 - 1}}$

25 $\left(\tan^{-1}\sqrt[3]{x}\right)' = \dfrac{\frac{1}{3}x^{-2/3}}{1 + \left(\sqrt[3]{x}\right)^2} = \dfrac{1}{3x^{2/3}(1 + x^{2/3})}$

27 $(x^2 \sec^{-1}\sqrt{x})' = x^2 \cdot \dfrac{1/(2\sqrt{x})}{\sqrt{x}\sqrt{(\sqrt{x})^2 - 1}} + 2x \sec^{-1}\sqrt{x}$

$= \dfrac{x}{2\sqrt{x - 1}} + 2x \sec^{-1}\sqrt{x}$

29 $(\sin 3x \sin^{-1} 3x)'$

$= \sin 3x \dfrac{3}{\sqrt{1 - (3x)^2}} + 3 \cos 3x \sin^{-1} 3x$

$= \dfrac{3 \sin 3x}{\sqrt{1 - 9x^2}} + 3 \cos 3x \sin^{-1} 3x$

31 $\left(\dfrac{x \sec^{-1} 3x}{e^{2x}}\right)' =$

$$\frac{e^{2x}\left(x \cdot \dfrac{3}{|3x|\sqrt{(3x)^2 - 1}} + \sec^{-1} 3x\right) - x \sec^{-1} 3x \cdot 2e^{2x}}{(e^{2x})^2}$$

$$= \frac{e^{2x}\left(\dfrac{x}{|x|\sqrt{9x^2 - 1}} + \sec^{-1} 3x - 2x \sec^{-1} 3x\right)}{e^{4x}}$$

$$= e^{-2x}\left(\frac{x}{|x|\sqrt{9x^2 - 1}} + (1 - 2x)\sec^{-1} 3x\right)$$

33 $(\arctan \sqrt{x})' = \dfrac{1/(2\sqrt{x})}{1 + (\sqrt{x})^2} = \dfrac{1}{2\sqrt{x}(1 + x)}$

35 $(\ln \sec^{-1} \sqrt{x})' =$

$$\frac{1}{\sec^{-1}\sqrt{x}} \cdot \frac{1/(2\sqrt{x})}{\sqrt{x}\sqrt{(\sqrt{x})^2 - 1}} = \frac{1}{2x\sqrt{x - 1}\sec^{-1}\sqrt{x}}$$

37 $\left(\dfrac{x}{\tan^{-1} 10^x}\right)'$

$$= \frac{(\tan^{-1} 10^x) - x \cdot \dfrac{1}{(10^x)^2 + 1} \cdot (\ln 10) \cdot 10^x}{(\tan^{-1} 10^x)^2}$$

$$= \frac{\tan^{-1} 10^x - \dfrac{(\ln 10)x \cdot 10^x}{10^{2x} + 1}}{(\tan^{-1} 10^x)^2}$$

$$= \frac{1}{\tan^{-1} 10^x} - \frac{(\ln 10)x \cdot 10^x}{(10^{2x} + 1)(\tan^{-1} 10^x)^2}$$

39 $\left(\sin^{-1} x - \sqrt{1 - x^2}\right)' = \dfrac{1}{\sqrt{1 - x^2}} - \dfrac{-x}{\sqrt{1 - x^2}}$

$$= \frac{1 + x}{\sqrt{1 - x^2}} = \frac{1 + x}{\sqrt{1 + x}\sqrt{1 - x}} = \frac{\sqrt{1 + x}}{\sqrt{1 - x}}$$

$$= \sqrt{\frac{1 + x}{1 - x}}$$

41 $[(\tan^{-1} 2x)^3]' = 3(\tan^{-1} 2x)^2 \dfrac{2}{1 + (2x)^2}$

$$= \frac{6(\tan^{-1} 2x)^2}{1 + 4x^2}$$

43 $\left[\dfrac{x}{2}\sqrt{2 - x^2} + \sin^{-1}(x/\sqrt{2})\right]'$

$$= \frac{x}{2} \cdot \frac{-x}{\sqrt{2 - x^2}} + \frac{1}{2}\sqrt{2 - x^2} + \frac{1/\sqrt{2}}{\sqrt{1 - (x/\sqrt{2})^2}}$$

$$= \frac{-x^2}{2\sqrt{2 - x^2}} + \frac{2 - x^2}{2\sqrt{2 - x^2}} + \frac{1}{\sqrt{2}\sqrt{1 - (x^2/2)}}$$

$$= \frac{2 - 2x^2}{2\sqrt{2 - x^2}} + \frac{1}{\sqrt{2 - x^2}}$$

$$= \frac{1 - x^2 + 1}{\sqrt{2 - x^2}} = \frac{2 - x^2}{\sqrt{2 - x^2}} = \sqrt{2 - x^2}$$

45 $\left[\dfrac{2}{3}\sec^{-1}\sqrt{3x^5}\right]' = \dfrac{2}{3} \cdot \dfrac{1}{\sqrt{3x^5}\sqrt{(\sqrt{3x^5})^2 - 1}} \cdot \dfrac{15x^4}{2\sqrt{3x^5}}$

$$= \frac{5x^4}{3x^5\sqrt{3x^5 - 1}} = \frac{5}{3x\sqrt{3x^5 - 1}}$$

47 $\left(\sqrt{1 + x}\sqrt{2 - x} - 3\sin^{-1}\sqrt{\dfrac{2 - x}{3}}\right)' =$

$$\sqrt{1 + x}\left(\frac{-1}{2\sqrt{2 - x}}\right) + \sqrt{2 - x} \cdot \frac{1}{2\sqrt{1 + x}} -$$

$$3 \cdot \frac{1}{\sqrt{1 - \left(\sqrt{\dfrac{2 - x}{3}}\right)^2}} \cdot \frac{-1/3}{2\sqrt{\dfrac{2 - x}{3}}}$$

$$= \frac{\sqrt{2 - x}}{2\sqrt{1 + x}} - \frac{\sqrt{1 + x}}{2\sqrt{2 - x}} + \frac{1}{\sqrt{1 - \dfrac{2 - x}{3}} \cdot 2\sqrt{\dfrac{2 - x}{3}}}$$

$$= \frac{(2-x)-(1+x)}{2\sqrt{1+x}\sqrt{2-x}} + \frac{1}{2\sqrt{\frac{1+x}{3}}\sqrt{\frac{2-x}{3}}}$$

$$= \frac{1-2x}{2\sqrt{1+x}\sqrt{2-x}} + \frac{3}{2\sqrt{1+x}\sqrt{2-x}}$$

$$= \frac{2-x}{\sqrt{1+x}\sqrt{2-x}} = \frac{\sqrt{2-x}}{\sqrt{1+x}} = \sqrt{\frac{2-x}{1+x}}$$

49 $\left(x(\sin^{-1}2x)^2 - 2x + \sqrt{1-4x^2}\sin^{-1}2x\right)'$

$$= x(2\sin^{-1}2x)\frac{2}{\sqrt{1-(2x)^2}} + (\sin^{-1}2x)^2 - 2 +$$

$$\sqrt{1-4x^2}\cdot\frac{2}{\sqrt{1-(2x)^2}} + \frac{-4x}{\sqrt{1-4x^2}}\cdot\sin^{-1}2x$$

$$= \frac{4x\sin^{-1}2x}{\sqrt{1-4x^2}} + (\sin^{-1}2x)^2 - 2 +$$

$$\frac{2\sqrt{1-4x^2}}{\sqrt{1-4x^2}} - \frac{4x\sin^{-1}2x}{\sqrt{1-4x^2}}$$

$$= (\sin^{-1}2x)^2 - 2 + 2 = (\sin^{-1}2x)^2$$

51 (a) $\left(\ln\left(x + \sqrt{x^2-9}\right)\right)'$

$$= \frac{1}{x+\sqrt{x^2-9}}\left(1 + \frac{x}{\sqrt{x^2-9}}\right)$$

$$= \frac{1}{x+\sqrt{x^2-9}}\cdot\frac{x+\sqrt{x^2-9}}{\sqrt{x^2-9}} = \frac{1}{\sqrt{x^2-9}}$$

(b) $\left(\sin^{-1}\frac{x}{3}\right)' = \frac{1/3}{\sqrt{1-(x/3)^2}} = \frac{1}{3\sqrt{1-x^2/9}}$

$$= \frac{1}{\sqrt{9-x^2}}$$

53 Since $\tan^{-1} 1 = \pi/4$, we could press [1], then [TAN^{-1}], and finally multiply by 4. This would give π to as many decimal places as the calculator offers.

55 (a) $d(\sin^{-1}x) = \frac{1}{\sqrt{1-x^2}}\cdot dx$

(b) Let $f(x) = \sin^{-1}x$. Note that $\sin^{-1}0.5 = \pi/6$. Then $f(0.47) = f(0.5 - 0.03) \approx$ $f'(0.5) + f'(0.5)(-0.03) =$

$$\frac{\pi}{6} + \frac{1}{\sqrt{1-(0.5)^2}}\cdot(-0.03) \approx 0.489.$$

57 $\tan\left(\tan^{-1}\frac{1}{2} + \tan^{-1}\frac{1}{3}\right) = \frac{\frac{1}{2}+\frac{1}{3}}{1-\frac{1}{2}\cdot\frac{1}{3}} = 1$, so

$$\tan^{-1}\frac{1}{2} + \tan^{-1}\frac{1}{3} = \frac{\pi}{4} + n\pi \text{ for some integer } n.$$

But $0 < \tan^{-1}\frac{1}{2} < \frac{\pi}{2}$ and $0 < \tan^{-1}\frac{1}{3} < \frac{\pi}{2}$,

so $0 < \tan^{-1}\frac{1}{2} + \tan^{-1}\frac{1}{3} < \pi$. Hence

$$\tan^{-1}\frac{1}{2} + \tan^{-1}\frac{1}{3} = \frac{\pi}{4}.$$

59 (a) $\left(\sin^{-1}\frac{x}{a}\right)' = \frac{1/a}{\sqrt{1-(x/a)^2}} = \frac{1}{a\sqrt{1-x^2/a^2}} =$

$$\frac{1}{\sqrt{a^2-x^2}}, \text{ so } \int\frac{dx}{\sqrt{a^2-x^2}} = \sin^{-1}\frac{x}{a} + C.$$

(b) Let $a = 5$. Then, by (a),

$$\int\frac{dx}{\sqrt{25-x^2}} = \sin^{-1}\frac{x}{5} + C.$$

(c) Let $a = \sqrt{5}$. Then $\int \frac{dx}{\sqrt{5 - x^2}}$

$= \sin^{-1} \frac{x}{\sqrt{5}} + C.$

61 (a) Using the fundamental theorem of calculus,

$\int_0^1 \frac{1}{1 + x^2}\, dx = \tan^{-1} x \big|_0^1 =$

$\tan^{-1} 1 - \tan^{-1} 0 = \frac{\pi}{4}.$

(b) $\int_0^1 \frac{1}{1 + x^2}\, dx \approx \frac{1/6}{3}\left[\frac{1}{1 + (0/6)^2} + \right.$

$\frac{4}{1 + (1/6)^2} + \frac{2}{1 + (2/6)^2} + \frac{4}{1 + (3/6)^2} +$

$\left.\frac{2}{1 + (4/6)^2} + \frac{4}{1 + (5/6)^2} + \frac{1}{1 + (6/6)^2}\right]$

$= \frac{1}{18}\left[\frac{36}{36} + 4 \cdot \frac{36}{37} + 2 \cdot \frac{36}{40} + 4 \cdot \frac{36}{45}\right.$

$\left. + 2 \cdot \frac{36}{52} + 4 \cdot \frac{36}{61} + \frac{36}{72}\right] \approx 0.7853979$

$\approx \frac{\pi}{4}$, so $\pi \approx 3.1415916$.

63 $\int_{\sqrt{2}}^{2} \frac{dx}{x\sqrt{x^2 - 1}} = \sec^{-1} x \big|_{\sqrt{2}}^{2} = \frac{\pi}{3} - \frac{\pi}{4} = \frac{\pi}{12}$

65 (a) Since the domain of arcsin x is $[-1, 1]$, x must be in this interval. For such x, $\sin(\arcsin x) = x$, by the definition of arcsin. So for all x in $[-1, 1]$, $\sin(\arcsin x) = x$.

(b) Since the range of arcsin x is $[-\pi/2, \pi/2]$, x must lie in this interval. For such x, $\arcsin(\sin x) = x$, by definition. So, for all x in $[-\pi/2, \pi/2]$, $\arcsin(\sin x) = x$.

6.7 The Differential Equation of Natural Growth and Decay

1 $\frac{dy}{dx} = \frac{x^2}{y^2}$, so separating the variables yields $y^2\, dy = x^3\, dx$. Integrating both sides gives

$\int y^2\, dy = \int x^3\, dx + C$, and $\frac{y^3}{3} = \frac{x^4}{4} + C.$

3 $\frac{dy}{dx} = \frac{y + 1}{x + 2}$, so separating the variables gives

$\frac{dy}{y + 1} = \frac{dx}{x + 2}$; then $\int \frac{dy}{y + 1} = \int \frac{dx}{x + 2} + C,$

so $\ln|y + 1| = \ln|x + 2| + C$. Thus $y + 1 = K(x + 2)$ where $K = \pm e^C$.

5 $\frac{dy}{dx} = \frac{\sin 3x}{\cos 2y}$, so $\cos 2y\, dy = \sin 3x\, dx$. Now

$\int \cos 2y\, dy = \int \sin 3x\, dx + C$, so $\frac{1}{2} \sin 2y$

$= -\frac{1}{3} \cos 3x + C.$

7 $\frac{dy}{dx} = \frac{e^y}{1 + x^2}$, so $e^{-y}\, dy = \frac{dx}{1 + x^2}$. Thus

$\int e^{-y}\, dy = \int \frac{dx}{1 + x^2} + C$ and

$-e^{-y} = \tan^{-1} x + C.$

9 Let $S(t)$ be the amount of the substance at time t. Then $S(t) = S(0)e^{kt}$ and $S(t + 1) = e^k S(t)$.

(a) The substance grows by 10% each hour, so $e^k = 1.1$, and $k = \ln 1.1 \approx 0.0953$. (Observe that the growth rate is small enough so that $k \approx 10\% = 0.10$ would be a good approximation.)

(b) From Eq. (13), $t_2 = \dfrac{\ln 2}{k} \approx 7.27$ hours.

11 (a) At $t = 0$, the initial amount is $10 \cdot 3^0 = 10$ grams.

(b) Let the mass of the bacteria after t hours be given by $M(t) = Me^{kt} = 10 \cdot 3^t$. Thus $e^k = 3$ and $k = \ln 3 \approx 1.099$.

(c) Since k is not small, we cannot use the approximation $M(t + 1) \approx M(t)(1 + k)$. However, note that $M(t + 1) = 3M(t) = M(t)(1 + 2)$. Hence the percent increase in any period of 1 hour is $2 \cdot 100 = 200$ percent.

13 Let $P(t) = P(0)e^{kt}$, where $t = 0$ corresponds to the year 1988. Thus $P(0) = 5.1$ (billion). If the population is increasing at a rate of 1.7 percent per year, then $k = 0.017$.

(a) $t_2 = \dfrac{\ln 2}{k} \approx 40.77$ years, so the population will double by the year 2028.

(b) The population will quadruple in twice its doubling time, in $2(40.77) = 81.54$ years, in the year 2069.

(c) Here $P(t) = 100 = 5.1e^{0.017t}$, so $e^{0.017t} \approx 19.608$ and $t \approx 175.05$ years. The population will reach 100 billion in 2163.

15 Measure time t with $t = 0$ corresponding to 1 P.M. Let $f(t)$ be the amount present after t hours. Thus, there are constants A and k such that $f(t) = Ae^{kt}$. Now 4:30 P.M. is $t = 3.5$, so $f(0) = 100$ and $f(3.5) = 250$. Therefore, $A = 100$ and $Ae^{k(3.5)} = 250$, so $e^{k(3.5)} = \dfrac{250}{100} = 2.5$. Then $k(3.5) = \ln 2.5$, so $k = \dfrac{\ln 2.5}{3.5} \approx 0.2618$.

(a) The culture weighs 400 grams when $100e^{kt} = 400$. We have $e^{kt} = 4$, so $kt = \ln 4$ and $t = \dfrac{\ln 4}{k} = \dfrac{3.5 \ln 4}{\ln 2.5} \approx 5.30$ hr $= 5$ hours and 18 minutes, which corresponds to 6:18 P.M.

(b) As shown, the growth constant is $k = \dfrac{\ln 2.5}{3.5} \approx 0.2618$.

17 (a) Let $A(t)$ be the mass of the substance at time t. Then $A(t) = A(0)e^{kt}$. When $A(0) = 10$ grams, we know that $A(1) = 10 - 0.05 = 9.95$ grams. Hence $A(1) = 9.95 = 10e^{k \cdot 1} \cdot 1$ and $k = \ln 0.995 \approx -0.005$. So when $A(0) = A$, we conclude that $A(t) \approx Ae^{-0.005t}$.

(b) The half-life is given by

$$t_{1/2} = \frac{-\ln 2}{k} = \frac{-\ln 2}{\ln 0.995} \approx 138.28 \text{ days.}$$

19 Let $U(t) = U(0)e^{k_1 t} = 246e^{k_1 t}$ be the population of the United States after t years (in millions of people) and let $M(t) = M(0)e^{k_2 t} = 87e^{k_2 t}$ be the population of Mexico after t years (in millions of people). The population of the U.S. grows at a rate of 0.7% per year, so $k_1 = 0.7/100 = 0.007$. Similarly, the population of Mexico grows at a rate of 1.8% per year, so $k_2 = 1.8/100 = 0.018$. We must find the time t when $U(t) = M(t)$, that is, when

$$246e^{0.007t} = 87e^{0.018t}$$

$$\ln 246 + 0.007t = \ln 87 + 0.018t$$

$$\ln \frac{247}{87} \approx 0.011t.$$

So $t = \dfrac{1}{0.011} \ln \dfrac{246}{87} \approx 94.5$ years. The populations would be the same during the year 2083.

21 If $A(t)$ is the concentration t years after death, then $A(t) = A_u e^{kt}$ for some constant k. Since the half-life is 5730 years, $k = \dfrac{\ln 2}{5730}$. If the age of the specimen is T, then $A_c = A(T) = A_u e^{kT}$, so $T =$
$$\frac{1}{k}\ln\left(\frac{A_c}{A_u}\right) = \frac{5730}{\ln 2}\ln\left(\frac{A_c}{A_u}\right) \approx 8300 \ln\left(\frac{A_c}{A_u}\right).$$

23 (a) If the interest rate is p percent, then $k = p/100$ and $t_2 = \dfrac{\ln 2}{k} = \dfrac{100 \ln 2}{p}$. So $100 \ln 2 \approx 69.315$ should be used instead of 72.

(b) 72 is used because it is relatively close to $100 \ln 2$ and has many factors such as 2, 3, 4, 6, 8, 9, and 12, making it easier to divide a given interest rate into it in one's head than $100 \ln 2$.

25 (a) $L\cdot\dfrac{di}{dt} + Ri = E$, so $L\cdot\dfrac{di}{dt} = E - Ri$ and
$$\frac{di}{dt} = \frac{E - Ri}{L}.$$

(b) $\dfrac{di}{dt} = \dfrac{E - Ri}{L}$, so $\dfrac{di}{E - Ri} = \dfrac{dt}{L}$, hence $\displaystyle\int \frac{di}{E - Ri} = \int \frac{dt}{L}$, so $-\dfrac{1}{R}\ln(E - Ri) = \dfrac{t}{L} + C$. When $t = 0$, $i = i_0$, so $-\dfrac{1}{R}\ln(E - Ri_0) = C$. Thus,
$$-\frac{1}{R}\ln(E - Ri) = \frac{t}{L} - \frac{1}{R}\ln(E - Ri_0),$$ and
$\ln(E - Ri) = \ln(E - Ri_0) - \dfrac{Rt}{L}$, so $E - Ri = (E - Ri_0)e^{-Rt/L}$; then $Ri = E - (E - Ri_0)e^{-Rt/L}$, so $i = \dfrac{1}{R}(E - (E - Ri_0)e^{-Rt/L})$.

27 (a) Let the investment at time t be $I(t)$. Then $I(0) = A$, the initial investment, and $I'(t) = f(t)$, since $f(t)$ is the rate of investment. Hence $I(t) = \int_0^t f(x)\,dx + I(0)$ by the first fundamental theorem of calculus. Thus $I(t) = A + \int_0^t f(x)\,dx$. The investment rate is proportional to the total investment at any time, so there exists a constant k such that $f(t) = kI(t) = k\left[A + \int_0^t f(x)\,dx\right]$.

(b) By differentiating the result in (a) we obtain $f'(t) = kf(t)$, so $f(t)$ is of the form Be^{kt}, where $B = f(0)$.

29 No. For example, suppose the population at time t of the western hemisphere is 2^t and that of the eastern is 4^t. If the world population is growing exponentially, we would have $2^t + 4^t = Pb^t$ for all t, where P and b are constants. Letting $t = 0$ gives $P = 2$. Letting $t = 1$ gives $6 = Pb = 2b$, so $b = 3$. But $2^2 + 4^2 = 20$ while $Pb^2 = 2\cdot 3^2 = 18$, a contradiction. (In fact, the sum of two exponentially growing quantities grows exponentially if and only if the two have the same growth constant.)

31 (a) Suppose h fish are harvested per unit time. Then, during a small period of time Δt, $h\Delta t$ fish are removed from the population. In the

same time, approximately $kP\Delta t$ fish are born, for some constant k. Hence ΔP, the increase in the population during the time Δt, is about $kP\Delta t - h\Delta t$.

(b) Dividing the equation in (a) by Δt gives $\frac{\Delta P}{\Delta t} \approx kP - h$. Taking the limit as $\Delta t \to 0$ gives $\frac{dP}{dt} = kP - h$. This is almost the differential equation for natural growth, but the right-hand side is $kP - h = k(P - h/k)$ rather than kP. This suggests looking at the derivative of $P - h/k$. Since h and k are constants, $\frac{d}{dt}\left(P - \frac{h}{k}\right) = \frac{dP}{dt} = k\left(P - \frac{h}{k}\right)$. Hence $P - h/k$ satisfies the differential equation for natural growth, so $P - h/k = Ae^{kt}$ for some A. Hence $P(t) = h/k + Ae^{kt}$. To express this in terms of $P(0)$, h, and k, let $t = 0$: $P(0) = \frac{h}{k} + A$. Hence $A = P(0) - \frac{h}{k}$, so $P(t) = \frac{h}{k} + \left(P(0) - \frac{h}{k}\right)e^{kt} = \frac{1}{k}[(kP(0) - h)e^{kt} + h]$.

(c) When $h = kP(0)$, $P(t) = \frac{1}{k}[(kP(0) - kP(0))e^{kt} + h] = h/k$. In this case, the population of fish remains constant. When $h > kP(0)$, $kP(0) - h < 0$ and $P(t) = 0$ for $t = \frac{1}{k}\ln\left(\frac{h}{h - kP(0)}\right)$. Thus the fish will eventually perish. When $h < kP(0)$, then $kP(0) - h > 0$, and $P(t)$ increases as t increases. So in this case the population of the fish will continue to grow until factors such as disease and low food supply stifle that growth.

33 $f'(x) = \frac{d}{dx}\left(3\int_0^x f(t)\,dt\right) = 3f(x)$, so $f(x) = Ce^{3x}$ for a constant C. But $C = f(0) = 3\int_0^0 f(t)\,dt = 0$. Hence $f(x) = 0$ for all x.

35 (a) If the investment continues to double in every five year period, by 1990 the amount should have been \$32,000.

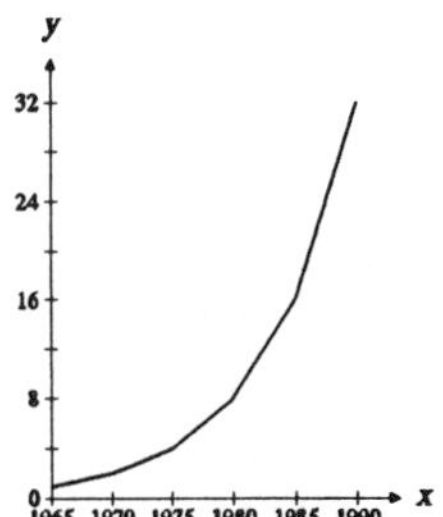

(b) The percentage increases during the five periods were 100%, 50%, $33\frac{1}{3}$%, 25%, and 40%. The first increase was the largest.

37 Since $k = \frac{\ln 2}{t_2}$, we have $P(t) = P(0)e^{\frac{\ln 2}{t_2}\cdot t} = P(0)2^{t/t_2}$.

39 Not only is the growth proportional to the population itself, as noted before, but it is also expected that the rate of population increase is proportional to the amount of room the population has left. As $P(t) \to M$, we expect $\frac{dP}{dt} \to 0$, hence the inclusion of the $(M - P(t))$ factor in the equation.

41 (a) Since $0 < P(t) < M$, both $P(t)$ and $M - P(t)$ are positive. If k is also positive, then $dP/dt = kP(t)[M - P(t)] > 0$.

(b) To find the maximum rate of change, we must set $\frac{d}{dt}\left(\frac{dP}{dt}\right) = \frac{d^2P}{dt^2} = 0$. Now, $\frac{d}{dt}\left(\frac{dP}{dt}\right) =$

$\frac{d}{dt}[kP(t)(M - P(t))]$

$= k\left[\frac{dP(t)}{dt}[M - P(t)] + \frac{d}{dt}(M - P(t))\cdot P(t)\right] =$

$k[kP(t)(M-P(t))[M-P(t)] - kP(t)(M-P(t))P(t)]$

$= k^2P(t)(M - P(t))[M - P(t) - P(t)]$.

So $\frac{d^2P}{dt^2} = 0$ when $M = 2P(t)$, or when $P(t)$

$= M/2$. Since $\frac{d^2P}{dt^2}$ changes from + to − at

$P(t) = M/2$, this represents the maximum rate of change of P.

(c) From (b), we know the graph of $P(t)$ has an inflection point at $P(t) = M/2$, as well.

6.8 l'Hôpital's Rule

Note: In the solutions below, the symbol $\underset{H}{=}$ (an "H" below an equal sign) is used to indicate an equality that follows from an application of l'Hôpital's rule.

1 $\lim_{x\to 2}(x^3 - 8) = 0 = \lim_{x\to 2}(x^2 - 4)$, so l'Hôpital's rule applies: $\lim_{x\to 2}\frac{x^3 - 8}{x^2 - 4} \underset{H}{=} \lim_{x\to 2}\frac{3x^2}{2x} = \frac{3\cdot 2^2}{2\cdot 2} = 3$.

3 $\lim_{x\to 0}\sin 3x = 0 = \lim_{x\to 0}\sin 2x$, so l'Hôpital's rule applies: $\lim_{x\to 0}\frac{\sin 3x}{\sin 2x} \underset{H}{=} \lim_{x\to 0}\frac{3\cos 3x}{2\cos 2x} = \frac{3}{2}$.

5 $\lim_{x\to\infty} e^x = \infty = \lim_{x\to\infty} x^3$, so l'Hôpital's rule applies; the result is another infinity-over-infinity form, so we continue to apply the rule until this is no longer the case: $\lim_{x\to\infty}\frac{x^3}{e^x} \underset{H}{=} \lim_{x\to\infty}\frac{3x^2}{e^x} \underset{H}{=} \lim_{x\to\infty}\frac{6x}{e^x} \underset{H}{=} \lim_{x\to\infty}\frac{6}{e^x}$

$= 0$. Observe that $\frac{6}{e^x} \to 0$ because $e^x \to \infty$ as $x \to \infty$; this was *not* obtained by l'Hôpital's rule because it does not apply in the last step.

7 $\lim_{x\to 0}(1 - \cos x) = 0 = \lim_{x\to 0} x^2$, so l'Hôpital's rule applies: $\lim_{x\to 0}\frac{1 - \cos x}{x^2} \underset{H}{=} \lim_{x\to 0}\frac{\sin x}{2x} =$

$\frac{1}{2}\lim_{x\to 0}\frac{\sin x}{x} = \frac{1}{2}\cdot 1 = \frac{1}{2}$.

9 $\lim_{x\to 0}\tan 3x = 0 = \lim_{x\to 0}\ln(1 + x)$, so l'Hôpital's rule applies: $\lim_{x\to 0}\frac{\tan 3x}{\ln(1 + x)} \underset{H}{=} \lim_{x\to 0}\frac{3\sec^2 3x}{1/(1 + x)} =$

$\frac{3\cdot 1^2}{1/(1 + 0)} = \frac{3}{1} = 3$.

11 $\lim_{x\to\infty}(\ln x)^2 = \infty = \lim_{x\to\infty} x$, so l'Hôpital's rule applies: $\lim_{x\to\infty}\frac{(\ln x)^2}{x} \underset{H}{=} \lim_{x\to\infty}\frac{2(\ln x)\cdot 1/x}{1} =$

$\lim_{x\to\infty}\frac{2\ln x}{x} \underset{H}{=} \lim_{x\to\infty}\frac{2/x}{1} = \lim_{x\to\infty}\frac{2}{x} = 0$.

13 Let $y = (1 - 2x)^{1/x}$. Then $\ln y = \frac{1}{x}\ln(1 - 2x)$

$= \frac{\ln(1 - 2x)}{x}$. Note that $\lim_{x\to 0} x = 0 =$

$\lim_{x\to 0}\ln(1 - 2x)$, so l'Hôpital's rule applies:

$$\lim_{x\to 0} \ln y = \lim_{x\to 0} \frac{\ln(1-2x)}{x} \underset{H}{=} \lim_{x\to 0} \frac{\frac{-2}{1-2x}}{1} =$$

$\lim_{x\to 0} \frac{-2}{1-2x} = \frac{-2}{1-0} = -2$. Thus $\lim_{x\to 0} \ln y =$

-2, so $\lim_{x\to 0} y = \lim_{x\to 0} e^{\ln y} = e^{-2}$.

15 Let $y = (\sin x)^{(e^x - 1)}$, so $\ln y = (e^x - 1)\ln \sin x$

$= \frac{\ln \sin x}{(e^x - 1)^{-1}}$. Note that $\lim_{x\to 0^+} \ln \sin x = -\infty$ and

$\lim_{x\to 0^+} \frac{1}{e^x - 1} = \infty$. So l'Hôpital's rule applies:

$$\lim_{x\to 0^+} \ln y = \lim_{x\to 0^+} \frac{\ln \sin x}{(e^x - 1)^{-1}}$$

$$\underset{H}{=} \lim_{x\to 0^+} \frac{(\cos x)/(\sin x)}{-(e^x - 1)^{-2}(e^x)} = \lim_{x\to 0^+} \frac{\cot x}{-e^x/(e^x - 1)^2}$$

$$= \lim_{x\to 0^+} \frac{1}{e^x} \lim_{x\to 0^+} \frac{(e^x - 1)^2}{-\tan x} = 1 \cdot \lim_{x\to 0^+} \frac{(e^x - 1)^2}{-\tan x}$$

$$\underset{H}{=} \lim_{x\to 0^+} \frac{2(e^x - 1)e^x}{-\sec^2 x} = \frac{2(1-1)\cdot 1}{-1} = 0. \text{ Thus}$$

$\ln y \to 0$, so $y \to 1$.

17 Let $y = (\tan x)^{\tan 2x}$. Then $\ln y = \tan 2x \ln \tan x =$

$\frac{\ln \tan x}{\cot 2x}$. Note that $\lim_{x\to 0+} \ln \tan x = -\infty$ and

$\lim_{x\to 0+} \cot 2x = \infty$, so l'Hôpital's rule applies. Thus

$$\lim_{x\to 0^+} \ln y = \lim_{x\to 0^+} \frac{\ln \tan x}{\cot 2x} \underset{H}{=} \lim_{x\to 0^+} \frac{\frac{1}{\tan x}\sec^2 x}{-2\csc^2 2x} =$$

$$\lim_{x\to 0^+} \frac{(\sin 2x)^2}{-2}\cdot\frac{\cos x}{\sin x}\cdot\frac{1}{\cos^2 x} = \lim_{x\to 0^+} \frac{(\sin 2x)^2}{-2\sin x \cos x}$$

$= \lim_{x\to 0^+} \frac{(\sin 2x)^2}{-\sin 2x} = \lim_{x\to 0^+} (-\sin 2x) = 0$. Hence,

$\lim_{x\to 0^+} \ln y = 0$, so $\lim_{x\to 0^+} y = e^0 = 1$.

19 $\lim_{x\to\infty} \frac{2^x}{3^x} = \lim_{x\to\infty} \left(\frac{2}{3}\right)^x = 0$, since $0 < 2/3 < 1$.

(l'Hôpital's rule applies, but is of little assistance.)

21 We saw in Sec. 6.1 that $\log_b x = \log_b c \log_c x$, so

$\log_2 x = \log_2 3 \log_3 x$. Thus $\frac{\log_2 x}{\log_3 x} = \log_2 3$, which

is a constant. Hence $\lim_{x\to\infty} \frac{\log_2 x}{\log_3 x} = \log_2 3$ (which

can also be expressed as $\frac{\ln 3}{\ln 2}$). Since $\lim_{x\to\infty} \log_2 x =$

$\infty = \lim_{x\to\infty} \log_3 x$, l'Hôpital's rule also applies, but

is not really necessary.

23 $\lim_{x\to\infty} \left(\frac{1}{x} - \frac{1}{\sin x}\right)$ does not exist, since $\lim_{x\to\infty} \frac{1}{x} = 0$

and $\lim_{x\to\infty} \frac{1}{\sin x}$ does not exist. (Note that

$\frac{1}{\sin x} = \pm 1$ whenever x is an odd multiple of $\pi/2$

and "blows up" to $\pm\infty$ at multiples of π.)

25 $$\lim_{x\to\infty} \frac{x^2 + 3\cos 5x}{x^2 - 2\sin 4x} = \lim_{x\to\infty} \frac{1 + 3\frac{\cos 5x}{x^2}}{1 - 2\frac{\sin 4x}{x^2}} =$$

$$\frac{1 + 3\cdot 0}{1 - 2\cdot 0} = 1$$

27 $$\lim_{x\to 0} \frac{3x^3 + x^2 - x}{5x^3 + x^2 + x} = \lim_{x\to 0} \frac{x(3x^2 + x - 1)}{x(5x^2 + x + 1)} =$$

$$\lim_{x\to 0} \frac{3x^2 + x - 1}{5x^2 + x + 1} = \frac{0 + 0 - 1}{0 + 0 + 1} = -1$$

29 $\lim_{x\to\infty} \frac{\sin x}{4 + \sin x}$ does not exist, since $\sin x$ varies periodically between -1 and 1 as $x \to \infty$.

31 $\lim_{x\to 1^+} (x - 1)\ln(x - 1) = \lim_{x\to 1^+} \frac{\ln(x - 1)}{(x - 1)^{-1}}$; now $\lim_{x\to 1^+} \ln(x - 1) = -\infty$ and $\lim_{x\to 1^+} \frac{1}{x - 1} = \infty$, so we apply l'Hôpital's rule to obtain $\lim_{x\to 1^+} \frac{\frac{1}{x - 1}}{-(x - 1)^{-2}}$

$= \lim_{x\to 1^+} (-(x - 1)) = -(1 - 1) = 0.$

33 Let $y = (\cos x)^{1/x}$. Then $\ln y = (1/x) \ln \cos x = \frac{\ln \cos x}{x}$. Now $\lim_{x\to 0} \ln \cos x = 0 = \lim_{x\to 0} x$, so l'Hôpital's rule applies. Thus $\lim_{x\to 0} \ln y =$

$$\lim_{x\to 0} \frac{\ln \cos x}{x} \underset{H}{=} \lim_{x\to 0} \frac{\frac{-\sin x}{\cos x}}{1} = \lim_{x\to 0} (-\tan x) = 0.$$

Hence $\lim_{x\to 0} \ln y = 0$, so $\lim_{x\to 0} y = e^0 = 1$.

35 $\lim_{x\to\infty} \frac{\sin 2x}{\sin 3x}$ does not exist; for $x = 2\pi n + \pi/3$, n an integer, we have $\frac{\sin 2x}{\sin 3x} = \frac{\sin\left(4n\pi + \frac{2\pi}{3}\right)}{\sin(6n\pi + \pi)} =$

$\frac{\sin 2\pi/3}{\sin \pi} = \frac{\sqrt{3}/2}{0}$, which is undefined.

37 $\lim_{x\to 0} \frac{xe^x(1 + x)^3}{e^x - 1} = \lim_{x\to 0} e^x(1 + x)^3 \lim_{x\to 0} \frac{x}{e^x - 1}$

$= 1 \cdot \lim_{x\to 0} \frac{x}{e^x - 1} \underset{H}{=} \lim_{x\to 0} \frac{1}{e^x} = 1$

39 $\lim_{x\to 0} (\csc x - \cot x) = \lim_{x\to 0} \left(\frac{1}{\sin x} - \frac{\cos x}{\sin x}\right) =$

$\lim_{x\to 0} \frac{1 - \cos x}{\sin x}$; now $\lim_{x\to 0} (1 - \cos x) = 0 = \lim_{x\to 0} \sin x$, so we apply l'Hôpital's rule to obtain

$$\lim_{x\to 0} \frac{1 - \cos x}{\sin x} \underset{H}{=} \lim_{x\to 0} \frac{\sin x}{\cos x} = \frac{0}{1} = 0.$$

41 $\lim_{x\to 0} (5^x - 3^x) = 0 = \lim_{x\to 0} \sin x$, so l'Hôpital's rule applies. Thus $\lim_{x\to 0} \frac{5^x - 3^x}{\sin x}$

$$\underset{H}{=} \lim_{x\to 0} \frac{(\ln 5)5^x - (\ln 3)3^x}{\cos x} = \frac{(\ln 5)\cdot 1 - (\ln 3)\cdot 1}{1}$$

$$= \ln 5 - \ln 3 = \ln \frac{5}{3}.$$

43 $\lim_{x\to 2} \frac{x^3 + 8}{x^2 + 5} = \frac{8 + 8}{4 + 5} = \frac{16}{9}$

45 $\lim_{x\to 0} \left(\frac{1}{1 - \cos x} - \frac{2}{x^2}\right) = \lim_{x\to 0} \frac{x^2 - 2 + 2\cos x}{x^2(1 - \cos x)}$, where $x^2 - 2 + 2\cos x \to 0$ and $x^2(1 - \cos x) \to 0$ as $x \to 0$, so l'Hôpital's rule applies. We thus obtain $\lim_{x\to 0} \frac{2x - 2\sin x}{x^2(\sin x) + 2x(1 - \cos x)}$, which is also a zero-over-zero form. A second application of l'Hôpital's rule produces

$$\lim_{x\to 0} \frac{2 - 2\cos x}{x^2 \cos x + 2x \sin x + 2x \sin x + 2(1 - \cos x)}$$

$$= \lim_{x\to 0} \frac{2(1 - \cos x)}{(x^2 - 2)\cos x + 4x \sin x + 2} \underset{H}{=}$$

$$\lim_{x\to 0} \frac{2\sin x}{(x^2-2)(-\sin x)+2x\cos x+4x\cos x+4\sin x}$$

$$= \lim_{x\to 0} \frac{2\sin x}{(6-x^2)\sin x + 6x\cos x} =$$

$$\lim_{x\to 0} \frac{2\cdot \dfrac{\sin x}{x}}{(6-x^2)\dfrac{\sin x}{x}+6\cos x} = \frac{2\cdot 1}{(6-0)\cdot 1+6\cdot 1}$$

$$= \frac{2}{6+6} = \frac{1}{6}.$$

47

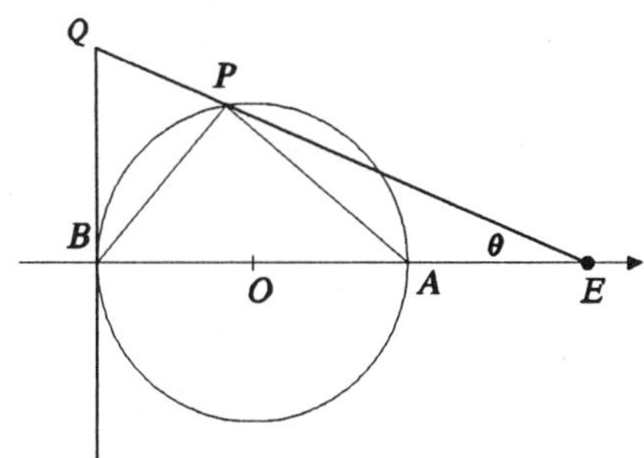

Label points as shown in the figure. Let $\theta = \angle BEQ$. Since EBQ is a right triangle, $\angle BQE = \pi/2 - \theta$. Since $\overline{BQ} = \overline{BP}$, $\angle BPQ = \angle BQP = \frac{\pi}{2} - \theta$. Also, $\angle BPA = \pi/2$, because AB is a diameter of a circle through P. Hence $\angle APE = \pi - \left(\frac{\pi}{2} - \theta\right) - \frac{\pi}{2} = \theta = \angle AEP$, so EAP is an isosceles triangle and $\overline{AE} = \overline{AP}$. As $Q \to B$, $P \to B$, so $\overline{AE} = \overline{AP} \to \overline{AB} = 2$ and E approaches the point (3, 0).

49 (b) $f(\theta) =$ Area of triangle $ABC = \frac{1}{2}(\sin\theta)(1-\cos\theta)$, and $g(\theta) =$ Area of sector OBC − Area of triangle OAC = $\theta/2 - 1/2\sin\theta\cos\theta$, so $\lim_{\theta\to 0}\frac{f(\theta)}{g(\theta)} =$

$$\lim_{\theta\to 0} \frac{\frac{1}{2}\sin\theta\,(1-\cos\theta)}{\frac{1}{2}\theta - \frac{1}{2}\sin\theta\cos\theta}$$

$$= \lim_{\theta\to 0} \frac{\sin\theta - \sin\theta\cos\theta}{\theta - \sin\theta\cos\theta}$$

$$\overset{H}{=} \lim_{\theta\to 0} \frac{\cos\theta + \sin^2\theta - \cos^2\theta}{1+\sin^2\theta - \cos^2\theta}$$

$$= \lim_{\theta\to 0} \frac{\cos\theta - \cos 2\theta}{1-\cos 2\theta}$$

$$\overset{H}{=} \lim_{\theta\to 0} \frac{-\sin\theta + 2\sin 2\theta}{2\sin 2\theta}$$

$$= \lim_{\theta\to 0}\left(1 - \frac{\sin\theta}{2\sin 2\theta}\right)$$

$$= \lim_{\theta\to 0}\left(1 - \frac{\sin\theta}{4\sin\theta\cos\theta}\right)$$

$$= \lim_{\theta\to 0}\left(1 - \frac{1}{4\cos\theta}\right) = 1 - \frac{1}{4} = \frac{3}{4}.$$

51 $\frac{y}{k} = \left(\alpha x_1^{-\rho} + (1-\alpha)x_2^{-\rho}\right)^{-1/\rho}$, so $\ln\frac{y}{k} =$

$$-\frac{1}{\rho}\ln\left(\alpha x_1^{-\rho} + (1-\alpha)x_2^{-\rho}\right) =$$

$$\frac{\ln\left(\alpha x_1^{-\rho} + (1-\alpha)x_2^{-\rho}\right)}{-\rho}.$$ As $\rho \to 0^+$,

$(\alpha x_1^{-\rho} + (1-\alpha)x_2^{-\rho}) \to (\alpha\cdot 1 + (1-\alpha)\cdot 1) = 1$, so $\ln(\alpha x_1^{-\rho} + (1-\alpha)x_2^{-\rho}) \to 0$. Thus, l'Hôpital's rule applies and $\lim_{\rho\to 0^+}\ln\frac{y}{k} =$

$$\lim_{\rho\to 0^+} \frac{\ln\left(\alpha x_1^{-\rho} + (1-\alpha)x_2^{-\rho}\right)}{-\rho} \underset{H}{=}$$

$$\lim_{\rho\to 0^+} \frac{\dfrac{(\ln x_1)\alpha x_1^{-\rho}(-1) + (\ln x_2)(1-\alpha)x_2^{-\rho}(-1)}{\alpha x_1^{-\rho} + (1-\alpha)x_2^{-\rho}}}{-1}$$

$$= \lim_{\rho\to 0^+} \frac{(\ln x_1)\alpha x_1^{-\rho} + (\ln x_2)(1-\alpha)x_2^{-\rho}}{\alpha x_1^{-\rho} + (1-\alpha)x_2^{-\rho}}$$

$$= \frac{(\ln x_1)\alpha\cdot 1 + (\ln x_2)(1-\alpha)\cdot 1}{\alpha\cdot 1 + (1-\alpha)\cdot 1}$$

$$= \frac{(\ln x_1)\alpha + (\ln x_2)(1-\alpha)}{1}$$

$$= (\ln x_1)\alpha + (\ln x_2)(1-\alpha).$$

Therefore $\lim_{\rho\to 0^+} \ln y/k = \ln(x_1^{\alpha}x_2^{1-\alpha})$, so $\lim_{\rho\to 0^+} y/k$

$= x_1^{\alpha}x_2^{1-\alpha}$; thus $\lim_{\rho\to 0^+} y = kx_1^{\alpha}x_2^{1-\alpha}$.

53 Let $y = \left(\dfrac{1+2^x}{2}\right)^{1/x}$. Then $\ln y = \dfrac{1}{x}\ln\left(\dfrac{1+2^x}{2}\right)$

$= \dfrac{\ln[(1+2^x)/2]}{x}$, where $\lim_{x\to 0} \ln\left(\dfrac{1+2^x}{2}\right) = 0 =$

$\lim_{x\to 0} x$, so l'Hôpital's rule applies. Thus $\lim_{x\to 0} \ln y =$

$$\lim_{x\to 0} \frac{\ln[(1+2^x)/2]}{x} \underset{H}{=} \lim_{x\to 0} \frac{\dfrac{2}{1+2^x}\cdot\dfrac{1}{2}(\ln 2)2^x}{1} =$$

$$\lim_{x\to 0} \frac{(\ln 2)2^x}{1+2^x} = \frac{(\ln 2)\cdot 1}{1+1} = \frac{\ln 2}{2}.$$ Therefore

$$\lim_{x\to 0} y = \lim_{x\to 0} e^{\ln y} = e^{(\ln 2)/2} = (e^{\ln 2})^{1/2} = 2^{1/2} = \sqrt{2}.$$

55 Let $y = x^x$ and note that at $(1, 1)$ lies on the graph.

Also, from Example 6 we know $\lim_{x\to 0^+} x^x = 1$. Now

$\ln y = x \ln x$ and applying logarithmic

differentiation yields $\dfrac{1}{y}\dfrac{dy}{dx} = x\cdot\dfrac{1}{x} + \ln x =$

$1 + \ln x$, so $\dfrac{dy}{dx} = x^x(1 + \ln x)$, which goes from

$-$ to $+$ at $x = e^{-1}$. Thus $(e^{-1}, (1/e)^{1/e})$ is a

minimum. Finally $\dfrac{d^2y}{dx^2} = x^x(1/x) + x^x(1 + \ln x)^2$

> 0 for $x > 0$, so x^x is concave up on $(0, 1]$.

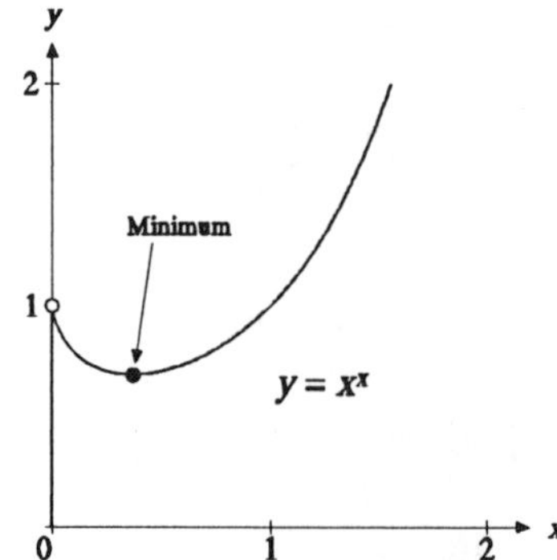

57 $y = x \ln x$ is only defined for $x > 0$. Now $\dfrac{dy}{dx} =$

$1 + \ln x$, which changes from $-$ to $+$ at $x = e^{-1}$.

Also, $\dfrac{d^2y}{dx^2} = \dfrac{1}{x} > 0$ for $x > 0$, so $x \ln x$ is

concave up for all $x > 0$. From Example 5, we

know $\lim_{x\to 0^+} x \ln x = 0$, and it is obvious that

$\lim_{x\to\infty} x \ln x = \infty$. Finally, $x \ln x = 0$ when $x = 1$.

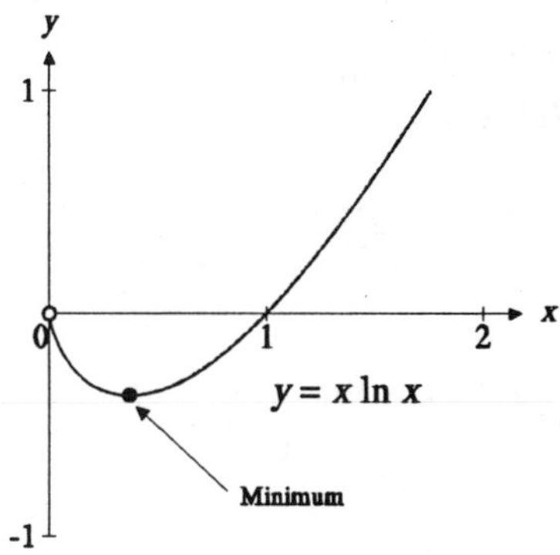

6.9 The Hyperbolic Functions and Their Inverses

1 $\cosh t + \sinh t = \dfrac{e^t + e^{-t}}{2} + \dfrac{e^t - e^{-t}}{2}$

$= \dfrac{e^t + e^{-t} + e^t - e^{-t}}{2} = \dfrac{2e^t}{2} = e^t$

3 $(\sinh t)' = \left(\dfrac{e^t - e^{-t}}{2}\right)' = \dfrac{e^t - (-e^{-t})}{2}$

$= \dfrac{e^t + e^{-t}}{2} = \cosh t$

5 $(\text{sech } t)' = \left(\dfrac{1}{\cosh t}\right)' = \dfrac{-(\cosh t)'}{\cosh^2 t} = \dfrac{-\sinh t}{\cosh^2 t}$

$= \dfrac{-1}{\cosh t} \cdot \dfrac{\sinh t}{\cosh t} = -\text{sech } t \tanh t$

7 $(\cosh 3x)' = 3 \sinh 3x$

9 $(\tanh \sqrt{x})' = (\text{sech}^2 \sqrt{x}) \cdot \dfrac{1}{2\sqrt{x}} = \dfrac{\text{sech}^2 \sqrt{x}}{2\sqrt{x}}$

11 $(e^{3x} \sinh x)' = e^{3x} \cosh x + 3e^{3x} \sinh x = e^{3x}(\cosh x + 3 \sinh x)$

13 $[(\cosh 4x)(\coth 5x)(\text{csch } x^2)]' =$

$(\cosh 4x)'(\coth 5x)(\text{csch } x^2) +$

$(\cosh 4x)(\coth 5x)'(\text{csch } x^2) +$

$(\cosh 4x)(\coth 5x)(\text{csch } x^2)'$

$= (4 \sinh 4x)(\coth 5x)(\text{csch } x^2) +$

$(\cosh 4x)(-5 \text{ csch}^2 5x)(\text{csch } x^2) +$

$(\cosh 4x)(\coth 5x)(-2x \text{ csch } x^2 \coth x^2)$

15 If we let $x = \cosh t = \dfrac{e^t + e^{-t}}{2}$, then $2x = e^t + \dfrac{1}{e^t}$ and $2xe^t = (e^t)^2 + 1$, so $(e^t)^2 - 2xe^t + 1 = 0$, which is a quadratic in the unknown e^t. Thus $e^t = \dfrac{2x \pm \sqrt{(2x)^2 - 4}}{2} = x \pm \sqrt{x^2 - 1}$. Recall that $\cosh t$ is an even function; to define $\cosh^{-1}$ unambiguously, consider $\cosh t$ for $t \geq 0$. But for $t \geq 0$ we must have $e^t \geq 1$. Since $\left(x - \sqrt{x^2 - 1}\right)\left(x + \sqrt{x^2 - 1}\right) = 1$, it follows that for $x > 1$, $x + \sqrt{x^2 - 1} > 1$ and $x - \sqrt{x^2 - 1} < 1$. We therefore choose the plus sign. Thus $e^t = x + \sqrt{x^2 - 1}$ and $\cosh^{-1} x = t = \ln\left(x + \sqrt{x^2 - 1}\right)$.

17 If we let $x = \coth t = \dfrac{e^t + e^{-t}}{e^t - e^{-t}} = \dfrac{e^{2t} + 1}{e^{2t} - 1}$, then $xe^{2t} - x = e^{2t} + 1, x + 1 = e^{2t}(x - 1), e^{2t} = \dfrac{x + 1}{x - 1}, 2t = \ln\left(\dfrac{x + 1}{x - 1}\right)$, and $t = \dfrac{1}{2} \ln\left(\dfrac{x + 1}{x - 1}\right)$.

Consequently $\coth^{-1} x = \dfrac{1}{2} \ln\left(\dfrac{x + 1}{x - 1}\right)$, where $|x| > 1$ (since $x = \coth t$ and $|\coth t| > 1$).

19 $(\cosh^{-1} x)' = \left(\ln\left(x + \sqrt{x^2 - 1}\right)\right)' = \dfrac{1 + \dfrac{x}{\sqrt{x^2 - 1}}}{x + \sqrt{x^2 - 1}}$

$= \dfrac{x + \sqrt{x^2 - 1}}{\sqrt{x^2 - 1}\left(x + \sqrt{x^2 - 1}\right)} = \dfrac{1}{\sqrt{x^2 - 1}}$, so

$\int \dfrac{dx}{\sqrt{x^2 - 1}} = \cosh^{-1} x + C.$

21 $(-\text{sech}^{-1} x)' = \left(-\ln\left(\dfrac{1 + \sqrt{1 - x^2}}{x}\right)\right)' =$

$\left(\ln x - \ln\left(1 + \sqrt{1 - x^2}\right)\right)' = \dfrac{1}{x} - \dfrac{-x/\sqrt{1 - x^2}}{1 + \sqrt{1 - x^2}}$

$= \dfrac{1}{x} + \dfrac{x}{\sqrt{1 - x^2}\left(1 + \sqrt{1 - x^2}\right)}$

$= \dfrac{\sqrt{1 - x^2}\left(1 + \sqrt{1 - x^2}\right) + x^2}{x\sqrt{1 - x^2}\left(1 + \sqrt{1 - x^2}\right)}$

$= \dfrac{\sqrt{1 - x^2} + (1 - x^2) + x^2}{x\sqrt{1 - x^2}\left(1 + \sqrt{1 - x^2}\right)}$

$= \dfrac{1 + \sqrt{1 - x^2}}{x\sqrt{1 - x^2}\left(1 + \sqrt{1 - x^2}\right)} = \dfrac{1}{x\sqrt{1 - x^2}}$. Thus

$\int \dfrac{dx}{x\sqrt{1 - x^2}} = -\text{sech}^{-1} x + C.$

23 (a) $\cosh x \cosh y + \sinh x \sinh y =$

$\left(\dfrac{e^x + e^{-x}}{2}\right)\left(\dfrac{e^y + e^{-y}}{2}\right) + \left(\dfrac{e^x - e^{-x}}{2}\right)\left(\dfrac{e^y - e^{-y}}{2}\right)$

$= \dfrac{e^{x+y} + e^{x-y} + e^{-x+y} + e^{-(x+y)} + e^{x+y} - e^{x-y} - e^{-x+y} + e^{-(x+y)}}{4}$

$= \dfrac{2e^{x+y} + 2e^{-(x+y)}}{4} = \dfrac{e^{x+y} + e^{-(x+y)}}{2}$

$= \cosh(x + y)$

(b) $\sinh x \cosh y + \cosh x \sinh y =$

$\left(\dfrac{e^x - e^{-x}}{2}\right)\left(\dfrac{e^y + e^{-y}}{2}\right) + \left(\dfrac{e^x + e^{-x}}{2}\right)\left(\dfrac{e^y - e^{-y}}{2}\right) =$

$\dfrac{e^{x+y} + e^{x-y} - e^{-x+y} - e^{-(x+y)} + e^{x+y} - e^{x-y} + e^{-x+y} - e^{-(x+y)}}{4}$

$= \dfrac{2e^{x+y} - 2e^{-(x+y)}}{4} = \dfrac{e^{x+y} - e^{-(x+y)}}{2}$

$= \sinh(x + y)$

25 (a) $\cosh x \cosh y - \sinh x \sinh y$

$= \left(\dfrac{e^x + e^{-x}}{2}\right)\left(\dfrac{e^y + e^{-y}}{2}\right) - \left(\dfrac{e^x - e^{-x}}{2}\right)\left(\dfrac{e^y - e^{-y}}{2}\right)$

$= \dfrac{e^{x+y} + e^{x-y} + e^{-x+y} + e^{-(x+y)} - (e^{x+y} - e^{x-y} - e^{-x+y} + e^{-(x+y)})}{4}$

$= \dfrac{2e^{x-y} + 2e^{-x+y}}{4} = \dfrac{e^{x-y} + e^{-x+y}}{2}$

$= \cosh(x - y)$

(b) $\sinh x \cosh y - \cosh x \sinh y$

$= \left(\dfrac{e^x - e^{-x}}{2}\right)\left(\dfrac{e^y + e^{-y}}{2}\right) - \left(\dfrac{e^x + e^{-x}}{2}\right)\left(\dfrac{e^y - e^{-y}}{2}\right)$

$= \dfrac{(e^{x+y} + e^{x-y} - e^{-x+y} - e^{-(x+y)}) - (e^{x+y} - e^{x-y} + e^{-x+y} - e^{-(x+y)})}{4}$

$= \dfrac{2e^{x-y} - 2e^{-x+y}}{4} = \dfrac{e^{x-y} - e^{-x+y}}{2}$

$= \sinh(x - y)$

27 (a) $2 \sinh^2(x/2) = 2\left(\dfrac{e^{x/2} - e^{-x/2}}{2}\right)^2 =$

$2\left(\dfrac{e^x - 2 + e^{-x}}{4}\right) = \dfrac{e^x - 2 + e^{-x}}{2}$

$$= \frac{e^x + e^{-x}}{2} - \frac{2}{2} = \cosh x - 1$$

(b) $2\cosh^2(x/2) = 2\left(\frac{e^{x/2} + e^{-x/2}}{2}\right)^2$

$$= 2\left(\frac{e^x + 2 + e^{-x}}{4}\right) = \frac{e^x + 2 + e^{-x}}{2}$$

$$= \frac{e^x + e^{-x}}{2} + \frac{2}{2} = \cosh x + 1$$

29 The graph crosses the axis at $x = 0$. Its slope there is $\frac{dy}{dx} = \operatorname{sech}^2 x = \operatorname{sech}^2 0 = 1$, so the angle is $\pi/4$.

31 (a)

t	-3	-2	-1
$\cosh t$	10.068	3.762	1.543
$\sinh t$	-10.018	-3.627	-1.175

t	0	1	2	3
$\cosh t$	1	1.543	3.762	10.068
$\sinh t$	0	1.175	3.627	10.018

(b)

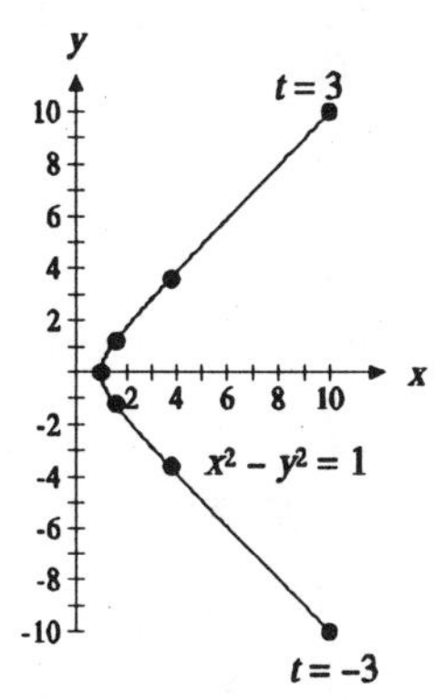

33 (a)

t	0	1	2	3
$\tanh t$	0	0.762	0.964	0.995

(b)

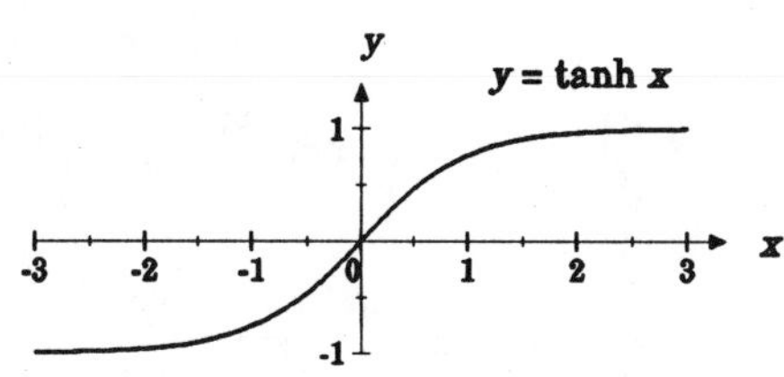

35 $y' = \operatorname{sech}^2 x$, $y'' = 2\operatorname{sech} x\,(-\operatorname{sech} x \tanh x) = -2\operatorname{sech}^2 x \tanh x$. Since $\operatorname{sech}^2 x = \frac{1}{\cosh^2 x} > 0$ for all x, y'' changes sign at x if and only if $\tanh x$ changes sign at x; that is, at $x = 0$. The only inflection point is $(0, 0)$.

37 (a) $v(t) = \frac{dx}{dt} = \frac{d}{dt}\left(\frac{V^2}{g}\ln\cosh\frac{gt}{V}\right)$

$$= \frac{V^2}{g}\cdot\frac{\left(\sinh\frac{gt}{V}\right)\frac{g}{V}}{\cosh\frac{gt}{V}} = V\tanh\frac{gt}{V}$$

(b) Note that $\lim\limits_{t\to\infty}\tanh kt = \lim\limits_{t\to\infty}\frac{e^{kt} - e^{-kt}}{e^{kt} + e^{-kt}} =$

$$\lim_{t\to\infty}\frac{1 - e^{-2kt}}{1 + e^{-2kt}} = \frac{1 - 0}{1 + 0} = 1, \text{ for } k > 0.$$

Since g and V are positive constants,

$\lim\limits_{t\to\infty} v(t) = \lim\limits_{t\to\infty} V\tanh\frac{gt}{V} = V\cdot 1 = V$, as claimed. (*Note*: g is positive because it is gravitational acceleration and is measured in the same direction as x, which increases as the object falls. We assume that V is positive also.)

(c) $\frac{dv}{dt} = \frac{d}{dt}\left(V \tanh \frac{gt}{V}\right) = V\left(\operatorname{sech}^2\left(\frac{gt}{V}\right)\right)\frac{g}{V}$

$= g \operatorname{sech}^2\left(\frac{gt}{V}\right)$

(d) $g - g\left(\frac{v}{V}\right)^2 = g - g \tanh^2 \frac{gt}{V}$

$= \frac{g}{\cosh^2 gt/V}\left(\cosh^2 \frac{gt}{V} - \sinh^2 \frac{gt}{V}\right)$

$= \frac{g}{\cosh^2 gt/V} = g \operatorname{sech}^2 \frac{gt}{V}$, which is the acceleration.

(e) The limit of the acceleration is

$\lim_{t\to\infty} g \operatorname{sech}^2 \frac{gt}{V} = \lim_{t\to\infty} \frac{g}{\cosh^2 gt/V}$. Since $\cosh gt/V \to \infty$ as $t \to \infty$, the limit of the acceleration is 0.

6.S Guide Quiz

1 (a) $\frac{d}{dx}(\sin^{-1}(x^3)) = \frac{1}{\sqrt{1 - (x^3)^2}} \cdot 3x^2$

$= \frac{3x^2}{\sqrt{1 - x^6}}$

(b) $\frac{d}{dx}(e^{\sin^{-1} 3x}) = e^{\sin^{-1} 3x} \cdot \frac{1}{\sqrt{1 - (3x)^2}} \cdot 3$

$= \frac{3e^{\sin^{-1} 3x}}{\sqrt{1 - 9x^2}}$

(c) $\frac{d}{dx}(\sin [(\tan^{-1} x)^2])$

$= \cos [(\tan^{-1} x)^2] \cdot 2 \tan^{-1} x \frac{1}{1 + x^2}$

$= \frac{2 \tan^{-1} x \cos [(\tan^{-1} x)^2]}{1 + x^2}$

(d) Let $y = (x^2 + 5)x^{\sqrt{x}}$. Then $\ln y = \ln(x^2 + 5) + \sqrt{x} \ln x$ and $\frac{1}{y}\frac{dy}{dx} = \frac{1}{x^2 + 5} \cdot 2x + \sqrt{x} \cdot \frac{1}{x} + \frac{1}{2\sqrt{x}} \ln x$, so $\frac{dy}{dx} = (x^2 + 5)x^{\sqrt{x}}\left[\frac{2x}{x^2 + 5} + \frac{1}{\sqrt{x}}\left(1 + \frac{1}{2} \ln x\right)\right]$.

(e) $\frac{d}{dx}(\ln(\sin x)) = \frac{1}{\sin x} \cdot \cos x = \cot x$

(f) $\frac{d}{dx}(7^{4x-1}) = (\ln 7)7^{4x-1} \cdot (4) = 4 \ln 7 \cdot 7^{4x-1}$

(g) $\left(\frac{\tan x}{\sec^{-1} 2x}\right)'$

$= \frac{\sec^{-1} 2x \sec^2 x - (\tan x)\frac{1}{|x|\sqrt{4x^2 - 1}}}{(\sec^{-1} 2x)^2}$

$= \frac{\sec^2 x}{\sec^{-1} 2x} - \frac{\tan x}{(\sec^{-1} 2x)^2 |x| \sqrt{4x^2 - 1}}$

(h) $\frac{d}{dx}(\ln(x^2 - 5x + 4)) = \frac{1}{x^2 - 5x + 4} \cdot (2x - 5) = \frac{2x - 5}{x^2 - 5x + 4}$

(i) Let $y = \left\{x^{-5}\left[\sqrt[3]{\cos 4x}\right]^8\right\}^{1/4}$. Then $\ln y =$

$\frac{1}{4}\left[-5 \ln x + \frac{8}{3} \ln(\cos 4x)\right]$ and $\frac{1}{y}\frac{dy}{dx} =$

$\frac{1}{4}\left[-\frac{5}{x} + \frac{8}{3} \cdot \frac{1}{\cos 4x} \cdot (-4 \sin 4x)\right] =$

$-\frac{5}{4x} - \frac{8}{3} \tan 4x$ and $\frac{dy}{dx} =$

$-\{x^{-5}[\sqrt[3]{\cos 4x}]^8\}^{1/4}\left[\frac{5}{4x} + \frac{8}{3} \tan 4x\right].$

2 (a) $\left[\frac{e^{ax}}{a^2 + b^2}(a \cos bx + b \sin bx)\right]'$

$= \frac{ae^{ax}}{a^2 + b^2}(a \cos bx + b \sin bx) +$

$\frac{e^{ax}}{a^2 + b^2}(-ab \sin bx + b^2 \cos bx)$

$= \frac{e^{ax}}{a^2 + b^2}[a^2 \cos bx + ab \sin bx -$

$ab \sin bx + b^2 \cos bx]$

$= \frac{e^{ax}}{a^2 + b^2}(a^2 + b^2) \cos bx = e^{ax} \cos bx.$

(b) $\left[x \sin^{-1} ax + \frac{1}{a}\sqrt{1 - a^2x^2}\right]'$

$= \sin^{-1} ax + x \cdot \frac{a}{\sqrt{1 - (ax)^2}} + \frac{1}{a} \cdot \frac{-a^2x}{\sqrt{1 - a^2x^2}}$

$= \sin^{-1} ax + \frac{ax}{\sqrt{1 - a^2x^2}} - \frac{ax}{\sqrt{1 - a^2x^2}}$

$= \sin^{-1} ax$

(c) $(x \tan^{-1} ax - \frac{1}{2a} \ln(1 + a^2x^2))'$

$= \tan^{-1} ax + x \cdot \frac{a}{1 + a^2x^2} - \frac{1}{2a} \cdot \frac{2a^2x}{1 + a^2x^2}$

$= \tan^{-1} ax + \frac{ax}{1 + a^2x^2} - \frac{ax}{1 + a^2x^2}$

$= \tan^{-1} ax$

(d) $\left(x \sec^{-1} ax - \frac{1}{a} \ln\left(ax + \sqrt{a^2x^2 - 1}\right)\right)' =$

$\sec^{-1} ax + x \cdot \frac{a}{|ax|\sqrt{(ax)^2 - 1}} - \frac{1}{a} \cdot \frac{a + (a^2x^2 - 1)^{-1/2}(a^2x)}{ax + \sqrt{a^2x^2 - 1}}$

$= \sec^{-1} ax + \frac{1}{\sqrt{a^2x^2 - 1}} - \frac{1}{a} \cdot \frac{a\left(\sqrt{a^2x^2 - 1} + ax\right)}{\sqrt{a^2x^2 - 1}\left(ax + \sqrt{a^2x^2 - 1}\right)}$

$= \sec^{-1} ax + \frac{1}{\sqrt{a^2x^2 - 1}} - \frac{1}{\sqrt{a^2x^2 - 1}}$

$= \sec^{-1} ax$

(Note that we must assume $ax > 0$.)

3 (a) $\frac{dy}{dx} = e^{-2y}x^3$, so $e^{2y}\, dy = x^3\, dx$, $\int e^{2y}\, dy$

$= \int x^3\, dx$, and $\frac{1}{2}e^{2y} = \frac{1}{4}x^4 + C.$

(b) $\frac{dy}{dx} = \frac{4y^2 + 1}{y}$, so $\frac{y}{4y^2 + 1}\, dy = dx$,

$\int \frac{y}{4y^2 + 1}\, dy = \int dx$, and $\frac{1}{8} \ln(4y^2 + 1)$

$= x + C.$

4 Let A be the initial amount and let $f(t) = Ab^t$ be the amount present after t days. $t_{1/2} = 3.825 = -\frac{\ln 2}{\ln b}$, so $\ln b = -\frac{\ln 2}{3.825}$. When only 10 percent of the original amount remains, $f(t) = \frac{1}{10}A = Ab^t$, so

$b^t = \frac{1}{10}$, $t \ln b = \ln 1/10 = -\ln 10$, and $t =$

$\frac{-\ln 10}{\ln b} = \frac{-\ln 10}{(-\ln 2)/3.825} \approx 12.706.$

Thus it takes 12.706 days. Since the half-life of radon is 3.875 days, it takes 3.875 days for the radon to diminish to 50% of its original amount.

5 (a) Both numerator and denominator approach zero, so l'Hôpital's rule applies:

$$\lim_{x\to 0} \frac{\sin 2x}{\tan^{-1} 3x} \underset{H}{=} \lim_{x\to 0} \frac{2\cos 2x}{3/(1 + 9x^2)} = \frac{2}{3}.$$

(b) $\lim_{x\to\infty} \frac{\sin 2x}{\tan^{-1} 3x}$ does not exist, since

$\lim_{x\to\infty} \tan^{-1} 3x = \pi/2$, but $\lim_{x\to\infty} \sin 2x$ does not exist; $\sin 2x$ varies periodically between -1 and 1 as $x \to \infty$.

(c) We rewrite the limit into a zero-over-zero form in order to apply l'Hôpital's rule:

$$\lim_{x\to\pi/2^-} (\sec x - \tan x) = \lim_{x\to\pi/2^-} \left(\frac{1}{\cos x} - \frac{\sin x}{\cos x}\right)$$

$$= \lim_{x\to\pi/2^-} \frac{1 - \sin x}{\cos x} \underset{H}{=} \lim_{x\to\pi/2^-} \frac{-\cos x}{-\sin x} = \frac{-0}{-1}$$

$= 0.$

(d) Let $y = (1 - \cos 2x)^x$. Then we have $\ln y =$ $x \ln(1 - \cos 2x) = \frac{\ln(1 - \cos 2x)}{1/x}$. By

l'Hôpital's rule, $\lim_{x\to 0} \ln y =$

$$\lim_{x\to 0} \frac{\ln(1 - \cos 2x)}{1/x} \underset{H}{=} \lim_{x\to 0} \frac{\frac{2\sin 2x}{1 - \cos 2x}}{-1/x^2}$$

$$= \lim_{x\to 0} \frac{-2x^2 \sin 2x}{1 - \cos 2x}$$

$$= \lim_{x\to 0} \frac{-2x^2(\sin 2x)(1 + \cos 2x)}{1 - \cos^2 2x}$$

$$= \lim_{x\to 0} \frac{-2x^2(\sin 2x)(1 + \cos 2x)}{\sin^2 2x}$$

$$= \lim_{x\to 0} -x\frac{2x}{\sin 2x}(1 + \cos 2x) = 0\cdot 1\cdot(1 + 1)$$

$= 0.$

Thus $\lim_{x\to 0} y = e^0 = 1.$

(e) Note that $\int \tan^2 \theta \, d\theta = \int (\sec^2 \theta - 1)\, d\theta$ $= \tan\theta - \theta + C$, so $\int_0^{\pi/2} \tan^2\theta \, d\theta =$ $(\tan\theta - \theta)\big|_0^{\pi/2} = \infty$. The limit is thus an infinity-over-infinity form, so by l'Hôpital's rule and the first fundamental theorem of

calculus, $\lim_{x\to\pi/2^-} \frac{\int_0^x \tan^2\theta\, d\theta}{\tan x} \underset{H}{=} \lim_{x\to\pi/2^-} \frac{\tan^2 x}{\sec^2 x}$

$$= \lim_{x\to\pi/2^-} \tan^2 x \cos^2 x = \lim_{x\to\pi/2^-} \sin^2 x = 1.$$

6 If $b > 1$, $\lim_{x\to\infty} \frac{b^x}{x^3} \underset{H}{=} \lim_{x\to\infty} \frac{(\ln b)b^x}{3x^2} \underset{H}{=}$

$\lim_{x\to\infty} \frac{(\ln b)^2 b^x}{6x} \underset{H}{=} \lim_{x\to\infty} \frac{(\ln b)^3 b^x}{6} = \infty$. Thus for large x, $x^3 < (1.001)^x < 2^x$, since $2 > 1.001$. If $b > 1$, $\lim_{x\to\infty} \frac{x^3}{\log_b x} \underset{H}{=} \lim_{x\to\infty} \frac{3x^2}{(\log_b e)/x} = \lim_{x\to\infty} \frac{3x^3}{\log_b e}$ $= \infty$. Thus, for large x, $\ln x < x^3$ and $\log_{10} x <$ x^3. Now $\log_{10} x = \frac{\ln x}{\ln 10}$, where $\ln 10 \approx 2.3 >$

1; thus, for positive x, $\log_{10} x < \ln x$. Therefore, for sufficiently large x, $\log_{10} x < \ln x < x^3 < (1.001)^x < 2^x$.

7 (a) $f(x) = (\ln x)/x$ is defined for $x > 0$.

(b) $f(x) = 0$ only for $x = 1$: $f(1) = (\ln 1)/1 = 0/1 = 0$.

(c) $$f'(x) = \frac{x \cdot \frac{1}{x} - (\ln x) \cdot 1}{x^2} = \frac{1 - \ln x}{x^2} = 0$$

when $1 - \ln x = 0$; that is, for $x = e$.

(d) $$f''(x) = \frac{x^2(-\frac{1}{x}) - (1 - \ln x) \cdot 2x}{x^4}$$

$$= \frac{-x - 2x + 2x \ln x}{x^4} = \frac{-3x + 2x \ln x}{x^4}$$

$$= \frac{-3 + 2 \ln x}{x^3}, \text{ so } f''(e) = \frac{-3 + 2 \ln e}{e^3}$$

$= -\frac{1}{e^3} < 0$, so $(e, 1/e)$ is a local maximum.

(e) $f'(x) > 0$ when $1 - \ln x > 0$; that is, when $0 < x < e$. Similarly $f'(x) < 0$ when $x > e$.

(f) From (e) we note that the graph of $f(x) = (\ln x)/x$ rises for $0 < x < e$, attains a maximum at $(e, 1/e)$, and falls thereafter. Thus $(e, 1/e)$ is a global maximum, while $\lim_{x \to 0^+} f(x) = -\infty$ shows that there is no global minimum.

(g) As noted in (f), $\lim_{x \to 0^+} f(x) = -\infty$, since $1/x \to \infty$ and $\ln x \to -\infty$ as $x \to 0^+$.

(h) $$\lim_{x \to \infty} \frac{\ln x}{x} \underset{H}{=} \lim_{x \to \infty} \frac{1/x}{1} = 0$$

(i)

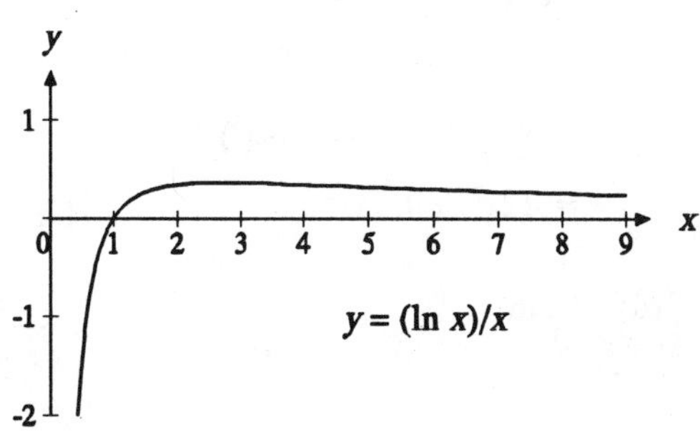

8 See Sec. 6.2.

9 Recall that $\log_2 3 = \dfrac{\log_{10} 3}{\log_{10} 2}$. So $\log_2 3$ can be found by applying the "log" key to 3, then to 2 and dividing.

10 $f'(x) = 2x - 2 = 2(x - 1)$

(a) Since $f(0) = 0 = f(2)$, f is not one-to-one on $[0, 2]$.

(b) Since $f'(x) > 0$ for x in $(1, 2]$, $f(x)$ is increasing for all x in $[1, 2]$ and is therefore one-to-one on $[1, 2]$. To find its inverse, set y equal to $x^2 - 2x$ and solve for x in terms of y. Thus, $x^2 - 2x - y = 0$ and by the quadratic formula, $x = \dfrac{2 \pm \sqrt{4 + 4y}}{2} = 1 \pm \sqrt{1 + y}$.

Since x lies in $[1, 2]$, we choose the positive root, and $f^{-1}(y) = 1 + \sqrt{1 + y}$.

(c) Since $f'(x) < 0$ for x in $[-1, 0]$, $f(x)$ is decreasing for all x in $[-1, 0]$ and is therefore one-to-one on $[-1, 0]$. From (b), we know that its inverse is of the form $x = 1 \pm \sqrt{1 + y}$. Here we choose the negative root because $x \le 0$, so $f^{-1}(y) = 1 - \sqrt{1 + y}$.

11 (a) $(\cosh 2x)' = 2 \sinh 2x$

(b) $(\sinh\sqrt{x})' = \cosh\sqrt{x}\cdot\frac{1}{2\sqrt{x}} = \frac{\cosh\sqrt{x}}{2\sqrt{x}}$

(c) $(\tanh 4x^3)' = \text{sech}^2(4x^3)\cdot 12x^2$

$= 12x^2\,\text{sech}^2(4x^3)$

(d) $(\sinh^{-1} e^x)' = \frac{1}{\sqrt{(e^x)^2+1}}e^x = \frac{e^x}{\sqrt{e^{2x}+1}}$

6.S Review Exercises

1 $(\sqrt{1+x^3})' = \frac{3x^2}{2\sqrt{1+x^3}}$

3 $(\sqrt{x})' = (x^{1/2})' = \frac{1}{2}x^{-1/2} = \frac{1}{2\sqrt{x}}$

5 $(\cos^2 3x)' = 2\cos 3x\,(-3\sin 3x)$

$= -6\cos 3x\sin 3x$

7 $\left(\sqrt{x^3}\right)' = (x^{3/2})' = \frac{3}{2}x^{1/2} = \frac{3}{2}\sqrt{x}$

9 $\left(\sqrt{\sin x}\right)' = \frac{\cos x}{2\sqrt{\sin x}}$

11 $(\cot x^2)' = -\csc^2 x^2\cdot 2x = -2x\csc^2 x^2$

13 $(x^{5/6}\sin^{-1}x)' = x^{5/6}\cdot\frac{1}{\sqrt{1-x^2}} + \frac{5}{6}x^{-1/6}\sin^{-1}x$

15 $(x^2e^{3x})' = x^2\cdot 3e^{3x} + 2xe^{3x} = x(3x+2)e^{3x}$

17 $(\ln(\sec 3x + \tan 3x))'$

$= \frac{3\sec 3x\tan 3x + 3\sec^2 3x}{\sec 3x + \tan 3x} =$

$\frac{3\sec 3x\,(\tan 3x + \sec 3x)}{\sec 3x + \tan 3x} = 3\sec 3x$

19 $\left(\cos\sqrt{x}\right)' = -\sin\sqrt{x}\cdot\frac{1}{2\sqrt{x}} = \frac{-\sin\sqrt{x}}{2\sqrt{x}}$

21 $(\ln(\sec x + \tan x))' = \frac{\sec x\tan x + \sec^2 x}{\sec x + \tan x}$

$= \frac{\sec x\,(\tan x + \sec x)}{\sec x + \tan x} = \sec x$

23 $\left(\frac{1}{\sqrt{6+3x^2}}\right)' = [(6+3x^2)^{-1/2}]' =$

$-\frac{1}{2}(6+3x^2)^{-3/2}\cdot 6x = -3x(6+3x^2)^{-3/2}$

25 $\left(\sqrt{\frac{2}{15}(5x+7)^3}\right)' = \left(\sqrt{\frac{2}{15}}(5x+7)^{3/2}\right)' =$

$\sqrt{\frac{2}{15}}\cdot\frac{3}{2}\left(\sqrt{5x+7}\right)\cdot 5 = \sqrt{\frac{15}{2}(5x+7)}$

27 $\left(\frac{x}{3} - \frac{4}{9}\ln(3x+4)\right)' = \frac{1}{3} - \frac{4}{9}\cdot\frac{3}{3x+4} =$

$\frac{1}{3} - \frac{4}{3(3x+4)} = \frac{3x}{3(3x+4)} = \frac{x}{3x+4}$

29 $((1+x^2)^5\sin 3x)' =$

$= (1+x^2)^5\,3\cos 3x + 5(1+x^2)^4\,2x\sin 3x$

$= (1+x^2)^4\,[3(1+x^2)\cos 3x + 10x\sin 3x]$

31 $((1+2x)^5\cos 3x)'$

$= (1+2x)^5(-3\sin 3x) + 5(1+2x)^4\,2\cos 3x$

$= (1+2x)^4[10\cos 3x - 3(1+2x)\sin 3x]$

33 $(\cos 3x\sin 4x)'$

$= \cos 3x\cdot 4\cos 4x - 3\sin 3x\sin 4x$

$= 4\cos 3x\cos 4x - 3\sin 3x\sin 4x$

35 $(\csc 3x^2)' = -\csc 3x^2\cot 3x^2\cdot 6x$

$= -6x\csc 3x^2\cot 3x^2$

37 $\left(x\left(\frac{x^2}{1+x}\right)^3\right)' = \left(\frac{x^7}{(1+x)^3}\right)'$

$= \frac{(1+x)^3\cdot 7x^6 - x^7\cdot 3(1+x)^2}{(1+x)^6}$

$= \frac{x^6}{(1+x)^4}[(1+x)\cdot 7 - 3x] = \frac{x^6(4x+7)}{(x+1)^4}$

39 $(x \sec 3x)' = x \cdot 3 \sec 3x \tan 3x + \sec 3x$

$= (\sec 3x)(3x \tan 3x + 1)$

41 $\left(\ln\left(x + \sqrt{x^2 + 1}\right)\right)' = \dfrac{1 + \frac{1}{2}(x^2 + 1)^{-1/2}(2x)}{x + \sqrt{x^2 + 1}}$

$= \dfrac{\sqrt{x^2 + 1} + x}{\sqrt{x^2 + 1}\left(x + \sqrt{x^2 + 1}\right)} = \dfrac{1}{\sqrt{x^2 + 1}}$

43 $(e^{-x} \tan^{-1} x^2)'$

$= e^{-x} \dfrac{2x}{1 + (x^2)^2} - e^{-x} \tan^{-1} x^2$

$= e^{-x}\left(\dfrac{2x}{1 + x^4} - \tan^{-1} x^2\right)$

45 $(2e^{\sqrt{x}}(\sqrt{x} - 1))'$

$= 2e^{\sqrt{x}} \cdot \dfrac{1}{2\sqrt{x}} + 2e^{\sqrt{x}} \cdot \dfrac{1}{2\sqrt{x}}(\sqrt{x} - 1)$

$= \dfrac{e^{\sqrt{x}}}{\sqrt{x}}(1 + (\sqrt{x} - 1)) = \dfrac{e^{\sqrt{x}}}{\sqrt{x}} \cdot \sqrt{x} = e^{\sqrt{x}}$

47 $\left(\dfrac{\ln x}{x^2}\right)' = \dfrac{x^2 \cdot \frac{1}{x} - \ln x \cdot 2x}{(x^2)^2} = \dfrac{x(1 - 2 \ln x)}{x^4}$

$= \dfrac{1 - 2 \ln x}{x^3}$

49 $\left(\dfrac{\sin^2 x}{\cos x}\right)' = \dfrac{\cos x \cdot 2 \sin x \cos x - \sin^2 x\,(-\sin x)}{\cos^2 x}$

$= \dfrac{\sin x}{\cos^2 x}(2 \cos^2 x + \sin^2 x) = \dfrac{\sin x}{\cos^2 x}(1 + \cos^2 x)$

$= (\sin x)(1 + \sec^2 x)$

51 Let $y = (\sec^{-1} 3x) \cdot x^2 \ln(1 + x^2)$. Then $\ln y =$ $\ln(\sec^{-1} 3x) + 2 \ln x + \ln(\ln(1 + x^2))$, so $\dfrac{1}{y} \cdot y' =$

$\dfrac{1}{\sec^{-1} 3x} \cdot \dfrac{3}{|3x|\sqrt{(3x)^2 - 1}} + \dfrac{2}{x} + \dfrac{1}{\ln(1 + x^2)} \cdot \dfrac{2x}{1 + x^2}$

$= \dfrac{1}{|x| \sec^{-1} 3x \sqrt{9x^2 - 1}} + \dfrac{2}{x} + \dfrac{2x}{(1 + x^2) \ln(1 + x^2)}$

and $y' = \dfrac{|x| \ln(1 + x^2)}{\sqrt{9x^2 - 1}} + 2x\,(\sec^{-1} 3x) \ln(1 + x^2)$

$+ \dfrac{2x^3}{1 + x^2} \sec^{-1} 3x$

53 $\left(x - 2 \ln(x - 1) + \dfrac{1}{x + 1}\right)'$

$= 1 - \dfrac{2}{x - 1} - \dfrac{1}{(x + 1)^2}$

55 $\left[\ln\left(\dfrac{1}{6x^2 + 3x + 1}\right)\right]' = [-\ln(6x^2 + 3x + 1)]'$

$= -\dfrac{12x + 3}{6x^2 + 3x + 1}$

57 $\left(\ln\left[\dfrac{(5x + 1)^3(6x + 1)^2}{(2x + 1)^4}\right]\right)$

$= (3 \ln(5x + 1) + 2 \ln(6x + 1) - 4 \ln(2x + 1))'$

$= 3 \cdot \dfrac{5}{5x + 1} + 2 \cdot \dfrac{6}{6x + 1} - 4 \cdot \dfrac{2}{2x + 1}$

$= \dfrac{15}{5x + 1} + \dfrac{12}{6x + 1} - \dfrac{8}{2x + 1}$

59 $\left[\dfrac{1}{2}\left(x\sqrt{9 - x^2} + 9 \sin^{-1}\left(\dfrac{x}{3}\right)\right)\right]'$

$= \dfrac{1}{2}\left[x \cdot \dfrac{-x}{\sqrt{9 - x^2}} + \sqrt{9 - x^2} + \dfrac{9}{3\sqrt{1 - (x/3)^2}}\right]$

$= \dfrac{1}{2}\left[\dfrac{-x^2 + (9 - x^2)}{\sqrt{9 - x^2}} + \dfrac{9}{\sqrt{9 - x^2}}\right]$

$$= \frac{1}{2} \cdot \frac{18 - 2x^2}{\sqrt{9 - x^2}} = \frac{9 - x^2}{\sqrt{9 - x^2}} = \sqrt{9 - x^2}$$

61 $(\tan 3x - 3x)' = 3 \sec^2 3x - 3$

$= 3(\sec^2 3x - 1) = 3 \tan^2 3x$

63 $$\left(\ln\left(x + \sqrt{x^2 + 25}\right)\right)' = \frac{1 + \frac{1}{2}(x^2 + 25)^{-1/2}(2x)}{x + \sqrt{x^2 + 25}}$$

$$= \frac{x + \sqrt{x^2 + 25}}{\sqrt{x^2 + 25}\left(x + \sqrt{x^2 + 25}\right)} = \frac{1}{\sqrt{x^2 + 25}}$$

65 $\left(x \sin^{-1} x + \sqrt{1 - x^2}\right)' =$

$$x \cdot \frac{1}{\sqrt{1 - x^2}} + \sin^{-1} x + \frac{-x}{\sqrt{1 - x^2}} = \sin^{-1} x$$

67 $\left[\frac{1}{3} \ln(\tan 3x + \sec 3x)\right]'$

$$= \frac{1}{3} \cdot \frac{3 \sec^2 3x + 3 \sec 3x \tan 3x}{\tan 3x + \sec 3x}$$

$$= \frac{1}{3} \cdot \frac{3 \sec 3x (\sec 3x + \tan 3x)}{\tan 3x + \sec 3x} = \sec 3x$$

69 $\left(-\frac{1}{3} \cos 3x + \frac{1}{9} \cos^3 3x\right)' =$

$$-\frac{1}{3}(-3 \sin 3x) + \frac{1}{9} \cdot 3 \cos^2 3x \,(-3 \sin 3x)$$

$= \sin 3x - \cos^2 3x \sin 3x = \sin 3x \,(1 - \cos^2 3x)$

$= \sin 3x \,(\sin^2 3x) = \sin^3 3x$

71 $\left[\frac{x}{2}\sqrt{4x^2 + 3} + \frac{3}{4} \ln\left(2x + \sqrt{4x^2 + 3}\right)\right]' =$

$$\frac{x}{2} \cdot \frac{4x}{\sqrt{4x^2 + 3}} + \frac{1}{2}\sqrt{4x^2 + 3} + \frac{3}{4} \cdot \frac{2 + 4x(4x^2 + 3)^{-1/2}}{2x + \sqrt{4x^2 + 3}}$$

$$= \frac{4x^2 + (4x^2 + 3)}{2\sqrt{4x^2 + 3}} + \frac{3}{2} \cdot \frac{\sqrt{4x^2 + 3} + 2x}{\sqrt{4x^2 + 3}\left(2x + \sqrt{4x^2 + 3}\right)}$$

$$= \frac{8x^2 + 3}{2\sqrt{4x^2 + 3}} + \frac{3}{2\sqrt{4x^2 + 3}} = \frac{8x^2 + 6}{2\sqrt{4x^2 + 3}}$$

$$= \frac{4x^2 + 3}{\sqrt{4x^2 + 3}} = \sqrt{4x^2 + 3}$$

73 $$\left(\frac{x^3(x^4 - x + 3)}{(x + 1)^2}\right)' = \left(\frac{x^7 - x^4 + 3x^3}{(x + 1)^2}\right)' =$$

$$\frac{(x + 1)^2(7x^6 - 4x^3 + 9x^2) - (x^7 - x^4 + 3x^3) \cdot 2(x + 1)}{((x + 1)^2)^2}$$

$$= \frac{(x + 1)(7x^6 - 4x^3 + 9x^2) - 2(x^7 - x^4 + 3x^3)}{(x + 1)^3} =$$

$$\frac{7x^7 + 7x^6 - 4x^4 - 4x^3 + 9x^3 + 9x^2 - 2x^7 + 2x^4 - 6x^3}{(x + 1)^3}$$

$$= \frac{5x^7 + 7x^6 - 2x^4 - x^3 + 9x^2}{(x + 1)^3}$$

$$= \frac{x^2}{(x + 1)^3}(5x^5 + 7x^4 - 2x^2 - x + 9)$$

75 Let $y = (1 + 3x)^{x^2}$. Then $\ln y = x^2 \ln(1 + 3x)$,

so $\frac{1}{y} \cdot y' = x^2 \cdot \frac{3}{1 + 3x} + 2x \ln(1 + 3x)$, hence

$$y' = (1 + 3x)^{x^2}\left(\frac{3x^2}{1 + 3x} + 2x \ln(1 + 3x)\right).$$

77 $\left(\frac{x \ln x}{(1 + x^2)^5}\right)'$

$$= \frac{(1 + x^2)^5[x \cdot \frac{1}{x} + \ln x] - x \ln x \cdot 5(1 + x^2)^4 \cdot 2x}{((1 + x^2)^5)^2}$$

$$= \frac{(1 + x^2)^4[(1 + x^2)(1 + \ln x) - 10x^2 \ln x]}{(1 + x^2)^{10}}$$

$$= \frac{1 + \ln x + x^2 - 9x^2 \ln x}{(1 + x^2)^6}$$

79 (a)

x	1/8	1/4	1/2	1	2	4	8
$\log_2 x$	−3	−2	−1	0	1	2	3

(b)

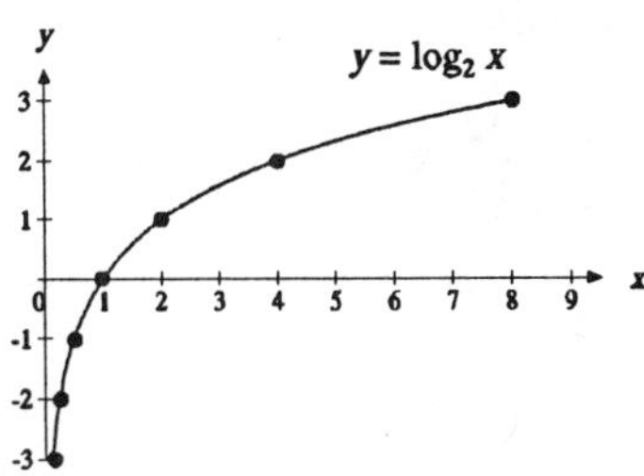

(c) As x grows large, so does $\log_2 x$: $\lim_{x\to\infty} \log_2 x = \infty$.

(d) As x approaches 0 from above, $\log_2 x$ grows large and negative: $\lim_{x\to 0^+} \log_2 x = -\infty$.

81 $$\left(\frac{x^2}{2}\ln ax - \frac{x^2}{4} + C\right)' =$$

$$\frac{x^2}{2}\cdot\frac{a}{ax} + \frac{2x}{2}\ln ax - \frac{2x}{4} = \frac{x}{2} + x\ln ax - \frac{x}{2}$$

$$= x \ln ax$$

83 $$(\ln(\ln ax) + C)' = \frac{1}{\ln ax}\cdot\frac{1}{ax}\cdot a = \frac{1}{x\ln ax}$$

85 $$\left(-\frac{1}{a}\ln|\cos ax| + C\right)' = -\frac{1}{a}\cdot\frac{1}{\cos ax}(-\sin ax)\cdot a$$

$$= \frac{\sin ax}{\cos ax} = \tan ax$$

87 $$\left(\frac{1}{a}\ln|\tan ax| + C\right)' = \frac{1}{a}\cdot\frac{1}{\tan ax}\sec^2 ax\cdot a$$

$$= \frac{\sec^2 ax}{\tan ax} = \frac{1}{\sin ax \cos ax}$$

89 $$\left(\frac{b}{a^2(ax+b)} + \frac{1}{a^2}\ln|ax+b| + C\right)'$$

$$= \frac{b}{a^2}\left(-\frac{1}{(ax+b)^2}\right)a + \frac{1}{a^2}\cdot\frac{1}{ax+b}\cdot a$$

$$= \frac{-b}{a(ax+b)^2} + \frac{ax+b}{a(ax+b)^2} = \frac{ax}{a(ax+b)^2}$$

$$= \frac{x}{(ax+b)^2}$$

91 (a) $$\left(\ln\left|\frac{1+x}{1-x}\right|\right)'$$

$$= \frac{1}{\left(\frac{1+x}{1-x}\right)}\cdot\frac{(1-x)\cdot 1 - (1+x)(-1)}{(1-x)^2}$$

$$= \frac{1-x}{1+x}\cdot\frac{1-x+1+x}{(1-x)^2} = \frac{2}{1-x^2}$$

(b) The area is $\int_0^{1/2}\frac{dx}{1-x^2}$. From (a),

$\frac{1}{2}\ln\left|\frac{1+x}{1-x}\right|$ is an antiderivative of $\frac{1}{1-x^2}$,

so Area $$= \left(\frac{1}{2}\ln\left|\frac{1+x}{1-x}\right|\right)\Bigg|_0^{1/2}$$

$$= \frac{1}{2}\ln\left|\frac{1+1/2}{1-1/2}\right| - \frac{1}{2}\ln\left|\frac{1+0}{1-0}\right|$$

$$= \frac{1}{2}\ln 3 - \frac{1}{2}\ln 1 = \frac{1}{2}\ln 3.$$

93 $$\left(\ln\left|\frac{\sqrt{ax+b} - b}{\sqrt{ax+b} + b}\right|\right)'$$

$$= (\ln|\sqrt{ax+b} - b|)' - (\ln|\sqrt{ax+b} + b|)'$$

$$= \frac{\frac{1}{2\sqrt{ax+b}}\cdot a}{\sqrt{ax+b} - b} - \frac{\frac{1}{2\sqrt{ax+b}}\cdot a}{\sqrt{ax+b} + b}$$

$$= \frac{a}{2\sqrt{ax+b}}\,\frac{(\sqrt{ax+b}+b) - (\sqrt{ax+b}-b)}{(\sqrt{ax+b}-b)(\sqrt{ax+b}+b)}$$

$$= \frac{a}{2\sqrt{ax+b}}\cdot\frac{2b}{ax+b-b^2}$$

$$= \frac{ab}{\sqrt{ax+b}(ax+b-b^2)}$$

95 (a) Since $(\ln|x^3 + x - 6|)' =$ $\frac{1}{x^3+x-6}(3x^2+1) = \frac{3x^2+1}{x^3+x-6}$, we have $\int \frac{3x^2+1}{x^3+x-6}\,dx =$ $\ln|x^3 + x - 6| + C.$

(b) Since $(\ln|\sin 2x|)' = \frac{1}{\sin 2x}\cdot 2\cos 2x =$ $2\cdot\frac{\cos 2x}{\sin 2x}$, we have $\int \frac{\cos 2x}{\sin 2x}\,dx =$ $\frac{1}{2}\ln|\sin 2x| + C.$

(c) Since $(\ln|5x+3|)' = \frac{1}{5x+3}\cdot 5$, we have $\int \frac{dx}{5x+3} = \frac{1}{5}\ln|5x+3| + C.$

(d) Since $\left(\frac{1}{5x+3}\right)' = ((5x+3)^{-1})' =$ $(-1)(5x+3)^{-2}\cdot 5 = \frac{-5}{(5x+3)^2}$, we have $\int \frac{dx}{(5x+3)^2} = -\frac{1}{5(5x+3)} + C.$

97 (a) $\ln\left(\frac{\sqrt[3]{\tan 4x}(1-2x)^5}{\sqrt[3]{(1+3x)^2}}\right) =$ $\frac{1}{3}\ln(\tan 4x) + 5\ln(1-2x) - \frac{2}{3}\ln(1+3x)$, so the derivative is

$$\frac{1}{3}\frac{\sec^2 4x}{\tan 4x}\cdot 4 + 5\cdot\frac{-2}{1-2x} - \frac{2}{3}\frac{3}{1+3x}$$

$$= \frac{4}{3}\frac{1}{\cos 4x \sin 4x} - \frac{10}{1-2x} - \frac{2}{1+3x}$$

$$= \frac{4}{3}\sec 4x \csc 4x - \frac{10}{1-2x} - \frac{2}{1+3x}.$$

(b) Let $y = \frac{(1+x^2)^3\sqrt{1+x}}{\sin 3x}$; then $\ln y =$ $3\ln(1+x^2) + \frac{1}{2}\ln(1+x) - \ln(\sin 3x).$

Hence $\frac{1}{y}\cdot y' =$ $3\cdot\frac{2x}{1+x^2} + \frac{1}{2}\frac{1}{1+x} - \frac{3\cos 3x}{\sin 3x} =$ $\frac{6x}{1+x^2} + \frac{1}{2(1+x)} - 3\cot 3x$ and $y' =$ $\frac{(1+x^2)^3\sqrt{1+x}}{\sin 3x}\left(\frac{6x}{1+x^2} + \frac{1}{2(1+x)} - 3\cot 3x\right).$

99 From Exercise 15 of Sec. 6.1, we have $\log_b a = \frac{1}{\log_a b}$.

(a) $\log_x 2 = \frac{1}{\log_2 x}$. As $x \to 1^+$, $\log_2 x \to 0$

through positive values, so $\log_x 2 \to \infty$. As $x \to 1^-$, $\log_2 x \to 0$ through negative values, so $\log_x 2 \to -\infty$. Hence $\lim_{x\to 1} \log_x 2$ does not exist.

(b) As $x \to \infty$, $\log_2 x \to \infty$, so $\log_x 2 = \frac{1}{\log_2 x} \to 0$. That is, $\lim_{x\to\infty} \log_x 2 = 0$.

101 We have $\frac{d}{dq}(M(q)) = -p\cdot\frac{1}{q} - (1-p)\frac{1}{1-q}(-1)$ $= -\frac{p}{q} + \frac{1-p}{1-q} = \frac{-p(1-q) + q(1-p)}{q(1-q)} = \frac{q-p}{q(1-q)}$. This quantity is negative for $q < p$, 0 for $q = p$, and positive for $q > p$, so $M(q)$ has a global minimum at $q = p$. Since $M(p) = H(p)$, the desired inequality follows.

103 Let $f(x) = \ln x$. Then $\lim_{x\to 0} \frac{\ln(2+x) - \ln 2}{x} = f'(2) = \frac{1}{2}$.

105 (a) For $h = 0.01$, $(1-2h)^{1/h} = 0.98^{100} \approx 0.13262$.

(b) $\lim_{h\to 0} (1-2h)^{1/h} = \left(\lim_{h\to 0} (1-2h)^{1/(-2h)}\right)^{-2}$ $= e^{-2} \approx 0.13534$.

107 Let $y = x^{1/\log_2 x}$. Then $\log_2 y = \frac{1}{\log_2 x}\cdot\log_2 x = 1$, so $y = 2^1 = 2$. Thus $\lim_{x\to\infty} y = \lim_{x\to\infty} 2 = 2$.

109 The limit equals the derivative of e^x at $x = 3$; namely, e^3.

111 By l'Hôpital's rule, $\lim_{x\to 0} \frac{\cos\sqrt{x} - 1}{\tan x} \underset{H}{=} \lim_{x\to 0} \frac{-\sin\sqrt{x}\cdot 1/(2\sqrt{x})}{\sec^2 x} = \lim_{x\to 0} \frac{-\sin\sqrt{x}}{2\sqrt{x}\sec^2\sqrt{x}} =$ $\lim_{x\to 0} \frac{-1}{2\sec^2 x}\cdot\frac{\sin\sqrt{x}}{\sqrt{x}} = -\frac{1}{2}\cdot 1 = -\frac{1}{2}$.

113 Let $y = (1+2x^2)^{1/x^2}$. Then $\ln y = \frac{1}{x^2}\ln(1+2x^2) = \frac{\ln(1+2x^2)}{x^2}$. By l'Hôpital's rule for zero-over-zero forms, $\lim_{x\to 0} \ln y = \lim_{x\to 0} \frac{\ln(1+2x^2)}{x^2} \underset{H}{=} \lim_{x\to 0} \frac{4x/(1+2x^2)}{2x} = \lim_{x\to 0} \frac{2}{1+2x^2} = \frac{2}{1+0} = 2$. Thus $\lim_{x\to 0} y = e^2$.

115 $\lim_{x\to-\infty} \frac{e^x - e^{-x}}{e^x + e^{-x}} = \lim_{x\to-\infty} \frac{e^{2x} - 1}{e^{2x} + 1} = \frac{0-1}{0+1} = -1$

117 By l'Hôpital's rule for a zero-over-zero form, $\lim_{x\to 0} \frac{1-\cos x}{x + \tan x} \underset{H}{=} \lim_{x\to 0} \frac{\sin x}{1 + \sec^2 x} = \frac{0}{1+1} = 0$.

119 By l'Hôpital's rule, $\lim_{x\to 0} \frac{\sin 2x}{e^{3x} - 1} \underset{H}{=} \lim_{x\to 0} \frac{2\cos 2x}{3e^{3x}} = \frac{2\cdot 1}{3\cdot 1} = \frac{2}{3}$.

121 $\lim_{x\to\infty} \frac{3x - \sin x}{x + \sqrt{x}} = \lim_{x\to\infty} \frac{3 - \frac{\sin x}{x}}{1 + \frac{1}{\sqrt{x}}} = \frac{3-0}{1+0} = 3$

123 By l'Hôpital's rule, $\lim_{x\to 2} \frac{x^3 - 8}{x^2 - 4} \underset{H}{=} \lim_{x\to 2} \frac{3x^2}{2x} = \frac{3\cdot 2^2}{2\cdot 2} = \frac{12}{4} = 3$.

125 $\lim_{x\to\pi/2} \dfrac{\sin x}{1+\cos x} = \dfrac{\sin\frac{\pi}{2}}{1+\cos\frac{\pi}{2}} = \dfrac{1}{1+0} = 1$

127 As $x \to 3^+$, $x - 3 \to 0$, $x - 3 > 0$, and $\ln(x-3) \to -\infty$, so $\lim_{x\to 3^+} \dfrac{\ln(x-3)}{x-3} = -\infty$.

129 $\lim_{x\to\infty} \dfrac{\cos x}{x} = 0$, since $\left|\dfrac{\cos x}{x}\right| \le \dfrac{1}{|x|}$ and $\dfrac{1}{x} \to 0$ as $x \to \infty$.

131 By l'Hôpital's rule, $\lim_{x\to 0} \dfrac{5^x - 3^x}{x} \underset{H}{=}$

$$\lim_{x\to 0} \frac{(\ln 5)\, 5^x - (\ln 3)\, 3^x}{1} = \frac{(\ln 5)\cdot 1 - (\ln 3)\cdot 1}{1} =$$

$\ln 5 - \ln 3 = \ln(5/3)$.

133 By l'Hôpital's rule, $\lim_{x\to\infty} \dfrac{\ln(x^2+1)}{\ln(x^2+8)} \underset{H}{=}$

$$\lim_{x\to\infty} \frac{\dfrac{2x}{x^2+1}}{\dfrac{2x}{x^2+8}} = \lim_{x\to\infty} \frac{x^2+8}{x^2+1} = 1.$$

135 Let $y = (2^x - x^{10})^{1/x}$. Then $\ln y = \dfrac{\ln(2^x - x^{10})}{x}$, which is an infinity-over-infinity form because 2^x grows more rapidly than x^{10}. Hence $\lim_{x\to\infty} \ln y =$

$$\lim_{x\to\infty} \frac{\ln(2^x - x^{10})}{x} \underset{H}{=} \lim_{x\to\infty} \frac{\dfrac{2^x \ln 2 - 10x^9}{2^x - x^{10}}}{1} =$$

$$\lim_{x\to\infty} \frac{\ln 2 - 10x^9/2^x}{1 - x^{10}/2^x} = \frac{\ln 2 - 0}{1 - 0} = \ln 2. \text{ Thus}$$

$\lim_{x\to\infty} y = e^{\ln 2} = 2.$

137 $\lim_{h\to 0} \dfrac{e^{3+h} - e^3}{1-h} = \dfrac{e^3 - e^3}{1-0} = \dfrac{0}{1} = 0$

139 (a) If x is measured in radians, then $\lim_{x\to 0} \dfrac{\sin x}{x} = 1$, from which it follows that $(\sin x)' = \cos x$. When x is measured in degrees, $(\mathrm{Sin}\, x)' = \dfrac{\pi}{180} \mathrm{Cos}\, x$, which is more cumbersome.

(b) $(\log_b x)' = \dfrac{\log_b e}{x}$, which is simplest when $b = e$, so that $\log_b e = 1$.

(c) $(b^x)' = (\log_e b)\cdot b^x$, which is simplest when $b = e$, so $\log_e b = 1$.

141 (a) Insufficient information. If $f(x) \ge 0$ near a, then the limit is $0^7 = 0$. But if $f(x) < 0$ for values of x arbitrarily close to a, then $f(x)^{g(x)}$ need not be defined for such x and the limit will not exist.

(b) $\lim_{x\to a} f(x)^{g(x)} = 2^0 = 1$

(c) Insufficient information. If $f(x) = 0$ and $g(x) = |x - a|$, then $\lim_{x\to a} f(x)^{g(x)} = 0$. If $f(x) = x - a$ and $g(x) = 0$, then $\lim_{x\to a} f(x)^{g(x)} = 1$.

(d) Insufficient information. If the limit exists, it must be 0. But if $f(x) < 0$ for values of x arbitrarily close to a, then $f(x)^{g(x)}$ may be undefined for such x and $\lim_{x\to a} f(x)^{g(x)}$ may not exist.

(e) Insufficient information. If $f(x) = \dfrac{1}{|x - a|}$ and $g(x) = 0$, then $\lim_{x\to a} f(x)^{g(x)} = 1$. If $f(x) = \dfrac{1}{|x - a|}$ and $g(x) = \dfrac{1}{\ln f(x)}$, then $f(x)^{g(x)} = e^{g(x)\ln f(x)} = e^1 = e$, so $\lim_{x\to a} f(x)^{g(x)} = e$.

(f) $\lim_{x\to a} f(x)^{g(x)} = 0$

143 See Theorem 1 of Sec. 6.8.

145 (a) $e^{\ln x} = x$, so e^x is the inverse of $\ln x$.

(b) $\ln e^x = x$, so $\ln x$ is the inverse of e^x.

(c) $\sqrt[3]{x^3} = x$, so $\sqrt[3]{x}$ is the inverse of x^3.

(d) $\frac{1}{3}(3x) = x$, so $\frac{1}{3}x$ is the inverse of $3x$.

(e) $\left(\sqrt[3]{x}\right)^3 = x$, so x^3 is the inverse of $\sqrt[3]{x}$.

(f) $\sin(\sin^{-1} x) = x$, so $\sin x$, restricted to $[-\pi/2, \pi/2]$, is the inverse of $\sin^{-1} x$.

147 (a) Let $f(t) = \log_{10} t$. Then $f'(t) = \dfrac{\log_{10} e}{t}$. If $t = 1$ and $\Delta t = x$, then $\log_{10}(1 + x) = f(t + \Delta t) \approx f(t) + \Delta f = f(1) + f'(1)\Delta t = \log_{10} 1 + \dfrac{\log_{10} e}{1}x = 0 + (\log_{10} e)x = (\log_{10} e)x \approx 0.434x$.

(b) Let $x = 0.05$ in (a). Then $\log_{10} 1.05 \approx (0.434)(0.05) = 0.0217$.

149 Differentiating both sides with respect to y gives

$$\frac{1}{1 + y} + x + \frac{dx}{dy}\cdot y = 0, \text{ so } \frac{dx}{dy} = -\frac{\frac{1}{1+y} + x}{y}$$

and $\dfrac{dy}{dx} = -\dfrac{y}{\frac{1}{1+y} + x}$. At $x = 0$, $y = 1$, so

$$y'(0) = -\frac{1}{\frac{1}{1+1} + 0} = -2.$$

Alternatively, we could differentiate the equation with respect to x, obtaining $\dfrac{1}{1 + y}\dfrac{dy}{dx} + x\dfrac{dy}{dx} + y = 0$. Hence, $\dfrac{dy}{dx} = -\dfrac{y}{\frac{1}{1+y} + x}$, as before.

151 Let $P = (x_0, y_0)$. P lies on the curve $y = e^x$, so $y_0 = e^{x_0}$. The line segment AP is tangent to the curve $y = e^x$, and $y' = e^x$, so the slope of the line through A and P is e^{x_0}. The equation of the line is $y - e^{x_0} = e^{x_0}(x - x_0)$, so $y = e^{x_0} + e^{x_0}(x - x_0) = e^{x_0}(x + (1 - x_0))$. The line crosses the x axis at A, so $y = 0$, which occurs when $x = -(1 - x_0) = x_0 - 1$. Thus $A = (x_0 - 1, 0)$ and $B = (x_0, 0)$, so the length of AB is $x_0 - (x_0 - 1) = 1$.

153 (a) Let $y = (1 + x)^{1/x}$. Then $\ln y = \dfrac{\ln(1 + x)}{x}$, so $\lim_{x\to 0} \ln y = \lim_{x\to 0} \dfrac{\frac{1}{x+1}}{1} = 1$. Thus $\lim_{x\to 0} f(x) = \lim_{x\to 0} y = e^1 = e = f(0)$, so f is continuous at $x = 0$.

(b) By definition, $f'(0) = \lim_{h\to 0} \dfrac{f(h) - f(0)}{h} =$

$\lim_{h \to 0} \frac{(1 + h)^{1/h} - e}{h}$. To apply l'Hôpital's rule to this zero-over-zero form, recall that $\ln y = \frac{\ln(1 + x)}{x}$, so $\frac{1}{y} \cdot y' = \frac{x \cdot \frac{1}{1 + x} - \ln(1 + x)}{x^2} = \frac{x - (1 + x)\ln(1 + x)}{x^2(1 + x)}$, and $y' = (1 + x)^{1/x} \cdot \frac{x - (1 + x)\ln(1 + x)}{x^2(1 + x)}$. Therefore

$$\lim_{h \to 0} \frac{(1 + h)^{1/h} - e}{h} \underset{H}{=} \lim_{h \to 0} \frac{(1 + h)^{1/h} \cdot \frac{h - (1 + h)\ln(1 + h)}{h^2(1 + h)} - 0}{1}$$

$$= \lim_{h \to 0} (1 + h)^{1/h} \cdot \frac{h - (1 + h)\ln(1 + h)}{h^2(1 + h)},$$

where $\lim_{h \to 0} (1 + h)^{1/h} = e$ and

$$\lim_{h \to 0} \frac{h - (1 + h)\ln(1 + h)}{h^2(1 + h)} \underset{H}{=} \lim_{h \to 0} \frac{1 - \frac{1 + h}{1 + h} - \ln(1 + h)}{2h(1 + h) + h^2}$$

$$= \lim_{h \to 0} \frac{-\ln(1 + h)}{3h^2 + 2h} \underset{H}{=} \lim_{h \to 0} \frac{-\frac{1}{1 + h}}{6h + 2} = \frac{-\frac{1}{1 + 0}}{0 + 2} = -\frac{1}{2}.$$

Therefore $f'(0) = e \cdot (-\frac{1}{2}) = -\frac{e}{2}$.

155 (a) If $\frac{ds}{dt} = ks$, then $s = Ae^{kt}$ for some A. (See Example 1 of Sec. 6.7.)

(b) $\frac{ds}{dt} = Ake^{kt}$

(c) At $t = 0$, $\frac{ds}{dt} = 0$, so $Ak = 0$. Hence, for all t, $\frac{ds}{dt} = 0e^{kt} = 0$.

157 (a) For $x > 0$, $(1^x + 2^x + 3^x)^{1/x} > 3^{1/x}$, so $\lim_{x \to 0^+} (1^x + 2^x + 3^x)^{1/x} = \infty$. For $-1 < x < 0$, $0 < (1^x + 2^x + 3^x)^{1/x} < (1^{-1} + 2^{-1} + 3^{-1})^{1/x} = (6/11)^{-1/x}$. Since $-1/x \to \infty$ as $x \to 0^-$, $(6/11)^{-1/x} \to 0$. Hence $\lim_{x \to 0^-} (1^x + 2^x + 3^x)^{1/x} = 0$, so $\lim_{x \to 0} (1^x + 2^x + 3^x)^{1/x}$ does not exist.

(b) For $x > 1$, $1^x + 2^x = 3^x\left[\left(\frac{1}{3}\right)^x + \left(\frac{2}{3}\right)^x\right] < 3^x\left[\frac{1}{3} + \frac{2}{3}\right] = 3^x$, so $3^x < 1^x + 2^x + 3^x < 2 \cdot 3^x$ and $3 < (1^x + 2^x + 3^x)^{1/x} < 3 \cdot 2^{1/x}$. Since $\lim_{x \to \infty} 2^{1/x} = 1$, $\lim_{x \to \infty} (1^x + 2^x + 3^x)^{1/x} = 3$.

(c) As $x \to -\infty$, $1^x + 2^x + 3^x \to 1 + 0 + 0 = 1$ and $\frac{1}{x} \to 0$, so $(1^x + 2^x + 3^x)^{1/x} \to 1^0 = 1$.

159 Because $f(a) = g(a) = 0$, we have, by the generalized mean-value theorem, $\frac{f(x)}{g(x)} =$

$\dfrac{f(x) - f(a)}{g(x) - g(a)} = \dfrac{f'(c)}{g'(c)}$ for some c in (a, x). As $x \to a^+$, $c \to a^+$; hence $\lim_{x\to a^+} \dfrac{f(x)}{g(x)} = \lim_{c\to a^+} \dfrac{f'(c)}{g'(c)}$ $= L$.

161 We can write $f(t)$ as $g(t)\cdot\dfrac{f(t)}{g(t)}$. Now $\lim_{t\to\infty} g(t) = \infty$, so for t large enough, $g(t) > 0$. Hence $\ln f(t)$ $= \ln g(t) + \ln\left(\dfrac{f(t)}{g(t)}\right)$, so $\lim_{t\to\infty} \dfrac{\ln f(t)}{\ln g(t)}$

$$= \lim_{t\to\infty} \frac{\ln g(t) + \ln\left(\frac{f(t)}{g(t)}\right)}{\ln g(t)}$$

$$= = \lim_{t\to\infty} \left(1 + \frac{\ln(f(t)/g(t))}{\ln g(t)}\right)$$

$= \lim_{t\to\infty} \left(1 + \dfrac{\ln(f(t)/g(t))}{\ln g(t)}\right)$. But, as $t \to \infty$, $\dfrac{f(t)}{g(t)} \to 3$, so $\ln(f(t)/g(t)) \to \ln 3$. Since $g(t) \to \infty$, $\dfrac{\ln(f(t)/g(t))}{\ln g(t)} \to 0$. Hence $\lim_{t\to\infty} \dfrac{\ln f(t)}{\ln g(t)} = 1 + 0$ $= 1$.

163 (a) The velocity is a constant, 10 m/sec, for $0 \le t \le 1$, so the particle moves 10 meters during [0, 1]. During [1, 2], the velocity drops from 10 m/sec to 0 m/sec; the average velocity during that interval is slightly larger than 5 m/sec, say, 6 m/sec. So the particle moves about 6 meters during [1, 2]. During [2, 4], the velocity drops from 0 to about -7 m/sec and rises back to 0; the average velocity is about -5 m/sec. So the particle moves approximately (2 sec)·(5 m/sec) = 10 meters during [2, 4]; since the velocity is negative, the particle travels 10 meters in the direction opposite to that of its initial motion.

(b)

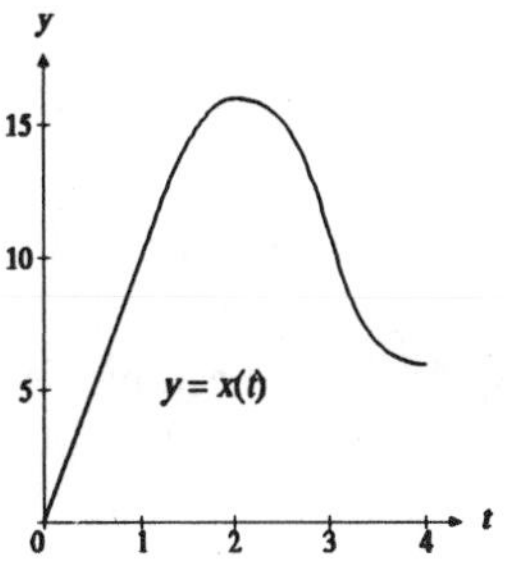

(c)

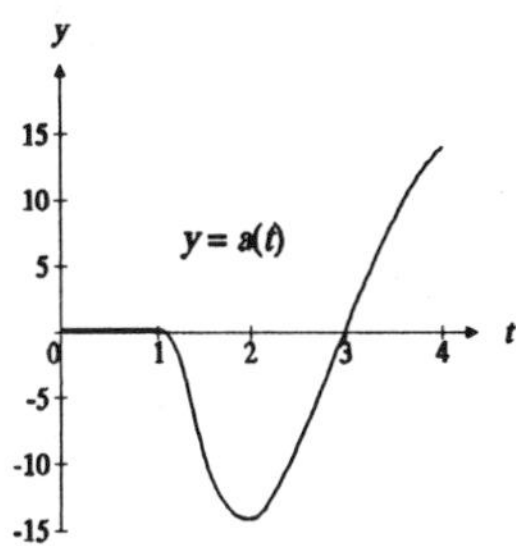

165 (a) With interest compounded n times per year at an annual rate of r, an initial deposit A grows to $A\left(1 + \dfrac{r}{n}\right)^n$. (See Eq. (1) of Sec. 6.2.) We wish to show that this increases as n increases. Let $x = 1/n$, so that the final amount is $A(1 + rx)^{1/x}$. If this is a decreasing function of x, then it diminishes as x grows and increases as x shrinks. But n grows as x shrinks, so it is equivalent to saying that the final amount increases as n grows.

(b) The derivative of $A(1 + rx)^{1/x}$ is

$A(1 + rx)^{1/x}\dfrac{\dfrac{rx}{1 + rx} - \ln(1 + rx)}{x^2}$, obtained by logarithmic differentiation. We will show that this is negative for $x > 0$. Let $f(x) = \frac{rx}{1 + rx} - \ln(1 + rx)$. Note that $f(0) = 0$ and

$$f'(x) = \frac{r}{(1 + rx)^2} - \frac{r}{1 + rx} = \frac{-r^2x}{(1 + rx)^2} <$$

0 for $x > 0$. Hence $f(x) < 0$ for $x > 0$, so

$(1 + rx)^{1/x} \frac{f(x)}{x^2}$ is negative for $x > 0$.

Therefore $(1 + rx)^{1/x}$ is decreasing for $x > 0$.

7 Computing Antiderivatives

7.1 Shortcuts, Integral Tables, and Machines

1 $\int 5x^3\,dx = 5\int x^3\,dx = 5\left(\frac{x^4}{4}\right) + C$

$= \frac{5}{4}x^4 + C$

3 $\int x^{1/3}\,dx = \frac{1}{\frac{1}{3}+1}x^{1/3+1} + C = \frac{3}{4}x^{4/3} + C$

5 $\int \frac{6\,dx}{x^2} = 6\int x^{-2}\,dx = 6\left[\frac{x^{-2+1}}{-2+1}\right] + C$

$= -6x^{-1} + C = -\frac{6}{x} + C$

7 From $(e^{-2x})' = -2e^{-2x}$, it follows that an antiderivative of e^{-2x} is $(-1/2)e^{-2x}$. Thus

$\int 5e^{-2x}\,dx = 5\int e^{-2x}\,dx = 5\left(-\frac{1}{2}e^{-2x}\right) + C$

$= -\frac{5}{2}e^{-2x} + C.$

9 $\int \frac{6\,dx}{|x|\sqrt{x^2-1}} = 6\int \frac{dx}{|x|\sqrt{x^2-1}}$

$= 6\sec^{-1} x + C$

11 Since the numerator is the derivative of the denominator, we see that $\int \frac{4x^3\,dx}{1+x^4} =$

$\ln|1+x^4| + C = \ln(1+x^4) + C$. The absolute value is unnecessary because $1 + x^4 > 0$.

13 Observe that $(1+\cos x)' = -\sin x$, so that except for the sign, the numerator is the derivative of the denominator. Hence $\int \frac{\sin x}{1+\cos x}\,dx =$

$-\int \frac{-\sin x}{1+\cos x}\,dx = -\ln|1+\cos x| + C =$

$-\ln(1+\cos x) + C$. (Since $1 + \cos x \geq 0$, the absolute value is unnecessary.)

15 Following the hint, $\int \frac{1+2x}{x^2}\,dx =$

$\int\left(\frac{1}{x^2} + \frac{2}{x}\right)dx = \int \frac{dx}{x^2} + 2\int \frac{dx}{x} =$

$\int x^{-2}\,dx + 2\int \frac{dx}{x} = \frac{x^{-2+1}}{-2+1} + 2\ln|x| + C =$

$-\frac{1}{x} + 2\ln|x| + C.$

17 $\int (x^2+3)^2\,dx = \int (x^4 + 6x^2 + 9)\,dx =$

$\frac{x^5}{5} + \frac{6}{3}x^3 + 9x + C = \frac{x^5}{5} + 2x^3 + 9x + C.$

19 $\int (1+3x)x^2\,dx = \int (x^2 + 3x^3)\,dx =$

$\frac{x^3}{3} + \frac{3}{4}x^4 + C.$

21 Note that $x^5\sqrt{1+x^2} = -\left[(-x)^5\sqrt{1+(-x)^2}\right]$, so $x^5\sqrt{1+x^2}$ is an odd function and Shortcut 1

applies: $\int_{-1}^{1} x^5\sqrt{1 + x^2}\,dx = 0.$

23 Note that $x^5\sqrt[4]{1 - x^2} = -\left[(-x)^5\sqrt[4]{1 - (-x)^2}\right]$, so $x^5\sqrt[4]{1 - x^2}$ is an odd function and Shortcut 1 applies: $\int_{-1}^{1} x^5\sqrt[4]{1 - x^2}\,dx = 0.$

25 Note that $\sqrt{9 - x^2} = \sqrt{9 - (-x)^2}$, so $\sqrt{9 - x^2}$ is an even function, and Shortcut 3 applies: $\int_{-3}^{3} \sqrt{9 - x^2}\,dx = 2\int_{0}^{3} \sqrt{9 - x^2}\,dx$. Now, $\sqrt{9 - x^2} = \sqrt{3^2 - x^2}$ is of the form $\sqrt{a^2 - x^2}$, so Shortcut 2 gives $2\int_{0}^{3} \sqrt{9 - x^2}\,dx = 2\left[\frac{1}{4}\pi 3^2\right] = \frac{9\pi}{2}$. Thus $\int_{-3}^{3} \sqrt{9 - x^2}\,dx = \frac{9\pi}{2}$.

(All formulas used in Exercises 27–31 are from the table of antiderivatives given in the endpapers of the text.)

27 (a) Using Formula 13 with $a = 3$ and $b = 2$ gives $\int \frac{dx}{(3x + 2)^2} = \frac{-1}{3(3x + 2)} + C.$

(b) Using Formula 15 with $a = 3$ and $b = 2$ gives $\int \frac{dx}{x(3x + 2)} = \frac{1}{2}\ln\left|\frac{x}{3x + 2}\right| + C.$

29 (a) Using Formula 22 with $a = 3$ and $b = -4$ gives $\int \frac{dx}{x\sqrt{3x - 4}} = \frac{2}{\sqrt{4}}\tan^{-1}\sqrt{\frac{3x - 4}{4}} + C$

$= \tan^{-1}\sqrt{\frac{3x - 4}{4}} + C.$

(b) Using Formula 23 with $a = 3$ and $b = -4$ gives $\int \frac{dx}{x^2\sqrt{3x - 4}} = -\frac{\sqrt{3x - 4}}{-4x} - \frac{3}{2(-4)}\int \frac{dx}{x\sqrt{3x - 4}}$, which from (a) reduces to $\int \frac{dx}{x^2\sqrt{3x - 4}} = \frac{\sqrt{3x - 4}}{4x} + \frac{3}{8}\tan^{-1}\sqrt{\frac{3x - 4}{4}} + C.$

31 (a) Using Formula 36 with $a = 1$, $b = 3$, and $c = 5$ (since $b^2 < 4ac$), we have

$$\int \frac{dx}{x^2 + 3x + 5}$$

$$= \frac{2}{\sqrt{4\cdot 1\cdot 5 - 3^2}}\tan^{-1}\frac{2\cdot 1x + 3}{\sqrt{4\cdot 1\cdot 5 - 3^2}} + C$$

$$= \frac{2}{\sqrt{11}}\tan^{-1}\frac{2x + 3}{\sqrt{11}} + C.$$

(b) Using Formula 36 with $a = 1$, $b = 2$, and $c = 5$, for the same reason, we have

$$\int \frac{dx}{x^2 + 2x + 5}$$

$$= \frac{2}{\sqrt{4\cdot 1\cdot 5 - 2^2}}\tan^{-1}\frac{2\cdot 1x + 2}{\sqrt{4\cdot 1\cdot 5 - 2^2}} + C$$

$$= \frac{1}{2}\tan^{-1}\frac{x + 1}{2} + C.$$

33 (a) If one sketches the graphs of $\sin^2\theta$ and $\cos^2\theta$ on $[0, \pi/2]$, the areas underneath the two curves and above $[0, \pi/2]$ appear to be equal.

(b) Note that $\int_0^{\pi/2} \sin^2\theta\,d\theta + \int_0^{\pi/2} \cos^2\theta\,d\theta$

$= \int_0^{\pi/2} (\sin^2\theta + \cos^2\theta)\,d\theta = \int_0^{\pi/2} 1\,d\theta$

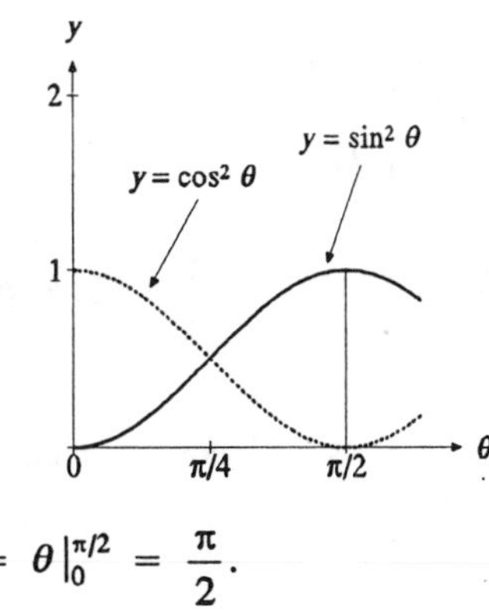

$= \theta\Big|_0^{\pi/2} = \dfrac{\pi}{2}.$

(c) Using the identity $\cos^2\theta = \dfrac{1}{2}(1 + \cos 2\theta)$,

we have $\int_0^{\pi/2} \cos^2\theta\, d\theta$

$= \dfrac{1}{2}\int_0^{\pi/2} (1 + \cos 2\theta)\, d\theta$

$= \dfrac{1}{2}\left(\theta + \dfrac{1}{2}\sin 2\theta\right)\Big|_0^{\pi/2} = \dfrac{1}{2}\left[\dfrac{\pi}{2}\right] = \dfrac{\pi}{4}.$

Since $\int_0^{\pi/2} \sin^2\theta\, d\theta + \int_0^{\pi/2} \cos^2\theta\, d\theta =$

$\pi/2$, we conclude that $\int_0^{\pi/2} \sin^2\theta\, d\theta = \pi/4$.

35 (a) Using Formula 40 with $a = 3$, $b = 1$ and

$c = 2$ gives $\displaystyle\int \frac{dx}{\sqrt{3x^2 + x + 2}} =$

$\dfrac{1}{\sqrt{3}} \ln\left|6x + 1 + 2\sqrt{3}\sqrt{3x^2 + x + 2}\right| + C.$

(b) Using Formula 41 with $a = -3$, $b = 1$, and

$c = 2$ gives $\displaystyle\int \frac{dx}{\sqrt{-3x^2 + x + 2}} =$

$\dfrac{1}{\sqrt{-(-3)}} \sin^{-1}\dfrac{-2(-3)x - 1}{\sqrt{1^2 - 4(-3)2}} + C =$

$\dfrac{1}{\sqrt{3}} \sin^{-1}\dfrac{6x - 1}{5} + C.$

7.2 The Substitution Method

1 Let $u = 1 + 3x$; then $du = 3\,dx$ and

$\displaystyle\int (1 + 3x)^5\, 3\, dx = \int u^5\, du = \frac{u^6}{6} + C =$

$\dfrac{(1 + 3x)^6}{6} + C.$

3 Let $u = 1 + x^2$; then $du = 2x\,dx$. When $x = 0$,

$u = 1$; when $x = 1$, $u = 2$. Hence, $\displaystyle\int_0^1 \frac{x\,dx}{\sqrt{1 + x^2}}$

$\displaystyle= \frac{1}{2}\int_0^1 \frac{2x\,dx}{\sqrt{1 + x^2}} = \frac{1}{2}\int_1^2 \frac{du}{\sqrt{u}} = \frac{1}{2}\cdot 2\sqrt{u}\Big|_1^2$

$= \sqrt{2} - 1.$

5 Let $u = 2x$; then $du = 2\,dx$ and $dx = du/2$, so

$\displaystyle\int \sin 2x\, dx = \int \sin u \frac{du}{2} = \frac{1}{2}\int \sin u\, du$

$= -\dfrac{1}{2}\cos u + C = -\dfrac{1}{2}\cos 2x + C.$

7 Let $u = 3x$; then $dx = \dfrac{1}{3}\,du$. When $x = -1$, $u =$

-3; when $x = 2$, $u = 6$. Hence $\int_{-1}^2 e^{3x}\, dx =$

$\displaystyle\int_{-3}^6 e^u \frac{1}{3}\, du = \frac{1}{3}e^u\Big|_{-3}^6 = \frac{1}{3}(e^6 - e^{-3}).$

9 Let $u = 3x$; then $du = 3\,dx$ and $dx = du/3$, so

$\displaystyle\int \frac{1}{\sqrt{1 - 9x^2}}\, dx = \int \frac{1}{\sqrt{1 - u^2}} \frac{du}{3}$

$\displaystyle= \frac{1}{3}\int \frac{du}{\sqrt{1 - u^2}} = \frac{1}{3}\sin^{-1} u + C$

$= \dfrac{1}{3}\sin^{-1} 3x + C.$

11 Let $u = \tan\theta$; then $du = \sec^2\theta\, d\theta$. When $\theta = \pi/6$, $u = \frac{1}{\sqrt{3}}$; when $\theta = \pi/4$, $u = 1$. Hence,

$$\int_{\pi/6}^{\pi/4} \tan\theta \sec^2\theta\, d\theta = \int_{1/\sqrt{3}}^{1} u\, du = \frac{1}{2}u^2\Big|_{1/\sqrt{3}}^{1}$$

$$= \frac{1}{2}\left(1 - \frac{1}{3}\right) = \frac{1}{3}.$$

13 Let $u = \ln x$; then $du = dx/x$, so $\int \frac{(\ln x)^4}{x}\, dx$

$$= \int (\ln x)^4 \frac{dx}{x} = \int u^4\, du = \frac{u^5}{5} + C$$

$$= \frac{(\ln x)^5}{5} + C.$$

15 Let $u = 1 - x^2$; then $du = -2x\, dx$, so

$$\int (1 - x^2)^5 x\, dx = -\frac{1}{2}\int (1 - x^2)^5(-2x\, dx)$$

$$= -\frac{1}{2}\int u^5\, du = -\frac{1}{12}u^6 + C$$

$$= -\frac{1}{12}(1 - x^2)^6 + C.$$

17 Let $u = 1 + x^2$; then $du = 2x\, dx$, so

$$\int \sqrt[3]{1 + x^2}\, x\, dx = \frac{1}{2}\int (1 + x^2)^{1/3}(2x\, dx)$$

$$= \frac{1}{2}\int u^{1/3}\, du = \frac{1}{2}\left(\frac{3}{4}u^{4/3}\right) + C$$

$$= \frac{3}{8}(1 + x^2)^{4/3} + C.$$

19 Let $u = \sqrt{t}$; then $du = \frac{dt}{2\sqrt{t}}$, so $\int \frac{e^{\sqrt{t}}}{\sqrt{t}}\, dt$

$$= 2\int e^{\sqrt{t}} \frac{dt}{2\sqrt{t}} = 2\int e^u\, du = 2e^u + C$$

$$= 2e^{\sqrt{t}} + C.$$

21 Let $u = 3\theta$; then $du = 3\, d\theta$, so $\int \sin 3\theta\, d\theta$

$$= \frac{1}{3}\int (\sin 3\theta)(3\, d\theta) = \frac{1}{3}\int \sin u\, du$$

$$= -\frac{1}{3}\cos u + C = -\frac{1}{3}\cos 3\theta + C.$$

23 Let $u = x - 3$; then $du = dx$, so

$$\int (x - 3)^{5/2}\, dx = \int u^{5/2}\, du = \frac{2}{7}u^{7/2} + C$$

$$= \frac{2}{7}(x - 3)^{7/2} + C.$$

25 Let $u = x^2 + 3x + 2$; then $du = (2x + 3)\, dx$, so

$$\int \frac{2x + 3}{x^2 + 3x + 2}\, dx = \int \frac{du}{u} = \ln|u| + C$$

$$= \ln|x^2 + 3x + 2| + C.$$

27 Let $u = 2x$; then $du = 2\, dx$, so $\int e^{2x}\, dx$

$$= \frac{1}{2}\int e^{2x}(2\, dx) = \frac{1}{2}\int e^u\, du = \frac{1}{2}e^u + C$$

$$= \frac{1}{2}e^{2x} + C.$$

29 Let $u = x^5$; then $du = 5x^4\, dx$, so $\int x^4 \sin x^5\, dx$

$$= \frac{1}{5}\int (\sin x^5)(5x^4\, dx) = \frac{1}{5}\int \sin u\, du$$

$$= -\frac{1}{5}\cos u + C = -\frac{1}{5}\cos x^5 + C.$$

31 Notice that the numerator is almost the derivative of x^2, while the denominator is $1 + (x^2)^2$. Let $u = x^2$; then $du = 2x\, dx$, so $\int \frac{x}{1 + x^4}\, dx =$

$$\frac{1}{2}\int \frac{2x\, dx}{1 + x^4} = \frac{1}{2}\int \frac{du}{1 + u^2} = \frac{1}{2}\tan^{-1} u + C$$

$= \frac{1}{2}\tan^{-1}x^2 + C.$

33 Let $u = 1 + x$; then $du = dx$ and $x = u - 1$, so

$$\int \frac{x\,dx}{(1+x)^3} = \int \frac{(u-1)\,du}{u^3} = \int\left(\frac{1}{u^2} - \frac{1}{u^3}\right)du$$

$$= \int (u^{-2} - u^{-3})\,du = -u^{-1} + \frac{1}{2}u^{-2} + C$$

$$= -\frac{1}{1+x} + \frac{1}{2(1+x)^2} + C.$$

35 Let $u = 3x$; then $du = 3\,dx$, so $\int \frac{\ln 3x}{x}\,dx =$

$\int \frac{\ln 3x}{3x}(3\,dx) = \int \ln u\left(\frac{du}{u}\right)$. Now with $w =$

$\ln u$, we have $dw = \frac{du}{u}$. Then $\int \ln u\left(\frac{du}{u}\right)$

$$= \int w\,dw = \frac{1}{2}w^2 + C = \frac{1}{2}(\ln u)^2 + C$$

$$= \frac{1}{2}(\ln 3x)^2 + C.$$

37 The area in question is represented by

$\int_1^2 e^{x^3}x^2\,dx$. Now let $u = x^3$, so $du = 3x^2\,dx$.

When $x = 1$, $u = 1$; when $x = 2$, $u = 2^3 = 8$.

Thus $\int_1^2 e^{x^3}x^2\,dx = \frac{1}{3}\int_1^2 e^{x^3}(3x^2\,dx)$

$$= \frac{1}{3}\int_1^8 e^u\,du = \frac{1}{3}(e^8 - e).$$

39 The area in question is represented by

$\int_0^1 \frac{x^2+3}{(x+1)^4}\,dx$. Now let $u = x + 1$, so $du = dx$

and $x = u - 1$. When $x = 0$, $u = 1$; when $x = 1$,

$u = 2$. Thus $\int_0^1 \frac{x^2+3}{(x+1)^4}\,dx = \int_1^2 \frac{(u-1)^2+3}{u^4}\,du$

$$= \int_1^2 \frac{u^2-2u+4}{u^4}\,du$$

$$= \int_1^2 (u^{-2} - 2u^{-3} + 4u^{-4})\,du$$

$$= \left(-u^{-1} + u^{-2} - \frac{4}{3}u^{-3}\right)\Big|_1^2$$

$$= \left(-\frac{1}{2} + \frac{1}{4} - \frac{4}{3}\cdot\frac{1}{8}\right) - \left(-1 + 1 - \frac{4}{3}\right) = \frac{11}{12}.$$

41 The area in question is represented by

$\int_1^e \frac{(\ln x)^3}{x}\,dx$. Now let $u = \ln x$, so $du = dx/x$.

When $x = 1$, $u = \ln 1 = 0$; when $x = e$, $u = \ln e$

$= 1$. Thus $\int_1^e \frac{(\ln x)^3}{x}\,dx = \int_1^e (\ln x)^3\,\frac{dx}{x}$

$$= \int_0^1 u^3\,du = \frac{u^4}{4}\Big|_0^1 = \frac{1}{4}.$$

43 Let $u = ax + b$, so $du = a\,dx$. Then $x = \frac{u-b}{a}$,

so $\int \frac{x^2\,dx}{ax+b} = \int \frac{1}{u}\left(\frac{u-b}{a}\right)^2\frac{du}{a}$

$$= \frac{1}{a^3}\int \frac{(u-b)^2}{u}\,du = \frac{1}{a^3}\int \frac{u^2-2bu+b^2}{u}\,du$$

$$= \frac{1}{a^3}\int\left(u - 2b + \frac{b^2}{u}\right)du$$

$$= \frac{1}{a^3}\left(\frac{1}{2}u^2 - 2bu + b^2\ln|u|\right) + K =$$

$$\frac{1}{a^3}\left(\frac{(ax+b)^2}{2} - 2b(ax+b) + b^2\ln|ax+b|\right) + K =$$

$$\frac{1}{a^3}\left(\frac{1}{2}(a^2x^2+2abx+b^2)-2abx-2b^2+b^2\ln|ax+b|\right)+K$$

$$= \frac{1}{a^3}\left(\frac{1}{2}a^2x^2 - abx + b^2\ln|ax+b|\right) + C,$$ where C

collects all of the constant terms; that is, $C = -\frac{3b^2}{2a^3} + K.$

45 Let $u = ax + b$; $du = a\,dx$. Then $x = \frac{u-b}{a}$ and

$$\int \frac{x^2\,dx}{(ax+b)^2} = \int \frac{1}{u^2}\left(\frac{u-b}{a}\right)^2 \frac{du}{a}$$

$$= \frac{1}{a^3}\int\left(1 - \frac{2b}{u} + \frac{b^2}{u^2}\right)du$$

$$= \frac{1}{a^3}\left(u - 2b\ln|u| - \frac{b^2}{u}\right) + K$$

$$= \frac{1}{a^3}\left((ax+b) - 2b\ln|ax+b| - \frac{b^2}{ax+b}\right) + K$$

$$= \frac{1}{a^3}\left(ax - 2b\ln|ax+b| - \frac{b^2}{ax+b}\right) + C,$$ with $C = \frac{b}{a^3} + K.$

47 Jack and Jill are both right. Both $-\cos^2\theta$ and $\sin^2\theta$ are antiderivatives of $2\cos\theta\sin\theta$. Notice that they differ by only a constant; in fact, $-\cos^2\theta + 1 = 1 - \cos^2\theta = \sin^2\theta$.

49 Jill is right. Jack made an error in substitution; $u = x^2$ implies that $x = \pm\sqrt{u}$. When $-2 \le x \le 0$, $x = -\sqrt{u}$, while when $0 \le x \le 1$, $x = \sqrt{u}$. His calculation should have been $\int_{-2}^{1} 2x^2\,dx$

$$= \int_{-2}^{0} x\cdot 2x\,dx + \int_0^1 x\cdot 2x\,dx$$

$$= \int_4^0 (-\sqrt{u})\,du + \int_0^1 \sqrt{u}\,du$$

$$= \int_0^4 \sqrt{u}\,du + \int_0^1 \sqrt{u}\,du = \frac{2}{3}u^{3/2}\Big|_0^4 + \frac{2}{3}u^{3/2}\Big|_0^1$$

$$= \frac{2}{3}(8-0) + \frac{2}{3}(1-0) = 6.$$

51 Let $u = -x$, so $du = -dx$. Then $\int_{-a}^{0} f(x)\,dx = \int_a^0 f(-u)(-du) = -\int_0^a f(u)(-du) = \int_0^a f(u)\,du$ $= \int_0^a f(x)\,dx$, since f is an even function and u is a dummy variable. Since $\int_{-a}^{0} f(x)\,dx = \int_0^a f(x)\,dx$, we have $\int_{-a}^{0} f(x)\,dx + \int_0^a f(x)\,dx = \int_0^a f(x)\,dx + \int_0^a f(x)\,dx$, and it follows that $\int_{-a}^{a} f(x)\,dx = 2\int_0^a f(x)\,dx$.

53 (a) Note that xe^{-x^2} is an odd function, so we consider only $x \ge 0$, and reflect the results across the line $y = -x$. Now, $y = xe^{-x^2} = 0$ when $x = 0$, so $(0, 0)$ is both the x intercept and the y intercept. Also $y' = e^{-x^2}(1 - 2x^2)$ changes from + to − at $x = \frac{1}{\sqrt{2}}$, so $(1/\sqrt{2}, 1/\sqrt{2e})$ is a maximum. Furthermore, $y'' = 2xe^{-x^2}(2x^2 - 3)$ changes from + to − at $x = 0$ and from − to + at $x = \sqrt{3/2}$, so $(0, 0)$ and $(\sqrt{3/2}, \sqrt{3/2}\,e^{-3/2})$ are inflection points. Finally, $\lim\limits_{x \to \infty} xe^{-x^2} = \lim\limits_{x \to \infty} \frac{x}{e^{x^2}} \underset{H}{=}$

$$\lim_{x \to \infty} \frac{1}{2xe^{x^2}} = 0.$$

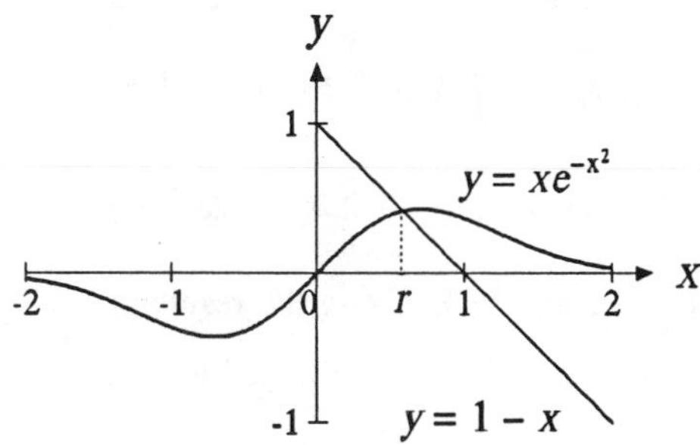

(b) From the figure, we see that the region bounded by $y = 1 - x$, $y = xe^{-x^2}$, and the x axis comprises two separate regions, whose total area is $\int_0^r xe^{-x^2}\,dx + \int_r^1 (1 - x)\,dx$, where xe^{-x^2} is the cross-sectional length of the region over $[0, r]$ and $1 - x$ is the cross-sectional length of the region over $[r, 1]$, where r remains to be determined. To find $\int_0^r xe^{-x^2}\,dx$, let $u = -x^2$, so $du = -2x\,dx$. When $x = 0$, $u = 0$; when $x = r$, $u = -r^2$. Thus, $\int_0^r xe^{-x^2}\,dx = -\frac{1}{2}\int_0^r e^{-x^2}(-2x\,dx) =$

$-\frac{1}{2}\int_0^{-r^2} e^u\,du = -\frac{1}{2}\left[e^{-r^2} - 1\right]$. Also,

$$\int_r^1 (1 - x)\,dx = \left(x - \frac{1}{2}x^2\right)\Big|_r^1 =$$

$\left(1 - \frac{1}{2}\right) - \left(r - \frac{1}{2}r^2\right) = \frac{1}{2} - r + \frac{1}{2}r^2$. So the entire area is given by

$$\int_0^r xe^{-x^2}\,dx + \int_r^1 (1 - x)\,dx =$$

$-\frac{1}{2}e^{-r^2} - r + \frac{1}{2}r^2 + 1$. To evaluate this result, we need to find r, the number for which $1 - x = xe^{-x^2}$. Applying Newton's method with $f(x) = 1 - x - xe^{-x^2}$ and $x_1 = 0.6$ yields $r \approx 0.584$ (to three decimal places). Upon substituting into our expression for the area, we find it to be approximately 0.231.

7.3 Integration by Parts

1 This exercise parallels Example 1. Let $u = x$ and $dv = e^{2x}\,dx$. Then $du = dx$ and $v = \frac{1}{2}e^{2x}$, so

$$\int xe^{2x}\,dx = x\left(\frac{1}{2}e^{2x}\right) - \int \frac{1}{2}e^{2x}dx =$$

$$\frac{1}{2}xe^{2x} - \frac{1}{4}e^{2x} + C = \frac{1}{4}e^{2x}(2x - 1) + C.$$

3 This problem is similar to Example 4. Letting $u = x$ and $dv = \sin 2x\,dx$, we have $du = dx$ and $v = -(1/2)\cos 2x$. Thus $\int x \sin 2x\,dx$

$$= x\left(-\frac{1}{2}\cos 2x\right) - \int\left(-\frac{1}{2}\cos 2x\right)dx$$

$$= -\frac{1}{2}x\cos 2x + \frac{1}{2}\int \cos 2x\,dx$$

$$= -\frac{1}{2}x\cos 2x + \frac{1}{4}\sin 2x + C.$$

5 This problem parallels Example 2. Use $u = \ln 3x$ and $dv = x\,dx$. Then $du = \frac{3}{3x}\,dx = \frac{dx}{x}$ and $v =$

$\frac{1}{2}x^2$. Hence $\int x \ln 3x \, dx$

$= \frac{1}{2}x^2 \ln 3x - \int \frac{1}{2}x^2 \frac{dx}{x}$

$= \frac{1}{2}x^2 \ln 3x - \frac{1}{2}\int x \, dx$

$= \frac{1}{2}x^2 \ln 3x - \frac{1}{4}x^2 + C$. An alternate approach to this problem directly uses the result of Example 2. Since $\ln 3x = \ln 3 + \ln x$, $\int x \ln 3x \, dx =$

$\int x \ln 3 \, dx + \int x \ln x \, dx$

$= (\ln 3)\frac{x^2}{2} + \frac{x^2}{2}\ln x - \frac{x^2}{4} + C$

$= \frac{x^2}{2}(\ln 3 + \ln x) - \frac{x^2}{4} + C$

$= \frac{x^2}{2}\ln 3x - \frac{x^2}{4} + C.$

7 This exercise is similar to Example 5. Let $u = x^2$ and $dv = e^{-x}\,dx$. Then $du = 2x\,dx$ and $v = -e^{-x}$. Hence $\int x^2e^{-x}\,dx = -x^2e^{-x} + \int 2xe^{-x}\,dx =$

$-x^2e^{-x} + 2\int xe^{-x}\,dx$. Another integration by parts is necessary to compute the new integral. Let $U = x$ and $dV = e^{-x}\,dx$. Then $dU = dx$ and $V =$ $-e^{-x}$. Thus $\int xe^{-x}\,dx = -xe^{-x} + \int e^{-x}\,dx =$

$-xe^{-x} - e^{-x} + K$. Therefore, $\int x^2e^{-x}\,dx =$ $-x^2e^{-x} + 2(-xe^{-x} - e^{-x} + K) = -x^2e^{-x} - 2xe^{-x} - 2e^{-x} + C = -e^{-x}(x^2 + 2x + 2) + C$, where $C = 2K$. Finally, we apply the fundamental theorem:

$\int_1^2 x^2e^{-x}\,dx = -e^{-x}(x^2 + 2x + 2)\Big|_1^2 =$

$-e^{-2}(4 + 4 + 2) + e^{-1}(1 + 2 + 2) = \frac{5}{e} - \frac{10}{e^2} =$

$\frac{5e - 10}{e^2}$. (This problem could also have been treated without integration by parts, with $u = -x$, $du = -dx$, and $\int x^2e^{-x}\,dx = \int u^2e^u\,(-du) =$ $-\int u^2e^u\,du = (u^2e^u - 2ue^u + 2e^u + K)$, by the result of Example 5. Using the resultant antiderivative, we evaluate the definite integral as above.)

9 Following the procedure of Example 3, let $u =$ $\sin^{-1} x$ and $dv = dx$. Then $du = \frac{dx}{\sqrt{1 - x^2}}$ and v $= x$. Hence, $\int \sin^{-1} x \, dx =$

$(\sin^{-1} x)x - \int \frac{x\,dx}{\sqrt{1 - x^2}}$. This new integral can be evaluated using substitution. Let $w = 1 - x^2$, $dw = -2x\,dx$. Then $\int \frac{x\,dx}{\sqrt{1 - x^2}} =$

$-\frac{1}{2}\int (1 - x^2)^{-1/2}(-2x\,dx) = -\frac{1}{2}\int w^{-1/2}\,dw =$

$-w^{1/2} + K = -\sqrt{1 - x^2} + K$. Thus, $\int \sin^{-1} x\,dx$

$= (\sin^{-1} x)x - \left[-\sqrt{1 - x^2} + K\right] =$

$x\sin^{-1} x + \sqrt{1 - x^2} + C$. Finally, $\int_0^1 \sin^{-1} x\,dx$

$= \left(x\sin^{-1} x + \sqrt{1 - x^2}\right)\Big|_0^1$

$= \left(1\sin^{-1} 1 + 0\right) - \left(0\sin^{-1} 0 + 1\right) = \sin^{-1} 1 - 1$

$= \frac{\pi}{2} - 1.$

11 Let $u = \ln x$ and $dv = x^2\,dx$; then $du = \dfrac{dx}{x}$ and

$v = \dfrac{x^3}{3}$. Thus $\int x^2 \ln x\,dx =$

$$\frac{1}{3}x^3 \ln x - \int \frac{x^3}{3}\cdot\frac{dx}{x} = \frac{1}{3}x^3 \ln x - \frac{1}{3}\int x^2\,dx$$

$$= \frac{1}{3}x^3 \ln x - \frac{1}{9}x^3 + C.$$

13 Let $u = (\ln x)^2$ and $dv = dx$. Then $du =$

$\dfrac{2\ln x}{x}\,dx$ and $v = x$, so $\int (\ln x)^2\,dx =$

$$x(\ln x)^2 - \int x\cdot\frac{2\ln x}{x}\,dx =$$

$x(\ln x)^2 - 2\int \ln x\,dx$. Another integration by parts, this time with $U = \ln x$, $dV = dx$, $dU = \dfrac{dx}{x}$, and $V = x$, evaluates this last integral:

$$\int \ln x\,dx = x\ln x - \int x\,\frac{dx}{x} = x\ln x - \int dx$$

$= x\ln x - x + K$. Hence $\int (\ln x)^2\,dx =$

$x(\ln x)^2 - 2x\ln x + 2x + C$ and

$$\int_2^3 (\ln x)^2\,dx = \left(x(\ln x)^2 - 2x\ln x + 2x\right)\Big|_2^3 =$$

$(3(\ln 3)^2 - 6\ln 3 + 6) - (2(\ln 2)^2 - 4\ln 2 + 4)$

$= 3(\ln 3)^2 - 6\ln 3 - 2(\ln 2)^2 + 4\ln 2 + 2$.

15 Let $u = \ln x$ and $dv = \dfrac{dx}{x^2}$. Then $du = \dfrac{dx}{x}$, $v =$

$-1/x$, and integration by parts yields $\int \dfrac{\ln x}{x^2}\,dx =$

$$-\frac{\ln x}{x} + \int \frac{1}{x}\cdot\frac{dx}{x} = -\frac{\ln x}{x} + \int \frac{dx}{x^2} =$$

$-\dfrac{\ln x}{x} - \dfrac{1}{x} + C$. Hence, $\int_1^e \dfrac{\ln x}{x^2}\,dx =$

$$\left(-\frac{\ln x}{x} - \frac{1}{x}\right)\Big|_1^e = \left(-\frac{1}{e} - \frac{1}{e}\right) - \left(-\frac{0}{1} - \frac{1}{1}\right)$$

$$= 1 - \frac{2}{e} = \frac{e-2}{e}.$$

17 This exercise parallels Example 8. Let $u = e^{3x}$ and $dv = \cos 2x\,dx$. Then $du = 3e^{3x}\,dx$ and $v =$

$\dfrac{1}{2}\sin 2x$, so $\int e^{3x}\cos 2x\,dx =$

$\dfrac{1}{2}e^{3x}\sin 2x - \dfrac{3}{2}\int e^{3x}\sin 2x\,dx$. Now let $U =$

$e^{3x}\,dx$ and $dV = \sin 2x\,dx$. This yields $dU = 3e^{3x}$

and $V = -\dfrac{1}{2}\cos 2x$. So $\int e^{3x}\cos 2x\,dx =$

$$\frac{1}{2}e^{3x}\sin 2x \;-$$

$$\frac{3}{2}\left(-\frac{1}{2}e^{3x}\cos 2x + \frac{3}{2}\int e^{3x}\cos 2x\,dx\right) =$$

$$\frac{1}{2}e^{3x}\sin 2x + \frac{3}{4}e^{3x}\cos 2x - \frac{9}{4}\int e^{3x}\cos 2x\,dx$$

and $\dfrac{13}{4}\int e^{3x}\cos 2x\,dx =$

$e^{3x}\left(\dfrac{1}{2}\sin 2x + \dfrac{3}{4}\cos 2x\right)$, from which it follows that

$$\int e^{3x}\cos 2x\,dx$$

$$= \frac{1}{13}e^{3x}(2\sin 2x + 3\cos 2x) + C.$$

19 Let $u = \ln(1 + x^2)$ and $dv = \dfrac{dx}{x^2}$. Then $du =$

$\dfrac{2x\,dx}{1 + x^2}$ and $v = -1/x$. Then $\int \dfrac{\ln(1 + x^2)}{x^2}\,dx =$

$$-\frac{\ln(1+x^2)}{x} + \int \frac{1}{x}\cdot\frac{2x\,dx}{1+x^2}$$

$$= -\frac{\ln(1+x^2)}{x} + 2\int \frac{dx}{1+x^2}$$

$$= -\frac{\ln(1+x^2)}{x} + 2\tan^{-1}x + C.$$

21 (a) Let $w = 3x + 7$ so that $x = \dfrac{w-7}{3}$ and $dx = \dfrac{dw}{3}$. Then $\int x\sqrt{3x+7}\,dx =$

$$\int \frac{w-7}{3} w^{1/2} \frac{dw}{3} = \frac{1}{9}\int (w^{3/2} - 7w^{1/2})\,dw$$

$$= \frac{1}{9}\left(\frac{2}{5}w^{5/2} - \frac{14}{3}w^{3/2}\right) + C$$

$$= \frac{2}{135}(3w - 35)w^{3/2} + C$$

$$= \frac{2}{135}(9x - 14)(3x+7)^{3/2} + C.$$

(b) Let $u = x$ and $dv = \sqrt{3x+7}\,dx = (3x+7)^{1/2}\,dx$. Then $du = dx$ and $v = \frac{2}{9}(3x+7)^{3/2}$. Thus $\int x\sqrt{3x+7}\,dx$

$$= \frac{2x}{9}(3x+7)^{3/2} - \frac{2}{9}\int (3x+7)^{3/2}\,dx$$

$$= \frac{2x}{9}(3x+7)^{3/2} - \frac{2}{9}\cdot\frac{2}{15}(3x+7)^{5/2} + C$$

$$= \frac{2x}{9}(3x+7)^{3/2} - \frac{4}{135}(3x+7)^{5/2} + C.$$ A little algebra will show that the answers for (a) and (b) agree. The integrals $\int (3x+7)^{1/2}\,dx$ and $\int (3x+7)^{3/2}\,dx$ were evaluated by using the substitution $w = 3x + 7$.

23 (a) Let $u = ax + b$, so that $x = \dfrac{w-b}{a}$ and $dx = \dfrac{dw}{a}$. Then $\int x(ax+b)^3\,dx$

$$= \int \frac{w-b}{a}\cdot w^3 \frac{dw}{a} = \frac{1}{a^2}\int (w^4 - bw^3)\,dw$$

$$= \frac{1}{a^2}\left(\frac{w^5}{5} - \frac{bw^4}{4}\right) + C$$

$$= \frac{1}{a^2}\left(\frac{(ax+b)^5}{5} - \frac{b(ax+b)^4}{4}\right) + C$$

$$= \frac{1}{20a^2}(ax+b)^4(4ax - b) + C.$$

(b) Let $u = x$ and $dv = (ax+b)^3\,dx$. Then $du = dx$ and $v = \dfrac{(ax+b)^4}{4a}$. Thus

$$\int x(ax+b)^3\,dx$$

$$= \frac{x(ax+b)^4}{4a} - \frac{1}{4a}\int (ax+b)^4\,dx$$

$$= \frac{x(ax+b)^4}{4a} - \frac{1}{4a}\cdot\frac{(ax+b)^5}{5a} + C$$

$$= \frac{1}{20a^2}(ax+b)^4(4ax - b) + C.$$

25 (a) Here $n = 2$ in the recursion of Example 7. Thus $\int \sin^2 x\,dx =$

$$-\frac{\sin x\cos x}{2} + \frac{1}{2}\int \sin^0 x\,dx$$

$$= -\frac{\sin x\cos x}{2} + \frac{1}{2}\int dx$$

$= -\dfrac{\sin x \cos x}{2} + \dfrac{1}{2}x + C$

$= \dfrac{1}{2}(x - \sin x \cos x) + C.$

(b) Letting $n = 4$ in the recursion formula,

$\int \sin^4 x\, dx = -\dfrac{\sin^3 x \cos x}{4} + \dfrac{3}{4}\int \sin^2 x\, dx$

$= -\dfrac{\sin^3 x \cos x}{4} + \dfrac{3}{8}(x - \sin x \cos x) + C$

from the result of (a).

(c) Using $n = 6$, $\int \sin^6 x\, dx$

$= -\dfrac{\sin^5 x \cos x}{6} + \dfrac{5}{6}\int \sin^4 x\, dx$

$= -\dfrac{\sin^5 x \cos x}{6} +$

$\dfrac{5}{6}\left(-\dfrac{\sin^3 x \cos x}{4} + \dfrac{3}{8}(x - \sin x \cos x)\right) + C$

$= -\dfrac{\sin^5 x \cos x}{6} - \dfrac{5 \sin^3 x \cos x}{24} +$

$\dfrac{5}{16}(x - \sin x \cos x) + C$, using the result of

(b).

27 Let $u = (\ln x)^{18}$ and $dv = x^{10}\, dx$, so $du =$

$18(\ln x)^{17}\dfrac{dx}{x}$ and $v = \dfrac{1}{11}x^{11}$. Hence

$\int x^{10}(\ln x)^{18}\, dx$

$= \dfrac{1}{11}x^{11}(\ln x)^{18} - \int \dfrac{1}{11}x^{11}\cdot 18\cdot(\ln x)^{17}\dfrac{dx}{x}$

$= \dfrac{1}{11}x^{11}(\ln x)^{18} - \dfrac{18}{11}\int x^{10}(\ln x)^{17}dx$. Now this

last integral has the same power of x in the integral but one less power of $\ln x$. Letting $u = (\ln x)^{17}$ and $dv = x^{10}\, dx$, and so on, will eventually give $\int x^{10}\, dx$, which can easily be evaluated.

29 (a) $y = e^x \sin x = 0$ when $x = 0$ and $x = \pi$, so $(0, 0)$ is the y intercept and $(0, 0)$ and $(\pi, 0)$ are the x intercepts. $y' = e^x(\cos x + \sin x)$ changes from + to − when $\tan x = -1$, or when $x = 3\pi/4 + k\pi$, where k is an integer, so

$\left(\dfrac{3\pi}{4}, \dfrac{\sqrt{2}}{2}e^{3\pi/4}\right)$ is a maximum on $[0, \pi]$. $y'' =$

$2e^x \cos x$ changes from + to − at $x = \pi/2$,

so $\left(\dfrac{\pi}{2}, e^{\pi/2}\right)$ is an inflection point.

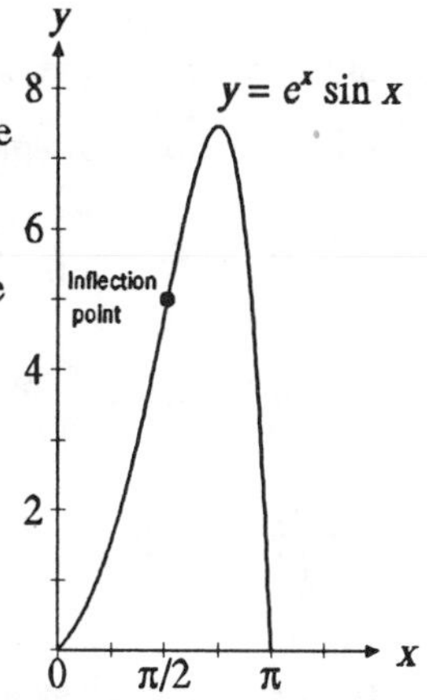

(b) We find $\int_0^\pi e^x \sin x\, dx$ by integration by

parts. Let $u = \sin x$ and $dv = e^x\, dx$. So $du =$

$\cos x\, dx$, $v = e^x$, and $\int e^x \sin x\, dx =$

$e^x \sin x - \int e^x \cos x\, dx$. Now let $U = \cos x$

and $dV = e^x\, dx$, so $dU = -\sin x\, dx$ and $V =$

e^x. Then $\int e^x \sin x\, dx =$

$e^x \sin x - \left[e^x \cos x - \int e^x(-\sin x)\, dx\right]$; by

solving for $\int e^x \sin x\, dx$, it follows that

$\int e^x \sin x\, dx = \dfrac{1}{2}e^x(\sin x - \cos x)$. Finally,

$$\int_0^\pi e^x \sin x\, dx = \frac{1}{2}e^x(\sin x - \cos x)\Big|_0^\pi$$

$$= \frac{1}{2}(e^\pi + 1).$$

31 Since volume is the definite integral of cross-sectional area, the volume is given by $\int_0^1 (xe^x)^2\, dx = \int_0^1 x^2e^{2x}\, dx$. Now let $u = 2x$, so $x = u/2$ and $du = 2\, dx$. Then $\int_0^1 x^2e^{2x}\, dx =$

$\int_0^2 \left(\frac{u}{2}\right)^2 e^u \frac{du}{2} = \frac{1}{8}\int_0^2 u^2e^u\, du$ (since when $x = 0$, $u = 0$ and when $x = 1$, $u = 2$). From Example 5 (or Exercise 28), we know $\int u^2e^u\, du =$ $(u^2 - 2u + 2)e^u$, so $\frac{1}{8}\int_0^2 u^2e^u\, du$

$$= \frac{1}{8}(u^2 - 2u + 2)e^u\Big|_0^2$$

$$= \frac{1}{8}[(4 - 4 + 2)e^2 - (0 - 0 + 2)e^0] = \frac{1}{8}(2e^2 - 2)$$

$$= \frac{1}{4}(e^2 - 1).$$

33 Let $w = \sqrt{x}$ so that $x = w^2$ and $dx = 2w\, dw$. Then

$$\int \sin\sqrt{x}\, dx = \int (\sin w)(2w\, dw)$$

$$= 2\int w \sin w\, dw = 2(-w\cos w + \sin w) + C$$

$= 2\left[\sin\sqrt{x} - \sqrt{x}\cos\sqrt{x}\right] + C$, by Example 4.

35 Let $w = \sqrt{x}$ so that $x = w^2$ and $dx = 2w\, dw$. Then $\int e^{\sqrt{x}}\, dx = 2\int we^w\, dw = 2(we^w - e^w) + C =$ $2\left(\sqrt{x}e^{\sqrt{x}} - e^{\sqrt{x}}\right) + C = 2e^{\sqrt{x}}\left(\sqrt{x} - 1\right) + C$, by Example 1.

37 Integrating by parts with $u = \ln x$ and $dv = \cos x\, dx$, we get $du = \frac{dx}{x}$, $v = \sin x$, and

$$\int (\cos x)(\ln x)\, dx = (\ln x)(\sin x) - \int \sin x \frac{dx}{x}.$$

Since the integral $\int \frac{\sin x}{x}\, dx$ is not elementary, $\int (\cos x)(\ln x)\, dx$ cannot be elementary either.

39 (a) $I_0 = \int_0^{\pi/2} \sin^0\theta\, d\theta = \int_0^{\pi/2} d\theta = \theta|_0^{\pi/2} = \frac{\pi}{2}$. $I_1 = \int_0^{\pi/2} \sin\theta\, d\theta = -\cos\theta|_0^{\pi/2} = -0 - (-1) = 1$.

(b) Using the recursion for $n \geq 2$, $I_n = \int_0^{\pi/2} \sin^n\theta\, d\theta =$

$$\left(-\frac{\sin^{n-1}\theta\cos\theta}{n}\right)\Big|_0^{\pi/2} + \frac{n-1}{n}\int_0^{\pi/2} \sin^{n-2}\theta\, d\theta$$

$$= \frac{n-1}{n}\int_0^{\pi/2} \sin^{n-2}\theta\, d\theta = \frac{n-1}{n}\cdot I_{n-2}.$$

(c) From (b), $I_2 = (1/2)I_0 = (1/2)(\pi/2) = \pi/4$ and $I_3 = (2/3)I_1 = 2/3$.

(d) $I_4 = \frac{3}{4}\cdot I_2 = \frac{3}{4}\cdot\frac{\pi}{4} = \frac{3\pi}{16}$, $I_5 = \frac{4}{5}\cdot I_3 = \frac{4}{5}\cdot\frac{2}{3} = \frac{8}{15}$.

(e) Since $I_3 = 2/3$, $I_5 = (2/3)(4/5)$, and there is a recursive relationship between I_n and I_{n-2} involving ratios of consecutive integers, it seems reasonable that $I_n = \frac{2}{3}\cdot\frac{4}{5}\cdot\frac{6}{7}\cdots\left(\frac{n-1}{n}\right)$

for $n \geq 3$ and odd.

(f) Since $I_2 = (1/2)(\pi/2)$ and $I_4 = (1/2)(3/4)(\pi/2)$, it seems plausible that $I_n = \frac{1}{2}\cdot\frac{3}{4}\cdot\frac{5}{6}\cdots\frac{(n-1)}{n}\cdot\frac{\pi}{2}$ for $n \geq 2$ and even for the same reason given in (e).

(g) Let $u = \pi/2 - \theta$ so $du = -d\theta$. When $\theta = 0$, $u = \pi/2$; when $\theta = \pi/2$, $u = 0$. Then $\int_0^{\pi/2} \cos^n\theta\, d\theta = \int_{\pi/2}^0 \cos^n\left(\frac{\pi}{2} - u\right)(-du) = \int_0^{\pi/2} \cos^n\left(\frac{\pi}{2} - u\right) du$. Recall that $\cos\left(\frac{\pi}{2} - u\right) = \sin u$, so $\int_0^{\pi/2} \cos^n\left(\frac{\pi}{2} - u\right) du = \int_0^{\pi/2} \sin^n u\, du = \int_0^{\pi/2} \sin^n\theta\, d\theta$ and $\int_0^{\pi/2} \cos^n\theta\, d\theta = \int_0^{\pi/2} \sin^n\theta\, d\theta$.

41 $u = x^n$ and $dv = e^{ax}\,dx$; then $du = nx^{n-1}\,dx$, $v = \frac{1}{a}e^{ax}$, and $\int x^n e^{ax}\,dx$

$$= x^n\cdot\frac{1}{a}e^{ax} - \int \frac{1}{a}e^{ax}\cdot nx^{n-1}\,dx$$

$$= \frac{1}{a}x^n e^{ax} - \frac{n}{a}\int x^{n-1}e^{ax}\,dx.$$

43 Let $u = x^n$ and $dv = \sin x\,dx$; then $du = nx^{n-1}\,dx$, $v = -\cos x$, and $\int x^n \sin x\,dx = x^n(-\cos x) - \int(-\cos x)nx^{n-1}\,dx = x^n\cos x + n\int x^{n-1}\cos x\,dx$. Now let $U = x^{n-1}$ and $dV = \cos x\,dx$ for a second integration by parts; then $dU = (n-1)x^{n-2}\,dx$, $V = \sin x$, and $\int x^n \sin x\,dx = x^n\cos x + n\int x^{n-1}\cos x\,dx = -x^n\cos x + n\left(x^{n-1}\sin x - \int(\sin x)(n-1)x^{n-2}\,dx\right)$

$$= -x^n\cos x + nx^{n-1}\sin x - n(n-1)\int x^{n-2}\sin x\,dx.$$

45 Since $x^3\sqrt{1 + x^{20}}$ is an odd function, $\int_{-1}^1 x^3\sqrt{1 + x^{20}}\,dx = 0$.

47 For the sake of argument, let $\int e^x\cos x\,dx = f(x) + K$, where K is a constant. Adding $C - K$ to both sides gives $\int e^x\cos x\,dx + (C - K) = f(x) + K + (C - K) = f(x) + C = \int e^x\cos x\,dx$. So $\int e^x\cos x\,dx$ and $\int e^x\cos x\,dx + C$, where C is an arbitrary constant, are equivalent forms. From Example 8, we have $\int e^x\cos x\,dx = e^x(\sin x + \cos x) - \int e^x\cos x\,dx$. From above, this can be rewritten as $\int e^x\cos x\,dx + C = e^x(\sin x + \cos x) - \left(\int e^x\cos x\,dx + C\right)$, so $2\int e^x\cos x\,dx + 2C = e^x(\sin x + \cos x)$ and $\int e^x\cos x\,dx = \frac{1}{2}e^x(\sin x + \cos x) + K$, where $K = -C$. Thus our missing constant has appeared and adding a constant at the end of the computation is justified.

7.4 How to Integrate Certain Rational Functions

1 Let $u = 3x - 4$, $du = 3\,dx$; then $\int \frac{dx}{3x - 4}$

$= \frac{1}{3}\int \frac{3\,dx}{3x - 4} = \frac{1}{3}\int \frac{du}{u} = \frac{1}{3}\ln|u| + C$

$= \frac{1}{3}\ln|3x - 4| + C.$

3 Let $u = 2x + 7$, $du = 2\,dx$; then $\int \frac{5\,dx}{(2x + 7)^2}$

$= \frac{5}{2}\int \frac{2\,dx}{(2x + 7)^2} = \frac{5}{2}\int \frac{du}{u^2} = -\frac{5}{2}\cdot\frac{1}{u} + C =$

$-\frac{5}{2(2x + 7)} + C.$

5 Motivated by Examples 2 and 3, let $x^2 = 9u^2$ or $x = 3u$. Then $dx = 3\,du$ and $\int \frac{dx}{x^2 + 9} =$

$\int \frac{du}{9u^2 + 9} = \frac{3}{9}\int \frac{du}{u^2 + 1} = \frac{1}{3}\tan^{-1}u + C =$

$\frac{1}{3}\tan^{-1}\frac{x}{3} + C.$

7 Note that the derivative of the denominator is $2x$.

Thus, $\int_0^3 \frac{x\,dx}{x^2 + 9} = \frac{1}{2}\int_0^3 \frac{2x\,dx}{x^2 + 9} =$

$\frac{1}{2}\ln(x^2 + 9)\Big|_0^3 = \frac{1}{2}\ln 18 - \frac{1}{2}\ln 9 = \frac{1}{2}\ln 2.$

9 $\int \frac{2x + 3}{x^2 + 9}\,dx = \int \frac{2x\,dx}{x^2 + 9} + \int \frac{3\,dx}{x^2 + 9} =$

$\ln(x^2 + 9) + 3\int \frac{dx}{x^2 + 9}$. By Exercise 5, we have

$\int \frac{dx}{x^2 + 9} = \frac{1}{3}\tan^{-1}\frac{x}{3} + K$, so $\int \frac{2x + 3}{x^2 + 9}\,dx$

$= \ln(x^2 + 9) + 3\left(\frac{1}{3}\tan^{-1}\frac{x}{3} + K\right)$

$= \ln(x^2 + 9) + \tan^{-1}\frac{x}{3} + C$, where $C = 3K$.

11 This problem is similar to Example 3. Let $16x^2 = 25u^2$; that is, let $4x = 5u$. Then $dx = \frac{5}{4}\,du$ and

$\int \frac{dx}{16x^2 + 25} = \int \frac{(5/4)\,du}{25(u^2 + 1)} = \frac{1}{20}\int \frac{du}{u^2 + 1}$

$= \frac{1}{20}\tan^{-1}u + C = \frac{1}{20}\tan^{-1}\frac{4x}{5} + C.$

13 The numerator is almost the derivative of the denominator. Hence $\int \frac{x\,dx}{16x^2 + 25} =$

$\frac{1}{32}\int \frac{32x\,dx}{16x^2 + 25} = \frac{1}{32}\ln(16x^2 + 25) + C.$

15 $\int \frac{x + 2}{9x^2 + 4}\,dx = \int \frac{x\,dx}{9x^2 + 4} + 2\int \frac{dx}{9x^2 + 4},$

where the two integrals on the right will be evaluated separately. First note that the derivative of $9x^2 + 4$ is $18x$. Multiplication and division by 18 in the first integral yields $\int \frac{x\,dx}{9x^2 + 4} =$

$\frac{1}{18}\int \frac{18x\,dx}{9x^2 + 4} = \frac{1}{18}\ln(9x^2 + 4) + C_1.$

Motivated by Example 3, we let $9x^2 = 4u^2$, or $3x = 2u$. Then $dx = \frac{2}{3}\,du$ and $\int \frac{dx}{9x^2 + 4} =$

$\int \frac{(2/3)\,du}{4u^2 + 4} = \frac{1}{6}\int \frac{du}{u^2 + 1} = \frac{1}{6}\tan^{-1}u + C_2$

$= \frac{1}{6}\tan^{-1}\frac{3x}{2} + C_2$. Combining these two results,

we have $\int \frac{x+2}{9x^2+4}\,dx = \frac{1}{18}\ln(9x^2+4) + C_1$

$+ 2\left(\frac{1}{6}\tan^{-1}\frac{3x}{2} + C_2\right) =$

$\frac{1}{18}\ln(9x^2+4) + \frac{1}{3}\tan^{-1}\frac{3x}{2} + C.$

17 As in Examples 2 and 3, choose u such that $2x^2 = 3u^2$; that is, $\sqrt{2}x = \sqrt{3}u$ and $\sqrt{2}\,dx = \sqrt{3}\,du$.

Then $\int \frac{dx}{2x^2+3} = \sqrt{\frac{3}{2}}\int \frac{du}{3u^2+3}$

$= \frac{1}{3}\sqrt{\frac{3}{2}}\int \frac{du}{u^2+1} = \frac{1}{\sqrt{6}}\tan^{-1}u + C$

$= \frac{1}{\sqrt{6}}\tan^{-1}\sqrt{\frac{2}{3}}x + C.$

19 Completing the square in the denominator yields

$\int \frac{dx}{x^2+2x+3} = \int \frac{dx}{(x+1)^2+2}$. Following

the procedure of Example 4, let u be such that

$(x+1)^2 = 2u^2$; that is, let $x + 1 = \sqrt{2}u$. Then dx

$= \sqrt{2}\,du$ and $\int \frac{dx}{(x+1)^2+2} = \int \frac{\sqrt{2}\,du}{2u^2+2}$

$= \frac{1}{\sqrt{2}}\int \frac{du}{u^2+1} = \frac{1}{\sqrt{2}}\tan^{-1}u + C$

$= \frac{1}{\sqrt{2}}\tan^{-1}\frac{x+1}{\sqrt{2}} + C.$

21 Completing the square in the denominator,

$\int \frac{dx}{x^2-2x+3} = \int \frac{dx}{(x-1)^2+2}$. Let $(x-1)^2$

$= 2u^2$, or $x - 1 = \sqrt{2}u$. Then $dx = \sqrt{2}\,du$ and

$\int \frac{dx}{(x-1)^2+2} = \int \frac{\sqrt{2}\,du}{2u^2+2} = \frac{1}{\sqrt{2}}\int \frac{du}{u^2+1}$

$= \frac{1}{\sqrt{2}}\tan^{-1}u + C = \frac{1}{\sqrt{2}}\tan^{-1}\frac{x-1}{\sqrt{2}} + C.$

23 Complete the square: $2x^2 + x + 3$

$= 2\left(x^2 + \frac{1}{2}x + \frac{1}{16}\right) - \frac{1}{8} + 3 =$

$2\left(x + \frac{1}{4}\right)^2 + \frac{23}{8}$. Let $2\left(x + \frac{1}{4}\right)^2 = \frac{23}{8}u^2$, so that

$u^2 = \frac{16}{23}\left(x + \frac{1}{4}\right)^2$ and $u = \frac{4}{\sqrt{23}}\left(x + \frac{1}{4}\right)$. Then du

$= \frac{4}{\sqrt{23}}\,dx$ and $\int \frac{dx}{2x^2+x+3} =$

$\int \frac{dx}{2\left(x+\frac{1}{4}\right)^2 + \frac{23}{8}} = \int \frac{\sqrt{23}\,\frac{du}{4}}{\frac{23}{8}(u^2+1)} =$

$\frac{2}{\sqrt{23}}\tan^{-1}u + C = \frac{2}{\sqrt{23}}\tan^{-1}\frac{4}{\sqrt{23}}\left(x + \frac{1}{4}\right) + C$

$= \frac{2}{\sqrt{23}}\tan^{-1}\frac{4x+1}{\sqrt{23}} + C.$

25 Completing the square: $x^2 + 4x + 7 =$

$(x+2)^2 + 3$. With $3u^2 = (x+2)^2$ or $\sqrt{3}u =$

$x + 2$, $\int \frac{dx}{x^2+4x+7} = \int \frac{dx}{(x+2)^2+3}$

$= \int \frac{\sqrt{3}\,du}{3u^2+3} = \frac{\sqrt{3}}{3}\int \frac{du}{u^2+1} =$

$\frac{1}{\sqrt{3}}\tan^{-1}u + C = \frac{1}{\sqrt{3}}\tan^{-1}\frac{x+2}{\sqrt{3}} + C.$

27 Completing the square, we have $2x^2 + 4x + 7 = 2(x^2 + 2x + 1) + 5 = 2(x + 1)^2 + 5$. The integral then becomes $\int \frac{dx}{2x^2 + 4x + 7} =$

$\int \frac{dx}{2(x+1)^2 + 5}$. Let $5u^2 = 2(x + 1)^2$ or $\sqrt{5}u = \sqrt{2}(x + 1)$. Then $dx = \frac{\sqrt{5}}{\sqrt{2}}\,du$ and

$$\int \frac{dx}{2(x+1)^2 + 5} = \int \frac{\sqrt{5}\,(du/\sqrt{2})}{5u^2 + 5}$$

$$= \frac{1}{\sqrt{10}} \int \frac{du}{u^2 + 1} = \frac{1}{\sqrt{10}} \tan^{-1} u + C$$

$$= \frac{1}{\sqrt{10}} \tan^{-1} \sqrt{\tfrac{2}{5}}(x + 1) + C$$

$$= \frac{1}{\sqrt{10}} \tan^{-1} \frac{2x + 2}{\sqrt{10}} + C.$$

29 Notice that the derivative of the denominator is $2x + 2$. With that in mind, rewrite the integral:

$$\int \frac{2x\,dx}{x^2 + 2x + 3} = \int \frac{(2x + 2) - 2}{x^2 + 2x + 3}\,dx$$

$$= \int \frac{(2x + 2)\,dx}{x^2 + 2x + 3} - 2 \int \frac{dx}{x^2 + 2x + 3}$$

$$= \ln(x^2 + 2x + 3) - \frac{2}{\sqrt{2}} \tan^{-1} \frac{x + 1}{\sqrt{2}} + C$$

$= \ln(x^2 + 2x + 3) - \sqrt{2} \tan^{-1} \frac{x + 1}{\sqrt{2}} + C$, where we used the result of Exercise 19 for the second integral.

31 This exercise parallels Example 6.

$$\int \frac{3x\,dx}{5x^2 + 3x + 2} = \frac{3}{10} \int \frac{10x\,dx}{5x^2 + 3x + 2}$$

$$= \frac{3}{10} \int \frac{(10x + 3 - 3)\,dx}{5x^2 + 3x + 2}$$

$$= \frac{3}{10} \int \frac{(10x + 3)\,dx}{5x^2 + 3x + 2} - \frac{9}{10} \int \frac{dx}{5\left(x^2 + \frac{3}{5}x + \frac{9}{100}\right) + \frac{155}{100}}$$

$$= \frac{3}{10} \ln(5x^2 + 3x + 2) - \frac{9}{10} \int \frac{dx}{5\left(x + \frac{3}{10}\right)^2 + \frac{31}{20}}.$$

If we let $5\left(x + \frac{3}{10}\right)^2 = \frac{31}{20}u^2$, or $x + \frac{3}{10} = \frac{\sqrt{31}}{10}u$, then the last integral is $\int \frac{\frac{\sqrt{31}}{10}\,du}{\frac{31}{20}u^2 + \frac{31}{20}}$

$$= \frac{2}{\sqrt{31}} \int \frac{du}{u^2 + 1} = \frac{2}{\sqrt{31}} \tan^{-1} u + K$$

$$= \frac{2}{\sqrt{31}} \tan^{-1} \frac{10}{\sqrt{31}}\left(x + \frac{3}{10}\right) + K$$

$= \frac{2}{\sqrt{31}} \tan^{-1} \frac{10x + 3}{\sqrt{31}} + K$. Combining these results, we have $\int \frac{3x\,dx}{5x^2 + 3x + 2} =$

$$\frac{3}{10} \ln(5x^2 + 3x + 2) - \frac{9}{5\sqrt{31}} \tan^{-1} \frac{10x + 3}{\sqrt{31}} + C.$$

33 Following the procedure of Example 6,

$$\int \frac{x + 1}{x^2 + x + 1}\,dx = \frac{1}{2} \int \frac{2x + 2}{x^2 + x + 1}\,dx$$

$$= \frac{1}{2} \int \frac{2x + 1 + 1}{x^2 + x + 1}\,dx$$

$$= \frac{1}{2}\int \frac{(2x+1)\,dx}{x^2+x+1} + \frac{1}{2}\int \frac{dx}{\left(x+\frac{1}{2}\right)^2+\frac{3}{4}}$$

$$= \frac{1}{2}\ln(x^2+x+1) + \frac{1}{2}\left(\frac{2}{\sqrt{3}}\tan^{-1}\frac{2x+1}{\sqrt{3}}\right) + C$$

$$= \frac{1}{2}\ln(x^2+x+1) + \frac{1}{\sqrt{3}}\tan^{-1}\frac{2x+1}{\sqrt{3}} + C,$$

where the second integral was evaluated by the substitution $(x+1/2)^2 = (3/4)u^2$, or $x + 1/2 = \frac{\sqrt{3}}{2}u$.

35 $\displaystyle\int \frac{3x+5}{3x^2+2x+1}\,dx = \frac{1}{2}\int \frac{6x+10}{3x^2+2x+1}\,dx$

$$= \frac{1}{2}\int \frac{(6x+2)+8}{3x^2+2x+1}\,dx$$

$$= \frac{1}{2}\int \frac{(6x+2)dx}{3x+2x+1} + 4\int \frac{dx}{3\left(x^2+\frac{2}{3}x+\frac{1}{9}\right)+\frac{2}{3}}$$

$$= \frac{1}{2}\ln(3x^2+2x+1) + 4\int \frac{dx}{3\left(x+\frac{1}{3}\right)^2+\frac{2}{3}}$$

$$= \frac{1}{2}\ln(3x^2+2x+1) + 4\cdot\frac{1}{\sqrt{2}}\tan^{-1}\frac{3}{\sqrt{2}}\left(x+\frac{1}{3}\right) + C$$

$$= \frac{1}{2}\ln(3x^2+2x+1) + 2\sqrt{2}\tan^{-1}\frac{3x+1}{\sqrt{2}} + C,$$

where the last integral was evaluated by the substitution $3\left(x+\frac{1}{3}\right)^2 = \frac{2}{3}u^2$, or $x + \frac{1}{3} = \frac{\sqrt{2}}{3}u$.

37 We seek $\displaystyle\int_0^1 \frac{x+1}{x^2+x+1}\,dx$. Recall from Exercise 33 that $\displaystyle\int \frac{x+1}{x^2+x+1}\,dx = \frac{1}{2}\ln(x^2+x+1) + \frac{1}{\sqrt{3}}\tan^{-1}\frac{2x+1}{\sqrt{3}} + C$. Thus

$$\int_0^1 \frac{x+1}{x^2+x+1}\,dx$$

$$= \left[\frac{1}{2}\ln(x^2+x+1) + \frac{1}{\sqrt{3}}\tan^{-1}\frac{2x+1}{\sqrt{3}}\right]\Bigg|_0^1 =$$

$$\frac{1}{2}\ln 3 + \frac{1}{\sqrt{3}}\tan^{-1}\sqrt{3} - \left(\frac{1}{2}\ln 1 + \frac{1}{\sqrt{3}}\tan^{-1}\frac{1}{\sqrt{3}}\right)$$

$$= \frac{1}{2}\ln 3 + \frac{1}{\sqrt{3}}\cdot\frac{\pi}{6}.$$

39 (a) Since $x = \dfrac{-b \pm \sqrt{b^2-4ac}}{2a}$ and $b^2 - 4ac < 0$, there can be no real solutions to the quadratic formula and thus no real roots.

(b) If $b^2 - 4ac > 0$, then by the quadratic formula $ax^2 + bx + c = 0$ has two distinct roots, $\dfrac{-b+\sqrt{b^2-4ac}}{2a}$ and $\dfrac{-b-\sqrt{b^2-4ac}}{2a}$.

(c) If $b^2 - 4ac = 0$, then by the quadratic formula $ax^2 + bx + c = 0$ has exactly one root, namely $\dfrac{-b}{2a}$.

41 (a) If $ax^2 + bx + c = 0$ is reducible, then it can be factored. This implies that when

$a\left(x^2 + \frac{b}{a}x + \frac{c}{a}\right) = 0$, there exist real numbers s_1 and s_2 such that $s_1 + s_2 = -b/a$ and $s_1s_2 = c/a$. It follows that $ax^2 + bx + c$ can be written as $a(x - s_1)(x - s_2)$.

(b) If $ax^2 + bx + c = a(x - s_1)(x - s_2) = 0$, then s_1 and s_2 are roots of $ax^2 + bx + c = 0$ by definition.

(c) Squaring both sides of $s_1 + s_2 = -b/a$ gives $s_1^2 + 2s_1s_2 + s_2^2 = \frac{b^2}{a^2}$. Now subtract $4s_1s_2$ from both sides, so $s_1^2 - 2s_1s_2 + s_2^2 = \frac{b^2}{a^2} - 4s_1s_2$. Then $(s_1 - s_2)^2 = \frac{b^2}{a^2} - 4s_1s_2 = \frac{b^2}{a^2} - 4\cdot\frac{c}{a} = \frac{b^2 - 4ac}{a^2}$. Since $(s_1 - s_2)^2 \geq 0$ for all real s_1 and s_2, $\frac{b^2 - 4ac}{a^2} \geq 0$. So $b^2 - 4ac \geq 0$ since $a^2 \geq 0$.

43 From Exercises 40 and 41, we know that a polynomial $ax^2 + bx + c$ is reducible if and only if $b^2 - 4ac \geq 0$.

(a) Here $b^2 - 4ac = 0^2 - 4(1)(-4) = 16 \geq 0$, so $x^2 - 4$ is reducible and $x^2 - 4 = (x + 2)(x - 2)$.

(b) Here $b^2 - 4ac = 0^2 - 4(1)(-3) = 12 \geq 0$, so $x^2 - 3$ is reducible and $x^2 - 3 = \left(x + \sqrt{3}\right)\left(x - \sqrt{3}\right)$.

(c) Since $b^2 - 4ac = 0^2 - 4(1)(3) = -12 < 0$, $x^2 + 3$ is irreducible.

(d) Since $b^2 - 4ac = 3^2 - 4(2)(1) = 1 \geq 0$, $2x^2 + 3x + 1$ is reducible. Hence $2x^2 + 3x + 1 = (2x + 1)(x + 1)$.

(e) Since $b^2 - 4ac = 3^2 - 4(2)(7) = -47 < 0$, so $2x^2 + 3x + 7$ is irreducible.

(f) Since $b^2 - 4ac = 3^2 - 4(2)(-7) = 65 \geq 0$, so $2x^2 + 3x - 7$ is irreducible. By Exercise 41, $2x^2 + 3x - 7 = 2(x - s_1)(x - s_2)$ where s_1 and s_2 are roots of $2x^2 + 3x - 7 = 0$. By the quadratic formula, $s_1 = \frac{-3 + \sqrt{65}}{4}$ and $s_2 = \frac{-3 - \sqrt{65}}{4}$. Thus $2x^2 + 3x - 7 = 2\left(x - \frac{-3 + \sqrt{65}}{4}\right)\left(x - \frac{-3 - \sqrt{65}}{4}\right) = 2\left(x + \frac{3 - \sqrt{65}}{4}\right)\left(x + \frac{3 + \sqrt{65}}{4}\right)$.

(g) Since $b^2 - 4ac = 0^2 - 4(49)(25) < 0$, $49x^2 + 25$ is irreducible.

45 We need to put $\int \frac{x\,dx}{(4x^2 + 8x + 13)^2}$ into a form where some of the formulas in the table of antiderivatives can be used. First note that if $u = 4x^2 + 8x + 13$, then $du = (8x + 8)\,dx = 8(x + 1)\,dx$. We can then write

$$\int \frac{x\,dx}{(4x^2 + 8x + 13)^2} = \int \frac{(x + 1 - 1)\,dx}{(4x^2 + 8x + 13)^2}$$

$$= \int \frac{(x + 1)\,dx}{(4x^2 + 8x + 13)^2} - \int \frac{dx}{(4x^2 + 8x + 13)^2}$$

$$= \frac{1}{8}\int \frac{8(x + 1)\,dx}{(4x^2 + 8x + 13)^2} - \int \frac{dx}{(4x^2 + 8x + 13)^2}$$

$$= \frac{1}{8}\int \frac{du}{u^2} - \int \frac{dx}{(4x^2 + 8x + 13)^2}$$

$$= -\frac{1}{8u} - \int \frac{dx}{(4x^2 + 8x + 13)^2}$$

$$= -\frac{1}{8(4x^2 + 8x + 13)} - \int \frac{dx}{(4x^2 + 8x + 13)^2}.$$

The problem is now reduced to finding

$\int \frac{dx}{(4x^2 + 8x + 13)^2}$, to which Formula 38 applies

with $n = 1$, $a = 4$, $b = 8$, and $c = 13$. Therefore

$$\int \frac{dx}{(4x^2 + 8x + 13)^2} =$$

$$\frac{8x + 8}{1(4 \cdot 4 \cdot 13 - 8^2)(4x^2 + 8x + 13)} +$$

$$\frac{2(2 - 1)4}{1(4 \cdot 4 \cdot 13 - 8^2)} \int \frac{dx}{4x^2 + 8x + 13} =$$

$\frac{x + 1}{18(4x^2 + 8x + 13)} + \frac{1}{18} \int \frac{dx}{4x^2 + 8x + 13}$. We

now apply Formula 36 to the remaining integral to obtain the further result

$$\frac{x + 1}{18(4x^2 + 8x + 13)} + \frac{1}{18} \cdot \frac{2}{12} \tan^{-1} \frac{8x + 8}{12} + C$$

$$= \frac{x + 1}{18(4x^2 + 8x + 13)} + \frac{1}{108} \tan^{-1} \frac{2(x + 1)}{3} + C.$$

Combining this with the earlier partial result, we have the antiderivative $-\frac{1}{8(4x^2 + 8x + 13)} -$

$$\frac{x + 1}{18(4x^2 + 8x + 13)} - \frac{1}{108} \tan^{-1} \frac{2(x + 1)}{3} + C =$$

$$\frac{-(4x + 13)}{72(4x^2 + 8x + 13)} - \frac{1}{108} \tan^{-1} \frac{2(x + 1)}{3} + C.$$

7.5 Integration of Rational Functions by Partial Fractions

1 $\frac{x + 3}{(x + 1)(x + 2)} = \frac{c_1}{x + 1} + \frac{c_2}{x + 2}$

3 $\frac{1}{(x - 1)^2(x + 2)} = \frac{c_1}{x - 1} + \frac{c_2}{(x - 1)^2} + \frac{c_3}{x + 2}$

5 $\frac{6x^2 - 2}{(x - 1)(x - 2)(2x - 3)} =$

$$\frac{c_1}{x - 1} + \frac{c_2}{x - 2} + \frac{c_3}{2x - 3}$$

7 $\frac{x}{(x + 1)(x^2 + x + 1)^2} =$

$$\frac{c_1}{x + 1} + \frac{c_2x + c_3}{x^2 + x + 1} + \frac{c_4x + c_5}{(x^2 + x + 1)^2}$$

9 $\frac{x + 3}{(x^2 - 1)(3x + 5)^3} =$

$$\frac{c_1}{3x + 5} + \frac{c_2}{(3x + 5)^2} + \frac{c_3}{(3x + 5)^3} + \frac{c_4x + c_5}{x^2 - 1}$$

11 Using long division, we obtain

$$\begin{array}{r} 1 \\ x^2 + x + 1 \overline{\smash{)}\, x^2} \\ \underline{x^2 + x + 1} \\ -x - 1 \end{array}$$

Therefore so $\frac{x^2}{x^2 + x + 1} = 1 + \frac{-x - 1}{x^2 + x + 1}$

$$= 1 - \frac{x + 1}{x^2 + x + 1}.$$

13 Here $\frac{x^5 - 2x + 1}{(x + 1)(x^2 + 1)} = \frac{x^5 - 2x + 1}{x^3 + x^2 + x + 1}$, and

using long division yields

$$\begin{array}{r} x^2 - x \\ x^3+x^2+x+1 \overline{\smash{)}\, x^5 - 2x + 1} \\ \underline{x^5 + x^4 + x^3 + x^2} \\ -x^4 - x^3 - x^2 - 2x \\ \underline{-x^4 - x^3 - x^2 - x} \\ -x + 1 \end{array}$$

Thus $\dfrac{x^5 - 2x + 1}{(x + 1)(x^2 + 1)} =$

$x^2 - x + \dfrac{1 - x}{(x + 1)(x^2 + 1)}$.

15 The degree of $A(x) = 4x - 3$ is 1, while the degree of $B(x) = x(x - 1)$ is 2, so long division is not required. Since x and $x - 1$ are irreducible, $B(x)$ is in factored form, so from Step 2 we know $\dfrac{4x - 3}{x(x - 1)} = \dfrac{c_1}{x} + \dfrac{c_2}{x - 1}$. Note that x and $x - 1$ are linear factors to the first power, so employ the procedure given in Example 3. Multiplying both sides by $x - 1$ and setting $x = 1$ gives $\dfrac{4x - 3}{x} =$ $(x - 1)\cdot\dfrac{c_1}{x} + c_2$, and $c_2 = \dfrac{4\cdot1 - 3}{1} = 1$. Now multiply both sides by x and set $x = 0$: $\dfrac{4x - 3}{x - 1}$ $= c_1 + x\cdot\dfrac{c_2}{(x - 1)}$, and $c_1 = \dfrac{4\cdot0 - 3}{0 - 1} = 3$. So $\dfrac{3}{x} + \dfrac{1}{x - 1}$ is the partial-fraction representation of $\dfrac{4x - 3}{x(x - 1)}$.

17 The degree of the numerator is less than the degree of the denominator, so long division is unnecessary. $B(x) = x^2(x - 1) = (x + 0)^2(x - 1)$ is in factored form, so from Step 2 we know $\dfrac{5x^2 - x - 1}{x^2(x - 1)} = \dfrac{c_1}{x} + \dfrac{c_2}{x^2} + \dfrac{c_3}{x - 1}$. Multiplying both sides of this equation by $x^2(x - 1)$ gives $5x^2 - x - 1 = c_1(x(x - 1)) + c_2(x - 1) + c_3x^2$. When $x = 1$, $c_3 = 5\cdot1 - 1 - 1 = 3$. When $x = 0$, $-c_2 = -1$, so $c_2 = 1$. Matching coefficients of x^2 produces $c_3 + c_1 = 5$, so $c_1 = 5 - 3 = 2$. Therefore, $\dfrac{5x^2 - x - 1}{x^2(x - 1)} = \dfrac{2}{x} + \dfrac{1}{x^2} + \dfrac{3}{x - 1}$.

19 The fraction is a proper rational function, and $B(x) = (x + 1)(x + 2)$ is in factored form, so by Step 2 $\dfrac{x}{(x + 1)(x + 2)} = \dfrac{c_1}{x + 1} + \dfrac{c_2}{x + 2}$. Since $x + 1$ and $x + 2$ are linear terms to the first power, we use the technique developed in Example 3. Multiplying both sides by $x + 1$ and setting $x = -1$ produces $c_1 = \left.\dfrac{x}{x + 2}\right|_{x=-1} = \dfrac{-1}{-1 + 2} = -1$, while multiplying the equation by $x + 2$ and letting $x = -2$ divulges $c_2 = \left.\dfrac{x}{x + 1}\right|_{x=-2} = \dfrac{-2}{-2 + 1} =$ 2. So $\dfrac{x}{(x + 1)(x + 2)} = -\dfrac{1}{x + 1} + \dfrac{2}{x + 2}$.

21 The fraction is a proper rational function, but $B(x) = x^2 - 1$ is reducible since $0^2 - 4\cdot1\cdot(-1) = 4 > 0$. Noting that $x^2 - 1 = (x + 1)(x - 1)$ and Step 2 applies, we have $\dfrac{2x}{x^2 - 1} = \dfrac{2x}{(x + 1)(x - 1)} =$ $\dfrac{c_1}{x + 1} + \dfrac{c_2}{x - 1}$. Now multiply the equation by

$x + 1$ and set $x = -1$, so $c_1 = \left.\dfrac{2x}{x-1}\right|_{x=-1} = \dfrac{2(-1)}{-1-1} = 1$. Finally, multiplying by $x - 1$ and setting $x = 1$ gives $c_2 = \left.\dfrac{2x}{x+1}\right|_{x=1} = \dfrac{2\cdot 1}{1+1} = 1$. Therefore, $\dfrac{2x}{x^2-1} = \dfrac{1}{x+1} + \dfrac{1}{x-1}$.

23 The fraction is a proper rational function, and $B(x) = x(x + 1)(x + 2)$ is in factored form, so by Step 2, $\dfrac{2x^2+3}{x(x+1)(x+2)} = \dfrac{c_1}{x} + \dfrac{c_2}{x+1} + \dfrac{c_3}{x+2}$.

Since x, $x + 1$, and $x + 2$ are linear terms to the first power, we follow the technique of Example 3. Multiplying the equation by x and letting $x = 0$ produces $c_1 = \left.\dfrac{2x^2+3}{(x+1)(x+2)}\right|_{x=0} = \dfrac{2\cdot 0^2+3}{(0+1)(0+2)} = \dfrac{3}{2}$. Multiplying the equation by $x + 1$ and setting $x = -1$ gives $c_2 = \left.\dfrac{2x^2+3}{x(x+2)}\right|_{x=-1} = \dfrac{2(-1)^2+3}{(-1)(-1+2)} = -5$. Finally, multiplying the equation by $x + 2$ and fixing $x = -2$ yields $c_3 = \left.\dfrac{2x^2+3}{x(x+1)}\right|_{x=-2} = \dfrac{2(-2)^2+3}{(-2)(-2+1)} = \dfrac{11}{2}$. Therefore, $\dfrac{2x^2+3}{x(x+1)(x+2)} = \dfrac{3/2}{x} - \dfrac{5}{x+1} + \dfrac{11/2}{x+2}$.

25 The fraction is a proper rational function, and expressing the denominator in factored form gives $B(x) = (x + 1)^2(x + 2)$. From Step 2 we have

$$\frac{6x^2-7x-1}{(x+1)^2(x+2)} = \frac{c_1}{x+1} + \frac{c_2}{(x+1)^2} + \frac{c_3}{x+2}.$$

Multiplying both sides of this equation by $(x + 1)^2(x + 2)$ yields $6x^2 - 7x - 1 = c_1(x + 1)(x + 2) + c_2(x + 2) + c_3(x + 1)^2$. When $x = -1$, $c_2 = 6(-1)^2 - 7(-1) - 1 = 12$. When $x = -2$, $c_3 = 6(-2)^2 - 7(-2) - 1 = 37$. Matching coefficients of x^2 reveals $c_1 + c_3 = 6$, so $c_1 = 6 - c_3 = 6 - 37 = -31$. Therefore,

$$\frac{6x^2-7x-1}{(x+1)^2(x+2)} = \frac{-31}{x+1} + \frac{12}{(x+1)^2} + \frac{37}{x+2}.$$

27 $\dfrac{5x^2+9x+6}{(x+1)(x^2+2x+2)}$ is a proper rational function, and $B(x) = (x + 1)(x^2 + 2x + 2)$ is in factored form because $x^2 + 2x + 2$ is irreducible ($b^2 - 4ac = 2^2 - 4\cdot 1\cdot 2 < 0$). From Steps 2 and 3, $\dfrac{5x^2+9x+6}{(x+1)(x^2+2x+2)} = \dfrac{c_1}{x+1} + \dfrac{c_2x+c_3}{x^2+2x+2}$, and $5x^2 + 9x + 6 = c_1(x^2 + 2x + 2) + (c_2x + c_3)(x + 1)$. Letting $x = -1$ in this equation gives $2 = c_1(1)$, or $c_1 = 2$. Now letting $x = 0$ gives $6 = 2c_1 + c_3$, so $c_3 = 6 - 2c_1 = 2$. Finally, letting $x = 1$ yields $20 = 5c_1 + 2c_2 + 2c_3$, so $c_2 = \dfrac{1}{2}(20 - 5c_1 - 2c_3) = 3$. Hence $\dfrac{5x^2+9x+6}{(x+1)(x^2+2x+2)} = \dfrac{2}{x+1} + \dfrac{3x+2}{x^2+2x+2}$.

29 Since the degree of the numerator is greater than the degree of the denominator, we use long division to get a polynomial and a proper rational function.

$$\begin{array}{r|l} & x - 1 \\ x^2 + x & x^3 \qquad + 5x + 1 \\ & \underline{x^3 + x^2} \\ & \quad -x^2 + 5x \\ & \quad \underline{-x^2 - x} \\ & \qquad\quad 6x + 1 \end{array}$$

So $\dfrac{x^3 + 5x + 1}{x(x + 1)} = x - 1 + \dfrac{6x + 1}{x(x + 1)}$. Now we carry out the procedure in Example 3 on $\dfrac{6x + 1}{x(x + 1)} = \dfrac{c_1}{x} + \dfrac{c_2}{x + 1}$. Multiplying the equation by x and setting $x = 0$ gives $c_1 = \left.\dfrac{6x + 1}{x + 1}\right|_{x=0} = 1$, while multiplying by $x + 1$ and setting $x = -1$ yields $c_2 = \left.\dfrac{6x + 1}{x}\right|_{x=-1} = 5$. Thus, $\dfrac{6x + 1}{x(x + 1)} = \dfrac{1}{x} + \dfrac{5}{x + 1}$ and $\dfrac{x^3 + 5x + 1}{x(x + 1)} = x - 1 + \dfrac{1}{x} + \dfrac{5}{x + 1}$.

31 The fraction is improper, so use long division:

$$\begin{array}{r|l} & 3 \\ x^3 + x & 3x^3 + 2x^2 + 3x + 1 \\ & \underline{3x^3 + \qquad\quad 3x} \\ & \qquad 2x^2 \qquad + 1 \end{array}$$

Thus, $\dfrac{3x^3 + 2x^2 + 3x + 1}{x(x^2 + 1)} = 3 + \dfrac{2x^2 + 1}{x(x^2 + 1)}$. Now find the partial-fraction expression for $\dfrac{2x^2 + 1}{x(x^2 + 1)}$.

First, $x^2 + 1$ is irreducible since $0^2 - 4\cdot 1\cdot 1 < 0$, so $B(x) = x(x^2 + 1)$ is in factored form, and Steps 2 and 3 state that $\dfrac{2x^2 + 1}{x(x^2 + 1)} = \dfrac{c_1}{x} + \dfrac{c_2x + c_3}{x^2 + 1}$.

Multiplying the equation by $x(x^2 + 1)$, we have $2x^2 + 1 = c_1(x^2 + 1) + (c_2x + c_3)x$. Now let $x = 0$ so $c_1 = 1$. Letting $x = 1$ gives $3 = 2c_1 + c_2 + c_3$ while letting $x = -1$ yields $3 = 2c_1 + c_2 - c_3$. Simplifying reveals two equations in two unknowns, $c_2 + c_3 = 1$ and $c_2 - c_3 = 1$. Since $c_2 + c_3 = 1 = c_2 - c_3$, we conclude that $c_3 = 0$ and $c_2 = 1$. So $\dfrac{2x^2 + 1}{x(x^2 + 1)} = \dfrac{1}{x} + \dfrac{x}{x^2 + 1}$ and

$$\frac{3x^3 + 2x^2 + 3x + 1}{x(x^2 + 1)} = 3 + \frac{1}{x} + \frac{x}{x^2 + 1}.$$

33 The fraction is proper, but $x^2 - x - 2$ is reducible $((-1)^2 - 4\cdot 1(-2) > 0)$, so factoring $x^2 - x - 2$ we have $x^2 - x - 2 = (x - 2)(x + 1)$. Now $\dfrac{x - 1}{(x - 2)(x + 1)} = \dfrac{c_1}{x - 2} + \dfrac{c_2}{x + 1}$ from Step 2, and $x + 1$ and $x - 2$ are linear terms to the first power, so the Example 3 procedure applies. Multiplying the equation above by $x - 2$ and setting $x = 2$ gives $c_1 = \left.\dfrac{x - 1}{x + 1}\right|_{x=2} = \dfrac{1}{3}$, and multiplying by $x + 1$ and letting $x = -1$ yields $c_2 = \left.\dfrac{x - 1}{x - 2}\right|_{x=-1} = \dfrac{2}{3}$. Therefore, $\dfrac{x - 1}{x^2 - x - 2} = \dfrac{1/3}{x - 2} + \dfrac{2/3}{x + 1}$.

35 The fraction is proper, but $x^2 + 3x - 4$ is reducible ($3^2 - 4(1)(-4) > 0$), and $x^2 + 3x - 4 = (x + 4)(x - 1)$. Now $\frac{2}{(x+4)(x-1)} = \frac{c_1}{x+4} + \frac{c_2}{x-1}$ from Step 2, and again the Example 3 procedure is valid since $x + 4$ and $x - 1$ are linear terms to the first power. So multiplying the equation above by $x + 4$ and letting $x = -4$ yields $c_1 = \left.\frac{2}{x-1}\right|_{x=-4} = -\frac{2}{5}$, and multiplying by $x - 1$ and letting $x = 1$ produces $c_2 = \left.\frac{2}{x+4}\right|_{x=1} = \frac{2}{5}$. Finally, $\frac{2}{x^2 + 3x - 4} = -\frac{2/5}{x+4} + \frac{2/5}{x-1}$.

37 We are given $3c_1 - 2c_2 = 6$, $c_1 + c_2 = 4$. Multiplying the bottom equation by 3 gives $3c_1 + 3c_2 = 12$, and subtracting this result from the top equation, we have $3c_1 - 2c_2 - 3c_1 - 3c_2 = 3 - 12$, and $-5c_2 = -9$ so $c_2 = 9/5$. Plugging this value into the bottom equation leaves $c_1 + 9/5 = 4$, so $c_1 = 11/5$. As a check, we see that $3\cdot\frac{11}{5} - 2\cdot\frac{9}{5} = \frac{15}{5} = 3$ and $\frac{9}{5} + \frac{11}{5} = \frac{20}{5} = 4$.

39 We are given $c_1 + 5c_2 = 6$, $2c_1 - 3c_2 = -2$. Solving for c_1 in the top equation, we have $c_1 = 6 - 5c_2$. Plugging this value into the bottom equation produces $2(6 - 5c_2) - 3c_2 = -2$, so $-13c_2 = -14$ and $c_2 = 14/13$. Thus $c_1 = 6 - 5\cdot\frac{14}{13} = \frac{8}{13}$. As a check, we see that $\frac{8}{13} + 5\cdot\frac{14}{13} = \frac{78}{13} = 6$ and $2\cdot\frac{8}{13} - 3\cdot\frac{14}{13} = -\frac{26}{13} = -2$.

41 We are given

$$c_1 + 2c_2 + c_3 = 9,$$
$$c_1 - c_2 = -1,$$
$$c_1 + c_3 = 3.$$

From the middle and bottom equations, we see that c_2 and c_3 can be easily solved in terms of c_1. Thus $c_2 = 1 + c_1$ and $c_3 = 3 - c_1$. Plugging these into the top equation gives $c_1 + 2(1 + c_1) + 3 - c_1 = 9$, so $2c_1 = 4$ and $c_1 = 2$. It follows that $c_2 = 1 + 2 = 3$ and $c_3 = 3 - 2 = 1$. As a check we see that $c_1 + 2c_2 + c_3 = 2 + 2\cdot3 + 1 = 9$, $c_1 - c_2 = 2 - 3 = -1$, and $c_1 + c_3 = 2 + 1 = 3$.

43 We are given

$$c_1 + c_3 = 4,$$
$$c_2 - c_3 = -6,$$
$$c_1 + c_2 + c_3 = -1.$$

From the top and middle equations, we see that c_1 and c_2 can be easily solved for in terms of c_3. Thus $c_1 = 4 - c_3$ and $c_2 = -6 + c_3$. Plugging these values into the bottom equation gives $(4 - c_3) + (-6 + c_3) + c_3 = -1$, so $c_3 = 1$, and it follows that $c_1 = 4 - 1 = 3$ and $c_2 = -6 + 1 = -5$. As a check we see that $c_1 + c_3 = 3 + 1 = 4$, $c_2 - c_3 = -5 - 1 = -6$, and $c_1 + c_2 + c_3 = 3 - 5 + 1 = -1$.

45 We wish to find $\int_1^2 \frac{dx}{x^3 + x} = \int_1^2 \frac{dx}{x(x^2 + 1)}$.

Note that the integrand is a proper rational function and can therefore be expressed in the form

$\frac{c_1}{x} + \frac{c_2x + c_3}{x^2 + 1}$. It follows that $\frac{1}{x(x^2 + 1)} =$

$\frac{c_1}{x} + \frac{c_2x + c_3}{x^2 + 1}$ and $1 = (x^2 + 1)c_1 + x(c_2x + c_3)$.

Letting $x = 0$ gives $c_1 = 1$. Letting $x = 1$ gives $1 = 2c_1 + c_2 + c_3$, and $c_2 + c_3 = -1$. Letting $x = -1$ gives $1 = 2c_1 + c_2 - c_3$, and $c_2 - c_3 = -1$. Since $c_2 + c_3 = -1 = c_2 - c_3$, we know that $c_3 = 0$ and $c_2 = -1$. So $\frac{1}{x(x^2 + 1)} =$

$\frac{1}{x} - \frac{x}{x^2 + 1}$. Now $\int_1^2 \frac{dx}{x(x^2 + 1)} =$

$$\int_1^2 \frac{dx}{x} - \int_1^2 \frac{x\,dx}{x^2 + 1} = \ln|x|\Big|_1^2 - \frac{1}{2}\int_1^2 \frac{2x\,dx}{x^2 + 1}$$

$$= \ln 2 - \frac{1}{2}\ln(x^2 + 1)\Big|_1^2 =$$

$$\ln 2 - \frac{1}{2}\ln 5 + \frac{1}{2}\ln 2 = \frac{3}{2}\ln 2 - \frac{1}{2}\ln 5.$$

47 When the region is revolved around the x axis, the cross-sections formed by planes perpendicular to the x axis are circles of radius $\frac{a + 2}{a^2 + a}$ at the x coordinate a. So the volume is $\int_1^2 \pi\left(\frac{x + 2}{x^2 + x}\right)^2 dx$

$= \pi \int_1^2 \frac{x^2 + 4x + 4}{x^2(x + 1)^2}\,dx$. To evaluate this integral, we must first find the partial-fraction representation of $\frac{x^2 + 4x + 4}{x^2(x + 1)^2} =$

$\frac{c_1}{x} + \frac{c_2}{x^2} + \frac{c_3}{x + 1} + \frac{c_4}{(x + 1)^2}$. Eliminating all denominators, we have $x^2 + 4x + 4 = x(x + 1)^2c_1 + (x + 1)^2c_2 + x^2(x + 1)c_3 + x^2c_4$. Letting $x = 0$ gives $c_2 = 4$. Letting $x = -1$ gives $c_4 = 1$. Now let $x = 1$, so that we have $9 = 4c_1 + 4c_2 + 2c_3 + c_4 = 4c_1 + 16 + 2c_3 + 1$, and $4c_1 + 2c_3 = -8$. Finally, let $x = -2$ so $0 = -2c_1 + c_2 - 4c_3 + 4c_4 = -2c_1 + 4 - 4c_3 + 4$, and $-2c_1 - 4c_3 = -8$. Solving this system of two equations in the two unknowns c_1 and c_3 yields $c_1 = -4$ and $c_3 = 4$. So $\frac{x^2 + 4x + 4}{x^2(x + 1)^2} =$

$-\frac{4}{x} + \frac{4}{x^2} + \frac{4}{x + 1} + \frac{1}{(x + 1)^2}$. Then

$$\pi \int_1^2 \frac{x^2 + 4x + 4}{x^2(x + 1)^2}\,dx$$

$$= \pi \int_1^2 \left(-\frac{4}{x} + \frac{4}{x^2} + \frac{4}{x + 1} + \frac{1}{(x + 1)^2}\right) dx$$

$$= \pi\left[-4\ln|x| - \frac{4}{x} + 4\ln|x + 1| - \frac{1}{x + 1}\right]\Bigg|_1^2$$

$$= \pi\left[\left(-4\ln 2 - \frac{4}{2} + 4\ln 3 - \frac{1}{3}\right) - \left(-4\ln 1 - \frac{4}{1} + 4\ln 2 - \frac{1}{2}\right)\right]$$

$$= \pi\left(-8\ln 2 + 4\ln 3 + \frac{13}{6}\right).$$

49 Before evaluating the integral, we need to find the partial-fraction representation of $\frac{1}{x^2 + 3x + 2} =$

$\frac{1}{(x + 2)(x + 1)} = \frac{c_1}{x + 2} + \frac{c_2}{x + 1}$. Multiplying the equation by $x + 2$ and letting $x = -2$ gives c_1

$= \left.\dfrac{1}{x+1}\right|_{x=-2} = -1$, while multiplying by $x + 1$ and letting $x = -1$ gives $c_2 = \left.\dfrac{1}{x+2}\right|_{x=-1} = 1$.

So $\dfrac{1}{x^2 + 3x + 2} = -\dfrac{1}{x+2} + \dfrac{1}{x+1}$ and

$$\int_0^1 \frac{dx}{x^2 + 3x + 2} = \int_0^1 \left(-\frac{1}{x+2} + \frac{1}{x+1}\right) dx$$

$= (-\ln|x+2| + \ln|x+1|)|_0^1$

$= -\ln 3 + \ln 2 - (-\ln 2 + \ln 1)$

$= 2\ln 2 - \ln 3 = \ln 4 - \ln 3 = \ln 4/3$.

51 The partial-fraction representation of $\dfrac{x^3}{x^3+1}$ needs to be found. Note that $x^3 + 1 = (x+1)(x^2 - x + 1)$, where $x^2 - x + 1$ is irreducible ($b^2 - 4ac < 0$), so $\dfrac{x^3}{x^3+1} = \dfrac{x^3+1-1}{x^3+1} = 1 - \dfrac{1}{x^3+1} = 1 - \dfrac{1}{(x+1)(x^2-x+1)}$. Now

$$\frac{1}{(x+1)(x^2-x+1)} = \frac{c_1}{x+1} + \frac{c_2x + c_3}{x^2-x+1}$$ for

appropriate c_1, c_2, and c_3. To find these, eliminate the denominators in the above equation. Thus $1 = (x^2 - x + 1)c_1 + (x+1)(c_2x + c_3)$. Now let $x = -1$ so $1 = 3c_1$ and $c_1 = 1/3$. Letting $x = 0$ gives $1 = c_1 + c_3$ so $c_3 = 2/3$. Letting $x = 1$ gives $1 = c_1 + 2c_2 + 2c_3$ so $c_2 = -1/3$. Then

$$\frac{1}{(x+1)(x^2-x+1)} = \frac{1/3}{x+1} + \frac{(-1/3)x + 2/3}{x^2-x+1}$$

and $\dfrac{x^3}{x^3+1} = 1 - \left(\dfrac{1/3}{x+1} + \dfrac{(-1/3)x + 2/3}{x^2-x+1}\right)$, so

$$\int \frac{x^3\,dx}{x^3+1} = \int\left(1 - \frac{1/3}{x+1} + \frac{(1/3)x - 2/3}{x^2-x+1}\right) dx =$$

$$\int dx - \frac{1}{3}\int\frac{dx}{x+1} + \int\frac{(1/3)x - 2/3}{x^2-x+1}\,dx =$$

$$\int dx - \frac{1}{3}\int\frac{dx}{x+1} + \int\frac{(1/3)x - 1/6}{x^2-x+1}\,dx +$$

$$\int\frac{-1/2}{x^2-x+1}\,dx = \int dx - \frac{1}{3}\int\frac{dx}{x+1} +$$

$$\frac{1}{6}\int\frac{2x-1}{x^2-x+1}\,dx - \frac{1}{2}\int\frac{dx}{x^2-x+1} =$$

$$x - \frac{1}{3}\ln|x+1| + \frac{1}{6}\ln|x^2-x+1| -$$

$\dfrac{1}{\sqrt{3}}\tan^{-1}\left(\dfrac{2x-1}{\sqrt{3}}\right) + C$, where the substitution $\dfrac{3}{4}u^2 = \left(x - \dfrac{1}{2}\right)^2$ was used for the last integral evaluation. Now

$$\int_1^2 \frac{x^3\,dx}{x^3+1} = \left[x - \frac{1}{3}\ln|x+1| + \frac{1}{6}\ln|x^2-x+1| - \frac{1}{\sqrt{3}}\tan^{-1}\left(\frac{2x-1}{\sqrt{3}}\right)\right]\Bigg|_1^2$$

$$= \left(2 - \frac{1}{3}\ln 3 + \frac{1}{6}\ln 3 - \frac{1}{\sqrt{3}}\tan^{-1}\sqrt{3}\right) -$$

$$\left(1 - \frac{1}{3}\ln 2 + \frac{1}{6}\ln 1 - \frac{1}{\sqrt{3}}\tan^{-1}\frac{1}{\sqrt{3}}\right)$$

$$= 1 + \frac{1}{3}\ln 2 - \frac{1}{6}\ln 3 - \frac{\pi}{3\sqrt{3}} + \frac{\pi}{6\sqrt{3}}$$

$$= \frac{1}{6}\left(6 + \ln\frac{4}{3} - \frac{\pi}{\sqrt{3}}\right).$$

53 (a) $x^2 + 6x + 5 = (x+5)(x+1)$

(b) $x^2 - 5 = (x + \sqrt{5})(x - \sqrt{5})$

(c) By the quadratic formula, the roots of

$2x^2 + 6x + 3$ are $\dfrac{-6 \pm \sqrt{(-6)^2 - 4(2)(3)}}{2\cdot 2} =$

$\dfrac{-3 \pm \sqrt{3}}{2}$. Let $r_1 = \dfrac{-3 + \sqrt{3}}{2}$ and $r_2 =$

$\dfrac{-3 - \sqrt{3}}{2}$. Then $2x^2 + 6x + 3 =$

$$2\left(x - \frac{-3 + \sqrt{3}}{2}\right)\left(x - \frac{-3 - \sqrt{3}}{2}\right)$$

$$= 2\left(x + \frac{3 - \sqrt{3}}{2}\right)\left(x + \frac{3 + \sqrt{3}}{2}\right).$$

55 (a) Here $a_n = a_2 = 1$ and $a_0 = -12$. If p/q is a root of $x^2 + x - 12$, then q must divide 1 and p must divide -12. So $q = 1$ ($q > 0$) and $p = \pm 1, \pm 2, \pm 3, \pm 4, \pm 6$, or ± 12. There are twelve combinations of p and q to check. A quick check shows that $p = -4$, $q = 1$ and $p = 3$, $q = 1$ produce the only rational roots. In fact, the roots of $x^2 + x - 12$ occur when $x^2 + x - 12 = (x + 4)(x - 3) = 0$, or when $x = -4$ or 3. Thus these rational roots are the only roots of $x^2 + x - 12$.

(b) Here $a_n = a_3 = 2$ and $a_0 = -6$. So q must divide 2 and p must divide -6. So $q = 1$ or $q = 2$ ($q > 0$) and $p = \pm 1, \pm 2, \pm 3$, or ± 6. When $q = 1$, there are eight p's to choose from. When $q = 2$, there are four p's to choose from, ± 1 and ± 3, since p/q is irreducible. The following table lists the results of trying all combinations with $p > 0$. So 2, 3, and 1/2 are all the rational roots of

q	p	p/q	$2(p/q)^3 - 11(p/q)^2 + 17(p/q) - 6$
1	+1	1	2
1	+2	2	0
1	+3	3	0
1	+6	6	132
2	+1	1/2	0
2	+3	3/2	3/2

$2x^3 - 11x^2 + 17x - 6$, since a cubic polynomial has at most three real roots.

(c) Here $a_n = a_4 = 1$ and $a_0 = 1$, so $q = 1$ and $p = \pm 1$. Neither $q = 1, p = 1$, or $q = 1, p = -1$ are roots of $x^4 + x^3 + x^2 + x + 1$, so no rational roots exist.

(d) Here $a_n = a_3 = 3$ and $a_0 = -1$, so $q = 1$ or 3 and $p = \pm 1$. Now $q = 1, p = 1$ does not yield a root because $3\cdot 1^3 - 2\cdot 1^2 - 4\cdot 1 - 1 = -4$. Also, $q = 1, p = -1$ does not yield a root because $3(-1)^3 - 2(-1)^2 - 4(-1) - 1 = -2$. Finally, $q = 3, p = 1$ yields a value of 50 and $q = 3, p = -1$ yields -88. So no rational roots exist.

57 If $x^4 + 1 = (x^2 + ax + 1)(x^2 + bx + 1)$, then multiplying out the right-hand side and comparing coefficients should reveal a and b. Thus $x^4 + 1 = x^4 + bx^3 + x^2 + ax^3 + abx^2 + ax + x^2 + bx + 1 = x^4 + (a + b)x^3 + (2 + ab)x^2 + (a + b)x + 1$. So $a + b = 0$ and $ab + 2 = 0$. Hence $a = -b$ and $ab = -b^2$, so $b = \pm\sqrt{2}$, and $a = \mp\sqrt{2}$. So

$$x^4 + 1 = (x^2 + \sqrt{2}x + 1)(x^2 - \sqrt{2}x + 1).$$

59 (a) From Exercise 57, $x^4 + x^2 + 1$

$= (x^2 + ax + 1)(x^2 + bx + 1) =$

$x^4 + (a + b)x^3 + (2 + ab)x^2 + (a + b)x + 1.$

So $a + b = 0$ and $2 + ab = 1$. Then $a = -b$ and $ab = -b^2 = -1$ so $b = \pm 1$ and a

$= \mp 1$. Therefore we have $x^4 + x^2 + 1 = (x^2 + x + 1)(x^2 - x + 1)$.

(b) The denominator $x^4 + x^2 + 1$ factors as $(x^2 + x + 1)(x^2 - x + 1)$, where both factors are irreducible. Thus, $\frac{1}{x^4 + x^2 + 1} =$

$\frac{c_1x + c_2}{x^2 + x + 1} + \frac{c_3x + c_4}{x^2 - x + 1}$, so $1 =$

$(c_1x + c_2)(x^2 - x + 1) + (c_3x + c_4)(x^2 + x + 1)$

$= c_1x^3 + (c_2 - c_1)x^2 + (c_1 - c_2)x + c_2 + c_3x^3 + (c_3 + c_4)x^2 + (c_3 + c_4)x + c_4$.

Equating coefficients of x^3 gives $0 = c_1 + c_3$, so $c_3 = -c_1$. Equating coefficients of x gives $0 = c_1 - c_2 + c_3 + c_4 = c_4 - c_2$, so $c_4 = c_2$. Equating coefficients of x^0 (the constants) gives $1 = c_2 + c_4$, so $c_2 = c_4 = 1/2$. Equating the coefficients of x^2 gives $0 = c_2 - c_1 + c_3 + c_4$, so $1 = c_1 - c_3$, $c_1 = 1/2$ and $c_3 = -1/2$. Thus $\frac{1}{x^4 + x^2 + 1} =$

$\frac{(1/2)x + 1/2}{x^2 + x + 1} + \frac{-(1/2)x + 1/2}{x^2 - x + 1}$. The integrands of these two quotients will be treated separately: $\int \frac{(1/2)x + 1/2}{x^2 + x + 1}\,dx$

$$= \frac{1}{4}\left[\int \frac{2x + 1}{x^2 + x + 1}\,dx + \int \frac{dx}{x^2 + x + 1}\right]$$

$$= \frac{1}{4}\left[\int \frac{2x + 1}{x^2 + x + 1}\,dx + \int \frac{dx}{x^2 + x + 1}\right] =$$

$$\frac{1}{4}\ln(x^2 + x + 1) + \frac{1}{4}\int \frac{dx}{(x + 1/2)^2 + 3/4}.$$

Now if we let $\frac{3}{4}u^2 = \left(x + \frac{1}{2}\right)^2$ and $\frac{\sqrt{3}}{2}u = x + \frac{1}{2}$, $\int \frac{dx}{\left(x + \frac{1}{2}\right)^2 + \frac{3}{4}} = \int \frac{\sqrt{3}/2\,du}{\frac{3}{4}(u^2 + 1)}$

$$= \frac{2}{\sqrt{3}}\tan^{-1} u + C_1 =$$

$\frac{2}{\sqrt{3}}\tan^{-1}\frac{2x + 1}{\sqrt{3}} + C_1$. Hence

$$\int \frac{(1/2)x + 1/2}{x^2 + x + 1}\,dx =$$

$$\frac{1}{4}\ln(x^2 + x + 1) + \frac{1}{2\sqrt{3}}\tan^{-1}\frac{2x + 1}{\sqrt{3}} + C_1.$$

Similarly, $\int \frac{-(1/2)x + 1/2}{x^2 - x + 1}\,dx =$

$$\frac{-1}{4}\ln(x^2 - x + 1) + \frac{1}{2\sqrt{3}}\tan^{-1}\frac{2x - 1}{\sqrt{3}} + C_2.$$

Putting these results together, we obtain

$$\int \frac{dx}{x^4 + x^2 + 1} = \frac{1}{4}\ln\frac{x^2 + x + 1}{x^2 - x + 1} +$$

$$\frac{1}{2\sqrt{3}}\tan^{-1}\frac{2x + 1}{\sqrt{3}} + \frac{1}{2\sqrt{3}}\tan^{-1}\frac{2x - 1}{\sqrt{3}} + C.$$

7.6 Special Techniques

1 Using the identity $\sin A \sin B = \frac{1}{2}\cos(A - B) - \frac{1}{2}\cos(A + B)$, we have

$$\int \sin 5x \sin 3x\, dx = \int \left(\frac{1}{2}\cos 2x - \frac{1}{2}\cos 8x\right) dx$$

$$= \frac{1}{4}\sin 2x - \frac{1}{16}\sin 8x + C.$$

3 Using the identity $\sin A \cos B = \frac{1}{2}\sin(A + B) + \frac{1}{2}\sin(A - B)$, we have

$$\int \cos 3x \sin 2x\, dx$$

$$= \int \left(\frac{1}{2}\sin 5x + \frac{1}{2}\sin(-x)\right) dx$$

$$= -\frac{1}{10}\cos 5x + \frac{1}{2}\cos x + C.$$

5 Using the identity $\sin^2 x = \frac{1}{2}(1 - \cos 2x)$, we have $\int \sin^2 3x\, dx = \int \left(\frac{1}{2} - \frac{1}{2}\cos 6x\right) dx$

$$= \frac{1}{2}x - \frac{1}{12}\sin 6x + C.$$

7 $\int (3\sin 2x + 4\sin^2 5x)\, dx$

$$= 3\int \sin 2x\, dx + 4\int \sin^2 5x\, dx$$

$$= -\frac{3}{2}\cos 2x + 4\int \left(\frac{1}{2} - \frac{1}{2}\cos 10x\right) dx$$

$= -\frac{3}{2}\cos 2x + 2x - \frac{1}{5}\sin 10x + C$, where the identity $\sin^2 x = \frac{1}{2}(1 - \cos 2x)$ was used.

9 $\int (3\sin^2 \pi x + 4\cos^2 \pi x)\, dx$

$$= \int (3\sin^2 \pi x + 3\cos^2 \pi x + \cos^2 \pi x)\, dx$$

$$= \int (3 + \cos^2 \pi x)\, dx$$

$$= \int \left(3 + \frac{1}{2}(1 + \cos 2\pi x)\right) dx$$

$$= \int \left(\frac{7}{2} + \frac{1}{2}\cos 2\pi x\right) dx$$

$$= \frac{7}{2}x + \frac{1}{2}\frac{\sin 2\pi x}{2\pi} + C$$

$= \frac{7}{2}x + \frac{1}{4\pi}\sin 2\pi x + C$. Note that the identity $\cos^2 x = \frac{1}{2}(1 + \cos 2x)$ was used.

11 Let $u = 2\theta$, so $du = 2\, d\theta$. Then $\int \tan 2\theta\, d\theta = \frac{1}{2}\int \tan u\, du = \frac{1}{2}(-\ln|\cos u|) + C = -\frac{1}{2}\ln|\cos 2\theta| + C$, where the result from Example 4 was used.

13 Let $u = 5x$, so $du = 5\, dx$. Then $\int \tan^2 5x\, dx$

$$= \frac{1}{5}\int \tan^2 u\, du = \frac{1}{5}\int (\sec^2 u - 1)\, du$$

$$= \frac{1}{5}(\tan u - u) + C = \frac{1}{5}(\tan 5x - 5x) + C.$$

15 Recall that $\cos(A \pm B) = \cos A \cos B \mp \sin A \sin B$. So $\frac{1}{2}\cos(A - B) = \frac{1}{2}(\cos A \cos B + \sin A \sin B)$ and $\frac{1}{2}\cos(A + B)$

$= \frac{1}{2}(\cos A \cos B - \sin A \sin B)$. Thus

$\frac{1}{2}\cos(A - B) - \frac{1}{2}\cos(A + B) = \frac{1}{2}\cos A \cos B$

$+ \frac{1}{2}\sin A \sin B -$

$\left(\frac{1}{2}\cos A \cos B - \frac{1}{2}\sin A \sin B\right) = \sin A \sin B.$

17 From the remark immediately following Example 3, we know $f(\alpha) = \int_0^\alpha \sec\theta\, d\theta =$

$\frac{1}{2}\ln\frac{1 + \sin\alpha}{1 - \sin\alpha}.$

When $\alpha = 20° = \pi/9$, $f(\pi/9) = \frac{1}{2}\ln\frac{1 + \sin\pi/9}{1 - \sin\pi/9}$

$\approx 0.3563785.$

When $\alpha = 40° = 2\pi/9$, $f(2\pi/9) =$

$\frac{1}{2}\ln\frac{1 + \sin 2\pi/9}{1 - \sin 2\pi/9} \approx 0.7629096.$

When $\alpha = 60° = \pi/3$, $f(\pi/3) = \frac{1}{2}\ln\frac{1 + \sin\pi/3}{1 - \sin\pi/3}$

$\approx 1.3169579.$

When $\alpha = 80° = 4\pi/9$, $f(4\pi/9) =$

$\frac{1}{2}\ln\frac{1 + \sin 4\pi/9}{1 - \sin 4\pi/9} \approx 2.4362461.$

19 Let $u = \sqrt{2x + 1}$, so $x = \frac{1}{2}(u^2 - 1)$ and $dx =$

$u\,du$. So $\int x^2\sqrt{2x + 1}\,dx$

$= \int \left[\frac{1}{2}(u^2 - 1)\right]^2 u(u\,du)$

$= \frac{1}{4}\int (u^4 - 2u^2 + 1)\,u^2\,du$

$= \frac{1}{4}\int (u^6 - 2u^4 + u^2)\,du$

$= \frac{1}{4}\left(\frac{u^7}{7} - \frac{2u^5}{5} + \frac{u^3}{3}\right) + C =$

$\frac{1}{4}\left(\frac{(\sqrt{2x + 1})^7}{7} - \frac{2(\sqrt{2x + 1})^5}{5} + \frac{(\sqrt{2x + 1})^3}{3}\right) + C$

$= \frac{\sqrt{(2x + 1)^3}}{4}\left(\frac{(2x + 1)^2}{7} - \frac{2(2x + 1)}{5} + \frac{1}{3}\right) + C$

$= \frac{\sqrt{(2x + 1)^3}}{4}\left(\frac{60x^2 - 24x + 8}{105}\right) + C$

$= \frac{(2x + 1)^{3/2}}{105}(15x^2 - 6x + 2) + C.$

21 Let $u = \sqrt{x}$, so that $x = u^2$ and $dx = 2u\,du$. Then

$\int \frac{dx}{\sqrt{x} + 3} = \int \frac{2u\,du}{u + 3} = 2\int \frac{(u + 3) - 3}{u + 3}\,du$

$= 2\int \left(1 - \frac{3}{u + 3}\right) du$

$= 2(u - 3\ln|u + 3|) + C$

$= 2(\sqrt{x} - 3\ln(\sqrt{x} + 3)) + C.$

23 Let $u = \sqrt[3]{3x + 2}$. Then $u^3 = 3x + 2$, so that

$x = \frac{u^3 - 2}{3}$ and $dx = u^2\,du$. Thus

$\int x\sqrt[3]{3x + 2}\,dx = \int \left(\frac{u^3 - 2}{3}\right) u(u^2\,du)$

$= \frac{1}{3}\int (u^6 - 2u^3)\,du = \frac{1}{3}\left(\frac{1}{7}u^7 - \frac{1}{2}u^4\right) + C$

$= \frac{(3x + 2)^{7/3}}{21} - \frac{(3x + 2)^{4/3}}{6} + C.$

25 Let $u = \sqrt{x}$. Then $\int \frac{x\,dx}{\sqrt{x} + 3} = \int \frac{u^2 \cdot 2u\,du}{u + 3}$

$= \int \left(2u^2 - 6u + 18 - \frac{54}{u + 3}\right) du$

$= \frac{2}{3}u^3 - 3u^2 + 18u - 54 \ln(u + 3) + C$

$= \frac{2}{3}x^{3/2} - 3x + 18\sqrt{x} - 54 \ln(\sqrt{x} + 3) + C.$

27 Let $u = x^{1/6}$. Then $\int \frac{dx}{\sqrt[3]{x} + \sqrt{x}} = \int \frac{6u^5\,du}{u^2 + u^3}$

$= \int \frac{6u^3}{1 + u}\,du$

$= \int \left(6u^2 - 6u + 6 - \frac{6}{u + 1}\right) du$

$= 2u^3 - 3u^2 + 6u - 6 \ln(u + 1) + C =$

$2x^{1/2} - 3x^{1/3} + 6x^{1/6} - 6 \ln(x^{1/6} + 1) + C.$

29 (a) Following the suggestion, $\int \tan^n \theta\,d\theta$

$= \int \tan^{n-2} \theta \tan^2 \theta\,d\theta$

$= \int \tan^{n-2} \theta (\sec^2 \theta - 1)\,d\theta$

$= \int \tan^{n-2} \theta \sec^2 \theta\,d\theta - \int \tan^{n-2} \theta\,d\theta$. To evaluate the first integral, let $u = \tan \theta$ so $du = \sec^2 \theta\,d\theta$. Then $\int \tan^{n-2} \theta \sec^2 \theta\,d\theta =$

$\int u^{n-2}\,du = \frac{u^{n-1}}{n - 1} + C = \frac{\tan^{n-1} \theta}{n - 1} + C.$

Finally, we have $\int \tan^n \theta\,d\theta =$

$\frac{\tan^{n-1} \theta}{n - 1} - \int \tan^{n-2} \theta\,d\theta.$

(b) From we know that (a), $\int \tan^3 \theta\,d\theta$

$= \frac{\tan^2 \theta}{2} - \int \tan \theta\,d\theta =$

$\frac{\tan^2\theta}{2} + \ln|\cos \theta| + C.$

(c) From (a), we know that $\int \tan^4 \theta\,d\theta =$ $\frac{\tan^3\theta}{3} - \int \tan^2 \theta\,d\theta$. From the same recursion, we have $\int \tan^2 \theta\,d\theta =$ $\tan \theta - \int d\theta = \tan \theta - \theta + K$. Thus,

$\int \tan^4 \theta\,d\theta = \frac{\tan^3\theta}{3} - (\tan \theta - \theta + K)$

$= \frac{\tan^3\theta}{3} - \tan \theta + \theta + C$, where $C = -K$.

31 (a) Observe that $\int \csc \theta\,d\theta = \int \frac{d\theta}{\sin\theta} =$ $\int \frac{\sin\theta}{\sin^2\theta}\,d\theta$. Now let $u = \cos \theta$ and $du =$ $-\sin \theta\,d\theta$, so $\int \frac{\sin \theta}{1 - \cos^2 \theta}\,d\theta =$

$-\int \frac{du}{1 - u^2} = -\frac{1}{2} \int \left(\frac{1}{1 + u} + \frac{1}{1 - u}\right) du =$

$-\frac{1}{2}[\ln(1 + u) - \ln(1 - u)] + C =$

$-\frac{1}{2} \ln \frac{1 + u}{1 - u} + C = \frac{1}{2} \ln \frac{1 - u}{1 + u} + C.$

Since $u = \cos \theta$, this becomes $\int \csc \theta\,d\theta =$ $\frac{1}{2} \ln \frac{1 - \cos \theta}{1 + \cos \theta} + C$. It can also be shown that $\int \csc \theta\,d\theta = \ln|\csc \theta - \cot \theta| + C.$

(b) Since $(\cot\theta)' = -\csc^2\theta$, $\int \csc^2\theta\, d\theta =$ $-\cot\theta + C$.

33 (a) Here $\int \sin^3\theta\cos^3\theta\, d\theta$

$= \int \sin^3\theta\cos^2\theta\cos\theta\, d\theta$

$= \int \sin^3\theta\,(1 - \sin^2\theta)\cos\theta\, d\theta$

$= \int \sin^3\theta\cos\theta\, d\theta - \int \sin^5\theta\cos\theta\, d\theta.$

For both integrals, let $u = \sin\theta$ so $du =$ $\cos\theta\, d\theta$. Then

$\int \sin^3\theta\cos\theta\, d\theta - \int \sin^5\theta\cos\theta\, d\theta =$

$\int u^3\, du - \int u^5\, du = \dfrac{u^4}{4} - \dfrac{u^6}{6} + C$

$= \dfrac{\sin^4\theta}{4} - \dfrac{\sin^6\theta}{6} + C.$

(b) Let $u = \sin\theta$ and $du = \cos\theta\, d\theta$. Then

$\int \sin^4\theta\cos\theta\, d\theta = \int u^4\, du = \dfrac{u^5}{5} + C$

$= \dfrac{\sin^5\theta}{5} + C.$

(c) Here $\int \sin^4\theta\cos^3\theta\, d\theta$

$= \int \sin^4\theta\cos^2\theta\cos\theta\, d\theta$

$= \int \sin^4\theta\,(1 - \sin^2\theta)\cos\theta\, d\theta$

$= \int \sin^4\theta\cos\theta\, d\theta - \int \sin^6\theta\cos\theta\, d\theta.$

Now let $u = \sin\theta$ so $du = \cos\theta\, d\theta$. Thus,

$\int \sin^4\theta\cos\theta\, d\theta - \int \sin^6\theta\cos\theta\, d\theta$

$= \int u^4\, du - \int u^6\, du = \dfrac{u^5}{5} - \dfrac{u^7}{7} + C$

$= \dfrac{\sin^5\theta}{5} - \dfrac{\sin^7\theta}{7} + C$. Therefore

$\int_0^{\pi/2} \sin^4\theta\cos\theta\, d\theta = \left.\dfrac{\sin^5\theta}{5} - \dfrac{\sin^7\theta}{7}\right|_0^{\pi/2}$

$= \left(\dfrac{1}{5} - \dfrac{1}{7}\right) - (0 - 0) = \dfrac{2}{35}.$

(d) Here $\int \cos^5\theta\, d\theta = \int \cos^4\theta\cos\theta\, d\theta$

$= \int (1 - \sin^2\theta)^2\cos\theta\, d\theta$

$= \int \cos\theta\, d\theta - 2\int \sin^2\theta\cos\theta\, d\theta +$

$\int \sin^4\theta\cos\theta\, d\theta$. Letting $u = \sin\theta$ and du $= \cos\theta\, d\theta$ for the last two integrals, we have

$\int \cos^5\theta\, d\theta$

$= \int \cos\theta\, d\theta - 2\int u^2\, du + \int u^4\, du$

$= \sin\theta - \dfrac{2}{3}u^3 + \dfrac{1}{5}u^5 + C$

$= \sin\theta - \dfrac{2}{3}\sin^3\theta + \dfrac{1}{5}\sin^5\theta + C.$

35 (a) Here $\int \cos^2\theta\sin^4\theta\, d\theta$

$= \int \left(\dfrac{1 + \cos 2\theta}{2}\right)\left(\dfrac{1 - \cos 2\theta}{2}\right)^2 d\theta =$

$\dfrac{1}{8}\int (1 + \cos 2\theta)(1 - 2\cos 2\theta + \cos^2 2\theta)\, d\theta$

$= \dfrac{1}{8}\int (1 - \cos 2\theta - \cos^2 2\theta + \cos^3 2\theta)\, d\theta =$

$\dfrac{1}{8}\int d\theta - \dfrac{1}{8}\int \cos 2\theta\, d\theta -$

$\dfrac{1}{8}\int \cos^2 2\theta\, d\theta + \dfrac{1}{8}\int \cos^3 2\theta\, d\theta$. Now

$\int \cos^2 2\theta \, d\theta = \int \frac{1}{2}(1 + \cos 4\theta) \, d\theta =$

$\frac{1}{2}\theta + \frac{1}{8} \sin 4\theta + C_1$, while $\int \cos^3 2\theta \, d\theta$

$= \int (1 - \sin^2 2\theta) \cos 2\theta \, d\theta$

$= \frac{1}{2} \sin 2\theta - \frac{1}{6} \sin^3 2\theta + C_2$, where the

substitution $u = \sin 2\theta$ was used. Finally,

$\int \cos^2 \theta \sin^4 \theta \, d\theta = \frac{1}{8}\theta - \frac{1}{8}\left(\frac{1}{2} \sin 2\theta\right) -$

$\frac{1}{8}\left(\frac{\theta}{2} + \frac{1}{8} \sin 4\theta\right) +$

$\frac{1}{8}\left(\frac{1}{2} \sin 2\theta - \frac{1}{6} \sin^3 2\theta\right) + C =$

$\frac{\theta}{16} - \frac{\sin 4\theta}{64} - \frac{\sin^3 2\theta}{48} + C.$

(b) Here $\int \cos^2 \theta \sin^2 \theta \, d\theta$

$= \int \left(\frac{1 + \cos 2\theta}{2}\right)\left(\frac{1 - \cos 2\theta}{2}\right) d\theta$

$= \frac{1}{4} \int (1 - \cos^2 2\theta) \, d\theta$

$= \frac{1}{4} \int d\theta - \frac{1}{4} \int \cos^2 2\theta \, d\theta$

$= \frac{1}{4}\theta - \frac{1}{4}\left(\frac{1}{2}\theta + \frac{1}{8} \sin 4\theta\right) + C$

$= \frac{1}{8}\theta - \frac{\sin 4\theta}{32} + C$, where the result from

(a) was used. Then $\int_0^{\pi/4} \cos^2 \theta \sin^2 \theta \, d\theta =$

$\left(\frac{1}{8}\theta - \frac{\sin 4\theta}{32}\right)\Big|_0^{\pi/4} = \frac{1}{8}\cdot\frac{\pi}{4} - 0 - (0 - 0) =$

$\frac{\pi}{32}.$

37 (a),(b)

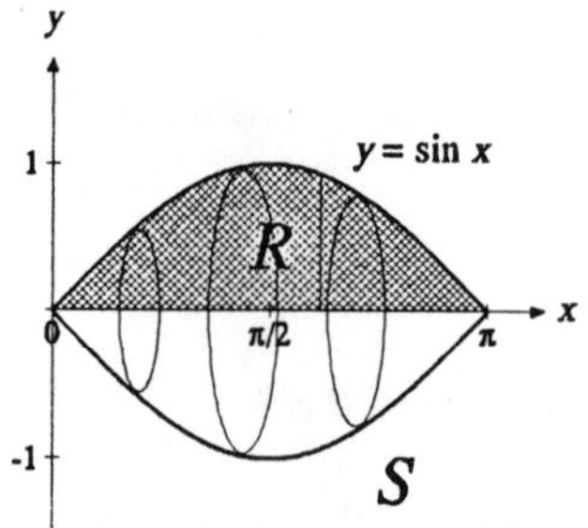

(c) Area of $R = \int_0^\pi \sin x \, dx$

(d) Volume of $S = \int_0^\pi \pi(\sin^2 x) \, dx$

$= \pi \int_0^\pi \sin^2 x \, dx$

(e) By the fundamental theorem of calculus,

$\int_0^\pi \sin x \, dx = -\cos x \big|_0^\pi = -(-1) - (-1) =$

2. Using the identity $\sin^2 x = \frac{1}{2}(1 - \cos 2x)$,

$\pi \int_0^\pi \sin^2 x \, dx = \pi \int_0^\pi \frac{1}{2}(1 - \cos 2x) \, dx$

$= \frac{\pi}{2}\left[x - \frac{1}{2} \sin 2x\right]\Big|_0^\pi$

$= \frac{\pi}{2}[(\pi - 0) - (0 - 0)] = \frac{\pi^2}{2}$. So the area

of R is 2 and the volume of S is $\pi^2/2$.

39 (a) $\frac{1}{\pi/2 - 0}\cdot\frac{1}{R} \int_0^{\pi/2} (M \sin t)^2 \, dt$

$= \frac{2M^2}{\pi R} \int_0^{\pi/2} \sin^2 t \, dt = \frac{2M^2}{\pi R}\cdot\frac{\pi}{4} = \frac{M^2}{2R}$

(b) $\frac{1}{\pi/2 - 0}\cdot\frac{1}{R} \int_0^{\pi/2} \left(\frac{M}{\sqrt{2}}\right)^2 dt = \frac{2}{\pi R}\cdot\frac{M^2}{2}\cdot\frac{\pi}{2}$

$$= \frac{M^2}{2R}$$

41 The integral involves a radical of the form $\sqrt{a^2 + x^2}$ with $a = 3$, so make the substitution $x = a \tan\theta = 3\tan\theta$. Then $dx = 3\sec^2\theta\, d\theta$ and $\sqrt{9 + x^2} = 3\sec\theta$. Therefore $\int \frac{dx}{\sqrt{9 + x^2}}$

$= \int \frac{3\sec^2\theta\, d\theta}{3\sec\theta} = \int \sec\theta\, d\theta = \ln|\sec\theta + \tan\theta| + C_1 = \ln\left|\frac{\sqrt{9 + x^2}}{3} + \frac{x}{3}\right| + C_1$

$$= \ln\left|\sqrt{9 + x^2} + x\right| - \ln 3 + C_1$$

$$= \ln\left|\sqrt{9 + x^2} + x\right| + C.$$

43 Since the integrand is a function of x and $\sqrt{a^2 - x^2}$, the integral satisfies Case 1 with $a = 1$. So let $x = a\sin\theta = \sin\theta$ and $dx = \cos\theta\, d\theta$. Then $\int x^3\sqrt{1 - x^2}\, dx$

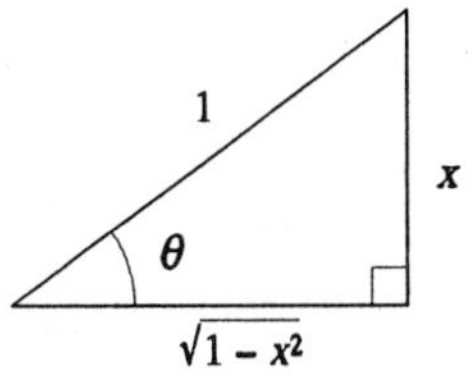

$$= \int \sin^3\theta \cos^2\theta\, d\theta$$

$$= \int \sin\theta\,(1 - \cos^2\theta)\cos^2\theta\, d\theta$$

$= \int \sin\theta\cos^2\theta\, d\theta - \int \sin\theta\cos^4\theta\, d\theta$. Now let $u = \cos\theta$, so $du = -\sin\theta\, d\theta$. Thus $\int \sin^3\theta\cos^2\theta\, d\theta = -\int u^2\, du + \int u^4\, du$

$$= -\frac{u^3}{3} + \frac{u^5}{5} + C = -\frac{\cos^3\theta}{3} + \frac{\cos^5\theta}{5} + C$$

$$= -\frac{\left(\sqrt{1 - x^2}\right)^3}{3} + \frac{\left(\sqrt{1 - x^2}\right)^5}{5} + C$$

$$= -\frac{1}{15}(1 - x^2)^{3/2}(3x^2 + 2) + C.$$

45 Make the trigonometric substitution $x = a\sin\theta$, $dx = a\cos\theta\, d\theta$. Then $\sqrt{a^2 - x^2} = a\cos\theta$ and

$$\int \sqrt{a^2 - x^2}\, dx = \int (a\cos\theta)(a\cos\theta\, d\theta)$$

$$= a^2\int \cos^2\theta\, d\theta = a^2\int \frac{1 + \cos 2\theta}{2}\, d\theta$$

$$= \frac{a^2}{2}\left(\theta + \frac{1}{2}\sin 2\theta\right) + C$$

$$= \frac{a^2}{2}(\theta + \sin\theta\cos\theta) + C$$

$$= \frac{a^2}{2}\left(\sin^{-1}\frac{x}{a} + \frac{x}{a}\frac{\sqrt{a^2 - x^2}}{a}\right) + C$$

$$= \frac{a^2}{2}\sin^{-1}\frac{x}{a} + \frac{x}{2}\sqrt{a^2 - x^2} + C.$$

47 Let $x = a\tan\theta$; then $\sqrt{a^2 + x^2} = a\sec\theta$, $dx = a\sec^2\theta\, d\theta$. Thus $\int \sqrt{a^2 + x^2}\, dx =$

$$\int (a\sec\theta)(a\sec^2\theta\, d\theta) = a^2\int \sec^3\theta\, d\theta$$

$$= a^2\left(\frac{\sec\theta\tan\theta}{2} + \frac{1}{2}\int \sec\theta\, d\theta\right)$$

$$= \frac{a^2}{2}(\sec\theta\tan\theta + \ln|\sec\theta + \tan\theta|) + C_1.$$

(We used the recursion for $\int \sec^n\theta\, d\theta$ from Exercise 30—Formula 56 from the table of antiderivatives.) Expressed in terms of x, this is

$$\frac{a^2}{2}\left(\frac{\sqrt{a^2 + x^2}}{a}\cdot\frac{x}{a} + \ln\left|\frac{\sqrt{a^2 + x^2}}{a} + \frac{x}{a}\right|\right) + C_1 =$$

$$\frac{1}{2}\left(x\sqrt{a^2 + x^2} + a^2\ln\left|\sqrt{a^2 + x^2} + x\right| - a^2\ln a\right) + C_1$$

$$= \frac{1}{2}\left(x\sqrt{a^2 + x^2} + a^2 \ln\left|\sqrt{a^2 + x^2} + x\right|\right) + C$$

$$= \frac{1}{2}\left[x\sqrt{a^2 + x^2} + a^2 \ln\left(\sqrt{a^2 + x^2} + x\right)\right] + C.$$

49 Let $25x^2 = 16 \sec^2 \theta$; that is, let $5x = 4 \sec \theta$ or $x = \frac{4}{5} \sec \theta$. Then $dx = \frac{4}{5} \sec \theta \tan \theta \, d\theta$ and $\sqrt{25x^2 - 16} = 4 \tan \theta$. Thus $\int \frac{dx}{\sqrt{25x^2 - 16}}$

$$= \int \frac{\frac{4}{5} \sec \theta \tan \theta \, d\theta}{4 \tan \theta} = \frac{1}{5} \int \sec \theta \, d\theta$$

$= \frac{1}{5} \ln |\sec \theta + \tan \theta| + C_1$. Expressed in terms of x, this becomes $\frac{1}{5} \ln\left|\frac{5x}{4} + \frac{\sqrt{25x^2 - 16}}{4}\right| + C_1$

$$= \frac{1}{5}\left[\ln\left|5x + \sqrt{25x^2 - 16}\right| - \ln 4\right] + C_1$$

$$= \frac{1}{5} \ln\left|5x + \sqrt{25x^2 - 16}\right| + C.$$

51 (a) The presence of $\sqrt{9 - x^2}$ in the integrand suggests the substitution $x = 3 \sin \theta$, so $dx = 3 \cos \theta \, d\theta$ and $\sqrt{9 - x^2} = 3 \cos \theta$. Thus,

$$\int \frac{x + \sqrt{9 - x^2}}{x^3} dx$$

$$= \int \frac{3 \sin \theta + 3 \cos \theta}{(3 \sin \theta)^3}(3 \cos \theta \, d\theta)$$

$$= \frac{1}{3} \int \frac{\sin \theta \cos \theta + \cos^2 \theta}{\sin^3 \theta}, \text{ and}$$

$\frac{\sin \theta \cos \theta + \cos^2 \theta}{\sin^3 \theta}$ is a rational function of $\cos \theta$ and $\sin \theta$.

(b) The presence of $\sqrt{5 - x^2}$ in the integrand suggests the substitution $x = \sqrt{5} \sin \theta$, so $dx = \sqrt{5} \cos \theta \, d\theta$ and $\sqrt{5 - x^2} = \sqrt{5} \cos \theta$.

Thus, $\int \frac{x^3\sqrt{5 - x^2}}{1 + \sqrt{5 - x^2}} dx =$

$$\int \frac{(5\sqrt{5} \sin^3 \theta)(\sqrt{5} \cos \theta)(\sqrt{5} \cos \theta \, d\theta)}{1 + \sqrt{5} \cos \theta}$$

$$= 25\sqrt{5} \int \frac{\sin^3 \theta \cos^2 \theta}{1 + \sqrt{5} \cos \theta} d\theta, \text{ and}$$

$\frac{\sin^3 \theta \cos^2 \theta}{1 + \sqrt{5} \cos \theta}$ is a rational function of $\cos \theta$ and $\sin \theta$.

53 (a) From the right triangle in the figure, it is clear that $\cos \frac{\theta}{2} = \frac{1}{\sqrt{1 + u^2}}$ and $\sin \frac{\theta}{2} = \frac{u}{\sqrt{1 + u^2}}$ by definition.

(b) Using the identity $\cos 2\theta = \cos^2 \theta - \sin^2 \theta$, we have $\cos \theta = \cos^2 \theta/2 - \sin^2 \theta/2 =$

$$\left(\frac{1}{\sqrt{1 + u^2}}\right)^2 - \left(\frac{u}{\sqrt{1 + u^2}}\right)^2 = \frac{1 - u^2}{1 + u^2}.$$

(c) Using the identity $\sin 2\theta = 2 \sin \theta \cos \theta$, we have $\sin \theta = 2 \sin \theta/2 \cos \theta/2 =$

$$2 \cdot \frac{u}{\sqrt{1 + u^2}} \cdot \frac{1}{\sqrt{1 + u^2}} = \frac{2u}{1 + u^2}.$$

(d) Since $\theta = 2 \tan^{-1} u$, it follows that $d\theta =$

$\dfrac{2\,du}{1+u^2}$.

55 Let $u = \tan(\theta/2)$. Then $\int_0^{\pi/2} \dfrac{d\theta}{4\sin\theta + 3\cos\theta}$

$$= \int_0^1 \frac{\dfrac{2\,du}{1+u^2}}{4\left(\dfrac{2u}{1+u^2}\right) + 3\left(\dfrac{1-u^2}{1+u^2}\right)} =$$

$$\int_0^1 \frac{2\,du}{8u + 3(1-u^2)} = 2\int_0^1 \frac{du}{-u^2 + 8u + 3}.$$

Now $-u^2 + 8u + 3$ is reducible since $8^2 - 4(-1)\cdot 3 > 0$, and its roots are

$\dfrac{-8 \pm \sqrt{(8)^2 - 4(-1)\cdot 3}}{2(-1)} = 4 \pm \sqrt{19}$. Therefore

$$-u^2 + 4u + 3 = -[u - (4+\sqrt{19})][u - (4-\sqrt{19})]$$

and $2\int_0^1 \dfrac{du}{-u^2 + 8u + 3} =$

$$-2\int_0^1 \frac{du}{[u - (4+\sqrt{19})][u - (4-\sqrt{19})]} =$$

$$-2\int_0^1 \frac{1}{2\sqrt{19}}\left(\frac{1}{u - (4+\sqrt{19})} - \frac{1}{u - (4-\sqrt{19})}\right) du$$

$$= -\frac{1}{\sqrt{19}}\left[\ln|u - (4+\sqrt{19})| - \ln|u - (4-\sqrt{19})|\right]_0^1$$

$$= -\frac{1}{\sqrt{19}}\left[\ln\left|\frac{u - (4+\sqrt{19})}{u - (4-\sqrt{19})}\right|\right]_0^1$$

$$= -\frac{1}{\sqrt{19}}\left[\ln\left|\frac{-3+\sqrt{19}}{-3-\sqrt{19}}\right| - \ln\left|\frac{4+\sqrt{19}}{4-\sqrt{19}}\right|\right]$$

$$= -\frac{1}{\sqrt{19}}\left[\ln\left|\frac{(-3+\sqrt{19})(4-\sqrt{19})}{(-3-\sqrt{19})(4+\sqrt{19})}\right|\right]$$

$$= -\frac{1}{\sqrt{19}}\ln\left[\frac{-31+7\sqrt{19}}{-31-7\sqrt{19}}\right].$$

57 (a) Since $\sec\theta + \tan\theta > 0$ for $0 \le \theta < \pi/2$, $|\sec\theta + \tan\theta| = \sec\theta + \tan\theta$ and

$$(\ln|\sec\theta + \tan\theta|)' = [\ln(\sec\theta + \tan\theta)]'$$

$$= \frac{\sec^2\theta + \sec\theta\tan\theta}{\sec\theta + \tan\theta}$$

$$= \frac{(\sec\theta)(\sec\theta + \tan\theta)}{(\sec\theta + \tan\theta)} = \sec\theta.$$

(b) There is no contradiction. Observe that

$$\ln(\sec\theta + \tan\theta) = \ln\left(\frac{1+\sin\theta}{\cos\theta}\right)$$

$$= \frac{1}{2}\ln\left[\left(\frac{1+\sin\theta}{\cos\theta}\right)^2\right]$$

$$= \frac{1}{2}\ln\left[\frac{(1+\sin\theta)^2}{\cos^2\theta}\right] = \frac{1}{2}\ln\left[\frac{(1+\sin\theta)^2}{1-\sin^2\theta}\right]$$

$$= \frac{1}{2}\ln\left[\frac{(1+\sin\theta)(1+\sin\theta)}{(1+\sin\theta)(1-\sin\theta)}\right]$$

$$= \frac{1}{2}\ln\left(\frac{1+\sin\theta}{1-\sin\theta}\right).$$

59 Let $u = \sqrt[n]{\dfrac{ax+b}{cx+d}}$. Then $u^n = \dfrac{ax+b}{cx+d}$, so

$(cx+d)u^n = ax + b$. Solving for x, $x =$

$\dfrac{b - u^n d}{u^n c - a}$; then $dx =$

$\dfrac{(u^n c - a)(-nu^{n-1}d) - (b - u^n d)(nu^{n-1}c)}{(u^n c - a)^2}\,du$. Any

rational function of x and $\sqrt[n]{\dfrac{ax+b}{cx+d}}$ is therefore a rational function of u alone. Hence it can be integrated using partial fractions.

7.7 What to Do in the Face of an Integral

1 Divide the numerator by the denominator, producing $x^{-2} + x^{-1}$. Now simply integrate using the power rule.

3 Factor the denominator as $x^2(x + 1)$ and use partial fractions.

5 Use integration by parts with $u = \tan^{-1} 2x$ and $dv = dx$.

7 Repeated integration by parts will work here. Let $u = x^{10}$ and $dv = e^x\,dx$ to begin with. Each application of integration by parts will reduce the exponent of x by 1, eventually leading to the integral of e^x.

9 Use substitution with $u = \tan\theta$ and $du = \sec^2\theta\,d\theta$. The integral is then $\int \frac{du}{u}$.

11 Make the substitution $u = \sqrt[3]{x + 2}$. Then $u^3 - 2 = x$ and $dx = 3u^2\,du$. This substitution will lead to an integrand that can be evaluated by several applications of the power rule.

13 Since $(x^2 + x + 1)' = 2x + 1$, use the substitution $u = x^2 + x + 1$. Then $du = (2x + 1)dx$ and the integral becomes $\int \frac{du}{u^5}$, which is easily evaluated using the power rule.

15 Use the trigonometric identity $\tan^2\theta = \sec^2\theta - 1$. Since $\int \sec^2\theta\,d\theta = \tan\theta + C$, this integrates immediately.

17 The substitution $w = \sqrt{x}$ transforms the integral to $2\int we^w\,dw$. Integration by parts with $u = w$ and $dv = e^w\,dw$ completes the problem.

19 Notice that $x^2 - 4x + 3 = (x - 3)(x - 1)$. Either use the recursion in Formula 38 of the table of antiderivatives, followed by partial fractions, or immediately apply the method of partial fractions.

21 Use polynomial division to write the integrand as $x^4 - x^3 + x^2 - x + 1 - \frac{1}{x + 1}$. These terms can be integrated immediately with the power rule and $\int \frac{du}{u} = \ln|u| + C$.

23 Making the substitution $u = e^x$, $du = e^x\,dx$, the integrand becomes $\int \frac{e^{3x}\,dx}{1 + e^x + e^{2x}} = \int \frac{e^{2x}(e^x\,dx)}{1 + e^x + e^{2x}} = \int \frac{u^2\,du}{1 + u + u^2}$. First carry out polynomial division to get $\int \left(1 - \frac{u + 1}{u^2 + u + 1}\right) du$. As in Sec. 7.4, write $\frac{1}{2}\left(\frac{(2u + 1) + 1}{u^2 + u + 1}\right) = \frac{1}{2}\left(\frac{2u + 1}{u^2 + u + 1}\right) + \frac{1}{2}\left(\frac{1}{u^2 + u + 1}\right)$. The first integrates to yield a logarithm. To integrate the second, complete the square in the denominator and make the appropriate substitution.

25 Simplifying, $\ln(e^x) = x \ln e = x$. Use the power rule.

27 Make the substitution $x = u - 2$, expand the

numerator, and divide. The resulting terms can be integrated by the power rule or $\int \frac{du}{u} = \ln|u| + C$.

29 Let $u = 3 + \sqrt{x}$. Then $du = \frac{du}{2\sqrt{x}}$ and the integral becomes $2\int \frac{du}{u^2}$. The power rule will complete this problem.

31 Use the substitution $u = 1 + \tan\theta$, $du = \sec^2\theta\, d\theta$. The resulting integral, $\int u^3\, du$, can be evaluated with the power rule.

33 The substitution $u = e^x$ transforms this into $\int \frac{u^2 + 1}{(u^2 - 1)u}\, du$, which can be evaluated by partial fractions.

35 The presence of $\sqrt[3]{x + 2}$ and $\sqrt{x + 2}$ in the integrand suggests using the substitution $u = \sqrt[6]{x + 2}$. The integral becomes $\int \frac{u^2 - 1}{u^3 + 1}(6u^5\, du)$. Employing long division and the techniques of Sec. 7.4 completes the problem.

37 The substitution $u = 3x + 2$ leads to an integral that can be evaluated using the power rule.

39 Recall that $a^x = e^{\ln a^x} = e^{x \ln a} = e^{kx}$ with $k = \ln a$. Rewrite the integrand as $\left(\frac{2}{4}\right)^x + \left(\frac{3}{4}\right)^x$ and apply this formula.

41 Break the integrand into two pieces: $\frac{x}{\sqrt{1 - x^2}}$ and $\frac{\sin^{-1} x}{\sqrt{1 - x^2}}$. The first piece can be integrated by the substitution $u = 1 - x^2$. The second can be integrated with the substitution $w = \sin^{-1} x$.

43 With $u = 1 + x^2$ and $du = 2x\, dx$, the integral becomes $\int x^2\sqrt{1 + x^2}\, x\, dx = \frac{1}{2}\int (u - 1)u^{1/2}\, du$. Multiply out and use the power rule.

45 Make the substitution $u = x^2 - 1$, $du = 2x\, dx$. The integral then becomes $\frac{1}{2}\int u^{-1/2}\, du$ and can be evaluated by using the power rule.

47 Let $u = x^2 - 9$. Then the integral becomes $\frac{1}{2}\int \frac{du}{u^{3/2}}$, which can be evaluated using the power rule.

49 Integrating by parts with $u = \tan^{-1} x$ and $dv = \frac{dx}{x^2}$ yields $-\frac{\tan^{-1} x}{x} + \int \frac{dx}{x(1 + x^2)}$, and this new integral can be evaluated using the method of partial fractions.

51 Use the substitution $u = \sin x$. Then the integral becomes $\int \ln u\, du$, which can be evaluated by integrating by parts.

53 Complete the square in the denominator:

$$\int \frac{dx}{x^2 + x + 5} = \int \frac{dx}{\left(x + \frac{1}{2}\right)^2 + \frac{19}{4}}.$$ Now use

the substitution $\left(x + \frac{1}{2}\right)^2 = \frac{19}{4}u^2$.

55 Let $u = x + 1$. The integral becomes $\int \frac{u+2}{u^5}\,du$

$= \int (u^{-4} + 2u^{-5})\,du$, which can be evaluated with the power rule.

57 The substitution $u = x^2 + 9$, $du = 2x\,dx$ yields

$\frac{1}{2}\int u^{10}\,du$. Now use the power rule.

59 Since the denominator is $(x + 1)^2(x - 2)^3$, the method of partial fractions will work.

61 The two cases where $\int (1 + x^2)^{1/n}\,dx$ can be evaluated are $n = 1$ and $n = 2$.

$n = 1$: $\int (1 + x^2)\,dx = x + \frac{1}{3}x^3 + C$

$n = 2$: Using the substitution $x = \tan\theta$ and $dx = \sec^2\theta\,d\theta$, we have $\int (1 + x^2)^{1/2}\,dx =$

$$\int \sec^3\theta\,d\theta = \frac{1}{2}\sec\theta\tan\theta + \frac{1}{2}\int \sec\theta\,d\theta$$

$$= \frac{1}{2}\sec\theta\tan\theta + \frac{1}{2}\ln|\sec\theta + \tan\theta| + C$$

$$= \frac{1}{2}x\sqrt{1 + x^2} + \frac{1}{2}\ln\left(x + \sqrt{1 + x^2}\right) + C.$$

63 $\int \frac{dx}{\sqrt{x+2} - \sqrt{x-2}}$

$$= \int \frac{\left(\sqrt{x+2} + \sqrt{x-2}\right)dx}{\left(\sqrt{x+2} - \sqrt{x-2}\right)\left(\sqrt{x+2} + \sqrt{x-2}\right)}$$

$$= \frac{1}{4}\int \left(\sqrt{x+2} + \sqrt{x-2}\right)dx$$

$$= \frac{1}{6}(x + 2)^{3/2} + \frac{1}{6}(x - 2)^{3/2} + C$$

65 Let $u = 9 - x^2$, so $du = -2x\,dx$, and

$$\int \frac{x\,dx}{(9 - x^2)^{5/2}} = -\frac{1}{2}\int \frac{du}{u^{5/2}} =$$

$$-\frac{1}{2}\left(-\frac{2}{3}u^{-3/2}\right) + C = \frac{1}{3}u^{-3/2} + C =$$

$$\frac{1}{3}(9 - x^2)^{-3/2} + C.$$

67 Let $x = 3\sin\theta$, so $dx = 3\cos\theta\,d\theta$. Thus

$$\int \frac{dx}{\sqrt{(9 - x^2)^5}}$$

$$= \int \frac{3\cos\theta\,d\theta}{(3\cos\theta)^5} = \frac{1}{3^4}\int \frac{d\theta}{\cos^4\theta}$$

$$= \frac{1}{3^4}\int \sec^4\theta\,d\theta$$

$$= \frac{1}{3^4}\left[\frac{\sec^2\theta\tan\theta}{3} + \frac{2}{3}\int \sec^2\theta\,d\theta\right]$$

$= \frac{1}{3^5}\sec^2\theta\tan\theta + \frac{2}{3^5}\tan\theta + C$, where the recursion formula for $\int \sec^n\theta\,d\theta$ given in Exercise 30 of Sec. 7.6 was used. (See also Formula 56 in the table of antiderivatives.) Finally,

$$\int \frac{dx}{\left(\sqrt{9 - x^2}\right)^5} =$$

$$\frac{1}{3^5}\left(\frac{3}{\sqrt{9 - x^2}}\right)^2\left(\frac{x}{\sqrt{9 - x^2}}\right) + \frac{2}{3^5}\left(\frac{x}{\sqrt{9 - x^2}}\right) + C$$

$$= \frac{1}{3^3}\cdot\frac{x}{(9 - x^2)^{3/2}} + \frac{2}{3^5}\cdot\frac{x}{(9 - x^2)^{1/2}} + C.$$

69 Let $u = x^2 + 9$, so $du = 2x\,dx$. Then $\int \frac{x\,dx}{\sqrt{x^2 + 9}}$

$= \int \frac{du}{2\sqrt{u}} = \sqrt{u} + C = \sqrt{x^2 + 9} + C.$

71 Break the integral in two: $\int (x^3 + x^2)\sqrt{x^2 - 5}\, dx$ $= \int x^3\sqrt{x^2 - 5}\, dx + \int x^2\sqrt{x^2 - 5}\, dx$. For the first integral, let $u = x^2 - 5$ so $du = 2x\, dx$ and $\int x^2\sqrt{x^2 - 5}\, x\, dx = \frac{1}{2}\int (u + 5)u^{1/2}\, du =$ $\frac{1}{2}\int (u^{3/2} + 5u^{1/2})\, du = \frac{1}{5}u^{5/2} + \frac{5}{3}u^{3/2} + C_1 =$ $\frac{1}{5}(x^2 - 5)^{5/2} + \frac{5}{3}(x^2 - 5)^{3/2} + C_1$. For the second integral, let $x = \sqrt{5}\sec\theta$ (and assume x varies through nonnegative values). Then $dx = \sqrt{5}\sec\theta \tan\theta\, d\theta$ and $\int x^2\sqrt{x^2 + 5}\, dx =$

$\int (5\sec^2\theta)(\sqrt{5}\tan\theta)(\sqrt{5}\sec\theta\tan\theta\, d\theta)$

$= 25\int \sec^3\theta\tan^2\theta\, d\theta = 25\int \frac{\sin^2\theta}{\cos^5\theta}\, d\theta$

$= 25\int \frac{1 - \cos^2\theta}{\cos^5\theta}\, d\theta$

$= 25\int (\sec^5\theta - \sec^3\theta)\, d\theta$

$= 25\left[\int \sec^5\theta\, d\theta - \int \sec^3\theta\, d\theta\right]$

$= 25\left[\frac{\sec^3\theta\tan\theta}{4} + \frac{3}{4}\int \sec^3\theta\, d\theta - \int \sec^3\theta\, d\theta\right]$

$= \frac{25}{4}\sec^3\theta\tan\theta - \frac{25}{4}\int \sec^3\theta\, d\theta$

$= \frac{25}{4}\sec^3\theta\tan\theta -$ $\frac{25}{4}\left[\frac{\sec\theta\tan\theta}{2} + \frac{1}{2}\int \sec\theta\, d\theta\right]$

$= \frac{25}{4}\sec^3\theta\tan\theta - \frac{25}{8}\sec\theta\tan\theta -$ $\frac{25}{8}\ln|\sec\theta + \tan\theta| + K_2$

$= \frac{25}{4}\left(\frac{x}{\sqrt{5}}\right)^3 \cdot \frac{\sqrt{x^2 - 5}}{\sqrt{5}} - \frac{25}{8}\frac{x}{\sqrt{5}} \cdot \frac{\sqrt{x^2 - 5}}{\sqrt{5}} -$ $\frac{25}{8}\ln\left|\frac{x}{\sqrt{5}} + \frac{\sqrt{x^2 - 5}}{\sqrt{5}}\right| + K_2$

$= \frac{x^3\sqrt{x^2 - 5}}{4} - \frac{5}{8}x\sqrt{x^2 - 5} -$ $\frac{25}{8}\ln|x + \sqrt{x^2 - 5}| + \frac{25}{8}\ln\sqrt{5} + K_2$

$= \frac{x^3\sqrt{x^2 - 5}}{4} - \frac{5}{8}x\sqrt{x^2 - 5} -$ $\frac{25}{8}\ln|x + \sqrt{x^2 - 5}| + C_2$. Finally,

$\int (x^3 + x^2)\sqrt{x^2 - 5}\, dx = \frac{1}{5}(x^2 - 5)^{5/2} +$ $\frac{5}{3}(x^2 - 5)^{3/2} + \frac{x^3(x^2 - 5)^{1/2}}{4} - \frac{5}{8}x(x^2 - 5)^{1/2} -$ $\frac{25}{8}\ln|x + (x^2 - 5)^{1/2}| + C.$

7.S Guide Quiz

1 The numerator is almost the derivative of the denominator. Thus $\int \frac{x^3\, dx}{1 + x^4} = \frac{1}{4}\int \frac{4x^3\, dx}{1 + x^4} =$

$\frac{1}{4}\ln(1 + x^4) + C.$

2 Use the substitution $u = 4x + 3$, so $du = 4\,dx$.

Then $\int \sqrt{4x + 3}\,dx = \frac{1}{4}\int \sqrt{u}\,du = \frac{1}{6}u^{3/2} + C$

$= \frac{1}{6}(4x + 3)^{3/2} + C.$

3 Rewrite $\int \sec^5 2x \tan 2x\,dx$ as

$\int \sec^4 2x \sec 2x \tan 2x\,dx$, and let $u = \sec 2x$.

Then $du = 2\sec 2x \tan 2x\,dx$ and

$\int \sec^4 2x \sec 2x \tan 2x\,dx = \frac{1}{2}\int u^4\,du$

$= \frac{1}{10}u^5 + C = \frac{1}{10}\sec^5 2x + C.$

4 Let $u = \tan x$ and $du = \sec^2 x\,dx$. So

$\int \tan^3 x \sec^2 x\,dx = \int u^3\,du = \frac{1}{4}u^4 + C$

$= \frac{1}{4}\tan^4 x + C.$

5 The partial-fraction representation of $\frac{1}{x^4 - 1}$ is

$-\frac{1}{2(x^2 + 1)} - \frac{1}{4(x + 1)} + \frac{1}{4(x - 1)}$. Thus

$\int \frac{dx}{x^4 - 1}$

$= -\frac{1}{2}\int \frac{dx}{1 + x^2} - \frac{1}{4}\int \frac{dx}{x + 1} + \frac{1}{4}\int \frac{dx}{x - 1}$

$= -\frac{1}{2}\tan^{-1}x - \frac{1}{4}\ln|x + 1| + \frac{1}{4}\ln|x - 1| + C$

$= -\frac{1}{2}\tan^{-1}x + \frac{1}{4}\ln\left|\frac{x - 1}{x + 1}\right| + C.$

6 Note $\frac{x^4}{x^4 - 1} = \frac{(x^4 - 1) + 1}{x^4 - 1} = 1 + \frac{1}{x^4 - 1}$

Therefore $\int \frac{x^4}{x^4 - 1}\,dx = \int \left(1 + \frac{1}{x^4 - 1}\right)dx$

$= x + \frac{1}{4}\ln\left|\frac{x - 1}{x + 1}\right| - \frac{1}{2}\tan^{-1}x + C$ from

Exercise 5.

7 Let $u = \ln x$, so $du = dx/x$. Then $\int \frac{\tan(\ln x)}{x}\,dx$

$= \int \tan u\,du = -\ln|\cos u| + C$

$= -\ln|\cos(\ln x)| + C.$

8 Let $u = x^3$, so $du = 3x^2\,dx$. Then $\int x^2 \cos^2 x^3\,dx$

$= \frac{1}{3}\int \cos^2 u\,du = \frac{1}{3}\left(\frac{u}{2} + \frac{\sin 2u}{4}\right) + C$

$= \frac{1}{6}x^3 + \frac{1}{12}\sin 2x^3 + C.$

9 Let $u = x^3$, so $du = 3x^2\,dx$. Then $\int x^2 \sec x^3\,dx$

$= \frac{1}{3}\int \sec u\,du = \frac{1}{3}\ln|\sec u + \tan u| + C$

$= \frac{1}{3}\ln|\sec x^3 + \tan x^3| + C.$

10 $\int \frac{x^4 - \sqrt{x}}{x^3}\,dx = \int (x - x^{-5/2})\,dx$

$= \frac{x^2}{2} + \frac{2}{3}x^{-3/2} + C$

11 Let $u = x + 3$, so $du = dx$. Then

$\int \frac{dx}{(x + 3)\sqrt{x + 3}} = \int \frac{dx}{(x + 3)^{3/2}} =$

$\int u^{-3/2}\,du = -2u^{-1/2} + C = -2(x + 3)^{-1/2} + C.$

12 From Exercise 51 of Sec. 7.5, $\int \frac{x^3\,dx}{x^3 + 1} =$

$x - \frac{1}{3}\ln|x + 1| + \frac{1}{6}\ln|x^2 - x + 1| -$

$\frac{1}{\sqrt{3}}\tan^{-1}\frac{2x - 1}{\sqrt{3}} + C$. So $\int_0^1 \frac{x^3}{x^3 + 1}\,dx =$

$x - \frac{1}{3}\ln|x + 1| + \frac{1}{6}\ln|x^2 - x + 1| -$

$\frac{1}{\sqrt{3}}\tan^{-1}\frac{2x - 1}{\sqrt{3}}\Big|_0^1 =$

$\left(1 - \frac{1}{3}\ln 2 + \frac{1}{6}\ln 1 - \frac{1}{\sqrt{3}}\tan^{-1}\frac{1}{\sqrt{3}}\right) -$

$\left(0 - \frac{1}{3}\ln 1 + \frac{1}{6}\ln 1 - \frac{1}{\sqrt{3}}\tan^{-1}\left(\frac{-1}{\sqrt{3}}\right)\right)$

$= 1 - \frac{1}{3}\ln 2 - \frac{2}{\sqrt{3}}\tan^{-1}\frac{1}{\sqrt{3}}$

$= 1 - \frac{1}{3}\ln 2 - \frac{\pi}{3\sqrt{3}}.$

13 We shall perform integration by parts with $u = x$ and $dv = \cos 3x\,dx$. So $du = dx$ and $v = \frac{1}{3}\sin 3x$. Thus $\int x\cos 3x\,dx$

$= \frac{x\sin 3x}{3} - \int \frac{1}{3}\sin 3x\,dx$

$= \frac{x\sin 3x}{3} + \frac{\cos 3x}{9} + C.$

14 Integrating by parts with $u = \tan^{-1} 5x$ and $dv = dx$, we have $du = \frac{5\,dx}{1 + 25x^2}$, $v = x$, and

$\int \tan^{-1} 5x\,dx = x\tan^{-1} 5x - \int \frac{5x\,dx}{1 + 25x^2}$

$= x\tan^{-1} 5x - \frac{1}{10}\ln(1 + 25x^2) + C.$

15 Note that the integral is almost of the form $\int \frac{f'(x)}{f(x)}\,dx = \ln|f(x)| + C$. Then $\int_{\sqrt{3}}^2 \frac{x}{7 - x^2}\,dx$

$= -\frac{1}{2}\int_{\sqrt{3}}^2 \frac{-2x}{7 - x^2}\,dx = -\frac{1}{2}\ln|7 - x^2|\Big|_{\sqrt{3}}^2$

$= -\frac{1}{2}\left[\ln|7 - 2^2| - \ln|7 - (\sqrt{3})^2|\right]$

$= -\frac{1}{2}(\ln 3 - \ln 4) = -\frac{1}{2}\ln\frac{3}{4} = \frac{1}{2}\ln\frac{4}{3}$

$= \ln\frac{2}{\sqrt{3}}.$

16 Let $7u^2 = x^2$, so $\sqrt{7}u = x$ and $\sqrt{7}\,du = dx$. Then $\int_0^{7/4} \frac{dx}{\sqrt{7 - x^2}}$

$= \int_0^{\sqrt{7}/4} \frac{\sqrt{7}\,du}{\sqrt{7 - 7u^2}} = \int_0^{\sqrt{7}/4} \frac{du}{\sqrt{1 - u^2}}$

$= (\sin^{-1} u)\Big|_0^{\sqrt{7}/4} = \sin^{-1}\frac{\sqrt{7}}{4} - \sin^{-1} 0 = \sin^{-1}\frac{\sqrt{7}}{4}.$

17 Make the substitution $u = x^4$, $du = 4x^3\,dx$. Then

$\int \frac{x^3\,dx}{1 + x^8} = \frac{1}{4}\int \frac{du}{1 + u^2} = \frac{1}{4}\tan^{-1} u + C$

$= \frac{1}{4}\tan^{-1} x^4.$

18 From Exercise 28(a) of Sec. 7.3, we know $\int P(x)e^x\,dx = [P(x) - P'(x) + P''(x) - \cdots]e^x + C$. To get the integrand in the form $P(x)e^x$, let $u =$

$-2x$, so $du = -2\,dx$. Then $\int x^3e^{-2x}\,dx =$

$\int\left(-\frac{u}{2}\right)^3 e^u\left(-\frac{du}{2}\right) = \frac{1}{16}\int u^3e^u\,du$. Here $P(u) =$

u^3, so $P'(u) = 3u^2$, $P''(u) = 6u$, and $P'''(u) = 6$

and $P^{(n)}(u) = 0$ for $n \geq 4$. Thus $\int u^3e^u\,du =$

$[u^3 - 3u^2 + 6u - 6]e^u + C$. Finally,

$\int x^3e^{-2x}\,dx$

$= \frac{1}{16}[(-2x)^3 - 3(-2x)^2 + 6(-2x) - 6]e^{-2x} + C =$

$\frac{1}{8}e^{-2x}[-4x^3 - 6x^2 - 6x - 3] + C$.

19 Given the form of the denominator, we shall use partial fractions. Now $\frac{x^2+2}{(x-3)(x+4)(x-1)} =$

$\frac{c_1}{x-3} + \frac{c_2}{x+4} + \frac{c_3}{x-1}$. So $c_1 =$

$\left.\frac{x^2+2}{(x+4)(x-1)}\right|_{x=3} = \frac{9+2}{7\cdot 2} = \frac{11}{14}$, $c_2 =$

$\left.\frac{x^2+2}{(x-3)(x-1)}\right|_{x=-4} = \frac{16+2}{(-7)(-5)} = \frac{18}{35}$, and c_3

$= \left.\frac{x^2+2}{(x-3)(x+4)}\right|_{x=1} = \frac{1+2}{(-2)(5)} = -\frac{3}{10}$. Thus

$\int \frac{x^2+2}{(x-3)(x+4)(x-1)}\,dx =$

$\int\left(\frac{11/14}{x-3} + \frac{18/35}{x+4} - \frac{3/10}{x-1}\right)dx$

$= \frac{11}{14}\ln|x-3| + \frac{18}{35}\ln|x+4| -$

$\frac{3}{10}\ln|x-1| + C$.

20 First, $\int \frac{x^3}{x^3-1}\,dx = \int \frac{(x^3-1)+1}{x^3-1}\,dx =$

$\int dx + \int \frac{dx}{x^3-1}$. In the last integral, note that $x^3 - 1 = (x-1)(x^2+x+1)$, where x^2+x+1 is irreducible. The method of partial fractions applies.

Thus $\frac{1}{x^3-1} = \frac{c_1}{x-1} + \frac{c_2x+c_3}{x^2+x+1}$, and $1 =$

$(x^2+x+1)c_1 + (x-1)(c_2x+c_3)$. Selecting $x =$ 1, 0, and -1 gives $c_1 = 1/3$, $c_2 = -1/3$, and $c_3 =$

$-2/3$. Hence $\int \frac{x^3}{x^3-1}\,dx$

$= \int dx + \int \frac{1/3}{x-1}\,dx - \int \frac{(1/3)x + 2/3}{x^2+x+1}\,dx$

$= x + \frac{1}{3}\ln|x-1| - \frac{1}{6}\int \frac{2x+4}{x^2+x+1}\,dx$

$= x + \frac{1}{3}\ln|x-1| - \frac{1}{6}\int \frac{2x+1}{x^2+x+1}\,dx -$

$\frac{1}{6}\int \frac{3\,dx}{x^2+x+1} = x + \frac{1}{3}\ln|x-1| -$

$\frac{1}{6}\ln|x^2+x+1| - \frac{1}{2}\int \frac{dx}{x^2+x+1}$

$= x + \frac{1}{3}\ln|x-1| - \frac{1}{6}\ln|x^2+x+1| -$

$\frac{1}{\sqrt{3}}\tan^{-1}\frac{2x+1}{\sqrt{3}} + C$, where the substitution $\frac{3}{4}u^2 = \left(x+\frac{1}{2}\right)^2$ was used for the evaluation of the last integral.

21 Let $9x^2 = 4\sin^2\theta$; that is, $3x = 2\sin\theta$ and $dx = (2/3)\cos\theta\,d\theta$. Then $\int \sqrt{4 - 9x^2}\,dx$

$$= \int (2\cos\theta)\left(\frac{2}{3}\cos\theta\right)d\theta = \frac{4}{3}\int \cos^2\theta\,d\theta$$

$$= \frac{4}{3}\int \frac{1 + \cos 2\theta}{2}\,d\theta = \frac{2}{3}\left(\theta + \frac{\sin 2\theta}{2}\right) + C$$

$$= \frac{2}{3}(\theta + \sin\theta\cos\theta) + C$$

$$= \frac{2}{3}\left(\sin^{-1}\frac{3x}{2} + \frac{3x}{2}\cdot\frac{\sqrt{4 - 9x^2}}{2}\right) + C$$

$$= \frac{2}{3}\sin^{-1}\frac{3x}{2} + \frac{x}{2}\sqrt{4 - 9x^2} + C.$$

22 Let $9x^2 = 16\tan^2\theta$; that is, $3x = 4\tan\theta$ and $dx = (4/3)\sec^2\theta\,d\theta$. Then $\int \frac{dx}{\sqrt{9x^2 + 16}} =$

$$\int \frac{4/3\,\sec^2\theta\,d\theta}{4\sec\theta} = \frac{1}{3}\int \sec\theta\,d\theta$$

$$= \frac{1}{3}\ln|\sec\theta + \tan\theta| + K$$

$$= \frac{1}{3}\ln\left|\frac{\sqrt{9x^2 + 16}}{4} + \frac{3x}{4}\right| + K$$

$$= \frac{1}{3}\ln\left(\sqrt{9x^2 + 16} + 3x\right) + C.$$

23 If we let $\theta = 3x$, we have $\int \frac{dx}{\sin^5 3x} =$

$\frac{1}{3}\int (\csc^5 3x)(3\,dx) = \frac{1}{3}\int \csc^5\theta\,d\theta$. Use integration by parts on $\int \csc^5\theta\,d\theta$ with $u = \csc^3\theta$ and $dv = \csc^2\theta\,d\theta$; then $du = -3\csc^3\theta\cot\theta\,d\theta$, $v = -\cot\theta$, and we have $\int \csc^5\theta\,d\theta =$

$$-\csc^3\theta\cot\theta - \int (\cot\theta)(3\csc^3\theta\cot\theta\,d\theta)$$

$$= -\csc^3\theta\cot\theta - 3\int \cot^2\theta\csc^3\theta\,d\theta$$

$$= -\csc^3\theta\cot\theta - 3\int (\csc^2\theta - 1)\csc^3\theta\,d\theta =$$

$$-\csc^3\theta\cot\theta - 3\int \csc^5\theta\,d\theta + 3\int \csc^3\theta\,d\theta.$$

Collecting occurrences of $\int \csc^5\theta\,d\theta$ on the left side of the equation and dividing through by 4, we obtain $\int \csc^5\theta\,d\theta = -\frac{1}{4}\csc^3\theta\cot\theta + \frac{3}{4}\int \csc^3\theta\,d\theta$. Now integrating $\int \csc^3\theta\,d\theta$ with $u = \csc\theta$ and $dv = \csc^2\theta\,d\theta$, we have $du = -\csc\theta\cot\theta\,d\theta$ and $v = -\cot\theta$, so $\int \csc^3\theta\,d\theta$

$$= -\csc\theta\cot\theta - \int (\cot\theta)(\csc\theta\cot\theta\,d\theta)$$

$$= -\csc\theta\cot\theta - \int \cot^2\theta\csc\theta\,d\theta$$

$$= -\csc\theta\cot\theta - \int (\csc^2\theta - 1)\csc\theta\,d\theta$$

$$= -\csc\theta\cot\theta - \int \csc^3\theta\,d\theta + \int \csc\theta\,d\theta.$$

Collecting occurrences of $\int \csc^3\theta\,d\theta$ on the left side of the equation and dividing through by 2 yields $\int \csc^3\theta\,d\theta =$

$$-\frac{1}{2}\csc\theta\cot\theta + \frac{1}{2}\int \csc\theta\,d\theta =$$

$$-\frac{1}{2}\csc\theta\cot\theta + \frac{1}{2}\ln|\csc\theta - \cot\theta| + C_1,$$

where we used the result of Exercise 31 of Sec. 7.6 for $\int \csc\theta\,d\theta$. Substituting back into the result for $\int \csc^5\theta\,d\theta$, we obtain $\int \csc^5\theta\,d\theta =$

$-\frac{1}{4}\csc^3\theta\cot\theta - \frac{3}{8}\csc\theta\cot\theta +$

$\frac{3}{8}\ln|\csc\theta - \cot\theta| + C_2$. Finally, then,

$$\int\frac{dx}{\sin^5 3x} = \frac{1}{3}\int\csc^5\theta\,d\theta$$

$$= -\frac{1}{12}\csc^3\theta\cot\theta - \frac{1}{8}\csc\theta\cot\theta + \frac{1}{8}\ln|\csc\theta - \cot\theta| + C$$

$$= -\frac{1}{12}\csc^3 3x\cot 3x - \frac{1}{8}\csc 3x\cot 3x + \frac{1}{8}\ln|\csc 3x - \cot 3x| + C.$$

24 If we make the trigonometric substitution $x = 3\sec\theta$, then $dx = 3\sec\theta\tan\theta\,d\theta$ and

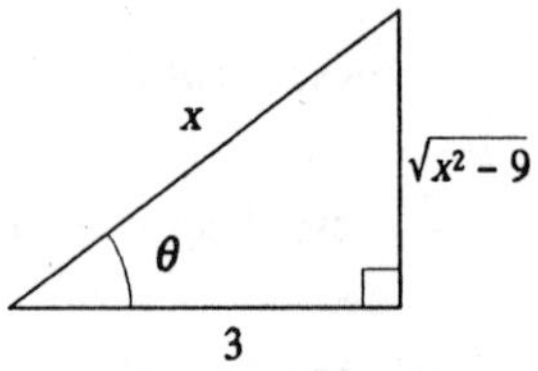

$$\int\frac{x^2\,dx}{\sqrt{x^2-9}} = \int\frac{(9\sec^2\theta)(3\sec\theta\tan\theta\,d\theta)}{3\tan\theta} =$$

$9\int\sec^3\theta\,d\theta$. The recursion formula of Exercise 30 of Sec. 7.6 (see also Formula 56 from the table of antiderivatives) yields $9\int\sec^3\theta\,d\theta$

$$= 9\left(\frac{\sec\theta\tan\theta}{2} + \frac{1}{2}\int\sec\theta\,d\theta\right)$$

$$= \frac{9}{2}(\sec\theta\tan\theta + \ln|\sec\theta + \tan\theta|) + C_1$$

$$= \frac{9}{2}\left(\frac{x}{3}\cdot\frac{\sqrt{x^2-9}}{3} + \ln\left|\frac{x}{3} + \frac{\sqrt{x^2-9}}{3}\right|\right) + C_1$$

$$= \frac{9}{2}\left(\frac{x\sqrt{x^2-9}}{9} + \ln\left|x + \sqrt{x^2-9}\right|\right) + C$$

$$= \frac{x}{2}\sqrt{x^2-9} + \frac{9}{2}\ln\left|x + \sqrt{x^2-9}\right| + C.$$

25 (See Exercise 53 of Sec. 7.6.) Let $u = \tan(x/2)$.

Then $$\int\frac{dx}{3+\cos x} = \int\frac{\frac{2\,du}{1+u^2}}{3 + \frac{1-u^2}{1+u^2}}$$

$$= \int\frac{2\,du}{3(1+u^2) + 1 - u^2} = 2\int\frac{du}{2u^2+4}$$

$$= \int\frac{du}{u^2+2} = \frac{1}{\sqrt{2}}\tan^{-1}\frac{u}{\sqrt{2}} + C$$

$$= \frac{1}{\sqrt{2}}\tan^{-1}\left(\frac{\tan x/2}{\sqrt{2}}\right) + C.$$

7.S Review Exercises

1 (a) Let $u = 1 + \cos\theta$, $du = -\sin\theta\,d\theta$. As θ varies from 0 to $\pi/2$, u goes from 2 to 1.

Thus $\int_0^{\pi/2}\sqrt{(1+\cos\theta)^3}\sin\theta\,d\theta =$

$$\int_2^1 u^{3/2}(-du) = \int_1^2 u^{3/2}\,du.$$

(b) $$\int_1^2 u^{3/2}\,du = \left.\frac{u^{5/2}}{5/2}\right|_1^2 = \frac{2}{5}(4\sqrt{2} - 1)$$

3 (a) $$\int_1^2(1+x^3)^2\,dx = \int_1^2(1 + 2x^3 + x^6)\,dx$$

$$= \left.\left(x + \frac{x^4}{2} + \frac{x^7}{7}\right)\right|_1^2$$

$$= \left(2 + \frac{16}{2} + \frac{128}{7}\right) - \left(1 + \frac{1}{2} + \frac{1}{7}\right) = \frac{373}{14}$$

(b) Rather than multiply out the integrand, let $u = 1 + x^3$ and $du = 3x^2\,dx$. Then

$$\int_1^2 (1 + x^3)^2 x^2\,dx = \int_2^9 u^2 \frac{du}{3} = \frac{u^3}{9}\Big|_2^9$$

$$= \frac{721}{9}.$$

5 (a) $\int \frac{dx}{x^3} = \int x^{-3}\,dx = \frac{x^{-2}}{-2} + C$

$$= -\frac{1}{2}x^{-2} + C$$

(b) Let $u = x + 1$. Then $du = dx$ and

$$\int \frac{dx}{\sqrt{x + 1}} = \int u^{-1/2}\,du = 2u^{1/2} + C$$

$$= 2\sqrt{x + 1} + C.$$

(c) Let $u = 1 + 5e^x$, $du = 5e^x\,dx$. Then

$$\int \frac{e^x\,dx}{1 + 5e^x} = \frac{1}{5}\int \frac{5e^x\,dx}{1 + 5e^x} = \frac{1}{5}\int \frac{du}{u}$$

$$= \frac{1}{5}\ln|u| + C = \frac{1}{5}\ln(1 + 5e^x) + C.$$

7 (a) $\int_0^3 \frac{x^3\,dx}{\sqrt{x + 1}} = \int_1^4 \frac{(u - 1)^3}{\sqrt{u}}\,du$

$$= \int_1^4 \frac{u^3 - 3u^2 + 3u - 1}{u^{1/2}}\,du$$

$$= \int_1^4 (u^{5/2} - 3u^{3/2} + 3u^{1/2} - u^{-1/2})\,du$$

$$= \left(\frac{2}{7}u^{7/2} - \frac{6}{5}u^{5/2} + 2u^{3/2} - 2u^{1/2}\right)\Big|_1^4 = \frac{388}{35}$$

(b) $\int_0^3 \frac{x^3}{\sqrt{x + 1}}\,dx = \int_1^2 \frac{(u^2 - 1)^3}{u} 2u\,du$

$$= 2\int_1^2 (u^6 - 3u^4 + 3u^2 - 1)\,du$$

$$= 2\left(\frac{1}{7}u^7 - \frac{3}{5}u^5 + u^3 - u\right)\Big|_1^2 = \frac{388}{35}$$

9 (a) Use integration by parts with $u = \ln(1 + x)$ and $dv = x^2\,dx$; then $du = \frac{dx}{1 + x}$, $v = \frac{x^3}{3}$,

and $\int x^2 \ln(1 + x)\,dx$

$$= \frac{x^3}{3}\ln(1 + x) - \frac{1}{3}\int \frac{x^3\,dx}{1 + x} =$$

$$\frac{x^3}{3}\ln(1 + x) - \frac{1}{3}\int\left(x^2 - x + 1 - \frac{1}{x + 1}\right)dx =$$

$$\frac{x^3}{3}\ln(1 + x) - \frac{1}{3}\left(\frac{x^3}{3} - \frac{x^2}{2} + x - \ln(1 + x)\right)$$

$$= \frac{x^3 + 1}{3}\ln(1 + x) - \frac{x^3}{9} + \frac{x^2}{6} - \frac{x}{3} + C.$$

(b) Make the substitution $u = 1 + x$, $du = dx$.

Then $\int x^2 \ln(1 + x)\,dx =$

$$\int (u - 1)^2 \ln u\,du =$$

$$\int u^2 \ln u\,du - 2\int u \ln u\,du + \int \ln u\,du.$$

Formula 66 from the table of antiderivatives

then gives $u^3\left[\frac{\ln u}{3} - \frac{1}{9}\right] - 2u^2\left[\frac{\ln u}{2} - \frac{1}{4}\right]$

$$+ u\left[\frac{\ln u}{1} - 1\right] + C_1 = \left(\frac{u^3}{3} - u^2 + u\right)\ln u$$

$$- \frac{u^3}{9} + \frac{u^2}{2} - u + C_1$$

$$= \left(\frac{(1 + x)^3}{3} - (1 + x)^2 + 1 + x\right)\ln(1 + x) -$$

$$\frac{(1 + x)^3}{9} + \frac{(1 + x)^2}{2} - 1 - x + C_1$$

$= \frac{1}{3}(x^3 + 1)\ln(1 + x) - \frac{1}{9}x^3 +$

$\frac{1}{6}x^2 - \frac{1}{3}x + C.$

11 Note that $x^3 - 1$ factors as $(x - 1)(x^2 + x + 1)$. Therefore, the partial-fraction representation of $\frac{2x^2 + 3x + 1}{x^3 - 1}$ has the form $\frac{c_1}{x - 1} + \frac{c_2x + c_3}{x^2 + x + 1}$; then $2x^2 + 3x + 1 = c_1(x^2 + x + 1) + (c_2x + c_3)(x - 1)$; when $x = 1$, we get $6 = 3c_1$, so $c_1 = 2$. When $x = 0$, we get $1 = c_1 - c_3 = 2 - c_3$, so $c_3 = 1$. Finally, when $x = -1$, we get $0 = c_1 - 2(-c_2 + c_3) = 2 + 2c_2 - 2 = 2c_2$. Hence $c_2 = 0$ and the partial-fraction representation is $\frac{2}{x - 1} + \frac{1}{x^2 + x + 1}$.

13 The denominator factors as $(x + 1)(x^2 - x + 1)$. The partial-fraction representation has the form $\frac{2x - 1}{x^3 + 1} = \frac{c_1}{x + 1} + \frac{c_2x + c_3}{x^2 - x + 1}$. Then $2x - 1 = c_1(x^2 - x + 1) + (c_2x + c_3)(x + 1)$. When $x = -1$, this yields $-3 = 3c_1$, so $c_1 = -1$. When $x = 0$, it yields $-1 = c_1 + c_3 = -1 + c_3$, so $c_3 = 0$. Finally, when $x = 1$, $1 = c_1 + 2(c_2 + c_3) = -1 + 2c_2$, so $c_2 = 1$. Hence $\frac{2x - 1}{x^3 + 1} =$

$\frac{-1}{x + 1} + \frac{x}{x^2 - x + 1}.$

15 $\frac{2x + 5}{x^2 + 3x + 2} = \frac{2x + 5}{(x + 1)(x + 2)}$

$= \frac{3}{x + 1} - \frac{1}{x + 2}$

17 Multiplied out, the denominator of $\frac{5x^3 + 6x^2 + 8x + 5}{(x^2 + 1)(x + 1)}$ is $x^3 + x^2 + x + 1$. By long division, we obtain $\frac{5x^3 + 6x^2 + 8x + 5}{x^3 + x^2 + x + 1} =$

$5 + \frac{x^2 + 3x}{x^3 + x^2 + x + 1} = 5 + \frac{x^2 + 3x}{(x^2 + 1)(x + 1)}.$

By partial fractions, we then have

$5 + \frac{x^2 + 3x}{(x^2 + 1)(x + 1)} = 5 + \frac{2x + 1}{x^2 + 1} - \frac{1}{x + 1}.$

19 (a) $\int \frac{x^3\,dx}{(x - 1)^2} = \int \left(x + 2 + \frac{3x - 2}{(x - 1)^2}\right) dx$

$= \int \left(x + 2 + \frac{3}{x - 1} + \frac{1}{(x - 1)^2}\right) dx$

$= \frac{1}{2}x^2 + 2x + 3\ln|x - 1| - \frac{1}{x - 1} + C$

(b) $\int \frac{x^3\,dx}{(x - 1)^2} = \int \frac{(u + 1)^3\,du}{u^2}$

$= \int \left(u + 3 + \frac{3}{u} + \frac{1}{u^2}\right) du$

$= \frac{1}{2}u^2 + 3u + 3\ln|u| - \frac{1}{u} + C$

$= \frac{1}{2}(x - 1)^2 + 3(x - 1) + 3\ln|x - 1| - \frac{1}{x - 1} + C$

21 (a) $\int x\sqrt[3]{x + 1}\,dx = \int (u^3 - 1)u \cdot 3u^2\,du$

$= \int (3u^6 - 3u^3)\,du = \frac{3}{7}u^7 - \frac{3}{4}u^4 + C$

$$= \frac{3}{7}(x+1)^{7/3} - \frac{3}{4}(x+1)^{4/3} + C$$

$$= \frac{3}{28}(x+1)^{4/3}(4x-3) + C$$

(b) $\int x\sqrt[3]{x+1}\,dx = \int (u-1)\sqrt[3]{u}\,du$

$$= \int (u^{4/3} - u^{1/3})\,du = \frac{3}{7}u^{7/3} - \frac{3}{4}u^{4/3} + C$$

$$= \frac{3}{7}(x+1)^{7/3} - \frac{3}{4}(x+1)^{4/3} + C$$

$$= \frac{3}{28}(x+1)^{4/3}(4x-3) + C$$

23 Let $u = x^4 + 1$, $du = 4x^3\,dx$; then $x^3\,dx = du/4$. As x goes from 0 to 1, u goes from 1 to 2, so

$$\int_0^1 (x^4+1)^5 x^3\,dx = \int_1^2 u^5 \cdot \frac{du}{4} = \frac{u^6}{24}\bigg|_1^2 =$$

$$\frac{2^6-1}{24} = \frac{21}{8}.$$

25 Let $u = 1 + \sin\theta$ and $du = \cos\theta\,d\theta$. As θ goes from 0 to $\pi/2$, u varies from 1 to 2. Thus

$$\int_0^{\pi/2} \frac{\cos\theta\,d\theta}{\sqrt{1+\sin\theta}} = \int_1^2 \frac{du}{\sqrt{u}} = 2\sqrt{u}\,\Big|_1^2$$

$$= 2(\sqrt{2} - 1).$$

27 (a) Not elementary

(b) The identity $\sqrt{1-\sin^2\theta} = \cos\theta$ is valid for $-\pi/2 \le \theta \le \pi/2$; for such values of θ, we have $\int \sqrt{4 - 4\sin^2\theta}\,d\theta =$

$$2\int \sqrt{1-\sin^2\theta}\,d\theta = 2\int \cos\theta\,d\theta$$

$$= 2\sin\theta + C.$$

(c) The identity $\sqrt{\cos^2(\theta/2)} = \cos\theta/2$ is valid for $-\pi/2 \le \theta/2 \le \pi/2$; that is, when $-\pi \le \theta \le \pi$. For such values of θ, $\int \sqrt{1+\cos\theta}\,d\theta$

$$= \int \sqrt{2\cos^2(\theta/2)}\,d\theta = \sqrt{2}\int \cos\frac{\theta}{2}\,d\theta$$

$$= 2\sqrt{2}\sin\frac{\theta}{2} + C.$$

29 Let $u = 5\theta$, $du = 5\,d\theta$. Then $\int \csc 5\theta\,d\theta =$

$$\frac{1}{5}\int \csc u\,du = \frac{1}{5}\ln|\csc u - \cot u| + C$$

$$= \frac{1}{5}\ln|\csc 5x - \cot 5x| + C.$$

31 (a) If $u = 1 + x^2$, then $x^2 = u - 1$ and $2x\,dx = du$. Thus $\int \frac{x^3\,dx}{(1+x^2)^4} = \frac{1}{2}\int \frac{(u-1)\,du}{u^4}$

$$= \frac{1}{2}\int (u^{-3} - u^{-4})\,du$$

$$= -\frac{1}{4}u^{-2} + \frac{1}{6}u^{-3} + C$$

$$= -\frac{1}{4}(1+x^2)^{-2} + \frac{1}{6}(1+x^2)^{-3} + C$$

$$= -\frac{1}{12}(1+3x^2)(1+x^2)^{-3} + C.$$

(b) If $x = \tan\theta$, then $dx = \sec^2\theta\,d\theta$, so $\int \frac{x^3\,dx}{(1+x^2)^4}$

$$= \int \frac{\tan^3\theta\sec^2\theta\,d\theta}{\sec^8\theta} = \int \sin^3\theta\cos^3\theta\,d\theta$$

$= \int \sin^3\theta\,(1-\sin^2\theta)\cos\theta\,d\theta$. Let $u = \sin\theta$, $du = \cos\theta\,d\theta$; the integral becomes

$$\int (u^3 - u^5)\, du = \frac{u^4}{4} - \frac{u^6}{6} + C_1$$

$$= \frac{1}{4}\sin^4\theta - \frac{1}{6}\sin^6\theta + C_1$$

$$= \frac{1}{4}\cdot\frac{x^4}{(1+x^2)^2} - \frac{1}{6}\cdot\frac{x^6}{(1+x^2)^3} + C_1$$

$$= \frac{1}{12}(1 + x^2)^{-3}(3x^4 + x^6) + C_1$$

$$= \frac{1}{12}(1+x^2)^{-3}[(1+x^2)^3 - 1 - 3x^2] + C_1$$

$$= \frac{1}{12} - \frac{1}{12}(1+3x^2)(1+x^2)^{-3} + C_1$$

$$= -\frac{1}{12}(1+3x^2)(1+x^2)^{-3} + C.$$

33 (a) If $u = \sqrt{1+x}$, then $u^2 = 1 + x$, $2u\,du = dx$, and $\int \frac{x^2}{\sqrt{1+x}}\,dx = \int \frac{(u^2-1)^2\,2u\,du}{u}$

$$= 2\int (u^2 - 1)^2\,du.$$

(b) If $u = 1 + x$, then $x = u - 1$, $dx = du$, and

$$\int \frac{x^2}{\sqrt{1+x}}\,dx = \int \frac{(u-1)^2\,du}{u^{1/2}}.$$

(c) If $x = \tan^2\theta$, then $dx = 2\tan\theta\sec^2\theta\,d\theta$, $\sqrt{1+x} = \sec\theta$, and $\int \frac{x^2}{\sqrt{1+x}}\,dx$

$$= \int \frac{\tan^4\theta\,(2\tan\theta\sec^2\theta\,d\theta)}{\sec\theta}$$

$$= 2\int \tan^5\theta\sec\theta\,d\theta.$$

(d) The integral in (a) is probably the easiest:

$$2\int (u^2-1)^2\,du = 2\int (u^4 - 2u^2 + 1)\,du$$

$$= 2\left(\frac{u^5}{5} - \frac{2u^3}{3} + u\right) + C =$$

$$\frac{2}{5}(x+1)^{5/2} - \frac{4}{3}(x+1)^{3/2} + 2(x+1)^{1/2} + C.$$

35 (a) $u = x^2$, $du = 2x\,dx$. The integral is

$$\int \sqrt{(1-u)^5}\cdot\frac{1}{2}\,du = \frac{1}{2}\int (1-u)^{5/2}\,du =$$

$$-\frac{1}{7}(1-u)^{7/2} + C = -\frac{1}{7}(1-x^2)^{7/2} + C.$$

(b) $u = 1 - x^2$, $du = -2x\,dx$. The integral is

$$\int \sqrt{u^5}\left(-\frac{1}{2}\,du\right) = -\frac{1}{2}\int u^{5/2}\,du$$

$$= -\frac{1}{7}u^{7/2} + C = -\frac{1}{7}(1-x^2)^{7/2} + C.$$

(c) Let $x = \sin\theta$, where $-\frac{\pi}{2} \le \theta \le \frac{\pi}{2}$. Then $\cos\theta > 0$, so $\sqrt{1-x^2} = \sqrt{1-\sin^2\theta}$ $= \cos\theta$. Hence $\sqrt{(1-x^2)^5} = \cos^5\theta$; also, $dx = \cos\theta\,d\theta$. The integral is therefore

$$\int \sin\theta\cos^5\theta\cos\theta\,d\theta$$

$$= -\int \cos^6\theta\,(-\sin\theta\,d\theta) = -\frac{1}{7}\cos^7\theta + C$$

$$= -\frac{1}{7}\left(\sqrt{1-x^2}\right)^7 + C = -\frac{1}{7}(1-x^2)^{7/2} + C.$$

37 Let $x = 3\sin\theta$, where $-\pi/2 \le \theta \le \pi/2$, so $\cos\theta > 0$. Then $dx = 3\cos\theta\,d\theta$ and $\int (9 - x^2)^{3/2}\,dx$

$$= \int 3\cos^3\theta\,(3\cos\theta\,d\theta) = 9\int \cos^4\theta\,d\theta.$$

Note that $\frac{1}{4}\sin^2 2\theta = \sin^2\theta\cos^2\theta$ and

$$9\int \cos^4\theta\,d\theta = 9\int \cos^2\theta\,(1 - \sin^2\theta)\,d\theta$$

$= 9\int \cos^2\theta\,d\theta - \frac{9}{4}\int \sin^2 2\theta\,d\theta$. Now let $u =$ 2θ so $du = 2\,d\theta$. Then the last integral becomes $\frac{1}{2}\int \sin^2 u\,du$. Thus

$$9\int \cos^2\theta\,d\theta - \frac{9}{8}\int \sin^2 u\,du$$

$$= 9\left(\frac{\theta}{2} + \frac{\sin 2\theta}{4}\right) - \frac{9}{8}\left(\frac{u}{2} - \frac{\sin 2u}{4}\right) + C$$

$$= \frac{9\theta}{2} + \frac{9}{4}\sin 2\theta - \frac{9u}{16} + \frac{9}{32}\sin 2u + C$$

$$= \frac{9\theta}{2} + \frac{9}{4}\sin 2\theta - \frac{18\theta}{16} + \frac{9}{32}\sin 4\theta + C$$

$= \frac{27\theta}{8} + \frac{9}{4}\sin 2\theta + \frac{9}{32}\sin 4\theta + C$. Since $\sin\theta$ $= x/3$ and $\cos\theta = \frac{\sqrt{9 - x^2}}{3}$, we have $\theta =$ $\sin^{-1}\frac{x}{3}$, $\sin 2\theta = 2\sin\theta\cos\theta = \frac{2}{9}x\sqrt{9 - x^2}$, $\cos 2\theta = \cos^2\theta - \sin^2\theta = \frac{9 - 2x^2}{9}$, and $\sin 4\theta$ $= 2\sin 2\theta\cos 2\theta = \frac{4}{81}x(9 - 2x^2)\sqrt{9 - x^2}$.

Therefore $\int (9 - x^2)^{3/2}\,dx = \frac{27}{8}\sin^{-1}\frac{x}{3} +$ $\frac{9}{4}\cdot\frac{2}{9}x\sqrt{9 - x^2} + \frac{9}{32}\cdot\frac{4}{81}x(9 - 2x^2)\sqrt{9 - x^2} + C =$ $\frac{27}{8}\sin^{-1}\frac{x}{3} + \frac{x\sqrt{9 - x^2}}{2} + \frac{x(9 - 2x^2)\sqrt{9 - x^2}}{72} + C.$

39 Let $4x^2 = 9\sec^2\theta$, so $x = 3/2\sec\theta$ and $dx =$ $3/2\sec\theta\tan\theta\,d\theta$. Then $\int \sqrt{4x^2 - 9}\,dx =$ $\int (3\tan\theta)\left(\frac{3}{2}\sec\theta\tan\theta\,d\theta\right)$

$$= \frac{9}{2}\int \tan^2\theta\sec\theta\,d\theta$$

$$= \frac{9}{2}\int (\sec^2\theta - 1)\sec\theta\,d\theta$$

$= \frac{9}{2}\int \sec^3\theta\,d\theta - \frac{9}{2}\int \sec\theta\,d\theta =$ $\frac{9}{2}\left(\frac{\sec\theta\tan\theta}{2} + \frac{1}{2}\int \sec\theta\,d\theta\right) - \frac{9}{2}\int \sec\theta\,d\theta$

$$= \frac{9}{4}\sec\theta\tan\theta - \frac{9}{4}\int \sec\theta\,d\theta$$

$$= \frac{9}{4}\sec\theta\tan\theta - \frac{9}{4}\ln|\sec\theta + \tan\theta| + C_1$$

$$= \frac{9}{4}\cdot\frac{2x}{3}\cdot\frac{\sqrt{4x^2 - 9}}{3} - \frac{9}{4}\ln\left|\frac{2x}{3} + \frac{\sqrt{4x^2 - 9}}{3}\right| + C_1$$

$$= \frac{x\sqrt{4x^2 - 9}}{2} - \frac{9}{4}\ln\left|2x + \sqrt{4x^2 - 9}\right| + C.$$

41 (a) $\frac{1}{2}x - \frac{1}{2}\sin x\cos x + C$

(b) Let $x = t^2$; then $\int \sin\sqrt{x}\,dx =$ $\int (\sin t)\,2t\,dt = 2(-t\cos t + \sin t) + C =$ $2\sin\sqrt{x} - 2\sqrt{x}\cos\sqrt{x} + C$ by Example 4 of Sec. 7.3.

43 $\int \frac{x^3\,dx}{(x^4 + 1)^3} = \frac{1}{4}\cdot\frac{(x^4 + 1)^{-2}}{-2}$

$$= -\frac{1}{8(x^4 + 1)^2} + C$$

45 $\int \frac{x^4 + x^2 + 1}{x^3}\,dx = \int (x + x^{-1} + x^{-3})\,dx$

$= \frac{x^2}{2} + \ln|x| - \frac{x^{-2}}{2} + C$

$= \frac{x^2}{2} - \frac{1}{2x^2} + \ln|x| + C$

47 $\int 10^x\,dx = \int e^{\ln 10^x}\,dx = \int e^{x \ln 10}\,dx$

$= \frac{e^{x \ln 10}}{\ln 10} + C = \frac{10^x}{\ln 10} + C$

49 Let $u = 3 + \cos x$, $du = -\sin x\,dx$. Then

$\int \frac{\sin x\,dx}{3 + \cos x} = \int \frac{-du}{u} = -\ln|u| + C$

$= -\ln(3 + \cos x) + C.$

51 With $u = x^3 - 1$ and $du = 3x^2\,dx$, we have

$\int x^2\sqrt{x^3 - 1}\,dx = \frac{1}{3}\int \sqrt{x^3 - 1}(3x^2\,dx)$

$= \frac{1}{3}\int u^{1/2}\,du = \frac{1}{3}\cdot\frac{u^{3/2}}{3/2} + C$

$= \frac{2}{9}(x^3 - 1)^{3/2} + C.$

53 Use integration by parts with $u = \tan^{-1} x$ and $dv = dx/x^2 = x^{-2}\,dx$. Then $du = \frac{dx}{1 + x^2}$, $v = -\frac{1}{x}$, and

$\int \frac{\tan^{-1} x}{x^2}\,dx = -\frac{\tan^{-1} x}{x} + \int \frac{dx}{x(x^2 + 1)}$. The partial-fraction representation of $\frac{1}{x(x^2 + 1)}$ has the form $\frac{c_1}{x} + \frac{c_2 x + c_3}{x^2 + 1}$, where $1 = c_1(x^2 + 1) + (c_2 x + c_3)x$. Substituting $x = 0$, 1, and -1, shows that $c_1 = 1$, $c_2 = -1$, and $c_3 = 0$. Then the integral becomes $\int \left(\frac{1}{x} - \frac{x}{x^2 + 1}\right) dx = \ln|x| - \frac{1}{2}\ln(x^2 + 1) + C$. Therefore

$\int \frac{\tan^{-1} x}{x^2}\,dx$

$= -\frac{\tan^{-1} x}{x} + \ln|x| - \frac{1}{2}\ln(x^2 + 1) + C.$

55 This requires integration by parts twice. Let $I = \int e^x \sin 3x\,dx$. Now choose $u = e^x$ and $dv = \sin 3x\,dx$, so that $du = e^x\,dx$ and $v = -\frac{1}{3}\cos 3x$ and $I = \frac{-e^x \cos 3x}{3} + \frac{1}{3}\int e^x \cos 3x\,dx$. Now let $U = e^x$ and $dv = \cos 3x\,dx$, so that $du = e^x\,dx$, $V = \frac{1}{3}\sin 3x$, and $I = -\frac{1}{3}e^x \cos 3x + \frac{1}{3}\left(\frac{1}{3}e^x \sin 3x - \frac{1}{3}\int e^x \sin 3x\,dx\right)$

$= -\frac{1}{3}e^x \cos 3x + \frac{1}{9}e^x \sin 3x - \frac{1}{9}(I)$. Hence $I = \frac{9}{10}\left(-\frac{1}{3}e^x \cos 3x + \frac{1}{9}e^x \sin 3x\right) + C$

$= \frac{1}{10}e^x(\sin 3x - 3\cos 3x) + C.$

57 Let $u = x^2 + 4$, $du = 2x\,dx$. Then

$\int x\sqrt{x^2 + 4}\,dx = \frac{1}{2}\int \sqrt{x^2 + 4}\,2x\,dx$

$= \frac{1}{2}\int u^{1/2}\,du = \frac{1}{2}\cdot\frac{u^{3/2}}{3/2} + C = \frac{1}{3}u^{3/2} + C$

$= \frac{1}{3}(x^2 + 4)^{3/2} + C.$

59 Let $u = x^3$, $du = 3x^2\,dx$. Then $\int \frac{x^2\,dx}{1 + x^6}$

$= \frac{1}{3}\int \frac{3x^2\,dx}{1 + (x^3)^2} = \frac{1}{3}\int \frac{du}{1 + u^2}$

$= \frac{1}{3}\tan^{-1}u + C = \frac{1}{3}\tan^{-1}x^3 + C.$

61 Let $u = x^3$ and $du = 3x^2\,dx$. Then $\int x^2 \sin x^3\,dx$

$= \frac{1}{3}\int \sin x^3\, 3x^2\,dx = \frac{1}{3}\int \sin u\,du$

$= -\frac{1}{3}\cos u + C = -\frac{1}{3}\cos x^3 + C.$

63 Use integration by parts with $u = \ln x$ and $dv =$ $x^4\,dx$. Then $du = \frac{dx}{x}$, $v = \frac{x^5}{5}$ and $\int x^4 \ln x\,dx$

$= \frac{x^5}{5}\ln x - \int \frac{x^5}{5}\cdot\frac{dx}{x} = \frac{x^5}{5}\ln x - \int \frac{x^4}{5}\,dx =$

$\frac{x^5}{5}\ln x - \frac{1}{25}x^5 + C.$

65 Let $u = \sqrt{x}$ and $du = \frac{dx}{2\sqrt{x}}$. Then $\int \frac{e^{\sqrt{x}}}{\sqrt{x}}\,dx$

$= 2\int e^{\sqrt{x}}\frac{dx}{2\sqrt{x}} = 2\int e^u\,du = 2e^u + C$

$= 2e^{\sqrt{x}} + C.$

67 Let $u = x^2 + 1$, $du = 2x\,dx$. Then $\int \frac{x\,dx}{\sqrt{(x^2 + 1)^3}}$

$= \frac{1}{2}\int u^{-3/2}\,du = -u^{-1/2} + C$

$= -\frac{1}{\sqrt{x^2 + 1}} + C.$

69 $\int \frac{dx}{\sqrt{(x + 1)^3}} = \int (x + 1)^{-3/2}\,dx$; letting $u = x +$ 1 and $du = dx$, we then have $\int u^{-3/2}\,du$

$= -2u^{-1/2} + C = \frac{-2}{\sqrt{x + 1}} + C.$

71 Let $u = x^2$, $du = 2x\,dx$. Then $\int \frac{x\,dx}{x^4 - 2x^2 - 3}$

$= \frac{1}{2}\int \frac{2x\,dx}{(x^2)^2 - 2x^2 - 3} = \frac{1}{2}\int \frac{du}{u^2 - 2u - 3}$

$= \frac{1}{2}\int \frac{du}{(u - 3)(u + 1)}$. Now $\frac{1}{(u - 3)(u + 1)}$ has a partial-fraction representation of the form

$\frac{c_1}{u - 3} + \frac{c_2}{u + 1}$, where $1 = c_1(u + 1) + c_2(u - 3)$.

Substituting $u = 3$ into this equation yields $c_1 = 1/4$, while $u = -1$ produces $c_2 = -1/4$.

Therefore $\frac{1}{2}\int \frac{du}{(u - 3)(u + 1)}$

$= \frac{1}{8}\int \left(\frac{1}{u - 3} - \frac{1}{u + 1}\right) du$

$= \frac{1}{8}(\ln|u - 3| - \ln|u + 1|) + C$

$= \frac{1}{8}\ln\left|\frac{u - 3}{u + 1}\right| + C = \frac{1}{8}\ln\left|\frac{x^2 - 3}{x^2 + 1}\right| + C.$

73 Let $u = \sqrt[3]{x - 1}$, so $x = u^3 + 1$ and $dx = 3u^2\,du$.

Then $\int \frac{x^2\,dx}{\sqrt[3]{x - 1}} = \int \frac{(u^3 + 1)^2\, 3u^2\,du}{u}$

$= 3\int (u^7 + 2u^4 + u)\,du$

$$= 3\left(\frac{u^8}{8} + \frac{2u^5}{5} + \frac{u^2}{2}\right) + C$$

$$= \frac{3}{8}(x - 1)^{8/3} + \frac{6}{5}(x - 1)^{5/3} + \frac{3}{2}(x - 1)^{2/3} + C.$$

75 Let $x = 2 \tan \theta$, $dx = 2 \sec^2 \theta \, d\theta$. Then $\sqrt{x^2 + 4} = 2 \sec \theta$ and $\int \frac{\sqrt{x^2 + 4}}{x} dx =$

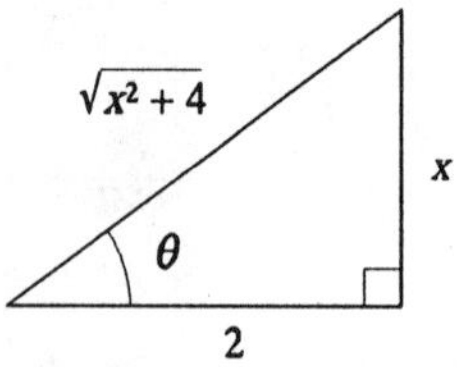

$$\int \frac{2 \sec \theta}{2 \tan \theta} 2 \sec^2 \theta \, d\theta = 2 \int \frac{\sec^3 \theta}{\tan \theta} d\theta$$

$$= 2 \int \frac{\sec \theta (\tan^2 \theta + 1)}{\tan \theta} d\theta$$

$$= 2 \int (\sec \theta \tan \theta + \csc \theta) \, d\theta$$

$$= 2(\sec \theta + \ln|\csc \theta - \cot \theta|) + C.$$

In terms of x, the answer is

$$2\left(\frac{\sqrt{x^2 + 4}}{2} + \ln\left|\frac{\sqrt{x^2 + 4}}{x} - \frac{2}{x}\right|\right) + C$$

$$= \sqrt{x^2 + 4} + 2 \ln\left|\frac{\sqrt{x^2 + 4} - 2}{x}\right| + C.$$

77 $\int \sec^5 \theta \tan \theta \, d\theta = \int \sec^4 \theta \, (\sec \theta \tan \theta \, d\theta)$

$$= \int u^4 \, du = \frac{u^5}{5} + C = \frac{1}{5} \sec^5 \theta + C$$

79 Let $x = 3 \tan \theta$, $dx = 3 \sec^2 \theta \, d\theta$. Then

$$\int \frac{dx}{x\sqrt{x^2 + 9}} = \int \frac{3 \sec^2 \theta \, d\theta}{(3 \tan \theta)(3 \sec \theta)}$$

$$= \frac{1}{3} \int \csc \theta \, d\theta = \frac{1}{3} \ln|\csc \theta - \cot \theta| + C$$

$$= \frac{1}{3}\left|\frac{\sqrt{x^2 + 9}}{x} - \frac{3}{x}\right| + C$$

$$= -\frac{1}{3} \ln\left|\frac{x}{\sqrt{x^2 + 9} - 3} \cdot \frac{\sqrt{x^2 + 9} + 3}{\sqrt{x^2 + 9} + 3}\right| + C$$

$$= -\frac{1}{3} \ln\left|\frac{\sqrt{x^2 + 9} + 3}{x}\right| + C.$$

81 $\int \frac{(1 - x)^2}{\sqrt[3]{x}} dx = \int \frac{1 - 2x + x^2}{x^{1/3}} dx$

$$= \int (x^{-1/3} - 2x^{2/3} + x^{5/3}) \, dx$$

$$= \frac{3}{2}x^{2/3} - \frac{6}{5}x^{5/3} + \frac{3}{8}x^{8/3} + C$$

83 Let $u = \sqrt{3} \cos x$. Then $\int \frac{\sin x \, dx}{1 + 3 \cos^2 x}$

$$= \int \frac{(-1/\sqrt{3}) \, du}{1 + u^2} = -\frac{1}{\sqrt{3}} \tan^{-1} u + C$$

$$= -\frac{1}{\sqrt{3}} \tan^{-1}(\sqrt{3} \cos x) + C.$$

85 $\int \left(e^x - \frac{1}{e^x}\right)^2 dx = \int (e^{2x} - 2 + e^{-2x}) \, dx$

$$= \frac{1}{2}e^{2x} - 2x - \frac{1}{2}e^{-2x} + C$$

87 Let $w = x^2$, $dw = 2x \, dx$. Then $\int x \sin^{-1} x^2 \, dx$

$$= \frac{1}{2} \int (\sin^{-1} x^2)(2x \, dx) = \frac{1}{2} \int \sin^{-1} w \, dw,$$

which, by Formula 67 from the table of antiderivatives with $a = 1$, equals

$$\frac{1}{2}\left(w \sin^{-1} w + \sqrt{1 - w^2}\right) + C$$

$= \frac{1}{2}\left(x^2 \sin^{-1} x^2 + \sqrt{1 - x^4}\right) + C.$

89 Let $u = e^x$; then $du = e^x\,dx = u\,dx$ and $dx = du/u$, so $\int \frac{dx}{e^{2x} + 5e^x} = \int \frac{dx}{(e^x)^2 + 5e^x} = \int \frac{du}{u(u^2 + 5u)} = \int \frac{du}{u^2(u + 5)}$. Now $\frac{1}{u^2(u + 5)}$

$= \frac{c_1}{u} + \frac{c_2}{u^2} + \frac{c_3}{u + 5}$, where $1 = c_1 u(u + 5) + c_2(u + 5) + c_3 u^2$. Substituting $u = 0, -5$, and 1 shows that $c_2 = 1/5$, $c_3 = 1/25$, and $c_1 = -1/25$, so $\int \frac{du}{u^2(u + 5)}$

$= \int \left(\frac{-1}{25u} + \frac{1}{5u^2} + \frac{1}{25(u + 5)}\right) du$

$= -\frac{1}{25} \ln|u| - \frac{1}{5u} + \frac{1}{25} \ln|u + 5| + C$

$= -\frac{1}{25} \ln e^x - \frac{1}{5e^x} + \frac{1}{25} \ln(e^x + 5) + C$

$= -\frac{x}{25} - \frac{1}{5}e^{-x} + \frac{1}{25} \ln(e^x + 5) + C.$

91 Let $u = \sqrt{3x + 2}$ so that $\frac{1}{3}(u^2 - 2) = x$ and $dx = \frac{2}{3}u\,du$. Then $\int (2x + 1)\sqrt{3x + 2}\,dx$

$= \int \left(\frac{2}{3}(u^2 - 2) + 1\right) u \left(\frac{2}{3}u\,du\right)$

$= \frac{2}{9} \int (2u^4 - u^2)\,du = \frac{2}{9}\left(\frac{2}{5}u^5 - \frac{1}{3}u^3\right) + C$

$= \frac{4}{45}(3x + 2)^{5/2} - \frac{2}{27}(3x + 2)^{3/2} + C$

$= \frac{1}{135}(36x + 14)(3x + 2)^{3/2} + C.$

93 Let $x = u + 1$, $dx = du$. Then $\int \frac{x^2\,dx}{(x - 1)^3}$

$= \int \frac{(u + 1)^2\,du}{u^3} = \int \frac{u^2 + 2u + 1}{u^3}\,du$

$= \int (u^{-1} + 2u^{-2} + u^{-3})\,du$

$= \ln|u| - 2u^{-1} - \frac{1}{2}u^{-2} + C$

$= \ln|x - 1| - \frac{2}{x - 1} - \frac{1}{2(x - 1)^2} + C.$

95 Let $u = e^x$. Then $du = e^x\,dx = u\,dx$, so $dx = du/u$ and $\int \frac{e^x + 1}{e^x - 1}\,dx = \int \frac{u + 1}{u - 1} \cdot \frac{du}{u} = \int \frac{(u + 1)\,du}{u(u - 1)}$. The partial-fraction representation of the integrand is $\frac{c_1}{u} + \frac{c_2}{u - 1}$, where $u + 1 = c_1(u - 1) + c_2 u$. Substituting $u = 0$ and 1 shows that $c_1 = -1$ and $c_2 = 2$. Thus $\int \frac{(u + 1)\,du}{u(u - 1)}$

$= \int \left(\frac{-1}{u} + \frac{2}{u - 1}\right) du$

$= -\ln|u| + 2\ln|u - 1| + C$

$= -\ln e^x + 2\ln|e^x - 1| + C$

$= -x + 2\ln|e^x - 1| + C.$

97 $\int (1 + 3x^2)^2\,dx = \int (1 + 6x^2 + 9x^4)\,dx$

$= x + 2x^3 + \frac{9}{5}x^5 + C$

99 Using the solution of Exercise 51 of Sec. 7.5, we

have $\int \frac{x^3}{x^3+1}\,dx = \int \left(1 - \frac{1}{x^3+1}\right) dx$

$= x - \frac{1}{3}\ln|x+1| + \frac{1}{6}\ln(x^2 - x + 1) -$

$\frac{1}{\sqrt{3}}\tan^{-1}\frac{2x-1}{\sqrt{3}} + C.$

101 Let $u = 2x + 1$, $du = 2\,dx$. Then $\int \frac{dx}{\sqrt{2x+1}}$

$= \frac{1}{2}\int \frac{du}{\sqrt{u}} = \frac{1}{2}(2\sqrt{u}) + C = \sqrt{2x+1} + C.$

103 $\int \sin^2 3x \cos^2 3x\,dx = \frac{1}{4}\int (2\sin 3x \cos 3x)^2\,dx$

$= \frac{1}{4}\int \sin^2 6x\,dx$; substituting $u = 6x$, $du =$

$6\,dx$, we have $\frac{1}{4}\int \sin^2 u\,\frac{du}{6}$

$= \frac{1}{24}\left(\frac{u}{2} - \frac{\sin u \cos u}{2}\right) + C$

$= \frac{1}{24}\left(3x - \frac{1}{2}\sin 6x \cos 6x\right) + C$

$= \frac{x}{8} - \frac{1}{48}\sin 6x \cos 6x + C =$

$\frac{x}{8} - \frac{\sin 12x}{96} + C.$

105 Let $u = 3\theta$, $du = 3\,d\theta$. Then $\int \tan^4 3\theta\,d\theta =$ $\frac{1}{3}\int \tan^4 u\,du$. By the recursion formula in Exercise 29 of Sec. 7.6 (see also Formula 53 in the table of antiderivatives), we have $\int \tan^4 u\,du =$ $\frac{1}{3}\tan^3 u - \tan u + u + C_1$. Hence $\int \tan^4 3\theta\,d\theta$

$= \frac{1}{3}\left(\frac{1}{3}\tan^3 u - \tan u + u + C_1\right)$

$= \frac{1}{9}\tan^3 3\theta - \frac{1}{3}\tan 3\theta + \theta + C.$

107 $\int \cos^2 x\,dx = \int \frac{1}{2}(1 + \cos 2x)\,dx$

$= \frac{1}{2}\left(x + \frac{\sin 2x}{2}\right) + C = \frac{1}{2}x + \frac{1}{4}\sin 2x + C$

109 Let $x = 2\sin\theta$. Then

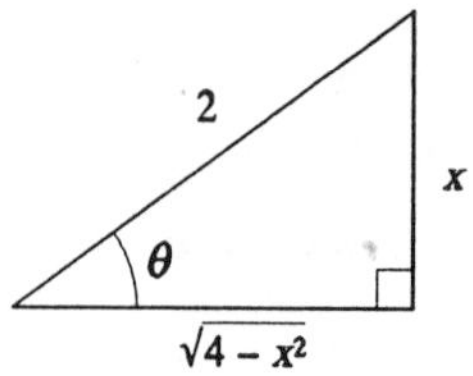

$dx = 2\cos\theta\,d\theta$,

$\sqrt{4 - x^2} = 2\cos\theta$,

and $\int \frac{dx}{(4 - x^2)^{3/2}} =$

$\int \frac{2\cos\theta\,d\theta}{(2\cos\theta)^3} = \int \frac{d\theta}{4\cos^2\theta} = \frac{1}{4}\int \sec^2\theta\,d\theta$

$= \frac{1}{4}\tan\theta + C = \frac{1}{4}\cdot\frac{x}{\sqrt{4 - x^2}} + C$

$= \frac{x}{4\sqrt{4 - x^2}} + C.$

111 Let $u = e^x$, $du = e^x\,dx$. Then $\int \frac{e^x\,dx}{1 + e^{2x}} =$

$\int \frac{du}{1 + u^2} = \tan^{-1} u + C = \tan^{-1}(e^x) + C.$

113 The partial-fraction representation of $\frac{1}{x^2 + 5x + 6}$

$= \frac{1}{(x+2)(x+3)}$ is of the form $\frac{c_1}{x+2} + \frac{c_2}{x+3}$,

where $1 = c_1(x + 3) + c_2(x + 2)$. Substituting $x = -2$ and -3 into this equation yields $c_1 = 1$ and $c_2 = -1$. Thus $\int \frac{dx}{x^2 + 5x + 6} =$

$$\int\left(\frac{1}{x+2} - \frac{1}{x+3}\right) dx =$$

$$\ln|x+2| - \ln|x+3| + C = \ln\left|\frac{x+2}{x+3}\right| + C.$$

115 $\int \frac{4x+10}{x^2+5x+6}\,dx = 2\int \frac{2x+5}{x^2+5x+6}\,dx$

$= 2\ln|x^2+5x+6| + C$

117 Note that $2x^2 + 5x + 6 = 2\left(x^2 + \frac{5}{2}x\right) + 6$

$$= 2\left(x^2 + \frac{5}{2}x + \frac{25}{16}\right) - \frac{25}{8} + 6$$

$$= 2\left(x + \frac{5}{4}\right)^2 + \frac{23}{8} = 2\left[\left(x + \frac{5}{4}\right)^2 + \frac{23}{16}\right].$$

Therefore $\int \frac{dx}{2x^2+5x+6}$

$$= \frac{1}{2}\int \frac{dx}{\left(x+\frac{5}{4}\right)^2 + \frac{23}{16}}$$

$$= \frac{1}{2}\cdot\frac{4}{\sqrt{23}}\tan^{-1}\left[\frac{4}{\sqrt{23}}\left(x+\frac{5}{4}\right)\right] + C$$

$= \frac{2}{\sqrt{23}}\tan^{-1}\frac{4x+5}{\sqrt{23}} + C$. (Use either Formula 9 or 36 from the text's table of antiderivatives.)

119 Note that $2x^2 + 5x - 6 =$

$$2\left(x^2 + \frac{5}{2}x + \frac{25}{16}\right) - \frac{73}{8} = 2\left[\left(x+\frac{5}{4}\right)^2 - \frac{73}{16}\right].$$

Thus $\int \frac{dx}{2x^2+5x-6} = \frac{1}{2}\int \frac{dx}{\left(x+\frac{5}{4}\right)^2 - \frac{73}{16}}$

$$= \frac{1}{2}\cdot\frac{2}{\sqrt{73}}\ln\left|\frac{x+\frac{5}{4}-\frac{\sqrt{73}}{4}}{x+\frac{5}{4}+\frac{\sqrt{73}}{4}}\right| + C$$

$= \frac{1}{\sqrt{73}}\ln\left|\frac{4x+5-\sqrt{73}}{4x+5+\sqrt{73}}\right| + C$. (We used Formula 25 from the text's table of antiderivatives, inverting the quantity inside the absolute value bars because of the difference in signs. Formula 35 is even more direct.)

121 $\int \frac{dx}{\sin^2 x} = \int \csc^2 x\,dx = -\cot x + C$

123 $u = \ln(x^2+5)$, $dv = dx$; $\int \ln(x^2+5)\,dx$

$$= [\ln(x^2+5)]x - \int x\cdot\frac{2x}{x^2+5}\,dx$$

$$= x\ln(x^2+5) - \int\left(2 - \frac{10}{x^2+5}\right)dx$$

$$= x\ln(x^2+5) - 2x + 2\sqrt{5}\tan^{-1}\frac{x}{\sqrt{5}} + C.$$

125 $\int \sqrt{(1+2x)(1-2x)}\,dx = \int \sqrt{1-4x^2}\,dx$; now let $2x = \sin\theta$, so $2\,dx = \cos\theta\,d\theta$ and

$$\int \sqrt{1-4x^2}\,dx = \frac{1}{2}\int \cos^2\theta\,d\theta$$

$$= \frac{1}{4}\int (1+\cos 2\theta)\,d\theta$$

$$= \frac{1}{4}\left(\theta + \frac{1}{2}\sin 2\theta\right) + C$$

$$= \frac{1}{4}(\theta + \sin\theta\cos\theta) + C$$

$= \frac{1}{4}\left(\sin^{-1} 2x + 2x\sqrt{1 - 4x^2}\right) + C.$

127 Let $u = x^2 + 1$, $du = 2x\,dx$. Then $\int \frac{2x\,dx}{\sqrt{x^2 + 1}}$

$= \int \frac{du}{\sqrt{u}} = \int u^{-1/2}\,du = 2u^{1/2} + C$

$= 2\sqrt{x^2 + 1} + C.$

129 The integrand is an improper fraction, so polynomial division is necessary:

$\frac{x^4 + 4x^3 + 6x^2 + 4x - 3}{x^4 - 1} = 1 + \frac{4x^3 + 6x^2 + 4x - 2}{x^4 - 1}$

$= 1 + \frac{4x^3 + 6x^2 + 4x - 2}{(x - 1)(x + 1)(x^2 + 1)}$. The partial-fraction representation of the remainder is

$\frac{c_1}{x - 1} + \frac{c_2}{x + 1} + \frac{c_3x + c_4}{x^2 + 1}$, where $4x^3 + 6x^2 + 4x - 2 = c_1(x + 1)(x^2 + 1) + c_2(x - 1)(x^2 + 1) + (c_3x + c_4)(x - 1)(x + 1)$. Substituting $x = -1$, 0, 1, and 2 into the equation shows that $c_1 = 3$, $c_2 = 1$, $c_3 = 0$, and $c_4 = 4$. Therefore

$\int \frac{x^4 + 4x^3 + 6x^2 + 4x - 3}{x^4 - 1}\,dx =$

$\int \left(1 + \frac{3}{x - 1} + \frac{1}{x + 1} + \frac{4}{x^2 + 1}\right) dx =$

$x + 3\ln|x - 1| + \ln|x + 1| + 4\tan^{-1} x + C.$

131 Use partial fractions: $\frac{12x^2 + 2x + 3}{4x^3 + x} =$

$\frac{c_1}{x} + \frac{c_2x + c_3}{4x^2 + 1}$, where $12x^2 + 2x + 3 = c_1(4x^2 + 1) + (c_2x + c_3)x$. Letting $x = 0$ yields $c_1 = 3$, while the values $x = \pm 1$ show that $c_2 = 0$ and $c_3 = 2$. Thus $\int \frac{12x^2 + 2x + 3}{4x^3 + x}\,dx =$

$\int \left(\frac{3}{x} + \frac{2}{4x^2 + 1}\right) dx = 3\ln|x| + \frac{1}{2}\int \frac{dx}{x^2 + 1/4}$

$= 3\ln|x| + \tan^{-1} 2x + C.$

133 Let $x = \tan\theta$. Then $\int \sqrt{\frac{1}{x^2} + \frac{1}{x^4}}\,dx$

$= \int \frac{\sqrt{x^2 + 1}}{x^2}\,dx = \int \frac{\sec\theta}{\tan^2\theta}\sec^2\theta\,d\theta$

$= \int \sec\theta\csc^2\theta\,d\theta = \int \sec\theta\,(1 + \cot^2\theta)\,d\theta$

$= \int \sec\theta\,d\theta + \int \csc\theta\cot\theta\,d\theta$

$= \ln|\sec\theta + \tan\theta| - \csc\theta + C$

$= \ln\left(\sqrt{1 + x^2} + x\right) - \frac{\sqrt{1 + x^2}}{x} + C.$

135 Let $3x = \sin\theta$, $3\,dx = \cos\theta\,d\theta$; then $\sqrt{1 - 9x^2} = \cos\theta$ and $\int \frac{dx}{\sqrt{1 - 9x^2}} = \int \frac{(1/3)\cos\theta\,d\theta}{\cos\theta}$

$= \frac{1}{3}\int d\theta = \frac{1}{3}\theta + C = \frac{1}{3}\sin^{-1} 3x + C.$

137 Use the trigonometric substitution $\sqrt{3}x = \sqrt{2}\tan\theta$; then $\sqrt{3}\,dx = \sqrt{2}\sec^2\theta\,d\theta$, $3x^2 + 2 = 2\sec^2\theta$, and $\int \frac{dx}{(3x^2 + 2)^{3/2}}$

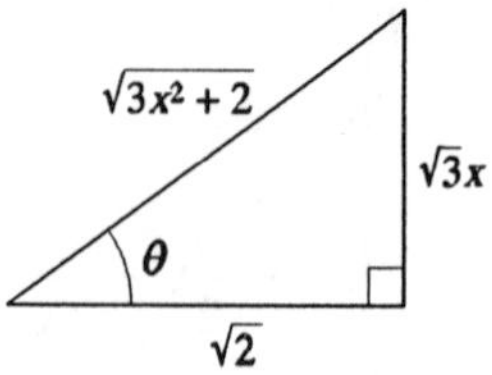

$= \frac{\sqrt{2}}{\sqrt{3}}\int \frac{\sec^2\theta\,d\theta}{2\sqrt{2}\sec^3\theta} = \frac{1}{2\sqrt{3}}\int \cos\theta\,d\theta$

$= \frac{1}{2\sqrt{3}} \sin\theta + C = \frac{1}{2\sqrt{3}} \cdot \frac{\sqrt{3}x}{\sqrt{3x^2 + 2}} + C$

$= \frac{x}{2\sqrt{3x^2 + 2}} + C.$

139 $\int \frac{dx}{\cos 4x} = \frac{1}{4} \int (\sec 4x)(4\ dx)$

$= \frac{1}{4} \ln|\sec 4x + \tan 4x| + C$

141 $\int e^x \sin^2 x\ dx = \int e^x\left(\frac{1 - \cos 2x}{2}\right) dx$

$= \frac{1}{2} \int (e^x - e^x \cos 2x)\ dx$

$= \frac{1}{2}e^x - \frac{1}{2} \int e^x \cos 2x\ dx$. The remaining integral can be done by integration by parts (twice) or with Formula 65 from the table of antiderivatives: $\int e^x \cos 2x\ dx = \frac{1}{5}(e^x \cos 2x + 2e^x \sin 2x) + C_1$. Therefore

$\int e^x \sin^2 2x\ dx$

$= \frac{1}{2}e^x - \frac{1}{10}e^x \cos 2x - \frac{1}{5}e^x \sin 2x + C.$

143 Let $u = \sqrt{x}$. Then $x = u^2$, $dx = 2u\ du$, and

$\int \sqrt{1 + \sqrt{1 + \sqrt{x}}}\ dx = \int \sqrt{1 + \sqrt{1 + u}}\ 2u\ du.$

Let $v = \sqrt{1 + u}$ so that $1 + u = v^2$, $u = v^2 - 1$, $du = 2v\ dv$, and thus $2 \int \sqrt{1 + \sqrt{1 + u}}\ u\ du$

$= 2 \int \sqrt{1 + v}(v^2 - 1)(2v)\ dv =$

$4 \int \sqrt{1 + v}(v^2 - 1)v\ dv$. Now let $w = \sqrt{1 + v}$; then $w^2 - 1 = v$, $2w\ dw = dv$, and

$4 \int \sqrt{1 + v}(v^2 - 1)v\ dv$

$= 4 \int w((w^2 - 1)^2 - 1)(w^2 - 1)\ 2w\ dw$

$= 8 \int w^2(w^2 - 1)(w^4 - 2w^2)\ dw$

$= 8 \int (w^8 - 3w^6 + 2w^4)\ dw$

$= 8\left(\frac{w^9}{9} - \frac{3w^7}{7} + \frac{2w^5}{5}\right) + C$

$= \frac{8}{315} w^5(35w^4 - 135w^2 + 126) + C$

$= \frac{8}{315}(1 + v)^{5/2}(35v^2 - 65v + 26) + C$

$= \frac{8}{315}\left(1 + \sqrt{1 + u}\right)^{5/2}\left(61 + 35u - 65\sqrt{1 + u}\right) + C$

$= \frac{8}{315}\left(1 + \sqrt{1 + \sqrt{x}}\right)^{5/2}\left(61 + 35\sqrt{x} - 65\sqrt{1 + \sqrt{x}}\right) + C.$

145 $\int \sec^4 x\ dx = \frac{1}{3} \sec^2 x \tan x + \frac{2}{3} \int \sec^2 x\ dx$

$= \frac{1}{3} \sec^2 x \tan x + \frac{2}{3} \tan x + C$

$= \frac{1}{3} \tan^3 x + \tan x + C$

147 $\int \cos^3 x\ dx = \int (1 - \sin^2 x) \cos x\ dx$

$= \sin x - \frac{1}{3} \sin^3 x + C$

149 $\int \frac{\ln x + \sqrt{x}}{x}\ dx = \int (\ln x)\frac{dx}{x} + \int x^{-1/2}\ dx$

$= \frac{1}{2}(\ln x)^2 + 2\sqrt{x} + C$

151 $\int e^{-x}\ dx = -e^{-x} + C$

153 $\int (e^{2x} + 1)e^{-x}\,dx = \int (e^x + e^{-x})\,dx$

$= e^x - e^{-x} + C$

155 Determine the partial-fraction representation of the integrand: $\dfrac{2x^2 + 4x + 3}{x^3 + 2x^2 + 3x} =$

$\dfrac{c_1}{x} + \dfrac{c_2x + c_3}{x^2 + 2x + 3}$, where $c_1(x^2 + 2x + 3) + (c_2x + c_3)x = 2x^2 + 4x + 3$. Substituting the values $x = -1$, 0, and 1 produces a system of equations whose solution is $c_1 = 1$, $c_2 = 1$, and $c_3 = 2$. Hence $\int \dfrac{2x^2 + 4x + 3}{x^3 + 2x^2 + 3x}\,dx$

$$= \int \left(\frac{1}{x} + \frac{x + 2}{x^2 + 2x + 3}\right) dx$$

$$= \ln|x| + \frac{1}{2}\int \frac{2x + 4}{x^2 + 2x + 3}\,dx =$$

$$\ln|x| + \frac{1}{2}\int \frac{2x + 2}{x^2 + 2x + 3}\,dx + \int \frac{dx}{(x + 1)^2 + 2} =$$

$$\ln|x| + \frac{1}{2}\ln(x^2 + 2x + 3) + \frac{1}{\sqrt{2}}\tan^{-1}\frac{x + 1}{\sqrt{2}} + C.$$

157 Recall that $\sqrt{\sec^2\theta - 1} = \tan\theta$, so let $u = \tan\theta$, $du = \sec^2\theta\,d\theta$. Then we have $\int \dfrac{\sec^2\theta\,d\theta}{\sqrt{\sec^2\theta - 1}} =$

$$\int \frac{du}{u} = \ln|u| + C = \ln|\tan\theta| + C.$$

159 Use integration by parts with $u = x$, $dv = \sin^2 x\,dx = \frac{1}{2}(1 - \cos 2x)\,dx$, $du = dx$, and $v = \frac{1}{2}\left(x - \frac{\sin 2x}{2}\right)$. Then $\int x\sin^2 x\,dx =$

$$\frac{x}{2}\left(x - \frac{\sin 2x}{2}\right) - \frac{1}{2}\int\left(x - \frac{\sin 2x}{2}\right) dx$$

$$= \frac{1}{2}\left(x^2 - \frac{x\sin 2x}{2} - \left(\frac{x^2}{2} + \frac{\cos 2x}{4}\right)\right) + C$$

$$= \frac{1}{8}(2x^2 - 2x\sin 2x - \cos 2x) + C.$$

161 Let $u = x$ and $dv = \tan^2 x\,dx = (\sec^2 x - 1)\,dx$. Then $du = dx$, $v = \tan x - x$, and $\int x\tan^2 x\,dx$

$$= x(\tan x - x) - \int (\tan x - x)\,dx$$

$$= x\tan x - x^2 + \ln|\cos x| + \frac{x^2}{2} + C$$

$$= x\tan x - \frac{x^2}{2} + \ln|\cos x| + C.$$

163 (a) The partial-fraction representation of the integrand is $\dfrac{1}{x^2 + 4x + 3} = \dfrac{1}{(x - 1)(x + 3)}$

$= \dfrac{c_1}{x + 1} + \dfrac{c_2}{x + 3}$, where $1 = c_1(x + 3) + c_2(x + 1)$. Substituting $x = -1$ shows that $c_1 = 1/2$, while $x = -3$ yields $c_2 = -1/2$. Thus

$$\int \frac{dx}{x^2 + 4x + 3} = \frac{1}{2}\int\left(\frac{1}{x + 1} - \frac{1}{x + 3}\right) dx$$

$$= \frac{1}{2}(\ln|x + 1| - \ln|x + 3|) + C$$

$$= \frac{1}{2}\ln\left|\frac{x + 1}{x + 3}\right| + C.$$

(b) $\displaystyle\int \frac{dx}{x^2 + 4x + 4} = \int \frac{dx}{(x + 2)^2}$

$$= -\frac{1}{x + 2} + C$$

(c) $\int \frac{dx}{x^2 + 4x + 5} = \int \frac{dx}{(x+2)^2 + 1}$

$= \tan^{-1}(x+2) + C$

(d) $x^2 + 4x - 2 = x^2 + 4x + 4 - 6$

$= (x+2)^2 - 6 = (x + 2 + \sqrt{6})(x + 2 - \sqrt{6})$, so the partial-fraction representation of the integrand is $\frac{1}{x^2 + 4x - 2} =$

$\frac{c_1}{x + 2 + \sqrt{6}} + \frac{c_2}{x + 2 - \sqrt{6}}$, where we also have $1 = c_1(x + 2 - \sqrt{6}) + c_2(x + 2 + \sqrt{6})$.

Substitution of $x = -2 \pm \sqrt{6}$ shows that $c_1 = -\frac{1}{2\sqrt{6}}$ and $c_2 = \frac{1}{2\sqrt{6}}$. Thus

$\int \frac{dx}{x^2 + 4x - 2}$

$= \frac{1}{2\sqrt{6}} \int \left[\frac{-1}{x + 2 + \sqrt{6}} + \frac{1}{x + 2 - \sqrt{6}}\right] dx$

$= \frac{1}{2\sqrt{6}}(-\ln|x + 2 + \sqrt{6}| + \ln|x + 2 - \sqrt{6}|) + C$

$= \frac{1}{2\sqrt{6}} \ln\left|\frac{x + 2 - \sqrt{6}}{x + 2 + \sqrt{6}}\right| + C$. (Formula 35 in the table of antiderivatives is a good alternative to doing this one by hand.)

165 Let $u = x^3$. Then $x = u^{1/3}$ and $dx = \frac{1}{3}u^{-2/3}\,du$, so

$\int \sqrt{1 - x^3}\,dx = \int \sqrt{1 - u}\left(\frac{1}{3}u^{-2/3}\right) du =$

$\frac{1}{3}\int u^{-2/3}(1 - u)^{1/2}\,du$. The integrand is of the form discussed in Exercise 164, with $p = -2/3$ and $q = 1/2$. Since none of p, q, and $p + q$ is an integer, the antiderivative is not elementary.

167 As suggested, let $u = \sin^2 x$; then $\sqrt{u} = \sin x$ and $x = \sin^{-1}\sqrt{u}$, so $dx = \frac{1}{\sqrt{1 - u}} \cdot \frac{1}{2\sqrt{u}}\,du =$

$\frac{1}{2}(1 - u)^{-1/2}u^{-1/2}\,du$. Therefore $\int \sqrt{\sin x}\,dx$

$= \int u^{1/4}\left[\frac{1}{2}(1 - u)^{-1/2}u^{-1/2}\right] du =$

$\frac{1}{2}\int u^{-1/4}(1 - u)^{-1/2}\,du$. The integrand is of the form discussed in Exercise 164, with $p = -1/4$ and $q = -1/2$. None of p, q, and $p + q$ is an integer, so the antiderivative is not elementary.

169 Let $u = \sin x$. Then $du = \cos x\,dx$ and

$\int \sin^p x \cos^q x\,dx = \int u^p(1 - u^2)^{(q-1)/2}\,du =$

$\int u^{p-1}(1 - u^2)^{(q-1)/2}\,u\,du$. Now let $v = u^2$, $dv = 2u\,du$; then the integral becomes

$\frac{1}{2}\int v^{(p-1)/2}(1 - v)^{(q-1)/2}\,dv$. By Exercise 164, this is elementary if and only if one of the following occurs:

(A) $\frac{p-1}{2}$ is an integer, say $\frac{p-1}{2} = n$; then $p = 2n + 1$ is an odd integer.

(B) $\frac{q-1}{2}$ is an integer, so q is an odd integer.

(C) $\frac{p-1}{2} + \frac{q-1}{2}$ is an integer, say, k; then $p + q = 2k + 2$ is an even integer.

171 (a) Let $u = x^n$; $x = u^{1/n}$, $dx = \frac{1}{n}u^{(1/n)-1}\,du$, and

$$\int \frac{x}{\sqrt{1+x^n}}\,dx = \int x(1+x^n)^{-1/2}\,dx$$

$$= \int u^{1/n}(1+u)^{-1/2}\frac{1}{n}u^{(1/n)-1}\,du$$

$$= \frac{1}{n}\int u^{(2/n)-1}(1+u)^{-1/2}\,du,$$ which is of the form discussed in Exercise 164, with $p = 2/n - 1$, $q = -1/2$, and $p + q = \frac{2}{n} - \frac{3}{2}$.

Since q is not an integer, the integral is elementary only if p is an integer or if $p + q$ is an integer. In the former case n must equal either 1 or 2. In the latter case, $\frac{2}{n} - \frac{3}{2} = \frac{4-3n}{2n}$ must be an integer. Note that $4 - 3n$ must be divisible by n; thus 4 is divisible by n since it equals the sum of $4 - 3n$ and $3n$, both of which are divisible by n. But the only positive divisors of 4 are 1, 2, and 4. Checking these values shows that $\frac{4-3n}{2n}$ is an integer only for $n = 4$. Thus the integral is elementary only for $n = 1$, 2, or 4.

(b) $n = 1$: The integral is $\int \frac{x\,dx}{\sqrt{1+x}}$. Let $u = \sqrt{1+x}$; then $x = u^2 - 1$ and $dx = 2u\,du$, so

$$\int \frac{x\,dx}{\sqrt{1+x}} = \int \frac{(u^2-1)\,2u\,du}{u}$$

$$= 2\int (u^2-1)\,du = 2\left(\frac{u^3}{3} - u\right) + C$$

$$= \frac{2}{3}(1+x)^{3/2} - 2(1+x)^{1/2} + C.$$

$n = 2$: The integral is $\int \frac{x\,dx}{\sqrt{1+x^2}} = \frac{1}{2}\cdot\frac{(1+x^2)^{1/2}}{1/2} + C = \sqrt{1+x^2} + C.$

$n = 4$: The integral is $\int \frac{x\,dx}{\sqrt{1+x^4}}$. Let $u = x^2$ and $du = 2x\,dx$. Then $\int \frac{x\,dx}{\sqrt{1+x^4}} = \frac{1}{2}\int \frac{du}{\sqrt{1+u^2}} = \frac{1}{2}\ln\left(u + \sqrt{1+u^2}\right) + C$

$= \frac{1}{2}\ln\left(x^2 + \sqrt{1+x^4}\right) + C$. (We used Formula 33 from the table of antiderivatives.)

173 (a) Let $u = x^4$; then $x = \pm u^{1/4}$ and $dx = \pm\frac{1}{4}u^{-3/4}\,du$, so $\int \frac{x^n}{\sqrt{1+x^4}}\,dx =$

$$\pm\int \frac{u^{n/4}}{\sqrt{1+u}}\frac{1}{4}u^{-3/4}\,du =$$

$\pm\frac{1}{4}\int u^{(n-3)/4}(1+u)^{-1/2}\,du$. Since 1/2 is not an integer, it follows from Exercise 164 that the integral is elementary if and only if either $\frac{n-3}{4}$ or $\frac{n-3}{4} + \left(-\frac{1}{2}\right) = \frac{n-5}{4}$ is an integer; that is, if and only if n is odd.

(b) $n = 3$: The integral is $\int \frac{x^3\,dx}{\sqrt{1 + x^4}}$. Let $u = 1 + x^4$; then $du = 4x^3\,dx$ and the integral becomes $\frac{1}{4}\int \frac{du}{u^{1/2}} = \frac{1}{2}u^{1/2} + C = \frac{1}{2}(1 + x^4)^{1/2} + C$.

$n = 5$: The integral is $\int \frac{x^5\,dx}{\sqrt{1 + x^4}}$. Let $u = \sqrt{1 + x^4}$; then $x^4 = u^2 - 1$, $4x^3\,dx = 2u\,du$, and $\int \frac{x^5\,dx}{\sqrt{1 + x^4}} = \frac{1}{4}\int \frac{x^2(4x^3\,dx)}{\sqrt{1 + x^4}}$

$= \frac{1}{4}\int \frac{\sqrt{u^2 - 1}(2u\,du)}{u} = \frac{1}{2}\int \sqrt{u^2 - 1}\,du$

$= \frac{1}{4}\left[u\sqrt{u^2 - 1} - \ln\left|u + \sqrt{u^2 - 1}\right|\right] + C$

$= \frac{1}{4}\left[x^2\sqrt{1 + x^4} - \ln\left(x^2 + \sqrt{1 + x^4}\right)\right] + C$.

(We used Formula 31 from the table of antiderivatives.)

175 (a) The fundamental theorem of calculus relies on the integrand having an elementary antiderivative, and $\int e^{x^2}\,dx$ is not elementary.

(b) Here $a = 0$, $b = 1$, $n = 6$, and $h = 1/6$, so Simpson's method gives $\int_0^1 e^{x^2}\,dx \approx$

$\frac{1/6}{3}\left[e^{0^2} + 4e^{(1/6)^2} + 2e^{(2/6)^2} + 4e^{(3/6)^2} + 2e^{(4/6)^2} + 4e^{(5/6)^2} + e^{1^2}\right] \approx 1.4628735$.

177 The integrand is an odd function and the interval of integration is centered at the origin, so the integral is 0.

179 (a) $x^3 - 1 = (x - 1)(x^2 + x + 1)$, so the partial-fraction representation of $\frac{x}{x^3 - 1}$ is $\frac{c_1}{x - 1} + \frac{c_2x + c_3}{x^2 + x + 1}$, where $x = c_1(x^2 + x + 1) + (c_2x + c_3)(x - 1)$. When $x = 1$, we get $c_1 = 1/3$; for $x = 0$, we obtain $0 = c_1 - c_3$, so $c_3 = c_1 = 1/3$; and when $x = -1$, we have $-1 = c_1 + (-c_2 + c_3)(-2) = 1/3 + (-c_2 + 1/3)(-2)$, so $c_2 = -1/3$. Thus

$$\frac{x}{x^3 - 1} = \frac{1}{3}\left(\frac{1}{x - 1} - \frac{x - 1}{x^2 + x + 1}\right).$$

(b) $x^3 - 3x + 2 = (x - 1)^2(x + 2)$, so the partial-fraction representation is $\frac{1}{x^3 + 3x + 2} = \frac{c_1}{x - 1} + \frac{c_2}{(x - 1)^2} + \frac{c_3}{x + 2}$, where $1 = c_1(x - 1)(x + 2) + c_2(x + 2) + c_3(x - 1)^2$. Substituting $x = -2$, 0, and 1 yields $c_1 = -1/9$, $c_2 = 1/3$, and $c_3 = 1/9$. Thus

$$\frac{1}{x^3 - 3x + 2} = \frac{1}{9}\left(\frac{-1}{x - 1} + \frac{3}{(x - 1)^2} + \frac{1}{x + 2}\right).$$

(c) Polynomial division yields $\frac{x^5}{x^3 + 1} = x^2 - \frac{x^2}{x^3 + 1}$. Now we have $x^3 + 1 = (x + 1)(x^2 - x + 1)$, so the partial-fraction representation is $\dfrac{x^2}{x^3+1}$

$= \dfrac{c_1}{x+1} + \dfrac{c_2x + c_3}{x^2 - x + 1}$, where $x^2 =$

$c_1(x^2 - x + 1) + (c_2x + c_3)(x + 1)$.
Substitution of $x = -1$, 0, and 1 yields c_1

$= 1/3$, $c_2 = 2/3$, and $c_3 = -1/3$, so $\dfrac{x^5}{x^3+1}$

$= x^2 - \dfrac{x^2}{x^3+1}$

$= x^2 - \dfrac{1}{3}\left(\dfrac{1}{x+1} + \dfrac{2x-1}{x^2-x+1}\right)$.

(d) Since $x^3 + 8 = (x + 2)(x^2 - 2x + 4)$, the partial-fraction representation is given by

$\dfrac{1}{x^3+8} = \dfrac{c_1}{x+2} + \dfrac{c_2x + c_3}{x^2 - 2x + 4}$, where 1

$= c_1(x^2 - 2x + 4) + (c_2x + c_3)(x + 2)$.
When $x = -2$, we obtain $c_1 = 1/12$; $x = 0$ yields $c_3 = 1/3$; and $x = 1$ yields $c_2 =$

$-1/12$. Thus $\dfrac{1}{x^3+8} =$

$\dfrac{1}{12}\left(\dfrac{1}{x+2} - \dfrac{x-4}{x^2-2x+4}\right)$.

181 The difference between the two purported antiderivatives is $a \ln \dfrac{\sqrt{x^2+a^2}+a}{x} +$

$a \ln \dfrac{\sqrt{x^2+a^2}-a}{x}$

$= a \ln\left(\dfrac{\sqrt{x^2+a^2}+a}{x}\cdot\dfrac{\sqrt{x^2+a^2}-a}{x}\right)$

$= a \ln\left(\dfrac{x^2+a^2-a^2}{x^2}\right) = a \ln 1 = 0$. The

antiderivatives are therefore equal. (It would have sufficed for our purposes had they differed by a constant.)

183 (a) Integrate by parts with $u = \tan x$ and $dv = x\,dx$. Then $du = \sec^2 x\,dx$, $v = \dfrac{x^2}{2}$, and

$\int x \tan x\,dx = (\tan x)\dfrac{x^2}{2} - \int \dfrac{x^2}{2}\sec^2 x\,dx$

$= \dfrac{1}{2}x^2 \tan x - \dfrac{1}{2}\int x^2 \sec^2 x\,dx$. So if

$\int x^2 \sec^2 x\,dx$ were elementary,

$\int x \tan x\,dx$ would be also.

(b) $\int x^2 \sec^2 x\,dx = \int x^2(1 + \tan^2 x)\,dx$

$= \int x^2\,dx + \int x^2 \tan^2 x\,dx$

$= \dfrac{1}{3}x^3 + \int x^2 \tan^2 x\,dx$. If $\int x^2 \tan^2 x\,dx$

were elementary, $\int x^2 \sec^2 x\,dx$ would also be elementary, contradicting (a).

(c) Let $x = 2u$. Then $\int \dfrac{x^2\,dx}{1+\cos x} =$

$8\int \dfrac{u^2\,du}{1+\cos 2u}$

$= 8\int \dfrac{u^2\,du}{1+(\cos^2 u - \sin^2 u)}$

$= 8\int \dfrac{u^2\,du}{2\cos^2 u} = 4\int u^2 \sec^2 u\,du$. Hence

$\int u^2 \sec^2 u\, du = \frac{1}{4}\int \frac{x^2\, dx}{1 + \cos x}$, so if

$\int \frac{x^2\, dx}{1 + \cos x}$ were elementary, then

$\int u^2 \sec^2 u\, du$ would be also, contradicting

(a).

185 (a) Let $u = 1/x$ and $dv = \sin x\, dx$. Then $du =$

$-\frac{dx}{x^2}$, $v = -\cos x$, and $\int \frac{\sin x}{x}\, dx$

$= \frac{1}{x}(-\cos x) - \int (-\cos x)\left(-\frac{1}{x^2}\, dx\right) =$

$-\frac{\cos x}{x} - \int \frac{\cos x}{x^2}\, dx$. Let $x = 2t$. Then

$\int \frac{\cos x}{x^2}\, dx = \int \frac{\cos 2t}{4t^2}(2\, dt)$

$= \frac{1}{2}\int \frac{\cos 2t}{t^2}\, dt = \frac{1}{2}\int \frac{2\cos^2 t - 1}{t^2}\, dt$

$= \int \frac{\cos^2 t}{t^2}\, dt - \frac{1}{2}\int t^{-2}\, dt$

$= \int \frac{\cos^2 t}{t^2}\, dt + \frac{1}{2t}$. Hence $\int \frac{\sin x}{x}\, dx =$

$-\frac{\cos x}{x} - \frac{1}{x} - \int \frac{\cos^2 t}{t^2}\, dt$. If $\int \frac{\cos^2 t}{t^2}\, dt$

were elementary, $\int \frac{\sin x}{x}\, dx$ would also be

elementary.

(b) $\int \frac{\cos^2 x}{x^2}\, dx = \int \frac{1 - \sin^2 x}{x^2}\, dx$

$= \int x^{-2}\, dx - \int \frac{\sin^2 x}{x^2}\, dx$

$= -\frac{1}{x} - \int \frac{\sin^2 x}{x^2}\, dx$. If $\int \frac{\sin^2 x}{x^2}\, dx$ were

elementary, then $\int \frac{\cos^2 x}{x^2}\, dx$ would be also,

contradicting (a).

(c) Let $x = e^u$. Then $\int \frac{\sin x}{x}\, dx$

$= \int \frac{\sin e^u}{e^u} e^u\, du = \int \sin e^u\, du$. If

$\int \sin e^u\, du$ were elementary, then

$\int \frac{\sin x}{x}\, dx$ would be also.

(d) Let $u = \sin x$ and $dv = dx/x$; then $du =$

$\cos x\, dx$, $v = \ln x$, and $\int \frac{\sin x}{x}\, dx =$

$\sin x \ln x - \int \ln x \cos x\, dx$. Thus if

$\int \cos x \ln x\, dx$ were elementary, $\int \frac{\sin x}{x}\, dx$

would also be.

187 Make the substitution $x = \sin\theta$, $dx = \cos\theta\, d\theta$.

Then we have $\int \frac{\sqrt{1 + ax^2}}{\sqrt{1 - x^2}}\, dx =$

$\int \frac{\sqrt{1 + a\sin^2\theta}}{\sqrt{1 - \sin^2\theta}} \cos\theta\, d\theta$

$= \int \frac{\sqrt{1 + a\sin^2\theta}}{\cos\theta} \cos\theta\, d\theta$

$= \int \sqrt{1 + a\sin^2\theta}\, d\theta$. Hence $\int \frac{\sqrt{1 + ax^2}}{\sqrt{1 - x^2}}\, dx$ is

elementary for the same values as

$\int \sqrt{1 + a \sin^2 \theta}\, d\theta$; namely $a = 0$ and $a = -1$.

189 Make the substitution $x = \cos \theta$, $dx = -\sin \theta\, d\theta$.

Then $\int \frac{\sqrt{1 + bx}}{\sqrt{1 - x^2}}\, dx =$

$$\int \frac{\sqrt{1 + b \cos \theta}}{\sqrt{1 - \cos^2 \theta}}(-\sin \theta\, d\theta)$$

$$= -\int \frac{\sqrt{1 + b \cos \theta}}{\sin \theta} \sin \theta\, d\theta$$

$= -\int \sqrt{1 + b \cos \theta}\, d\theta$. Hence $\int \frac{\sqrt{1 + bx}}{\sqrt{1 - x^2}}\, dx$ is

elementary for the same values of b as

$\int \sqrt{1 + b \cos \theta}\, d\theta$ is, namely $b = -1, 0,$ and 1.

8 Applications of the Definite Integral

8.1 Computing Area by Parallel Cross Sections

1 (a)

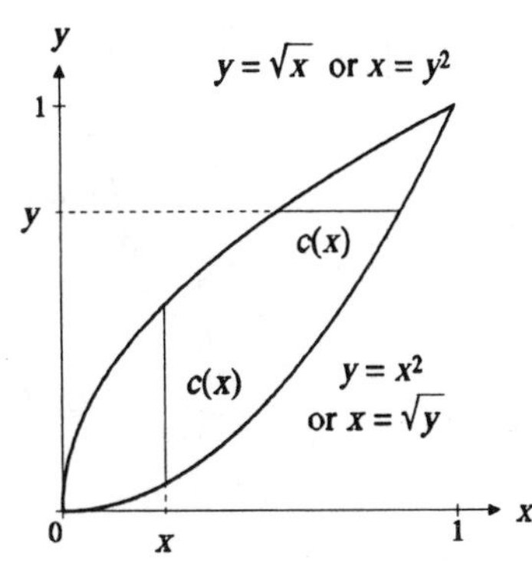

(b) From the figure in (a), $c(x) = \sqrt{x} - x^2$.

(c) From the figure in (a), $c(y) = \sqrt{y} - y^2$.

3 (a)

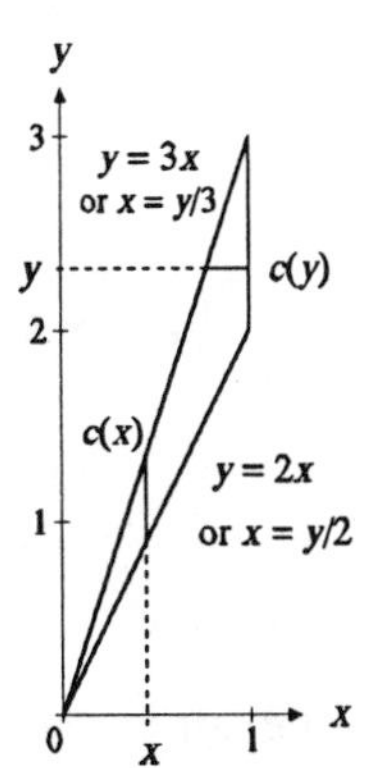

(b) From the figure in (a), $c(x) = 3x - 2x = x$.

(c) From the figure, it is apparent that $c(y)$ does not have the same formula throughout $[0, 3]$. Indeed, for y in $[0, 2]$, $c(y) = y/2 - y/3 = y/6$, while for y in $[2, 3]$, $c(y) = 1 - y/3$.

Thus $c(y) = \begin{cases} y/6 & 0 \le y \le 2 \\ 1 - y/3 & 2 < y \le 3 \end{cases}$.

5 (a)

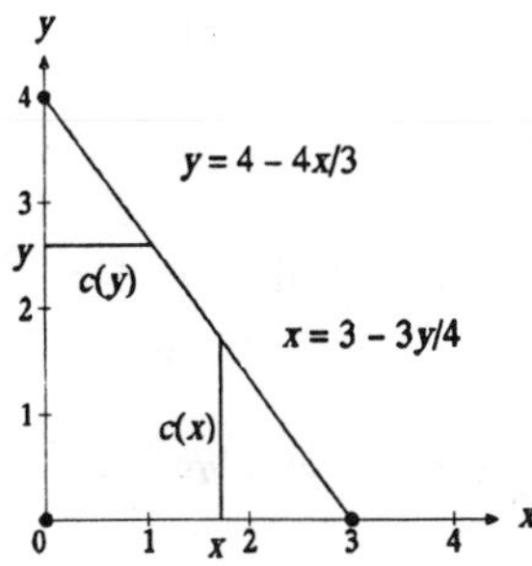

(b) The line representing the hypotenuse of the triangle is given by $\frac{x}{3} + \frac{y}{4} = 1$ since the x intercept is 3 and the y intercept is 4. So $y = 4 - (4/3)x$. We conclude from the figure that $c(x) = 4 - (4/3)x - 0 = 4 - (4/3)x$.

(c) Note that $x = 3 - (3/4)y$, so from the figure $c(y) = 3 - (3/4)y - 0 = 3 - (3/4)y$.

7 The two curves intersect when $x^2 = 3x - 2$, or $0 = x^2 - 3x + 2 = (x - 2)(x - 1)$. Thus, the two points of intersection are $(1, 1)$ and $(2, 4)$. The region is shown in the accompanying figure.

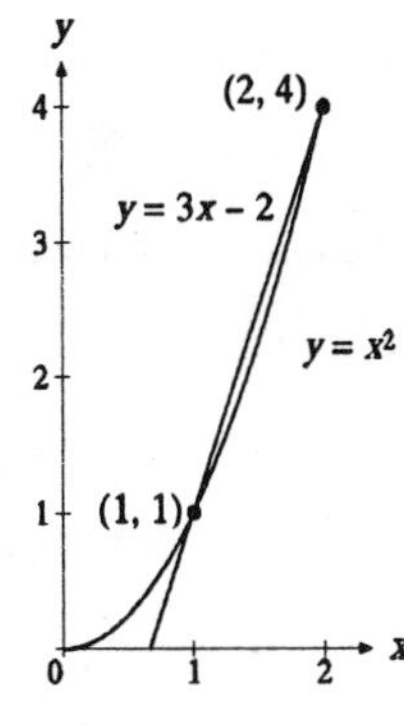

(a) Each vertical cross section is bounded above by the line $y = 3x - 2$ and below by the parabola $y = x^2$. Hence $c(x) = 3x - 2 - x^2$, so the area is $\int_1^2 (3x - 2 - x^2)\, dx =$

$$\int_1^2 (3x - 2 - x^2)\, dx = \left(\frac{3x^2}{2} - 2x - \frac{x^3}{3}\right)\Bigg|_1^2$$

$$= \left(6 - 4 - \frac{8}{3}\right) - \left(\frac{3}{2} - 2 - \frac{1}{3}\right) = \frac{1}{6}.$$

(b) Each horizontal cross section is bounded on the right by the parabola $y = x^2$ (or $x = \sqrt{y}$) and on the left by the line $y = 3x - 2$, or $x = \frac{y+2}{3}$. Therefore $c(y) = \sqrt{y} - \frac{y+2}{3}$

and Area $= \int_1^4 c(y)\, dy$

$$= \int_1^4 \left(y^{1/2} - \frac{y}{3} - \frac{2}{3}\right) dy$$

$$= \left[\frac{2}{3}y^{3/2} - \frac{y^2}{6} - \frac{2}{3}y\right]\Bigg|_1^4$$

$$= \left(\frac{16}{3} - \frac{8}{3} - \frac{8}{3}\right) - \left(\frac{2}{3} - \frac{1}{6} - \frac{2}{3}\right) = \frac{1}{6}.$$

9 The two curves intersect when $4x = 2x^2$; thus $0 = 2x^2 - 4x = 2x(x - 2)$, so $x = 0$ or 2. The region is sketched in the figure.

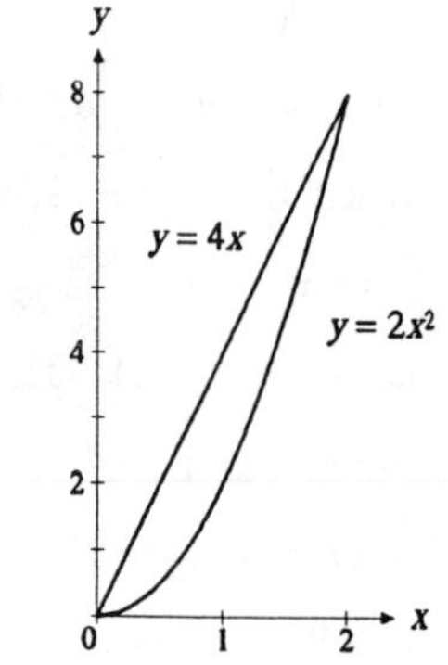

(a) Each vertical cross section is bounded above by the line $y = 4x$ and below by the parabola $y = 2x^2$. Thus, $c(x) = 4x - 2x^2$, and Area $= \int_0^2 (4x - 2x^2)\, dx =$

$$\left(2x^2 - \frac{2}{3}x^3\right)\Bigg|_0^2 = 8 - \frac{16}{3} = \frac{8}{3}.$$

(b) Each horizontal cross section is bounded on the right by the parabola and on the left by the line. Solving the equations of these two curves for x in terms of y, we get $x = \sqrt{y/2} = \frac{1}{\sqrt{2}}y^{1/2}$ on the parabola and $x = y/4$ on the line. Hence $c(y) = \frac{1}{\sqrt{2}}y^{1/2} - \frac{y}{4}$ and

$$\text{Area} = \int_0^8 \left[\frac{1}{\sqrt{2}}y^{1/2} - \frac{y}{4}\right] dy$$

$$= \left[\frac{1}{\sqrt{2}}\cdot\frac{2}{3}y^{3/2} - \frac{y^2}{8}\right]\Bigg|_0^8 = \frac{\sqrt{2}}{3}(8)^{3/2} - 8$$

$$= \frac{\sqrt{2}}{3}(16\sqrt{2}) - 8 = \frac{32}{3} - 8 = \frac{8}{3}.$$

11 The region is indicated in the figure.

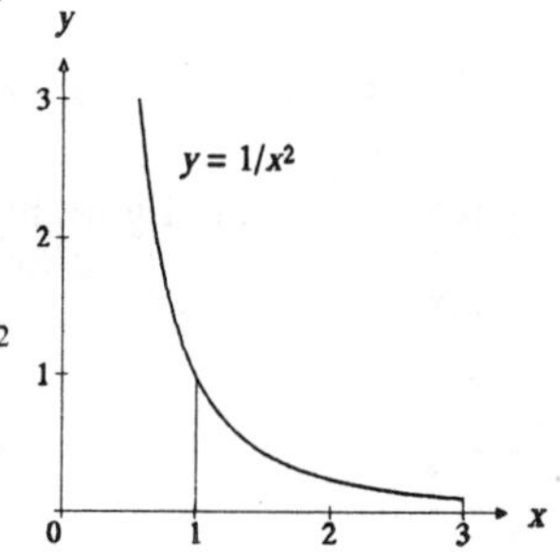

(a) Each vertical cross section has length $c(x) = 1/x^2 = x^{-2}$, so the area is

$$\int_1^3 x^{-2}\, dx = -\frac{1}{x}\Bigg|_1^3 = -1/3 - (-1) = 2/3.$$

(b) For $0 \le y \le 1/9$, the horizontal cross sections are bounded on the right by the line $x = 3$ and on the left by the line $x = 1$. Hence they all have length 2.

For $1/9 \le y \le 1$, the horizontal cross sections are bounded on the right by the curve $y = 1/x^2$, or $x = y^{-1/2}$, and on the left by the line $x = 1$. Hence, their length is given by $y^{-1/2} - 1$. Thus

$$\text{Area} = \int_0^{1/9} 2\,dy + \int_{1/9}^{1} (y^{-1/2} - 1)\,dy$$

$$= 2y\Big|_0^{1/9} + (2y^{1/2} - y)\Big|_{1/9}^{1}$$

$$= \frac{2}{9} + [(2 - 1) - (\frac{2}{3} - \frac{1}{9})] = \frac{2}{3}.$$

13 The region is shown in the accompanying diagram. The cross-sectional length $c(x)$ is $c(x) = x^2 - x^3$, so Area =

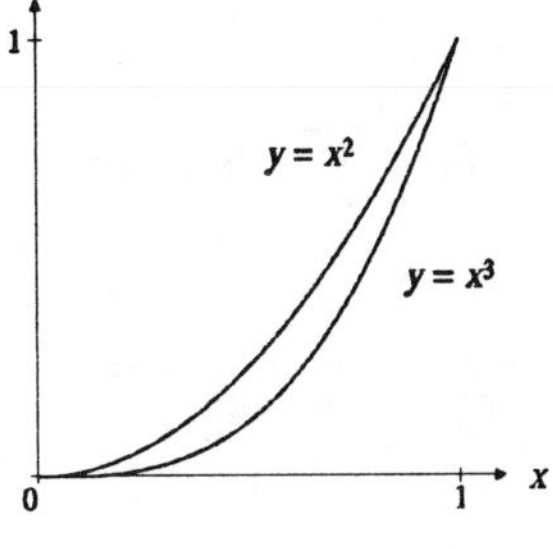

$$\int_0^1 (x^2 - x^3)\,dx$$

$$= \left(\frac{x^3}{3} - \frac{x^4}{4}\right)\Big|_0^1 = \left(\frac{1}{3} - \frac{1}{4}\right) - (0 - 0) = \frac{1}{12}.$$

15 As seen in the figure, $c(x) = \sqrt{x} - x^2 = x^{1/2} - x^2$, so Area =

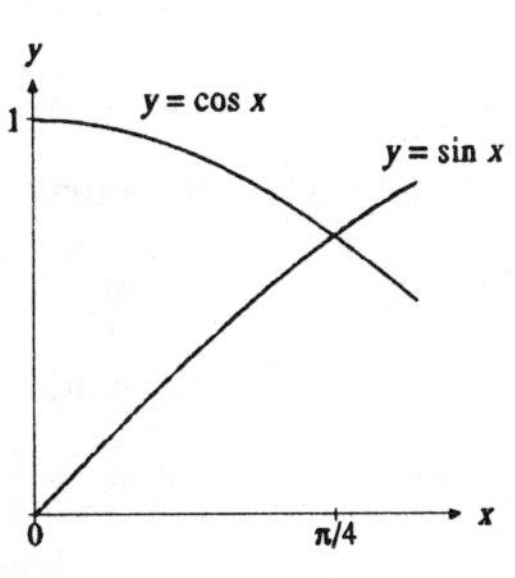

$$\int_0^1 (x^{1/2} - x^2)\,dx =$$

$$\left(\frac{2}{3}x^{3/2} - \frac{x^3}{3}\right)\Big|_0^1$$

$$= \left(\frac{2}{3} - \frac{1}{3}\right) - (0 - 0) = \frac{1}{3}.$$

17 Here we have $c(x) = \cos x - \sin x$, so Area

$$= \int_0^{\pi/4} (\cos x - \sin x)\,dx$$

$$= (\sin x + \cos x)\Big|_0^{\pi/4}$$

$$= \left(\frac{1}{\sqrt{2}} + \frac{1}{\sqrt{2}}\right) - (0 + 1)$$

$$= \frac{2}{\sqrt{2}} - 1 = \sqrt{2} - 1.$$

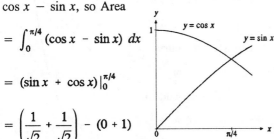

19 Each vertical cross section is bounded above by $y = x^3$ and bounded below by $y = \sqrt[3]{2x - 1}$, so $c(x) = x^3 - \sqrt[3]{2x - 1}$. Thus the area is

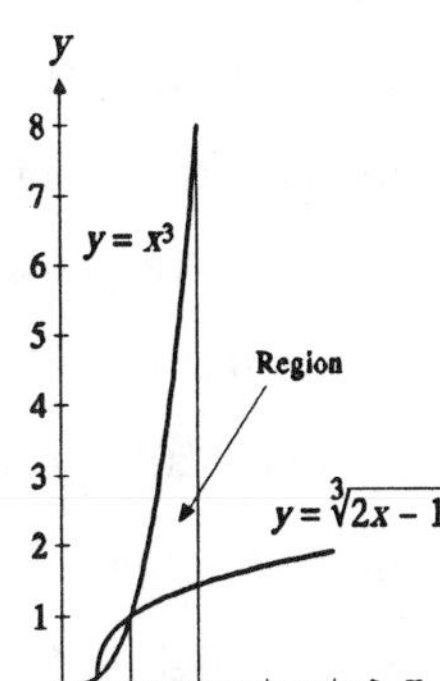

$$\int_1^2 \left(x^3 - \sqrt[3]{2x - 1}\right) dx$$

$$= \int_1^2 x^3\,dx - \int_1^2 \sqrt[3]{2x - 1}\,dx.$$

For the last integral, let $u = 2x - 1$ so $du = 2\,dx$; When $x = 1$, $u = 1$ and when $x = 2$, $u = 3$.

Hence Area $= \int_1^2 x^3\,dx + \frac{1}{2}\int_1^3 u^{1/3}\,du =$

$$\frac{1}{4}x^4\Big|_1^2 + \frac{1}{2}\left[\frac{3}{4}u^{4/3}\right]\Big|_1^3 = \left(4 - \frac{1}{4}\right) - \frac{1}{2}\left(\frac{3}{4}\cdot 3^{4/3} - \frac{3}{4}\right)$$

$$= \frac{33}{8} - \frac{3}{8}\cdot 3^{4/3} = \frac{3}{8}(11 - 3^{4/3}).$$

21 The area is $\int_{1/2}^{1/\sqrt{3}} \tan^{-1} 2x\,dx =$

$$\left(x \tan^{-1} 2x - \frac{1}{4}\ln(1 + 4x^2)\right)\Big|_{1/2}^{1/\sqrt{3}} =$$

$$\left(\frac{1}{\sqrt{3}} \tan^{-1}\frac{2}{\sqrt{3}} - \frac{1}{4}\ln\frac{7}{3}\right) - \left(\frac{1}{2}\tan^{-1} 1 - \frac{1}{4}\ln 2\right)$$

$$= \frac{1}{\sqrt{3}}\tan^{-1}\frac{2}{\sqrt{3}} - \frac{\pi}{8} - \frac{1}{4}\ln\frac{7}{6}.$$

23 At an intersection, $-7x + 29 = \dfrac{8}{x^2 - 8}$, so $0 =$

$$(x^2 - 8)(7x - 29) + 8 = 7x^3 - 29x^2 - 56x + 240.$$

We find that 3 and 4 are roots of this equation (by plotting points or using a calculator). Factoring out $(x - 3)(x - 4)$ shows that the other root is $-20/7$. The only intersections in the first quadrant are (3, 8) and (4, 1). The graph shows that the region we want is given by $3 \le x \le 4$, $\frac{8}{x^2 - 8} \le y \le 29 - 7x$. Its area is

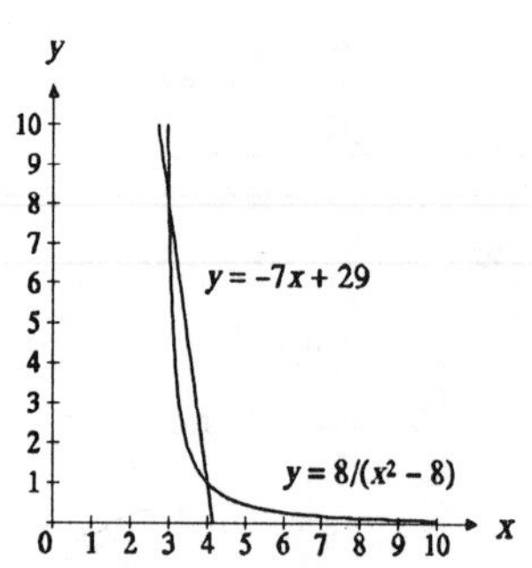

$$\int_3^4 \left(29 - 7x - \frac{8}{x^2 - 8}\right) dx$$

$$= \left(29x - \frac{7}{2}x^2\right)\Big|_3^4 + 8\int_3^4 \frac{dx}{8 - x^2}$$

$$= \frac{9}{2} + \left(8 \cdot \frac{1}{2\sqrt{8}} \ln\left|\frac{\sqrt{8} + x}{\sqrt{8} - x}\right|\right)\Big|_3^4$$

$$= \frac{9}{2} + \sqrt{2}\left(\ln\frac{4 + \sqrt{8}}{4 - \sqrt{8}} - \ln\frac{3 + \sqrt{8}}{3 - \sqrt{8}}\right)$$

$$= \frac{9}{2} + \sqrt{2}\ \ln\left[\frac{(4 + \sqrt{8})(3 - \sqrt{8})}{(4 - \sqrt{8})(3 + \sqrt{8})}\right]$$

$= \frac{9}{2} - \sqrt{2}\ \ln(3 + \sqrt{8})$. (We used the Formula 25 from the text's table of antiderivatives.)

25 $\int_1^2 \frac{x}{x^2 + 5x + 6}\, dx = \int_1^2 \left[\frac{3}{x + 3} - \frac{2}{x + 2}\right] dx$

$$= 3\ln\frac{5}{4} - 2\ln\frac{4}{3} = \ln\frac{1125}{1024}$$

27

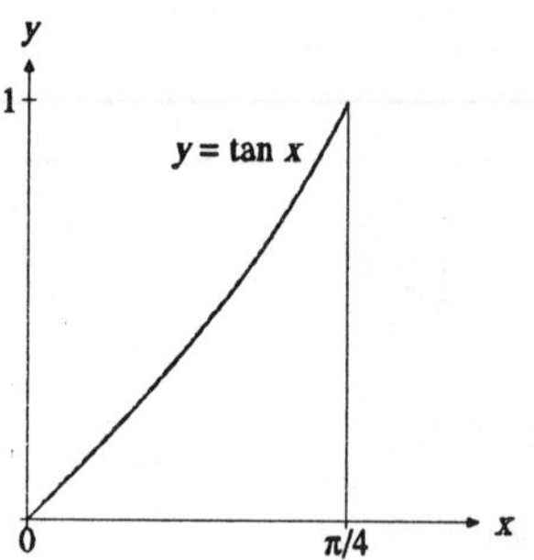

(a) $\int_0^{\pi/4} \tan x\, dx = \ln|\sec x|\,\Big|_0^{\pi/4} = \frac{1}{2}\ln 2$

(b) $\int_0^1 (\frac{\pi}{4} - \tan^{-1} y)\, dy$

$$= \left(\frac{\pi}{4}y - y\tan^{-1} y + \frac{1}{2}\ln(1 + y^2)\right)\Big|_0^1$$

$$= \frac{1}{2}\ln 2$$

29 By horizontal cross sections, $A(t) =$

$\int_0^{t^2} \left(\sqrt{y} + \sqrt{\frac{y}{2}}\right) dy = \left(1 - \frac{1}{\sqrt{2}}\right)\frac{2}{3}y^{3/2}\Big|_0^{t^2} =$

$\frac{1}{3}(2 - \sqrt{2})t^3$. Since $R(t) = t^3$, $A(t)/R(t)$ is constant, $\frac{1}{3}(2 - \sqrt{2})$. Hence both limits equal $\frac{1}{3}(2 - \sqrt{2})$.

31 The vertical cross sections are bounded above by the line $y = x + 2$ and bounded below by the curve $y = x^3$, so $c(x) = (x + 2) - x^3$ and the area is

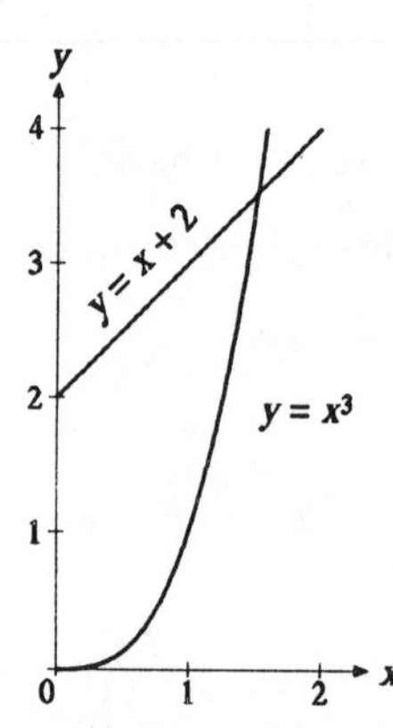

$\int_0^r (x + 2 - x^3)\, dx$, where r is the value for

which $c(r) = 0$ (see the figure). To estimate r, we apply Newton's method with $f(x) = x + 2 - x^3$ and $x_1 = 3/2$. This gives, to three decimal places, $r \approx 1.521$. Now Area $= \int_0^r (x + 2 - x^3)\,dx =$

$$\left(\frac{1}{2}x^2 + 2x - \frac{1}{4}x^4\right)\Big|_0^r = \frac{1}{2}r^2 + 2r - \frac{1}{4}r^4 \approx$$

2.861.

33 $\frac{1}{a \cdot a^4}\int_0^a x^4\,dx = \frac{1}{a^5}\cdot\frac{a^5}{5} = \frac{1}{5}$

35 Let $u = f(t)$. By Fig. 12,

$\int_0^t f(x)\,dx + \int_0^{f(t)} f^{-1}(y)\,dy = t \cdot f(t)$; that is,

$\int_0^u f^{-1}(y)\,dy = uf^{-1}(u) - \int_0^{f^{-1}(u)} f(x)\,dx$. Hence,

if both f^{-1} and $\int f(x)\,dx$ are elementary, so is

$\int_0^u f^{-1}(y)\,dy$.

37 Exploration problems are in the *Instructor's Manual.*

39 Center the square at the origin of the x and y axes, as in the figure. We shall consider only the one-eighth of the square shown (given the symmetry of the problem) and will multiply the determined area of the section of R by eight. For a point on the boundary of R, we have $\sqrt{x^2 + y^2} = \frac{a}{2} - x$. Squaring and simplifying gives $x = \frac{a}{4} - \frac{1}{a}y^2$. Therefore the horizontal cross section at height y has length $\left(\frac{a}{4} - \frac{1}{a}y^2\right) - y$. The boundary meets the diagonal $x = y$ when $\sqrt{y^2 + y^2} = \frac{a}{2} - y$; that is, $\frac{a}{2} = (1 + \sqrt{2})y$. Thus $y =$

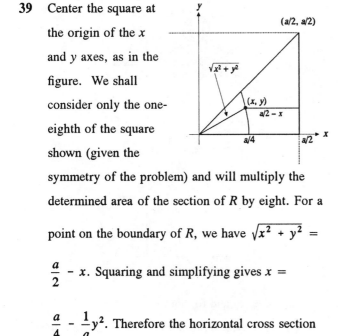

$$\frac{a}{2(1 + \sqrt{2})} = \frac{\sqrt{2} - 1}{2}a = ka, \text{ where we let } k = \frac{\sqrt{2} - 1}{2},$$

for convenience. So the area of R is

$$8\int_0^{ka}\left(\frac{a}{4} - \frac{1}{a}y^2 - y\right)dy$$

$$= 8\left(\frac{a}{4}y - \frac{1}{3a}y^3 - \frac{1}{2}y^2\right)\Big|_0^{ka}$$

$$= 2ka^2 - \frac{8k^3a^3}{3a} - 4k^2a^2;$$

plugging in the value of k and simplifying yields the area $\frac{4\sqrt{2} - 5}{3}a^2$.

41 (a)

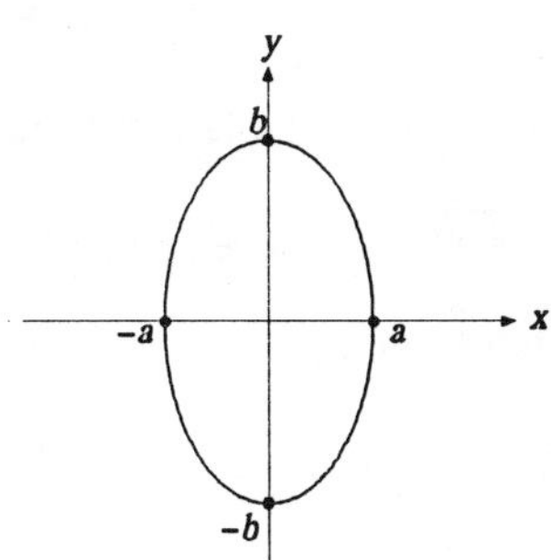

(b) Solving for y in terms of x gives $y =$

$$\sqrt{b^2 - \frac{b^2}{a^2}x^2} = b\sqrt{1 - \left(\frac{x}{a}\right)^2}$$

for the upper half of the ellipse and $y = -b\sqrt{1 - \left(\frac{x}{a}\right)^2}$ for the lower half. Thus $c(x) =$

$$b\sqrt{1 - \left(\frac{x}{a}\right)^2} - \left(-b\sqrt{1 - \left(\frac{x}{a}\right)^2}\right)$$

$$= 2b\sqrt{1 - \left(\frac{x}{a}\right)^2}.$$

(c) Solving for x in terms of y gives

$x = \sqrt{a^2 - \frac{a^2}{b^2}y^2} = a\sqrt{1 - (y/b)^2}$ for the right half of the ellipse and $x = -a\sqrt{1 - (y/b)^2}$ for the left half. Thus $c(y) = 2a\sqrt{1 - (y/b)^2}$.

(d) Using vertical cross sections, the area is

$$\int_{-a}^{a} c(x)\,dx = \int_{-a}^{a} 2b\sqrt{1 - \left(\frac{x}{a}\right)^2}\,dx$$

$$= 2b\int_{-a}^{a} \sqrt{1 - \left(\frac{x}{a}\right)^2}\,dx$$

$$= \frac{2b}{a}\int_{-a}^{a} \sqrt{a^2 - x^2}\,dx = \frac{2b}{a}\cdot\frac{\pi a^2}{2} = ab\pi,$$

where we used the fact that the area under $\sqrt{a^2 - x^2}$ and above $[-a, a]$ is that of half a circle of radius a.

8.2 Some Pointers on Drawing

1 (a)

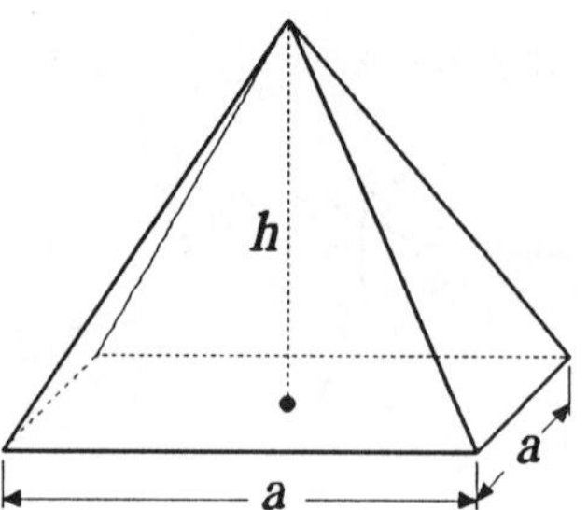

(b)

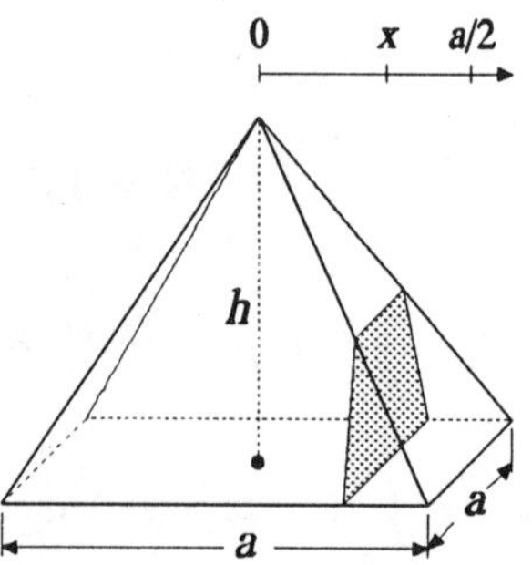

(c)

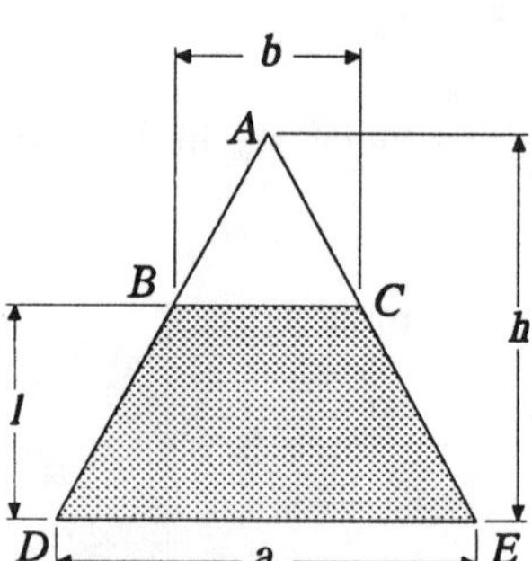

(d)

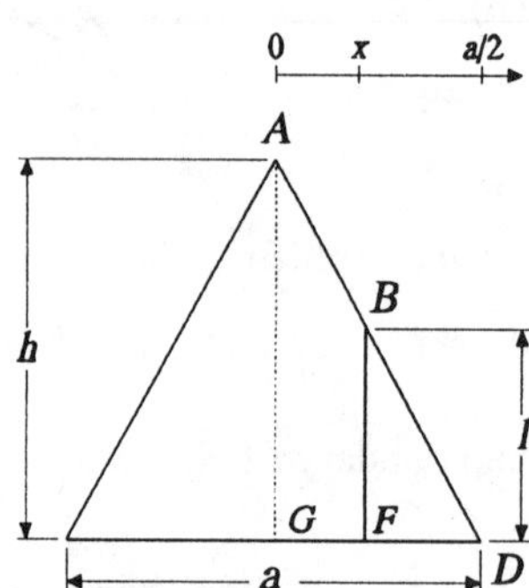

(e) See (c) and (d) above.

(f) The typical cross section is a trapezoid with

bases a and b and height l, and its area is $\frac{1}{2}(a + b)l$. To get l in terms of h, a, and b, note that triangles ABC and ADE in the figure in (c) are similar. Since corresponding parts of similar triangles are proportional, we see in particular that $\frac{h - l}{h} = \frac{b}{a}$, so $l = h(1 - b/a)$. Also, triangles AGD and BFD in the figure in (d) are similar. Hence for the same reason given above we may conclude that $\frac{h}{l} = \frac{a/2}{a/2 - x}$. Thus $(a/2 - x)h = h\left(1 - \frac{b}{a}\right)\frac{a}{2}$, and it follows that $b = 2x$. Now that we have l in terms of h, a, and b and b in terms of x, we can find $A(x)$ as follows:

$$A(x) = \frac{1}{2}(a + b)l = \frac{1}{2}(a + 2x)h\left(1 - \frac{2x}{a}\right)$$

$$= \frac{h}{2}\left(a - \frac{4x^2}{a}\right).$$ As a check, note that $A(0) = \frac{1}{2}ah$ (the area of a triangle of base a and height h) and $A(a/2) = 0$ in concurrence with the figures in (b) and (c).

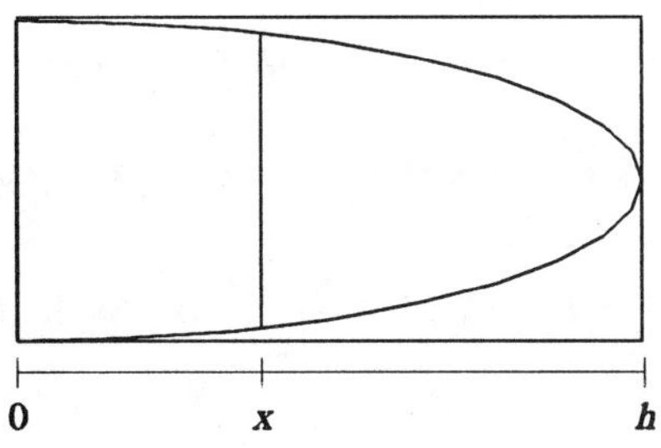

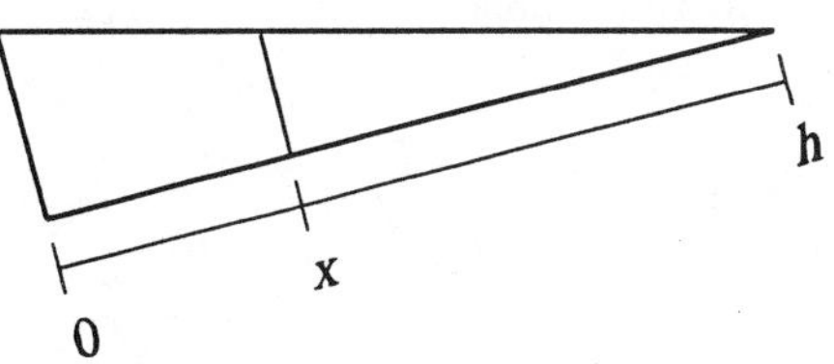

3

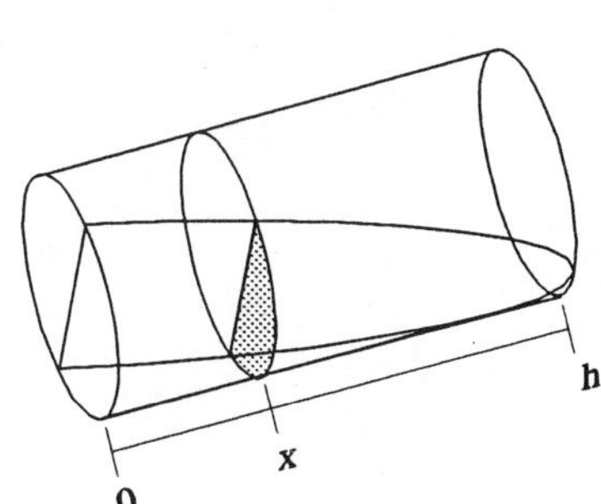

5 (a)

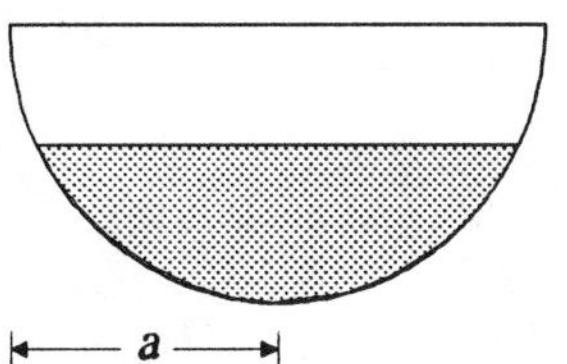

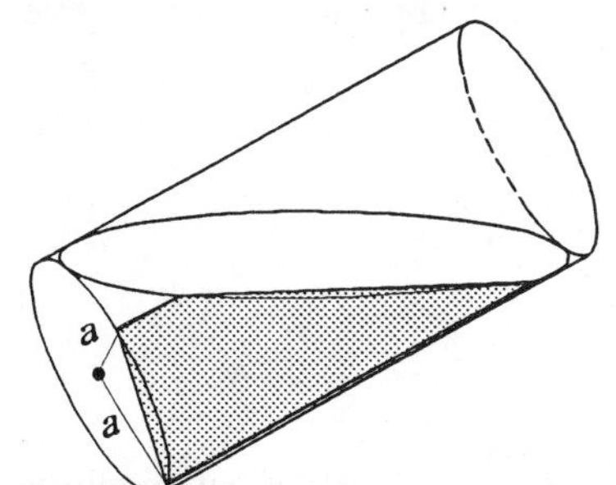

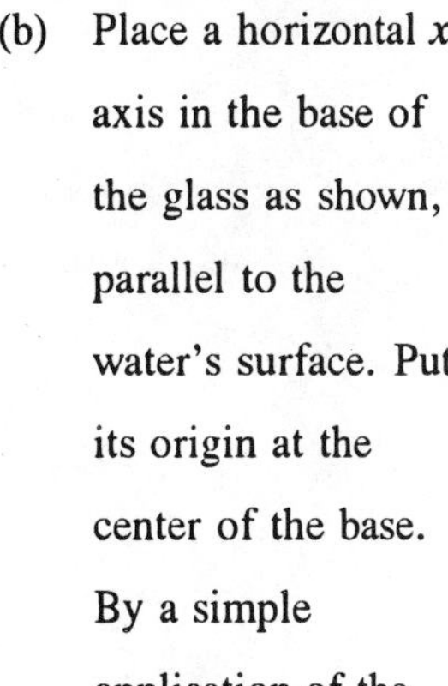

(b) Place a horizontal x axis in the base of the glass as shown, parallel to the water's surface. Put its origin at the center of the base. By a simple application of the

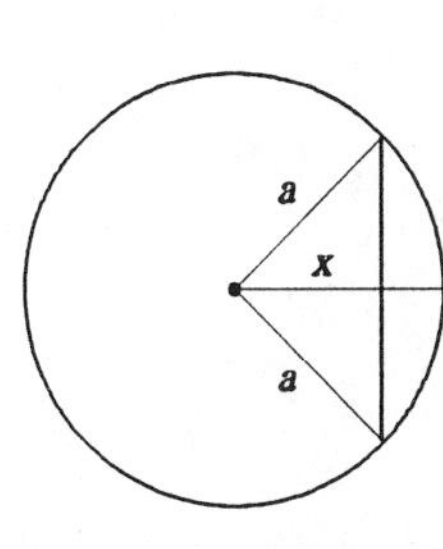

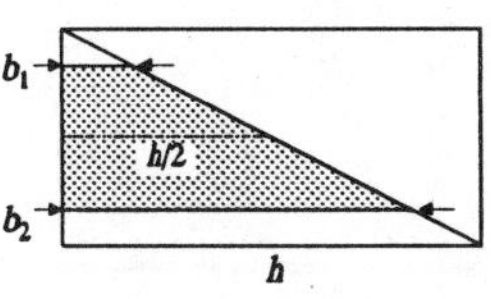

Pythagorean theorem, the trapezoidal cross section with coordinate x has height $H = 2\sqrt{a^2 - x^2}$. Now the area of a trapezoid is $\frac{1}{2}H(b_1 + b_2) = \frac{b_1 + b_2}{2}H$. As the figure shows, however, $\frac{b_1 + b_2}{2} = \frac{h}{2}$, so the trapezoid's area is $\frac{h}{2}\cdot H = \frac{h}{2}\cdot 2\sqrt{a^2 - x^2} = h\sqrt{a^2 - x^2}$.

7

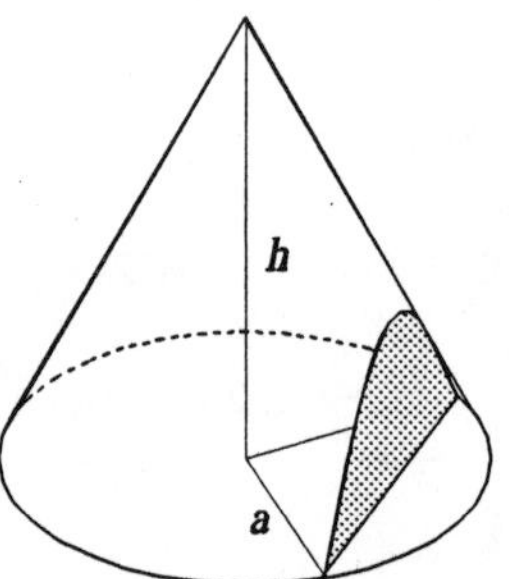

9 (a) A horizontal plane cuts out a circle.

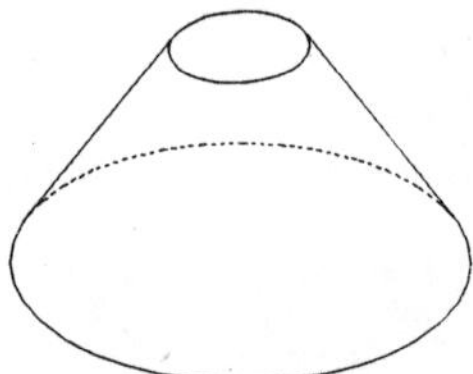

(b) A tilted plane can cut out an ellipse.

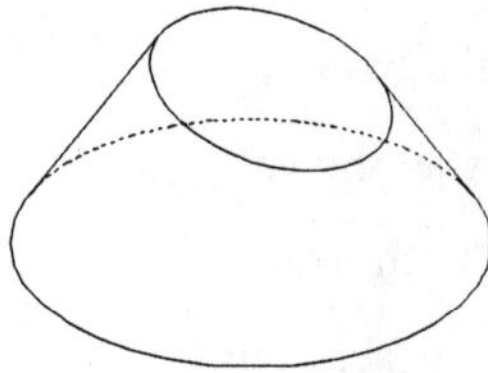

(c) A tilted plane that is parallel to the cone's edge cuts out a parabola.

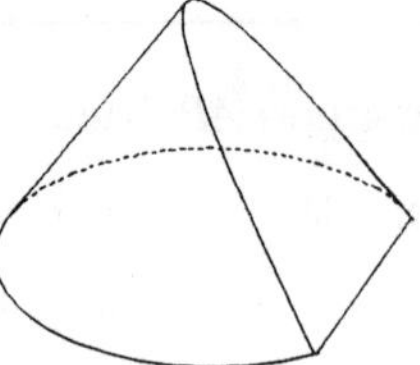

(d) A vertical plane cuts out a hyperbola.

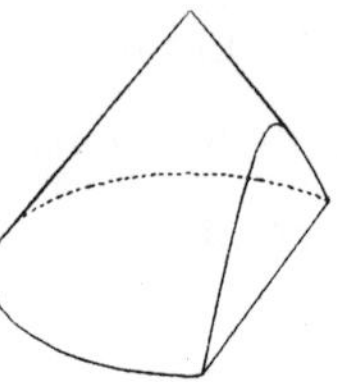

11 (a)

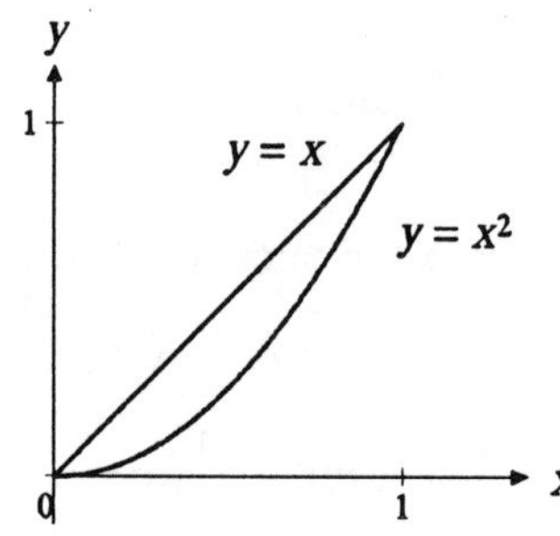

(b),(c)

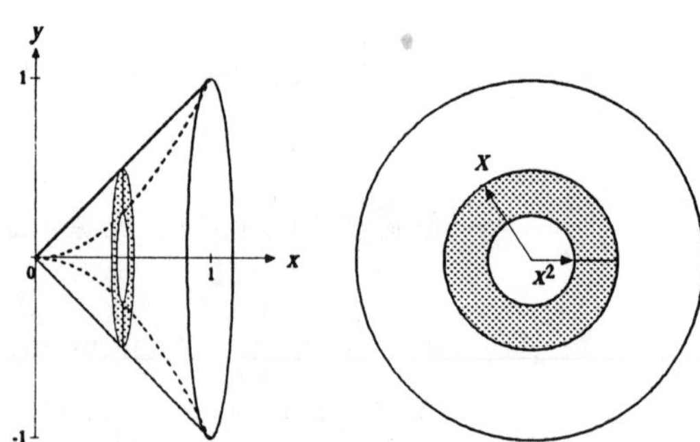

(d) From the figures in (c), Area $= \pi x^2 - \pi(x^2)^2 = \pi x^2(1 - x^2)$.

8.3 Setting Up a Definite Integral

1

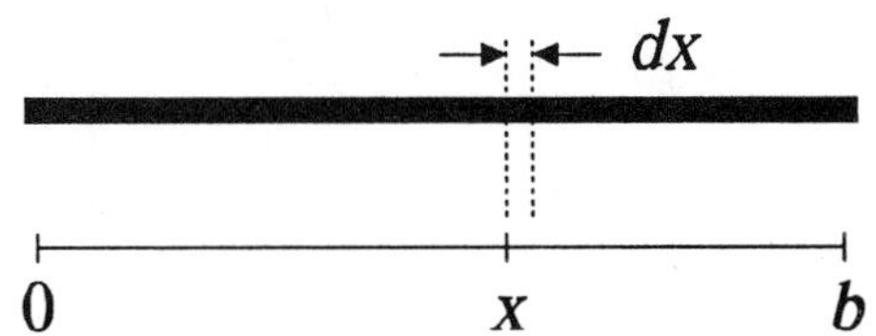

Select a narrow piece of wire a distance x from the left end of the wire, and let its width be dx. Then the mass of this piece is approximately $f(x)\,dx$. Since the endpoints of the wire are 0 and b, we conclude from the informal style that the total mass of the wire is $\int_0^b f(x)\,dx$.

3 This problem is very similar to Example 2.

(a) The depth of the rain that falls between r and $r + dr$ feet from the center of the storm is approximately $g(r)$ feet throughout the region. The area of this ring is approximately its inner circumference times its thickness dr. Therefore the volume is approximately $g(r)\cdot 2\pi r\,dr$ cubic feet.

(b) The total volume of rain between 1000 and 2000 feet is $\int_{1000}^{2000} 2\pi r g(r)\,dr$ cubic feet, which follows from (a) and the use of the informal approach.

5 (a) The area of the ring is equal to the difference of the area A_2 of the larger circle and the area A_1 of the smaller circle: $A_2 - A_1$

$= \pi(x + dx)^2 - \pi x^2$

$= \pi(x^2 + 2x\,dx + dx^2 - x^2)$

$= \pi(2x\,dx + dx^2) = 2\pi x\,dx + \pi(dx)^2.$

(b) The area of a trapezoid is $\frac{1}{2}(b_1 + b_2)h$. In this case, the area is $\frac{1}{2}[2\pi x + 2\pi(x + dx)]\,dx$

$= \pi[x + (x + dx)]\,dx = \pi(2x + dx)\,dx$

$= 2\pi x\,dx + \pi(dx)^2.$

7 (a)

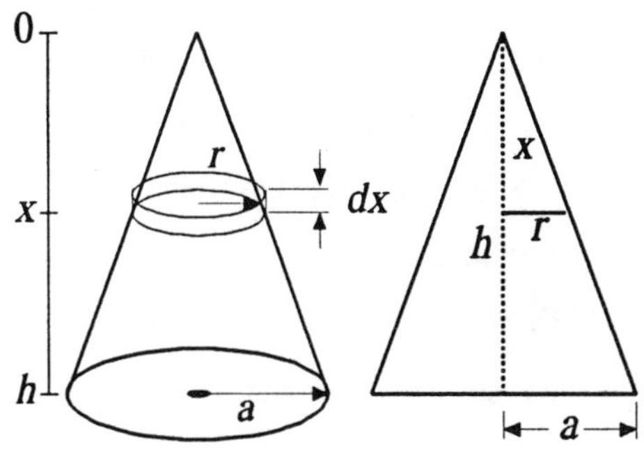

(b) The local approximation is a disk of radius r and width dx. Its volume is $\pi r^2\,dx$. To get r in terms of x, note from the side view that $\frac{x}{r} = \frac{h}{a}$, so $r = \frac{ax}{h}$. Now the local approximation of volume is given by $\pi\left(\frac{ax}{h}\right)^2 dx$, so the volume of the cone is $\int_0^h \pi\left(\frac{ax}{h}\right)^2 dx$, from the informal approach.

(c) $\int_0^h \pi\left(\frac{ax}{h}\right)^2 dx = \frac{\pi a^2}{h^2}\int_0^h x^2\,dx$

$= \frac{\pi a^2}{h^2}\frac{x^3}{3}\bigg|_0^h = \frac{1}{3}\pi a^2 h$

9 (a)

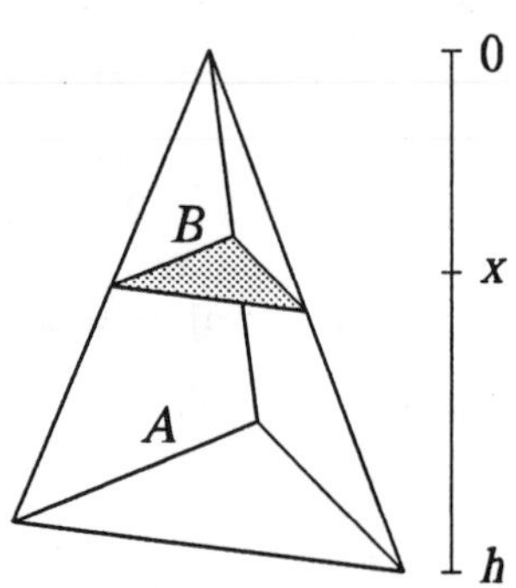

(b) Set up an x axis as in (a), where $x = 0$ corresponds to the peak of the pyramid and $x = h$ to the base. Let the triangular cross section at a distance x from the peak have area B; this triangle is similar to the base triangle, which has area A. Since corresponding parts of similar triangles are proportional, we know that $b_B = kb_A$ and $h_B = kh_A$, where b_B and b_A are the bases of the B and A triangles, respectively, and h_B and h_A are the heights. (It doesn't matter which sides we pick for the bases as long as we choose corresponding parts of the two triangles.) To find the constant of proportionality k, note that $k = \frac{b_B}{b_A} = \frac{x}{h}$. Then the area B is related to the area A by the equation $B = \frac{1}{2}b_B h_B = \frac{1}{2}(kb_A)(kh_A) = k^2\left(\frac{1}{2}b_A h_A\right) = k^2 A = A\left(\frac{x}{h}\right)^2$. Since the approximating slab has height dx, its volume is approximately $A\left(\frac{x}{h}\right)^2 dx$. Therefore total volume is $\int_0^h A\left(\frac{x}{h}\right)^2 dx$ by the informal approach.

(c) $\int_0^h A\left(\frac{x}{h}\right)^2 dx = \frac{A}{h^2}\int_0^h x^2\, dx = \frac{A}{h^2}\cdot\frac{h^3}{3}$

$= \frac{1}{3}Ah$

11 (a) $F(150) = 0$, barring any future dramatic medical advances between now and 2050.

(b) $F(0) = 1$ because all people are alive 0 years after their (successful) birth.

(c)

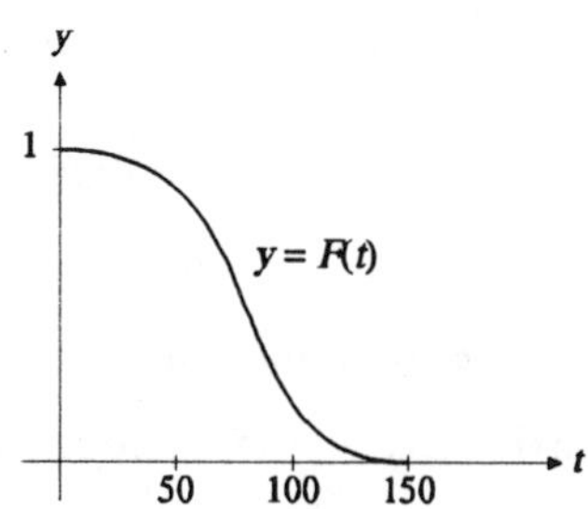

(d) $f(t) = F'(t)$ is and remains negative so long as the technology to bring the dead back to life does not exist.

(e) The fraction of people born in 1900 that die within t years is $1 - F(t)$. So the fraction of people who die during the time interval $(t, t + dt]$ is $(1 - F(t + dt)) - (1 - F(t)) = F(t) - F(t + dt)$.

(f) Since $f(t) = F'(t)$, we have $\int_t^{t+dt} f(x)\, dx = F(t + dt) - F(t)$, so $-\int_t^{t+dt} f(x)\, dx$ is the desired quantity. (We choose the dummy variable x for the integrand because t is being used in the limits of integration.)

(g) $\int_0^{150} f(t)\, dt = F(t)\big|_0^{150} = F(150) - F(0)$

$= -1$

(h) If dt is small, then the fraction of people who die during the time interval $[t, t + dt]$ is approximately $-f(t)\,dt$. Let N be the number of people. Now the sum of the life spans of all the people who die during $[t, t + dt]$ is about $-N \cdot t \cdot f(t)\,dt$. By the informal approach, the sum of the life spans of all the people born in 1900 is $-\int_0^{150} Ntf(t)\,dt$. Dividing by the number of people, N, gives the average life span of the people born in 1900:

$$-\int_0^{150} tf(t)\,dt.$$

13 (a) $c(t)e(t)\,dt$ dollars

(b) $\int_0^{24} c(t)e(t)\,dt$ dollars

15 (a) $\int_a^b f(x)\,dx$

(b) $\int_a^b f(x)m(x)\,dx$

17 (a) The area of the narrow ring of inner radius r and width dr shown in the figure is approximately $2\pi r\,dr$ square meters, so its mass is about $3 \cdot 2\pi r\,dr = 6\pi r\,dr$ kilograms. The ring is spun around the axis through the center of the disk 5 times per second; thus every point on the ring travels about $2\pi r \cdot 5 = 10\pi r$ meters per second. Hence the kinetic energy of the ring is approximately

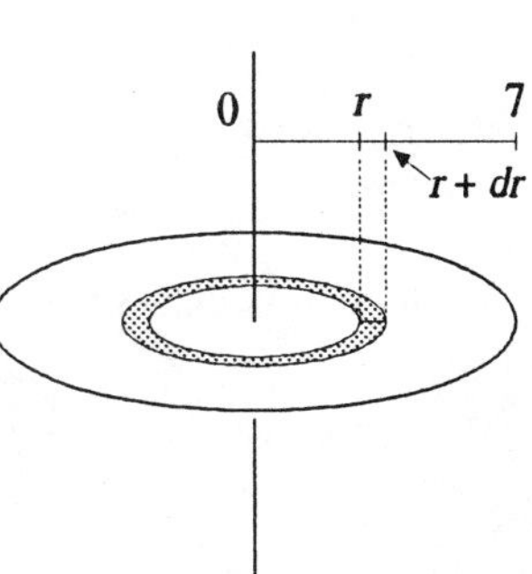

$$\frac{1}{2}(6\pi r\,dr)(10\pi r)^2 = 300\pi^3 r^3\,dr \text{ joules.}$$

(Note: A joule has the units of $\frac{kg\,m^2}{s^2}$; that is, 1 joule $= 10^7$ ergs.)

(b) The total kinetic energy is $\int_0^7 300\pi^3 r^3\,dr$ joules.

(c) $\int_0^7 300\pi^3 r^3\,dr = 300\pi^3 \int_0^7 r^3\,dr =$

$$300\pi^3 \cdot \frac{r^4}{4}\bigg|_0^7 = 300\pi^3 \cdot \frac{7^4}{4} = 180075\pi^3 \text{ joules}$$

19 (a) It would be best to use a cross-section of the square that is parallel to the edge about which the square is rotating, for all points contained in this cross-section would be moving at about the same velocity (thus simplifying the ensuing integration).

(b) The area of the narrow rectangle in the figure above is $a\,dx$ square meters. Since the square is homogeneous, its density is given by $\frac{M}{a^2}$ kilograms per square meter. Hence the mass of the rectangle is $a\,dx\left(\frac{M}{a^2}\right) = \frac{M\,dx}{a}$ kilograms. Because the square is rotating around its edge 5 times per second, the velocity of every point on the rectangle is approximately $5(2\pi x) = 10\pi x$ meters per second. Therefore, the kinetic energy of the rectangle is about

$$\frac{1}{2}\left(\frac{M\,dx}{a}\right)(10\pi x)^2 = \frac{50M\pi^2x^2}{a}\,dx \text{ joules.}$$

(c) The total kinetic energy of the square is

$$\int_0^a \frac{50\,M\pi^2x^2}{a}\,dx \text{ joules.}$$

(d) $$\int_0^a \frac{50M\pi^2x^2}{a}\,dx = \frac{50M\pi^2}{a}\int_0^a x^2\,dx =$$

$$\frac{50M\pi^2}{a}\frac{x^3}{3}\bigg|_0^a = \frac{50Ma^2\pi^2}{3} \text{ joules}$$

21 Using a narrow ring concentric with the disk would be best because every point on that circle would be moving at the same velocity.

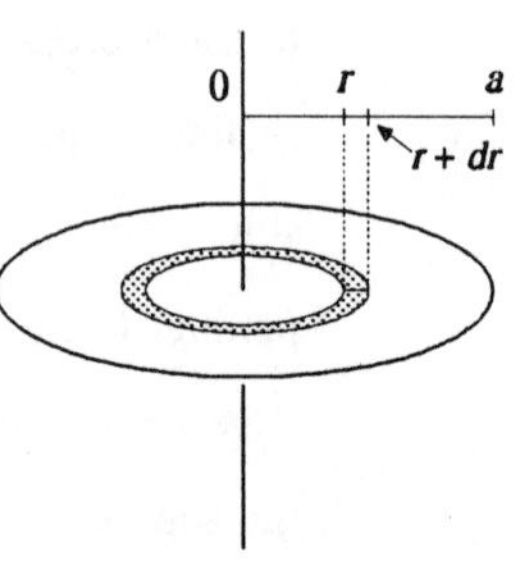

(b) The area of the narrow ring in the figure above is about $2\pi r\,dr$ square meters. Since the circle is homogeneous, its density is $\dfrac{M}{\pi a^2}$ kilograms per square meter. Hence the mass of the ring is $(2\pi r\,dr)\left(\dfrac{M}{\pi a^2}\right) = \dfrac{2Mr\,dr}{a^2}$.

Now because there are 2π radians in one revolution of a circle, the circle is rotating at the rate of $\omega\,\dfrac{\text{rad}}{\text{sec}}\cdot\dfrac{1\text{ rev}}{2\pi\text{ rad}} = \dfrac{\omega}{2\pi}\,\dfrac{\text{rev}}{\text{sec}}$. For the ring, one revolution is approximately $2\pi r$, so the velocity of the ring is about $\left(\dfrac{\omega}{2\pi}\right)2\pi r$ $= \omega r$ meters per second. Therefore, the kinetic energy of the ring is about

$$\frac{1}{2}\left(\frac{2Mr\,dr}{a^2}\right)(\omega r)^2 = \frac{M\omega^2r^3\,dr}{a^2} \text{ joules.}$$

(c) The total kinetic energy of the disk is

$$\int_0^a \frac{M\omega^2r^3}{a^2}\,dr \text{ joules.}$$

(d) $$\int_0^a \frac{M\omega^2r^3}{a^2}\,dr = \frac{M\omega^2}{a^2}\int_0^a r^3\,dr =$$

$$\frac{M\omega^2}{a^2}\frac{r^4}{4}\bigg|_0^a = \frac{M\omega^2}{a^2}\cdot\frac{a^4}{4} = \frac{1}{4}M\omega^2a^2 \text{ joules}$$

23 The situation is given in the figure. Now the mass of the region is M, while its area A is

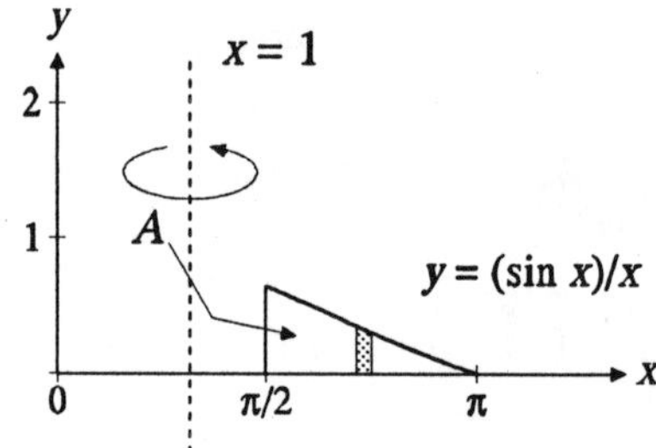

$\displaystyle\int_{\pi/2}^{\pi} \frac{\sin x}{x}\,dx$. Its density is $\dfrac{M}{A}$. Since the area of the vertical strip between x and $x + dx$ is approximately $\dfrac{\sin x}{x}\,dx$, its mass is $\dfrac{M\sin x}{Ax}\,dx$.

This strip has distance about $x - 1$ from the axis of rotation. By the discussion in Exercise 21, the velocity of a point in the strip is about $\omega(x - 1)$. Hence the kinetic energy of the strip is about

$$\frac{1}{2}\left(\frac{M\sin x}{Ax}\,dx\right)[\omega(x-1)]^2 =$$

$\dfrac{M\omega^2}{2A}\dfrac{(x-1)^2}{x}\sin x\,dx$. So the total kinetic energy of the region is $\int_{\pi/2}^{\pi} \frac{M\omega^2}{2A} \frac{(x-1)^2}{x} \sin x \, dx =$

$$\frac{M\omega^2}{2A}\left(\int_{\pi/2}^{\pi} x \sin x \, dx - 2\int_{\pi/2}^{\pi} \sin x \, dx + \int_{\pi/2}^{\pi} \frac{\sin x}{x} \, dx\right).$$

From Formula 57 at the front of the text, we have (for $a = 1$) $\int x \sin x \, dx = \sin x - x \cos x + C.$

Then $\int_{\pi/2}^{\pi} x \sin x \, dx = (\sin x - x \cos x)\Big|_{\pi/2}^{\pi} =$

$(0 - \pi(-1)) - (1 - 0) = \pi - 1$. Also,

$\int_{\pi/2}^{\pi} \sin x \, dx = -\cos x \Big|_{\pi/2}^{\pi} = 1$ and $\int_{\pi/2}^{\pi} \frac{\sin x}{x} \, dx$

$= A$, so the total kinetic energy is

$$\frac{M\omega^2}{2A}(\pi - 3 + A) = \left(\frac{\pi - 3}{2A} + \frac{1}{2}\right)M\omega^2.$$

(a) Using the trapezoidal method to find A, we have $\int_{\pi/2}^{\pi} \frac{\sin x}{x} \, dx \approx \frac{(\pi - \pi/2)/6}{2}\left[\frac{\sin \pi/2}{\pi/2}\right.$

$$+ 2\cdot\frac{\sin(\pi/2 + \pi/12)}{\pi/2 + \pi/12} + 2\cdot\frac{\sin(\pi/2 + \pi/6)}{\pi/2 + \pi/6}$$

$$+ 2\cdot\frac{\sin(\pi/2 + \pi/4)}{\pi/2 + \pi/4} + 2\cdot\frac{\sin(\pi/2 + \pi/3)}{\pi/2 + \pi/3} +$$

$$\left. 2\cdot\frac{\sin(\pi/2 + 5\pi/12)}{\pi/2 + 5\pi/12} + \frac{\sin \pi}{\pi}\right] =$$

$$\frac{\pi}{24}\left[\frac{2}{\pi} + \frac{24}{7\pi}\sin\frac{7\pi}{12} + \frac{3}{\pi}\sin\frac{2\pi}{3} + \right.$$

$$\left.\frac{8}{3\pi}\sin\frac{3\pi}{4} + \frac{12}{5\pi}\sin\frac{5\pi}{6} + \frac{24}{11\pi}\sin\frac{11\pi}{12}\right]$$

≈ 0.4817, so the total kinetic energy is about

$$\left(\frac{\pi - 3}{2(0.4817)} + \frac{1}{2}\right)M\omega^2 \approx 0.6470M\omega^2.$$

(b) Using Simpson's method to find A, we have

$$\int_{\pi/2}^{\pi} \frac{\sin x}{x} \, dx \approx \frac{(\pi - \pi/2)/6}{3}\left[\frac{\sin \pi/2}{\pi/2} + \right.$$

$$4\cdot\frac{\sin(\pi/2 + \pi/12)}{\pi/2 + \pi/12} + 2\cdot\frac{\sin(\pi/2 + \pi/6)}{\pi/2 + \pi/6} +$$

$$4\cdot\frac{\sin(\pi/2 + \pi/4)}{\pi/2 + \pi/4} + 2\cdot\frac{\sin(\pi/2 + \pi/3)}{\pi/2 + \pi/3} +$$

$$\left.4\cdot\frac{\sin(\pi/2 + 5\pi/12)}{\pi/2 + 5\pi/12} + \frac{\sin \pi}{\pi}\right]$$

$$= \frac{\pi}{36}\left[\frac{2}{\pi} + \frac{48}{7\pi}\sin\frac{7\pi}{12} + \frac{3}{\pi}\sin\frac{2\pi}{3} + \right.$$

$$\left.\frac{16}{3\pi}\sin\frac{3\pi}{4} + \frac{12}{5\pi}\sin\frac{5\pi}{6} + \frac{48}{11\pi}\sin\frac{11\pi}{12}\right]$$

≈ 0.4812, so the total kinetic energy is about

$$\left(\frac{\pi - 3}{2(0.4812)} + \frac{1}{2}\right)M\omega^2 \approx 0.6471M\omega^2.$$

25 The region has density

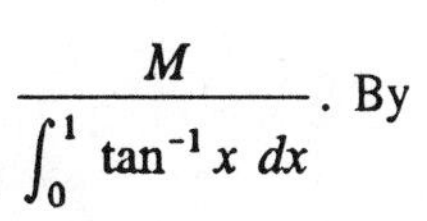

. By Formula 71 from the table of antiderivatives,

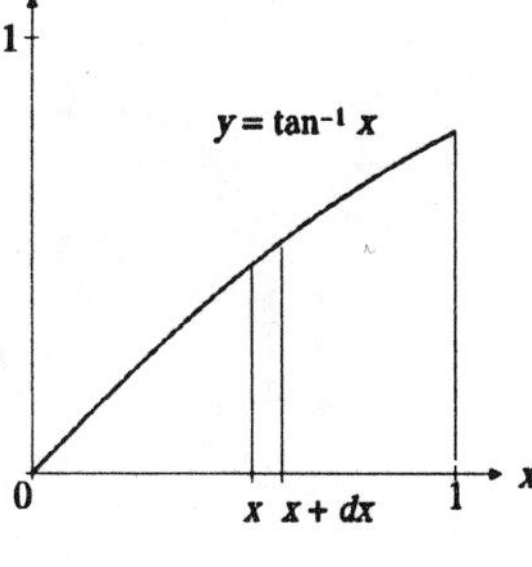

we have $\int_0^1 \tan^{-1} x \, dx$

$$= \left(x \tan^{-1} x - \frac{1}{2}\ln(1 + x^2)\right)\Big|_0^1 = \frac{\pi}{4} - \frac{1}{2}\ln 2,$$

so the density is $\frac{M}{\frac{\pi}{4} - \frac{1}{2}\ln 2}$. Thus, since the area of the strip between x and $x + dx$ is $\tan^{-1} x \, dx$, its mass is $\frac{M}{\frac{\pi}{4} - \frac{1}{2}\ln 2}\tan^{-1} x \, dx$. We know its velocity to be ωx, so the local estimate for the kinetic energy is $\frac{1}{2}\cdot\frac{M\tan^{-1}x\,dx}{\frac{\pi}{4}-\frac{1}{2}\ln 2}(\omega x)^2 =$

$\frac{M\omega^2}{\frac{\pi}{2}-\ln 2}x^2\tan^{-1}x\,dx$. Hence the total kinetic

energy is $\frac{M\omega^2}{\frac{\pi}{2}-\ln 2}\int_0^1 x^2\tan^{-1}x\,dx$. Now let u

$= \tan^{-1}x$ and $dv = x^2\,dx$ so that $\int_0^1 x^2\tan^{-1}x\,dx$

$= \frac{x^3}{3}\tan^{-1}x\Big|_0^1 - \frac{1}{3}\int_0^1 \frac{x^3}{1+x^2}\,dx$. Here

$\int_0^1 \frac{x^3}{1+x^2}\,dx = \int_0^1\left(x - \frac{x}{1+x^2}\right)dx$

$= \left[\frac{x^2}{2} - \frac{1}{2}\ln(1+x^2)\right]\Big|_0^1 = \frac{1}{2} - \frac{1}{2}\ln 2$. Hence

$\int_0^1 x^2\tan^{-1}x\,dx = \frac{x^3\tan^{-1}x}{3}\Big|_0^1 - \frac{1}{3}\left(\frac{1}{2} - \frac{1}{2}\ln 2\right)$

$= \frac{1}{3}\cdot\frac{\pi}{4} - \frac{1}{6} + \frac{1}{6}\ln 2 = \frac{1}{6}\left(\frac{\pi}{2} + \ln 2 - 1\right)$.

Finally, the total kinetic energy is

$\frac{M\omega^2}{\pi/2 - \ln 2}\cdot\frac{1}{6}\left(\frac{\pi}{2} + \ln 2 - 1\right)$

$= \frac{(\pi/2 + \ln 2 - 1)M\omega^2}{3\pi - 6\ln 2}$.

27 The region has density $\frac{M}{\int_0^2\sqrt{1+x^2}\,dx}$. If we let $x = \tan\theta$, then $dx = \sec^2\theta\,d\theta$ and

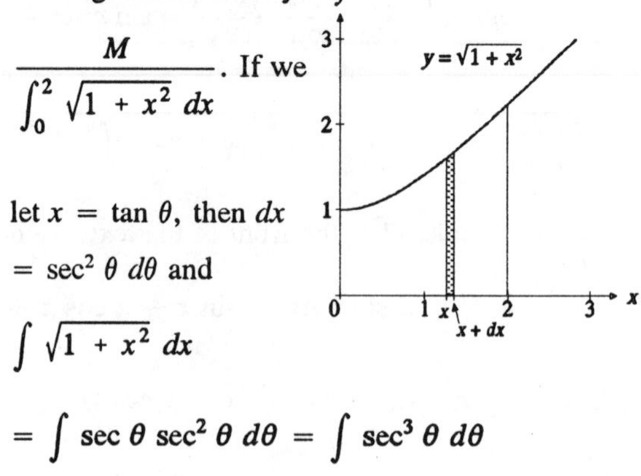

$\int\sqrt{1+x^2}\,dx$

$= \int\sec\theta\sec^2\theta\,d\theta = \int\sec^3\theta\,d\theta$

$= \frac{\sec\theta\tan\theta}{2} + \frac{1}{2}\ln|\sec\theta + \tan\theta| + C$

$= \frac{x\sqrt{1+x^2}}{2} + \frac{1}{2}\ln\left(x + \sqrt{1+x^2}\right) + C$, where

Formula 55 from the table of antiderivatives was used. Hence $\int_0^2\sqrt{1+x^2}\,dx =$

$\left(\frac{x\sqrt{1+x^2}}{2} + \frac{1}{2}\ln\left(x+\sqrt{1+x^2}\right)\right)\Big|_0^2$

$= \sqrt{5} + \frac{1}{2}\ln(2+\sqrt{5})$ and the density of the homogeneous region is $\frac{M}{\sqrt{5} + \frac{1}{2}\ln(2+\sqrt{5})}$. Thus

since the area of the strip between x and $x + dx$ is $\sqrt{1+x^2}\,dx$, its mass is $\frac{M\sqrt{1+x^2}\,dx}{\sqrt{5}+\frac{1}{2}\ln(2+\sqrt{5})}$. We know its velocity is ωx, so the local estimate of the kinetic energy is $\frac{1}{2}\frac{M\sqrt{1+x^2}\,dx}{\sqrt{5}+\frac{1}{2}\ln(2+\sqrt{5})}(\omega x)^2$

$= \dfrac{M\omega^2}{2\sqrt{5} + \ln(2 + \sqrt{5})} x^2\sqrt{1 + x^2}\, dx$. Therefore

the total kinetic energy is

$\dfrac{M\omega^2}{2\sqrt{5} + \ln(2 + \sqrt{5})} \int_0^2 x^2\sqrt{1 + x^2}\, dx$. Again,

letting $x = \tan\theta$ gives $\int x^2\sqrt{1 + x^2}\, dx =$

$\int \tan^2\theta \sec\theta\,(\sec^2\theta\, d\theta) = \int \sec^3\theta \tan^2\theta\, d\theta$

$= \int \sec^3\theta\,(\sec^2\theta - 1)\, d\theta$

$= \int (\sec^5\theta - \sec^3\theta)\, d\theta$

$= \dfrac{\sec^3\theta\tan\theta}{4} + \dfrac{3}{4}\int \sec^3\theta\, d\theta - \int \sec^3\theta\, d\theta$

$= \dfrac{\sec^3\theta\tan\theta}{4} - \dfrac{1}{4}\int \sec^3\theta\, d\theta$, by Formula 56

from the table of antiderivates. Hence

$\int_0^2 x^2\sqrt{1 + x^2}\, dx$

$= \left[\dfrac{x(1+x^2)^{3/2}}{4} - \dfrac{1}{4}\left(\dfrac{x\sqrt{1+x^2}}{2} + \dfrac{1}{2}\ln\left(x + \sqrt{1+x^2}\right)\right)\right]\Bigg|_0^2$

$= \dfrac{5^{3/2}}{2} - \dfrac{1}{4}\left(\sqrt{5} + \dfrac{1}{2}\ln(2 + \sqrt{5})\right)$

$= \dfrac{5}{2}\sqrt{5} - \dfrac{1}{4}\sqrt{5} - \dfrac{1}{8}\ln(2 + \sqrt{5})$

$= \dfrac{9}{4}\sqrt{5} - \dfrac{1}{8}\ln(2 + \sqrt{5})$. Finally, the total kinetic

energy is

$\dfrac{M\omega^2}{2\sqrt{5} + \ln(2 + \sqrt{5})}\left(\dfrac{9}{4}\sqrt{5} - \dfrac{1}{8}\ln(2 + \sqrt{5})\right)$

$= \dfrac{18\sqrt{5} - \ln(2 + \sqrt{5})}{16\sqrt{5} + 8\ln(2 + \sqrt{5})} M\omega^2$.

29 The local approximation is a cylindrical shell of radius x, height $2\sqrt{a^2 - x^2}$, and width dx. (See the

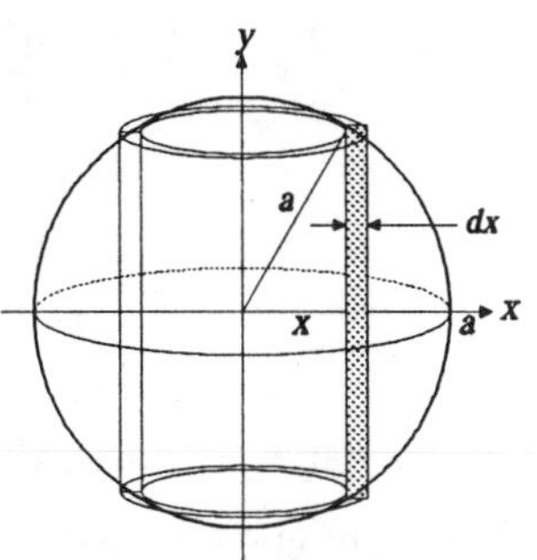

figures.) Hence its volume is

$2\pi x \cdot 2\sqrt{a^2 - x^2}\, dx =$

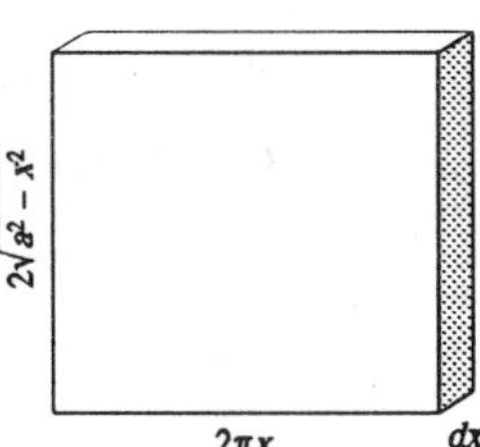

$4\pi x\sqrt{a^2 - x^2}\, dx$. Recall

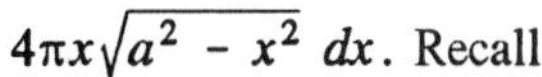

that the volume of a sphere with radius a is $4\pi a^3/3$. Thus the density of the homogeneous sphere is $\dfrac{M}{4\pi a^3/3} = \dfrac{3M}{4\pi a^3}$.

Consequently, the mass of the shell is

$4\pi x\sqrt{a^2 - x^2}\, dx \cdot \dfrac{3M}{4\pi a^3} = \dfrac{3M}{a^3} x\sqrt{a^2 - x^2}\, dx$.

Since its velocity is ωx (all points on the cylinder are equidistant from the axis of rotation), its kinetic energy is $\dfrac{1}{2}\dfrac{3M}{a^3} x\sqrt{a^2 - x^2}\, dx\,(\omega x)^2 =$

$\dfrac{3M\omega^2}{2a^3} x^3\sqrt{a^2 - x^2}\, dx$. Therefore the total kinetic

energy of the sphere is $\dfrac{3M\omega^2}{2a^3}\int_0^a x^3\sqrt{a^2 - x^2}\, dx$.

Let $y = a^2 - x^2$, so $dy = -2x\, dx$. When $x = 0$, $y = a^2$, and when $x = a$, $y = 0$, so

$\int_0^a x^3\sqrt{a^2 - x^2}\, dx = \int_{a^2}^0 (a^2 - y)x\sqrt{y}\,\dfrac{dy}{-2x}$

$$= \frac{1}{2}\int_0^{a^2} (a^2y^{1/2} - y^{3/2})\, dy$$

$$= \frac{1}{2}\left(\frac{2}{3}a^2y^{3/2} - \frac{2}{5}y^{5/2}\right)\Bigg|_0^{a^2} = \frac{2a^5}{15}.$$ Finally, the

total kinetic energy is $\frac{3M\omega^2}{2a^5}\cdot\frac{2a^5}{15} = \frac{1}{5}M\omega^2a^2$.

31 (a) The velocity at every point within the ring is approximately $g(r)$ centimeters per second. The area of the ring is approximately $2\pi r\cdot dr$; that is, the circumference of the inner circle times the thickness of the ring. Hence the flow is approximately $g(r)\cdot 2\pi r\cdot dr$ cubic centimeters per second.

(b) The total flow is $\int_0^b 2\pi rg(r)\, dr$.

(c) If $g(r) = k(b^2 - r^2)$, then the flow of blood is

$$\int_0^b 2\pi rg(r)\, dr = 2\pi\int_0^b r[k(b^2 - r^2)]\, dr$$

$$= 2\pi k\int_0^b (b^2r - r^3)\, dr$$

$$= 2\pi k\left(\frac{b^2r^2}{2} - \frac{r^4}{4}\right)\Bigg|_0^b = 2\pi k\cdot\frac{b^4}{4}$$

$= \frac{\pi k}{2}b^4$ cm^3/sec. This is proportional to b^4, the fourth power of the radius of the artery.

8.4 Computing volumes

1 (a)

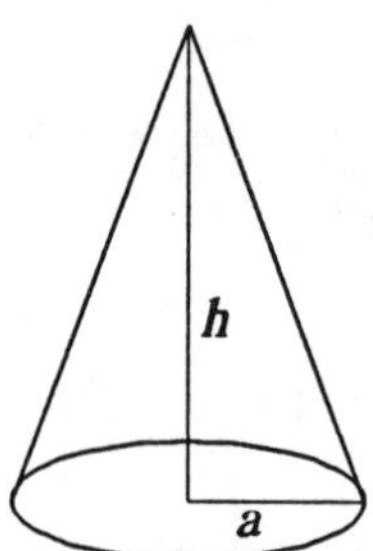

(b)

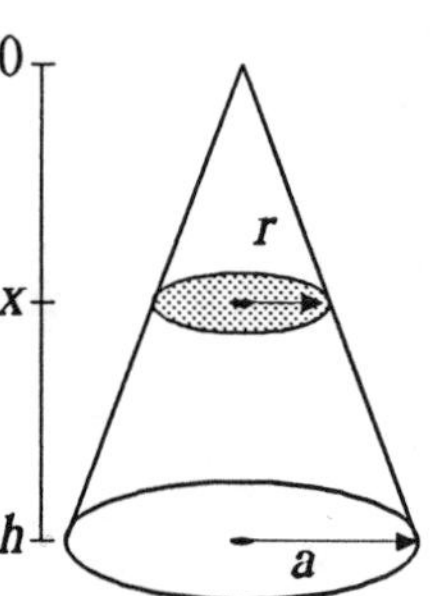

(c)

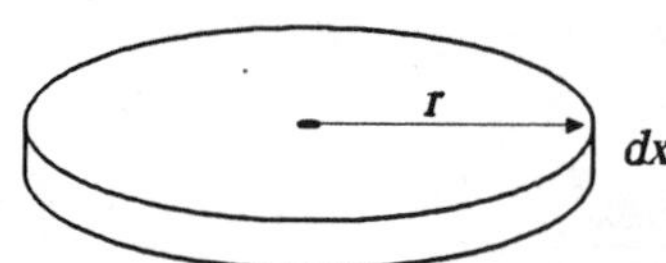

(d) The volume of the local approximation is $\pi r^2\, dx$. From the similar triangles in (b), $\frac{r}{x} = \frac{a}{h}$ and $r = \frac{ax}{h}$. So the volume in terms of x is $\pi\left(\frac{ax}{h}\right)^2 dx$.

(e) Since x varies from 0 to h, the volume of the cone is $\int_0^h \pi\left(\frac{ax}{h}\right)^2 dx$.

(f) $\int_0^h \pi\left(\frac{ax}{h}\right)^2 dx = \frac{\pi a^2}{h^2}\int_0^h x^2\, dx$

$$= \frac{\pi a^2}{h^2} \cdot \frac{x^3}{3}\bigg|_0^h = \frac{1}{3}\pi a^2 h.$$

3 (a) The solid is shown in Fig. 14 of the text.

(b) The typical cross section is shown in Fig. 14 of the text.

(c)

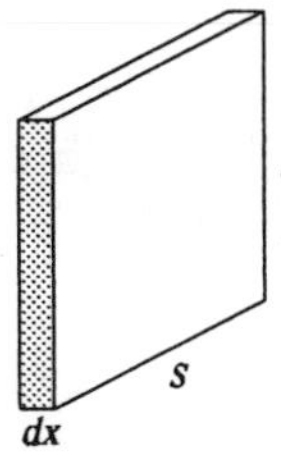

(d) The volume of the local approximation is $s^2\,dx$. But from the figure, $s = 2\sqrt{3^2 - x^2} = 2\sqrt{9 - x^2}$. So the volume in terms of x is $\left(2\sqrt{9 - x^2}\right)^2 dx = 4(9 - x^2)\,dx$.

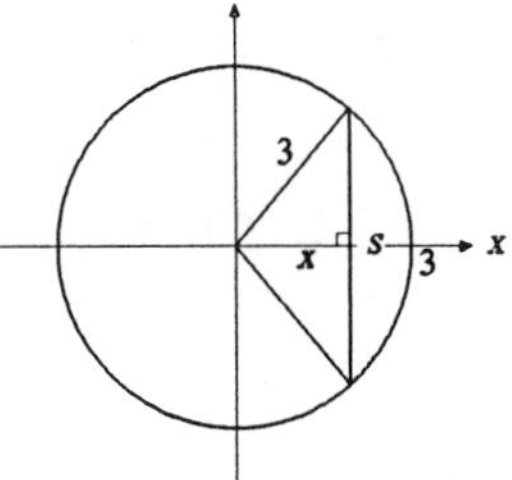

(e) Since x varies from -3 to 3, the volume of the solid is $\int_{-3}^{3} 4(9 - x^2)\,dx$.

(f) $\int_{-3}^{3} 4(9 - x^2)\,dx = 4\int_{-3}^{3} (9 - x^2)\,dx$

$$= 4\left(9x - \frac{x^3}{3}\right)\bigg|_{-3}^{3} = 4 \cdot 36 = 144$$

5 (a)

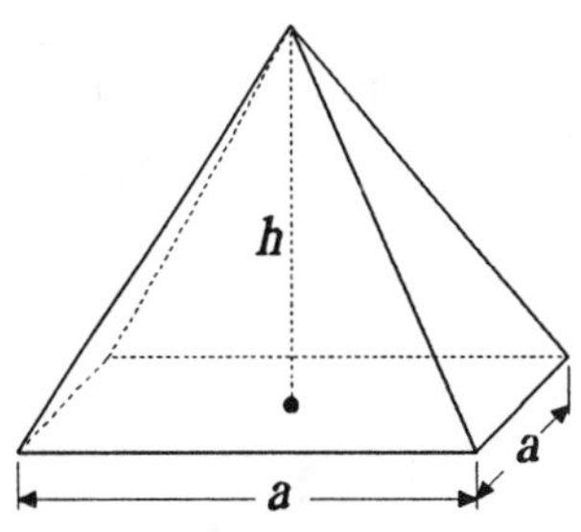

(b)

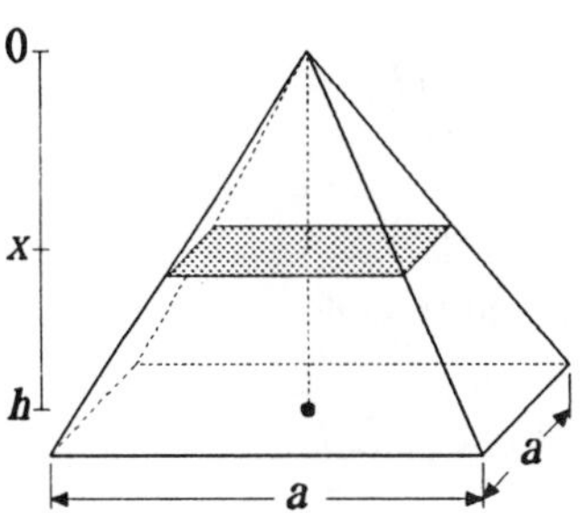

(c)

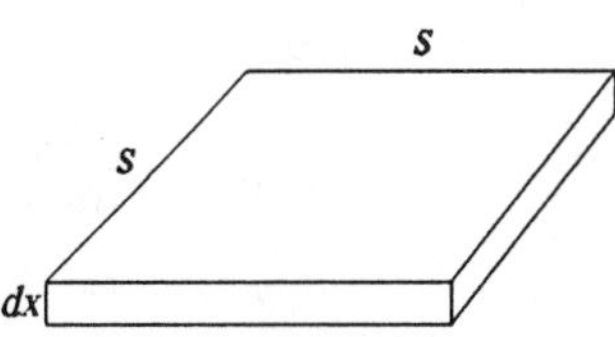

(d) The volume of the local approximation is $s^2\,dx$. But from the figure, $\frac{s/2}{x} = \frac{a/2}{h}$ so $s = \frac{ax}{h}$. The volume in terms of x is $\left(\frac{ax}{h}\right)^2 dx$.

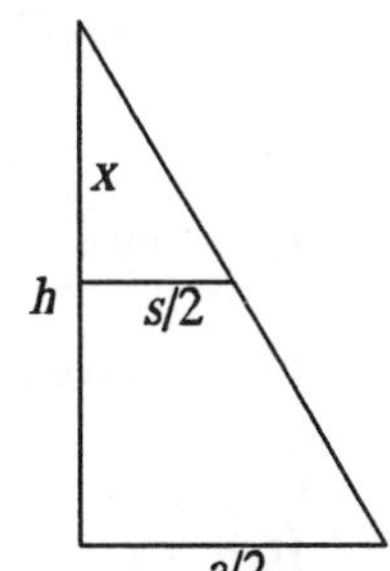

(e) Since x varies from 0 to h, the volume of the pyramid is $\int_0^h \left(\frac{ax}{h}\right)^2 dx$.

(f) $\int_0^h \left(\frac{ax}{h}\right)^2 dx = \frac{a^2}{h^2}\int_0^h x^2\,dx = \frac{a^2}{h^2}\frac{x^3}{3}\bigg|_0^h$

$= \frac{1}{3}a^2h$

7 (a) The solid is shown in Fig. 15 of the text.

(b) The typical cross section is shown in Fig. 15 of the text.

(c)

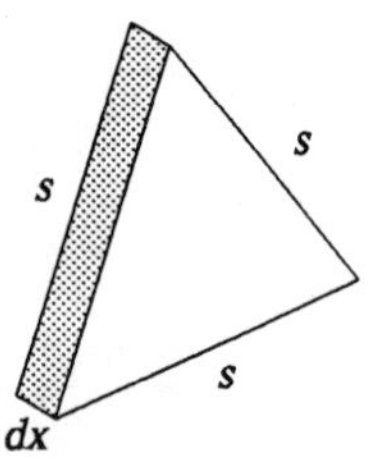

(d) The area of an equilateral triangle with side s is $\frac{s^2\sqrt{3}}{4}$, so the volume of the local approximation is $\frac{s^2\sqrt{3}}{4}\,dx$. From the figure, $s = 2\sqrt{5^2 - x^2}$ $= 2\sqrt{25 - x^2}$. Thus the volume in terms of x is $\frac{\left(2\sqrt{25 - x^2}\right)^2\sqrt{3}}{4}\,dx = \sqrt{3}(25 - x^2)\,dx$.

(e) Since x varies from -5 to 5, the volume of the solid is $\int_{-5}^{5} \sqrt{3}(25 - x^2)\,dx$.

(f) $\int_{-5}^{5} \sqrt{3}(25 - x^2)\,dx = 2\sqrt{3}\int_0^5 (25 - x^2)\,dx$

$$= 2\sqrt{3}\left[25x - \frac{x^3}{3}\right]\Bigg|_0^5 = 2\sqrt{3}\left[125 - \frac{125}{3}\right]$$

$$= \frac{500}{\sqrt{3}}$$

9 Volume $= \int_1^2 \pi(\sqrt{x})^2\,dx = \int_1^2 \pi x\,dx = \frac{\pi x^2}{2}\Bigg|_1^2$

$$= \frac{\pi}{2}(4 - 1) = \frac{3\pi}{2}$$

11 Volume $= \int_1^2 \pi\left[\left(\frac{1}{\sqrt{x}}\right)^2 - \left(\frac{1}{x}\right)^2\right] dx$

$$= \pi\int_1^2 \left(\frac{1}{x} - \frac{1}{x^2}\right) dx = \pi\left(\ln x + \frac{1}{x}\right)\Bigg|_1^2$$

$$= \pi\left(\ln 2 - \frac{1}{2}\right)$$

13 Volume $= \int_0^{\pi/4} \pi\,[(\tan x)^2 - (\sin x)^2]\,dx$

$$= \pi\int_0^{\pi/4} (\tan^2 x - \sin^2 x)\,dx$$

$$= \pi\int_0^{\pi/4} \left(\sec^2 x - 1 - \frac{1 - \cos 2x}{2}\right) dx$$

$$= \pi\left(\tan x - \frac{3}{2}x + \frac{\sin 2x}{4}\right)\Bigg|_0^{\pi/4}$$

$$= \pi\left[\left(1 - \frac{3\pi}{8} + \frac{1}{4}\right) - 0\right] = \frac{\pi}{8}(10 - 3\pi)$$

15 (a) The solid region is shown in Fig. 17 of the text.

(b)

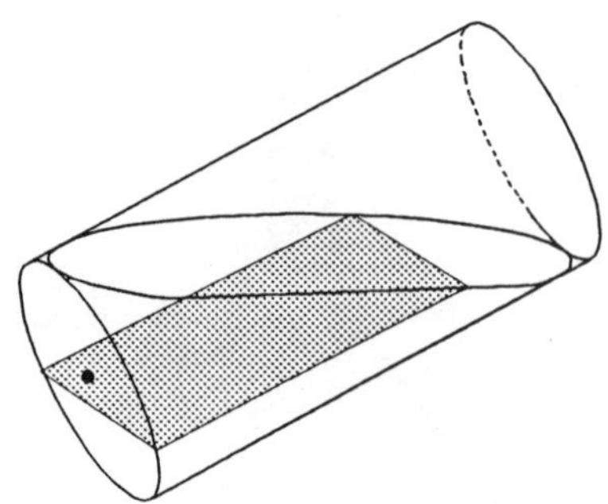

(c)

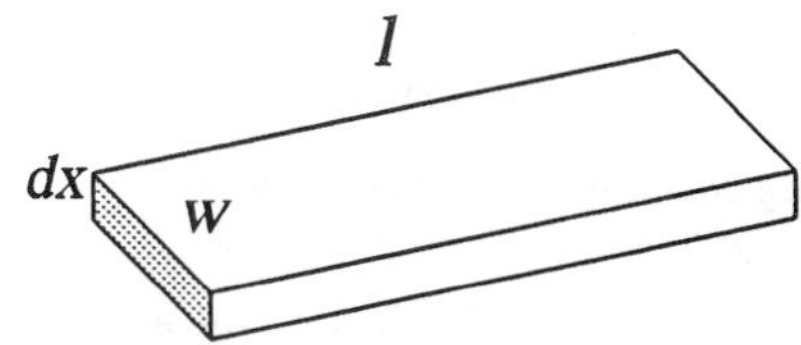

(d) From the figure with the bottom of the glass illustrated, we see $w = 2\sqrt{a^2 - x^2}$. From the

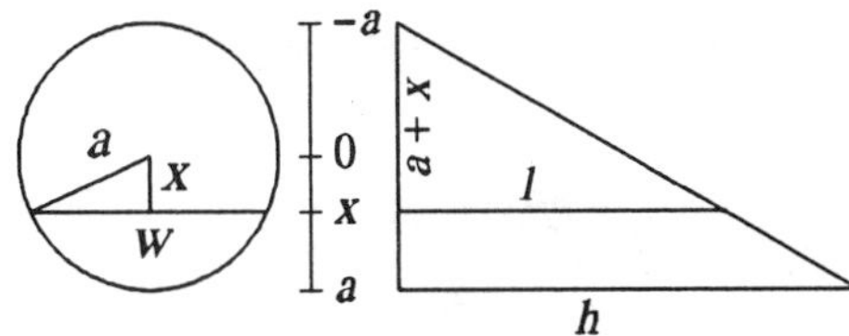

similar triangles in the side view, it follows that $l = \frac{h}{2a}(a + x)$. Thus the volume of the local approximation is $wl\,dx =$

$\frac{h}{a}(a + x)\sqrt{a^2 - x^2}\,dx.$

(e) Since x varies from $-a$ to a, Volume =

$\int_{-a}^{a} \frac{h}{a}(a + x)\sqrt{a^2 - x^2}\,dx.$

(f) $\int_{-a}^{a} \frac{h}{a}(a + x)\sqrt{a^2 - x^2}\,dx =$

$h\int_{-a}^{a} \sqrt{a^2 - x^2}\,dx + \frac{h}{a}\int_{-a}^{a} x\sqrt{a^2 - x^2}\,dx$

$= h\cdot\frac{\pi}{2}a^2 + 0 = \frac{\pi}{2}a^2h$, since the second integral is the area of a half disk and the third integrand is an odd function.

17 Note that the water takes up exactly half the volume of the glass. Since the volume of the glass (a cylinder of radius a and height h) is πa^2h, the volume of the water must be $\frac{1}{2}(\pi a^2h) = \frac{\pi}{2}a^2h.$

19 Place an x axis in the base of the glass, perpendicular to the intersection of the base with the water's surface. Let its origin be at the center of the base. The rectangle with coordinate x has width $2\sqrt{a^2 - x^2}$. Its length is $\frac{h}{a}x$. Hence its area is $\frac{2h}{a}\cdot x\sqrt{a^2 - x^2}$ So the volume of the water is

$V = \int_0^a \frac{2h}{a}x\sqrt{a^2 - x^2}\,dx$. Let $u = a^2 - x^2$. Then

$V = \frac{h}{a}\int_{a^2}^{0} \sqrt{u}\,(-du) = \frac{h}{a}\int_0^{a^2} u^{1/2}\,du$

$= \frac{h}{a}\cdot\frac{2}{3}u^{3/2}\Big|_0^{a^2} = \frac{2}{3}a^2h.$

21 (a)

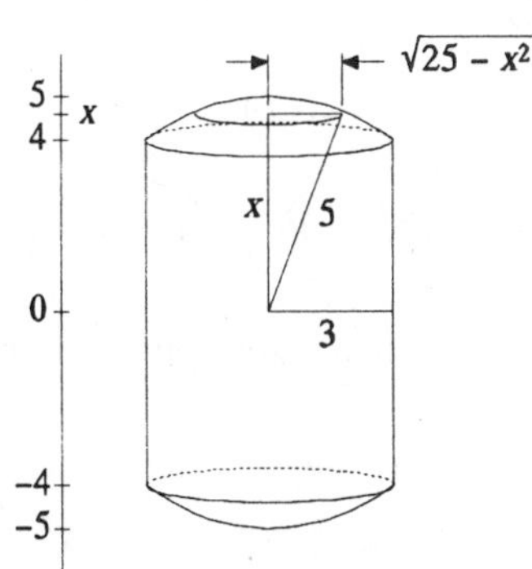

(b) For $-4 \le x \le 4$, cross sections are circles with radius 3. However, for $-5 \le x \le -4$ and $4 \le x \le 5$, the circular cross sections do not all have the same radius. As seen in the figure, the radius of such a cross section is $\sqrt{25 - x^2}$. Therefore,

$$A(x) = \begin{cases} 9\pi, & -4 \le x \le 4 \\ \pi(25 - x^2), & -5 \le x \le -4 \text{ or } 4 \le x \le 5. \end{cases}$$

(c) The volume removed from the center of the sphere is $\int_{-5}^{5} A(x)\,dx = \int_{-5}^{4} \pi(25 - x^2)\,dx +$

$$\int_{-4}^{4} 9\pi\, dx + \int_{4}^{5} \pi(25 - x^2)\, dx$$

$$= \left(25x - \frac{x^3}{3}\right)\Bigg|_{-5}^{-4} + 9\pi x\Bigg|_{-4}^{4} + \pi\left(25x - \frac{x^3}{3}\right)\Bigg|_{4}^{5}$$

$$= \pi\left[\left(-100 + \frac{64}{3}\right) - \left(-125 + \frac{125}{3}\right)\right]$$

$$+ 9\pi(4 - (-4))$$

$$+ \pi\left[\left(125 - \frac{125}{3}\right) - \left(100 - \frac{64}{3}\right)\right] = \frac{244\pi}{3}.$$

23 The typical cross section perpendicular to the y axis is a ring with inner

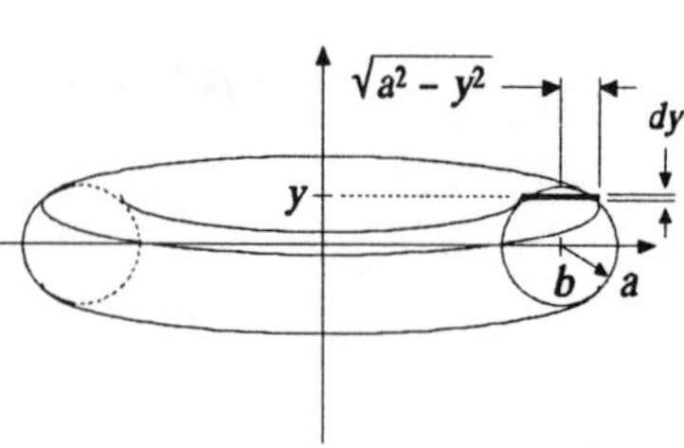

radius $b - \sqrt{a^2 - y^2}$ and outer radius $b + \sqrt{a^2 - y^2}$. Its area is $\pi\left[\left(b + \sqrt{a^2 - y^2}\right)^2 - \left(b - \sqrt{a^2 - y^2}\right)^2\right] = 4\pi b\sqrt{a^2 - y^2}$. Hence the volume of the torus is $\int_{-a}^{a} 4\pi b\sqrt{a^2 - y^2}\, dy = 4\pi b \int_{-a}^{a} \sqrt{a^2 - y^2}\, dy = 4\pi b \cdot \frac{\pi a^2}{2} = 2\pi^2 a^2 b$ (since the last integral is the area of a semicircle of radius a).

25 Let $f(y)$ and $g(y)$ be defined as in the figure, so that the area of the region R is $A = \int_c^d [f(y) - g(y)]\, dy$. By the washer technique, the volume of the solid of revolution formed by revolving R about the y axis is $V = \pi \int_c^d [(f(x))^2 - (g(y))^2]\, dy$. If we revolve R about the line $x = -k$, we instead obtain a solid of revolution whose volume is $V^* = \pi \int_c^d [(f(y) + k)^2 - (g(y) + k)^2]\, dy =$

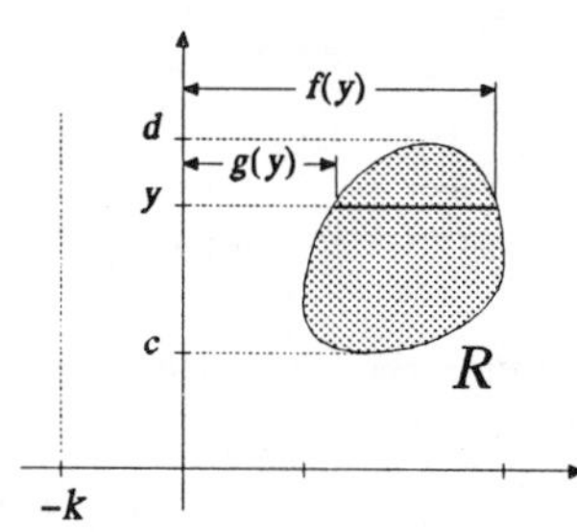

$$\pi \int_c^d [(f(y))^2 - 2f(y)k + k^2 - (g(y))^2 + 2g(y)k - k^2]\, dy$$

$$= \pi \int_c^d [(f(y))^2 - (g(y))^2 + 2k(f(y) - g(y))]\, dy =$$

$$\pi \int_c^d [(f(y))^2 - (g(y))^2]\, dy + 2\pi k \int_c^d [f(y) - g(y)]\, dy$$

$$= V + 2\pi k A.$$

8.5 The Shell Method

1 By inspection, we see that $c(x) = x$ and $R(x) = x$ (since the trapezoid is revolved about the line $x =$

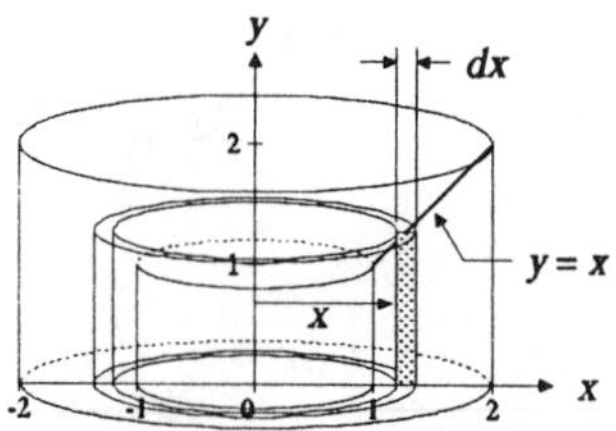

0.) Thus the local approximation of the volume is $2\pi R(x)c(x)\, dx = 2\pi x^2\, dx$. Now x varies from 1 to 2, so the total volume of the solid is $\int_1^2 2\pi x^2\, dx$

$$= 2\pi \int_1^2 x^2\, dx = 2\pi \frac{x^3}{3}\Bigg|_1^2 = \frac{14\pi}{3}.$$

3 Clearly $R(x) = x$ since the triangle is revolved around the line $x = 0$. To find $c(x)$, we must find a

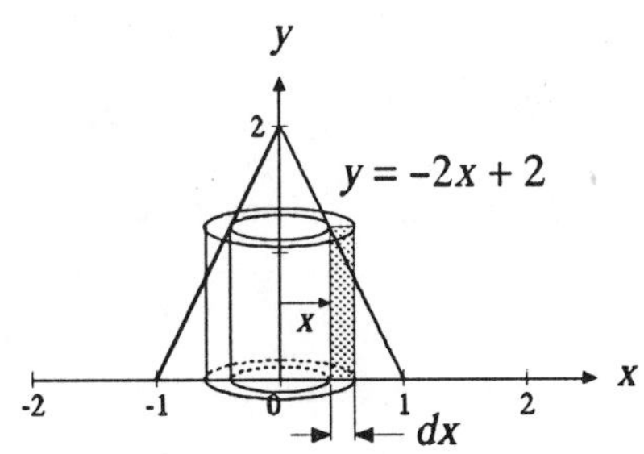

formula for the line between (0, 2) and (1, 0). The slope of the line is $\frac{2-0}{0-1} = -2$, and its y intercept is 2, so $y = -2x + 2$. Thus $c(x) = -2x + 2$ and the local approximation of the volume is $2\pi x(-2x + 2)\,dx = 4\pi(x - x^2)\,dx$. Here x varies from 0 to 1, so the total volume of the solid is

$$4\pi \int_0^1 (x - x^2)\,dx = 4\pi\left[\frac{x^2}{2} - \frac{x^3}{3}\right]\Bigg|_0^1 = 4\pi \cdot \frac{1}{6}$$

$= \frac{2\pi}{3}$. (Note that the solid of revolution is actually a right circular cone of radius 1 and height 2.)

5 Note that $x^2 > x^3$, for all x in (0, 1), so $c(x) = x^2 - x^3$, and $R(x) = x$ by inspection. Hence the

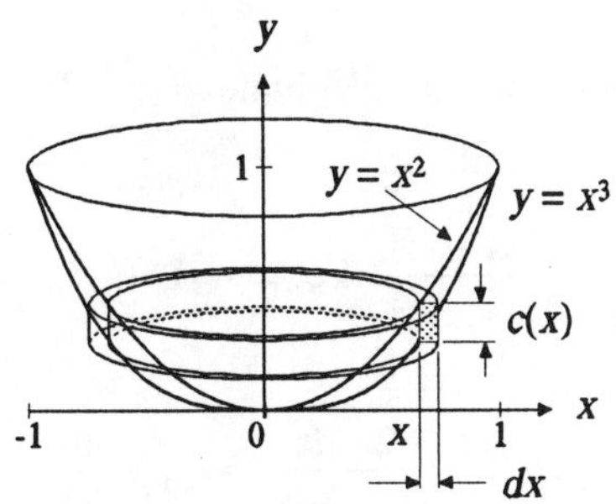

local estimate of the volume is $2\pi x(x^2 - x^3)\,dx$, and the total volume is thus $2\pi \int_0^1 (x^3 - x^4)\,dx$

$= 2\pi\left[\frac{x^4}{4} - \frac{x^5}{5}\right]\Bigg|_0^1 = \frac{\pi}{10}$. (Observe that this result matches that of Exercise 5. The problems are equivalent, with x and y interchanged.)

7 Since the region is revolved about the x axis, we must integrate with respect to y. Now $y^2 > y^3$ for all y in (0, 1), so $c(y) = y^2 - y^3$. By inspection, $R(y) = y$; thus the local approximation for the volume is $2\pi y(y^2 - y^3)\,dy = 2\pi(y^3 - y^4)\,dy$. Because y varies from 0 to 1, the total volume of the solid is $2\pi \int_0^1 (y^3 - y^4)\,dy =$

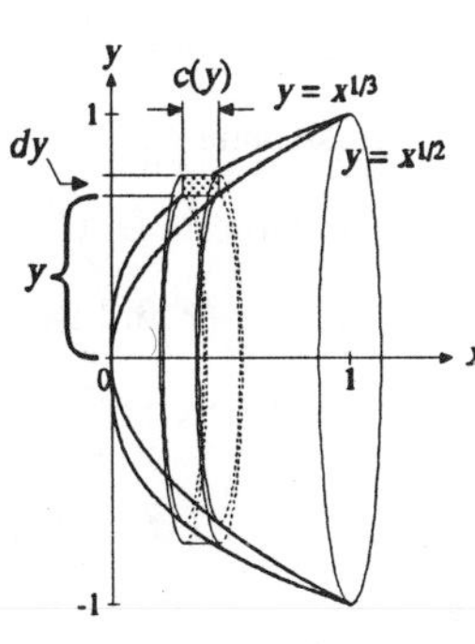

$$2\pi\left[\frac{y^4}{4} - \frac{y^5}{5}\right]\Bigg|_0^1 = \frac{\pi}{10}.$$

9 The right circular cone can be formed by revolving a right triangle with legs of length a and h around the leg of length h (see the figure). By similar triangles, $\frac{c(y)}{a - y} = \frac{h}{a}$, so $c(y) = h - (h/a)y$. The radius of revolution is $R(y) = y$ and the local estimate of the volume is $2\pi y(h - (h/a)y)\,dy = 2\pi(hy - (h/a)y^2)\,dy$. Now y varies from 0 to a, so the volume of the cone is

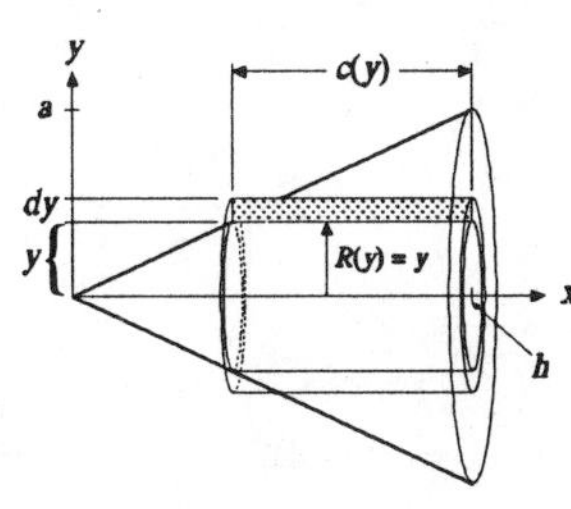

$$2\pi \int_0^a \left(hy - \frac{h}{a}y^2\right)dy = 2\pi\left[\frac{h}{2}y^2 - \frac{h}{3a}y^3\right]\Bigg|_0^a$$

$= \frac{1}{3}\pi a^2 h$.

11 From the figure, we see that $c(x) = x + x^3$ and x varies from 1 to 2.

(a) Here $R(x) = x$, so the local estimate of the volume is $2\pi x(x + x^3)\,dx$ and the total volume is

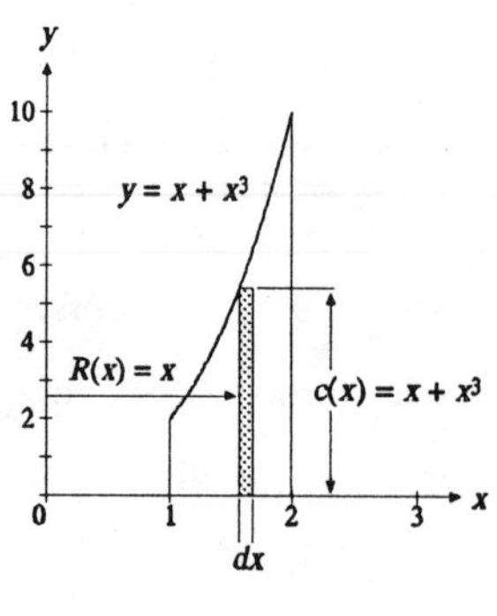

$$2\pi \int_1^2 (x^2 + x^4)$$

$$= 2\pi\left[\frac{x^3}{3} + \frac{x^5}{5}\right]\Bigg|_1^2$$

$$= 2\pi\left[\frac{8}{3} + \frac{32}{5} - \left(\frac{1}{3} + \frac{1}{5}\right)\right] = \frac{256\pi}{15}.$$

(b) Here $R(x) = 3 - x$, so the local estimate is $2\pi(3 - x)(x + x^3)\,dx$ and the total volume is

$$2\pi \int_1^2 (-x^4 + 3x^3 - x^2 + 3x)\,dx$$

$$= 2\pi\left[-\frac{x^5}{5} + \frac{3x^4}{4} - \frac{x^3}{3} + \frac{3x^2}{2}\right]\Bigg|_1^2$$

$$= 2\pi\left[-\frac{32}{5} + 12 - \frac{8}{3} + 6 - \left(-\frac{1}{5} + \frac{3}{4} - \frac{1}{3} + \frac{3}{2}\right)\right]$$

$$= 2\pi \cdot \frac{433}{60} = \frac{433\pi}{30}.$$

13 The region to be revolved is shown in the figure.

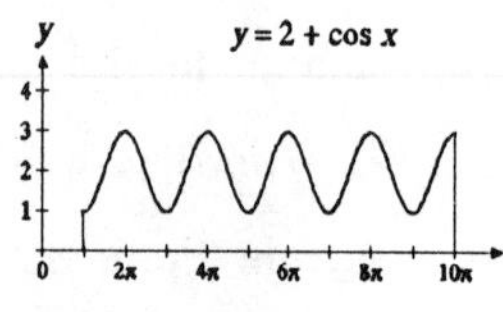

(a) The method of cylindrical shells is called for. Here $R(x) = x$, $c(x) = 2 + \cos x$, and the volume is

$$\int_\pi^{10\pi} 2\pi R(x)c(x)\,dx$$

$$= 2\pi \int_\pi^{10\pi} x(2 + \cos x)\,dx$$

$$= 2\pi(x^2 + x\sin x + \cos x)\Big|_\pi^{10\pi}$$

$$= 2\pi[(100\pi^2 + 0 + 1) - (\pi^2 + 0 - 1)]$$

$= 2\pi(99\pi^2 + 2)$. (Integration by parts was used to evaluate $\int x\cos x\,dx$.)

(b) Parallel cross sections should be used. A typical cross section is a circular disk of radius $R(x) = 2 + \cos x$. The volume is therefore $\int_\pi^{10\pi} \pi(2 + \cos x)^2\,dx =$

$$\pi \int_\pi^{10\pi} (4 + 4\cos x + \cos^2 x)\,dx$$

$$= \pi \int_\pi^{10\pi} \left(4 + 4\cos x + \frac{1 + \cos 2x}{2}\right) dx$$

$$= \pi \int_\pi^{10\pi} \left(\frac{9}{2} + 4\cos x + \frac{\cos 2x}{2}\right) dx$$

$$= \pi\left(\frac{9}{2}x + 4\sin x + \frac{\sin 2x}{4}\right)\Bigg|_\pi^{10\pi}$$

$$= \pi\left[(45\pi + 0 + 0) - \left(\frac{9\pi}{2} + 0 + 0\right)\right]$$

$$= \frac{81\pi^2}{2}.$$

15 Using the shell technique with $R(x) = x$ and $c(x) = \frac{1}{(1 + x^2)^2}$, we have Volume $= \int_0^1 2\pi x \cdot \frac{dx}{(1 + x^2)^2}$ $= \pi \int_0^1 \frac{2x\,dx}{(1 + x^2)^2}$. Let $u = 1 + x^2$, so $du = 2x\,dx$. Then $\int_0^1 \frac{2x\,dx}{(1 + x^2)^2} = \int_1^2 \frac{du}{u^2} = -\frac{1}{u}\Big|_1^2 = -\frac{1}{2} + 1 = \frac{1}{2}$, so the total volume is $\pi/2$.

17 Two problems arise if we try to use parallel cross sections. First, such a cross section may have up to 6 separate pieces (which may not be clear from the

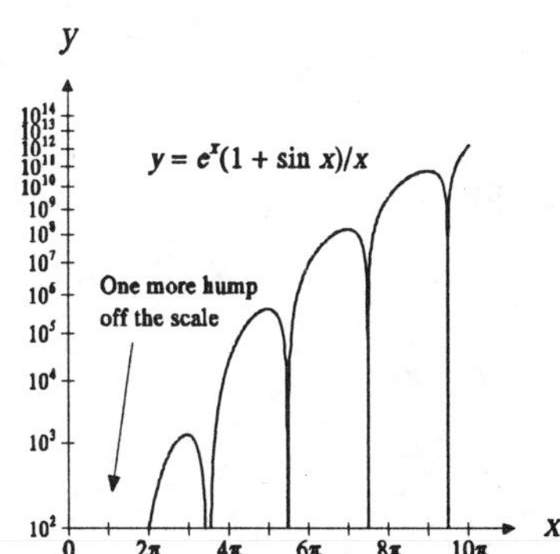

figures because the y values vary so greatly). Second, to find where the line $y = h$ intersects the curve $\dfrac{e^x(1 + \sin x)}{x}$ involves solving the equation $\dfrac{e^x(1 + \sin x)}{x} = h$. (I certainly can't do it; can you?) On the other hand, the shell technique is easy: The volume is $\displaystyle\int_\pi^{10\pi} 2\pi x\left(\frac{e^x(1 + \sin x)}{x}\right) dx$

$$= 2\pi \int_\pi^{10\pi} e^x(1 + \sin x)\, dx$$

$$= \pi\left(e^x + \frac{1}{2}e^x(\sin x - \cos x)\right)\Big|_\pi^{10\pi}$$

$= \pi(e^{10\pi} - 3e^\pi)$. (The antiderivative of $e^x \sin x$ can be found in by the method of Example 8 in Sec. 7.3 or with Formula 64 from the table of antiderivatives.)

19 For $0 \le x \le 1$, $R(x) = x - (-2) = x + 2$ and $c(x) = \dfrac{1}{x^2 + 4x + 1}$, so the total volume is

$$\int_0^1 2\pi(x + 2)\cdot\frac{dx}{x^2 + 4x + 1} =$$

$$\pi \int_0^1 \frac{2x + 4}{x^2 + 4x + 1}\, dx = \pi \ln(x^2 + 4x + 1)\Big|_0^1$$

$= \pi \ln 6$. (Note that the derivative of the denominator of the integrand is equal to the numerator of the integrand.)

21 In the first case, we know $R(x) = x$ so the volume is given by

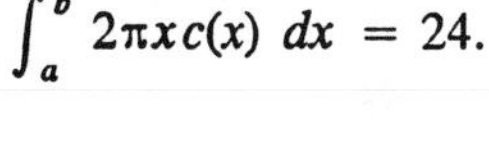
$$\int_a^b 2\pi x c(x)\, dx = 24.$$

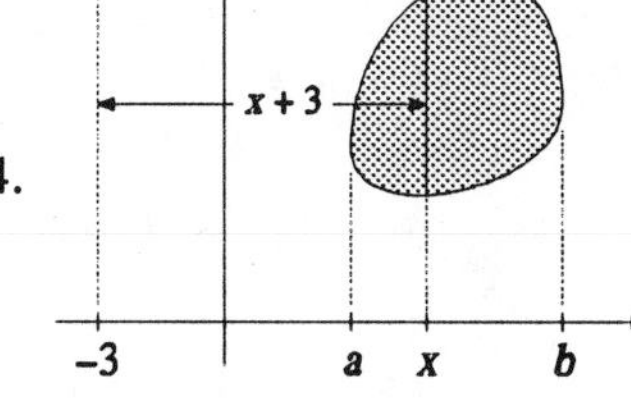

In the second case, $R(x) = x - (-3) = x + 3$, so the volume is $\displaystyle\int_a^b 2\pi x(x + 3)\, dx =$ $\displaystyle\int_a^b 2\pi x c(x)\, dx + 6\pi \int_a^b c(x)\, dx = 82$. Now the area of R, A, is $\displaystyle\int_a^b c(x)\, dx$, so the above equation becomes $\displaystyle\int_a^b 2\pi x c(x)\, dx + 6\pi A = 82$. Since $\displaystyle\int_a^b 2\pi x c(x)\, dx = 24$, this reduces to $24 + 6\pi A = 82$, and $A = 29/(3\pi)$.

23 Let R be the region below $y = f(x)$ and above the interval $[a, b]$. Assume that R lies in the first quadrant. The solid formed by revolving R about the x axis has volume $\displaystyle\int_a^b \pi(f(x))^2\, dx$. The solid formed by revolving R about the y axis has volume $\displaystyle\int_a^b 2\pi x f(x)\, dx$. So we want $\displaystyle\int (f(x))^2\, dx$ to be nonelementary while $\displaystyle\int x f(x)\, dx$ is elementary. The function $f(x) = e^{x^2}$ will work. While $\displaystyle\int (e^{x^2})^2\, dx = \int e^{2x^2}\, dx = \frac{1}{\sqrt{2}} \int e^{u^2}\, du$ (where $u = \sqrt{2}x$) is nonelementary, $\displaystyle\int x e^{x^2}\, dx =$

$\frac{1}{2}\int e^v\,dv = \frac{1}{2}e^v + C$ (where $v = x^2$) is elementary. To specify R completely, we may choose any a and b with $0 \le a < b$; for example, $a = 0$ and $b = 1$.

25 (a) Let $u = x^2$, $dv = f'(x)\,dx$; $\int x^2 f'(x)\,dx$ $= x^2f(x) - \int f(x)\cdot 2x\,dx =$ $x^2f(x) - 2\int xf(x)\,dx$. Since $x^2f(x)$ is elementary, $\int x^2f'(x)\,dx$ is elementary if and only if $\int xf(x)\,dx$ is.

(b) By shells, $V = \int_0^a 2\pi xf(x)\,dx$ $= 2\pi\int_0^a xf(x)\,dx$. By cross sections, $V =$ $\int_0^b \pi[f^{-1}(y)]^2\,dy$. Let $y = f(x)$. Then $V =$ $\pi\int_a^0 x^2f'(x)\,dx$.

8.6 The Centroid of a Plane Region

1 The area of R is $\int_{-2}^2 (4 - x^2)\,dx$ $= 2\int_0^2 (4 - x^2)\,dx$ $= 2\left(4x - \frac{x^3}{3}\right)\Big|_0^2$ $= \frac{32}{3}$. Note, by symmetry, that $\bar{x} = 0$. Finally,

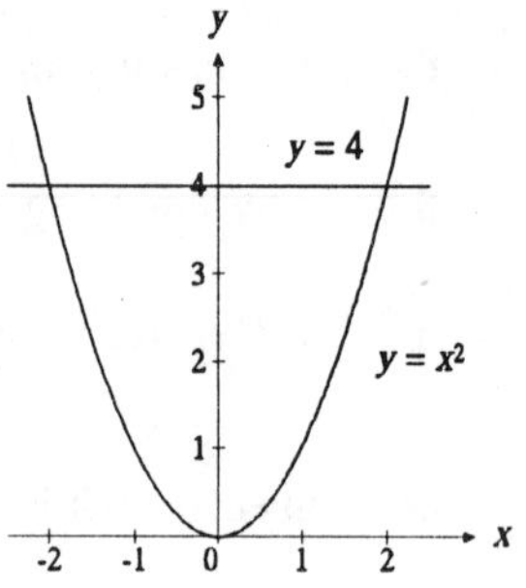

$$\bar{y} = \frac{\int_c^d yc(y)\,dy}{\text{Area of } R} = \frac{3}{32}\int_0^4 y(2\sqrt{y})\,dy = \frac{3}{16}\int_0^4 y^{3/2}\,dy = \frac{3}{16}\cdot\frac{2}{5}y^{5/2}\Big|_0^4 = \frac{12}{5}, \text{ so } (\bar{x}, \bar{y})$$

$= (0, 12/5)$.

3 The area of R is $\int_0^4 (4x - x^2)\,dx$ $= \left(2x^2 - \frac{x^3}{3}\right)\Big|_0^4$ $= \frac{32}{3}$. Then $\bar{x}$

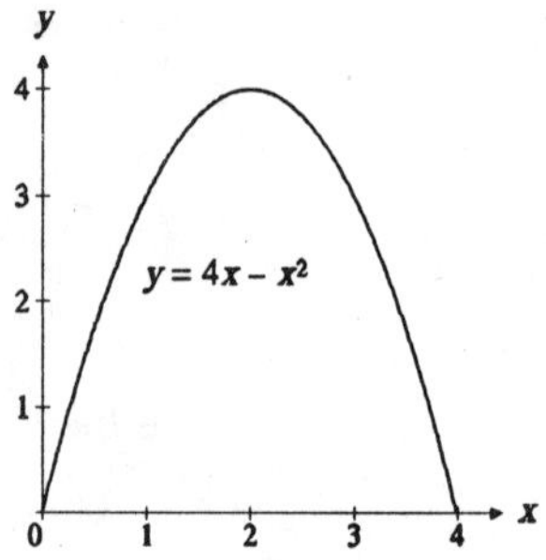

$$= \frac{\int_a^b xc(x)\,dx}{\text{Area of } R} = \frac{3}{32}\int_0^4 x(4x - x^2)\,dx$$

$$= \frac{3}{32}\int_0^4 (4x^2 - x^3)\,dx = \frac{3}{32}\left(\frac{4}{3}x^3 - \frac{1}{4}x^4\right)\Big|_0^4$$

$= 2$. (One could argue this from parabolic symmetry as well.) Finally, since it is difficult to derive a formula for x in terms of y, we use formula (10); thus $\bar{y} = \dfrac{\int_a^b \frac{1}{2}(f(x))^2}{\text{Area of } R}$

$$= \frac{3}{64}\int_0^4 (4x - x^2)^2\,dx$$

$$= \frac{3}{64}\int_0^4 (16x^2 - 8x^3 + x^4)\,dx$$

$$= \frac{3}{64}\left(\frac{16}{3}x^3 - 2x^4 + \frac{1}{5}x^5\right)\Big|_0^4 = \frac{8}{5}. \text{ So } (\bar{x}, \bar{y})$$

$= (2, 8/5)$.

5 Let A be the area of the region. Then

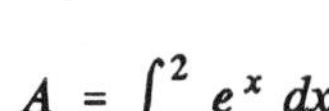

$$A = \int_1^2 e^x\,dx$$

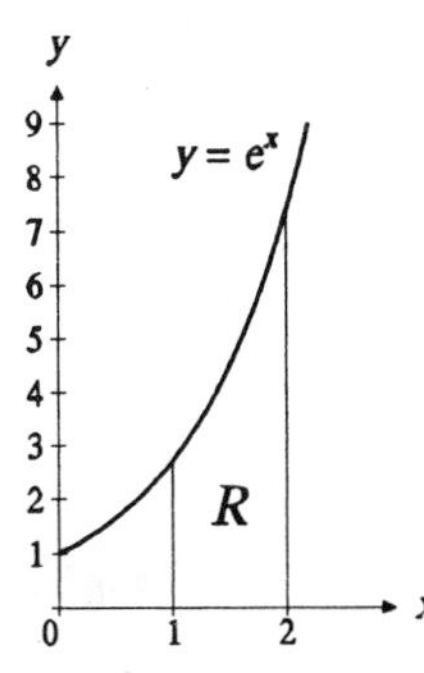

$= e^2 - e$; $\bar{x} = \dfrac{1}{A}\displaystyle\int_1^2 xe^x\,dx = \dfrac{1}{A}(xe^x - e^x)\Big|_1^2 = \dfrac{2e^2 - e^2 - e + e}{e(e - 1)}$

$= \dfrac{e}{e - 1}$; $\bar{y} = \dfrac{1}{A}\displaystyle\int_1^2 \frac{1}{2}(e^x)^2\,dx = \dfrac{1}{2A}\left(\dfrac{1}{2}e^{2x}\right)\Big|_1^2$

$= \dfrac{e^4 - e^2}{4e(e - 1)} = \dfrac{e^2 + e}{4}$. Hence the centroid is

$\left(\dfrac{e}{e - 1}, \dfrac{e^2 + e}{4}\right)$.

7 The area of the region is $A = \displaystyle\int_0^3 \sqrt{1 + x}\,dx$

$= \dfrac{(1 + x)^{3/2}}{3/2}\Big|_0^3$

$= \dfrac{14}{3}$. Then

$\bar{x} = \dfrac{1}{A}\displaystyle\int_0^3 x\sqrt{1 + x}\,dx$. If we let $u = 1 + x$, then

$x = u - 1$ and $dx = du$, so

$\bar{x} = \dfrac{3}{14}\displaystyle\int_1^4 (u - 1)\sqrt{u}\,du$

$= \dfrac{3}{14}\displaystyle\int_1^4 (u^{3/2} - u^{1/2})\,du$

$= \dfrac{3}{14}\left[\dfrac{2}{5}u^{5/2} - \dfrac{2}{3}u^{3/2}\right]\Big|_1^4$

$= \dfrac{3}{14}\left[\left(\dfrac{64}{5} - \dfrac{16}{3}\right) - \left(\dfrac{2}{5} - \dfrac{2}{3}\right)\right] = \dfrac{58}{35}$. Finally, $\bar{y}$

$= \dfrac{1}{A}\displaystyle\int_0^3 \frac{1}{2}(\sqrt{1 + x})^2\,dx = \dfrac{1}{2A}\displaystyle\int_0^3 (1 + x)\,dx$

$= \dfrac{3}{28}\left(x + \dfrac{x^2}{2}\right)\Big|_0^3 = \dfrac{45}{56}$, so the centroid is

(58/35, 45/56).

9 (a) Set up coordinate axes as in the figure, where the line L corresponds to the y axis. Then

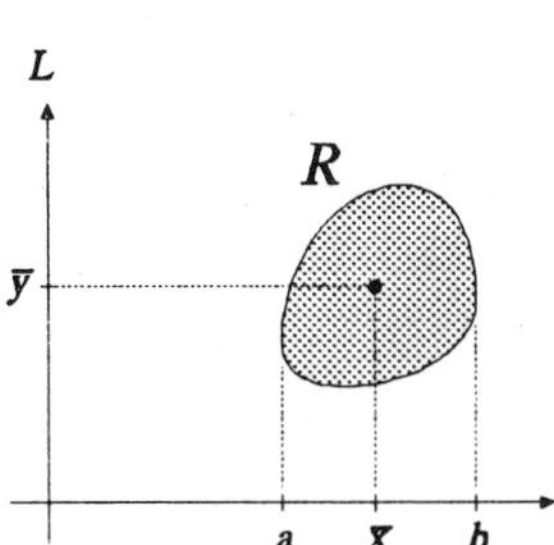

the volume V of the solid produced when R is revolved about L is $V = \displaystyle\int_a^b 2\pi x c(x)\,dx =$

$A\cdot 2\pi \dfrac{\int_a^b xc(x)\,dx}{A} = A\cdot 2\pi\bar{x}$. Since $2\pi\bar{x}$ is the distance the centroid of R is revolved and A is the area of R, we have proved Pappus's theorem.

(b) The distance between the centroid of the circle (its center) and the axis of revolution is 5 inches. The circle's area is $\pi\cdot 3^2 = 9\pi$ square inches. By the theorem of Pappus, the volume of the torus is $9\pi\cdot 2\pi(5) = 90\pi^2$ cubic inches.

11 From the figure, we see that revolving the right triangle about the x axis yields a right circular

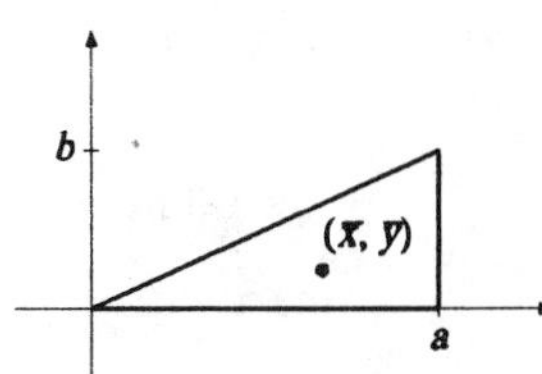

cone of radius b, height a, and volume $(1/3)\pi b^2 a$. By Pappus's theorem we have $(1/3)\pi b^2 a =$

$2\pi\bar{y}\cdot(1/2)ab$, so $\bar{y} = \frac{b}{3}$. Revolving the right triangle about the line $x = a$ gives a right circular cone of radius a, height b, and volume $(1/3)\pi a^2 b$. The centroid travels a distance $2\pi(a - \bar{x})$. By Pappus's theorem, $\frac{1}{3}\pi a^2 b = 2\pi(a - \bar{x})\cdot\frac{1}{2}ab$, so

$\bar{x} = a - \frac{a}{3} = \frac{2a}{3}$ Finally, the centroid of the right triangle is $(\bar{x}, \bar{y}) = \left(\frac{2a}{3}, \frac{b}{3}\right)$.

12 (a) We already know that the moment of R about the y axis is $\int_a^b xc(x)\,dx$. In this case, $c(x) = f(x) - g(x)$, so the moment is $\int_a^b x[f(x) - g(x)]\,dx$.

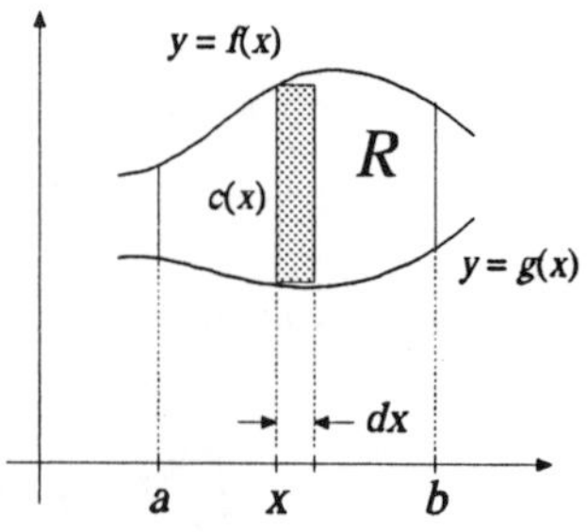

(b) The following procedure is similar to that used in establishing equation (10). Consider the portion of the region R between x and $x + dx$. This strip is approximately a rectangle with area $(f(x) - g(x))\,dx$. The y coordinate of the centroid of this region is the midpoint of its height, $\frac{f(x) + g(x)}{2}$.

According to equation (8), the moment about the x is the product of the y coordinate of the rectangle's centroid with its area. Hence the moment of the rectangle about the x axis $\frac{f(x) + g(x)}{2}(f(x) - g(x))\,dx =$ $\frac{1}{2}[(f(x))^2 - (g(x))^2]\,dx$. Integrating over $[a, b]$ to find the total moment of R about the x axis, we obtain $\frac{1}{2}\int_a^b [(f(x))^2 - (g(x))^2]\,dx$.

13 Refer to the solution to Exercise 12. In the present case we have $f(x) = x^2$ and $g(x) = x^3$, where $a = 0$ and $b = 1$.

(a) Hence the moment about the y axis is

$$\int_a^b x[f(x) - g(x)]\,dx = \int_0^1 x(x^2 - x^3)\,dx$$

$$= \frac{1}{20}.$$

(b) The moment about the x axis is

$$\frac{1}{2}\int_a^b [(f(x))^2 - (g(x))^2]\,dx =$$

$$\frac{1}{2}\int_0^1 [(x^2)^2 - (x^3)^2]\,dx = \frac{1}{35}.$$

(c) The area of R is $\int_0^1 (x^2 - x^3)\,dx = \frac{1}{12}$.

(d) $\bar{x} = \frac{1/20}{1/12} = \frac{3}{5}$

(e) $\bar{y} = \frac{1/35}{1/12} = \frac{12}{35}$

15 Refer to the solution for Exercise 12.

(a) $\int_1^e x(3^x - 2^x)\,dx \approx 18.9738$. (This result can be computed exactly using the formula

$$\int xa^x\,dx = \frac{xa^x}{\ln a} - \frac{a^x}{(\ln a)^2} + C,$$ but the results are too messy to be worthwhile.)

(b) $\frac{1}{2}\int_1^e \left[(3^x)^2 - (2^x)^2\right] dx =$

$\frac{1}{2}\int_1^e (3^{2x} - 2^{2x})\, dx \approx 73.1042$

(c) $\int_1^e (3^x - 2^x)\, dx \approx 8.6950$

(d) $\bar{x} \approx \frac{18.9738}{8.6950} \approx 2.1821$

(e) $\bar{y} \approx \frac{73.1042}{8.6950} \approx 8.4076$

17 Exercise 11 showed that $\bar{x} = a/3$ for the triangle of Figure 16. The line $x = a/3$ divides the triangle into two pieces. Verify for yourself that one part has area $ab/18$ while the other has area $4ab/9$, which are certainly not equal. You should be able to come up with other examples as well.

19 (b) The area of R^* is $\frac{1}{2}\cdot 1\cdot 1 = \frac{1}{2}$. Then $\bar{y}$ for R^* is

$$\frac{\int_0^1 \frac{1}{2}x^2\, dx}{1/2} = \frac{x^3}{3}\Big|_0^1 = \frac{1}{3}.$$

The area of R is

$$\frac{1}{2}\cdot 1\cdot 1 + \frac{1}{2}\cdot\frac{1}{3}\cdot\frac{1}{3} = \frac{5}{9},$$

found by adding the areas of the two triangles.

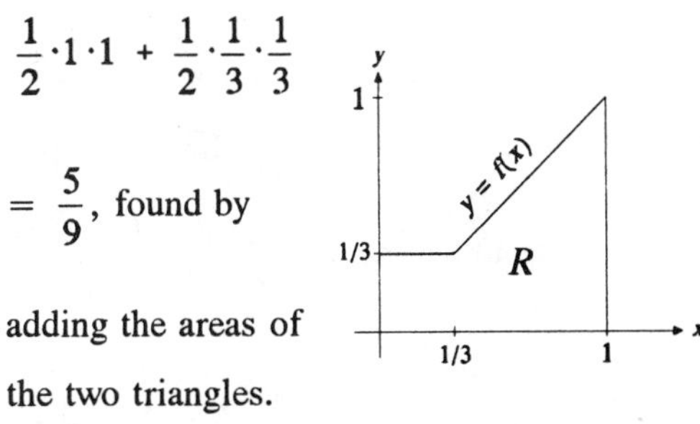

Thus $\bar{y}$ for R is

$$\frac{\int_0^{1/3} \frac{1}{2}(1/9)\, dx + \int_{1/3}^1 \frac{1}{2}x^2\, dx}{5/9}$$

$$= \frac{9}{5}\left[\frac{1}{18}\cdot\frac{1}{3} + \frac{1}{6} - \frac{1}{6}\left(\frac{1}{3}\right)^3\right] = \frac{29}{90}.$$

The centroid of R^* has a larger y coordinate than the centroid of R.

(c) The area of the region R is given by

$$\frac{1}{2}\cdot 1\cdot 1 + \frac{1}{2}\cdot a\cdot a = \frac{1}{2}(a^2 + 1).$$

Thus

$$\bar{y} = \frac{\int_0^a \frac{1}{2}a^2\, dx + \int_a^1 \frac{1}{2}x^2\, dx}{(a^2+1)/2}$$

$$= \frac{a^3 + \left(\frac{1}{3}x^3\right)\Big|_a^1}{a^2+1} = \frac{a^3 + \frac{1}{3} - \frac{1}{3}a^3}{a^2+1}$$

$$= \frac{2a^3 + 1}{3(a^2+1)}.$$

(d) Let $h(a) = \frac{2a^3+1}{3(a^2+1)}$. We wish to minimize h. First, find critical values of h by taking its derivative and setting it equal to 0: $h'(a) =$

$$\frac{3(a^2+1)(6a^2) - (2a^3+1)(6a)}{9(a^2+1)^2}$$

$$= \frac{6a(3a^3 + 3a - 2a^3 - 1)}{9(a^2+1)^2}$$

$$= \frac{6a(a^3 + 3a - 1)}{9(a^2+1)^2}.$$

So $h'(a) = 0$ when $a = 0$ or $a^3 + 3a - 1 = 0$; that is, when a is a root of $x^3 + 3x - 1 = 0$. When $a = 0$, $h(a) = 1/3$, already shown in (b) not to be a minimum. When $a = 1$, $h(a) = 1/2$, also not a minimum. Hence the minimum value for $h(a)$ occurs at a root of $x^3 + 3x - 1 = 0$.

(e) Let $p(x) = x^3 + 3x - 1$. Then $p'(x) = 3x^2 + 3 = 3(x^2 + 1)$. So $p'(x)$ is always strictly greater than 0 and $p(x)$ is an increasing function. Hence it can cross the x axis at most once. Further, since $p(0) = -1 < 0$, $p(1) = 3 > 0$, and $p(x)$ is a continuous function, it must cross the x axis at least once. Therefore the graph of $p(x)$ crosses the x axis exactly once, and $x^3 + 3x - 1 = 0$ has exactly one real root, which we call q.

(f) Employing Newton's method (or a calculator's root-finder) with $f(x) = x^3 + 3x - 1$ and $x_1 = 0.3$, we have $x_2 = x_1 - \dfrac{f(x_1)}{f'(x_1)} =$

$$0.3 - \frac{(0.3)^3 + 3\cdot(0.3) - 1}{3[(0.3)^2 + 1]} \approx 0.32232,\ x_3 =$$

$$x_2 - \frac{f(x_2)}{f'(x_2)} \approx 0.32219, \text{ and } x_4 = x_3 - \frac{f(x_3)}{f'(x_3)}$$

≈ 0.32219. Thus, to two decimal places, $q \approx 0.3222$.

8.7 Work

1 We know that $F = 5$ when $x = 0.2$, so $F = kx$, and $k = \dfrac{F}{x} = \dfrac{5}{0.2} = 25$.

(a) Following Example 1, the work accomplished in stretching the spring 0.2 meter from its rest length is $\int_0^{0.2} kx\,dx = \int_0^{0.2} 25x\,dx$

$$= \frac{25}{2}x^2\Big|_0^{0.2} = \frac{1}{2} \text{ joule.}$$

(b) Similarly, the work accomplished is

$$\int_0^{0.3} kx\,dx = \int_0^{0.3} 25x\,dx = \frac{25}{2}x^2\Big|_0^{0.3} =$$

$\dfrac{9}{8}$ joules.

3 The work done in stretching the spring from x to $x + dx$ is Force × Distance $\approx 3x^2\,dx$ joules. Thus the total work done in stretching the spring 0.80 meter from its rest length is $\int_0^{0.8} 3x^2\,dx = x^3\Big|_0^{0.8}$ $= 0.512$ joule.

5 Following Example 2, we see that the work done is

$$\int_{4000}^{8000} \left[\frac{4000}{r}\right]^2 dr = -\frac{4000^2}{r}\Bigg|_{4000}^{8000}$$

$$= -4000^2\left[\frac{1}{8000} - \frac{1}{4000}\right]$$

$$= \frac{4000^2}{8000} = 2000 \text{ mi-lb.}$$

7 The work accomplished in lifting the one-pound payload from a distance r to a distance $r + dr$ (relative to the center of the earth) is Force × Distance $\approx \left(\dfrac{4000}{r}\right)^3 dr$ joules. Since the moon is 240,000 miles from the center of the earth, the total work accomplished in lifting the payload from the earth to the moon is $\int_{4000}^{240,000} \left[\dfrac{4000}{r}\right]^3 dr =$

$$-\frac{4000^3}{2r^2}\Bigg|_{4000}^{240,000} = -\frac{4000^3}{2}\left[\frac{1}{(240,000)^2} - \frac{1}{(4000)^2}\right]$$

$$= \left(-\frac{5}{9} + 2000\right) \text{mi-lb} \approx 1999 \text{ mi-lb.}$$

9 (a) Let the x axis point down and have its origin at the top.

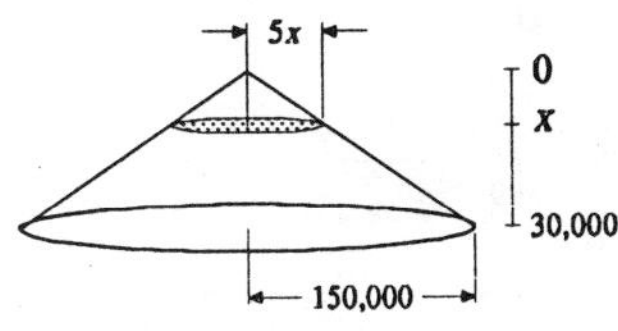

Points at level x have been raised $30000 - x$ feet. The radius at level x is proportional to x and equals 150,000 when $x = 30{,}000$; thus it equals $5x$. Hence the area at level x is $\pi(5x)^2 = 25\pi x^2$, so Work =

$$200 \int_0^{30,000} (30{,}000 - x)25\pi x^2 \, dx$$

$$= 5000\pi \int_0^{30,000} (30{,}000x^2 - x^3) \, dx$$

$$= 5000\pi \left(10000x^3 - \frac{1}{4}x^4\right)\Bigg|_0^{30,000}$$

$$= 5000\pi(30{,}000)^3\left(10{,}000 - \frac{1}{4}\cdot 30{,}000\right)$$

$$= 3.375 \times 10^{20}\pi \text{ ft-lbs.}$$

(b) $$\frac{3.375 \cdot 10^{20}\pi}{3 \cdot 10^{14}} = 1{,}125{,}000\pi \approx 3{,}500{,}000.$$

Thus the energy needed to raise Mt. Everest is equivalent to that released by about 3.5 million 1-megaton H-bombs.

11 The volume of the water that lies between x and $x + dx$ is approximately $A(x)\,dx$ cubic feet. Hence its weight is $62.4\,A(x)\,dx$. Since the water lies $b - x$ feet below the top of the container, the work required to pump the water outside the container is $(b - x)\cdot 62.4\,A(x)\,dx = 62.4(b - x)A(x)\,dx$. Thus the total work required to empty the container is

$$62.4 \int_a^b (b - x)A(x) \, dx.$$

13 Here $A(x) = c(x)h$, so from Exercise 11 the work required to empty the tank is

$$62.4 \int_a^b (b - x)A(x) \, dx =$$

$62.4h \int_a^b (b - x)c(x) \, dx$. Now let A be the area of R. Then $62.4h \int_a^b (b - x)c(x) \, dx =$

$$62.4h\left[b \int_a^b c(x) \, dx - \int_a^b xc(x) \, dx\right]$$

$= 62.4h[bA - \bar{x}A] = 62.4hA[b - \bar{x}]$. But $62.4hA$ is the weight of water in the tank and $b - \bar{x}$ is the distance that the centroid is from the top of the tank. So the work is indeed the same as if all the water were at the same depth as the centroid of R.

8.8 Improper Integrals

1 $$\int_1^\infty \frac{dx}{x^3} = \lim_{b\to\infty} \int_1^b \frac{dx}{x^3} = \lim_{b\to\infty}\left(-\frac{1}{2}x^{-2}\Big|_1^b\right) =$$

$\lim_{b\to\infty}\left(-\frac{1}{2b^2} + \frac{1}{2}\right) = \frac{1}{2}$. The improper integral is convergent.

3 Convergent: $\int_0^\infty e^{-x}\,dx = \lim_{b\to\infty} -e^{-x}\Big|_0^b$

$$= \lim_{b\to\infty}(-e^{-b} + e^0) = 0 + 1 = 1.$$

5 $$\int_0^\infty \frac{x^3\,dx}{x^4 + 1} = \lim_{b\to\infty} \int_0^b \frac{x^3\,dx}{x^4 + 1} =$$

$$\lim_{b\to\infty} \frac{1}{4}\ln(x^4 + 1)\Big|_0^b = \lim_{b\to\infty}\frac{1}{4}\ln(b^4 + 1) = \infty.$$

The improper integral is divergent.

7 $$\int_0^\infty \frac{dx}{(x + 2)^3} = \lim_{b\to\infty}\int_0^b \frac{dx}{(x + 2)^3}$$

$= \lim_{b\to\infty} \frac{(x+2)^{-2}}{-2}\Big|_0^b$

$= -\frac{1}{2}\lim_{b\to\infty}\left(\frac{1}{(b+2)^2} - \frac{1}{(0+2)^2}\right) = -\frac{1}{2}\left(0 - \frac{1}{4}\right)$

$= \frac{1}{8}$. The improper integral is convergent.

9 Divergent. $\int_1^\infty x^{-0.99}\,dx = \lim_{b\to\infty}\int_1^b x^{-0.99}\,dx =$

$\lim_{b\to\infty} 100x^{0.01}\Big|_1^b = \lim_{b\to\infty} 100(b^{0.01} - 1) = \infty$.

11 The integral is improper for two reasons: The integrand is undefined at 0 and the upper limit of integration is infinite. So we break the integral into two pieces: $\int_0^c \frac{\sin x}{x^2}\,dx + \int_c^\infty \frac{\sin x}{x^2}\,dx$, where we are free to choose any convenient $c > 0$. Since $\lim_{x\to 0}\frac{\sin x}{x} = 1$, there is some $c > 0$ such that $\frac{\sin x}{x} > \frac{1}{2}$ for $0 < x < c$. By the analog of Theorem 1 for the unbounded integrands, if $\int_0^c \frac{\sin x}{x^2}\,dx = \int_0^c \frac{\sin x}{x}\cdot\frac{1}{x}\,dx$ were convergent, then $\int_0^c \frac{1}{2}\cdot\frac{1}{x}\,dx$ would be also. But $\int_0^c \frac{1}{2}\cdot\frac{1}{x}\,dx$

$= \lim_{t\to 0^+}\frac{1}{2}\int_t^c \frac{dx}{x} = \frac{1}{2}\lim_{t\to 0^+}(\ln x)\Big|_t^c =$

$\frac{1}{2}\lim_{t\to 0^+}(\ln c - \ln t) = \infty$. Hence the given integral diverges.

13 $\int_1^\infty \frac{\ln x\,dx}{x} = \lim_{b\to\infty}\int_1^b \frac{\ln x\,dx}{x} = \lim_{b\to\infty}\frac{1}{2}(\ln x)^2\Big|_1^b$

$= \lim_{b\to\infty}\frac{1}{2}(\ln b)^2 = \infty$. The improper integral is divergent.

15 $\int_0^\infty \frac{x\,dx}{x^4+1} = \lim_{b\to\infty}\frac{1}{2}\int_0^b \frac{2x\,dx}{(x^2)^2+1} =$

$\frac{1}{2}\lim_{b\to\infty}\tan^{-1}(x^2)\Big|_0^{b^2} = \frac{1}{2}\lim_{b\to\infty}\tan^{-1}(b^4) = \frac{\pi}{4}$, so

$\int_0^\infty \frac{x\,dx}{x^4+1}$ converges.

17 Since the integrand is undefined at $x = 0$ and $x = 1$, split up the integral: $\int_0^1 \frac{dx}{\sqrt{x}\sqrt{1-x}} =$

$\lim_{t\to 0^+}\int_t^{1/2}\frac{dx}{\sqrt{x}\sqrt{1-x}} + \lim_{s\to 1^-}\int_{1/2}^s \frac{dx}{\sqrt{x}\sqrt{1-x}}$. Now let $x = \sin^2\theta$ and $dx = 2\sin\theta\cos\theta\,d\theta$. Then

$\int \frac{dx}{\sqrt{x}\sqrt{1-x}} = \int \frac{2\sin\theta\cos\theta\,d\theta}{\sin\theta\cos\theta} = 2\theta + C$

$= 2\sin^{-1}\sqrt{x} + C$. Hence

$\lim_{t\to 0^+}\int_t^{1/2}\frac{dx}{\sqrt{x}\sqrt{1-x}} + \lim_{s\to 1^-}\int_{1/2}^s \frac{dx}{\sqrt{x}\sqrt{1-x}}$

$= \lim_{t\to 0^+} 2\sin^{-1}\sqrt{x}\Big|_t^{1/2} + \lim_{s\to 1^-} 2\sin^{-1}\sqrt{x}\Big|_{1/2}^s$

$= 2\lim_{t\to 0^+}\left(\sin^{-1}\frac{1}{\sqrt{2}} - \sin^{-1}\sqrt{t}\right) +$

$2\lim_{s\to 1^-}\left(\sin^{-1}\sqrt{s} - \sin^{-1}\frac{1}{\sqrt{2}}\right)$

$= 2\left[\frac{\pi}{4} - 0 + \frac{\pi}{2} - \frac{\pi}{4}\right] = \pi$. So $\int_0^1 \frac{dx}{\sqrt{x}\sqrt{1-x}}$ is convergent and equals π.

19 The integrand is undefined at $x = 0$, hence we

examine $\lim_{t\to 0^+} \int_t^1 \frac{dx}{\sqrt[3]{x}}$. Now $\int \frac{dx}{\sqrt[3]{x}} = \int x^{-1/3}\, dx$

$= \frac{3}{2}x^{2/3} + C$, so $\lim_{t\to 0^+} \int_t^1 \frac{dx}{\sqrt[3]{x}} = \lim_{t\to 0^+} \frac{3}{2}x^{2/3}\Big|_t^1$

$= \lim_{t\to 0^+} \left[\frac{3}{2} - \frac{3}{2}t^{2/3}\right] = \frac{3}{2}$. Thus $\int_0^1 \frac{dx}{\sqrt[3]{x}}$ converges

to 3/2.

21 The integral equals $\int_0^\infty \frac{e^{-\sqrt{x}}}{\sqrt{x}}\, dx = \int_0^1 \frac{e^{-\sqrt{x}}}{\sqrt{x}}\, dx +$

$\int_1^\infty \frac{e^{-\sqrt{x}}}{\sqrt{x}}\, dx$. For $0 < x \le 1$, $e^{-x} < 1$ so $\frac{e^{-x}}{\sqrt{x}} <$

$\frac{1}{\sqrt{x}}$. But $\int_0^1 \frac{1}{\sqrt{x}}\, dx = \lim_{t\to 0+} \int_t^1 \frac{1}{\sqrt{x}}\, dx =$

$\lim_{t\to 0+} 2\sqrt{x}\Big|_t^1 = 2 \lim_{t\to 0+} (1 - \sqrt{t})\Big|_t^1 = 2$; by the

comparison test, $\int_0^1 \frac{1}{\sqrt{x}}\, dx$ converges. For $x \ge 1$,

$x \ge \sqrt{x}$, so $e^{-x} \le e^{-\sqrt{x}}$ and $\frac{e^{-x}}{\sqrt{x}} \le \frac{e^{-\sqrt{x}}}{\sqrt{x}}$. But

$\int_1^\infty \frac{e^{-\sqrt{x}}}{\sqrt{x}}\, dx = \lim_{t\to\infty} \int_1^t \frac{e^{-\sqrt{x}}}{\sqrt{x}}\, dx = \lim_{t\to\infty} \left(-2e^{-\sqrt{x}}\right)\Big|_1^t$

$= 2 \lim_{t\to\infty} \left(e^{-1} - e^{-\sqrt{t}}\right) = 2e^{-1}$. By the comparison

test, $\int_1^\infty \frac{e^{-\sqrt{x}}}{\sqrt{x}}\, dx$ converges. Hence $\int_0^\infty \frac{e^{-\sqrt{x}}}{\sqrt{x}}\, dx$

converges.

23 The result is finite: Area $= \int_1^\infty \left(\frac{1}{x} - \frac{1}{x+1}\right) dx$

$= \lim_{b\to\infty} \ln\left(\frac{x}{x+1}\right)\Big|_1^b = \ln 2.$

25 One method involves writing $\int_0^\infty e^{-x^2}\, dx =$

$\int_0^a e^{-x^2}\, dx + \int_a^\infty e^{-x^2}\, dx$, where $a > 0$ will be

chosen in a moment. For $x \ge a$, we have $x^2 \ge$

ax, so $e^{-x^2} \le e^{-ax}$ and $0 \le \int_a^\infty e^{-x^2}\, dx \le$

$\int_a^\infty e^{-ax}\, dx = \lim_{b\to\infty} \left(-\frac{1}{a}e^{-ax}\right)\Big|_a^b = \frac{1}{a}e^{-a^2}$. Pick a

so large that $\frac{1}{a}e^{-a^2} < 0.01$. (For example, $a =$

2.) Thus $0 < \int_0^\infty e^{-x^2}\, dx - \int_0^2 e^{-x^2}\, dx < 0.01$,

so $\left|\int_0^\infty e^{-x^2}\, dx - \left(0.005 + \int_0^2 e^{-x^2}\, dx\right)\right| < 0.005$.

If we can compute $\int_0^2 e^{-x^2}\, dx$ with an error less

than 0.005, then we will know $\int_0^\infty e^{-x^2}\, dx$ with an

error less than $0.005 + 0.005 = 0.01$. To apply Simpson's rule, we need to know the fourth

derivative of $y = e^{-x^2}$. We have $y' = -2xe^{-x^2}$, y''

$= (4x^2 - 2)e^{-x^2}$, $y^{(3)} = (12x - 8x^3)e^{-x^2}$, and $y^{(4)}$

$= (12 - 48x^2 + 16x^4)e^{-x^2}$. The function $12 -$

$48x^2 + 16x^4$ is decreasing for $0 \le x \le \sqrt{3/2}$ and

increasing for $\sqrt{3/2} \le x \le 2$. Evaluating it for x

$= 0$, $\sqrt{3/2}$, and 2 shows that its maximum absolute

value is $12 - 48\cdot 2^2 + 16\cdot 2^4 = 76$. Also, for $0 \le$

$x \le 2$, $|e^{-x^2}| \le 1$. Hence $|y^{(4)}| \le 76$. Hence

the error in Simpson's method, using n steps of

size $2/n$ is at most $\dfrac{(2-0)\cdot 76\cdot(2/n)^4}{180} = \dfrac{608}{45n^4}$.

Pick n so large that $\dfrac{608}{45n^4} < 0.005$; $n = 8$ works.

27 Note that $f(x)$ (hence $[f(x)]^2$) is continuous. Therefore, $\int_{-\infty}^{\infty} [f(x)]^2\,dx = \int_{-\infty}^{-1} [f(x)]^2\,dx + \int_{-1}^{1} [f(x)]^2\,dx + \int_{1}^{\infty} [f(x)]^2\,dx$, where the middle integral is proper. Substituting $u = -x$ in the integral over $(\infty, -1)$, $\int_{-\infty}^{-1} [f(x)]^2\,dx = \int_{1}^{\infty} [f(x)]^2\,dx$, so $\int_{-\infty}^{\infty} [f(x)]^2\,dx = 2\int_{1}^{\infty} [f(x)]^2\,dx + \int_{-1}^{1} [f(x)]^2\,dx$. For $x \ge 1$, $\dfrac{\sin^2 x}{x^2} \le \dfrac{1}{x^2}$, so $0 \le \int_1^\infty \dfrac{\sin^2 x}{x^2}\,dx \le \int_1^\infty \dfrac{1}{x^2}\,dx = \lim_{b\to\infty} \dfrac{1}{x}\Big|_1^b = 1$, a finite number. Hence, by Theorem 1, $\int_1^\infty [f(x)]^2\,dx$ is convergent. It follows that $\int_{-\infty}^{\infty} [f(x)]^2\,dx$ is convergent.

29 For some $c > 0$ and all $x \ge c$, $e^x - 1 > x^5$, so $\dfrac{x^3}{e^x-1} < \dfrac{1}{x^2}$. Thus $\int_c^\infty \dfrac{x^3\,dx}{e^x-1}$ converges by comparison with $\int_c^\infty \dfrac{dx}{x^2}$. Also, $\lim_{x\to 0} \dfrac{x^3}{e^x-1} \underset{H}{=} \lim_{x\to 0}\dfrac{3x^2}{e^x} = 0$, so $\dfrac{x^3}{e^x-1}$ is bounded for $0 < x < c$. Hence $\int_0^c \dfrac{x^3}{e^x-1}\,dx$ converges.

31 Assume $p \ne 1$, so $\int_1^\infty \dfrac{dx}{x^p} = \lim_{b\to\infty}\int_1^b \dfrac{dx}{x^p} = \lim_{b\to\infty}\dfrac{x^{1-p}}{1-p}\Big|_1^b = \lim_{b\to\infty}\left(\dfrac{b^{1-p}}{1-p} - \dfrac{1}{1-p}\right)$. If $p < 1$, $\lim_{b\to\infty}\left(\dfrac{b^{1-p}}{1-p} - \dfrac{1}{1-p}\right) = \infty$. If $p > 1$, then $\lim_{b\to\infty}\left(\dfrac{b^{1-p}}{1-p} - \dfrac{1}{1-p}\right) = \lim_{b\to\infty}\left(\dfrac{1}{b^{p-1}(1-p)} - \dfrac{1}{1-p}\right) = \dfrac{1}{p-1}$. If $p = 1$, then $\int_1^\infty \dfrac{dx}{x^p} = \lim_{b\to\infty} \ln x\Big|_1^b = \lim_{b\to\infty}\ln b = \infty$. We conclude that $\int_1^\infty \dfrac{dx}{x^p}$ converges for $p > 1$ and otherwise diverges.

33 (a) From equation (4) in Example 6, the present value is $\int_0^\infty e^{-rt} f(t)\,dt = \int_0^\infty ke^{-rt}\,dt = \lim_{b\to\infty}\int_0^b ke^{-rt}\,dt = \lim_{b\to\infty}\left(-\dfrac{k}{r}e^{-rt}\right)\Big|_0^b = \lim_{b\to\infty}\left(-\dfrac{k}{r}e^{-rb} + \dfrac{k}{r}\right) = \dfrac{k}{r}$.

35 The integral $\int_0^1 \dfrac{1}{x}\,dx$ diverges. Therefore you cannot really write equations using it. (Looking at it another way, once you see that it diverges to infinity, there is no special contradiction in saying that it equals twice itself: "multiplying" infinity by 2 doesn't change it.)

37 Using integration by parts, with $u = t^2$, $dv = e^{-rt}\,dt$, we have $du = 2t\,dt$, $v = -e^{-rt}/r$, and $P(r)$

$= \int_0^\infty e^{-rt} t^2\, dt$

$= \lim_{b\to\infty}\left(t^2\left(-\frac{1}{r}e^{-rt}\right)\Big|_0^b\right) - \int_0^\infty \left(-\frac{1}{r}e^{-rt}\right)\cdot 2t\, dt$

$= \frac{2}{r}\int_0^\infty te^{-rt}\, dt$. Using parts again, this time with $U = t$, $dV = e^{-rt}\, dt$, we have $dU = dt$, $V = -e^{-rt}/r$, and $P(r) =$

$\frac{2}{r}\left[\lim_{b\to\infty} t\left(-\frac{1}{r}e^{-rt}\right)\Big|_0^b - \int_0^\infty \left(-\frac{1}{r}e^{-rt}\right) dt\right]$

$= \frac{2}{r^2}\int_0^\infty e^{-rt}\, dt = \lim_{b\to\infty} \frac{2}{r^2}\left(-\frac{1}{r}e^{-rt}\right)\Big|_0^b = \frac{2}{r^3}.$

39 $P(r) = \int_0^\infty e^{-rt}\sin t\, dt$

$= \lim_{b\to\infty}\left(-\frac{e^{-rt}}{r^2+1}(r\sin t + \cos t)\Big|_0^b\right) = \frac{1}{r^2+1}$

41 Use integration by parts with $u = e^{-rt}$, $dv = f'(t)\, dt$; then $du = -re^{-rt}\, dt$, $v = f(t)$, and $Q(r)$ $= \int_0^\infty e^{-rt}f'(t)\, dt =$

$\lim_{b\to\infty}(e^{-rt}f(t)\,|_0^b) - \int_0^\infty f(t)(-re^{-rt}\, dt) =$

$0 - f(0) + r\int_0^\infty e^{-rt}f(t)\, dt = -f(0) + rP(r).$

(Note that $\lim_{t\to\infty} e^{-rt}f(t) = 0$ because $\int_0^\infty e^{-rt}f(t)\, dt$ converges.)

8.S Guide Quiz

1 (a)

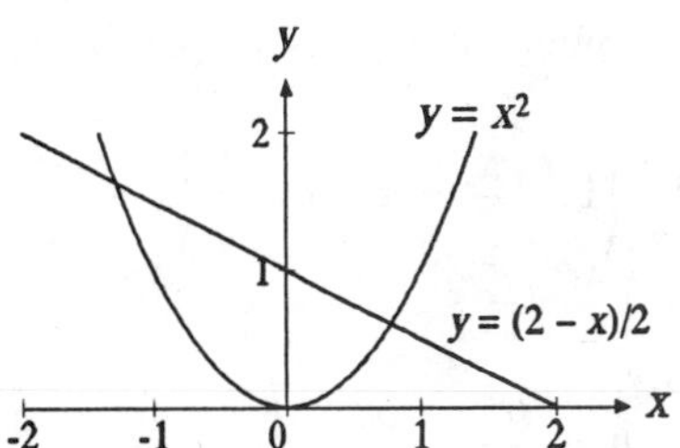

(b) The curves intersect when $x^2 = \frac{2-x}{2}$; hence $2x^2 = 2 - x$, $2x^2 + x - 2 = 0$, and $x = \frac{-1 \pm \sqrt{17}}{4}$. The cross-sectional height is given by $c(x) = \frac{2-x}{2} - x^2 =$ $1 - \frac{x}{2} - x^2$, so the area is

$\int_{(-1-\sqrt{17})/4}^{(-1+\sqrt{17})/4}\left(1 - \frac{x}{2} - x^2\right) dx.$

(c) Writing the curves in terms of y gives us $x = 2 - 2y$ and $x = \pm\sqrt{y}$. Hence, at the intersections, $(2 - 2y)^2 = y$, $4 - 8y + 4y^2 = y$, $4y^2 - 9y + 4 = 0$, and $y = \frac{9 \pm \sqrt{17}}{8}$.

For $0 \le y \le \frac{9 - \sqrt{17}}{8}$, the cross-sectional length is $c(y) = \sqrt{y} - (-\sqrt{y}) = 2\sqrt{y}$, while for $\frac{9-\sqrt{17}}{8} \le y \le \frac{9+\sqrt{17}}{8}$, the cross-sectional length is $c(y) = (2 - 2y) - (-\sqrt{y})$

$= 2 - 2y + \sqrt{y}$. Therefore the area is

$$\int_0^{(9-\sqrt{17})/8} 2\sqrt{y}\,dy + \int_{(9-\sqrt{17})/8}^{(9+\sqrt{17})/8} (2 - 2y + \sqrt{y})\,dy.$$

(d) The integral in (b) is simpler:

$$\int_a^b \left(1 - \frac{x}{2} - x^2\right) dx = \left(x - \frac{x^2}{4} - \frac{x^3}{3}\right)\Bigg|_a^b$$

$$= b - \frac{b^2}{4} - \frac{b^3}{3} - \left(a - \frac{a^2}{4} - \frac{a^3}{3}\right).$$

Substituting $a = \dfrac{-1 - \sqrt{17}}{4}$ and $b = \dfrac{-1 + \sqrt{17}}{4}$ and simplifying, we find that the area is $\dfrac{17\sqrt{17}}{48}$.

2 See the figures in the answer section of the text or examine the results of Exercise 9 of Sec. 8.2.

3

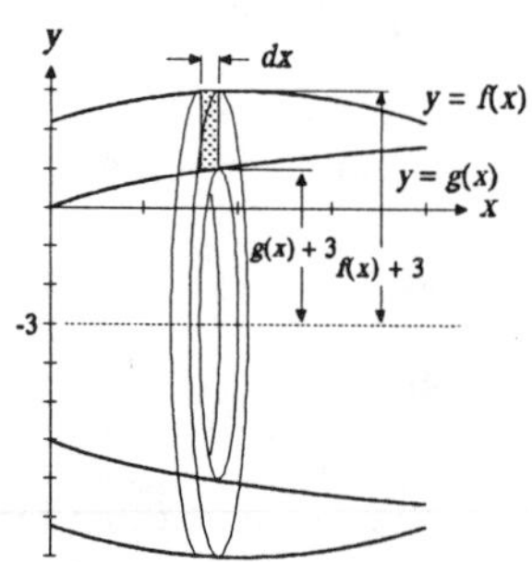

From the figure, we see that the local approximation is $\pi[(f(x)+3)^2 - (g(x)+3)^2]\,dx$. Thus the total volume of S is

$$\int_a^b \pi[(f(x) + 3)^2 - (g(x) + 3)^2]\,dx.$$

4 (a) Assume the material is homogeneous with density equal to 1. Then the local estimate of the moment about the x axis is Lever × Mass $= y \cdot 1 \cdot c(y)\,dy = y\,c(y)\,dy$. Thus the total moment is

$$\int_0^k y\,c(y)\,dy.$$

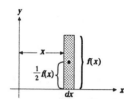

(b) Recall that the moment of a rectangle about the x axis can be alternatively calculated by multiplying the total mass of the rectangle and the distance of its centroid from the x axis. The region between x and $x + dx$ is approximately a rectangle of area $f(x)\,dx$ with centroid of height $f(x)/2$, so if $f(x)$ is the height of the rectangle, its moment is $\frac{1}{2}f(x) \cdot f(x)\,dx$. Thus the total moment about the x axis is $\int_a^b \frac{1}{2}[f(x)]^2\,dx$.

5 The equation of the line through (2, 1) and (3, 0) is $y = \dfrac{1 - 0}{2 - 3}(x - 3) = -x + 3$. Now the area of

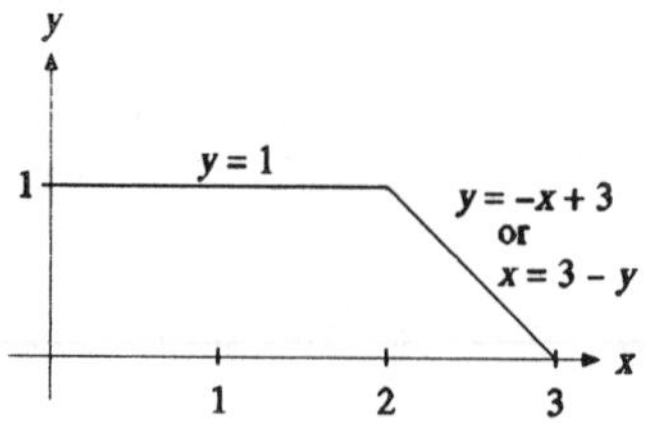

the trapezoid is $(1/2)(2 + 3) \cdot 1 = 5/2$. The moment of the trapezoid about the x axis is $\int_0^1 y(3 - y)\,dy$

$$= \left(\frac{3}{2}y^2 - \frac{y^3}{3}\right)\Bigg|_0^1 = \frac{3}{2} - \frac{1}{3} = \frac{7}{6}. \text{ Thus}$$

$\bar{y} = \dfrac{7/6}{5/2} = \dfrac{7}{15}$. The moment of the trapezoid about the y axis is $\int_0^2 x\cdot 1\,dx + \int_2^3 x(-x+3)\,dx$

$$= \frac{x^2}{2}\bigg|_0^2 + \left(-\frac{x^3}{3} + \frac{3}{2}x^2\right)\bigg|_2^3$$

$$= 2 + \left(-\frac{27}{3} + \frac{27}{2} + \frac{8}{3} - \frac{12}{2}\right) = \frac{19}{6}, \text{ hence}$$

$$\bar{x} = \frac{19/6}{5/2} = \frac{19}{15}. \text{ So } (\bar{x}, \bar{y}) = \left(\frac{19}{15}, \frac{7}{15}\right).$$

6 Assuming that the force required to stretch a spring is proportional to the distance stretched, we have $F = kx$. Further, the work required to stretch the spring 6 cm from its rest position is $\int_0^6 kx\,dx =$

$$\frac{k}{2}x^2\bigg|_0^6 = 18k = W, \text{ so } k = \frac{W}{18}. \text{ The work}$$

required to stretch the spring 3 cm from its rest position is $\int_0^3 kx\,dx = \frac{k}{2}x^2\bigg|_0^3 = \frac{9}{2}k = \frac{9}{2}\cdot\frac{W}{18}$

$$= \frac{W}{4}.$$

7 (a) $y = 1/x^2$, so $x = \frac{1}{\sqrt{y}}$. Now y varies from 0 to 1, so the integral is $\int_0^1 \left(\frac{1}{\sqrt{y}} - 1\right) dy$.

(b) The integrand in (a) is undefined at $y = 0$; hence $\int_0^1 \left(\frac{1}{\sqrt{y}} - 1\right) dy$

$$= \lim_{t\to 0^+} \int_t^1 \left(\frac{1}{\sqrt{y}} - 1\right) dy$$

$$= \lim_{t\to 0^+} \int_t^1 (y^{-1/2} - 1)\,dy = \lim_{t\to 0^+} [2y^{1/2} - y]\big|_t^1$$

$$= \lim_{t\to 0^+} [2 - 2t^{1/2} - 1 + t] = 1.$$

(c) $y = 1/x^2$ and x varies from 1 to ∞, so the area of R is $\int_1^\infty \frac{dx}{x^2}$.

(d) $$\int_1^\infty \frac{dx}{x^2} = \lim_{b\to\infty} \int_1^b \frac{dx}{x^2} = \lim_{b\to\infty} \left[-\frac{1}{x}\right]\bigg|_1^b$$

$$= \lim_{b\to\infty} \left[1 - \frac{1}{b}\right] = 1$$

(e) The volume of the local approximation is $\pi\left(\frac{1}{x^2}\right)^2 dx$, so the volume of S is given by

$$\int_1^\infty \pi\left(\frac{1}{x^2}\right)^2 dx.$$

(f) $$\int_1^\infty \pi\left(\frac{1}{x^2}\right)^2 dx = \lim_{b\to\infty} \int_1^b \pi\cdot\frac{dx}{x^4}$$

$$= \pi \lim_{b\to\infty} \int_1^b \frac{dx}{x^4} = \pi \lim_{b\to\infty} \left[-\frac{1}{3x^3}\right]\bigg|_1^b$$

$$= \pi \lim_{b\to\infty} \left[-\frac{1}{3b^3} + \frac{1}{3}\right] = \frac{\pi}{3}$$

(g) Here $R(y) = y$ and $c(y) = \frac{1}{\sqrt{y}} - 1$, so the volume of S is $\int_0^1 2\pi y\left(\frac{1}{\sqrt{y}} - 1\right) dy$.

(h) $$\int_0^1 2\pi y\left(\frac{1}{\sqrt{y}} - 1\right) dy = 2\pi \int_0^1 (\sqrt{y} - y)\,dy$$

$$= 2\pi\left[\frac{2}{3}y^{3/2} - \frac{y^2}{2}\right]\bigg|_0^1 = 2\pi\left[\frac{2}{3} - \frac{1}{2}\right] = \frac{\pi}{3}$$

8 (a) See the proof of Theorem 3 in Sec. 8.8.

(b) $\int_a^\infty f(x)\,dx$ is between -15 and 15.

8.S Review Exercises

1 (a) $c(x) = 2x - x^2$, so the area is

$$\int_0^2 (2x - x^2)\,dx.$$

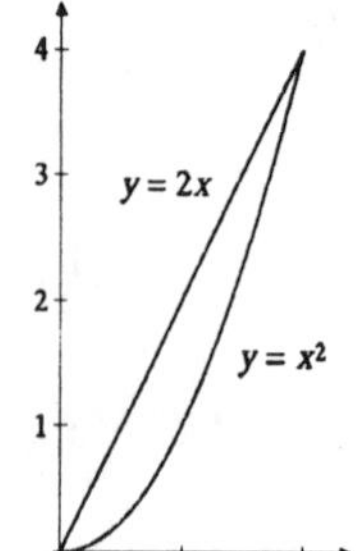

(b) $y = x^2$ implies that $x = \sqrt{y}$, since $x > 0$.

$y = 2x$ implies that $x = y/2$. Thus the area is $\int_0^4 \left(\sqrt{y} - \frac{y}{2}\right) dy$.

3 The equation of the line containing the points (2, 0) and (3, 2) is $y = 2x - 4$ or $x = \frac{1}{2}(y + 4)$. The

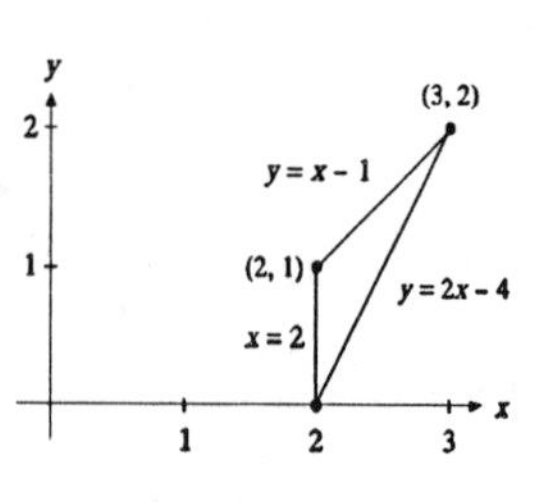

line joining (2, 1) and (3, 2) has equation $y = x - 1$ or $x = y + 1$. The line containing (2, 0) and (2, 1) has equation $x = 2$. Each cylindrical shell has radius $R(y) = y$. For $0 \le y \le 1$, $c(y) = \frac{y + 4}{2} - 2 = \frac{y}{2}$. For $1 \le y \le 2$, $c(y) = \frac{y + 4}{2} - (y + 1) = 1 - \frac{y}{2}$. Hence the volume is $\int_0^2 2\pi R(y) c(y)\,dy$

$$= \int_0^1 2\pi y\left(\frac{y}{2}\right) dy + \int_1^2 2\pi y\left(1 - \frac{y}{2}\right) dy$$

$$= \int_0^1 \pi y^2\,dy + \int_1^2 \pi y(2 - y)\,dy.$$

5 (a) The area is

$$\int_1^2 c(x)\,dx$$

$$= \int_1^2 \frac{x\,dx}{1 + x}$$

$$= \int_1^2 \left(1 - \frac{1}{x + 1}\right) dx = [x - \ln(x + 1)]\Big|_1^2$$

$$= (2 - \ln 3) - (1 - \ln 2) = 1 + \ln\frac{2}{3}.$$

(b) Using parallel cross sections, the volume of the solid formed by revolving R about the x axis is $\int_1^2 \pi(c(x))^2\,dx$

$$= \pi \int_1^2 \left(1 - \frac{1}{1 + x}\right)^2 dx$$

$$= \pi \int_1^2 \left(1 - \frac{2}{1 + x} + \frac{1}{(1 + x)^2}\right) dx$$

$$= \pi\left(x - 2\ln(1 + x) - \frac{1}{1 + x}\right)\Bigg|_1^2$$

$$= \pi\left[\left(2 - 2\ln 3 - \frac{1}{3}\right) - \left(1 - 2\ln 2 - \frac{1}{2}\right)\right]$$

$$= \pi\left(\frac{7}{6} + 2\ln\frac{2}{3}\right).$$

(c) Using cylindrical shells the volume formed by revolving R about the y axis is

$$\int_1^2 2\pi x\left(\frac{x}{x + 1}\right) dx$$

$$= 2\pi \int_1^2 \left(x - 1 + \frac{1}{x + 1}\right) dx$$

$$= 2\pi\left[\frac{x^2}{2} - x + \ln(1 + x)\right]\Bigg|_1^2$$

$$= 2\pi\left[(2 - 2 + \ln 3) - (\tfrac{1}{2} - 1 + \ln 2)\right]$$

$$= 2\pi\left(\frac{1}{2} + \ln\frac{3}{2}\right).$$

(d) If R is revolved about the line $y = -1$, a typical cross section is the region between two concentric circles (a "washer"). Hence the volume is $\int_1^2 \pi\left[\left(1 + \frac{x}{1+x}\right)^2 - 1^2\right] dx$

$$= \pi \int_1^2 \left[3 - \frac{4}{1+x} + \frac{1}{(1+x)^2}\right] dx$$

$$= \pi\left(3x - 4\ln(1+x) - \frac{1}{1+x}\right)\Bigg|_1^2$$

$$= \pi\left[\left(6 - 4\ln 3 - \frac{1}{3}\right) - \left(3 - 4\ln 2 - \frac{1}{2}\right)\right]$$

$$= \pi\left(\frac{19}{6} + 4\ln\frac{2}{3}\right).$$

7 (a) The area of R is

$$\int_0^{\pi/2} \sin 2x\, dx$$

$$= -\frac{1}{2}\cos 2x\Bigg|_0^{\pi/2}$$

$$= -\frac{1}{2}\cos\pi + \frac{1}{2}\cos 0 = -\frac{1}{2}(-1) + \frac{1}{2}(1)$$

$$= 1.$$

(b) With parallel cross sections, the volume of the solid formed by revolving R about the x axis is $\int_0^{\pi/2} \pi(\sin 2x)^2\, dx$

$$= \pi\int_0^{\pi/2} \frac{1 - \cos 4x}{2}\, dx$$

$$= \frac{\pi}{2}\left(x - \frac{1}{4}\sin 4x\right)\Bigg|_0^{\pi/2}$$

$$= \frac{\pi}{2}\left[\left(\frac{\pi}{2} - 0\right) - (0 - 0)\right] = \frac{\pi^2}{4}.$$

(c) By the method of cylindrical shells, the volume is $\int_0^{\pi/2} 2\pi x(\sin 2x)\, dx$

$$= 2\pi\int_0^{\pi/2} x\sin 2x\, dx$$

$$= 2\pi\left[\frac{1}{4}\sin 2x - \frac{1}{2}x\cos 2x\right]\Bigg|_0^{\pi/2}$$

$$= 2\pi\left[\left(\frac{1}{4}\cdot 0 - \frac{\pi}{4}(-1)\right) - \left(\frac{1}{4}\cdot 0 - 0\cdot 1\right)\right]$$

$$= \frac{\pi^2}{2}.$$

(d) When R is revolved about the line $y = -1$, parallel cross sections consist of the area between two concentric circles. Hence the volume is $\int_0^{\pi/2} \pi[(\sin 2x + 1)^2 - 1]\, dx$

$$= \pi\int_0^{\pi/2} (\sin^2 2x + 2\sin 2x)\, dx$$

$$= \pi\int_0^{\pi/2} \sin^2 2x\, dx + 2\pi\int_0^{\pi/2} \sin 2x\, dx$$

$= \frac{\pi^2}{4} + 2\pi\cdot 1 = \frac{\pi^2}{4} + 2\pi$, making use of the results of (a) and (b).

9 (a) The area of R is

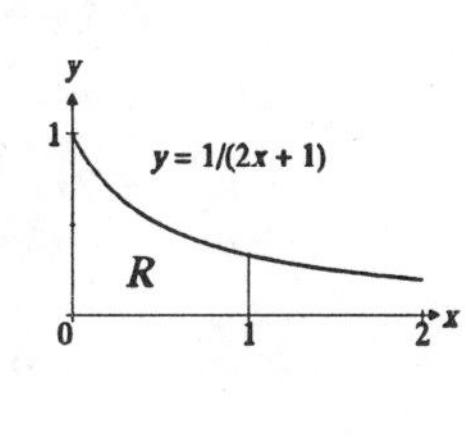

$$\int_0^1 \frac{dx}{2x+1}$$

$$= \frac{1}{2}\ln(2x+1)\Bigg|_0^1$$

$$= \frac{1}{2}\ln 3 - \frac{1}{2}\ln 1 = \frac{1}{2}\ln 3.$$

(b) The volume of the solid formed by revolving R about the x axis is, by parallel cross sections, $\int_0^1 \frac{\pi}{(2x+1)^2}\,dx$

$$= \frac{\pi}{2}\int_0^1 \frac{2\,dx}{(2x+1)^2} = \frac{\pi}{2}\left(\frac{-1}{2x+1}\right)\Big|_0^1$$

$$= \frac{\pi}{2}\left(-\frac{1}{3}+1\right) = \frac{\pi}{3}.$$

(c) Using the method of cylindrical shells, the volume is $\int_0^1 2\pi x\left(\frac{1}{2x+1}\right)dx$

$$= \pi\int_0^1\left(1 - \frac{1}{2x+1}\right)dx$$

$$= \pi\left(x - \frac{1}{2}\ln(2x+1)\right)\Big|_0^1 = \pi\left(1 - \frac{1}{2}\ln 3\right)$$

$$= \frac{\pi}{2}(2 - \ln 3).$$

(d) When R is revolved about the line $y = -1$, parallel cross sections consist of the area between two concentric circles ("washers"). The volume is $\int_0^1 \pi\left[\left(\frac{1}{2x+1}+1\right)^2 - 1\right]dx$

$$= \pi\int_0^1\left[\frac{1}{(2x+1)^2} + \frac{2}{2x+1}\right]dx$$

$$= \frac{\pi}{3} + 2\pi\left(\frac{1}{2}\ln 3\right) = \frac{\pi}{3}(1 + 3\ln 3),$$

making use of the results of (a) and (b).

11 The moment of R about the y axis is

$$\int_{\pi/2}^{\pi} x\left(\frac{\sin x}{x}\right)dx = \int_{\pi/2}^{\pi}\sin x\,dx = -\cos x\Big|_{\pi/2}^{\pi}$$

$$= -(-1) - 0 = 1.$$

13 The moment of R about the y axis is

$$\int_0^1 x\cdot\frac{1}{\sqrt{x^2+1}}\,dx = \frac{1}{2}\int_0^1\frac{2x\,dx}{\sqrt{x^2+1}}$$

$$= \frac{1}{2}\left(2\sqrt{x^2+1}\right)\Big|_0^1 = \sqrt{2} - 1.$$

15

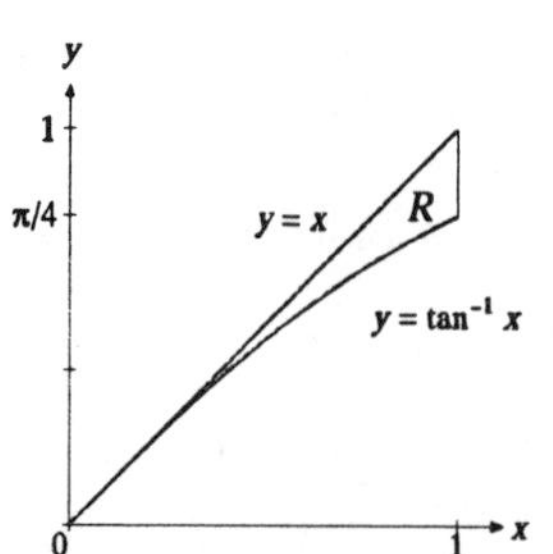

The area of R is $\int_0^1 (x - \tan^{-1}x)\,dx$

$$= \left[\frac{x^2}{2} - \left(x\tan^{-1}x - \frac{1}{2}\ln(1+x^2)\right)\right]\Big|_0^1$$

$$= \left[\frac{1}{2} - \frac{\pi}{4} + \frac{1}{2}\ln 2\right] - 0 = \frac{1}{4}(2 + 2\ln 2 - \pi).$$

(The antiderivative $\int \tan^{-1}x\,dx$ was found in Example 3 of Sec. 7.3.)

17 The area is $\int_0^{\pi/4}(\cos^3 x - \sin^3 x)\,dx =$

$$\int_0^{\pi/4}[(\cos x)(1-\sin^2 x) - (\sin x)(1-\cos^2 x)]\,dx =$$

$$\int_0^{\pi/4}(1-\sin^2 x)\cos x\,dx + \int_0^{\pi/4}(1-\cos^2 x)(-\sin x\,dx)$$

$$= \left(\sin x - \frac{\sin^3 x}{3}\right)\Big|_0^{\pi/4} + \left(\cos x - \frac{\cos^3 x}{3}\right)\Big|_0^{\pi/4}$$

$$= \left[\frac{\sqrt{2}}{2} - \frac{1}{3}\left(\frac{\sqrt{2}}{2}\right)^3\right] + \left[\frac{\sqrt{2}}{2} - \frac{1}{3}\left(\frac{\sqrt{2}}{2}\right)^3 - \left(1 - \frac{1}{3}\right)\right]$$

$$= \sqrt{2} - \frac{2}{3}\left(\frac{\sqrt{2}}{2}\right)^3 - \frac{2}{3} = \frac{5\sqrt{2}}{6} - \frac{2}{3}.$$

19

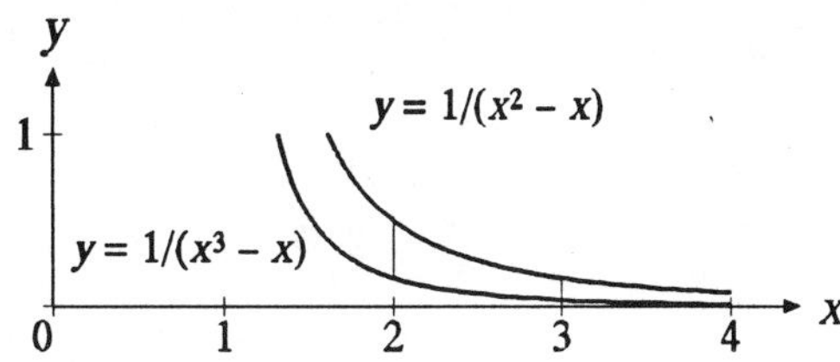

(a) The area is $\int_2^3 \left(\frac{1}{x^2 - x} - \frac{1}{x^3 - x}\right) dx$

$$= \int_2^3 \frac{1}{x(x-1)}\left(1 - \frac{1}{x+1}\right) dx$$

$$= \int_2^3 \frac{x\,dx}{x(x-1)(x+1)}$$

$$= \int_2^3 \frac{dx}{(x-1)(x+1)}$$

$$= \frac{1}{2}\int_2^3 \left(\frac{1}{x-1} - \frac{1}{x+1}\right) dx$$

$$= \frac{1}{2}[\ln(x-1) - \ln(x+1)]\Big|_2^3$$

$$= \frac{1}{2}\ln\left(\frac{x-1}{x+1}\right)\Big|_2^3 = \frac{1}{2}\left[\ln\frac{2}{4} - \ln\frac{1}{3}\right]$$

$$= \frac{1}{2}\ln\frac{3}{2}.$$

(b) The area is $\int_3^\infty \left(\frac{1}{x^2 - x} - \frac{1}{x^3 - x}\right) dx$

$$= \lim_{b\to\infty}\int_3^b \frac{1}{2}\left(\frac{1}{x-1} - \frac{1}{x+1}\right) dx$$

$$= \lim_{b\to\infty}\frac{1}{2}\ln\left(\frac{x-1}{x+1}\right)\Big|_3^b$$

$$= \lim_{b\to\infty}\left[\frac{1}{2}\ln\left(\frac{b-1}{b+1}\right) - \frac{1}{2}\ln\frac{2}{4}\right] = \frac{1}{2}\ln 2.$$

21

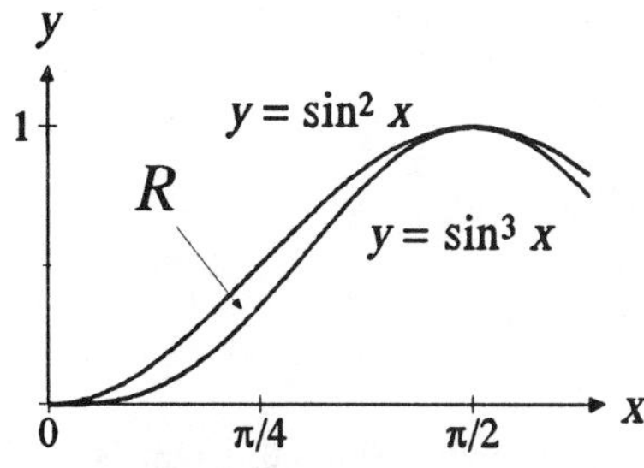

(a) The area of R is $\int_0^{\pi/2} (\sin^2 x - \sin^3 x)\,dx$

$$= \frac{\pi}{4} - \frac{2}{3} = \frac{3\pi - 8}{12}.$$

(b) The moment of R about the x axis is $\int_0^{\pi/2} \frac{1}{2}[(\sin^2 x)^2 - (\sin^3 x)^2]\,dx$. (This formula was derived in Exercise 12 of Sec. 8.6. The moment is therefore

$$\frac{1}{2}\int_0^{\pi/2} (\sin^4 x - \sin^6 x)\,dx$$

$$= \frac{1}{2}\left[\frac{1\cdot 3}{2\cdot 4}\cdot\frac{\pi}{2} - \frac{1\cdot 3\cdot 5}{2\cdot 4\cdot 6}\cdot\frac{\pi}{2}\right] = \frac{\pi}{64}.$$ (We used Formula 73 from the text's table of integrals.)

(c) $\bar{y} = \dfrac{\text{Moment of } R \text{ about the } x \text{ axis}}{\text{Area of } R}$

$$= \frac{\pi/64}{(3\pi - 8)/12} = \frac{3\pi}{16(3\pi - 8)}$$

23 The ring is formed by revolving the region shown

about the y axis. The region is described by $a \le x \le b$, $-\sqrt{b^2 - x^2} \le y \le \sqrt{b^2 - x^2}$. By the shell technique, the volume of the ring is $\int_a^b 2\pi x \cdot 2\sqrt{b^2 - x^2}\, dx$

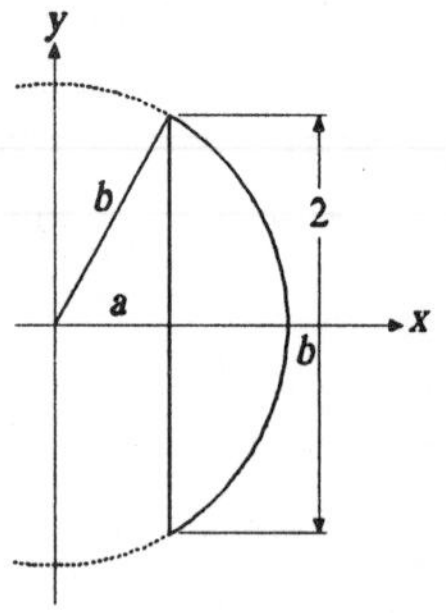

$$= 2\pi \int_a^b \sqrt{b^2 - x^2}\, 2x\, dx$$

$$= 2\pi\left(-\frac{2}{3}(b^2 - x^2)^{3/2}\right)\Big|_a^b = \frac{4\pi}{3}(b^2 - a^2)^{3/2}.$$

The height of the ring is, by the Pythagorean theorem, $2\sqrt{b^2 - a^2}$. Since this equals 2 inches, $b^2 - a^2 = 1$. It follows that the volume is $\frac{4\pi}{3}$ cubic inches. (Note that this does not depend on the size of the sphere or drill; only the height of the ring matters.)

25 (a) Area $= \int_0^\infty e^{-x}\, dx = \lim_{b\to\infty} \int_0^b e^{-x}\, dx$

$$= \lim_{b\to\infty} (-e^{-x})\Big|_0^b = \lim_{b\to\infty} (1 - e^{-b}) = 1$$

(b) The typical cross section is a disk of radius e^{-x}. Hence, Volume $= \int_0^\infty \pi(e^{-x})^2\, dx$

$$= \lim_{b\to\infty} \pi \int_0^b e^{-2x}\, dx = \pi \lim_{b\to\infty}\left(-\frac{1}{2}e^{-2x}\right)\Big|_0^b$$

$$= \frac{\pi}{2}\lim_{b\to\infty}(1 - e^{-2b}) = \frac{\pi}{2}.$$

(c) By cylindrical shells, the volume is

$$\int_0^\infty 2\pi x e^{-x}\, dx = 2\pi \lim_{b\to\infty}\int_0^b x e^{-x}\, dx$$

$$= 2\pi \lim_{b\to\infty}(-(x+1)e^{-x})\Big|_0^b$$

$$= 2\pi \lim_{b\to\infty}(1 - (b+1)e^{-b}) = 2\pi.$$

27 See Sec. 8.5.

29 $\int_0^\infty e^{-rx}\cos ax\, dx = \lim_{b\to\infty}\int_0^b e^{-rx}\cos ax\, dx$

$$= \lim_{b\to\infty}\left[\frac{e^{-rx}}{a^2 + r^2}(a\sin ax - r\cos ax)\right]\Bigg|_0^b$$

$$= \frac{1}{a^2 + r^2}\lim_{b\to\infty}[e^{-rb}(a\sin ab - r\cos ab) - (-r)]$$

$$= \frac{1}{a^2 + r^2}(0 + r) = \frac{r}{a^2 + r^2},$$ since, as $b \to \infty$, $e^{-rb} \to 0$ and $a\sin ab - r\cos ab$ is bounded.

31 Consider the region in the first quadrant below the curve $y = \frac{1}{1 + x^2}$ and above the x axis. By vertical cross sections its area is $\int_0^\infty \frac{dx}{1 + x^2}$. Solving $y = \frac{1}{1 + x^2}$ for x, we get

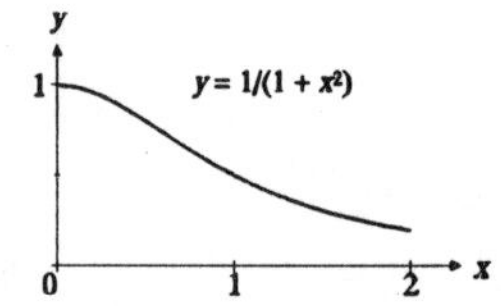

$$1 + x^2 = \frac{1}{y},\ x^2 = \frac{1}{y} - 1 = \frac{y - 1}{y},$$

$x = \sqrt{\frac{y - 1}{y}}$. So, by horizontal cross sections, the area is $\int_0^1 \sqrt{\frac{y - 1}{y}}\, dy$.

33 In the first integral, we make the substitution $u = x^2$, $du = 2x\, dx$. As x goes from 0 to ∞, so does u, so $\int_0^\infty \frac{\sin x^2}{x}\, dx = \frac{1}{2}\int_0^\infty \frac{\sin x^2}{x^2}\, 2x\, dx$

$= \frac{1}{2}\int_0^{\infty} \frac{\sin u}{u}\,du = \frac{1}{2}\int_0^{\infty} \frac{\sin x}{x}\,dx.$

35 Let $u = 2x + 3$, $x = \frac{u-3}{2}$, $dx = \frac{1}{2}\,du$. As x goes from 0 to ∞, u goes from 3 to ∞, so $\int_0^{\infty} e^{-x}\sin(2x+3)\,dx = \int_3^{\infty} e^{-(u-3)/2}\sin u\,\frac{du}{2}$

$= \frac{1}{2}e^{3/2}\int_3^{\infty} e^{-u/2}\sin u\,du$. Apply Formula 64 from the table of antiderivatives with $a = -1/2$ and $b = 1$. Then $\int_0^{\infty} e^{-x}\sin(2x+3)\,dx$

$= \frac{1}{2}e^{3/2}\cdot\frac{e^{-u/2}}{\left(-\frac{1}{2}\right)^2 + 1^2}\left(-\frac{1}{2}\sin u - \cos u\right)\Big|_3^{\infty}$

$= \frac{1}{2}e^{3/2}\cdot\frac{4}{5}e^{-3/2}\left(\frac{1}{2}\sin 3 + \cos 3\right)$

$= \frac{1}{5}(\sin 3 + 2\cos 3)$. (The notation $f(x)\big|_a^{\infty}$ is short for $\lim_{b\to\infty} f(x)\big|_a^b$.)

37 Apply integration by parts with $u = e^x$, $dv = dx/x$, $du = e^x\,dx$, $v = \ln x$. Then $\int \frac{e^x}{x}\,dx$

$= e^x \ln x - \int e^x \ln x\,dx$. If $\int e^x \ln x\,dx$ were elementary, then $\int \frac{e^x}{x}\,dx$ would be also.

39 The fundamental theorem of calculus requires the function whose definite integral is being computed to be continuous throughout the interval of integration. But $1/x^2$ is undefined at $x = 0$.

41 Apply integration by parts with $u = \ln x$, $dv = x^4\,dx$, $du = \frac{dx}{x}$, $v = \frac{x^5}{5}$: $\int x^4 \ln x\,dx$

$= \frac{x^5}{5}\ln x - \int \frac{x^5}{5}\cdot\frac{1}{x}\,dx = \frac{x^5}{5}\ln x - \int \frac{x^4}{5}\,dx$

$= \frac{x^5}{5}\ln x - \frac{x^5}{25} + C$. So $\int_0^1 x^4 \ln x\,dx$

$= \lim_{t\to 0^+}\int_t^1 x^4 \ln x\,dx$

$= \lim_{t\to 0^+}\left[-\frac{1}{25} - \left(\frac{t^5}{5}\ln t - \frac{t^5}{25}\right)\right] = -\frac{1}{25}$. We used l'Hôpital's rule here: $\lim_{t\to 0^+} t^5 \ln t$

$= \lim_{t\to 0^+}\frac{\ln t}{t^{-5}} \underset{H}{=} \lim_{t\to 0^+}\frac{1/t}{-5t^{-6}} = -\frac{1}{5}\lim_{t\to 0^+} t^5 = 0.$

43 Make the substitution $u = -\ln x$. Then $x = e^{-u}$, so $dx = -e^{-u}\,du$. As x increases from 0 to 1, u decreases from ∞ to 0, so $\int_0^1 (-\ln x)^3\,dx$

$= \int_{\infty}^0 u^3(-e^{-u}\,du) = \int_0^{\infty} u^3 e^{-u}\,du$

$= \int_0^{\infty} x^3 e^{-x}\,dx.$

45 The parabola and line intersect when $x^2 = 3x - 2$; that is, when $x^2 - 3x + 2 = 0$, $(x-1)(x-2) = 0$, $x = 1$ or 2. Thus the region is described by $1 \le x \le 2$, $x^2 \le y \le 3x - 2$. The region's area is

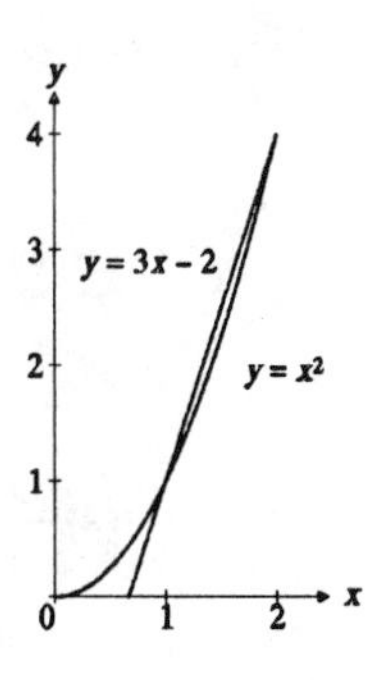

$A = \int_1^2 [(3x-2) - x^2]\,dx$

$= \left(\frac{3}{2}x^2 - 2x - \frac{1}{3}x^3\right)\Big|_1^2$

$= \left(6 - 4 - \frac{8}{3}\right) - \left(\frac{3}{2} - 2 - \frac{1}{3}\right) = \frac{1}{6}$. The x coordinate of the centroid is

$\bar{x} = \frac{1}{A}\int_1^2 x[3x - 2 - x^2]\,dx$

$= 6\int_1^2 (3x^2 - 2x - x^3)\,dx$

$= 6\left(x^3 - x^2 - \frac{1}{4}x^4\right)\Big|_1^2$

$= 6\left[(8 - 4 - 4) - \left(1 - 1 - \frac{1}{4}\right)\right] = \frac{3}{2}$. To compute the y coordinate, we use the moment formula from Exercise 12 of Sec. 8.6 and divide by the area: $\bar{y}$

$= \frac{1}{A}\int_1^2 \frac{1}{2}[(3x - 2)^2 - (x^2)^2]\,dy$

$= 3\int_1^2 (9x^2 - 12x + 4 - x^4)\,dx$

$= 3\left(3x^3 - 6x^2 + 4x - \frac{x^5}{5}\right)\Big|_1^2$

$= 3\left[\left(24 - 24 + 8 - \frac{32}{5}\right) - \left(3 - 6 + 4 - \frac{1}{5}\right)\right] = \frac{12}{5}$.

The centroid is $\left(\frac{3}{2}, \frac{12}{5}\right)$.

47 For $x \geq 1/5$, $\left|\frac{x^2 - 5x^3}{x^6 + 1}\sin 3x\right|$

$= \frac{5x^3 - x^2}{x^6 + 1}|\sin 3x| \leq \frac{5x^3}{x^6}\cdot 1 = \frac{5}{x^3}$. Since

$\int_1^\infty \frac{5}{x^3}\,dx = \lim_{b\to\infty}\left(-\frac{5}{2}x^{-2}\right)\Big|_1^b = \lim_{b\to\infty}\frac{5}{2}(1 - b^{-2})$

$= \frac{5}{2}$ is convergent, so is $\int_1^\infty \frac{x^2 - 5x^3}{x^6 + 1}\sin 3x\,dx$.

But $\int_0^1 \frac{x^2 - 5x^3}{x^6 + 1}\sin 3x\,dx$ is proper, so

$\int_0^\infty \frac{x^2 - 5x^3}{x^6 + 1}\sin 3x\,dx$ is convergent.

49 (a) $\Gamma(1) = \int_0^\infty e^{-x}\,dx = \lim_{b\to\infty}(-e^{-x})\big|_0^b = 1$

(b) Let $u = x^n$ and $dv = e^{-x}\,dx$; then $du = nx^{n-1}\,dx$, $v = -e^{-x}$, and $\Gamma(n + 1)$

$= \lim_{t\to\infty}\int_0^t x^n e^{-x}\,dx$

$= \lim_{t\to\infty}\left(x^n(-e^{-x})\big|_0^t - \int_0^t (-e^{-x})nx^{n-1}\,dx\right)$

$= \lim_{t\to\infty}(-t^n e^{-t}) + n\lim_{t\to\infty}\int_0^t e^{-x}x^{n-1}\,dx$

$= n\Gamma(n)$.

(c) 1, 2, 6, 24

(d) $\Gamma(n + 1) = n!$

51 The volume, when the depth is y, is $V = \pi r^2 y$.

Thus $\frac{d}{dt}(\pi r^2 y) = -\sqrt{y}$; that is, $\pi r^2\cdot\frac{dy}{dt} = -\sqrt{y}$.

Separating variables, we get $dt = -\pi r^2\frac{dy}{\sqrt{y}}$. Hence

$t = \int dt = -\int \pi r^2\frac{dy}{\sqrt{y}} = -2\pi r^2\sqrt{y} + C$. To evaluate the constant C, recall that $y = h$ when $t = 0$, so $C = 2\pi r^2\sqrt{h}$. Hence $t = 2\pi r^2(\sqrt{h} - \sqrt{y})$.

(a) When $y = h/2$, $t = 2\pi r^2(\sqrt{h} - \sqrt{h/2})$

$= \pi r^2\sqrt{h}(2 - \sqrt{2})$ seconds.

(b) When $y = 0$, $t = 2\pi r^2(\sqrt{h} - \sqrt{0}) = 2\pi r^2\sqrt{h}$ seconds.

53 (a) $G(0) = \int_0^\infty \frac{dx}{2(1+x^2)} = \lim_{b\to\infty} \frac{1}{2}\tan^{-1}x\Big|_0^b$

$= \frac{1}{2}\left(\frac{\pi}{2} - 0\right) = \frac{\pi}{4}.$

$G(1) = \int_0^\infty \frac{dx}{(1+x)(1+x^2)}$; the partial-fraction representation of $\frac{1}{(1+x)(1+x^2)}$ is of the form $\frac{c_1}{1+x} + \frac{c_2x + c_3}{1+x^2}$, where $1 = c_1(1+x^2) + (c_2x + c_3)(1+x)$. Setting $x = -1, 0,$ and 1 gives $c_1 = \frac{1}{2}$, $c_3 = \frac{1}{2}$, and $c_2 = -\frac{1}{2}$. Hence $\frac{1}{(1+x)(1+x^2)} = \frac{1}{2}\left(\frac{1}{1+x} - \frac{x-1}{1+x^2}\right)$, so $\int \frac{dx}{(1+x)(1+x^2)}$

$= \frac{1}{2}\int \frac{dx}{1+x} - \frac{1}{2}\int \frac{x-1}{1+x^2}\,dx =$

$\frac{1}{2}\ln|1+x| - \frac{1}{4}\ln(1+x^2) + \frac{1}{2}\tan^{-1}x + C$

$= \frac{1}{4}\ln\frac{(1+x)^2}{1+x^2} + \frac{1}{2}\tan^{-1}x + C.$

Therefore, $G(1)$

$= \left[\frac{1}{4}\ln\frac{(1+x)^2}{1+x^2} + \frac{1}{2}\tan^{-1}x\right]\Bigg|_0^\infty$

$= \frac{1}{4}\ln 1 + \frac{1}{2}\cdot\frac{\pi}{2} - \left(\frac{1}{4}\ln 1 + \frac{1}{2}\cdot 0\right) = \frac{\pi}{4}.$

$G(2) = \int_0^\infty \frac{dx}{(1+x^2)^2}$. By Formula 38 from the text's list of antiderivatives, with $a = 1$, $b = 0$, $c_2 = 1$, and $n = 1$, we obtain

$\int \frac{dx}{(1+x^2)^2} = \frac{2x}{4(1+x^2)} + \frac{2}{4}\int \frac{dx}{1+x^2}$

$= \frac{x}{2(1+x^2)} + \frac{1}{2}\tan^{-1}x + C.$

Hence, $G(2) = \left[\frac{x}{2(1+x^2)} + \frac{1}{2}\tan^{-1}x\right]\Bigg|_0^\infty$

$= \left(0 + \frac{1}{2}\cdot\frac{\pi}{2}\right) - \left(0 + \frac{1}{2}\cdot 0\right) = \frac{\pi}{4}.$

(b) With $x = 1/y$, we have $dx = -\frac{dy}{y^2}$. As x increases from 0 to ∞, y decreases from ∞ to 0. Hence $G(a) =$

$\int_\infty^0 \frac{1}{(1+y^{-a})(1+y^{-2})}\left(-\frac{dy}{y^2}\right)$

$= \int_0^\infty \frac{y^a\,dy}{(1+y^a)(1+y^2)}$

$= \int_0^\infty \frac{x^a\,dx}{(1+x^a)(1+x^2)}.$

(c) Adding the equations in (a) and (b), we get

$2G(a) = \int_0^\infty \frac{dx}{(1+x^a)(1+x^2)} +$

$\int_0^\infty \frac{x^a\,dx}{(1+x^a)(1+x^2)}$

$= \int_0^\infty \frac{(1+x^a)\,dx}{(1+x^a)(1+x^2)} = \int_0^\infty \frac{dx}{1+x^2}$

$= \tan^{-1}x\,\Big|_0^\infty = \frac{\pi}{2}.$

Hence $G(a) = \pi/4$.

55 Place the circles in an xy coordinate system as shown. The two circles intersect when $(x-1)^2 = x^2$ or $x^2 - 2x + 1 = x^2$. This occurs when $x = 1/2$.

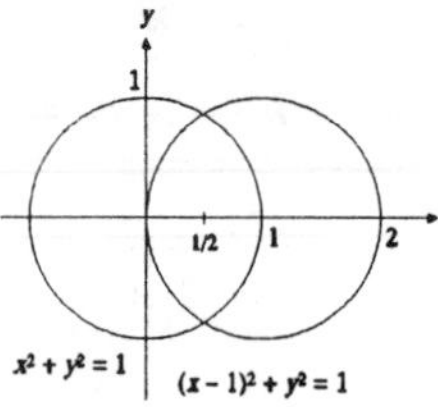

(a) By symmetry, the area is 4 times the area of the region $1/2 \le x \le 1$, $0 \le y \le \sqrt{1-x^2}$.

Thus the area is $4\int_{1/2}^{1}\sqrt{1-x^2}\,dx$. Let $x = \sin\theta$, so $dx = \cos\theta\,d\theta$ and

$$4\int_{1/2}^{1}\sqrt{1-x^2}\,dx = 2\int_{\pi/6}^{\pi/2} 2\cos^2\theta\,d\theta$$

$$= 2\int_{\pi/6}^{\pi/2}(1+\cos 2\theta)\,d\theta$$

$$= 2\left(\theta + \frac{\sin 2\theta}{2}\right)\Bigg|_{\pi/6}^{\pi/2}$$

$$= 2\left[\left(\frac{\pi}{2}+0\right) - \left(\frac{\pi}{6}+\frac{\sqrt{3}}{4}\right)\right] = \frac{2\pi}{3} - \frac{\sqrt{3}}{2}.$$

(b) Horizontal cross sections are bounded on the right by the circle $x^2 + y^2 = 1$ (so $x = \sqrt{1-y^2}$) and on the left by the circle $(x-1)^2 + y^2 = 1$ (so $x - 1 = -\sqrt{1-y^2}$, $x = 1 - \sqrt{1-y^2}$). Hence, $c(y) = \sqrt{1-y^2} - (1-\sqrt{1-y^2}) = 2\sqrt{1-y^2} - 1$, and the area is

$$\int_{-\sqrt{3}/2}^{\sqrt{3}/2}\left(2\sqrt{1-y^2} - 1\right)dy$$

$$= \int_{-\sqrt{3}/2}^{\sqrt{3}/2} 2\sqrt{1-y^2}\,dy - \sqrt{3}.$$ Making the trigonometric substitution $y = \sin\theta$ as in (a),

$$\int_{-\sqrt{3}/2}^{\sqrt{3}/2} 2\sqrt{1-y^2}\,dy = \int_{-\pi/3}^{\pi/3} 2\cos^2\theta\,d\theta$$

$$= \left(\theta + \frac{\sin 2\theta}{2}\right)\Bigg|_{-\pi/3}^{\pi/3}$$

$$= \left[\frac{\pi}{3} + \frac{\sqrt{3}/2}{2}\right] - \left[-\frac{\pi}{3} + \frac{-\sqrt{3}/2}{2}\right]$$

$= \dfrac{2\pi}{3} + \dfrac{\sqrt{3}}{2}$, so the area is

$$\left(\frac{2\pi}{3} + \frac{\sqrt{3}}{2}\right) - \sqrt{3} = \frac{2\pi}{3} - \frac{\sqrt{3}}{2}.$$

(c) The area of the region is twice the area of the indicated region in the accompanying diagram. The area of the shaded region is found by subtracting the area of the triangle AOB from the area of the circular section AOB. The area of the circular section is $\frac{1}{3}(\pi \cdot 1^2) = \frac{\pi}{3}$. The area of the triangle is $\frac{1}{2}\left(2\frac{\sqrt{3}}{2}\right)\cdot\frac{1}{2} = \frac{\sqrt{3}}{4}$.

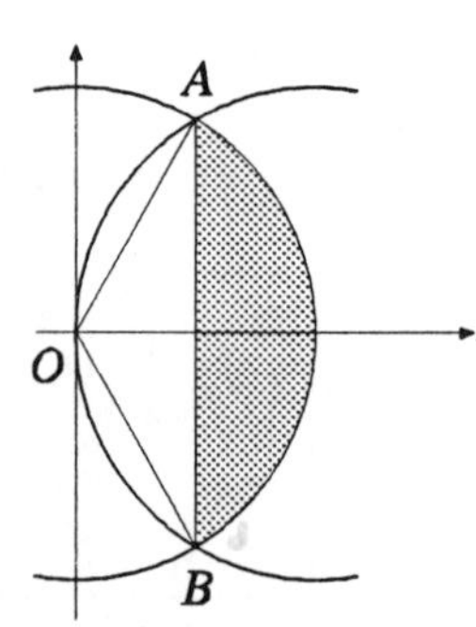

Therefore the area of the shaded region is $\frac{\pi}{3} - \frac{\sqrt{3}}{4}$. Hence the area of the original region is $2\left(\frac{\pi}{3} - \frac{\sqrt{3}}{4}\right) = \frac{2\pi}{3} - \frac{\sqrt{3}}{2}$.

57 The substitution $u = 1/x$ is not valid. The function u is not defined at $x = 0$.

59 The trignometric identity $\sin(x + a)$

$= \sin x \cos a + \cos x \sin a$ gives

$\int \sin x \sin(x + a)\,dx = \cos a \int \sin^2 x\,dx +$

$\sin a \int \sin x \cos x\,dx = \cos a\left(\frac{x}{2} - \frac{\sin 2x}{4}\right) +$

$\sin a \int \frac{1}{2}\sin 2x\,dx = \frac{1}{4}(\cos a)(2x - \sin 2x) -$

$\frac{1}{4}\sin a \cos 2x + C$. (We used the result of Example 2 in Sec. 7.6. See also Formula 44 in the table of antiderivatives.) Hence

$\int_0^{2\pi} \sin x \sin(x + a)\,dx$

$= \frac{1}{4}(\cos a)[(4\pi - 0) - (0 - 0)] - \frac{1}{4}(\sin a)[1 - 1]$

$= \pi \cos a$, which is maximal when $\cos a = 1$; that is, when $a = 0$ or 2π. The maximum value of the integral is π.

61 (a) Make the substitution $x = \sqrt{2}\,ku$, $dx = \sqrt{2}\,k\,du$. As x goes from $-\infty$ to ∞, so does u. Hence

$\int_{-\infty}^{\infty} f(x)\,dx = \frac{1}{\sqrt{2\pi}k}\int_{-\infty}^{\infty} e^{-x^2/(2k^2)}\,dx$

$= \frac{1}{\sqrt{2\pi}k}\int_{-\infty}^{\infty} e^{-u^2}\sqrt{2}k\,du$

$= \frac{1}{\sqrt{\pi}}\int_{-\infty}^{\infty} e^{-u^2}\,du$

$= \frac{1}{\sqrt{\pi}}\left(\int_{-\infty}^{0} e^{-u^2}\,du + \int_0^{\infty} e^{-u^2}\,du\right) =$

$\frac{2}{\sqrt{\pi}}\left(\int_0^{\infty} e^{-u^2}\,du\right)$, since the integrand is an even function. By assumption, this equals

$\frac{2}{\sqrt{\pi}}\frac{\sqrt{\pi}}{2} = 1.$

(b) $\mu = \frac{1}{\sqrt{2\pi}k}\int_{-\infty}^{\infty} xe^{-x^2/(2k^2)}\,dx$. Make the substitution $u = -x^2/(2k^2)$, so that $du = -x\,dx/k^2$ and $x\,dx = -k^2\,du$. Then

$\int xe^{-x^2/(2k^2)}\,dx = \int e^u(-k^2\,du)$

$= -k^2e^u + C = -k^2e^{-x^2/(2k^2)} + C.$

As $x \to \infty$ or $x \to -\infty$, $-k^2e^{-x^2/(2k^2)} \to 0$, so

$\mu = \frac{1}{\sqrt{2\pi}k}(0 - 0) = 0.$

(c) $\mu_2 = \int_{-\infty}^{\infty} (x - 0)^2 \cdot \frac{1}{\sqrt{2\pi}\,k} e^{-x^2/(2k^2)}\,dx$

$= \frac{1}{\sqrt{2\pi}\,k}\int_{-\infty}^{\infty} x^2e^{-x^2/(2k^2)}\,dx$. Let $x = \sqrt{2}\,ku$, $dx = \sqrt{2}\,k\,du$. Then $\int x^2e^{-x^2/(2k^2)}\,dx$

$= \int 2k^2u^2e^{-u^2}(\sqrt{2}\,k\,du)$

$= 2\sqrt{2}k^3\int u^2e^{-u^2}\,du$. Use integration by parts with $U = u$, $dV = ue^{-u^2}\,du$, $dU = du$, and $V = -\frac{1}{2}e^{-u^2}$ Then $\int u^2e^{-u^2}\,du$

$= \int U\,dV = UV - \int V\,dU$

$= -\frac{u}{2}e^{-u^2} + \frac{1}{2}\int e^{-u^2}\,du.$

Hence μ_2

$= \frac{1}{\sqrt{2\pi}k}2\sqrt{2}k^3\left(-\frac{u}{2}e^{-u^2} + \frac{1}{2}\int e^{-u^2}\,du\right)\Big|_{-\infty}^{\infty}$

$= \frac{k^2}{\sqrt{\pi}}\left[-ue^{-u^2}\Big|_{-\infty}^{\infty} + \int_{-\infty}^{\infty} e^{-u^2}\,du\right]$. As $u \to$

$-\infty$ or $u \to \infty$, $e^{-u^2} \to 0$. Also, as shown in

(a), $\int_{-\infty}^{\infty} e^{-u^2}\,du = \sqrt{\pi}$, so μ_2

$= \frac{k^2}{\sqrt{\pi}}[(0 - 0) + \sqrt{\pi}] = k^2$.

(d) $\sigma = \sqrt{\mu_2} = \sqrt{k^2} = k$

63 For $v \geq \sqrt{k/2}$, we have $-1 \leq 1 - \frac{k}{v^2} \leq 1$, so

$\left|1 - \frac{k}{v^2}\right| \leq 1$. Let b be the larger of $\sqrt{k/2}$ and a.

Then, for $v \geq b$, $\left|\left(1 - \frac{k}{v^2}\right)e^{-v^2}\right| \leq e^{-v^2}$. By

Example 2 of Sec. 8.8, $\int_b^{\infty} e^{-v^2}\,dv$ converges. By

Theorems 1 and 3 of Sec. 8.8, therefore, so does

$\int_b^{\infty}\left(1 - \frac{k}{v^2}\right)e^{-v^2}\,dv$. Hence $\int_a^{\infty}\left(1 - \frac{k}{v^2}\right)e^{-v^2}\,dv$

$= \int_a^{b}\left(1 - \frac{k}{v^2}\right)e^{-v^2}\,dv + \int_b^{\infty}\left(1 - \frac{k}{v^2}\right)e^{-v^2}\,dv$

converges.

65 Let $x = u/a$. Then $\int x^n e^{-ax}\,dx = \int \frac{u^n}{a^n}e^{-u}\frac{du}{a}$

$= \frac{1}{a^{n+1}}\int u^n e^{-u}\,du$. Hence $\int_0^{\infty} x^n e^{-ax}\,dx$

$= \lim_{b\to\infty}\int_0^{b} x^n e^{-ax}\,dx = \lim_{b\to\infty}\frac{1}{a^{n+1}}\int_0^{ab} u^n e^{-u}\,du$

$= \frac{1}{a^{n+1}}\int_0^{\infty} u^n e^{-u}\,du = \frac{n!}{a^{n+1}}$.

9 Plane Curves and Polar Coordinates

9.1 Polar Coordinates

1

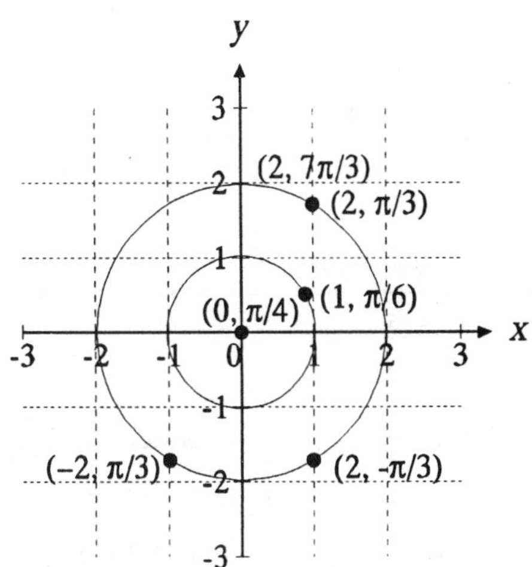

3 (a) Changing the angle θ by increments of 2π,

$$\left(3, \frac{\pi}{4}\right) = \left(3, \frac{9\pi}{4}\right) = \left(3, -\frac{7\pi}{4}\right) =$$

$$\left(3, \frac{\pi}{4} + 2\pi k\right) \text{ for any integer } k.$$

(b) $$\left(3, \frac{\pi}{4}\right) = \left(-3, \frac{5\pi}{4}\right) = \left(-3, -\frac{3\pi}{4}\right) =$$

$$\left(-3, \frac{13\pi}{4}\right) = \left(-3, \frac{5\pi}{4} + 2\pi k\right) \text{ for any}$$

integer k.

5 Multiplying the equation $r = \sin\theta$ by r, we have $r^2 = r\sin\theta$ or $x^2 + y^2 = y$.

7 The given equation is $r = \dfrac{3}{4\cos\theta + 5\sin\theta}$.

Multiplying through by the denominator yields $r(4\cos\theta + 5\sin\theta) = 3$. Then $4r\cos\theta + 5r\sin\theta = 3$. But $x = r\cos\theta$ and $y = r\sin\theta$, so this becomes $4x + 5y = 3$.

9 Since $x = r\cos\theta$, this equation becomes $r\cos\theta = -2$ or $r = \dfrac{-2}{\cos\theta}$. Hence $r = -2\sec\theta$.

11 Substituting $x = r\cos\theta$ and $y = r\sin\theta$ into the equation $xy = 1$ yields $r^2\sin\theta\cos\theta = 1$ or $r^2 = \dfrac{2}{2\sin\theta\cos\theta} = \dfrac{2}{\sin 2\theta}$.

13 First make a table for $r = 1 + \sin\theta$.

θ	0	$\pi/4$	$\pi/2$	$3\pi/4$
r	1	$1 + 1/\sqrt{2}$	2	$1 + 1/\sqrt{2}$

θ	π	$5\pi/4$	$3\pi/2$	$7\pi/4$	2π
r	1	$1 - 1/\sqrt{2}$	0	$1 - 1/\sqrt{2}$	1

Note that the graph starts to repeat itself. It is periodic with period 2π; that is, $1 + \sin(\theta + 2\pi) = 1 + \sin\theta$. This is the graph of Example 2 rotated by 90°.

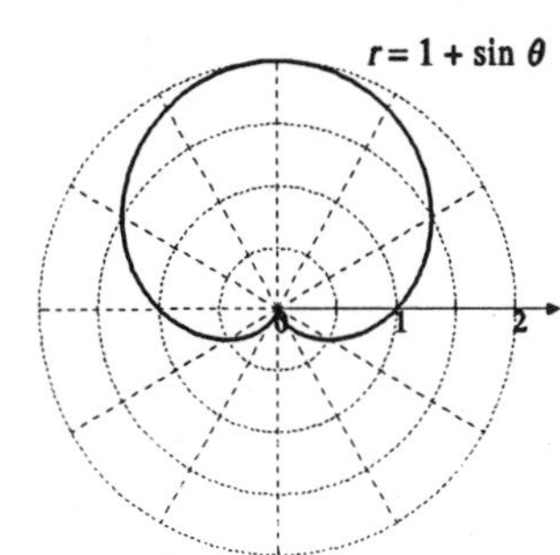

15 $r = 2^{-\theta/\pi}$

θ	0	$\pi/4$	$\pi/2$	$3\pi/4$	π
r	1	$2^{-1/4} \approx$ 0.841	$2^{-1/2} \approx$ 0.707	$2^{-3/4} \approx$ 0.595	$2^{-1} =$ 0.5

Note that $\lim_{\theta\to\infty} r = \lim_{\theta\to\infty} 2^{-\theta/\pi} = 0$, and $\lim_{\theta\to-\infty} r = \lim_{\theta\to-\infty} 2^{-\theta/\pi} = \infty$.

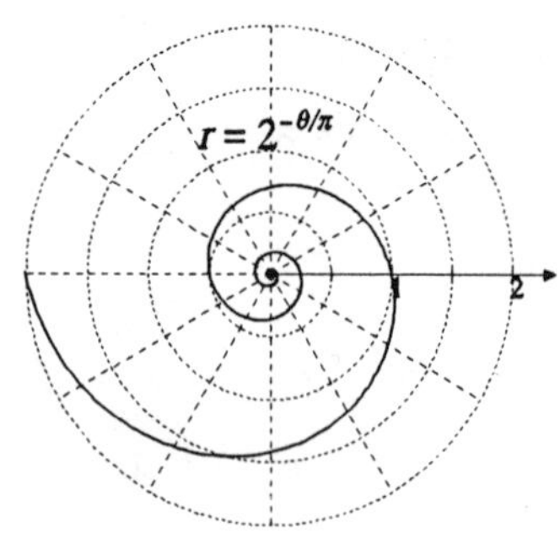

17 $r = \cos 3\theta$. Note that r is 0 when 3θ is $\pm\pi/2$, $\pm 3\pi/2$, $\pm 5\pi/2$, ... (any odd multiple of $\pi/2$). Thus $r = 0$ when θ is $\pm\pi/6$, $\pm\pi/2$, $\pm 5\pi/6$ etc. So one loop is formed for θ in $[-\pi/6, \pi/6]$, another for θ in $[\pi/6, \pi/2]$ (these values of θ yielding negative r), and so on. Since $|r|$ is at most 1, the graph must be the indicated three-leaved rose. The graph is similar to the one in Example 4.

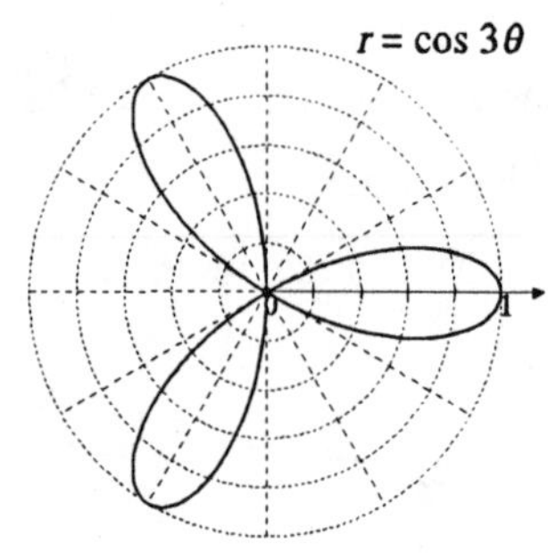

19 Regardless of the choice of θ, $r = 2$; hence the graph is a circle of radius 2 (centered at the pole).

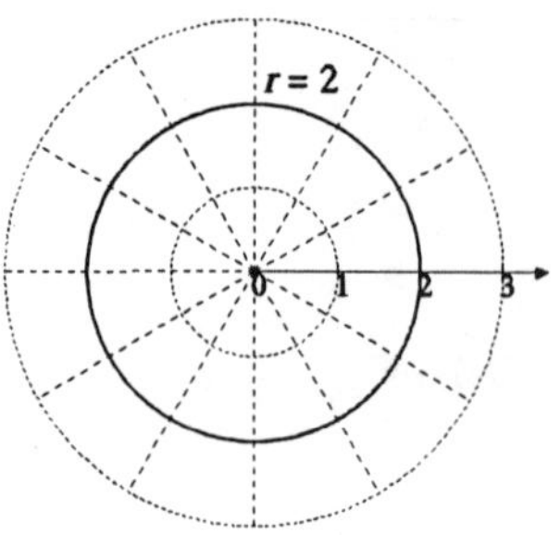

21 $r = 3 \sin \theta$

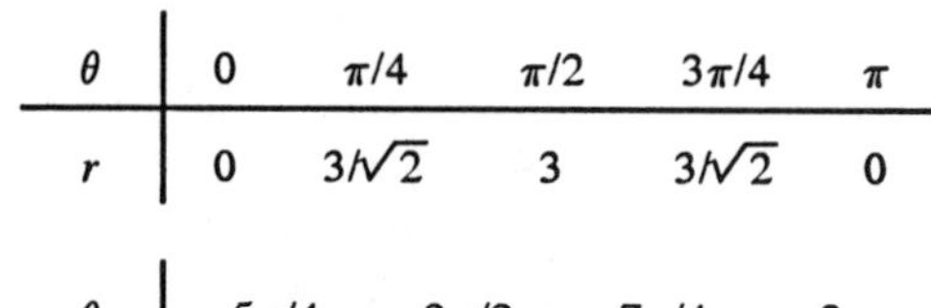

θ	0	$\pi/4$	$\pi/2$	$3\pi/4$	π
r	0	$3/\sqrt{2}$	3	$3/\sqrt{2}$	0

θ	$5\pi/4$	$3\pi/2$	$7\pi/4$	2π
r	$-3/\sqrt{2}$	-3	$-3/\sqrt{2}$	0

The graph of $r = 3 \sin \theta$ is periodic with period π. (It retraces itself for θ in $[\pi, 2\pi]$.) The graph is a circle of radius 3/2 centered at $(3/2, \pi/2)$.

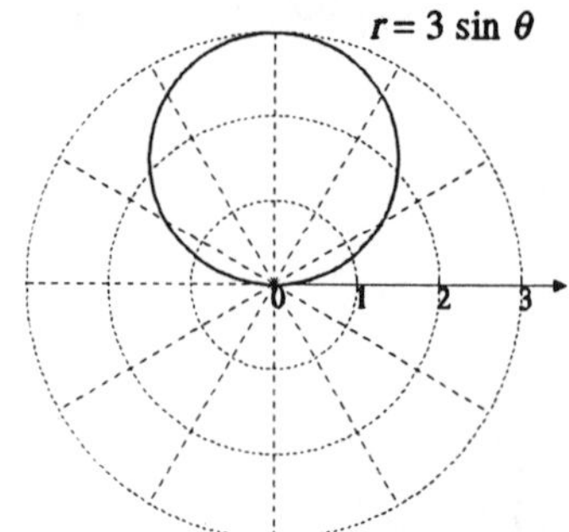

23 (a)

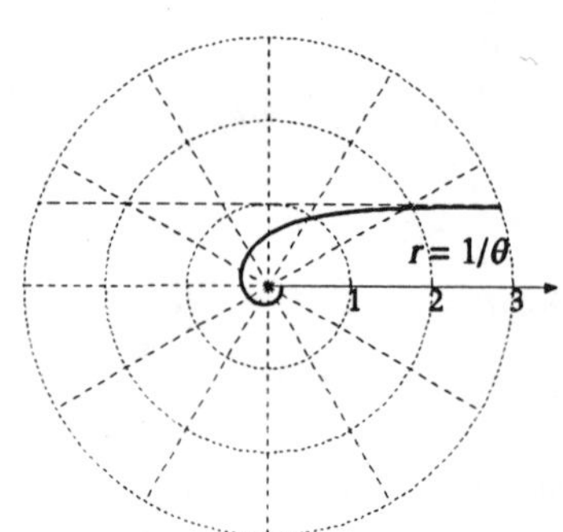

(b) From (a), $y = r \sin \theta = \dfrac{\sin \theta}{\theta}$. Hence $\lim_{\theta\to\infty} y$

$= \lim_{\theta \to \infty} \frac{\sin \theta}{\theta} = 0$ (because $|\sin \theta| \le 1$ for all θ). Thus the y coordinate approaches 0 as $\theta \to \infty$. (As $\theta \to 0$, $y \to 1$, which is a help in sketching the graph in (a).)

25 If (r, θ) satisfies the equation $r = 1 + \cos \theta$, then $-r = -1 - \cos \theta = \cos(\theta + \pi) - 1$, so $(-r, \theta + \pi)$ satisfies the equation $r = \cos \theta - 1$. But (r, θ) and $(-r, \theta + \pi)$ represent the same point, so every point of the first curve is also on the second. Similarly, every point of the second is on the first. The two curves coincide.

27 First note that the origin is on both curves, since $(0, 0)$ satisfies the first equation and $(0, \pi/6)$ satisfies the second. Suppose that (r, θ) satisfies the first equation, $r \ne 0$, and (r, θ) is on the second curve. Then either $r = \cos 3(\theta + 2n\pi)$ or $-r = \cos 3(\theta + (2n + 1)\pi)$ for some integer n. But since $\cos 3(\theta + 2n\pi) = \cos(3\theta + 6n\pi) = \cos 3\theta$ and $\cos 3(\theta + (2n + 1)\pi) = \cos(3\theta + \pi + (6n + 2)\pi) = \cos(3\theta + \pi) = -\cos 3\theta$, each of these equations implies that $r = \cos 3\theta$. Therefore, it suffices to solve $r = \sin 3\theta$ and $r = \cos 3\theta$ simultaneously. These equations imply that $\sin 3\theta = \cos 3\theta$, so $\tan 3\theta = 1$ and therefore $3\theta = \pi/4 + n\pi$ for some integer n. Hence $\theta = \frac{\pi}{12} + n \cdot \frac{\pi}{3}$ and the point is either $\left(\frac{1}{\sqrt{2}}, \frac{\pi}{12}\right)$, $\left(-\frac{1}{\sqrt{2}}, \frac{5\pi}{12}\right)$, $\left(\frac{1}{\sqrt{2}}, \frac{9\pi}{12}\right)$, $\left(-\frac{1}{\sqrt{2}}, \frac{13\pi}{12}\right)$, $\left(\frac{1}{\sqrt{2}}, \frac{17\pi}{12}\right)$, or $\left(-\frac{1}{\sqrt{2}}, \frac{21\pi}{12}\right)$.

Eliminating duplications and including the origin, we find four intersections: $(0, 0)$, $\left(\frac{1}{\sqrt{2}}, \frac{\pi}{12}\right)$, $\left(\frac{1}{\sqrt{2}}, \frac{3\pi}{4}\right)$, and $\left(\frac{1}{\sqrt{2}}, \frac{17\pi}{12}\right)$.

29 The origin is an intersection, since $(0, 0)$ is on the first curve and $(0, \pi/4)$ is on the second. For $r \ne 0$, suppose that $r = \sin \theta$ and that (r, θ) is on the second curve. Then either $r = \cos 2(\theta + 2n\pi)$ or $-r = \cos 2(\theta + (2n + 1)\pi)$ for some integer n. In either case, $\pm r = \cos 2\theta = 1 - 2\sin^2 \theta = 1 - 2r^2$. Hence $2r^2 \pm r - 1 = 0$ and $r = \frac{\pm 1 \pm \sqrt{(\pm 1)^2 - 4 \cdot 2 \cdot (-1)}}{2 \cdot 2} = \frac{\pm 1 \pm 3}{4}$ (four possible choices of sign). Thus $r = -1, -1/2, 1/2$, or 1, so θ is either $\pm\frac{\pi}{6}$, $\pm\frac{\pi}{2}$, $\pm\frac{5\pi}{6}$, or an angle differing from one of these by a multiple of 2π. So the point is either $\left(\frac{1}{2}, \frac{\pi}{6}\right)$, $\left(-\frac{1}{2}, -\frac{\pi}{6}\right)$, $\left(1, \frac{\pi}{2}\right)$, $\left(-1, -\frac{\pi}{2}\right)$, $\left(\frac{1}{2}, \frac{5\pi}{6}\right)$ or $\left(-\frac{1}{2}, -\frac{5\pi}{6}\right)$. Eliminating duplications and including the origin, there are four intersections: $(0, 0)$, $\left(\frac{1}{2}, \frac{\pi}{6}\right)$, $\left(1, \frac{\pi}{2}\right)$, and $\left(\frac{1}{2}, \frac{5\pi}{6}\right)$.

31 First make a table.
Since the period of $1 + 2\cos \theta$ is 2π, these values will suffice to sketch the graph.

θ	0	$\pm\pi/6$	$\pm\pi/4$	$\pm\pi/3$	$\pm\pi/2$
r	3	$1+\sqrt{3}$ ≈ 2.7	$1+\sqrt{2}$ ≈ 2.4	2	1

θ	$\pm 2\pi/3$	$\pm 3\pi/4$	$\pm 5\pi/6$	π
r	0	$1-\sqrt{2}$ ≈ -0.4	$1-\sqrt{3}$ ≈ -0.7	-1

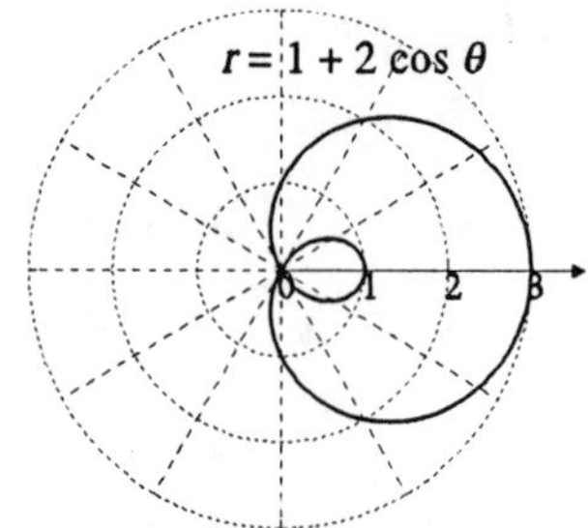

33 (a)

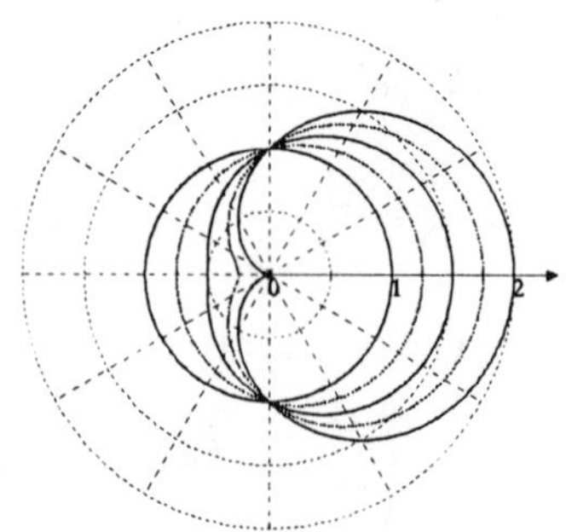

(b) For a point on the curve, we have $x = r\cos\theta = \cos\theta + a(\cos\theta)^2$. To minimize this is equivalent to minimizing $c + ac^2$ for $-1 \le c \le 1$. But $\frac{d}{dc}(c + ac^2) = 1 + 2ac$, which is negative for $c < -\frac{1}{2a}$ and positive for $c > -\frac{1}{2a}$. Since $-1 \le -\frac{1}{2a} \le 1$ for $a \ge 1/2$, the minimum for $-1 \le c \le 1$ occurs at $c = -\frac{1}{2a}$; $x = \left(-\frac{1}{2a}\right) + a\left(-\frac{1}{2a}\right)^2 = -\frac{1}{4a}$. This accounts for two values of θ, namely, $\cos^{-1}\left(-\frac{1}{2a}\right)$ and $-\cos^{-1}\left(-\frac{1}{2a}\right)$. Thus there are two points on the curve at which x is minimal; their rectangular coordinates are $\left(-\frac{1}{4a}, y\right)$ and $\left(-\frac{1}{4a}, -y\right)$ for some y. If the curve were convex, then their midpoint, $\left(-\frac{1}{4a}, 0\right)$, would have to lie *inside* the curve; hence $r(\pi)$ would have to be at least $\frac{1}{4a}$. We have $r(\pi) = 1 + a\cos\pi = 1 - a$. If $1 - a \ge \frac{1}{4a}$, then $4a - 4a^2 \ge 1$ and $0 \ge 1 - 4a + 4a^2 = (1 - 2a)^2$. Since $a \ne 1/2$, this is impossible. Hence the curve is not convex.

35 $y = r\sin\theta = \cos 2\theta \sin\theta = (1 - 2\sin^2\theta)\sin\theta = (\sin\theta) - 2(\sin\theta)^3$. The top half of the right-hand leaf is traced out as θ goes from 0 to $\pi/4$, so we wish to maximize $(\sin\theta) - 2(\sin\theta)^3$ for θ in this range. This is equivalent to maximizing $s - 2s^3$ for $0 \le s \le \frac{1}{\sqrt{2}}$. Since $\frac{d}{ds}(s - 2s^3) = 1 - 6s^2$ is positive for $0 \le s \le \frac{1}{\sqrt{6}}$ and negative for $\frac{1}{\sqrt{6}} \le s \le \frac{1}{\sqrt{2}}$, the maximum occurs at $s = \frac{1}{\sqrt{6}}$. We have $y = \frac{1}{\sqrt{6}} - 2\left(\frac{1}{\sqrt{6}}\right)^3 = \frac{\sqrt{6}}{9}$.

37 The origin is on both curves. For $r \ne 0$, suppose that $r = 2\theta$ and that (r, θ) lies on the first curve. Then either $r = \theta + 2n\pi$ or $-r = \theta + (2n + 1)\pi$

for some integer n. In the first case, $2\theta = r = \theta + 2n\pi$, so $\theta = 2n\pi$ and $r = 4n\pi$. Since we are taking $\theta \geq 0$ and $r \neq 0$, this gives the points $(4n\pi, 0)$ for the integers $n > 0$. In the second case, $-2\theta = \theta + (2n + 1)\pi$. But since we are considering only $\theta \geq 0$ for both curves, we have both $\theta \geq 0$ and $\theta + (2n + 1)\pi \geq 0$. So $-2\theta \leq 0$ and $-2\theta \geq 0$; hence $\theta = 0$ and $n = -1/2$. But n is supposed to be an integer, so the second case cannot occur. The intersections are $(4n\pi, 0)$ for integers $n \geq 0$.

39 (a)

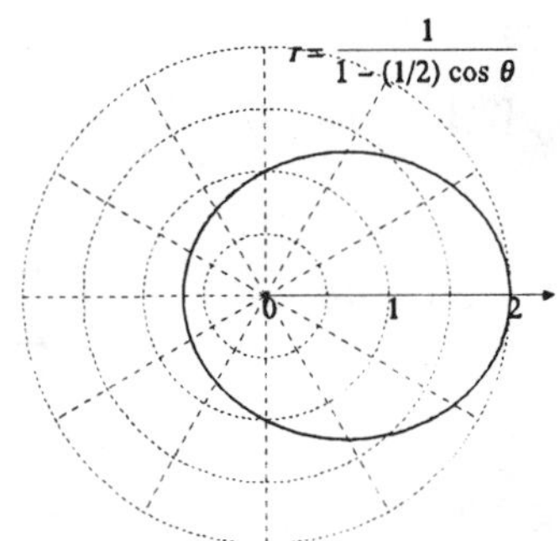

(b) We have $r - \frac{1}{2}r\cos\theta = 1$, so

$\sqrt{x^2 + y^2} - \frac{1}{2}x = 1$ and $2\sqrt{x^2 + y^2} = 2 + x$. Squaring both sides and collecting like terms, we obtain $3x^2 + 4y^2 - 4x - 4 = 0$.

9.2 Area in Polar Coordinates

1 The area is $\int_0^{\pi/2} \frac{1}{2}(2\theta)^2\, d\theta = 2\int_0^{\pi/2} \theta^2\, d\theta$

$$= \frac{2}{3}\theta^3\Big|_0^{\pi/2} = \frac{\pi^3}{12}.$$

3 The area is $\int_{\pi/4}^{\pi/2} \frac{1}{2}\left(\frac{1}{1 + \theta}\right)^2 d\theta =$

$$\frac{1}{2}\int_{\pi/4}^{\pi/2} \frac{d\theta}{(1 + \theta)^2} = \frac{1}{2}\cdot\frac{-1}{1 + \theta}\Big|_{\pi/4}^{\pi/2} =$$

$$-\frac{1}{2}\left(\frac{1}{1 + \pi/2} - \frac{1}{1 + \pi/4}\right) = \frac{\pi}{(4 + \pi)(2 + \pi)}.$$

5 The area is $\int_0^{\pi/4} \frac{1}{2}(\tan\theta)^2\, d\theta =$

$$\frac{1}{2}\int_0^{\pi/4} (\sec^2\theta - 1)\, d\theta = \frac{1}{2}(\tan\theta - \theta)\Big|_0^{\pi/4}$$

$$= \frac{1}{2}\left(1 - \frac{\pi}{4}\right) = \frac{4 - \pi}{8}.$$

7

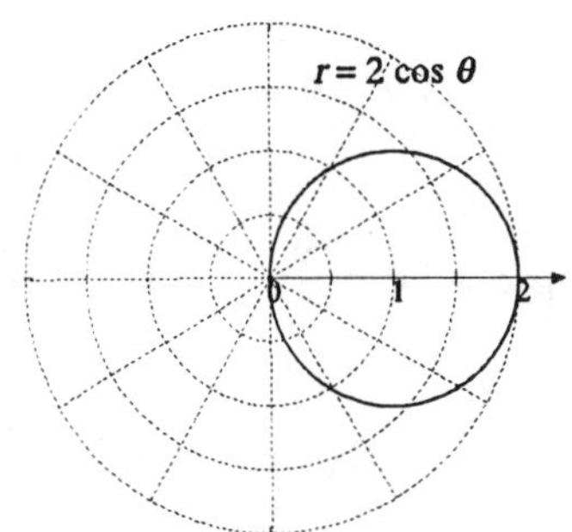

The graph of $r = 2\cos\theta$ is a circle of radius 1, hence its area is $\pi\cdot 1^2 = \pi$. Alternatively, the entire circle is completed as θ varies from 0 to π, so its area is $\int_0^{\pi} \frac{1}{2}(2\cos\theta)^2\, d\theta = 2\int_0^{\pi} \cos^2\theta\, d\theta =$

$2\cdot\frac{\pi}{2} = \pi$. (Recall that $\int_0^{n\pi/2} \cos^2\theta\, d\theta = \frac{n\pi}{4}$, where n is a positive integer.)

9

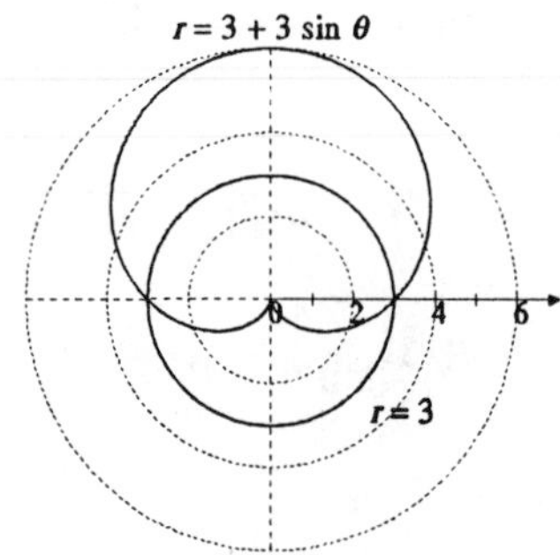

The graphs of the two curves are shown and the region is shaded. Both areas of concern are swept out as θ varies from 0 to π, so the area is

$$\frac{1}{2}\int_0^\pi [(3 + 3\sin\theta)^2 - 3^2]\, d\theta$$

$$= \frac{1}{2}\int_0^\pi (18\sin\theta + 9\sin^2\theta)\, d\theta$$

$$= \frac{1}{2}\int_0^\pi \left(18\sin\theta + \frac{9}{2}(1 - \cos 2\theta)\right) d\theta$$

$$= \frac{1}{2}\left(-18\cos\theta + \frac{9}{2}\theta - \frac{9}{4}\sin 2\theta\right)\Bigg|_0^\pi$$

$$= \frac{1}{2}\left(36 + \frac{9}{2}\pi\right) = 18 + \frac{9\pi}{4}.$$

11 See the graph accompanying Example 4 of Sec. 9.1. The area is $\int_0^{\pi/3} \frac{1}{2}(\sin^2 3\theta)\, d\theta =$

$$\frac{1}{4}\int_0^{\pi/3} (1 - \cos 6\theta)\, d\theta = \frac{1}{4}\left(\theta - \frac{\sin 6\theta}{6}\right)\Bigg|_0^{\pi/3}$$

$$= \frac{\pi}{12}.$$

13

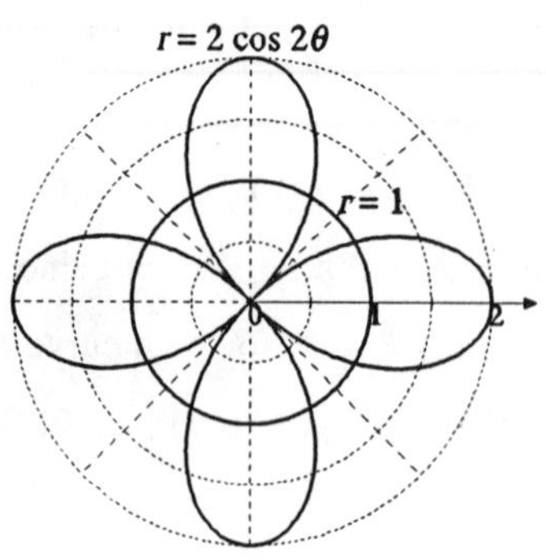

The region is shown in the figure. The graphs intersect when $1 = 2\cos 2\theta$; that is, when $\cos 2\theta = 1/2$. The area is $\frac{1}{2}\int_{-\pi/6}^{\pi/6} (4\cos^2 2\theta - 1)\, d\theta$

$$= \frac{1}{2}\int_{-\pi/6}^{\pi/6} [2(1 + \cos 4\theta) - 1]\, d\theta$$

$$= \int_0^{\pi/6} (1 + 2\cos 4\theta)\, d\theta = \left(\theta + \frac{1}{2}\sin 4\theta\right)\Bigg|_0^{\pi/6}$$

$$= \frac{\pi}{6} + \frac{1}{2}\sin\frac{2\pi}{3} = \frac{\pi}{6} + \frac{\sqrt{3}}{4}.$$

15

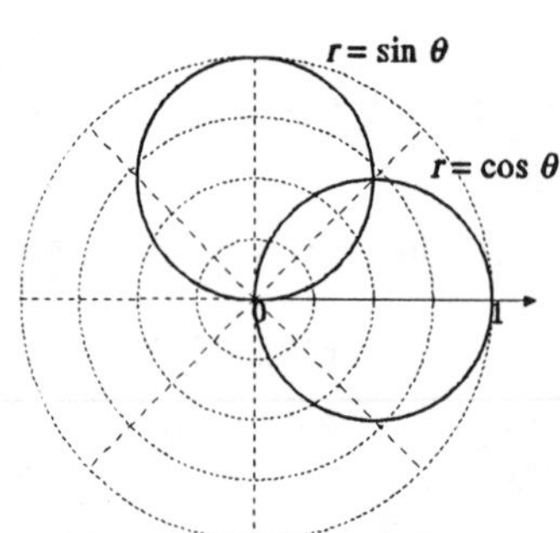

The graphs of $r = \sin\theta$ and $r = \cos\theta$ are circles of radius 1/2. The area of the region under consideration can be found by subtracting the area of their intersection from the area of the circle $r = \sin\theta$. The intersection is divided into two pieces of equal area by the ray $\theta = \pi/4$. Thus its area is

$$2\left(\int_0^{\pi/4} \frac{1}{2}\sin^2\theta\, d\theta\right) = \int_0^{\pi/4} \sin^2\theta\, d\theta =$$

$$\frac{1}{2}\int_0^{\pi/4}(1 - \cos 2\theta)\,d\theta = \frac{1}{2}\left(\theta - \frac{1}{2}\sin 2\theta\right)\Big|_0^{\pi/4}$$

$$= \frac{1}{2}\left(\frac{\pi}{4} - \frac{1}{2}\right) = \frac{\pi - 2}{8}.$$

17

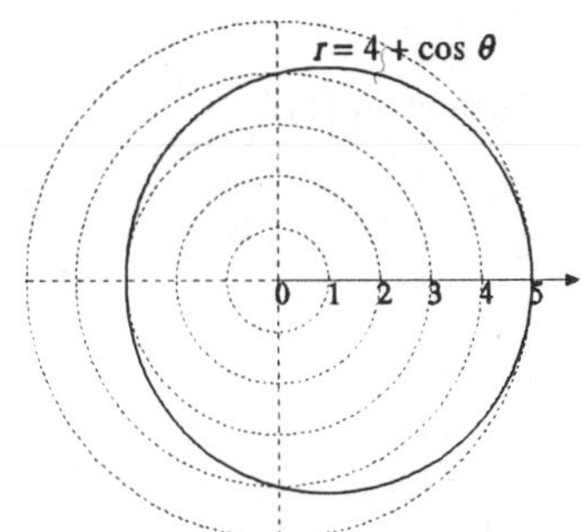

To see if $r = 4 + \cos\theta$ is a circle, express the graph in terms of rectangular coordinates and check whether the resulting formula is of the form $(x - a)^2 + (y - b)^2 = r^2$ for some constants a, b, and r. Observe that $r = 4 + \cos\theta$ implies $r^2 = 4r + r\cos\theta$, so $x^2 + y^2 = 4\sqrt{x^2 + y^2} + x$ and $x^2 - x + y^2 = 4\sqrt{x^2 + y^2}$. Squaring both sides yields $(x^2 - x)^2 + 2y^2(x^2 - x) + y^4 = 16(x^2 + y^2)$, which is clearly not of the form given above. Therefore the graph of $r = 4 + \cos\theta$ is not a circle.

19 Let $r = \overline{AD}$. The square has area r^2. The crescent is obtained by deleting a quarter disk of radius $\overline{BD} = r\sqrt{2}$ from the union of triangle BDE and a half disk of radius r. Hence its area is

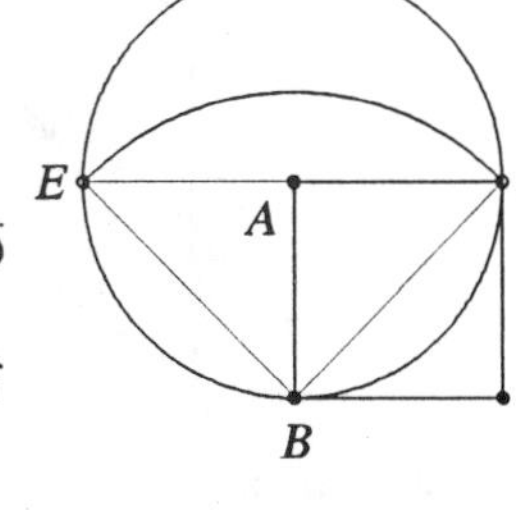

$$\left(\frac{1}{2}\cdot r\cdot 2r + \frac{1}{2}\pi r^2\right) - \frac{1}{4}\pi\left(r\sqrt{2}\right)^2 = r^2.$$

21 (a)

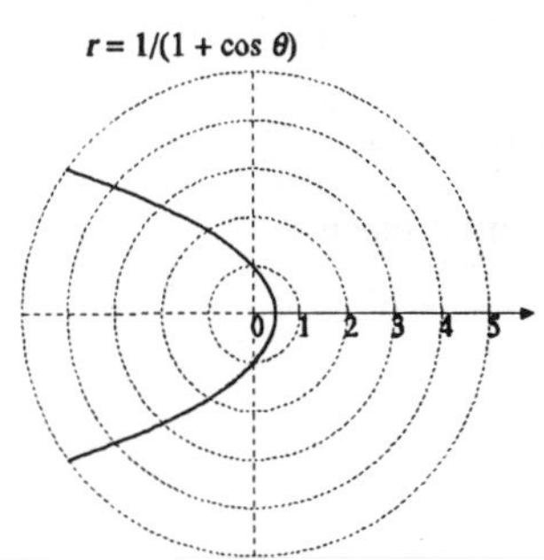

(b) $r = \dfrac{1}{1 + \cos\theta}$, so $r + r\cos\theta = 1$ and therefore $\sqrt{x^2 + y^2} = 1 - x$ and thus $x^2 + y^2 = (1 - x)^2$. Upon multiplying out and collecting like terms, we obtain $x = \frac{1}{2}(1 - y^2)$, which is the equation of a parabola.

(c) $$\int_0^{3\pi/4}\frac{1}{2}\left(\frac{1}{1 + \cos\theta}\right)^2 d\theta$$

$= \frac{1}{8}\int_0^{3\pi/4}\sec^4\frac{\theta}{2}\,d\theta$; if we let $u = \theta/2$, then $du = d\theta/2$ and we obtain

$$\frac{1}{8}\int_0^{3\pi/8}\sec^4 u\,(2\,du)$$

$$= \frac{1}{4}\int_0^{3\pi/8}\sec^2 u\,(1 + \tan^2 u)\,du$$

$$= \frac{1}{4}\left(\tan u + \frac{\tan^3 u}{3}\right)\Big|_0^{3\pi/8}$$

$$= \frac{1}{12}\tan\frac{3\pi}{8}\left(3 + \tan^2\frac{3\pi}{8}\right) = \frac{1}{6}\left(5 + 4\sqrt{2}\right).$$

(Note that $\tan\frac{3\pi}{8} = 1 + \sqrt{2}$.)

(d) The line $\theta = \frac{3\pi}{4}$ is the line $y = -x$, which meets the parabola $x = \frac{1}{2}(1 - y^2)$ when

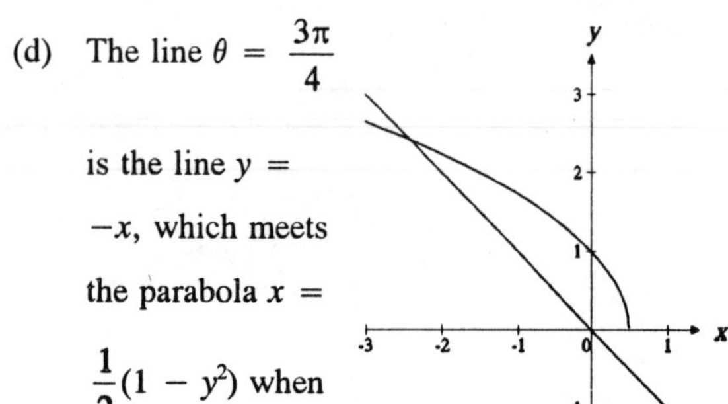

$y > 0$ and $-y = \frac{1}{2}(1 - y^2)$. Thus $y = 1 + \sqrt{2}$, so the region is described by $0 \le y \le 1 + \sqrt{2}$, $-y \le x \le \frac{1}{2}(1 - y^2)$. Its area is $\int_0^{1+\sqrt{2}} \left[\frac{1}{2}(1 - y^2) - (-y)\right] dy$

$= \frac{1}{6}(5 + 4\sqrt{2})$.

23 The area is $\int_0^{\pi/2} \frac{1}{2}\left(\sqrt[3]{1 + \theta^2}\right)^2 d\theta = \frac{1}{2}\int_0^{\pi/2} (1 + \theta^2)^{2/3}\, d\theta$. Using Simpson's method with $n = 4$ $(h = \pi/8)$ yields

$$\frac{1}{2}\int_0^{\pi/2} (1 + \theta^2)^{2/3}\, d\theta = \frac{1}{2}\frac{\pi}{24}\left[(1 + 0^2)^{2/3} + 4\left(1 + \left(\frac{\pi}{8}\right)^2\right)^{2/3} + 2\left(1 + \left(\frac{\pi}{4}\right)^2\right)^{2/3} + 4\left(1 + \left(\frac{3\pi}{8}\right)^2\right)^{2/3} + \left(1 + \left(\frac{\pi}{2}\right)^2\right)^{2/3}\right] \approx 1.1514920.$$

25 Exploration problems are in the *Instructor's Manual*.

9.3 Parametric Equations

1 (a)

t	-2	-1	0	1	2
x	-3	-1	1	3	5
y	-3	-2	-1	0	1

(b) See the graph in (c).

(c)

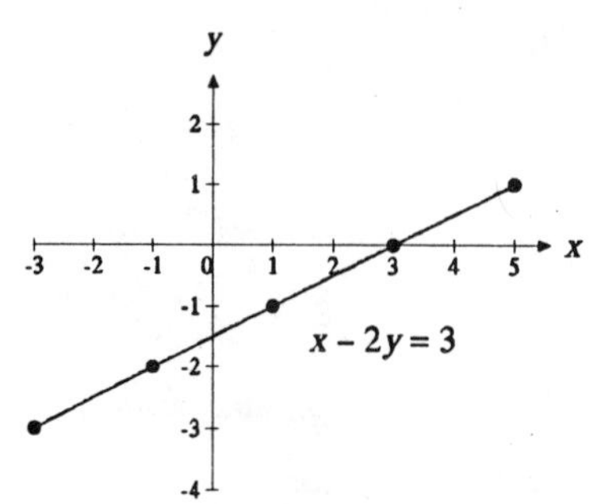

(d) $t = y + 1$, so $x = 2(y + 1) + 1 = 2y + 3$. Thus $x - 2y = 3$.

3 (a)

t	-3	-2	-1	0	1	2	3
x	9	4	1	0	1	4	9
y	6	2	0	0	2	6	12

(b) See the graph in (c).

(c)

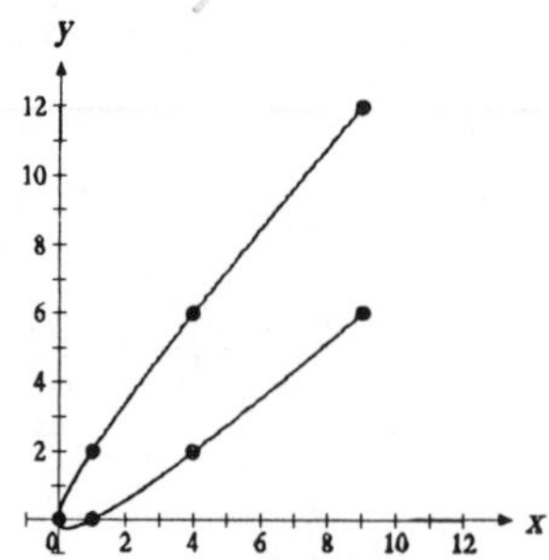

(d) Since $y = t^2 + t$, $t = y - t^2 = y - x$. Then $x = t^2 = (y - x)^2 = y^2 - 2xy + x^2$. In terms of x and y, the equation of the curve is therefore $x^2 + y^2 - 2xy - x = 0$.

5 Let $x = t$. Then $y = \sqrt{1 + x^3} = \sqrt{1 + t^3}$. Thus we have $\begin{cases} x = t \\ y = \sqrt{1 + t^3} \end{cases}$.

7 With $\theta = t$, we have $r = \cos 2t$, so $\begin{cases} \theta = t \\ r = \cos 2t \end{cases}$.

9 Recall that $\frac{dy}{dx} = \frac{dy/dt}{dx/dt}$. Hence $\frac{dy}{dx} = \frac{7t^6 + 1}{3t^2 + 1}$.

Now $\frac{d^2y}{dx^2} = \frac{d}{dx}\left(\frac{dy}{dx}\right) = \frac{\frac{d}{dt}\left(\frac{dy}{dx}\right)}{dx/dt} = \frac{\frac{d}{dt}\left(\frac{7t^6 + 1}{3t^2 + 1}\right)}{\frac{d}{dt}(t^3 + t)}$

$= \frac{1}{3t^2 + 1} \cdot \frac{(3t^2 + 1)(42t^5) - (7t^6 + 1)(6t)}{(3t^2 + 1)^2}$

$= \frac{84t^7 + 42t^5 - 6t}{(3t^2 + 1)^3}$.

11 Note $\frac{dx}{dt} = \frac{d}{dt}(1 + \ln t) = \frac{1}{t}$ and $\frac{dy}{dt} = \frac{d}{dt}(t \ln t) = t \cdot \frac{1}{t} + \ln t = 1 + \ln t$. Hence $\frac{dy}{dx} = \frac{dy/dt}{dx/dt} = \frac{1 + \ln t}{1/t} = t + t \ln t$. Now $\frac{d^2y}{dx^2}$

$= \frac{d}{dx}\left(\frac{dy}{dx}\right) = \frac{\frac{d}{dt}\left(\frac{dy}{dx}\right)}{dx/dt} = \frac{\frac{d}{dt}(t + t \ln t)}{1/t}$

$= t\left(1 + t \cdot \frac{1}{t} + \ln t\right) = 2t + t \ln t$.

13 The following solution shows how this problem can be done by manual calculations. It's messy enough to show the benefits of using a symbolic math system. Wolfram Research's Mathematica, for example, gives a solution for $\frac{d^2y}{dx^2}$ in the form

$\frac{64 \cos^6 \theta - 96 \cos^4 \theta + 36 \cos^2 \theta - 9}{(4 \sin^3 \theta \cos^3 \theta)(8 \cos^2 \theta - 3)^3}$; because of trigonometric identities, there are many different forms for this answer. In the solution below, we find a second form of $\frac{d^2y}{dx^2}$, as well as computing the answer for $\frac{dy}{dx}$.

If $r = \cos 3\theta$, then $x = r \cos \theta = \cos 3\theta \cos \theta$ and $y = r \sin \theta = \cos 3\theta \sin \theta$. Then $\frac{dy}{dx} = \frac{dy/d\theta}{dx/d\theta} =$

$\frac{\cos 3\theta \cos \theta - 3 \sin 3\theta \sin \theta}{-\cos 3\theta \sin \theta - 3 \sin 3\theta \cos \theta} =$

$\frac{3 \sin 3\theta \sin \theta - \cos 3\theta \cos \theta}{3 \sin 3\theta \cos \theta + \cos 3\theta \sin \theta}$. Now $\frac{d^2y}{dx^2} =$

$\frac{d}{dx}\left(\frac{dy}{dx}\right) = \frac{\frac{d}{d\theta}\left(\frac{dy}{dx}\right)}{dx/d\theta}$. Differentiating our expression for $\frac{dy}{dx}$ with respect to θ will involve both the quotient and product rules. Some trigonometric identities let us avoid the product rule. Note that $3 \sin 3\theta \sin \theta - \cos 3\theta \cos \theta$

$= 2 \sin 3\theta \sin \theta - 2 \cos 3\theta \cos \theta + \sin 3\theta \sin \theta$

$+ \cos 3\theta \cos \theta = -2 \cos 4\theta + \cos 2\theta =$

$\cos 2\theta - 2 \cos 4\theta$, and

$3 \sin 3\theta \cos \theta + \cos 3\theta \sin \theta = 2 \sin 3\theta \cos \theta$

$+ 2 \cos 3\theta \sin \theta + \sin 3\theta \cos \theta - \cos 3\theta \sin \theta$

$= 2 \sin 4\theta + \sin 2\theta$.

Thus $\frac{dy}{dx} = \frac{\cos 2\theta - 2\cos 4\theta}{2\sin 4\theta + \sin 2\theta}$ and $\frac{d}{d\theta}\left(\frac{dy}{dx}\right) =$

$\frac{vu' - uv'}{v^2}$, where $v^2 = (2\sin 4\theta + \sin 2\theta)^2$ and

$vu' - uv' =$

$(2\sin 4\theta + \sin 2\theta)(-2\sin 2\theta + 8\sin 4\theta) -$

$(\cos 2\theta - 2\cos 4\theta)(8\cos 4\theta + 2\cos 2\theta)$

$= -4\sin 4\theta \sin 2\theta - 2\sin^2 2\theta + 16\sin^2 4\theta +$

$8\sin 2\theta \sin 4\theta - 8\cos 2\theta \cos 4\theta + 16\cos^2 4\theta$

$- 2\cos^2 2\theta + 4\cos 4\theta \cos 2\theta$

$= 4\sin 4\theta \sin 2\theta - 4\cos 2\theta \cos 4\theta + 14$

$= -4\cos 6\theta + 14 = 4(1 - \cos 6\theta) + 10$

$= 4(2\sin^2 3\theta) + 10 = 8\sin^2 3\theta + 10.$

Therefore, $\frac{d^2y}{dx} =$

$-\frac{8\sin^2 3\theta + 10}{(3\sin 3\theta \cos\theta + \cos 3\theta \sin\theta)^3}$. [If you thought that this approach was too involved, try it the "straightforward" way. First, get lots of paper.]

15 $\frac{dy}{dx} = \frac{dy/dt}{dx/dt} = \frac{\frac{d}{dt}(t^5 + t)}{\frac{d}{dt}(t^3 + t^2)} = \frac{5t^4 + 1}{3t^2 + 2t}$, and by inspection $t = 1$ gives $x = y = 2$; hence $\left.\frac{dy}{dx}\right|_{(2,2)}$

$= \left.\frac{dy}{dx}\right|_{t=1} = \frac{5\cdot 1^4 + 1}{3\cdot 1^2 + 2\cdot 1} = \frac{6}{5}$. So the slope of the tangent line to the curve at (2, 2) is 6/5. By the point-slope equation of a line, we find that an equation of the tangent line at (2, 2) is $y =$ $\frac{6}{5}x - \frac{2}{5}$.

17 $\frac{dy}{dx} = \frac{dy/dt}{dx/dt} = \frac{2t + 1}{3t^2 + 1}$, so $\frac{d^2y}{dx^2} = \frac{\frac{d}{dt}(dy/dx)}{dx/dt}$

$$= \frac{\frac{(3t^2 + 1)(2) - (2t + 1)(6t)}{(3t^2 + 1)^2}}{3t^2 + 1} = \frac{-6t^2 - 6t + 2}{(3t^2 + 1)^3}.$$

19 From Exercise 17, we have $\frac{d^2y}{dx^2} =$ $\frac{-6t^2 - 6t + 2}{(3t^2 + 1)^3}$. Now $3t^2 + 1 > 0$ for all t, and $(3t^2 + 1)^3 > 0$ as a consequence. Note that $-(6t^2 + 6t - 2) = -2(3t^2 + 3t - 1)$; also, $3t^2 + 3t - 1 = 0$ when $t = \frac{-3 \pm \sqrt{3^2 - 4(3)(-1)}}{2\cdot 3}$ $= -\frac{1}{2} \pm \frac{\sqrt{21}}{6}$. Hence $3t^2 + 3t - 1$ changes from $+$ to $-$ at $-\frac{1}{2} - \frac{\sqrt{21}}{6}$ and from $-$ to $+$ at $-\frac{1}{2} + \frac{\sqrt{21}}{6}$; so $\frac{d^2y}{dx^2} = \frac{-2(3t^2 + 3t - 1)}{(3t^2 + 1)^3}$ goes from $-$ to $+$ at $t = -\frac{1}{2} - \frac{\sqrt{21}}{6}$ and from $+$ to $-$ at $t = -\frac{1}{2} + \frac{\sqrt{21}}{6}$. We conclude that the curve given by $\begin{cases} x = t^3 + t + 1 \\ y = t^2 + t + 2 \end{cases}$ is concave down for t in $\left(-\infty, -\frac{1}{2} - \frac{\sqrt{21}}{6}\right)$, concave up for t in

$\left(-\frac{1}{2} - \frac{\sqrt{21}}{6}, -\frac{1}{2} + \frac{\sqrt{21}}{6}\right)$, and concave down for t in $\left(-\frac{1}{2} + \frac{\sqrt{21}}{6}, \infty\right)$.

21 $x = \sin 3\theta \cos \theta$, so it follows that $\frac{dx}{d\theta} = 3 \cos 3\theta \cos \theta - \sin 3\theta \sin \theta$; $y = \sin 3\theta \sin \theta$, so $\frac{dy}{d\theta} = 3 \cos 3\theta \sin \theta + \sin 3\theta \cos \theta$; then $\frac{dy}{dx} = \frac{dy/d\theta}{dx/d\theta} = \frac{3 \cos 3\theta \sin \theta + \sin 3\theta \cos \theta}{3 \cos 3\theta \cos \theta - \sin 3\theta \sin \theta}$. At $\theta = \frac{\pi}{12}$, we have $\cos 3\theta = \sin 3\theta = \frac{1}{\sqrt{2}}$. Also, $\sin \theta = \sin\left(\frac{\pi}{4} - \frac{\pi}{6}\right) = \frac{\sqrt{3} - 1}{2\sqrt{2}}$ and $\cos \theta = \cos\left(\frac{\pi}{4} - \frac{\pi}{6}\right) = \frac{\sqrt{3} + 1}{2\sqrt{2}}$. Hence $\frac{dy}{dx} = \frac{2\sqrt{3} - 1}{\sqrt{3} + 2} = 5\sqrt{3} - 8$.

23 Select an arbitrary point P on the circle at some angle θ. The x coordinate of P is $x = h + a \cos \theta$, and the y coordinate of P is $y = k + a \sin \theta$. So the circle is given in parametric form as

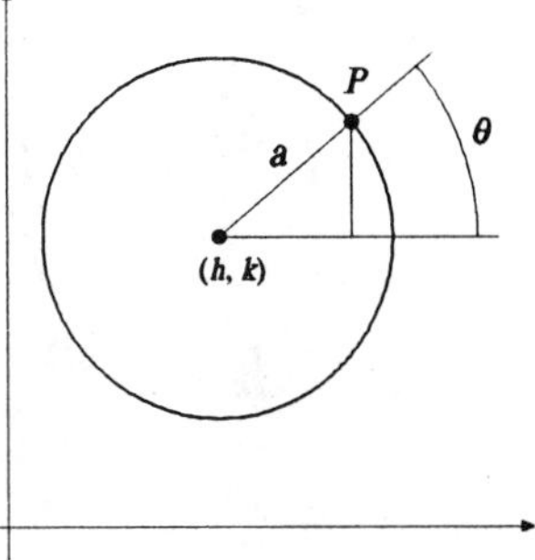

$\begin{cases} x = h + a \cos \theta \\ y = k + a \sin \theta \end{cases}$. As a check, note that $x - h = a \cos \theta$ and $y - k = a \sin \theta$. Hence $(x - h)^2 = a^2 \cos^2 \theta$, $(y - k)^2 = a^2 \sin^2 \theta$, and it follows that $(x - h)^2 + (y - k)^2 = a^2(\cos^2 \theta + \sin^2 \theta) = a^2$.

As expected, $(x - h)^2 + (y - k)^2 = a^2$ is the formula for a circle of radius a centered at (h, k).

25 (a) Since $x = (v_0 \cos \alpha)t$, $t = \frac{x}{v_0 \cos \alpha}$ and so

$$y = (v_0 \sin \alpha)t - 16t^2$$

$$= (v_0 \sin \alpha)\left(\frac{x}{v_0 \cos \alpha}\right) - 16\left(\frac{x}{v_0 \cos \alpha}\right)^2$$

$$= x \tan \alpha - \frac{16x^2}{v_0^2 \cos^2 \alpha}.$$

(b) The ball's path follows part of a parabola.

(c) We maximize the y coordinate by setting $\frac{dy}{dx} = 0$. Hence $\frac{dy}{dx} = \frac{d}{dx}\left[x \tan \alpha - \frac{16x^2}{v_0^2 \cos^2 \alpha}\right] = \tan \alpha - \frac{32x}{v_0^2 \cos^2 \alpha} = 0$ when $x = \frac{v_0^2}{32} \sin \alpha \cos \alpha$. At this value of x, $y =$

$$\left(\frac{v_0^2}{32} \sin \alpha \cos \alpha\right) \tan \alpha - \frac{16}{v_0^2 \cos^2 \alpha}\left(\frac{v_0^2}{32} \sin \alpha \cos \alpha\right)^2$$

$$= \frac{v_0^2}{32} \sin^2 \alpha - \frac{v_0^2}{64} \sin^2 \alpha = \frac{v_0^2}{64} \sin^2 \alpha.$$

Now $\frac{d^2y}{dx^2} = \frac{-32}{v_0^2 \cos^2 \alpha} < 0$ for all x, and the y coordinate is maximal at this x. Thus the coordinates of the ball's highest point are

$$\left(\frac{v_0^2}{32}\sin\alpha\cos\alpha, \frac{v_0^2}{64}\sin^2\alpha\right).$$

27 The slope of the tangent line of $r = \theta$ (at any θ) is $\frac{dy}{dx}$

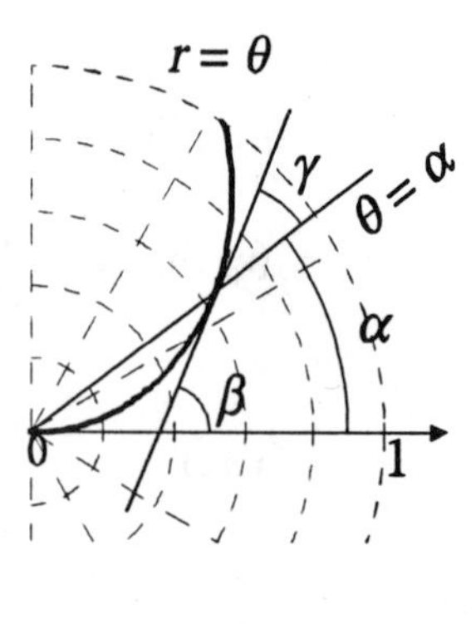

$$= \frac{dy/d\theta}{dx/d\theta}$$

$$= \frac{\frac{d}{d\theta}[\theta\sin\theta]}{\frac{d}{d\theta}[\theta\cos\theta]}$$

$= \frac{\sin\theta + \theta\cos\theta}{\cos\theta - \theta\sin\theta}$. So the angle of inclination, or the angle the tangent line makes with the positive x axis is $\beta = \tan^{-1}\left(\frac{\sin\theta + \theta\cos\theta}{\cos\theta - \theta\sin\theta}\right)$ and the angle the ray $\theta = \alpha$ makes with the positive x axis is α; hence the angle γ between the spiral and the ray is

$$\gamma = |\beta - \alpha| = \left|\tan^{-1}\left(\frac{\sin\theta + \theta\cos\theta}{\cos\theta - \theta\sin\theta}\right) - \alpha\right|.$$

When $r = \theta = \alpha + 2\pi n$, this becomes

$$\left|\tan^{-1}\left(\frac{\sin(\alpha+2\pi n) + (\alpha+2\pi n)\cos(\alpha+2\pi n)}{\cos(\alpha+2\pi n) - (\alpha+2\pi n)\sin(\alpha+2\pi n)}\right) - \alpha\right|.$$

Now $\sin(\alpha + 2\pi n) = \sin\alpha$ and $\cos(\alpha + 2\pi n) = \cos\alpha$, so

$$\lim_{n\to\infty}\left|\tan^{-1}\left(\frac{\sin\alpha + (\alpha+2\pi n)\cos\alpha}{\cos\alpha - (\alpha+2\pi n)\sin\alpha}\right) - \alpha\right| =$$

$$\lim_{n\to\infty}\left|\tan^{-1}\left(\frac{\sin\alpha + \alpha\cos\alpha + 2\pi n\cos\alpha}{\cos\alpha - \alpha\sin\alpha - 2\pi n\sin\alpha}\right) - \alpha\right|$$

$$= \lim_{n\to\infty}\left|\tan^{-1}\left(\frac{\frac{\sin\alpha}{n} + \frac{\alpha\cos\alpha}{n} + 2\pi\cos\alpha}{\frac{\cos\alpha}{n} - \frac{\alpha\sin\alpha}{n} - 2\pi\sin\alpha}\right) - \alpha\right|$$

$= |\tan^{-1}(-\cot\alpha) - \alpha| = |-1||\tan^{-1}(\cot\alpha) + \alpha|$ $= |\tan^{-1}(\cot\alpha) + \alpha|$. From the figure we see that $\cot\alpha = a/b$ and $\tan^{-1}(\cot\alpha) = \tan^{-1}(a/b) = \pi/2 - \alpha$.

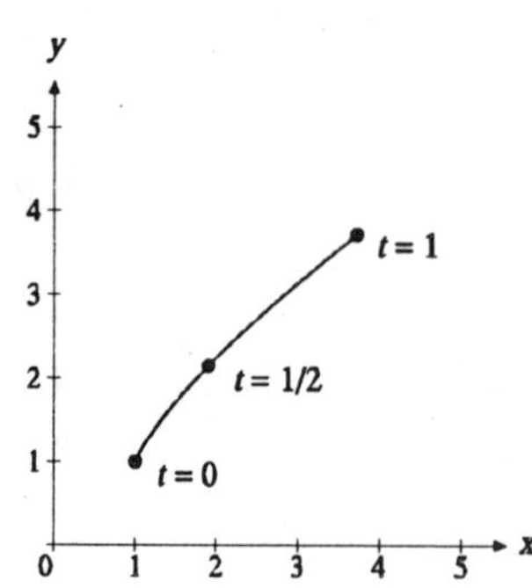

Therefore, as $n \to \infty$, the angle between the spiral and the ray approaches $|\tan^{-1}(\cot\alpha) + \alpha|$ $= |(\pi/2 - \alpha) + \alpha| = \pi/2$.

29 (a)

(b) $\frac{dy}{dx} = \frac{1 + e^t}{2t + e^t}$, which equals 2 when $t = 0$.

(c) $$\int_1^{e+1} y\,dx = \int_0^1 y\frac{dx}{dt}\,dt$$

$$= \int_0^1 (t + e^t)(2t + e^t)\,dt$$

$$= \int_0^1 (2t^2 + 3te^t + e^{2t})\,dt$$

$$= \left[\frac{2}{3}t^3 + 3(t-1)e^t + \frac{1}{2}e^{2t}\right]\Big|_0^1 = \frac{19 + 3e^2}{6}.$$

31 The region bounded above by the parametric equations $x = a\theta - a\sin\theta$, $y = a - a\cos\theta$ for θ in the interval $[0, 2\pi]$ is revolved about the x axis. Using circular slices, we see that the volume is

$\int_0^{2\pi a} \pi y^2 \, dx$

$= \pi \int_0^{2\pi} (a - a\cos\theta)^2[(a - a\cos\theta)\, d\theta]$

$= \pi a^3 \int_0^{2\pi} (1 - \cos\theta)^3 \, d\theta =$

$\pi a^3 \int_0^{2\pi} (1 - 3\cos\theta + 3\cos^2\theta - \cos^3\theta)\, d\theta =$

$\pi a^3\left(\theta - 3\sin\theta + \frac{3}{2}\theta + \frac{3}{4}\sin 2\theta - \sin\theta + \frac{1}{3}\sin^3\theta\right)\Big|_0^{2\pi}$

$= \pi a^3(5\pi) = 5\pi^2 a^3.$

33 Consider the parameterized curve $x = g(t)$, $y = f(t)$ and let P be the point $(g(t), f(t))$. Then $\dfrac{f(t)}{g(t)}$ is the slope of the line joining P to the origin, while $\dfrac{f'(t)}{g'(t)}$ is the slope of the tangent line at P. Now l'Hôpital's rule implies that if the curve passes through the origin at $t = 0$ and the slope at P approaches a limit as $t \to 0$, then the slope of the line OP approaches the same limit.

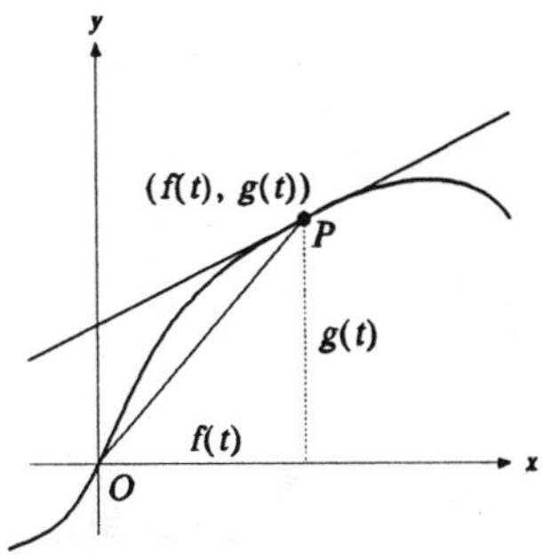

35 The curve is defined by $x = a\cos^3\theta$ and $y = a\sin^3\theta$. Let $P = (a\cos^3\beta, a\sin^3\beta)$. Then the slope of the tangent line at P is $\dfrac{dy}{dx}\Big|_P = \dfrac{dy/d\theta}{dx/d\theta}\Big|_{\theta=\beta}$

$= \dfrac{3a\sin^2\beta\cos\beta}{-3a\cos^2\beta\sin\beta} = -\tan\beta$. Thus the equation of the tangent line is $y - a\sin^3\beta = -\tan\beta\,(x - a\cos^3\beta)$. Successively letting $x = 0$ and $y = 0$, the points of intersection of the tangent line with the coordinate axes are found to be $(0, a\sin\beta)$ and $(a\cos\beta, 0)$. Finally, by the distance formula, the length of the line segment is

$\sqrt{a^2\cos^2\beta + a^2\sin^2\beta} = a.$

9.4 Arc Length and Speed on a Curve

1 The arc length is $\int_1^2 \sqrt{1 + \left(\frac{3}{2}\sqrt{x}\right)^2}\, dx =$

$\int_1^2 \sqrt{1 + \frac{9}{4}x}\, dx$. To evaluate this integral, let $u = 1 + \frac{9}{4}x$, $du = \frac{9}{4}\,dx$. The arc length is

$\frac{4}{9}\int_{13/4}^{11/2} u^{1/2}\, du = \frac{4}{9}\cdot\frac{2}{3}u^{3/2}\Big|_{13/4}^{11/2} =$

$\frac{8}{27}\left[\left(\frac{11}{2}\right)^{3/2} - \left(\frac{13}{4}\right)^{3/2}\right] = \frac{22\sqrt{22} - 13\sqrt{13}}{27}.$

3 Note that $y = (e^x + e^{-x})/2 = \cosh x$ and $\dfrac{dy}{dx} = \sinh x$, so the arc length is $\int_0^b \sqrt{1 + \sinh^2 x}\, dx =$

$\int_0^b \cosh x\, dx = \sinh x\big|_0^b = \sinh b = \dfrac{e^b - e^{-b}}{2}.$

5 Arc length $= \int_0^{\pi/2} \sqrt{\left(\frac{dx}{dt}\right)^2 + \left(\frac{dy}{dt}\right)^2}\, dt$

$= \int_0^{\pi/2} \sqrt{(-3\cos^2 t\sin t)^2 + (3\sin^2 t\cos t)^2}\, dt$

$= \int_0^{\pi/2} \sqrt{9\cos^4 t\sin^2 t + 9\sin^4 t\cos^2 t}\, dt$

$= \int_0^{\pi/2} \sqrt{(9\cos^2 t\sin^2 t)(\cos^2 t + \sin^2 t)}\, dt$

$$= \int_0^{\pi/2} 3 \sin t \cos t \, dt = \frac{3}{2} \sin^2 t \Big|_0^{\pi/2} = \frac{3}{2}.$$

7 Arc length $= \int_0^{\pi} \sqrt{r^2 + (r')^2} \, d\theta$

$$= \int_0^{\pi} \sqrt{(1 + \cos\theta)^2 + (-\sin\theta)^2} \, d\theta$$

$$= \int_0^{\pi} \sqrt{1 + 2\cos\theta + \cos^2\theta + \sin^2\theta} \, d\theta$$

$$= \int_0^{\pi} \sqrt{2 + 2\cos\theta} \, d\theta$$

$$= \int_0^{\pi} 2\sqrt{\frac{1}{2}(1 + \cos\theta)} \, d\theta = \int_0^{\pi} 2\cos\frac{\theta}{2} \, d\theta$$

$$= 4 \sin\frac{\theta}{2}\Big|_0^{\pi} = 4.$$

9 The speed is $\frac{ds}{dt} = \sqrt{\left(\frac{dx}{dt}\right)^2 + \left(\frac{dy}{dt}\right)^2}$

$$= \sqrt{(50^2 + (-32t)^2} = \sqrt{2500 + 1024t^2}$$

$$= 2\sqrt{625 + 256t^2}.$$

11 The speed is $\frac{ds}{dt} = \sqrt{\left(\frac{dx}{dt}\right)^2 + \left(\frac{dy}{dt}\right)^2}$

$$= \sqrt{(1 - \sin t)^2 + (2 - \cos t)^2}$$

$$= \sqrt{1 - 2\sin t + \sin^2 t + 4 - 4\cos t + \cos^2 t}$$

$$= \sqrt{6 - 2\sin t - 4\cos t}.$$

13 (a)

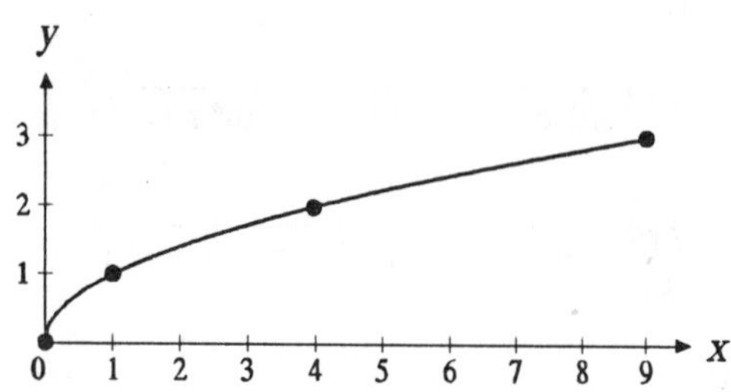

(b)

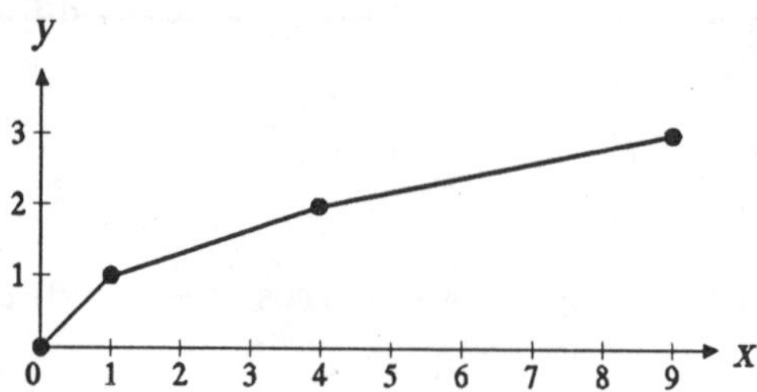

The inscribed polygonal arc contains three line segments, for a total length of

$$\sqrt{(1-0)^2 + (1-0)^2} + \sqrt{(4-1)^2 + (2-1)^2}$$

$$+ \sqrt{(9-4)^2 + (3-2)^2} = \sqrt{2} + \sqrt{10} + \sqrt{26}$$

$$\approx 9.6755.$$

(c) $\frac{dx}{dt} = 2t$ and $\frac{dy}{dt} = 1$, so the arc length is

$$\int_0^3 \sqrt{\left(\frac{dx}{dt}\right)^2 + \left(\frac{dy}{dt}\right)^2} \, dt = \int_0^3 \sqrt{4t^2 + 1} \, dt.$$

Since $y = t$, this integral could also be expressed in terms of y as $\int_0^3 \sqrt{4y^2 + 1} \, dy$. (The dummy variable of integration is really not important.)

(d) Arc length $\approx \frac{1}{2}\Big[\sqrt{4\cdot 0^2 + 1} + 2\sqrt{4\cdot 1^2 + 1}$

$$+ 2\sqrt{4\cdot 2^2 + 1} + \sqrt{4\cdot 3^2 + 1}\Big] =$$

$$\frac{1}{2}\left[1 + 2\sqrt{5} + 2\sqrt{17} + \sqrt{37}\right] \approx 9.9006$$

(e) Arc length $\approx \frac{1/2}{3}\Big[1 + 4\sqrt{2} + 2\sqrt{5} + 4\sqrt{10}$

$$+ 2\sqrt{17} + 4\sqrt{26} + \sqrt{37}\Big] \approx 9.7505$$

(f) $\int_0^3 \sqrt{4t^2 + 1} \, dt \approx 9.7471$, according to a

Hewlett-Packard 48SX calculator.

15 Arc length $= \int_0^\infty \sqrt{(e^{-3\theta})^2 + (-3e^{-3\theta})^2}\, d\theta =$

$$\sqrt{10}\int_0^\infty e^{-3\theta}\, d\theta = \sqrt{10}\cdot\left(-\frac{1}{3}e^{-3\theta}\right)\Big|_0^\infty = \frac{\sqrt{10}}{3}$$

17 The parameter here is y. Let $y = t$ and $x = f(t)$. Then t varies from c to d and arc length is

$$\int_c^d \sqrt{(dx/dt)^2 + (dy/dt)^2}\, dt = \int_c^d \sqrt{\left(\frac{dx}{dt}\right)^2 + 1}\, dt.$$

Now since $y = t$, $\dfrac{dx}{dt} = \dfrac{dx}{dy}$ and $dt = dy$. Thus

the integral is $\displaystyle\int_c^d \sqrt{1 + \left(\frac{dx}{dy}\right)^2}\, dy$.

19 (a)

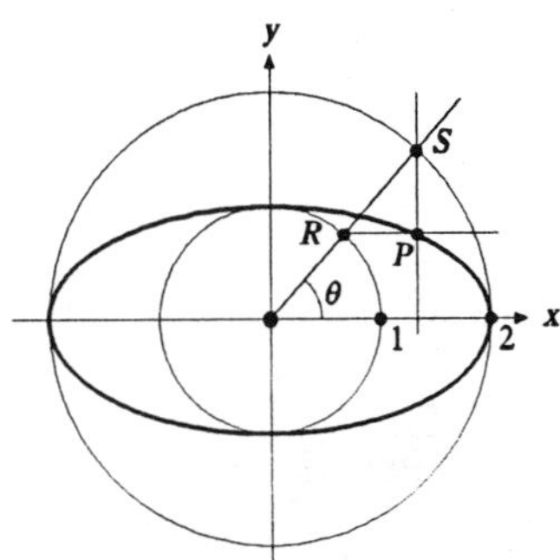

(b) Label the points R and S as in the figure. Observe that the coordinates of R are $(\cos\theta, \sin\theta)$, while the coordinates of S are $(2\cos\theta, 2\sin\theta)$. Since P has the same x coordinate as S and the same y coordinate as R, we have $P = (2\cos\theta, \sin\theta)$, as claimed.

(c) $\displaystyle\int_0^{2\pi} \sqrt{\left(\frac{dx}{dt}\right) + \left(\frac{dy}{dt}\right)^2}\, dt$

$$= \int_0^{2\pi} \sqrt{(-2\sin\theta)^2 + (\cos\theta)^2}\, d\theta$$

$$= \int_0^{2\pi} \sqrt{1 + 3\sin^2\theta}\, d\theta$$

(d) $\dfrac{x^2}{4} + \dfrac{y^2}{1} = \left(\dfrac{x}{2}\right)^2 + y^2$

$$= (\cos\theta)^2 + (\sin\theta)^2 = 1$$

21 (a) Here $x = \cos \pi t$ and $y = \sin \pi t$, so distance

$$= \int_1^2 \sqrt{(-\pi \sin \pi t)^2 + (\pi \cos \pi t)^2}\, dt$$

$$= \pi \int_1^2 dt = \pi.$$

(b) Speed $= \sqrt{(-\pi \sin \pi t)^2 + (\pi \cos \pi t)^2} = \pi$

(c)

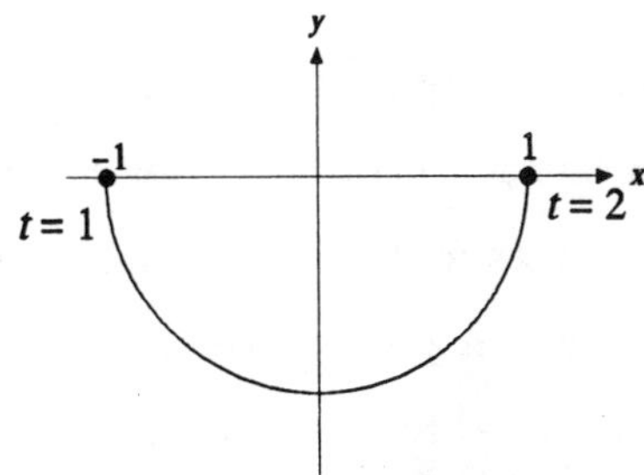

(d) $x^2 + y^2 = (\cos \pi t)^2 + (\sin \pi t)^2 = 1$

23 (a) The average value is $\dfrac{\int_0^\pi (1 + \cos\theta)\, d\theta}{\pi - 0} =$

$$\frac{1}{\pi}(\theta + \sin\theta)\Big|_0^\pi = 1.$$

(b) Note that $ds = \sqrt{r^2 + (r')^2}\, d\theta$

$$= \sqrt{(1 + \cos\theta)^2 + (-\sin\theta)^2}\, d\theta$$

$= \sqrt{2 + 2\cos\theta}\, d\theta$. Hence the average is

$$\frac{\int_a^b f(s)\, ds}{\int_a^b ds} = \frac{\int_0^\pi (1 + \cos\theta)\sqrt{2 + 2\cos\theta}\, d\theta}{\int_0^\pi \sqrt{2 + 2\cos\theta}\, d\theta}$$

$$= \frac{\int_0^\pi (1 + \cos\theta)^{3/2}\, d\theta}{\int_0^\pi (1 + \cos\theta)^{1/2}\, d\theta}$$

$$= \frac{\int_0^\pi \left(2\cos^2\frac{\theta}{2}\right)^{3/2} d\theta}{\int_0^\pi \left(2\cos^2\frac{\theta}{2}\right)^{1/2} d\theta} = \frac{2\int_0^\pi \cos^3\frac{\theta}{2}\,d\theta}{\int_0^\pi \cos\frac{\theta}{2}\,d\theta}$$

$$= \frac{4\int_0^\pi \left(\cos^3\frac{\theta}{2}\right)\frac{1}{2}\,d\theta}{2\sin\frac{\theta}{2}\Big|_0^\pi}$$

$$= 2\left(\sin\frac{\theta}{2} - \frac{1}{3}\sin^3\frac{\theta}{2}\right)\Big|_0^\pi = \frac{4}{3}.$$ (See Formula 49 from the table of antiderivatives or Exercise 147 of the Chapter 7 Review Exercises for the antiderivative of $\cos^3 x$.)

25 (a) The arc length of the curve $r = f(\theta)$ for θ in the interval $[a, t]$ is $s(t) =$

$$\int_a^t \sqrt{[f(\theta)]^2 + [f'(\theta)]^2}\,d\theta.$$

(b) Using the first fundamental theorem of calculus, we have $\frac{ds}{dt}$

$$= \frac{d}{dt}\left(\int_a^t \sqrt{[f(\theta)]^2 + [f'(\theta)]^2}\,d\theta\right)$$

$$= \sqrt{[f(t)]^2 + [f'(t)]^2}.$$

29 Assume $m \neq 1$. Arc length $=$

$\int \sqrt{1 + (mx^{m-1})^2}\,dx = \int \sqrt{1 + m^2x^{2m-2}}\,dx$. Let $t = m^2x^{2m-2}$. Then the integral becomes

$\int \sqrt{1 + t}\cdot c\cdot t^{\frac{1}{2m-2}-1}\,dt$, where $c =$

$\frac{1}{2(m-1)m^{1/(m-1)}}$. By Chebyshev's theorem, this is elementary if and only if either $\frac{1}{2m-2}$ or

$\frac{1}{2m-2} + \frac{1}{2}$ is an integer; that is, if and only if

$\frac{1}{m-1}$ is an integer. Thus if $m \neq 1$, m must have the form $1 + 1/n$ for some integer n. If $m = 1$, then $y = x$ and arc length involves the antiderivative $\int \sqrt{1 + 1^2}\,dx$, which is easily evaluated.

9.5 The Area of a Surface of Revolution

1 $R = y = x^3$ and $ds =$

$\sqrt{1 + \left(\frac{dy}{dx}\right)^2}\,dx =$

$\sqrt{1 + 9x^4}\,dx$, so the surface area is $\int_a^b 2\pi x^3\,ds =$

$\int_1^2 2\pi x^3\sqrt{1 + 9x^4}\,dx.$

3 $R = x = y^{1/3}$ and $ds =$

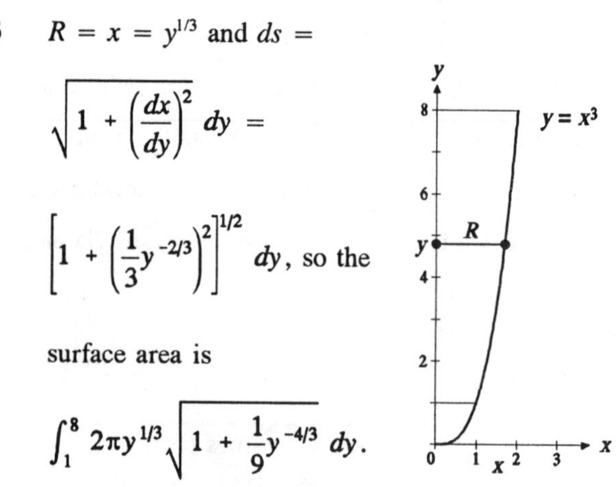

$\sqrt{1 + \left(\frac{dx}{dy}\right)^2}\,dy =$

$\left[1 + \left(\frac{1}{3}y^{-2/3}\right)^2\right]^{1/2} dy$, so the surface area is

$\int_1^8 2\pi y^{1/3}\sqrt{1 + \frac{1}{9}y^{-4/3}}\,dy.$

5 $R = y = e^x$ and $ds = \sqrt{1 + (e^x)^2}\,dx$; the surface

area is

$\int_0^1 2\pi e^x\sqrt{1+(e^x)^2}\,dx$. Let

$u = e^x$; then $du = e^x\,dx$.

When $x = 0$, $u = 1$; when

$x = 1$, $u = e$. Then the

surface area is

$$2\pi\int_1^e \sqrt{1+u^2}\,du =$$

$$2\pi\cdot\frac{1}{2}\left[u\sqrt{1+u^2} + \ln\left|u+\sqrt{1+u^2}\right|\right]\Big|_1^e =$$

$$\pi\left[e\sqrt{1+e^2} + \ln\left(e+\sqrt{1+e^2}\right) - \sqrt{2} - \ln\left(1+\sqrt{2}\right)\right].$$

7

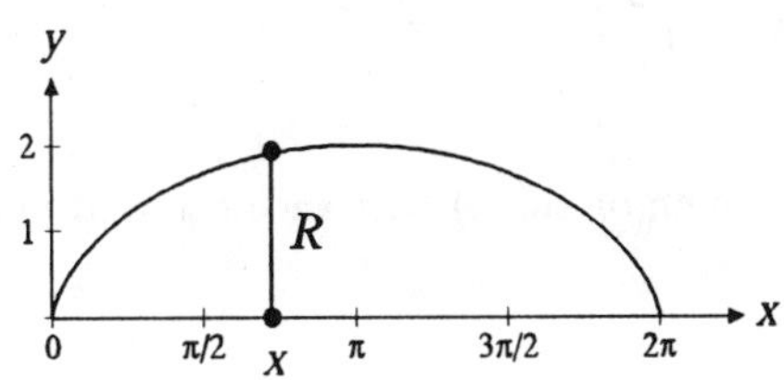

Since the cycloid is being revolved about the x axis, $R = y = 1 - \cos\theta$. With θ as parameter, ds

$$= \sqrt{\left(\frac{dx}{d\theta}\right)^2 + \left(\frac{dy}{d\theta}\right)^2}\,d\theta =$$

$$\sqrt{(1-\cos\theta)^2 + (\sin\theta)^2}\,d\theta = \sqrt{2 - 2\cos\theta}\,d\theta$$

$= \sqrt{2}\sqrt{1-\cos\theta}\,d\theta$. The surface area is then

$$\int_0^{2\pi} 2\pi(1-\cos\theta)\sqrt{2}\sqrt{1-\cos\theta}\,d\theta =$$

$$2\sqrt{2}\pi\int_0^{2\pi}\left(2\sin^2\frac{\theta}{2}\right)^{3/2} d\theta = 8\pi\int_0^{2\pi}\sin^3\frac{\theta}{2}\,d\theta$$

$$= 8\pi\left(-2\cos\frac{\theta}{2} + \frac{2}{3}\cos^3\frac{\theta}{2}\right)\Big|_0^{2\pi} = \frac{64\pi}{3}.$$

9 $\int_0^1 2\pi\cdot 2x^3\sqrt{1+(6x^2)^2}\,dx =$

$4\pi\int_0^1 x^3\sqrt{1+36x^4}\,dx$. Let $u = 1 + 36x^4$.

11 $\int_1^2 2\pi\cdot x^2\sqrt{1+(2x)^2}\,dx =$

$2\pi\int_1^2 x^2\sqrt{1+4x^2}\,dx$. Let $x = \frac{1}{2}\tan\theta$.

13 $\int_1^8 2\pi(x^{2/3}-1)\sqrt{1+\left(\frac{2}{3}x^{-1/3}\right)^2}\,dx =$

$2\pi\int_1^8 (x^{2/3}-1)\sqrt{1+\frac{4}{9}x^{-2/3}}\,dx$. Let $x = u^{3/2}$ and

$u = v - 4/9$.

15 $\frac{dy}{dx} = x^2 - \frac{1}{4x^2}$, so $\frac{ds}{dx} = \sqrt{1+\left(x^2-\frac{1}{4x^2}\right)^2}$

$= x^2 + \frac{1}{4x^2}$, and the area is

$\int_1^2 2\pi\left(\frac{1}{3}x^3 + \frac{1}{4x} + 1\right)\left(x^2 + \frac{1}{4x^2}\right)dx$. Expand

and use the power rule.

17 Let a be the radius of the sphere.

(a) The volume of the sphere is $\frac{4}{3}\pi a^3$. Since the diameter and the height of the cylinder are $2a$, the volume of the cylinder is $\pi r^2 h = \pi a^2\cdot 2a = 2\pi a^3$. The volume of the sphere is 2/3 that of the cylinder.

(b) The surface area of the sphere is $4\pi a^2$. The surface area of the can, excluding the top and bottom is $2\pi rh = 2\pi\cdot a\cdot 2a = 4\pi a^2$. The areas are equal.

19 $\int_1^3 2\pi x^{1/4}\left(1+\left(\frac{1}{4}x^{-3/4}\right)^2\right)^{1/2} dx$

$= \int_1^3 2\pi x^{1/4} \cdot \frac{1}{4}(16 + x^{-3/2})^{1/2}\, dx$

$= \frac{\pi}{2} \int_1^3 x^{1/4}(16 + x^{-3/2})^{1/2}\, dx \approx 15.0065$, by calculator.

21 Since the curve is being revolved about the polar axis (that is, the x axis), $R = y = r \sin\theta$. As shown in Sec. 9.4, $ds = \sqrt{r^2 + (r')^2}\, d\theta$. Thus

$\int_a^b 2\pi R\, ds = \int_\alpha^\beta 2\pi r \sin\theta \sqrt{r^2 + (r')^2}\, d\theta.$

23 $r = a$, $\alpha = 0$, and $\beta = \pi$, so the area is

$\int_0^\pi 2\pi a \sin\theta \sqrt{a^2 + 0^2}\, d\theta = 2\pi a^2 \int_0^\pi \sin\theta\, d\theta$

$= 4\pi a^2.$

25 $r = \sin 2\theta$, so $r' = 2\cos 2\theta$ and $r^2 + (r')^2 = \sin^2 2\theta + 4\cos^2 2\theta = 1 + 3\cos^2 2\theta$. By Exercise 21, the area is

$\int_0^{\pi/2} 2\pi \sin 2\theta \sin\theta \sqrt{1 + 3\cos^2 2\theta}\, d\theta.$

27 The curve $y = \frac{b}{a}\sqrt{a^2 - x^2}$ for $-a \le x \le a$ is revolved about the x axis. Since $\frac{dy}{dx} = -\frac{bx}{a\sqrt{a^2 - x^2}}$, $\frac{ds}{dx} = \frac{\sqrt{a^4 - (a^2 - b^2)x^2}}{a\sqrt{a^2 - x^2}}$ and the area is $\int_{-a}^a 2\pi \cdot \frac{b}{a}\sqrt{a^2 - x^2} \cdot \frac{\sqrt{a^4 - (a^2 - b^2)x^2}}{a\sqrt{a^2 - x^2}}\, dx$

$= \frac{2\pi b}{a^2} \int_{-a}^a \sqrt{a^4 - (a^2 - b^2)x^2}\, dx =$

$2\pi\left[b^2 + \frac{a^2 b}{\sqrt{a^2 - b^2}} \sin^{-1}\frac{\sqrt{a^2 - b^2}}{a}\right]$. As $b \to a^-$,

$\frac{a}{\sqrt{a^2 - b^2}} \sin^{-1}\frac{\sqrt{a^2 - b^2}}{a} \to 1$, so the area approaches $2\pi[a^2 + a^2] = 4\pi a^2$.

29 The area of the sector is $\frac{\theta}{2\pi} \cdot \pi l^2 = \frac{2\pi r}{l} \cdot \frac{l^2}{2} = \pi r l$. So the (lateral) surface of the cone is $\pi r l$.

31 Since $\frac{d}{dr}\left(\frac{4}{3}\pi r^3\right) = 4\pi r^2$, this implies that the increase (or decrease) in the volume of the sphere is directly proportional to its surface area. This is clearly not a coincidence.

33 Area $= \int_a^b 2\pi y\, ds = s \cdot 2\pi \cdot \frac{\int_a^b y\, ds}{s} = s \cdot 2\pi \bar{y}$

= (Length of curve)·(Distance centroid moves)

35

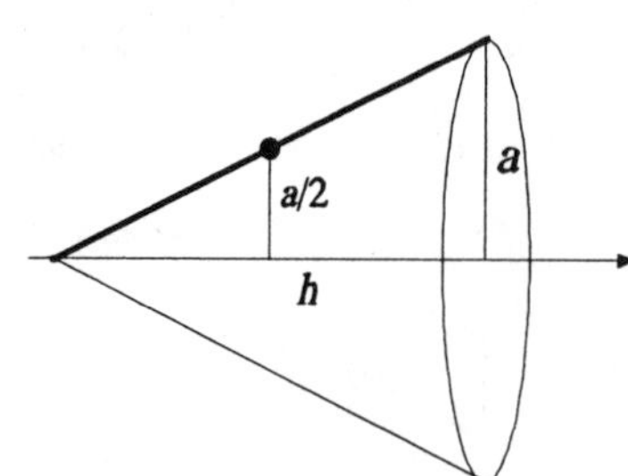

The cone is obtained by revolving a line segment of length $\sqrt{a^2 + h^2}$ about an axis passing through one end of the segment and at a distance a from the other end. The centroid of the line segment (its midpoint) is at a distance $a/2$ from the axis, so the area is $\sqrt{a^2 + h^2} \cdot 2\pi \cdot \frac{a}{2} = \pi a\sqrt{a^2 + h^2}$.

9.6 Curvature

1 Here $y = x^2$, $y' = 2x$, and $y'' = 2$. The curvature at (1, 1) is $\kappa = \left.\frac{|y''|}{[1 + (y')^2]^{3/2}}\right|_{x=1} = \frac{2}{(1 + 4)^{3/2}} = \frac{2}{5\sqrt{5}}$; the radius of curvature is $\frac{1}{\kappa} = \frac{5\sqrt{5}}{2}$.

3 We have $y = e^{-x}$, $y' = -e^{-x}$, and $y'' = e^{-x}$. The curvature at $(1, 1/e)$ is $\kappa = \left.\frac{|y''|}{[1 + (y')^2]^{3/2}}\right|_{x=1} = \frac{e^{-1}}{(1 + e^{-2})^{3/2}}$. The radius of curvature is $\frac{1}{\kappa} = \frac{(1 + e^{-2})^{3/2}}{e^{-1}} = e(1 + e^{-2})^{3/2}$.

5 Here we have $y = \tan x$, $y' = \sec^2 x$, and $y'' = 2 \sec x (\sec x \tan x) = 2 \sec^2 x \tan x$. The curvature at $(\pi/4, 1)$ is $\kappa = \left.\frac{|y''|}{[1 + (y')^2]^{3/2}}\right|_{x=\pi/4} = \frac{2 \cdot 2}{(1 + 2^2)^{3/2}} = \frac{4}{5\sqrt{5}}$. The radius of curvature is $1/\kappa = \frac{5\sqrt{5}}{4}$.

7 We have $x = 2 \cos 3t$, $\dot{x} = -6 \sin 3t$, and $\ddot{x} = -18 \cos 3t$; $y = 2 \sin 3t$, $\dot{y} = 6 \cos 3t$, and $\ddot{y} = -18 \sin 3t$. The curvature when $t = 0$ is $\left.\frac{|\dot{x}\ddot{y} - \dot{y}\ddot{x}|}{[(\dot{x})^2 + (\dot{y})^2]^{3/2}}\right|_{t=0} = \frac{|0 - 6(-18)|}{[0^2 + 6^2]^{3/2}} = \frac{6 \cdot 18}{(36)^{3/2}} = \frac{1}{2}$. (This answer is not surprising, since the curve is a circle of radius 2.)

9 $x = e^{-t} \cos t$

$\dot{x} = -e^{-t} \cos t - e^{-t} \sin t$

$\ddot{x} = e^{-t} \cos t + e^{-t} \sin t + e^{-t} \sin t - e^{-t} \cos t = 2e^{-t} \sin t$

$y = e^{-t} \sin t$

$\dot{y} = -e^{-t} \sin t + e^{-t} \cos t$

$\ddot{y} = e^{-t} \sin t - e^{-t} \cos t - e^{-t} \cos t - e^{-t} \sin t = -2e^{-t} \cos t$

$\dot{x}\ddot{y} - \ddot{x}\dot{y} = (-e^{-t} \cos t - e^{-t} \sin t)(-2e^{-t} \cos t) - (2e^{-t} \sin t)(-e^{-t} \sin t + e^{-t} \cos t) = 2e^{-2t}$

$\dot{x}^2 + \dot{y}^2 = e^{-2t}(\cos t + \sin t)^2 + e^{-2t}(\cos t - \sin t)^2 = 2e^{-2t}$

The curvature at $t = \pi/6$ is $\left.\frac{2e^{-2t}}{(2e^{-2t})^{3/2}}\right|_{t=\pi/6} = \left.\frac{e^{t}}{\sqrt{2}}\right|_{t=\pi/6} = \frac{e^{\pi/6}}{\sqrt{2}}$.

11 (a) $y = \frac{e^x + e^{-x}}{2} = \cosh x$, $y' = \frac{e^x - e^{-x}}{2} = \sinh x$, $y'' = \frac{e^x + e^{-x}}{2} = \cosh x$. Now the curvature is $\kappa = \frac{|y''|}{[1 + (y')^2]^{3/2}} = \frac{\cosh x}{[1 + \sinh^2 x]^{3/2}} = \frac{\cosh x}{[\cosh^2 x]^{3/2}} = \frac{1}{\cosh^2 x} = \frac{4}{(e^x + e^{-x})^2}$. Then the radius of curvature is $\cosh^2 x = \frac{(e^x + e^{-x})^2}{4}$.

(b) As we saw in (a), the radius of curvature is

$\cosh^2 x = y^2$.

13 Since $y = \frac{dy}{dx} = \frac{d^2y}{dx^2} = e^x$, the curvature is $\kappa =$

$\frac{|e^x|}{(1 + (e^x)^2)^{3/2}} = \frac{y}{(1 + y^2)^{3/2}}$ and the radius of

curvature is $\frac{1}{\kappa} = \frac{(1 + y^2)^{3/2}}{y}$. Hence $\frac{d}{dy}\left(\frac{1}{\kappa}\right) =$

$$\frac{y \cdot \frac{3}{2}(1 + y^2)^{1/2} \cdot 2y - (1 + y^2)^{3/2} \cdot 1}{y^2} =$$

$\frac{(2y^2 - 1)(1 + y^2)^{1/2}}{y^2}$ is negative for $0 < y < \frac{1}{\sqrt{2}}$

and positive for $y > \frac{1}{\sqrt{2}}$. The radius of curvature

is minimal for $y = \frac{1}{\sqrt{2}}$; that is, for $x = \ln \frac{1}{\sqrt{2}} =$

$-\frac{1}{2} \ln 2$.

15 (a) If the tangent at some point is parallel to the x axis, then $y' = 0$ at that point; hence $\kappa =$

$$\frac{|y''|}{[1 + (y')^2]^{3/2}} = \frac{|y''|}{1^{3/2}} = |y''|.$$

(b) Since $1 + (y')^2 \geq 1$, $\frac{1}{[1 - (y')^2]^{3/2}} \leq 1$.

Hence $\kappa = \frac{|y''|}{[1 + (y')^2]^{3/2}} = |y''|\frac{1}{[1 + (y')^2]^{3/2}}$

$\leq |y''|$.

17 Notice that at the point (1000, 0) the tangent line is vertical. Hence the curvature cannot be calculated using the formula $\frac{|y''|}{[1 + (y')^2]^{3/2}}$. With y as the parameter (using Theorem 3 of Sec. 9.6), the curvature is $\frac{|-x''|}{[1 + (x')^2]^{3/2}}$, where derivatives are with respect to y. Now, $\frac{x^2}{1000^2} + \frac{y^2}{500^2} = 1$, so $\frac{2xx'}{1000^2} + \frac{2y}{500^2} = 0$ by implicit differentiation. At (1000, 0), we see that $x' = 0$. Once again using implicit differentiation, $\frac{2[xx'' + (x')^2]}{1000^2} + \frac{2}{500^2} = 0$. At (1000, 0) then, $x'' = -\frac{1}{250}$. The curvature at (1000, 0) is thus $\frac{1/250}{1} = \frac{1}{250}$ and the radius of curvature is $R = 250$. Hence $\tan A = \frac{88^2}{32 \cdot 250}$ $= 0.968$ and $A = \tan^{-1} 0.968 \approx 0.77$ radians ($\approx$ 44°). At (0, 500), the standard formula for curvature applies: $\frac{x^2}{1000^2} + \frac{y^2}{500^2} = 1$, so $\frac{2x}{1000^2} + \frac{2yy'}{500^2} = 0$, and $\frac{2}{1000^2} + \frac{2[yy'' + (y')^2]}{500^2}$ $= 0$, so $y' = 0$ and $y'' = -\frac{1}{2000}$ at (0, 500). The curvature is then $\frac{1/2000}{1}$ and the radius of curvature is $R = 2000$. Hence $\tan A = \frac{88^2}{32 \cdot 2000}$

$= 0.121$ and $A = \tan^{-1} 0.121 \approx 0.12$ radians ($\approx 7°$).

19 $x = a\cos\theta$, $\dot{x} = -a\sin\theta$, $\ddot{x} = -a\cos\theta$, $y = b\sin\theta$, $\dot{y} = b\cos\theta$, $\ddot{y} = -b\sin\theta$, $\dot{x}\ddot{y} - \ddot{x}\dot{y} = ab$, $\dot{x}^2 + \dot{y}^2 = a^2\sin^2\theta + b^2\cos^2\theta = \dfrac{a^4y^2 + b^4x^2}{a^2b^2}$, so the radius of curvature is

$$\frac{(a^4y^2 + b^4x^2)^{3/2}}{a^4b^4}.$$

21 Using the formula from Exercise 19 with $a = 3$ and $b = 2$, at $(\pm 3, 0)$ we have $R = \dfrac{(3^4\cdot 0^2 + 2^4\cdot 3^2)^{3/2}}{3^4 2^4} = \dfrac{2^6 3^3}{3^4 2^4} = 4/3$, while at $(0, \pm 2)$, $R = \dfrac{(3^4\cdot 2^2 + 2^4\cdot 0^2)^{3/2}}{3^4 2^4} = \dfrac{3^6 2^3}{3^4 2^4} = 9/2$.

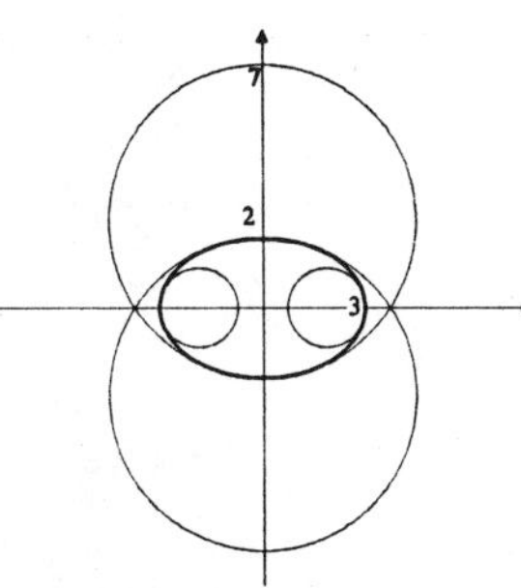

23 Here $x = r\cos\theta = (1 + \cos\theta)\cos\theta = \cos\theta + \cos^2\theta$ and $y = r\sin\theta = (1 + \cos\theta)\sin\theta = \sin\theta + \sin\theta\cos\theta$. Therefore $\dfrac{dx}{d\theta} = -\sin\theta - 2\cos\theta\sin\theta = -\sin\theta - \sin 2\theta$, $\dfrac{d^2x}{d\theta^2} = -\cos\theta - 2\cos 2\theta$, $\dfrac{dy}{d\theta} = \cos\theta + (\cos^2\theta - \sin^2\theta) = \cos\theta + \cos 2\theta$, and $\dfrac{d^2y}{d\theta^2} = -\sin\theta - 2\sin 2\theta$. Using Theorem 3 with parameter θ, we have $\kappa = \dfrac{|\dot{x}\ddot{y} - \dot{y}\ddot{x}|}{[(\dot{x})^2 + (\dot{y})^2]^{3/2}} =$

$$\frac{|(-\sin\theta - \sin 2\theta)(-\sin\theta - 2\sin 2\theta) - (\cos\theta + \cos 2\theta)(-\cos\theta - 2\cos 2\theta)|}{[(-\sin\theta - \sin 2\theta)^2 + (\cos\theta + \cos 2\theta)^2]^{3/2}}$$

$$= \frac{|\sin^2\theta + 3\sin 2\theta\sin\theta + 2\sin^2 2\theta + \cos^2\theta + 3\cos 2\theta\cos\theta + 2\cos^2 2\theta|}{[\sin^2\theta + 2\sin 2\theta\sin\theta + \sin^2 2\theta + \cos^2\theta + 2\cos 2\theta\cos\theta + \cos^2 2\theta]^{3/2}}$$

$$= \frac{3\,|1 + \sin 2\theta\sin\theta + \cos 2\theta\cos\theta|}{2^{3/2}\,|1 + \sin 2\theta\sin\theta + \cos 2\theta\cos\theta|^{3/2}}$$

$$= \frac{3\sqrt{2}}{4(1 + 2\sin^2\theta\cos\theta + \cos^3\theta - \sin^2\theta\cos\theta)^{1/2}}$$

$$= \frac{3\sqrt{2}}{4(1 + \sin^2\theta\cos\theta + \cos^3\theta)^{1/2}}$$

$$= \frac{3\sqrt{2}}{4(1 + \cos\theta)^{1/2}} = \frac{3\sqrt{2}}{4\sqrt{r}}.$$

25 Since $\dfrac{dy}{dx} = y^3$, $\dfrac{d^2y}{dx^2} = 3y^2\cdot\dfrac{dy}{dx} = 3y^5$ and $\kappa = \dfrac{|3y^5|}{[1 + (y^3)^2]^{3/2}} = \dfrac{3|y|^5}{(1 + y^6)^{3/2}}$.

27 The wheel isn't really rotating about its point of contact with the ground. Compare the motion of pivoting the wheel about its point of contact with the motion that actually occurs. The wheel cannot pivot into the ground.

9.7 The Reflection Properties of the Conic Sections

1 The angle between the two lines is $\dfrac{3\pi}{4} - \dfrac{\pi}{4} = \dfrac{\pi}{2}$.

3 Here $m = -3$ and $m' = 2$, so $\tan\theta = \dfrac{m - m'}{1 + mm'}$

$= \dfrac{-3 - 2}{1 + (-3)(2)} = \dfrac{-5}{-5} = 1.$

5 Here $m = -2$ and $m' = -3$, so $\tan\theta =$

$\dfrac{m - m'}{1 + mm'} = \dfrac{-2 - (-3)}{1 + (-2)(-3)} = \dfrac{1}{7}.$

7 The angle between the curves at $\left(\dfrac{\pi}{4}, \dfrac{\sqrt{2}}{2}\right)$ is the angle between the two tangent lines at that point. All that is necessary is to compute the slopes of the tangent lines and apply the formula. The slope of a tangent line at a point is simply $\dfrac{dy}{dx}$ evaluated at that point. For $y = \sin x$, $\dfrac{dy}{dx} = \cos x$. Hence the slope of the tangent line is $\cos\dfrac{\pi}{4} = \dfrac{\sqrt{2}}{2}$. For $y = \cos x$, $\dfrac{dy}{dx} = -\sin x$ and the slope of the tangent line is $-\sin\dfrac{\pi}{4} = -\dfrac{\sqrt{2}}{2}$. Since a line of slope $-\dfrac{\sqrt{2}}{2}$ has a larger angle of inclination than a line of slope $\dfrac{\sqrt{2}}{2}$, $m = -\dfrac{\sqrt{2}}{2}$, and $m' = \dfrac{\sqrt{2}}{2}$; hence $\tan\theta =$

$\dfrac{m - m'}{1 + mm'} = -2\sqrt{2}.$

9 The curve $y = e^x$ has slope at (0, 1) given by $y'(0) = e^0 = 1$. The curve $y = e^{-x}$ has slope at (0, 1) given by $y'(0) = -e^{-0} = -1$. Then $\tan\theta =$

$\dfrac{(-1) - 1}{1 + (-1)(1)}$, which is undefined, so $\theta = \pi/2$; the curves are perpendicular.

11 We have $\tan\alpha = \dfrac{2(cx)^{1/2}x^{1/2} - c^{1/2}x + c^{3/2}}{x^{3/2} - cx^{1/2} + 2(cx)^{1/2}c^{1/2}}$

$= \dfrac{2c^{1/2}x - c^{1/2}x + c^{3/2}}{x^{3/2} - cx^{1/2} + 2cx^{1/2}} = \dfrac{c^{1/2}x + c^{3/2}}{x^{3/2} + cx^{1/2}}$

$= \dfrac{c^{1/2}(x + c)}{x^{1/2}(x + c)} = \dfrac{c^{1/2}}{x^{1/2}}$, as was to be shown.

13

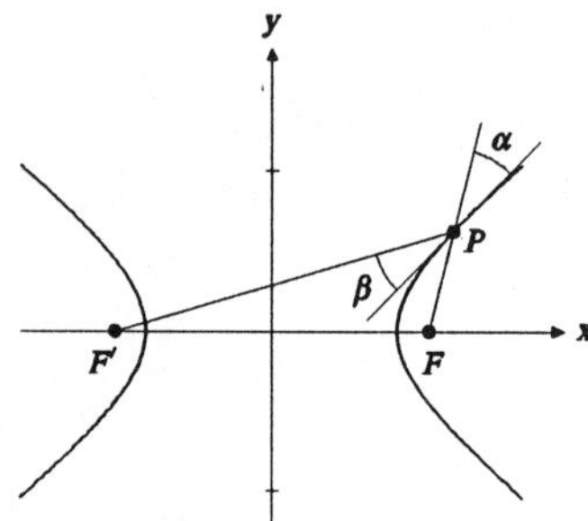

Position the axes as in the figure. The hyperbola has the equation $\dfrac{x^2}{a^2} - \dfrac{y^2}{b^2} = 1$. Its foci are $F = (c, 0)$ and $F' = (-c, 0)$, where $c^2 = a^2 + b^2$; its vertices are $V = (a, 0)$ and $V' = (-a, 0)$. Without loss of generality, suppose $P = (x, y)$ is on the right half of the hyperbola. Then the slopes of FP and $F'P$ are $\dfrac{y}{x - c}$ and $\dfrac{y}{x + c}$, respectively. To find the slope of the tangent at P, differentiate implicitly: $\dfrac{2x}{a^2} - \dfrac{2y\dfrac{dy}{dx}}{b^2} = 0$; so the slope is $\dfrac{dy}{dx} = \dfrac{b^2x}{a^2y}$. By equation (1), $\tan\alpha = \dfrac{\dfrac{y}{x - c} - \dfrac{b^2x}{a^2y}}{1 + \dfrac{y}{x - c}\dfrac{b^2x}{a^2y}}$

$= \dfrac{ya^2y - b^2x(x-c)}{a^2y(x-c) + b^2xy} = \dfrac{b^2cx - (b^2x^2 - a^2y^2)}{(a^2+b^2)xy - a^2cy} =$

$\dfrac{b^2cx - a^2b^2}{c^2xy - a^2cy} = \dfrac{b^2(cx - a^2)}{cy(cx - a^2)} = \dfrac{b^2}{cy}$. Similarly,

$$\tan\beta = \frac{\dfrac{b^2x}{a^2y} - \dfrac{y}{x+c}}{1 + \dfrac{b^2x}{a^2y}\dfrac{y}{x+c}} = \frac{b^2x^2 - a^2y^2 + b^2cx}{a^2xy + a^2cy + b^2xy}$$

$= \dfrac{a^2b^2 + b^2cx}{c^2xy + a^2cy} = \dfrac{b^2(a^2 + cx)}{cy(a^2 + cx)} = \dfrac{b^2}{cy}$. Hence

$\tan\alpha = \tan\beta$, so $\alpha = \beta$.

15 $f(\theta) = f'(\theta) = e^\theta$, so $\tan\gamma = e^\theta/e^\theta = 1$ and $\gamma = \pi/4$.

17 If $\gamma = \theta$, then $\tan\gamma = \dfrac{r}{dr/d\theta} = \tan\theta$. Thus $r' = r\cot\theta$ and $r' - r\cot\theta = 0$. Multiplying this equation through by $\csc\theta$, $r'\csc\theta - r\csc\theta\cot\theta = 0$ and $(r\csc\theta)' = 0$. Hence $r\csc\theta = a$, so $r = a\sin\theta$ for some constant a.

19 Since $\tan\gamma = \dfrac{f(\theta)}{f'(\theta)}$, $f(\theta) = (\tan\gamma)f'(\theta)$. If γ is constant, then so is $k = \tan\gamma$. Hence $f(\theta)$ satisfies the linear differential equation $f(\theta) = kf'(\theta)$, so $f(\theta) = Ce^{k\theta}$, for some constant C.

9.S Guide Quiz

1 The curve $r = \cos 5\theta$ is a five-leaved rose. One loop lies between $\theta = -\dfrac{\pi}{10}$ and $\theta = \dfrac{\pi}{10}$. Its area is $\int_{-\pi/10}^{\pi/10} \dfrac{1}{2}r^2\, d\theta = \dfrac{1}{2}\int_{-\pi/10}^{\pi/10} \cos^2 5\theta\, d\theta$.

2 The curve length is $\int_1^2 \sqrt{1 + \left(\dfrac{dy}{dx}\right)^2}\, dx =$

$\int_1^2 \sqrt{1 + (4x^3)^2}\, dx = \int_1^2 \sqrt{1 + 16x^6}\, dx$.

3 $R = 20 - y = 20 - x^4$, so the surface area is

$\int_a^b 2\pi R\, ds = \int_1^2 2\pi(20 - x^4)\sqrt{1 + 16x^6}\, dx$.

4 If the curve of Exercise 2 is revolved about the line $x = 1$, then $R = x - 1$ and the surface area is

$\int_a^b 2\pi R\, ds = \int_1^2 2\pi(x-1)\sqrt{1 + 16x^6}\, dx$.

5 The curve length is $\int_0^{2\pi} \sqrt{r^2 + (r')^2}\, d\theta$

$= \int_0^{2\pi} \sqrt{(2 + \sin\theta)^2 + \cos^2\theta}\, d\theta$

$= \int_0^{2\pi} \sqrt{4 + 4\sin\theta + \sin^2\theta + \cos^2\theta}\, d\theta$

$= \int_0^{2\pi} \sqrt{5 + 4\sin\theta}\, d\theta$.

6 Here $r = y = r\sin\theta = (2 + \sin\theta)\sin\theta$. Hence the surface area is $\int_a^b 2\pi R\, ds =$

$\int_0^\pi 2\pi(2 + \sin\theta)\sin\theta\sqrt{5 + 4\sin\theta}\, d\theta$. (Note that ds was computed in Exercise 5.)

7 The curve length is $\int_0^1 \sqrt{\left(\dfrac{dx}{dt}\right)^2 + \left(\dfrac{dy}{dt}\right)^2}\, dt$

$= \int_0^1 \sqrt{(10t)^2 + \left(\dfrac{1}{2}t^{-1/2}\right)^2}\, dt$

$= \int_0^1 \sqrt{100t^2 + \dfrac{1}{4}t^{-1}}\, dt$.

8 Speed $= \sqrt{\left(\dfrac{dx}{dt}\right)^2 + \left(\dfrac{dy}{dt}\right)^2} = \sqrt{12^2 + (-32t + 5)^2}$.

Thus when $t = 1$, the speed is $\sqrt{144 + 27^2} = \sqrt{873}$.

9 Since $\sec^2 t = \tan^2 t + 1$, the equation is $y^2 = x^2 + 1$.

10 (a) We have $x = r\cos\theta = \cos 2\theta \cos\theta$, so $\frac{dx}{d\theta} = (\cos 2\theta)(-\sin\theta) + (\cos\theta)(-2\sin 2\theta) = -\cos 2\theta \sin\theta - 2\sin 2\theta\cos\theta$ and $y = r\sin\theta = \cos 2\theta \sin\theta$, so $\frac{dy}{d\theta} = (\cos 2\theta)(\cos\theta) + (\sin\theta)(-2\sin 2\theta) = \cos 2\theta\cos\theta - 2\sin 2\theta\sin\theta$. Hence $\frac{dy}{dx} = \frac{dy/d\theta}{dx/d\theta} = -\frac{\cos 2\theta\cos\theta - 2\sin 2\theta\sin\theta}{\cos 2\theta\sin\theta + 2\sin 2\theta\cos\theta}$.

At $\theta = \pi/8$, we have $\frac{dy}{dx} = -\frac{\cos\pi/4\cos\pi/8 - 2\sin\pi/4\sin\pi/8}{\cos\pi/4\sin\pi/8 + 2\sin\pi/4\cos\pi/8}$. Since $\cos\pi/4 = \sin\pi/4$, we can cancel out this quantity, producing $-\frac{\cos\pi/8 - 2\sin\pi/8}{\sin\pi/8 + 2\cos\pi/8}$.

Upon dividing through by $\cos\pi/8$, we obtain $\frac{dy}{dx} = -\frac{1 - 2\tan\pi/8}{\tan\pi/8 + 2} \approx -0.0711$. Hence $\tan\alpha = -0.0711$, where α is the angle of inclination. Since α must lie between 0 and π, we have $\alpha = \pi + \tan^{-1}(-0.0711) \approx 3.07 \approx 176°$.

As an alternative, we could have used the approach of Exercise 14 of Sec. 9.7. First find γ: $\tan\gamma = \frac{r}{dr/d\theta} = \frac{\cos 2\theta}{-2\sin 2\theta} = -\frac{1}{2}\cot 2\theta$. When $\theta = \frac{\pi}{8}$, $\tan\gamma = -\frac{1}{2}\cot\frac{\pi}{4} = -\frac{1}{2}$. Thus γ is between $\pi/2$ and π and has a tangent of $-\frac{1}{2}$. So $\gamma = \pi - \tan^{-1}\frac{1}{2}$. Finally, the angle of inclination is $\theta + \gamma = \frac{\pi}{8} + \left(\pi - \tan^{-1}\frac{1}{2}\right) = \frac{9\pi}{8} - \tan^{-1}\frac{1}{2} \approx 3.07 \text{ radians} \approx 176°$.

(b) Note that at $\theta = \pi/8$ the tangent is very nearly horizontal, as expected from the result in (a).

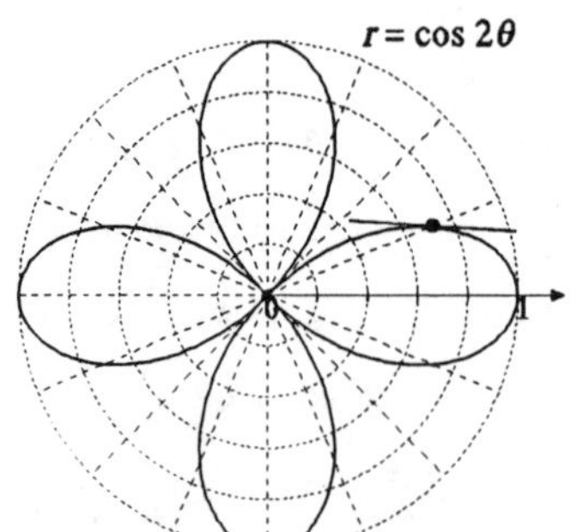

11 $y = 1/x = x^{-1}$, $y' = -x^{-2}$, and $y'' = 2x^{-3}$, so the curvature of the curve at (1, 1) is $\left.\frac{|y''|}{[1 + (y')^2]^{3/2}}\right|_{x=1} = \frac{2}{(1+1)^{3/2}} = \frac{1}{\sqrt{2}}$. Therefore the radius of curvature, that is, the radius of the circle that best approximates the curve at that point is $\sqrt{2}$.

12 See Sec. 9.7.

9.S Review Exercises

1 (a) See Sec. 9.2.

(b) See Figure 6 in Sec. 9.2.

3 (a) See Sec. 9.5.

(b) See Figure 8 in Sec. 9.5.

5 (a) See p. 551.

(b) See the proof of Theorem 2 in Sec. 9.6.

7 Arc length $= \int_0^{\pi} \sqrt{1 + \left(\frac{dy}{dx}\right)^2}\,dx =$

$\int_0^{\pi} \sqrt{1 + \cos^2 x}\,dx$. If the curve is revolved about the x axis, then $R = y = \sin x$ and the surface area is $\int_a^b 2\pi R\,ds = \int_0^{\pi} 2\pi \sin x \sqrt{1 + \cos^2 x}\,dx$.

Making the substitution $u = \cos x$, $du = -\sin x\,dx$, the surface area is

$-2\pi \int_1^{-1} \sqrt{1 + u^2}\,du = 4\pi \int_0^1 \sqrt{1 + u^2}\,du$

$= 4\pi\left[\frac{1}{2}\left(u\sqrt{1 + u^2} + \ln\left|u + \sqrt{1 + u^2}\right|\right)\right]\Big|_0^1$

$= 2\pi\left[\sqrt{2} + \ln\left(1 + \sqrt{2}\right) - (0 + \ln 1)\right]$

$= 2\pi\left[\sqrt{2} + \ln\left(1 + \sqrt{2}\right)\right]$.

If the curve is revolved about the y axis, then $R = x$ and the surface area is $\int_a^b 2\pi R\,ds =$

$\int_0^{\pi} 2\pi x\sqrt{1 + \cos^2 x}\,dx$.

9 The y coordinate is $y = r \sin\theta =$ $(1 + \cos\theta)\sin\theta = \sin\theta + \sin\theta\cos\theta$. To find where y is maximum, compute $\frac{dy}{d\theta}$ and set it equal to 0: $\frac{dy}{d\theta} = \cos\theta + \cos^2\theta - \sin^2\theta = \cos\theta +$ $\cos^2\theta - (1 - \cos^2\theta) = 2\cos^2\theta + \cos\theta - 1 =$ $(2\cos\theta - 1)(\cos\theta + 1)$. Thus $\frac{dy}{d\theta} = 0$ when $\cos\theta = 1/2$ or $\cos\theta = -1$; that is, when $\theta = \pm\frac{\pi}{3}$ or $\theta = \pi$. The maximum value of y is $\frac{3\sqrt{3}}{4}$ and it occurs when $\theta = \frac{\pi}{3}$.

11 Since $r = \frac{1}{1 + \theta}$, we have $r' = -\frac{1}{(1 + \theta)^2}$, and

$r^2 + (r')^2 = \frac{1}{(1 + \theta)^2} + \frac{1}{(1 + \theta)^4} =$

$\frac{(1 + \theta)^2 + 1}{(1 + \theta)^4}$. Hence the total length of the curve is $\int_0^{\infty} \sqrt{r^2 + (r')^2}\,d\theta = \int_0^{\infty} \frac{\sqrt{(1 + \theta)^2 + 1}}{(1 + \theta)^2}\,d\theta =$

$\int_1^{\infty} \frac{\sqrt{u^2 + 1}}{u^2}\,du$, where $u = 1 + \theta$. Note that

$\sqrt{u^2 + 1} > u$, so $\int_1^{\infty} \frac{\sqrt{u^2 + 1}}{u^2}\,du \geq \int_1^{\infty} \frac{u}{u^2}\,du$

$= \int_1^{\infty} \frac{du}{u} = \ln u\big|_1^{\infty}$, which diverges to infinity. It follows that the length of the given curve is infinite.

13 (a)

(b) $x = r\cos\theta = \sin 2\theta\cos\theta = 2\sin\theta\cos^2\theta$,

so $\frac{dx}{d\theta} = -4\sin^2\theta\cos\theta + 2\cos^3\theta$, which is 0 when $2\cos^3\theta = 4\sin^2\theta\cos\theta$. Hence either $\cos\theta = 0$, which clearly corresponds to a minimum, or $\cos^2\theta = 2\sin^2\theta$; that is,

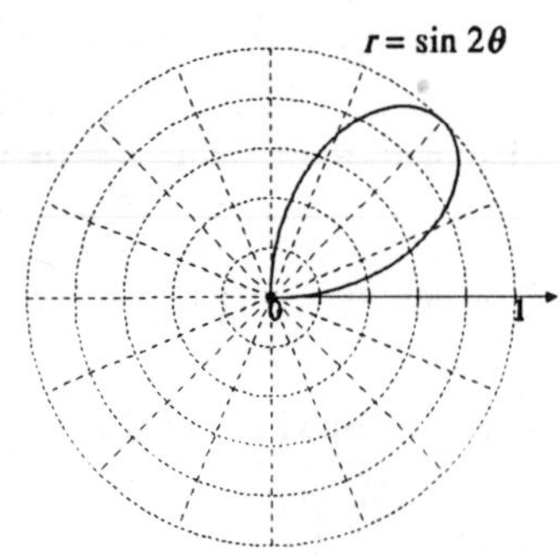

$\tan^2\theta = \frac{1}{2}$ or $\tan\theta = \frac{1}{\sqrt{2}}$. (Since θ should be a first-quadrant angle, we can ignore $-\frac{1}{\sqrt{2}}$.) So $\theta = \tan^{-1}\frac{1}{\sqrt{2}} \approx 0.615 \approx 35.26°$ and $r = \sin 2\theta \approx 0.943$. The point is approximately $(0.943, 35.26°)$.

(c) $y = r\sin\theta = \sin 2\theta \sin\theta = 2\sin^2\theta\cos\theta$, so $\frac{dy}{d\theta} = -2\sin^3\theta + 4\sin\theta\cos^2\theta = 0$ when $2\sin^3\theta = 4\sin\theta\cos^2\theta$, or when $\tan\theta = \sqrt{2}$. So $\theta = \tan^{-1}\sqrt{2} \approx 0.955 \approx 54.74°$ and $r = \sin 2\theta \approx 0.943$. The point is approximately $(0.943, 54.74°)$. Note that the similarity between the answers in (b) and (c) is a direct result of the symmetry of the leaf about the line $\theta = \pi/4 = 45°$.

(d) $\frac{dr}{d\theta} = \frac{d}{d\theta}(\sin 2\theta) = 2\cos 2\theta = 0$ when $\theta = \pi/4$. The point with maximal r is $(1, \pi/4)$.

15 $x = \cos 2t$, $\dot{x} = -2\sin 2t$, $\ddot{x} = -4\cos 2t$, $y = \sin 3t$, $\dot{y} = 3\cos 3t$, and $\ddot{y} = -9\sin 3t$, so $\frac{dy}{dx} = \frac{dy/dt}{dx/dt} = \frac{\dot{y}}{\dot{x}} = -\frac{3\cos 3t}{2\sin 2t}$, and, by Review Exercise 14, $\frac{d^2y}{dx^2} = \frac{\dot{x}\ddot{y} - \ddot{x}\dot{y}}{(\dot{x})^3} =$

$$\frac{18\sin 2t\sin 3t + 12\cos 2t\cos 3t}{-8\sin^3 2t}.$$

17 (a)

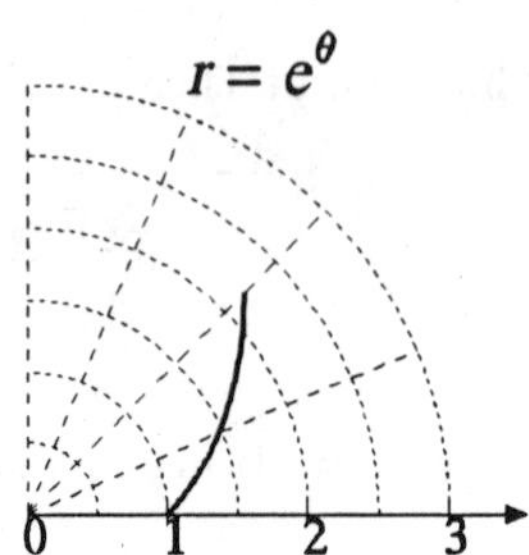

(b) The area of R is given by $\int_a^b y\,dx$, where $a = 1$, $b = \frac{e^{\pi/4}}{\sqrt{2}}$, and y is the cross-sectional height. Substituting in terms of θ, we have $y = r\sin\theta = e^\theta\sin\theta$, $x = r\cos\theta = e^\theta\cos\theta$, and $dx = (e^\theta\cos\theta - e^\theta\sin\theta)\,d\theta$. Then the area is $\int_0^{\pi/4}(e^\theta\sin\theta)e^\theta(\cos\theta - \sin\theta)\,d\theta$.

(c) In terms of circular cross sections, the volume is given by $\int_a^b \pi y^2\,dx$. Using the calculation from (b), we have

$$\int_0^{\pi/4}\pi e^{3\theta}(\sin^2\theta)(\cos\theta - \sin\theta)\,d\theta.$$

(d) In terms of cylindrical shells, the volume is

$$\int_a^b 2\pi R c(x)\,dx = \int_a^b 2\pi xy\,dx =$$

$$\int_0^{\pi/4} 2\pi(e^\theta\cos\theta)(e^\theta\sin\theta)e^\theta(\cos\theta - \sin\theta)\,d\theta$$

$$= \int_0^{\pi/4} 2\pi e^{3\theta}(\sin\theta\cos\theta)(\cos\theta - \sin\theta)\,d\theta.$$

(e) Here $R = y = r\sin\theta = e^\theta\sin\theta$ and $ds =$

$\sqrt{r^2 + (r')^2}\, d\theta = \sqrt{e^{2\theta} + e^{2\theta}}\, d\theta =$

$\sqrt{2}e^{\theta}\, d\theta$. The surface area is $\int_a^b 2\pi R\, ds =$

$\int_0^{\pi/4} 2\sqrt{2}\pi e^{2\theta} \sin\theta\, d\theta$.

(f) $R = x = r\cos\theta = e^{\theta}\cos\theta$, $ds = \sqrt{2}\, e^{\theta}\, d\theta$ as in (e). The surface area is $\int_0^{\pi/4} 2\sqrt{2}\pi e^{2\theta}\cos\theta\, d\theta$. Observe that the integrals in (b) through (d) involve the exponential function multiplied by products of sines and cosines. The products of sines and cosines can be written as sums and differences of sines and cosines using the equations at the top of page 440. The resulting integrals can be evaluated by Formulas 64 and 65 from the text's table of antiderivatives, as can the integrals in (e) and (f).

19 The slope of the tangent line to a point on $r = \theta$ is

$$\frac{dy}{dx} = \frac{dy/d\theta}{dx/d\theta} = \frac{\frac{d}{d\theta}(\theta\sin\theta)}{\frac{d}{d\theta}(\theta\cos\theta)} =$$

$$\frac{\theta\cos\theta + \sin\theta}{-\theta\sin\theta + \cos\theta}.$$ At $\theta = \pi/3$, $\frac{dy}{dx} =$

$$\frac{\pi/3 \cdot 1/2 + \sqrt{3}/2}{-\pi/3\cdot\sqrt{3}/2 + 1/2} = \frac{\pi + 3\sqrt{3}}{3 - \sqrt{3}\pi}.$$ Now the point $(\pi/3, \pi/3)$ in polar coordinates corresponds to the point $(\pi/3\cos\pi/3, \pi/3\sin\pi/3) = (\pi/6, \sqrt{3}\pi/6)$ in rectangular coordinates. By the point-slope formula, the equation of the tangent line is $y - \frac{\sqrt{3}\pi}{6} =$

$$\frac{\pi + 3\sqrt{3}}{3 - \sqrt{3}\pi}\left(x - \frac{\pi}{6}\right).$$

21 Let some point 0 enclosed by the convex curve be the pole of a polar coordinate system; suppose the polar axis intersects the curve at the point P. Recall that curvature is defined as $\kappa = \frac{d\phi}{ds}$, where $\phi(s)$ is the angle that the tangent line makes with the polar axis. Suppose that $s = 0$ at P and $\phi(0) = \alpha$. As one traverses the curve counterclockwise from P, $\phi(s)$ increases. When $s = L$, the length of the curve, one is back at P and $\phi(L) = \alpha + 2\pi$. We now see that the average of curvature with respect to arc length is $\frac{1}{L}\int_0^L \kappa\, ds = \frac{1}{L}\int_0^L \frac{d\phi}{ds}\, ds =$

$$\frac{1}{L}\cdot\phi(s)\Big|_0^L = \frac{1}{L}[\phi(L) - \phi(0)] = \frac{1}{L}[\alpha + 2\pi - \alpha]$$

$= \frac{2\pi}{L}$, as claimed.

23 The length of the curve is $1\cdot\frac{\pi}{6} = \frac{\pi}{6}$. By symmetry, the centroid lies on the ray $\theta = \pi/12$. Now the moment of the curve $r = 1$ $(0 \le \theta \le \pi/6)$ about the y axis is $\int_a^b x\, ds =$

$$\int_0^{\pi/6} (r\cos\theta)\sqrt{r^2 + (r')^2}\, d\theta = \int_0^{\pi/6}\cos\theta\, d\theta =$$

$\sin\theta\big|_0^{\pi/6} = \frac{1}{2}$. Thus $\bar{x} = \frac{\int_a^b x\, ds}{\text{Length of curve}} =$

$\frac{1/2}{\pi/6} = \frac{3}{\pi}$. Since $(\bar{x}, \bar{y})$ lies on the ray $\theta = \pi/12$, we see $\bar{r} = \dfrac{3}{\pi \cos \pi/12}$; so $\bar{y} = \dfrac{3}{\pi} \tan \dfrac{\pi}{12} =$

$\dfrac{3}{\pi}(2 - \sqrt{3}) = \dfrac{6 - 3\sqrt{3}}{\pi}$. So the centroid of the

curve is $(\bar{x}, \bar{y}) = \left(\dfrac{3}{\pi}, \dfrac{6 - 3\sqrt{3}}{\pi}\right)$.

25 (a) The generalized mean value theorem says that

$\dfrac{f(b) - f(a)}{g(b) - g(a)} = \dfrac{f'(T)}{g'(T)}$ for some T in (a, b).

Note that $\dfrac{f(b) - f(a)}{g(b) - g(a)}$ is the slope of the

chord joining $(g(a), f(a))$ and $(g(b), f(b))$.

Also, $\dfrac{f'(T)}{g'(T)} = \left.\dfrac{dy/dt}{dx/dt}\right|_{t=T} = \left.\dfrac{dy}{dx}\right|_{t=T}$ is the

slope of the tangent line at $(g(T), f(T))$. So the theorem says that the slope of the chord joining two points on the curve equals the slope of the tangent line at some point on the curve between them.

(b) The equation of the line through $(g(a), f(a))$ and $(g(b), f(b))$ is $y - f(a) =$

$\dfrac{f(b) - f(a)}{g(b) - g(a)}(x - g(a))$. Thus the ordinate of

the point on the line with abscissa $g(t)$ is

$f(a) + \dfrac{f(b) - f(a)}{g(b) - g(a)}(g(t) - g(a))$. Now $h(t) =$

$f(t) - \left[f(a) + \dfrac{f(b) - f(a)}{g(b) - g(a)}(g(t) - g(a))\right]$, so

$h(a) =$

$f(a) - \left[f(a) + \dfrac{f(b) - f(a)}{g(b) - g(a)}(g(a) - g(a))\right] = 0$

and $h(b) =$

$f(b) - \left[f(a) + \dfrac{f(b) - f(a)}{g(b) - g(a)}(g(b) - g(a))\right] =$

$f(b) - f(a) - f(b) + f(a) = 0$. Hence, by Rolle's theorem, there is a number T in (a, b) such that $h'(T) = 0$. But $h'(t) =$

$f'(t) - \dfrac{f(b) - f(a)}{g(b) - g(a)}g'(t)$, so this means that

there is a T in (a, b) with $f'(T) =$

$\dfrac{f(b) - f(a)}{g(b) - g(a)}g'(T)$ or $\dfrac{f'(T)}{g'(T)} = \dfrac{f(b) - f(a)}{g(b) - g(a)}$

(assuming $g'(T) \neq 0$).

27 Here $R = y = x^a$ and $ds = \sqrt{1 + \left(\dfrac{dy}{dx}\right)^2}\, dx =$

$\sqrt{1 + (ax^{a-1})^2}\, dx$, so the surface area is

$\int_1^2 2\pi x^a(1 + (ax^{a-1})^2)^{1/2}\, dx$

$= 2\pi \int_1^2 x^a(1 + a^2x^{2a-2})^{1/2}\, dx$. Now let $u =$

a^2x^{2a-2}; so $x = \left(\dfrac{u}{a^2}\right)^{\frac{1}{2a-2}}$. Then $du =$

$a^2(2a - 2)x^{2a-3}\, dx$ and $dx = \dfrac{du}{a^2(2a - 2)x^{2a-3}} =$

$\left(\dfrac{u}{a^2}\right)^{-\frac{2a-3}{2a-2}} \cdot \dfrac{du}{a^2(2a - 2)}$ assuming $a \neq 1$. Also

when $x = 1$, $u = a^2$; when $x = 2$, $u = a^2 2^{2a-2} = a^2 4^{a-1}$. Hence the integral becomes

$2\pi \int_{a^2}^{a^2 4^{a-1}} \left(\dfrac{u}{a^2}\right)^{\frac{a}{2a-2}} (1 + u)^{1/2} \left(\dfrac{u}{a^2}\right)^{-\frac{2a-3}{2a-2}} \dfrac{du}{a^2(2a - 2)}$

$= C \int_{a^2}^{a^2 4^{a-1}} u^{\frac{3-a}{2a-2}}(1 + u)^{1/2}\, du$, where C is the product of all constant terms. Here $p = \frac{3 - a}{2a - 2}$ and $q = 1/2$. Clearly both are rational for all rational a. Therefore, the integral is elementary if either $\frac{3 - a}{2a - 2}$ or $\frac{3 - a}{2a - 2} + \frac{1}{2}$ is an integer.

Suppose $\frac{3 - a}{2a - 2} = k$ for some integer k; then $3 - a = 2ak - 2k$ and $a = \frac{2k + 3}{2k + 1} = 1 + \frac{2}{2k + 1}$, where $2k + 1$ is a nonzero integer.

Now assume $\frac{3 - a}{2a - 2} + \frac{1}{2} = \frac{1}{a - 1} = l$ for some nonzero integer l. (We know that l cannot be 0, because $\frac{3 - a}{2a - 2} + \frac{1}{2} = 0$ implies that $\frac{3 - a}{a - 1} = -1$ or $3 - a = 1 - a$, which is impossible.) Solving $\frac{1}{a - 1} = l$ for a yields $a = 1 + \frac{1}{l} = 1 + \frac{2}{2l} = 1 + \frac{2}{n}$, where n is not zero. Thus, $a = 1 + 2/n$ for some nonzero integer n. When $a = 1$, we have $\int_1^2 2\pi x\sqrt{1 + 1^2}\, dx = 2\pi\sqrt{2} \int_1^2 x\, dx = 15\pi\sqrt{2}$; thus the case $a = 1$ yields an elementary integral as well.

29 (a) Let A and B be the initial and current points of contact, as shown. Since the smaller circle rolls without slipping, the arc lengths $\widehat{AB}$ and $\widehat{BP}$ must be equal. But $\widehat{AB} = a\theta$ and $\widehat{BP} = b\phi$, so $b\phi = a\theta$.

(b) Let C be the center of the smaller circle. The line segment OC has length $a - b$ and makes an angle θ with the x axis, so C has coordinates $((a - b) \cos \theta, (a - b) \sin \theta))$. Also, the segment CP has length b and makes an angle $-(\phi - \theta)$ with the x axis. Hence P has coordinates $(x, y) =$

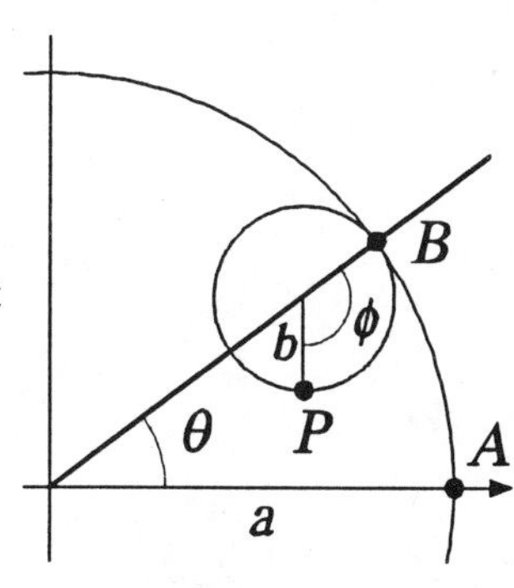

$$((a - b) \cos \theta + b \cos(-(\phi - \theta)),$$
$$(a - b) \sin \theta + b \sin(-(\phi - \theta)))$$
$$= ((a - b) \cos \theta + b \cos(\phi - \theta),$$
$$(a - b) \sin \theta - b \sin(\phi - \theta)),$$ so the desired equations follow. From part (a), we get

$$\phi - \theta = \frac{a\theta}{b} - \theta = \frac{a - b}{b}\theta.$$ Hence, in terms of the parameter θ, we have

$$x = (a - b) \cos \theta + b \cos\left(\frac{a - b}{b}\theta\right),$$

$$y = (a - b) \sin \theta - b \sin\left(\frac{a - b}{b}\theta\right).$$

10 Series

10.1 An Informal Introduction to Series

1 (a) $e^{1/3} \approx 1 + \frac{1}{3} + \frac{\left(\frac{1}{3}\right)^2}{2!} + \frac{\left(\frac{1}{3}\right)^3}{3!} = \frac{113}{81}$

≈ 1.3950617

(b) $e^{1/3} \approx 1 + \left(\frac{1}{3}\right) + \frac{\left(\frac{1}{3}\right)^2}{2!} + \frac{\left(\frac{1}{3}\right)^3}{3!} + \frac{\left(\frac{1}{3}\right)^4}{4!}$

$= \frac{2713}{1944} \approx 1.3955761$

3

x	$x - x^3/6$	$\sin x$	$\sin x - (x - x^3/6)$
0.1	0.0998333	0.0998334	0.0000001
0.2	0.1986666	0.1986693	0.0000027
0.3	0.2955000	0.2955202	0.0000202
0.4	0.3893333	0.3894183	0.0000850
0.5	0.4791666	0.4794255	0.0002589

5 (a) $\int_{1/2}^{1} \frac{e^x - 1}{x}\, dx$

$\approx \int_{1/2}^{1} \frac{\left(1 + x + \frac{x^2}{2!}\right) - 1}{x}\, dx$

$= \int_{1/2}^{1} \left(1 + \frac{x}{2}\right) dx = \left(x + \frac{1}{4}x^2\right)\Big|_{1/2}^{1} = \frac{11}{16}$

$= 0.6875$

(b) $\int_{1/2}^{1} \frac{e^x - 1}{x}\, dx$

$\approx \int_{1/2}^{1} \frac{\left(1 + x + \frac{x^2}{2!} + \frac{x^3}{3!}\right) - 1}{x}\, dx$

$= \int_{1/2}^{1} \left(1 + \frac{x}{2} + \frac{x^2}{6}\right) dx =$

$\left(x + \frac{1}{4}x^2 + \frac{1}{18}x^3\right)\Big|_{1/2}^{1} = \frac{53}{72} \approx 0.7361111$

7 (a) Let $S_n =$

$\frac{1}{2} - \frac{(1/2)^2}{2} + \frac{(1/2)^3}{3} - \dots + (-1)^{n-1}\frac{(1/2)^n}{n}$.

n	S_n	Decimal
1	1/2	0.5
2	1/2 − 1/8 = 3/8	0.375
3	1/2 − 1/8 + 1/24 = 5/12	0.4166667
4	1/2 − 1/8 + 1/24 − 1/64 = 77/192	0.4010417
5	1/2 − 1/8 + 1/24 − 1/64 + 1/160 = 391/960	0.4072917

(b) Calculator: $\ln(1.5) \approx 0.4054651$

Error: $|0.4054651 - 0.4072917|$

≈ 0.0018266

9 (a) Calculator: $\ln 2 \approx 0.6931471$

Estimate 1: $1 - 1/2 + 1/3 - 1/4 + 1/5$

≈ 0.7833333

Estimate 2: $-\left[-\frac{1}{2} - \frac{(-1/2)^2}{2} + \frac{(-1/2)^3}{3} - \frac{(-1/2)^4}{4} + \frac{(-1/2)^5}{5}\right] \approx -(-0.6885417)$

$= 0.6885417$

(b) Estimate 2 is better than estimate 1, probably because it uses integer powers of $-1/2$, while estimate 1 uses integer powers of 1. Thus, successive terms in estimate 2 get smaller faster than those in estimate 1. This suggests that estimate 2 becomes a better estimate of ln 2 faster than estimate 1.

11 (a) $\frac{1}{\sqrt{1}} + \frac{1}{\sqrt{2}} \approx 1.7071068 > 1.4142136 \approx \sqrt{2}$

(b) $\frac{1}{\sqrt{1}} + \frac{1}{\sqrt{2}} + \frac{1}{\sqrt{3}} + \frac{1}{\sqrt{4}} \approx 2.7844571 > 2 = \sqrt{4}$

13 (a) $\frac{1}{1+t} = 1 - t + t^2 - t^3 + \cdots + (-1)^{n-1}t^{n-1} + \frac{(-1)^n t^n}{1+t}$, so $\int_0^x \frac{1}{1+t}\,dt = \int_0^x \left[1 - t + t^2 - t^3 + \cdots + (-1)^{n-1}t^{n-1}\right] dt + \int_0^x \frac{(-1)^n t^n}{1+t}\,dt$, from which we obtain

$\ln(1+t)\big|_0^x = \left[t - \frac{1}{2}t^2 + \frac{1}{3}t^3 - \cdots + \frac{(-1)^{n-1}}{n}t^n\right]\Big|_0^x + (-1)^n \int_0^x \frac{t^n}{1+t}\,dt$ and therefore $\ln(1+x) = x - \frac{1}{2}x^2 + \frac{1}{3}x^3 - \frac{1}{4}x^4 + \cdots + \frac{(-1)^{n-1}}{n}x^n + (-1)^n \int_0^x \frac{t^n}{1+t}\,dt.$

(b) For $0 \le x \le 1$ we have $0 \le \int_0^x \frac{t^n}{1+t}\,dt \le \int_0^x \frac{t^n}{1+0}\,dt = \frac{t^{n+1}}{n+1}\Big|_0^x = \frac{x^{n+1}}{n+1} \le \frac{1}{n+1}$, which approaches 0 as $n \to \infty$. By comparison it follows that $\int_0^x \frac{t^n}{1+t}\,dt \to 0$ as $n \to \infty$.

15 Exploration exercises are in the *Instructor's Manual.*

10.2 Sequences

1 $\{0.999^n\}$: 0.999, 0.998001, 0.997002999, $\cdots$; the sequence converges with $\lim_{n\to\infty} 0.999^n = 0$ since $-1 < 0.999 < 1$ and $r^n \to 0$ whenever $|r| < 1$.

3 $\{1^n\}$: 1, 1, 1, $\cdots$; the sequence converges with $\lim_{n\to\infty} 1^n = 1$.

5 $\{n!\}$: 1, 2, 6, $\cdots$; clearly the sequence diverges since $n!$ grows without bound as n increases.

7 $\left\{\frac{3n+5}{5n-3}\right\}$: 4, $\frac{11}{7}$, $\frac{7}{6}$, $\cdots$ the sequence converges

with $\lim_{n\to\infty} \frac{3n+5}{5n-3} = \lim_{n\to\infty} \frac{3+5/n}{5-3/n} = \frac{3}{5}$.

9 $\left\{\frac{\cos n}{n}\right\}$: 0.5403023, −0.2080734, −0.3299975, ⋯; the sequence converges with $\lim_{n\to\infty} \frac{\cos n}{n} = 0$ since $-\frac{1}{n} \le \frac{\cos n}{n} \le \frac{1}{n}$ and both $\lim_{n\to\infty} \frac{-1}{n} = 0$ and $\lim_{n\to\infty} \frac{1}{n} = 0$.

11 $\left\{\left(1+\frac{2}{n}\right)^n\right\}$: 3, 4, $\frac{125}{27}$, ⋯; the sequence converges with $\lim_{n\to\infty}\left(1+\frac{2}{n}\right)^n = \lim_{n\to\infty}\left[\left(1+\frac{2}{n}\right)^{n/2}\right]^2 = e^2$.

13 (a)

n	$6^n/n! = a_n$
1	6
2	18
3	36
4	54
5	64.8
6	64.8
7	55.543
8	41.657

(b)

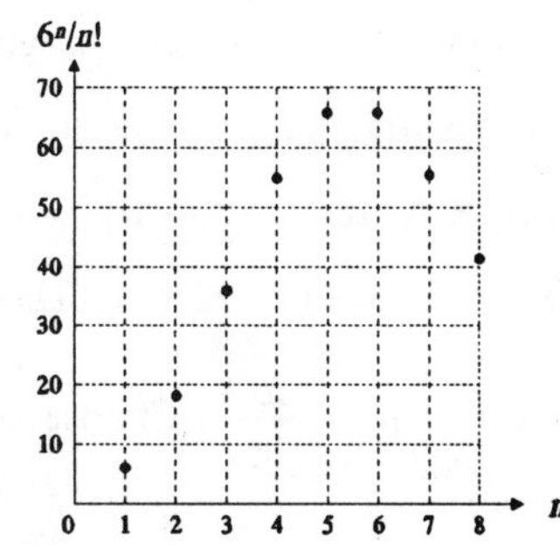

(c) The largest value of $\frac{6^n}{n!}$ is 64.8, which occurs as the sixth (or fifth) term, a_6, in the sequence. Since $a_7 = \left(\frac{6}{7}\right)a_6$, $a_8 = \left(\frac{6}{8}\right)\left(\frac{6}{7}\right)a_6$, $a_9 = \left(\frac{6}{9}\right)\left(\frac{6}{8}\right)\left(\frac{6}{7}\right)a_6$, ⋯, all succeeding terms are smaller than a_6.

(d) $\lim_{n\to\infty} \frac{6^n}{n!} = 0$ by Example 3.

15 (a) $(0.999)^{10{,}000} = 0.000045173 < 0.0001$

(b) We want $(0.999)^n = 0.0001$, which implies that $\ln(0.999)^n = \ln(0.0001)$. But then we have $n \ln(0.999) = \ln(0.0001)$, so $n = \frac{\ln(0.0001)}{\ln(0.999)} \approx 9205.734$. Therefore $n =$ 9206 is large enough.

17 The value of the dollar after n years of 5% annual loss is $a_n = (0.95)^n$.

(a) $a_4 = (0.95)^4 = 0.8145063 \approx 0.81$

(b) $\lim_{n\to\infty} a_n = \lim_{n\to\infty} 0.95^n = 0$ since $-1 < 0.95 < 1$.

19 (a) $F_7 = F_6 + F_5 = 8 + 5 = 13$

$F_8 = F_7 + F_6 = 13 + 8 = 21$

$F_9 = F_8 + F_7 = 21 + 13 = 34$, and

$F_{10} = F_9 + F_8 = 34 + 21 = 55$

(b)

n	F_{n+1}/F_n
1	$1/1 = 1$
2	$2/1 = 2$
3	$3/2 = 1.5$
4	$5/3 \approx 1.6666667$
5	$8/5 = 1.6$
6	$13/8 = 1.625$
7	$21/13 \approx 1.6153846$
8	$34/21 \approx 1.6190476$
9	$55/34 \approx 1.6176471$
10	$89/55 \approx 1.6181818$

(c) Conjecture: $\lim_{n\to\infty} \dfrac{F_{n+1}}{F_n} \approx 1.618$

21 Exploration problems are in the *Instructor's Manual.*

23 Divide the interval [0, 1] into n equal parts, each of length $1/n$. Let $x_i = i/n$ and $f(x) = x^2$. Then

$$\lim_{n\to\infty} \sum_{i=1}^{n} \left(\frac{i}{n}\right)^2 \frac{1}{n} = \sum_{i=1}^{n} f(x_i)\frac{1}{n} = \int_0^1 f(x)\,dx = \frac{1}{3}.$$

25 (a) The graphs intersect near $x = \pi/4 \approx 0.7853982$.

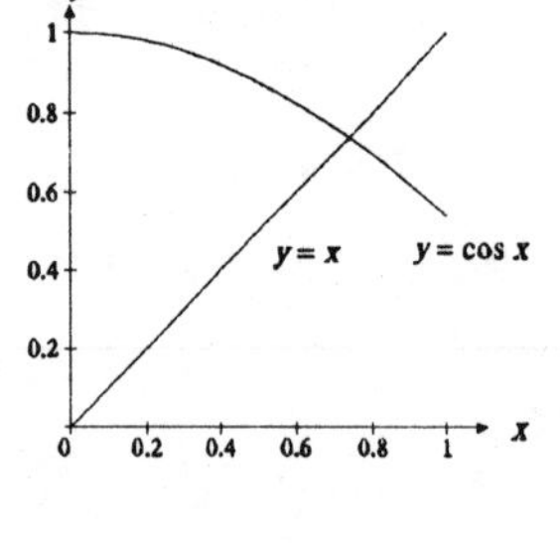

(b) Newton's method stipulates that x_{n+1}

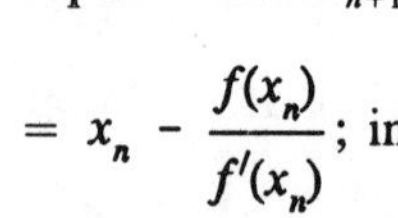

$= x_n - \dfrac{f(x_n)}{f'(x_n)}$; in this case, $f(x) = x - \cos x$ and $f'(x) = 1 + \sin x$, so that $x_{n+1} = x_n - \dfrac{x_n - \cos x_n}{1 + \sin x_n}$.

Starting with $x_1 = 0.7853982$, we obtain the values in the following table.

n	x_n
1	0.7853982
2	0.7395361
3	0.7390852
4	0.7390852

Our result suggests that $r \approx 0.7390852$.

(c)

n	$x_{n+1} = \cos x_n$
1	0.7853982
2	0.7071068
3	0.7602446
4	0.7246675
5	0.7487199
6	0.7325608
7	0.7434642
8	0.7361283
9	0.7410737
10	0.7377442
11	0.7399878
12	0.7384768
13	0.7394948

Our result suggests that $r \approx 0.739$.

(d) Newton's method in part (b) requires more work than the calculator method in part (c). However, the method in part (c) is much slower in approaching a limit.

27 (a) If r satisfies $0 < r < 1$, then $\lim_{n\to\infty} r^n = 0$.

Proof: Let $\epsilon > 0$ be given. Assume also that $\epsilon < 1$. We now determine an integer N so that if $n \ge N$, then $|r^n - 0| = r^n < \epsilon$. We have $r^n < \epsilon$, so $\ln r^n < \ln \epsilon$. Then $n \ln r < \ln \epsilon$ and thus $n > \dfrac{\ln \epsilon}{\ln r}$. Choose an integer $N > \dfrac{\ln \epsilon}{\ln r}$. Thus, if $n \ge N$, then $r^n < \epsilon$ for

$0 < \epsilon < 1$. (For $\epsilon \geq 1$, choose $N = 1$.)

(b) If r satisfies $-1 < r < 0$, then $\lim_{n\to\infty} r^n = 0$.

Proof: Since $-1 < r < 0$, it follows that $0 < -r < 1$ and $\lim_{n\to\infty} (-r)^n = 0$ by part (a).

Then $-(-r)^n \leq r^n \leq (-r)^n$ and $\lim_{n\to\infty} \left(-(-r)^n\right) = -\lim_{n\to\infty} (-r)^n = -1\cdot 0 = 0$, so that $\lim_{n\to\infty} r^n = 0$.

29 We show that $\lim_{n\to\infty} \frac{3}{n^2} = 0$.

Proof: Let $\epsilon > 0$ be given. Determine an integer N so that if $n \geq N$, then $\left|\frac{3}{n^2} - 0\right| = \frac{3}{n^2} < \epsilon$. We have $\frac{3}{n^2} < \epsilon$, so $\frac{3}{\epsilon} < n^2$ and $\sqrt{\frac{3}{\epsilon}} < n$. Choose $N > \sqrt{\frac{3}{\epsilon}}$; then $n \geq N$ implies $\frac{3}{n^2} < \epsilon$.

10.3 Series

1 (a) False. Consider $a_n = 1/n$, where $a_n \to 0$, but $\sum_{n=1}^{\infty} a_n$ diverges.

(b) False. Consider $a_n = 1/n$, where $a_n \to 0$, but $\sum_{n=1}^{\infty} a_n \neq 0$.

(c) False. Consider $a_n = (1/2)^n$, where $a_n \to 0$, but $\sum_{n=1}^{\infty} a_n$ has a sum of 1.

(d) True. Consider $a_n = 1/n$ and $a_n = (1/2)^n$, where the sum of the first diverges, but the sum of the second converges.

3 (a) False. If $S_n \to 3$, then the series converges.

(b) False. If $S_n \to 3$, then the series converges, so $a_n \to 0$.

(c) True. This is the contrapositive of the nth-term test.

(d) True. Series convergence is defined by the convergence of the sequence of partial sums, S_n.

(e) True. The sum of the series is defined to be the limit of the sequence of partial sums.

(f) False. See part (c).

5 $1 + \frac{1}{2} + \frac{1}{4} + \frac{1}{8} + \cdots$ is a geometric series with initial term $a = 1$ and ratio $r = 1/2$; by Theorem 1, the sum is $\frac{a}{1 - r} = \frac{1}{1 - \frac{1}{2}} = 2$.

7 $\sum_{n=1}^{\infty} 10^{-n}$ is a geometric series with initial term $a = 10^{-1} = \frac{1}{10}$ and ratio $r = \frac{1}{10}$. By Theorem 1, the sum is $\frac{a}{1 - r} = \frac{\frac{1}{10}}{1 - \frac{1}{10}} = \frac{1}{9}$.

9 $\sum_{n=1}^{\infty} 5(0.99)^n$ is a geometric series with initial term $a = 5(0.99) = 4.95$ and ratio $r = 0.99$. By Theorem 1, the sum is $\frac{a}{1 - r} = \frac{4.95}{1 - 0.99} = 495$.

11 $\sum_{n=1}^{\infty} 4\left(\frac{2}{3}\right)^n$ is geometric with initial term $a = 4\left(\frac{2}{3}\right)^1 = \frac{8}{3}$ and ratio $r = \frac{2}{3}$. By Theorem 1, the sum is $\frac{a}{1-r} = \frac{8/3}{1-2/3} = 8$.

13 $-5 + 5 - 5 + 5 - \cdots = \sum_{n=1}^{\infty} (-1)^n 5$ diverges since $\lim_{n\to\infty} (-1)^n 5 \neq 0$.

15 $\sum_{n=1}^{\infty} \frac{2}{n}$ diverges since its terms are twice those of the harmonic series. The partial sums of the harmonic series diverge, so sums twice as large must also diverge.

17 $\sum_{n=1}^{\infty} 6\left(\frac{4}{5}\right)^n = \frac{6\left(\frac{4}{5}\right)}{1-\frac{4}{5}} = 24$

19 $\sum_{n=1}^{\infty} (2^{-n} + 3^{-n}) = \sum_{n=1}^{\infty} 2^{-n} + \sum_{n=1}^{\infty} 3^{-n}$

$= \sum_{n=1}^{\infty} \left(\frac{1}{2}\right)^n + \sum_{n=1}^{\infty} \left(\frac{1}{3}\right)^n = \frac{1/2}{1-1/2} + \frac{1/3}{1-1/3}$

$= 1 + \frac{1}{2} = \frac{3}{2}$

21 The ball initially falls 1 meter and rebounds 1(0.6) meter; in the second descent it falls the distance it has just risen, 1(0.6), and rebounds to 1(0.6)(0.6) $= 1(0.6)^2$. This pattern continues, so that the total distance traveled is $1 + 2(0.6) + 2(0.6)^2 + 2(0.6)^3 + \cdots = 1 + \sum_{n=1}^{\infty} 2(0.6)^n = 1 + \frac{2(0.6)}{1-(0.6)} = 1 + 3 = 4$ ft.

23 $3.171717\cdots = 3 + \frac{17}{100} + \frac{17}{100^2} + \frac{17}{100^3} + \cdots$

$= 3 + \frac{17/100}{1 - 1/100} = 3 + \frac{17}{99} = 3\frac{17}{99}$

25 $4.1256256256\cdots = 4 + \frac{1}{10} + \frac{256}{10,000} + \frac{256}{10,000,000} + \frac{256}{10,000,000,000} + \cdots$

$= 4 + \frac{1}{10} + \frac{\frac{256}{10,000}}{1 - \frac{1}{1000}} = 4\frac{251}{1998}$

27 $P = g(t)(e^{-rt}) + g(t)(e^{-rt})^2 + g(t)(e^{-rt})^3 + \cdots$

$= g(t)[(e^{-rt}) + (e^{-rt})^2 + (e^{-rt})^3 + \cdots]$; observe that the series in the brackets is geometric with initial term $a = e^{-rt}$ and ratio $s = e^{-rt}$. If rt is positive, then e^{-rt} is less than 1, so Theorem 1 says the sum is $\frac{a}{1-s} = \frac{e^{-rt}}{1-e^{-rt}}$. Then $P = \frac{g(t)e^{-rt}}{1-e^{-rt}}$, as claimed.

29 The money supply begins with the initial deposit of \$1000. The bank then loans 80% of the deposit, (0.8)(1000) = 800, to a second party, who could then deposit it in a bank account. The bank has now recorded \$1800 in deposits as a result of the initial \$1000. The bank can then lend out 80% of the new \$800 deposit, $(0.8)(800) = (0.8)^2(1000) = 640$; if the \$640 is also deposited, then 80%, $(0.8)(640) = (0.8)^3(1000) = 512$, can be loaned

out. If this continues, total bank deposits will be $1000 + (0.8)1000 + (0.8)^2 1000 + \cdots =$ $1000[1 + 0.8 + (0.8)^2 + \cdots] = 1000 \cdot \frac{1}{1 - 0.8} =$ $1000(5) = 5000$. The initial deposit has grown to \$5000.

31 In Sec. 4.3, we learned that the height of a freely falling object is $y = \frac{a}{2}t^2 + v_0 t + y_0$, where a is gravitational acceleration, v_0 is initial velocity, and y_0 is initial height. Using $a = -32$ ft/sec², $v_0 =$ 0 ft/sec, and $y_0 = h$ ft, we find that $y = 0$ when $0 = -16t^2 + h$, so $t = \frac{\sqrt{h}}{4}$, giving the time t it takes to fall from a height h. The fall from 6 ft therefore requires $\frac{\sqrt{6}}{4}$ seconds. The time for the first complete bounce of height 6(0.9) (since the ball takes as long to rise as to fall) is $2\frac{\sqrt{6(0.9)}}{4} =$ $\frac{\sqrt{6}(0.9)^{1/2}}{2}$ sec. Then the next bounce takes $\frac{\sqrt{6}(0.9)^{2/2}}{2}$ seconds, while the one after that takes $\frac{\sqrt{6}(0.9)^{3/2}}{2}$ seconds. The total time is therefore

$$\frac{\sqrt{6}}{4} + \frac{\sqrt{6}}{2}[(0.9)^{1/2} + (0.9)^{2/2} + (0.9)^{3/2} + \cdots]$$

$$= \frac{\sqrt{6}}{4} + \frac{\sqrt{6}}{2}\left[\frac{0.9^{1/2}}{1 - 0.9^{1/2}}\right] \approx 23.25 \text{ seconds.}$$

33 (a) If we simply add up the area of the rectangular regions as drawn in Fig. 4, we

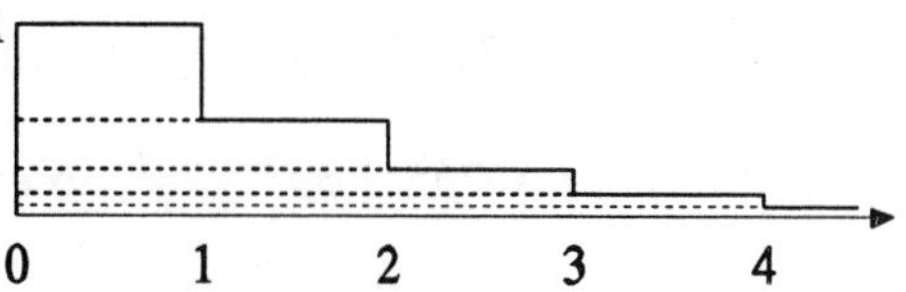

obtain $1 + \frac{1}{2} + \frac{1}{4} + \frac{1}{8} + \cdots$. But if we view the area in terms of horizontal cross sections instead, as shown in the figure, we obtain $\frac{1}{2} + \frac{2}{4} + \frac{3}{8} + \cdots$. These two views of the same area show that $1 + \frac{1}{2} + \frac{1}{4} + \frac{1}{8} + \cdots$ $= \frac{1}{2} + \frac{2}{4} + \frac{3}{8} + \cdots$, as claimed.

(b) By part (a), we have $\sum_{n=1}^{\infty} \frac{n}{2^n} =$

$$1 + \left(\frac{1}{2}\right) + \left(\frac{1}{2}\right)^2 + \cdots = \frac{1}{1 - 1/2} = 2.$$

(c) Constructing a graph analogous to Figure 4, using p in place of 1/2, with $0 < p < 1$, we

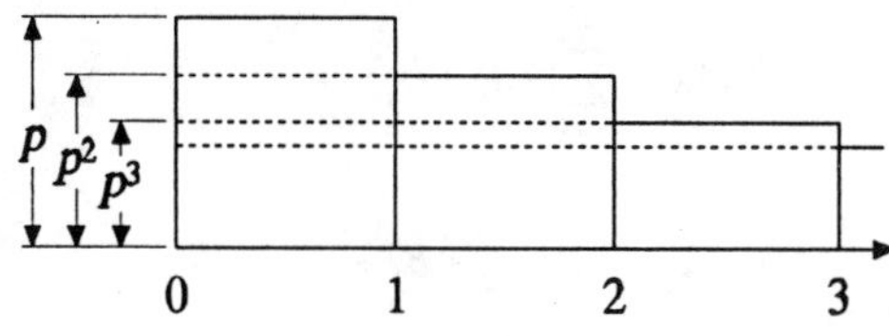

see that $p + p^2 + p^3 + \cdots =$ $1(p - p^2) + 2(p^2 - p^3) + 3(p^3 - p^4) + \cdots$. Since the left-hand side of the equation is geometric, we sum that series to obtain $\frac{p}{1 - p}$. The right-hand side can be factored to produce

$1(p - p^2) + 2(p^2 - p^3) + 3(p^3 - p^4) + \cdots$

$= p(1 - p) + 2p^2(1 - p) + 3p^3(1 - p) + \cdots$

$= (1 - p)[p + 2p^2 + 3p^3 + \cdots]$, so our equation has become $\dfrac{p}{1 - p} =$ $(1 - p)[p + 2p^2 + 3p^3 + \cdots]$. Hence $p + 2p^2 + 3p^3 + \cdots = \dfrac{p}{(1 - p)^2}$.

35 (a) Since $S_n = a_1 + a_2 + \cdots + a_n$ and $\sum_{n=1}^{\infty} a_n =$ L, $\lim_{n\to\infty} S_n = L$. Let $T_n = ca_1 + ca_2 + \cdots +$ $ca_n = cS_n$, then $\sum_{n=1}^{\infty} ca_n = \lim_{n\to\infty} T_n = \lim_{n\to\infty} cS_n$ $= c \lim_{n\to\infty} S_n = cL$.

(b) Let $S_n = a_1 + a_2 + \cdots + a_n$ and $T_n = b_1 +$ $b_2 + \cdots + b_n$. Since $\sum_{n=1}^{\infty} a_n = L$ and $\sum_{n=1}^{\infty} b_n$ $= M$, we know $\lim_{n\to\infty} S_n = L$ and $\lim_{n\to\infty} T_n =$ M. Then $\sum_{n=1}^{\infty} (a_n + b_n) =$

$\lim_{n\to\infty} (a_1 + b_1 + a_2 + b_2 + \cdots + a_n + b_n) =$

$\lim_{n\to\infty} (S_n + T_n) = \lim_{n\to\infty} S_n + \lim_{n\to\infty} T_n = L + M.$

10.4 The Integral Test

1 Let $f(x) = \dfrac{1}{x^{1.1}}$, which is continuous, positive, and decreasing for $x \geq 1$; since $\int_1^{\infty} \dfrac{1}{x^{1.1}}\,dx = \dfrac{x^{-0.1}}{-0.1}\Big|_1^{\infty}$ $= \dfrac{-10}{x^{0.1}}\Big|_1^{\infty} = 0 - \dfrac{-10}{1} = 10$, $\sum_{n=1}^{\infty} \dfrac{1}{n^{1.1}}$ converges.

3 Let $f(x) = \dfrac{x}{x^2 + 1}$, which is continuous, positive, and decreasing for $x \geq 1$; since $\int_1^{\infty} \dfrac{x}{x^2 + 1}\,dx =$ $\dfrac{1}{2} \ln(x^2 + 1)\Big|_1^{\infty} = \infty$, $\sum_{n=1}^{\infty} \dfrac{n}{n^2 + 1}$ diverges.

5 Let $f(x) = \dfrac{1}{x \ln x}$, which is continuous, positive, and decreasing for $x \geq 2$; since $\int_2^{\infty} \dfrac{1}{x \ln x}\,dx =$ $\ln|\ln x|\,\big|_2^{\infty} = \infty$, $\sum_{n=2}^{\infty} \dfrac{1}{n \ln n}$ diverges.

7 Let $f(x) = \dfrac{\ln x}{x}$, which is continuous, positive, and decreasing for $x \geq e$; since $\int_3^{\infty} \dfrac{\ln x}{x}\,dx =$ $\dfrac{1}{2} (\ln x)^2\Big|_3^{\infty} = \infty$, $\sum_{n=3}^{\infty} \dfrac{\ln n}{n}$ diverges and hence $\sum_{n=1}^{\infty} \dfrac{\ln n}{n}$ diverges.

9 $\sum_{n=1}^{\infty} \dfrac{1}{n^{1/3}}$ diverges since $p = 1/3 \leq 1$.

11 $\sum_{n=1}^{\infty} \dfrac{1}{n^{1/2}}$ diverges since $p = 1/2 \leq 1$.

13 (a) If $p > 1$, then $\int_1^\infty \frac{1}{x^p}\,dx = \int_1^\infty x^{-p}\,dx =$

$$\left.\frac{x^{1-p}}{1-p}\right|_1^\infty = \left.\frac{1}{1-p}\cdot\frac{1}{x^{p-1}}\right|_1^\infty = 0 - \frac{1}{1-p} =$$

$\frac{1}{p-1}$; by the integral test, $\sum_{n=1}^\infty \frac{1}{n^p}$

converges.

(b) Obviously $\sum_{n=1}^\infty \frac{1}{n^p} > 1$. Theorem 3 gives an

upper bound: $R_1 < \int_1^\infty \frac{dx}{x^p} = \frac{1}{p-1}$, so

$$\sum_{n=1}^\infty \frac{1}{n^p} = 1 + R_1 < 1 + \frac{1}{p-1} = \frac{p}{p-1}.$$

15 (a) $S_4 = 1 + \frac{1}{8} + \frac{1}{27} + \frac{1}{64} \approx 1.1777$

(b) $\frac{1}{50} = \int_5^\infty \frac{1}{x^3}\,dx < R_4 < \int_4^\infty \frac{1}{x^3}\,dx = \frac{1}{32}$

(c) $1.1976 = 1.1776 + 1/50 < \sum_{n=1}^\infty \frac{1}{n^3} <$

$1.1777 + 1/32 < 1.2090$

17 (a) $S_4 = \frac{1}{2} + \frac{1}{5} + \frac{1}{10} + \frac{1}{17} \approx 0.8588$

(b) Since $\int \frac{1}{x^2+1}\,dx = \tan^{-1} x + C$, we have

$0.1973 < 0.1974 \approx \frac{\pi}{2} - \tan^{-1} 5$

$= \int_5^\infty \frac{1}{x^2+1}\,dx < R_4 < \int_4^\infty \frac{1}{x^2+1}\,dx$

$= \frac{\pi}{2} - \tan^{-1} 4 < 0.2450.$

(c) $1.0561 = 0.8588 + 0.1973 < \sum_{n=1}^\infty \frac{1}{n^2+1}$

$< 0.8589 + 0.2450 = 1.1039$

19 (a) $\int_{101}^\infty \frac{1}{x^2}\,dx < R_{100} < \int_{100}^\infty \frac{1}{x^2}\,dx$; that is,

$\frac{1}{101} < R_{100} < \frac{1}{100}$.

(b) If $n \geq 10{,}000$ then $R_n < \int_n^\infty \frac{1}{x^2}\,dx \leq$

$\int_{10{,}000}^\infty \frac{1}{x^2}\,dx = 0.0001.$

21 (a) $R_n < \int_n^\infty \frac{1}{x^4}\,dx = \frac{1}{3n^3}$, which is less than

0.0001 for $n > \left(\frac{10{,}000}{3}\right)^{1/3} \approx 14.9$. Thus we

should use 15 terms.

(b) By part (a), S_{15} will give the prescribed

accuracy: $S_{15} = \frac{1}{1^4} + \frac{1}{2^4} + \frac{1}{3^4} + \cdots + \frac{1}{15^4}$

$\approx 1.082.$

23 $\int_1^\infty k^x\,dx = \left.\frac{k^x}{\ln k}\right|_1^\infty = \frac{-k}{\ln k}$, so $\sum_{n=1}^\infty k^n$ converges

for $0 < k < 1$.

25 Let $f(x) = x^3 e^{-x} = x^3/e^x$; by the quotient rule,

$f'(x) = \frac{e^x \cdot 3x^2 - x^3 \cdot e^x}{e^{2x}} = \frac{3x^2 - x^3}{e^x} =$

$\frac{x^2(3-x)}{e^x}$, which is negative for $x > 3$. Hence f

is decreasing for $x > 3$. By Formula 63 in the

table of antiderivatives, $\int_3^\infty x^3 e^{-x}\,dx =$

$\lim_{b\to\infty} [(-e^{-x})(x^3 + 3x^2 + 6x + 6)]\Big|_3^b = e^{-3}(78)$, so

$\sum_{n=1}^{\infty} n^3 e^{-n}$ converges.

27 From the two diagrams it follows that $\int_n^{2n+1} \frac{1}{x}\,dx$

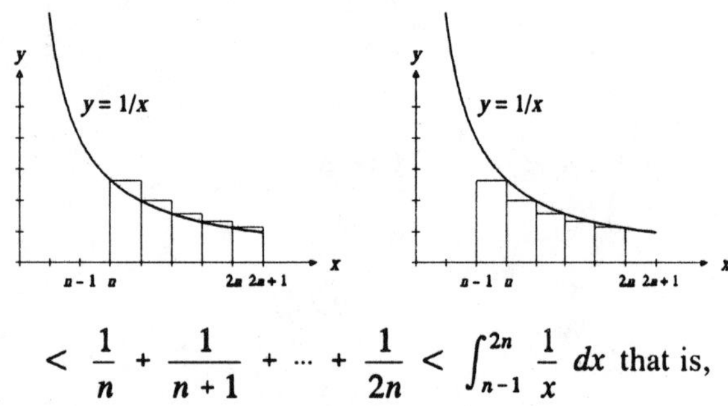

$< \frac{1}{n} + \frac{1}{n+1} + \cdots + \frac{1}{2n} < \int_{n-1}^{2n} \frac{1}{x}\,dx$ that is,

$\ln(2n+1) - \ln n < \frac{1}{n} + \frac{1}{n+1} + \cdots + \frac{1}{2n} <$

$\ln 2n - \ln(n-1)$. We can simplify this to

$\ln\left(\frac{2n+1}{n}\right) < \frac{1}{n} + \frac{1}{n+1} + \cdots + \frac{1}{2n} <$

$\ln\left(\frac{2n}{n-1}\right)$, an inequality we now use in (a) and (b).

(a) If $n = 100$ then $\ln\frac{201}{100} <$

$\frac{1}{100} + \frac{1}{101} + \cdots + \frac{1}{200} < \ln\frac{200}{99}$.

(b) Taking the limit of the inequality as $n \to \infty$

yields $\ln 2 \le \lim_{n\to\infty} \sum_{i=n}^{2n} \frac{1}{i} \le \ln 2$. Therefore

we must have $\lim_{n\to\infty} \sum_{i=n}^{2n} \frac{1}{i} = \ln 2$.

29 $\lim_{n\to\infty} \frac{\frac{1}{n^{1.01}}}{\frac{1}{n}} = \lim_{n\to\infty} \frac{n}{n^{1.01}} = \lim_{n\to\infty} \frac{1}{n^{.01}} = 0$

31 (a) Let $f(x) = 1 + x - e^x$. Note that $f(0) = 0$ and $f'(x) = 1 - e^x$, so $f'(x) \le 0$ for $x \ge 0$. Hence $f(x) \le 0$ for $x \ge 0$; that is, $1 + x - e^x \le 0$, so $1 + x \le e^x$ for $x \ge 0$.

It follows that $1 + a_i \le e^{a_i}$ for $a_i \ge 0$.

(b) If $\sum_{n=1}^{\infty} a_n = L < \infty$, then $\prod_{i=1}^{n} (1 + a_i) \le$

$\prod_{i=1}^{n} e^{a_i} = e^{a_1 + a_2 + \cdots + a_n} = \exp\left(\sum_{i=1}^{n} a_i\right) \le$

$\exp\left(\lim_{n\to\infty} \sum_{i=1}^{n} a_i\right) = \exp\left(\sum_{n=1}^{\infty} a_n\right) = e^L$. Since

$S_n = \prod_{i=1}^{n} (1 + a_i)$ is a bounded increasing

sequence, it follows that $\lim_{n\to\infty} S_n =$

$\lim_{n\to\infty} \prod_{i=1}^{n} (1 + a_i)$ exists and is less than or

equal to e^L.

10.5 Comparison Tests

1 $\sum_{n=1}^{\infty} \frac{1}{n^2+3}$ converges since $\frac{1}{n^2+3} \le \frac{1}{n^2}$ and

$\sum_{n=1}^{\infty} \frac{1}{n^2}$ converges (p-series with $p = 2$).

3 $\sum_{n=1}^{\infty} \frac{\sin^2 n}{n^2}$ converges since $\frac{\sin^2 n}{n^2} \le \frac{1}{n^2}$ and

$\sum_{n=1}^{\infty} \frac{1}{n^2}$ converges (p-series with $p = 2$).

5 $\sum_{n=1}^{\infty} \frac{5n+1}{(n+2)n^2}$ converges since $\lim_{n\to\infty} \frac{\frac{5n+1}{(n+2)n^2}}{1/n^2} =$

$\lim_{n\to\infty} \frac{5n+1}{n+2} = 5$ and $\sum_{n=1}^{\infty} \frac{1}{n^2}$ converges (p-series

with $p = 2$).

7 $\sum_{n=1}^{\infty} \frac{n+1}{(5n+2)\sqrt{n}}$ diverges since $\lim_{n\to\infty} \frac{\frac{n+1}{(5n+2)\sqrt{n}}}{1/\sqrt{n}}$

$= \lim_{n\to\infty} \frac{n+1}{5n+2} = \frac{1}{5}$ and $\sum_{n=1}^{\infty} \frac{1}{\sqrt{n}}$ diverges

(p-series with $p = 1/2$).

9 $\sum_{n=1}^{\infty} \frac{n^2}{3^n}$ converges since $\frac{n^2}{3^n} < \frac{2.9^n}{3^n}$ and $\sum_{n=1}^{\infty} \frac{2.9^n}{3^n}$

$= \sum_{n=1}^{\infty} \left(\frac{2.9}{3}\right)^n$ converges (geometric series with

ratio less than 1).

11 $\sum_{n=1}^{\infty} \frac{1}{n^n}$ converges since $\frac{1}{n^n} \le \frac{1}{n^2}$ and $\sum_{n=1}^{\infty} \frac{1}{n^2}$

converges (p-series with $p = 2$).

13 $\sum_{n=1}^{\infty} \frac{4n+1}{(2n+3)n^2}$ converges since $\lim_{n\to\infty} \frac{\frac{4n+1}{(2n+3)n^2}}{1/n^2}$

$= \lim_{n\to\infty} \frac{4n+1}{2n+3} = 2$ and $\sum_{n=1}^{\infty} \frac{1}{n^2}$ converges

(p-series with $p = 2$).

15 $\sum_{n=1}^{\infty} \frac{1+\cos n}{n^2}$ converges since $\frac{1+\cos n}{n^2} \le$

$\frac{1+1}{n^2} = \frac{2}{n^2}$ and $\sum_{n=1}^{\infty} \frac{2}{n^2}$ converges (p-series

with $p = 2$).

17 $\sum_{n=1}^{\infty} \frac{\ln n}{n^2}$ converges since $\frac{\ln n}{n^2} \le \frac{\sqrt{n}}{n^2} = \frac{1}{n^{3/2}}$ and

$\sum_{n=1}^{\infty} \frac{1}{n^{3/2}}$ converges (p-series with $p = 3/2$).

19 $\sum_{n=1}^{\infty} \frac{2^n}{n!}$ converges since $\frac{2^n}{n!} \le 2\left(\frac{2}{3}\right)^{n-2}$ for $n \ge 2$

and $\sum_{n=2}^{\infty} 2\left(\frac{2}{3}\right)^{n-2}$ converges (geometric series).

21 $\sum_{n=1}^{\infty} \frac{e^n}{\pi^n} = \sum_{n=1}^{\infty} \left(\frac{e}{\pi}\right)^n$ converges (geometric series).

23 $\sum_{n=1}^{\infty} \frac{3k+1}{2k+10}$ diverges since $\lim_{k\to\infty} \frac{3k+1}{2k+10} = \frac{3}{2} \ne$

0 (nth-term test for divergence).

25 $\sum_{n=2}^{\infty} \frac{1}{\ln n}$ diverges since $\ln n < \sqrt{n}$; thus $\frac{1}{\sqrt{n}} <$

$\frac{1}{\ln n}$. But $\sum_{n=2}^{\infty} \frac{1}{\sqrt{n}}$ diverges (p-series with $p =$

$1/2$), so $\sum_{n=2}^{\infty} \frac{1}{\ln n}$ also diverges.

27 $\sum_{n=1}^{\infty} \left(\frac{n+1}{n+3}\right)^n$ diverges since $\lim_{n\to\infty} \left(\frac{n+1}{n+3}\right)^n =$

$\lim_{n\to\infty} \left(\frac{1+1/n}{1+3/n}\right)^n = \lim_{n\to\infty} \frac{(1+1/n)^n}{(1+3/n)^n} = \frac{e}{e^3} = e^{-2}$

$\ne 0$ (nth-term test for divergence).

29 The result merely implies that $\sum_{n=1}^{\infty} a_n$ is less than the divergent series $\sum_{n=1}^{\infty} b_n$. No conclusion can be drawn.

31 If $\sum_{n=1}^{\infty} b_n$ is convergent, then $\sum_{n=1}^{\infty} 3b_n$ and $\sum_{n=1}^{\infty} 5b_n$ are convergent. If $3b_n \le a_n \le 5b_n$, then $\sum_{n=1}^{\infty} a_n$ is also convergent.

33 If $\sum_{n=1}^{\infty} b_n$ is convergent, then $\lim_{n\to\infty} b_n = 0$. Thus, there is a number N such that $b_n < 1$ for $n \ge N$. Then $a_n < b_n^2 < b_n$ for $n \ge N$, so $\sum_{n=N}^{\infty} a_n$ is convergent. Hence $\sum_{n=1}^{\infty} a_n$ is convergent.

35 If $x > 0$ then $\sum_{n=1}^{\infty} \frac{x^n}{n2^n} = \sum_{n=1}^{\infty} \frac{1}{n}\left(\frac{x}{2}\right)^n$ converges for $0 < x < 2$ since $\frac{1}{n}\left(\frac{x}{2}\right)^n \le \left(\frac{x}{2}\right)^n$ and $\sum_{n=1}^{\infty} \left(\frac{x}{2}\right)^n$ converges (geometric series); and diverges for $x \ge 2$ since $\frac{1}{n} \le \frac{1}{n}\left(\frac{x}{2}\right)^n$ and $\sum_{n=1}^{\infty} \frac{1}{n}$ diverges (harmonic series).

37 If $\lim_{n\to\infty} \frac{p_n}{d_n} = L > 0$, then there is an integer N so that $\frac{p_n}{d_n} > \frac{L}{2}$ for $n \ge N$. Thus, $\frac{L}{2}d_n < p_n$ for $n \ge N$. Since $\sum_{n=N}^{\infty} d_n$ diverges, so do $\sum_{n=N}^{\infty} \frac{L}{2}d_n$ and $\sum_{n=N}^{\infty} p_n$. Thus $\sum_{n=1}^{\infty} p_n$ diverges. If $\lim_{n\to\infty} \frac{p_n}{d_n} = \infty$, then there is an integer N such that $\frac{p_n}{d_n} > 1$ for $n \ge N$, that is, $d_n < p_n$ for $n \ge N$. Since $\sum_{n=N}^{\infty} d_n$ diverges, so do $\sum_{n=N}^{\infty} p_n$ and $\sum_{n=1}^{\infty} p_n$.

39 (a) $\sum_{n=1}^{\infty} \frac{1}{1 + 2^n}$ converges since $\frac{1}{1 + 2^n} \le \frac{1}{2^n} = \left(\frac{1}{2}\right)^n$ and $\sum_{n=1}^{\infty} \left(\frac{1}{2}\right)^n$ converges (geometric series).

(b) $S_3 = 1/3 + 1/5 + 1/9 \approx 0.644$ and $\sum_{n=4}^{\infty} \frac{1}{1 + 2^n} \le \sum_{n=4}^{\infty} \left(\frac{1}{2}\right)^n = \left(\frac{1}{2}\right)^4\left[1 + \left(\frac{1}{2}\right) + \left(\frac{1}{2}\right)^2 + \cdots\right] = 0.125$, so $0.644 \le \sum_{n=1}^{\infty} \frac{1}{1 + 2^n} \le 0.644 + 0.125 = 0.769$.

41 Since $\lim_{n\to\infty} c_n = 0$ there is some integer N such that $0 \le c_n < 1$ for $n \ge N$. Then $a_n c_n^2 \le a_n c_n$ for

$n \ge N$. Since $\sum_{n=N}^{\infty} a_n c_n$ converges, so do $\sum_{n=N}^{\infty} a_n c_n^2$ and $\sum_{n=1}^{\infty} a_n c_n^2$.

43 Let $f(x) = \dfrac{x+2}{(x+1)x^3}$ which is continuous, positive, and decreasing for $x > 0$. Then $S_5 = \sum_{n=1}^{5} \dfrac{n+2}{n+1} \cdot \dfrac{1}{n^3} = \dfrac{3}{2} + \dfrac{1}{6} + \dfrac{5}{108} + \dfrac{3}{160} + \dfrac{7}{750} \approx 1.741$. By Theorem 3 of Sec. 10.4, $R_n < \int_n^{\infty} \dfrac{x+2}{(x+1)x^3}\,dx = \int_n^{\infty} \left[\dfrac{-1}{x+1} + \dfrac{1}{x} + \dfrac{-1}{x^2} + \dfrac{2}{x^3}\right] dx = \left[-\ln(x+1) + \ln x + \dfrac{1}{x} - \dfrac{1}{x^2}\right]\Bigg|_n^{\infty} = \dfrac{1}{n^2} - \dfrac{1}{n} - \ln\dfrac{n}{n+1}$, so $R_5 < 0.023$. Thus, $\sum_{n=1}^{\infty} \dfrac{n+2}{(n+1)n^3} < 1.741 + 0.023 = 1.764$.

10.6 Ratio Tests

1 $\lim_{n\to\infty} \dfrac{a_{n+1}}{a_n} = \lim_{n\to\infty} \dfrac{(n+1)^2}{3^{n+1}} \cdot \dfrac{3^n}{n^2} = \lim_{n\to\infty} \dfrac{1}{3} \cdot \left(\dfrac{n+1}{n}\right)^2 = \dfrac{1}{3} \cdot 1^2 = \dfrac{1}{3} < 1$, so $\sum_{n=1}^{\infty} \dfrac{n^2}{3^n}$ converges.

3 $\lim_{n\to\infty} \dfrac{a_{n+1}}{a_n} = \lim_{n\to\infty} \dfrac{(n+1)\ln(n+1)}{3^{n+1}} \cdot \dfrac{3^n}{n \ln n} = \lim_{n\to\infty} \dfrac{n+1}{n} \cdot \dfrac{\ln(n+1)}{\ln n} \cdot \dfrac{1}{3} = \dfrac{1}{3} \lim_{n\to\infty} \dfrac{\ln(n+1)}{\ln n} \underset{H}{=} \dfrac{1}{3} \lim_{n\to\infty} \dfrac{\frac{1}{n+1}}{\frac{1}{n}} = \dfrac{1}{3} < 1$, so $\sum_{n=1}^{\infty} \dfrac{n \ln n}{3^n}$ converges.

5 $\lim_{n\to\infty} \dfrac{a_{n+1}}{a_n} = \lim_{n\to\infty} \dfrac{(2(n+1)+1)(2^{n+1}+1)}{3^{n+1}+1} \cdot \dfrac{3^n+1}{(2n+1)(2^n+1)} = \lim_{n\to\infty} \dfrac{2n+3}{2n+1} \cdot \dfrac{2 + \frac{1}{2^n}}{1 + \frac{1}{2^n}} \cdot \dfrac{1 + \frac{1}{3^n}}{3 + \frac{1}{3^n}} = 1 \cdot 2 \cdot \dfrac{1}{3} = \dfrac{2}{3} < 1$, so $\sum_{n=1}^{\infty} \dfrac{(2n+1)(2^n+1)}{3^n+1}$ converges.

7 (a) $\lim_{n\to\infty} \dfrac{a_{n+1}}{a_n} = \lim_{n\to\infty} \dfrac{2^{n+1}}{((n+1)+1)^{n+1}} \cdot \dfrac{(n+1)^n}{2^n} = \lim_{n\to\infty} \dfrac{2}{n+2} \cdot \left(\dfrac{n+1}{n+2}\right)^n = \lim_{n\to\infty} \dfrac{2}{n+2} \cdot \left(\dfrac{1+1/n}{1+2/n}\right)^n = \lim_{n\to\infty} \dfrac{2}{n+2} \cdot \dfrac{(1+1/n)^n}{(1+2/n)^n} = 0 \cdot \dfrac{e^1}{e^2} = 0 < 1$, so $\sum_{n=1}^{\infty} \dfrac{2^n}{(n+1)^n}$ converges.

(b) $\lim_{n\to\infty} (a_n)^{1/n} = \lim_{n\to\infty} \dfrac{2}{n+1} = 0 < 1$, so

$\sum_{n=1}^{\infty} \frac{2^n}{(n+1)^n}$ converges.

(c) $\frac{2^n}{(n+1)^n} < \frac{2^n}{3^n}$ for $n \geq 2$ and $\sum_{n=2}^{\infty} \frac{2^n}{3^n} =$

$\sum_{n=2}^{\infty} \left(\frac{2}{3}\right)^n$ is a convergent geometric series, so

$\sum_{n=2}^{\infty} \frac{2^n}{(n+1)^n}$ and $\sum_{n=1}^{\infty} \frac{2^n}{(n+1)^n}$ converge.

9 $\lim_{n\to\infty} \frac{a_{n+1}}{a_n} = \lim_{n\to\infty} \frac{(n+1)x^{n+1}}{nx^n} = \lim_{n\to\infty} \left(1 + \frac{1}{n}\right)x =$

x, permitting us to state the following two conclusions.

(a) For $0 < x < 1$ the series $\sum_{n=1}^{\infty} nx^n$ converges.

(b) For $x > 1$ the series $\sum_{n=1}^{\infty} nx^n$ diverges. If

$x = 1$ then $\sum_{n=1}^{\infty} nx^n = \sum_{n=1}^{\infty} n$ diverges by the

nth-term test for divergence.

11 Note that $\lim_{n\to\infty} \frac{a_{n+1}}{a_n} = \lim_{n\to\infty} \frac{x^{n+1}}{2^{n+1}} \cdot \frac{2^n}{x^n} = \frac{x}{2}$.

(a) For $0 < x < 2$ the series $\sum_{n=1}^{\infty} \frac{x^n}{2^n}$ converges.

(b) For $x > 2$ the series $\sum_{n=1}^{\infty} \frac{x^n}{2^n}$ diverges. If $x =$

2 then $\sum_{n=1}^{\infty} \frac{x^n}{2^n} = \sum_{n=1}^{\infty} 1$ diverges by nth-term

test for divergence.

13 $\lim_{n\to\infty} (a_n)^{1/n} = \lim_{n\to\infty} \left(\frac{n^n}{3^{n^2}}\right)^{1/n} = \lim_{n\to\infty} \frac{n}{3^n} \underset{H}{=}$

$\lim_{n\to\infty} \frac{1}{3^n \ln 3} = 0 < 1$, so $\sum_{n=1}^{\infty} \frac{n^n}{3^{n^2}}$ converges.

15 $n^2 < 1.5^n$ for $n \geq 13$ so that $\sum_{n=13}^{\infty} \frac{n^2}{2^n} <$

$\sum_{n=13}^{\infty} \frac{1.5^n}{2^n} = \sum_{n=13}^{\infty} (0.75)^n = \frac{(0.75)^{13}}{1 - 0.75} < 0.095.$

We have $S_{13} = \frac{1}{2} + \frac{4}{4} + \frac{9}{8} + \frac{16}{16} + \frac{25}{32} +$

$\frac{36}{64} + \cdots + \frac{169}{8192} \approx 5.972$, so $\sum_{n=1}^{\infty} \frac{n^2}{2^n} < 5.972$

$+ 0.095 = 6.067$.

17 Note that $\frac{6^3}{6!} = \frac{6 \cdot 6 \cdot 6}{6 \cdot 5 \cdot 4 \cdot 3 \cdot 2} = 0.3$, $\frac{7^3}{7!} \approx 0.068$

$< (0.3)^2$, $\frac{8^3}{8!} \approx 0.0127 < (0.3)^3$, $\frac{9^3}{9!} \approx 0.002$

$< (0.3)^4$, …, so that $\sum_{n=6}^{\infty} \frac{n^3}{n!} < \sum_{n=6}^{\infty} (0.3)^{n-1} =$

$\frac{0.3^5}{1 - 0.3} < 0.0035$. Then $S_5 =$

$\frac{1}{1} + \frac{8}{2!} + \frac{27}{3!} + \frac{64}{4!} + \frac{125}{5!} \approx 13.2083$, so that

$\sum_{n=1}^{\infty} \frac{n^3}{n!} < 13.2083 + 0.0035 = 13.2118$.

19 The function $f(x) = \frac{1}{x^2 - 1}$ is continuous,

positive, and decreasing for $x > 1$ so that $R_n <$

$\int_n^{\infty} \frac{1}{x^2 - 1}\, dx = \int_n^{\infty} \left[\frac{1/2}{x - 1} + \frac{-1/2}{x + 1}\right] dx =$

$\left[\frac{1}{2}\ln(x - 1) - \frac{1}{2}\ln(x + 1)\right]\Big|_n^{\infty} =$

$\lim_{b\to\infty} \frac{1}{2}\ln\left(\frac{x - 1}{x + 1}\right)\Big|_n^b = -\frac{1}{2}\ln\frac{n - 1}{n + 1}$. If $n = 4$ then

$R_4 < 0.255$ and $\sum_{n=2}^{\infty} \frac{1}{n^2 - 1} < \frac{1}{3} + \frac{1}{8} + \frac{1}{15} +$

$0.255 = 0.780$.

21 The function $f(x) = \frac{\ln x}{x}$ is continuous, positive and decreasing for $x > e$ so that the nth partial sum $S_m = f(1) + f(2) + \sum_{n=3}^{m} f(k) > f(1) + f(2) + \int_3^{m+1} f(x)\, dx$, that is, $S_m > \frac{\ln 2}{2} + \int_3^{m+1} \frac{\ln x}{x}\, dx$

$= \frac{\ln 2}{2} + \frac{1}{2}(\ln x)^2\Big|_3^{m+1} = \frac{\ln 2}{2} +$

$\frac{1}{2}(\ln(m + 1))^2 - \frac{1}{2}(\ln 3)^2 \approx -0.2569 +$

$\frac{1}{2}(\ln(m + 1))^2$; that is, $S_m > -0.2569 + \frac{1}{2}(\ln(m + 1))^2$. To make S_m exceed 1000, choose $-0.2569 + \frac{1}{2}(\ln(m + 1))^2 > 1000$. This implies that $(\ln(m + 1))^2 > 2001$; hence $\ln(m + 1) > (2001)^{1/2}$, which leads to $m + 1 > e^{\sqrt{2001}}$. We should choose $m > e^{\sqrt{2001}} - 1$.

23 As noted in Appendix C (see p. S-24 of the text), the mth partial sum of the geometric series $a + ax + ax^2 + \cdots$ is $\frac{a(1 - x^m)}{1 - x}$. For the series $\sum_{n=1}^{\infty} (1.01)^n$ we have both $a = 1.01$ and $x = 1.01$, so the mth partial sum is $S_m = \frac{(1.01)^{m+1} - 1.01}{0.01}$.

If we want $S_m > 1000$, then $\frac{(1.01)^{m+1} - 1.01}{0.01} >$ 1000, so

$$(1.01)^{m+1} - 1.01 > 10,$$
$$(1.01)^{m+1} > 11.01,$$
$$(m + 1)\ln 1.01 > \ln 11.01,$$
$$m + 1 > \frac{\ln(11.01)}{\ln(1.01)}.$$

We therefore need to choose $m > \frac{\ln 11.01}{\ln 1.01} - 1$ ≈ 240.08, so $m = 241$ will do.

25 By Example 2 the series $\sum_{n=1}^{\infty} \frac{x^n}{n!}$ converges for $x > 0$. Since the series converges, $\lim_{n\to\infty} \frac{x^n}{n!} = 0$.

27 Since $a_n = 2^n/n$, $\lim_{n\to\infty} \sqrt[n]{a_n} = \lim_{n\to\infty} \left(\frac{2^n}{n}\right)^{1/n} =$

$\lim_{n\to\infty} \frac{2}{n^{1/n}} = 2 > 1$, so $\sum_{n=1}^{\infty} \frac{2^n}{n}$ diverges by the root test.

29 (a) Since $(p_n)^{1/n} < r$ for $n > N$ we have $p_n < r^n$ for $n > N$. Then $p_{N+1} < r^{N+1}$, $p_{N+2} < r^{N+2}$, $p_{N+3} < r^{N+3}$, $\cdots$ so that $\sum_{n=N+1}^{\infty} p_n = p_{N+1} +$

$p_{N+2} + p_{N+3} + \cdots < r^{N+1} + r^{N+2} + r^{N+3} + \cdots = \dfrac{r^{N+1}}{1 - r} < \infty$. Therefore $\sum_{n=1}^{\infty} p_n$ converges.

(b) Since $(p_n)^{1/n} > r > 1$ for $n > N$ we have $p_n > 1^n = 1$ for $n > N$. Thus, $\lim_{n\to\infty} p_n \geq 1$ and therefore $\sum_{n=1}^{\infty} p_n$ diverges by the nth-term test for divergence.

10.7 Tests for Series with Both Positive and Negative Terms

1 $\sum_{n=1}^{\infty} (-1)^{n+1}\dfrac{n}{n+1}$ diverges since $\lim_{n\to\infty} (-1)^{n+1}\dfrac{n}{n+1} \neq 0$ (nth-term test for divergence).

3 $\sum_{n=1}^{\infty} (-1)^{n+1}\dfrac{1}{\sqrt{n}}$ converges since $a_n = \dfrac{1}{\sqrt{n}}$ is a positive, decreasing sequence with $a_n \to 0$ as $n \to \infty$. (Observe that without the alternating signs, this series would be a divergent p-series.)

5 $\dfrac{3}{\sqrt{1}} - \dfrac{2}{\sqrt{1}} + \dfrac{3}{\sqrt{2}} - \dfrac{2}{\sqrt{2}} + \dfrac{3}{\sqrt{3}} - \dfrac{2}{\sqrt{3}} + \cdots = \dfrac{1}{\sqrt{1}} + \dfrac{1}{\sqrt{2}} + \dfrac{1}{\sqrt{3}} + \cdots = \sum_{n=1}^{\infty} \dfrac{1}{\sqrt{n}}$ diverges (p-series with $p = 1/2$).

7 $\sum_{n=1}^{\infty} (-1)^{n+1}\dfrac{n}{2n+1}$ diverges since $\lim_{n\to\infty} (-1)^{n+1}\dfrac{n}{2n+1} \neq 0$ (nth-term test for divergence).

9 (a) $S_5 = \dfrac{1}{1} - \dfrac{1}{2} + \dfrac{1}{3} - \dfrac{1}{4} + \dfrac{1}{5} = \dfrac{47}{60} \approx 0.78333$ and $S_6 = S_5 - \dfrac{1}{6} = \dfrac{47}{60} - \dfrac{1}{6} = \dfrac{37}{60} \approx 0.61667$.

(b) Recall that the sum S is between S_n and S_{n+1} for every n. Since S_5 is larger than S_6 and S must lie between them, S_5 must be larger than S.

(c) As noted in (b), the sum of the series is between S_5 and S_6, that is, between 0.78333 and 0.61667.

11 $\sum_{n=1}^{\infty} \dfrac{(-1)^n}{\sqrt[3]{n^2}} = \sum_{n=1}^{\infty} \dfrac{(-1)^n}{n^{2/3}}$ converges conditionally: it is a decreasing alternating series whose terms approach 0, so $\sum_{n=1}^{\infty} \dfrac{(-1)^n}{n^{2/3}}$ converges; but $\sum_{n=1}^{\infty} \dfrac{1}{n^{2/3}}$ diverges (p-series with $p = 2/3$).

13 $\sum_{n=2}^{\infty} \dfrac{(-1)^n}{n \ln n}$ is conditionally convergent: it is a decreasing alternating series whose terms approach 0, so $\sum_{n=2}^{\infty} \dfrac{(-1)^n}{n \ln n}$ converges; but $\int_2^{\infty} \dfrac{1}{x \ln x}\, dx = \ln(\ln x)\big|_2^{\infty} = \infty$, so $\sum_{n=2}^{\infty} \dfrac{1}{n \ln n}$ diverges.

15 $\sum_{n=1}^{\infty} \left(1 - \cos\dfrac{\pi}{n}\right)$ converges absolutely since

$$\lim_{n\to\infty} \frac{\left|1 - \cos\frac{\pi}{n}\right|}{1/n^2} = \lim_{n\to\infty} \frac{1 - \cos\frac{\pi}{n}}{1/n^2} \underset{H}{=}$$

$$\lim_{n\to\infty} \frac{\left(\sin\frac{\pi}{n}\right)\frac{-\pi}{n^2}}{\frac{-2}{n^3}} = \lim_{n\to\infty} \frac{\pi}{2} \cdot \frac{\sin\frac{\pi}{n}}{\frac{1}{n}} =$$

$$\lim_{n\to\infty} \frac{\pi^2}{2} \cdot \frac{\sin\frac{\pi}{n}}{\frac{\pi}{n}} = \frac{\pi^2}{2} \text{ and } \sum_{n=1}^{\infty} \frac{1}{n^2} \text{ converges}$$

(p-series with $p = 2$).

17 $\sum_{n=1}^{\infty} \sin\frac{\pi}{n^2}$ converges absolutely since $\lim_{n\to\infty} \frac{\left|\sin\frac{\pi}{n^2}\right|}{\frac{1}{n^2}}$

$$= \lim_{n\to\infty} \frac{\sin\frac{\pi}{n^2}}{\frac{1}{n^2}} = \pi \lim_{n\to\infty} \frac{\sin\frac{\pi}{n^2}}{\frac{\pi}{n^2}} = \pi, \text{ and } \sum_{n=1}^{\infty} \frac{1}{n^2}$$

converges (p-series with $p = 2$).

19 $\frac{1}{1^2} + \frac{1}{2^2} - \frac{1}{3^2} - \frac{1}{4^2} + \frac{1}{5^2} + \frac{1}{6^2} - \cdots$ converges absolutely since the series of the absolute value of the terms is $\sum_{n=1}^{\infty} \frac{1}{n^2}$, which is convergent ($p$-series with $p = 2$).

21 $\sum_{n=1}^{\infty} \frac{\cos n\pi}{2n+1} = \sum_{n=1}^{\infty} \frac{(-1)^n}{2n+1}$, which converges conditionally. Note that it is a decreasing alternating series whose terms approach 0, so $\sum_{n=1}^{\infty} \frac{(-1)^n}{2n+1}$ converges; but $\sum_{n=1}^{\infty} \frac{1}{2n+1}$ diverges since $\frac{1}{2n+1} > \frac{1}{2n+2}$ and $\sum_{n=1}^{\infty} \frac{1}{2n+2} =$

$\frac{1}{2} \sum_{n=1}^{\infty} \frac{1}{n+1}$ diverges.

23 $\sum_{n=1}^{\infty} \frac{(-9)^n}{10^n + n}$ converges absolutely since $\left|\frac{(-9)^n}{10^n + n}\right|$

$= \frac{9^n}{10^n + n} < \frac{9^n}{10^n} = \left(\frac{9}{10}\right)^n$; thus $\sum_{n=1}^{\infty} \frac{(-9)^n}{10^n + n}$

converges absolutely by comparison to a convergent geometric series.

25 $\sum_{n=1}^{\infty} \frac{(-1.01)^n}{n!}$ converges absolutely since

$$\lim_{n\to\infty} \left|\frac{\frac{(-1.01)^{n+1}}{(n+1)!}}{\frac{(-1.01)^n}{n!}}\right| = \lim_{n\to\infty} \frac{1.01}{n+1} = 0 < 1.$$

27 $\lim_{n\to\infty} \left|\frac{a_{n+1}}{a_n}\right| = \lim_{n\to\infty} \left|\frac{\frac{x^{n+1}}{(n+1)!}}{x^n/n!}\right| = \lim_{n\to\infty} \frac{|x|}{n+1} = 0 <$

1 for all x values so that $\sum_{n=1}^{\infty} \frac{x^n}{n!}$ converges

absolutely for all x values.

29 (a) $|R_n| = \left|\sum_{k=n+1}^{\infty} \frac{\sin k}{k^2}\right| \le \sum_{k=n+1}^{\infty} \frac{|\sin k|}{k^2} <$

$\sum_{k=n+1}^{\infty} \frac{1}{k^2} < \int_n^{\infty} \frac{1}{x^2}\,dx = \left.\frac{-1}{x}\right|_n^{\infty} = \frac{1}{n}$. To

make sure $|R_n| < 0.005$ choose $\frac{1}{n} < 0.005$.

Thus $n > 200$, so we may choose $n = 201$.

(b) The partial sum S_{201} will estimate $\sum_{n=1}^{\infty} \frac{\sin n}{n^2}$ to two decimal places; $S_{201} =$

$$\frac{\sin 1}{1^2} + \frac{\sin 2}{2^2} + \frac{\sin 3}{3^2} + \cdots + \frac{\sin 201}{201^2}$$

≈ 1.01. (Experimenting with your calculator will show that we can obtain good two-decimal approximations for much smaller values of n. Since $\sin n$ is negative as often as it is positive, convergence is faster than indicated by our analysis above.)

31 (a) By Example 2 of Sec. 10.6, $\sum_{n=1}^{\infty} \frac{2^n}{n!}$ converges.

(b) If we use the sum of the first n terms to approximate the series, the error is $\sum_{k=n+1}^{\infty} \frac{2^k}{k!}$

$$= \sum_{k=n+1}^{\infty} \frac{2^k}{k(k-1)\cdots(n+1)n!} \le$$

$$\sum_{k=n+1}^{\infty} \frac{2^k}{(n+1)^{k-n}n!} = \frac{1}{n!}\sum_{k=n+1}^{\infty} \frac{2^k}{(n+1)^{k-n}} =$$

$$\frac{1}{n!}\frac{\frac{2^k}{(n+1)}}{1-\frac{2}{n+1}} = \frac{2^{n+1}}{n!(n+1-2)} = \frac{2^{n+1}}{n!(n-1)},$$

which is approximately 0.00181 when $n = 8$. Hence $S_8 = \frac{2012}{315} \approx 6.39$ is an estimate accurate to two decimal places.

33 First, consider the even-numbered partial sums for $\sum_{n=1}^{\infty} (-1)^{n+1}p_n$. We have $S_2 = p_1 - p_2$,

$S_4 = (p_1 - p_2) + (p_3 - p_4) = S_2 + (p_3 - p_4)$, $S_6 = (p_1 - p_2) + (p_3 - p_4) + (p_5 - p_6) = S_4 + (p_5 - p_6), \cdots$; since p_n is a decreasing sequence, $S_2, S_4, S_6, S_8, \cdots$ is an increasing sequence. Also, $S_2 = p_1 + (-p_2) < p_1$, $S_4 = p_1 + (-p_2 + p_3) - p_4 < p_1$, $S_6 = p_1 + (-p_2 + p_3) + (-p_4 + p_5) - p_6 < p_1, \cdots$; that is, $S_{2n} < p_1$ for $n = 1, 2, 3, \cdots$. It follows from Theorem 2 of Sec. 10.2 that $\lim_{n\to\infty} S_{2n}$ exists, say $\lim_{n\to\infty} S_{2n} = S$. Next consider $S_1 = p_1$,

$S_3 = (p_1 - p_2) + p_3 = S_2 + p_3$,
$S_5 = (p_1 - p_2) + (p_3 - p_4) + p_5 = S_4 + p_5, \cdots$,
and in general, $S_{2n+1} = S_{2n} + p_{n+1}$ for $n = 0, 1, 2, 3, \cdots$. Hence $\lim_{n\to\infty} S_{2n+1} = \lim_{n\to\infty} S_{2n} + \lim_{n\to\infty} p_{n+1} = S + 0 = S$. We have shown that the sequence of partial sums $S_1, S_2, S_3, S_4, \cdots$ converges to S, so that $\sum_{n=1}^{\infty} (-1)^{n+1}p_n$ converges.

35 $$\lim_{n\to\infty} \left| \frac{\frac{n+1}{2(n+1)+1}x^{n+1}}{\frac{n}{2n+1}x^n} \right| = \lim_{n\to\infty} \frac{n+1}{n}\cdot\frac{2n+1}{2n+3}|x| =$$

$|x|$. If $|x| < 1$, that is, if $-1 < x < 1$, then $\sum_{n=1}^{\infty} \frac{nx^n}{2n+1}$ converges absolutely. And if $|x| > 1$, that is, if $x < -1$ or $x > 1$, then the series diverges. If $x = 1$, then $\sum_{n=1}^{\infty} \frac{nx^n}{2n+1} = \sum_{n=1}^{\infty} \frac{n}{2n+1}$

diverges since $\lim_{n\to\infty} \frac{n}{2n+1} = \frac{1}{2} \neq 0$ (nth-term test for divergence). If $x = -1$, then $\sum_{n=1}^{\infty} \frac{nx^n}{2n+1} = \sum_{n=1}^{\infty} (-1)^n \frac{n}{2n+1}$ diverges since $\lim_{n\to\infty} (-1)^n \frac{n}{2n+1} \neq 0$ (nth-term test for divergence).

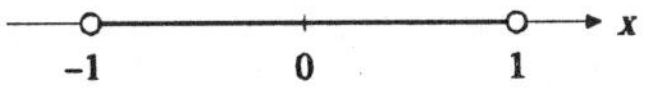

37 This argument is not correct. The series $1 - 1/2 + 1/3 - 1/4 + 1/5 - \cdots$ is conditionally convergent; hence, rearrangement of its terms may change the value of the original convergent series.

39 (a) By comparing the graphs of $y = x$ and $y = \sin x$ for $x \geq 0$ we see that $0 \leq \sin x \leq x$, so that $0 \leq \sin a_n \leq a_n$ and $\sum_{n=1}^{\infty} \sin a_n$ converges by the comparison test since $\sum_{n=1}^{\infty} a_n$ converges.

(b) $\sum_{n=1}^{\infty} a_n$ converges so that $a_n \to 0$ as $n \to \infty$. Thus $\sum_{n=1}^{\infty} \cos a_n$ diverges since $\lim_{n\to\infty} \cos a_n = \cos\left(\lim_{n\to\infty} a_n\right) = \cos 0 = 1 \neq 0$ (nth-term test for divergence).

10.S Guide Quiz

2 Consider $\sum_{n=1}^{\infty} \frac{(-1)^{n+1}}{\sqrt{n}}$, which converges by the alternating-series test. However, $\sum_{n=1}^{\infty} \frac{1}{n}$ is divergent, although its terms are the squares of the terms of the first series.

3 (a) Since $|R_n| < |a_{n+1}| = \frac{1}{(n+1)^2}$, choosing $\frac{1}{(n+1)^2} < 0.0005$ ensures that S_n will estimate $\sum_{n=1}^{\infty} \frac{(-1)^{n+1}}{n^2}$ to three places. Thus we must have $(n+1)^2 > 2000$, so $n + 1 > 44.72$ and $n > 43.72$. We see that $n = 44$ suffices.

(b) $|R_k| = \left| \sum_{n=k+1}^{\infty} \frac{(-1)^{n+1}}{n^2} \right| \leq \sum_{n=k+1}^{\infty} \frac{1}{n^2} < \int_k^{\infty} \frac{1}{x^2}\,dx = \frac{-1}{x}\Big|_k^{\infty} = \frac{1}{k}$, so choosing $1/k < 0.0005$ ensures that S_k will estimate $\sum_{n=1}^{\infty} \frac{(-1)^{n+1}}{n^2}$ to three places. Hence $k > 2000$ suffices.

4 Review the statement of the convergence tests to determine the series to which each applies.

5 (a) $R_n < \int_n^{\infty} \frac{x}{e^x}\,dx = \int_n^{\infty} xe^{-x}\,dx = \frac{n+1}{e^n}$, where we used Formula 62 from the table of antiderivatives (with $a = -1$). Hence choosing $\frac{n+1}{e^n} < 0.005$ ensures that S_n approximates the series to two decimal places. Finding that $n = 8$ suffices (use your calculator), we compute $S_8 = 1/e + 2/e^2 + \cdots + 8/e^8 \approx 0.919$ as our approximation of $\sum_{n=1}^{\infty} \frac{n}{e^n}$ to two places.

(b) Since $n < 1.5^n$ (other choices are possible), we have $|R_k| = \sum_{n=k+1}^{\infty} \frac{n}{e^n} < \sum_{n=k+1}^{\infty} \frac{1.5^n}{e^n} =$

$$\sum_{n=k+1}^{\infty} \left(\frac{1.5}{e}\right)^n = \frac{(1.5/e)^{k+1}}{1 - 1.5/e},$$ which we find is approximately equal to 0.0032 when $k = 10$. Hence this choice of k ensures that S_k estimates the series to two decimal places.

Upon evaluating $S_{10} = \frac{1}{e} + \frac{2}{e^2} + \cdots + \frac{10}{e^{10}} \approx 0.920$, we see that this estimate agrees to two decimal places with the one computed in (a).

6 (a) $$\lim_{n\to\infty} \frac{a_{n+1}}{a_n} = \lim_{n\to\infty} \frac{(n+1)!}{(n+1)^{n+1}} \cdot \frac{n^n}{n!} =$$

$$\lim_{n\to\infty} \left(\frac{n}{n+1}\right)^n = \lim_{n\to\infty} \left(\frac{1}{1+1/n}\right)^n =$$

$$\lim_{n\to\infty} \frac{1}{(1+1/n)^n} = \frac{1}{e} < 1,$$ so $\sum_{n=1}^{\infty} \frac{n!}{n^n}$ converges.

(b) $\sum_{n=1}^{\infty} \tan \frac{1}{n}$ diverges by limit-comparison with the harmonic series: $\lim_{n\to\infty} \frac{\tan 1/n}{1/n} = \lim_{x\to 0} \frac{\tan x}{x} = 1.$

(c) By the ratio test, we have

$$\lim_{n\to\infty} \left| \frac{\frac{(n+1)^2+1}{n+1} \cdot \frac{x^{n+1}}{(n+1)+1}}{\frac{n^2+1}{n} \cdot \frac{x^n}{n+1}} \right| =$$

$$\lim_{n\to\infty} \frac{n^2+2n+2}{n+1} \cdot \frac{n}{n^2+1} \cdot \frac{n+1}{n+2} |x| = |x|.$$ If $|x| < 1$, that is, if $-1 < x < 1$, then the series converges. If $|x| > 1$, that is, if $x < -1$ or $x > 1$, then the series diverges. If $x = 1$ or $x = -1$, then the series diverges since

$$\lim_{n\to\infty} \left| \frac{n^2+1}{n} \cdot \frac{x^n}{n+1} \right| = \lim_{n\to\infty} \frac{n^2+1}{n(n+1)} = 1 \neq 0$$

(nth-term test for divergence).

(d) By the root test, $\sum_{n=1}^{\infty} \frac{2^n}{(1 + 1/n)^{n^2}}$ converges,

since $\lim_{n\to\infty} (a_n)^{1/n} = \lim_{n\to\infty} \left[\frac{2^n}{(1+1/n)^{n^2}}\right]^{\frac{1}{n}} =$

$$\lim_{n\to\infty} \frac{2}{(1+1/n)^n} = \frac{2}{e} < 1.$$

7 (a) False. Let B be the largest value of $|b_n|$ for $n = 1, 2, 3, \ldots$. and let $a_n = -B$ for all n. Then $a_n \le b_n$ but $\sum_{n=1}^{\infty} a_n = -\sum_{n=1}^{\infty} B$ clearly diverges.

(b) False. See part (a).

(c) False. Let $a_n = b_n$ for $n = 1, 2, 3, \cdots$; then $a_n \le b_n$ and $\sum_{n=1}^{\infty} a_n = \sum_{n=1}^{\infty} b_n$ converges.

(d) True. See parts (a) and (c).

8 (a) If $|a_n| \le \left(\frac{1}{2}\right)^n$ for $n \ge 1$, then $\sum_{n=1}^{\infty} |a_n| \le \sum_{n=1}^{\infty} \left(\frac{1}{2}\right)^n = \frac{1/2}{1 - 1/2} = 1$.

(b) By (a), $\sum_{n=1}^{\infty} |a_n|$ converges, so $\sum_{n=1}^{\infty} a_n$ must also converge. Also by part (a) it follows that

$$-1 \le -\sum_{n=1}^{\infty} |a_n| \le \sum_{n=1}^{\infty} a_n \le \sum_{n=1}^{\infty} |a_n| \le 1.$$

9 (a) No. The error is $\sum_{n=10}^{\infty} \frac{1}{n^2}$, which is certainly larger than its first term, $1/10^2$. However, $R_n < \int_n^{\infty} \frac{1}{x^2}\,dx = \left.\frac{-1}{x}\right|_n^{\infty} = \frac{1}{n}$ so $R_9 < 1/9$.

(b) Since $\sum_{n=1}^{\infty} \frac{(-1)^n}{n^n}$ is a strictly decreasing alternating series, $|R_n| < |a_{n+1}|$ so that $|R_9| < |a_{10}| = 1/10^2$.

10.S Review Exercises

1 $\sum_{n=1}^{\infty} \frac{(-1)^n}{n^2}$ converges.

Alternating series test: $\frac{1}{n^2}$ is a positive, decreasing sequence with $\frac{1}{n^2} \to 0$ as $n \to \infty$.

Absolute convergence test: $\sum_{n=1}^{\infty} \left|\frac{(-1)^n}{n^2}\right| = \sum_{n=1}^{\infty} \frac{1}{n^2}$ converges (p-series with $p = 2$).

3 $\sum_{k=1}^{\infty} \frac{\sqrt{k}}{k^2 + 1}$ converges.

Comparison test: $\frac{\sqrt{k}}{k^2 + 1} < \frac{\sqrt{k}}{k^2} = \frac{1}{k^{3/2}}$ and $\sum_{k=1}^{\infty} \frac{1}{k^{3/2}}$ converges (p-series with $p = 3/2$).

Limit comparison test: $\lim_{k\to\infty} \frac{\frac{\sqrt{k}}{k^2 + 1}}{1/k^{3/2}} = \lim_{k\to\infty} \frac{k^2}{k^2 + 1} = 1$ and $\sum_{k=1}^{\infty} \frac{1}{k^{3/2}}$ converges.

5 $\sum_{n=1}^{\infty} \left[\frac{3 + 1/n}{2 + 1/n}\right]^n$ diverges.

nth-term test for divergence: $\lim_{n\to\infty} a_n = \lim_{n\to\infty} \left[\frac{3 + 1/n}{2 + 1/n}\right]^n = \lim_{n\to\infty} \left[\frac{3[1 + (1/3)/n]}{2[1 + (1/2)/n]}\right]^n =$

$$\left(\lim_{n\to\infty} \frac{3^n}{2^n}\right) \lim_{n\to\infty} \frac{\left(1 + \frac{1/3}{n}\right)^n}{\left(1 + \frac{1/2}{n}\right)^n} = \left(\lim_{n\to\infty} \frac{3^n}{2^n}\right) \frac{e^{1/3}}{e^{1/2}} = \infty.$$

Root test: $\lim_{n\to\infty} (a_n)^{1/n} = \lim_{n\to\infty} \left(\left[\frac{3 + 1/n}{2 + 1/n}\right]^n\right)^{1/n} =$

$$\lim_{n\to\infty} \frac{3 + 1/n}{2 + 1/n} = \frac{3}{2} > 1.$$

Limit-comparison test: Compare with $(3/2)^n$.

7 $\sum_{n=1}^{\infty} \frac{1}{2^n - 3}$ converges.

Limit-comparison test: $\lim_{n\to\infty} \frac{\frac{1}{2^n - 3}}{1/2^n} =$

$\lim_{n\to\infty} \frac{2^n}{2^n - 3} = \lim_{n\to\infty} \frac{1}{1 - 3/2^n} = \frac{1}{1 - 0} = 1$ and

$\sum_{n=1}^{\infty} \frac{1}{2^n} = \sum_{n=1}^{\infty} \left(\frac{1}{2}\right)^n$ converges (geometric series);

Ratio test: $\lim_{n\to\infty} \frac{a_{n+1}}{a_n} = \lim_{n\to\infty} \frac{1}{2^{n+1} - 3} \cdot \frac{2^n - 3}{1} =$

$\lim_{n\to\infty} \frac{1 - 3/2^n}{2 - 3/2^n} = \frac{1 - 0}{2 - 0} = \frac{1}{2} < 1.$

9 $\sum_{k=1}^{\infty} \frac{\cos^2 k}{2^k}$ converges.

Comparison test: $\frac{\cos^2 k}{2^k} \le \frac{1}{2^k} = \left(\frac{1}{2}\right)^k$ and

$\sum_{k=1}^{\infty} \left(\frac{1}{2}\right)^k$ converges (geometric series).

11 $\sum_{n=1}^{\infty} \frac{n^n}{n!}$ diverges.

Ratio test: $\lim_{n\to\infty} \frac{(n+1)^{n+1}}{(n+1)!} \cdot \frac{n!}{n^n} = \lim_{n\to\infty} \frac{(n+1)^n}{n^n} =$

$\lim_{n\to\infty} \left(1 + \frac{1}{n}\right)^n = e > 1.$

nth-term test for divergence: $\lim_{n\to\infty} a_n = \lim_{n\to\infty} \frac{n^n}{n!} =$

$\infty \ne 0.$

Comparison test: $\frac{n^n}{n!} \ge n$ for $n = 1, 2, 3, \cdots$ and

$\sum_{n=1}^{\infty} n$ diverges (nth-term test for divergence).

13 $\sum_{n=1}^{\infty} \frac{1}{n\sqrt{n}}$ converges.

p-series test: $\frac{1}{n\sqrt{n}} = \frac{1}{n^{3/2}}$ so $p = 3/2$.

Integral test: $\int_1^{\infty} \frac{1}{x^{3/2}}\,dx = \int_1^{\infty} x^{-3/2}\,dx =$

$-2x^{-1/2}\Big|_1^{\infty} = \frac{-2}{\sqrt{x}}\Big|_1^{\infty} = 2.$

15 $\sum_{n=1}^{\infty} (-1)^n \ln\left(\frac{n+1}{n}\right)$ converges.

Alternating-series test: $p_n = \ln\left(\frac{n+1}{n}\right)$ is a positive, decreasing sequence with $p_n \to 0$ as $n \to \infty$.

17 $\sum_{n=1}^{\infty} n \sin\frac{1}{n}$ diverges.

nth-term test for divergence: $\lim_{n\to\infty} a_n =$

$\lim_{n\to\infty} n \sin\frac{1}{n} = \lim_{n\to\infty} \frac{\sin 1/n}{1/n} = 1 \ne 0.$

19 $\sum_{n=0}^{\infty} \frac{5n^3 + 6n + 1}{n^5 + n^3 + 2}$ converges.

Limit-comparison test: $\lim_{n\to\infty} \dfrac{\dfrac{5n^3 + 6n + 1}{n^5 + n^3 + 2}}{1/n^2} =$

$\lim_{n\to\infty} \dfrac{5n^5 + 6n^3 + n^2}{n^5 + n^3 + 2} = \lim_{n\to\infty} \dfrac{5 + 6/n^2 + 1/n^3}{1 + 1/n^2 + 2/n^5} =$

$\dfrac{5}{1} = 5$ and $\sum_{n=1}^{\infty} \dfrac{1}{n^2}$ converges (p-series with $p =$ 2);

Comparison test: $\dfrac{5n^3 + 6n + 1}{n^5 + n^3 + 2} <$

$\dfrac{5n^3 + 6n^3 + n^3}{n^5 + 0 + 0} = \dfrac{12}{n^2}$ and $\sum_{n=1}^{\infty} \dfrac{1}{n^2}$ converges

(p-series with $p = 2$).

21 $\sum_{n=1}^{\infty} \dfrac{2^{-n}}{n}$ converges.

Comparison test: $\dfrac{2^{-n}}{n} = \dfrac{1}{2^n n} \le \dfrac{1}{2^n} = \left(\dfrac{1}{2}\right)^n$ and

$\sum_{n=1}^{\infty} \left(\dfrac{1}{2}\right)^n$ converges (geometric series).

Limit-comparison test: $\lim_{n\to\infty} \dfrac{2^{-n}/n}{(1/2)^n} =$

$\lim_{n\to\infty} \dfrac{(1/2)^n \cdot 1/n}{(1/2)^n} = \lim_{n\to\infty} \dfrac{1}{n} = 0$ and $\sum_{n=1}^{\infty} \left(\dfrac{1}{2}\right)^n$

converges (geometric series).

Ratio test: $\lim_{n\to\infty} \dfrac{a_{n+1}}{a_n} = \lim_{n\to\infty} \dfrac{1}{2^{n+1}(n+1)} \cdot \dfrac{2^n n}{1} =$

$\lim_{n\to\infty} \dfrac{1}{2} \cdot \dfrac{n}{n+1} = \dfrac{1}{2} < 1.$

Root test: $\lim_{n\to\infty} (a_n)^{1/n} = \lim_{n\to\infty} \left(\dfrac{1}{2^n n}\right)^{1/n} =$

$\lim_{n\to\infty} \dfrac{1}{2} \cdot \dfrac{1}{n^{1/n}} = \dfrac{1}{2} \cdot \dfrac{1}{1} = \dfrac{1}{2} < 1.$

23 $\sum_{n=0}^{\infty} \dfrac{(-1)^n (1/2)^n}{n!}$ converges.

Alternating-series test: $p_n = \dfrac{(1/2)^n}{n!}$ is a positive,

decreasing sequence with $p_n \to 0$ as $n \to \infty$.

Absolute-convergence test: $|a_n| = \dfrac{(1/2)^n}{n!} \le \left(\dfrac{1}{2}\right)^n$

and $\sum_{n=1}^{\infty} \left(\dfrac{1}{2}\right)^n$ converges (geometric series).

Absolute-ratio test: $\lim_{n\to\infty} \left|\dfrac{a_{n+1}}{a_n}\right| =$

$\lim_{n\to\infty} \dfrac{(1/2)^{n+1}}{(n+1)!} \cdot \dfrac{n!}{(1/2)^n} = \lim_{n\to\infty} \dfrac{1/2}{n+1} = 0 < 1.$

25 $\sum_{n=0}^{\infty} \dfrac{n+2}{n+1}\left(\dfrac{2}{3}\right)^n$ converges.

Comparison test: $\dfrac{n+2}{n+1}\left(\dfrac{2}{3}\right)^n \le 2\left(\dfrac{2}{3}\right)^n$ and

$\sum_{n=0}^{\infty} 2\left(\dfrac{2}{3}\right)^n$ converges (geometric series).

Limit-comparison test: $\lim_{n\to\infty} \dfrac{\dfrac{n+2}{n+1}\left(\dfrac{2}{3}\right)^n}{(2/3)^n} =$

$\lim_{n\to\infty} \dfrac{n+2}{n+1} = 1$ and $\sum_{n=0}^{\infty} \left(\dfrac{2}{3}\right)^n$ converges (geometric

series).

Ratio test: $\lim_{n\to\infty} \frac{a_{n+1}}{a_n} =$

$$\lim_{n\to\infty} \frac{n+3}{n+2}\left(\frac{2}{3}\right)^{n+1} \cdot \frac{n+1}{n+2}\left(\frac{3}{2}\right)^n =$$

$$\lim_{n\to\infty} \frac{n^2+4n+3}{n^2+4n+4}\left(\frac{2}{3}\right) = \frac{1}{1}\left(\frac{2}{3}\right) = \frac{2}{3} < 1.$$

27 $\sum_{n=1}^{\infty} \frac{n \cos n}{1+n^4}$ converges.

Absolute-convergence test: $|a_n| = \frac{n\,|\cos n|}{1+n^4} \le$

$\frac{n}{1+n^4} < \frac{n}{n^4} = \frac{1}{n^3}$ and $\sum_{n=1}^{\infty} \frac{1}{n^3}$ converges

(p-series with $p = 3$).

29 $\sum_{n=1}^{\infty} \frac{n-3}{n\sqrt{n}}$ diverges.

Limit comparison test: $\lim_{n\to\infty} \frac{\frac{n-3}{n\sqrt{n}}}{1/\sqrt{n}} = \lim_{n\to\infty} \frac{n-3}{n}$

$= 1$ and $\sum_{n=1}^{\infty} \frac{1}{\sqrt{n}}$ diverges (p-series with $p = 1/2$).

Integral test: $f(x) = \frac{x-3}{x\sqrt{x}}$ is positive, continuous,

and decreasing for $x \ge 9$ and $\int_9^{\infty} \frac{x-3}{x\sqrt{x}}\,dx =$

$$\int_9^{\infty} \left(\frac{1}{\sqrt{x}} - \frac{3}{x^{3/2}}\right) dx = \left(2\sqrt{x} + \frac{6}{\sqrt{x}}\right)\Bigg|_9^{\infty} = \infty.$$

31 $\sum_{n=0}^{\infty} \frac{10^n}{n!}$ converges.

Ratio test: $\lim_{n\to\infty} \frac{a_{n+1}}{a_n} = \lim_{n\to\infty} \frac{10^{n+1}}{(n+1)!}\frac{n!}{10^n} =$

$$\lim_{n\to\infty} \frac{10}{n+1} = 0.$$

Limit-comparison test: $\lim_{n\to\infty} \frac{10^n/n!}{1/2^n} = \lim_{n\to\infty} \frac{20^n}{n!} =$

0 and $\sum_{n=0}^{\infty} \frac{1}{2^n}$.

33 By Theorem 1 in Sec. 10.3, the series $\sum_{k=n+1}^{\infty} \left(\frac{1}{3}\right)^k$

is equal to $\frac{1/3}{1 - 1/3} = \frac{1}{2} = 0.5$.

35 $|R_n| < |a_{n+1}| = \frac{n+1}{5^{n+1}}$, so if we want $\frac{n+1}{5^{n+1}} <$

0.005, then $n \ge 4$. Choosing $n = 4$ will ensure

that $S_4 = \frac{-1}{5} + \frac{2}{25} + \frac{-3}{125} + \frac{4}{625} = -0.1376$

≈ -0.14 estimates $\sum_{n=1}^{\infty} (-1)^n \frac{n}{5^n}$ to two places.

Alternatively, the solution to part (c) of Exercise 33 in Sec. 10.3 also applies when $-1 < p < 0$; in

this case, $p = -1/5$. Therefore $\sum_{n=1}^{\infty} (-1)^n \frac{n}{5^n} =$

$$\frac{-1/5}{(1 - (-1/5))^2} = -\frac{5}{36} \approx -0.1388 \approx -0.14.$$

37 (a) If $\frac{a_{n+1}}{a_n} \le \frac{b_{n+1}}{b_n}$, then $\frac{a_{n+1}}{b_{n+1}} \le \frac{a_n}{b_n}$ for $n =$

1, 2, 3, $\cdots$, so that $\frac{a_2}{b_2} \le \frac{a_1}{b_1}$, $\frac{a_3}{b_3} \le \frac{a_2}{b_2}$,

$\frac{a_4}{b_4} \le \frac{a_3}{b_3}, \cdots$, that is, $\frac{a_1}{b_1} \ge \frac{a_2}{b_2} \ge \frac{a_3}{b_3} \ge$

$\frac{a_4}{b_4} \ge \cdots \ge \frac{a_n}{b_n} \ge \cdots > 0$. Thus, $\frac{a_n}{b_n} \le \frac{a_1}{b_1}$

for $n \ge 1$, that is, $a_n \le \frac{a_1}{b_1} b_n$ for $n \ge 1$.

Since $\sum_{n=1}^{\infty} \frac{a_1}{b_1} b_n = \frac{a_1}{b_1} \sum_{n=1}^{\infty} b_n$ converges,

$\sum_{n=1}^{\infty} a_n$ converges by the comparison test.

(b) Since $\lim_{n\to\infty} \frac{a_{n+1}}{a_n} = r < 1$, there is an integer

N satisfying $\frac{a_{n+1}}{a_n} \le \frac{r+1}{2} < 1$ for $n \ge N$.

Let $b_n = \left(\frac{r+1}{2}\right)^n$ so that for $n \ge N$, $\frac{a_{n+1}}{a_n}$

$\le \frac{b_{n+1}}{b_n} = \frac{r+1}{2}$. Since $\sum_{n=1}^{\infty} b_n =$

$\sum_{n=1}^{\infty} \left(\frac{r+1}{2}\right)^n$ converges (geometric series) it

follows from part (a) that $\sum_{n=1}^{\infty} a_n$ converges.

39 (a) If $\sum_{n=1}^{\infty} a_n$ converges, then $a_n \to 0$ as $n \to \infty$.

Thus, there is an integer N so that $a_n < 1$ for

$n \ge N$. Hence, $a_n^2 \le a_n$ for $n \ge N$, so that

$\sum_{n=N}^{\infty} a_n^2$ and $\sum_{n=1}^{\infty} a_n^2$ converge by the

comparison test.

(b) Let $a_n = (-1)^{n+1} \frac{1}{\sqrt{n}}$. Then $\sum_{n=1}^{\infty} a_n =$

$\sum_{n=1}^{\infty} \frac{(-1)^{n+1}}{\sqrt{n}}$ converges by the alternating

series test, but $\sum_{n=1}^{\infty} a_n^2 = \sum_{n=1}^{\infty} \frac{1}{n}$ diverges

(harmonic series).

41 (a) Assume that $\sum_{n=1}^{\infty} a_n^2$ and $\sum_{n=1}^{\infty} b_n^2$ converge, so

that $\sum_{n=1}^{\infty} \left(a_n^2 + b_n^2\right)$ converges. We have $0 \le$

$(|a_n| - |b_n|)^2 = a_n^2 - 2|a_n b_n| + b_n^2$, that is,

$0 \le |a_n b_n| \le \frac{1}{2}\left(a_n^2 + b_n^2\right)$. Since

$\frac{1}{2} \sum_{n=1}^{\infty} \left(a_n^2 + b_n^2\right)$ converges, $\sum_{n=1}^{\infty} |a_n b_n|$

converges by the comparison test, so that

$\sum_{n=1}^{\infty} a_n b_n$ converges by the absolute-

convergence test.

(b) Since $\sum_{n=1}^{\infty} a_n^2$ and $\sum_{n=1}^{\infty} \left(\frac{1}{n}\right)^2$ converges, it

follows from part (a) that $\sum_{n=1}^{\infty} a_n\left(\frac{1}{n}\right)$

converges.

43 (a) If $y = \ln x$, then for $x \ge 1$, $y \ge 0$, $y' = 1/x$

> 0, and $y'' = -\frac{1}{x^2} < 0$. From Exercise

42(b) it follows that $a_1 + a_2 + a_3 + \cdots + a_n$

$$= \left(\int_1^2 \ln x\,dx - \frac{\ln 1 + \ln 2}{2}\right) +$$

$$\left(\int_2^3 \ln x\,dx - \frac{\ln 2 + \ln 3}{2}\right) +$$

$$\left(\int_3^4 \ln x - \frac{\ln 3 + \ln 4}{2}\right) + \cdots +$$

$$\left(\int_{n-1}^n \ln x\,dx - \frac{\ln(n-1) + \ln n}{2}\right)$$ has a limit

as $n \to \infty$, that is, $\lim_{n\to\infty} \left\{\int_1^n \ln x\,dx - \right.$

$$\left[\frac{\ln 1 + \ln 2}{2} + \frac{\ln 2 + \ln 3}{2} + \cdots + \right.$$

$$\left.\left.\frac{\ln(n-1) + \ln n}{2}\right]\right\}$$ exists, call it C.

(b) Rewriting the last limit in part (a) gives

$$\lim_{n\to\infty} \left\{(x\ln x - x)\Big|_1^n - \right.$$

$$\left.\left[\ln 2 + \ln 3 + \cdots + \ln(n-1) + \ln n - \frac{\ln n}{2}\right]\right\} =$$

$$\lim_{n\to\infty} \left[n\ln n - n + 1 - \ln(2\cdot 3\cdot 4\cdots(n-1)\cdot n) + \frac{1}{2}\ln n\right]$$

$$= \lim_{n\to\infty} \left[n\ln n - n + 1 - \ln n! + \ln\sqrt{n}\right] =$$

C.

45 (a) $I_0 = \int_0^{\pi/2} \sin^0 x\,dx = \int_0^{\pi/2} 1\,dx = x\big|_0^{\pi/2} =$

$\frac{\pi}{2}$ and $I_1 = \int_0^{\pi/2} \sin^1 x\,dx = -\cos x\big|_0^{\pi/2} =$

1.

(b) By Exercise 39 in Sec. 7.3, $I_{2n} =$

$$\frac{2n-1}{2n}\cdot\frac{2n-3}{2n-2}\cdots\frac{5}{6}\cdot\frac{3}{4}\cdot\frac{1}{2}\cdot\frac{\pi}{2}$$ and $I_{2n+1} =$

$$\frac{2n}{2n+1}\cdot\frac{2n-2}{2n-1}\cdots\frac{6}{7}\cdot\frac{4}{5}\cdot\frac{2}{3}$$ for $n = 1, 2, 3, \cdots$.

(c) $I_7 = \frac{6}{7}\cdot\frac{4}{5}\cdot\frac{2}{3}$ and $I_6 = \frac{5}{6}\cdot\frac{3}{4}\cdot\frac{1}{2}\cdot\frac{\pi}{2}$, so

$$\frac{I_7}{I_6} = \frac{6}{7}\cdot\frac{6}{5}\cdot\frac{4}{5}\cdot\frac{4}{3}\cdot\frac{2}{3}\cdot\frac{2}{1}\cdot\frac{2}{\pi}.$$

(d) From part (b) we have $\frac{I_{2n+1}}{I_n} =$

$$\frac{\frac{2n}{2n+1}\cdot\frac{2n-2}{2n-1}\cdot\frac{2n-4}{2n-3}\cdots\frac{6}{7}\cdot\frac{4}{5}\cdot\frac{2}{3}}{\frac{2n-1}{2n}\cdot\frac{2n-3}{2n-2}\cdot\frac{2n-5}{2n-4}\cdots\frac{5}{6}\cdot\frac{3}{4}\cdot\frac{1}{2}\cdot\frac{\pi}{2}}$$

$$= \frac{2n}{2n+1}\cdot\frac{2n}{2n-1}\cdot\frac{2n-2}{2n-1}\cdot\frac{2n-2}{2n-3}\cdots\frac{6}{7}\cdot\frac{6}{5}\cdot\frac{4}{5}\cdot\frac{4}{3}\cdot\frac{2}{3}\cdot\frac{2}{1}\cdot\frac{2}{\pi}.$$

(e) Since $I_n = \int_0^{\pi/2} \sin^n x\,dx$ and $0 < \sin x < 1$ for $0 < x < \pi/2$, it follows that $I_k < I_m$ if $k > m$, since $\sin^k x < \sin^m x$ for $0 < x < \pi/2$.

Thus, $I_{2n} < I_{2n-1}$, so $\frac{2n}{2n+1}I_{2n} <$

$$\frac{2n}{2n+1}I_{2n-1} =$$

$$\frac{2n}{2n+1}\cdot\frac{2n-2}{2n-1}\cdot\frac{2n-4}{2n-3}\cdots\frac{6}{7}\cdot\frac{4}{5}\cdot\frac{2}{3} = I_{2n+1} <$$

I_{2n}, that is, $\frac{2n}{2n+1}I_{2n} < I_{2n+1} < I_{2n}$. Hence

$\frac{2n}{2n+1} < \frac{I_{2n+1}}{I_{2n}} < 1$, so $\lim_{n\to\infty}\frac{2n}{2n+1} \le$

$\lim_{n\to\infty}\frac{I_{2n+1}}{I_{2n}} \le \lim_{n\to\infty} 1$ and therefore $1 \le$

$\lim_{n\to\infty}\frac{I_{2n+1}}{I_{2n}} \le 1$ and $\lim_{n\to\infty}\frac{I_{2n+1}}{I_{2n}} = 1$, as claimed.

(f) Combining parts (d) and (e) gives $1 =$

$$\lim_{n\to\infty} \frac{I_{2n+1}}{I_{2n}}$$

$$= \lim_{n\to\infty} \frac{2}{\pi}\cdot\frac{2}{1}\cdot\frac{2}{3}\cdot\frac{4}{3}\cdot\frac{4}{5}\cdot\frac{6}{5}\cdot\frac{6}{7}\cdots\frac{2n}{2n-1}\cdot\frac{2n}{2n+1}$$

$$= \frac{2}{\pi}\lim_{n\to\infty} \frac{2\cdot 2}{1\cdot 3}\cdot\frac{4\cdot 4}{3\cdot 5}\cdot\frac{6\cdot 6}{5\cdot 7}\cdots\frac{2n\cdot 2n}{(2n-1)(2n+1)},$$

$$\text{so } \lim_{n\to\infty} \frac{2\cdot 2}{1\cdot 3}\cdot\frac{4\cdot 4}{3\cdot 5}\cdot\frac{6\cdot 6}{5\cdot 7}\cdots\frac{2n\cdot 2n}{(2n-1)(2n+1)}$$

$$= \frac{\pi}{2}.$$

47 (a) From Exercise 44 it follows that

$$\lim_{n\to\infty} \frac{n!}{(n/e)^n\sqrt{n}} = k \text{ and } \lim_{n\to\infty} \frac{(2n)!}{(2n/e)^{2n}\sqrt{2n}} = k.$$

From Exercise 46(c), $\sqrt{\dfrac{\pi}{2}}$

$$= \lim_{n\to\infty} \frac{(n!)^2\, 4^n}{(2n)!\,\sqrt{2n+1}} =$$

$$\lim_{n\to\infty} \left(\frac{n!}{(n/e)^n\sqrt{n}}\right)^2 \cdot \frac{(n/e)^{2n}\cdot n}{\dfrac{(2n)!}{(2n/e)^{2n}\cdot\sqrt{2n}}\left(\dfrac{2n}{e}\right)^{2n}\sqrt{2n}} \cdot \frac{4^n}{\sqrt{2n+1}}$$

$$= \frac{k^2}{k}\lim_{n\to\infty} \frac{(n/e)^{2n}\, n}{(2n/e)^{2n}\sqrt{2n}}\cdot\frac{4^n}{\sqrt{2n+1}}$$

$$= k\lim_{n\to\infty} \left(\frac{n}{e}\cdot\frac{e}{2n}\right)^{2n}\cdot 4^n\cdot\frac{n}{\sqrt{4n^2+2n}}$$

$$= k\lim_{n\to\infty} \left(\frac{1}{2}\right)^{2n} 4^n \sqrt{\frac{n^2}{4n^2+2n}}$$

$$= k\lim_{n\to\infty} \left(\frac{1}{4}\right)^{n} 4^n \sqrt{\frac{1}{4+2/n}}$$

$$= k\lim_{n\to\infty} \sqrt{\frac{1}{4+2/n}} = k\cdot\frac{1}{2}, \text{ so } \sqrt{\frac{\pi}{2}} = \frac{k}{2}$$

and $k = \sqrt{2\pi}$.

(b) By calculator $20! \approx 2.432902008 \times 10^{18}$ and for $n = 20$ we have $\dfrac{\sqrt{2\pi n}\,(n/e)^n}{n!} \approx$ 0.995842347.

49 (a) $f(x) = \lim_{n\to\infty} \dfrac{x^{2n}}{1+x^{2n}}$ so that $f\left(\dfrac{1}{2}\right) =$

$$\lim_{n\to\infty} \frac{(1/2)^{2n}}{1+(1/2)^{2n}} = \lim_{n\to\infty} \frac{(1/4)^n}{1+(1/4)^n}\cdot\frac{4^n}{4^n} =$$

$$\lim_{n\to\infty} \frac{1}{4^n+1} = 0;\ f(2) = \lim_{n\to\infty} \frac{2^{2n}}{1+2^{2n}} =$$

$$\lim_{n\to\infty} \frac{4^n}{1+4^n} = \lim_{n\to\infty} \frac{1}{\dfrac{1}{4^n}+1} = \frac{1}{0+1} = 1;$$

$$f(1) = \lim_{n\to\infty} \frac{1^{2n}}{1+1^{2n}} = \frac{1}{1+1} = \frac{1}{2}.$$

(b) This function is defined for all values of x:

$$f(x) = \begin{cases} 0, & -1 < x < 1 \\ 1, & x > 1,\ x < -1 \\ \dfrac{1}{2}, & x = 1,\ x = -1 \end{cases}$$

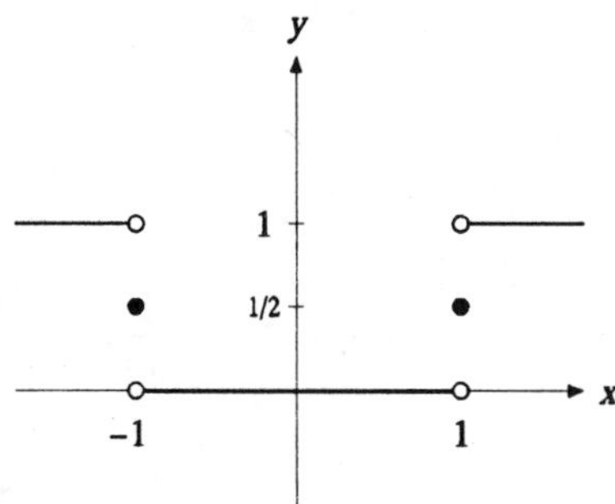

(c) The function f is continuous everywhere except at $x = 1$ and $x = -1$.

51 (a) If $q > 1$, then $\sum_{n=1}^{\infty} \frac{(\ln n)^p}{n^q}$ converges for all $p > 0$: By l'Hôpital's rule, $\lim_{x\to\infty} \frac{\ln x}{x} \underset{H}{=} \lim_{x\to\infty} \frac{1/x}{1} = 0$. Consequently, for any $k > 0$,

$$\lim_{x\to\infty} \frac{\ln x}{x^k} = \lim_{x\to\infty} \frac{\frac{1}{k}\ln x^k}{x^k} = \frac{1}{k}\lim_{y\to\infty} \frac{\ln y}{y} =$$

0, where $y = x^k$. Therefore, for each number $r > 0$, there is an integer N such that $\ln n < n^r$ for $n \geq N$. Hence $\frac{(\ln n)^p}{n^q} < \frac{(n^r)^p}{n^q} = n^{rp-q}$. We need to choose r so that $\sum_{n=N}^{\infty} n^{rp-q}$ converges, because then $\sum_{n=N}^{\infty} \frac{(\ln n)^p}{n^q}$ converges by the comparison test. We must therefore have $rp - q < -1$; solving this inequality for r, we obtain $r < (q - 1)/p$. Choosing $r = (q - 1)/(2p)$ suffices, showing that $\sum_{n=N}^{\infty} \frac{(\ln n)^p}{n^q}$ converges, so $\sum_{n=1}^{\infty} \frac{(\ln n)^p}{n^q}$ converges as well.

If $0 < q \leq 1$, then $\sum_{n=1}^{\infty} \frac{(\ln n)^p}{n^q}$ diverges for all $p > 0$: For $n \geq 3$, $\frac{1}{n^q} < \frac{(\ln n)^p}{n^q}$. Since $\sum_{n=3}^{\infty} \frac{1}{n^q}$ diverges, so does $\sum_{n=3}^{\infty} \frac{(\ln n)^p}{n^q}$ by the comparison test. Hence, $\sum_{n=1}^{\infty} \frac{(\ln n)^p}{n^q}$ diverges.

(b) The series $\sum_{n=1}^{\infty} \frac{n^q}{(\ln n)^p}$ diverges for all values of $p > 0$ and $q > 0$. As shown in part (a), for any $r > 0$ there is an N such that $\ln n < n^r$ for $n \geq N$. As in (a), let $r = \frac{q-1}{2p}$. Then, for $n \geq N$, $\frac{n^q}{(\ln n)^p} > \frac{n^q}{(n^r)^p} = n^{q-rp} = n^{(q+1)/2}$, which approaches infinity as $n \to \infty$. Hence $\lim_{n\to\infty} \frac{n^q}{(\ln n)^p} = \infty$, so the series diverges by the nth-term test for divergence.

53 $\sum_{n=1}^{\infty} \frac{e^{1/n} - 1}{n}$ converges since $\lim_{n\to\infty} \frac{\frac{e^{1/n} - 1}{n}}{1/n^2} =$

$$\lim_{n\to\infty} \frac{e^{1/n} - 1}{1/n} \underset{H}{=} \lim_{n\to\infty} \frac{e^{1/n} \cdot \frac{-1}{n^2}}{-\frac{1}{n^2}} = \lim_{n\to\infty} e^{1/n} = e^0$$

$= 1$ and $\sum_{n=1}^{\infty} \frac{1}{n^2}$ converges.

11 Power Series and Complex Numbers

11.1 Taylor Series

1 $f(x) = \dfrac{1}{1 + x}$, $f'(x) = \dfrac{-1}{(1 + x)^2}$, and $f''(x) = \dfrac{2}{(1 + x)^3}$, so $a_0 = f(0) = 1$, $a_1 = f'(0) = -1$, $a_2 = \dfrac{f''(0)}{2} = 1$ and $P_1(x; 0) = 1 - x$, $P_2(x; 0) = 1 - x + x^2$.

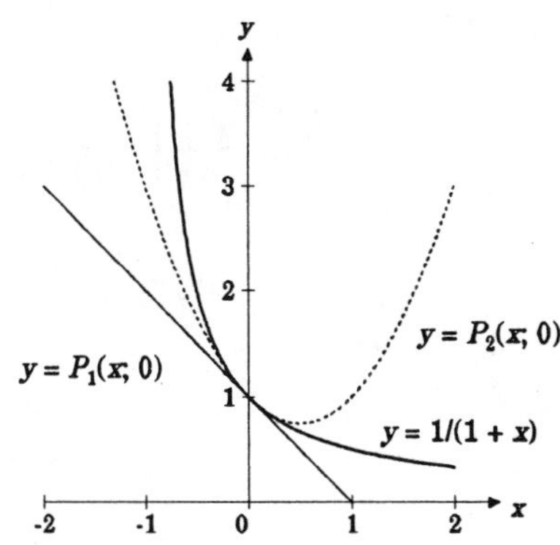

3 $f(x) = \ln(1 + x)$, $f'(x) = \dfrac{1}{1 + x}$, $f''(x) = \dfrac{-1}{(1 + x)^2}$, and $f'''(x) = \dfrac{2}{(1 + x)^3}$, so $a_0 = f(0) = 0$, $a_1 = f'(0) = 1$, $a_2 = \dfrac{f''(0)}{2} = -\dfrac{1}{2}$, $a_3 = \dfrac{f'''(0)}{3!} = \dfrac{1}{3}$, $P_1(x; 0) = x$, $P_2(x; 0) = x - \dfrac{1}{2}x^2$, $P_3(x; 0) = x - \dfrac{1}{2}x^2 + \dfrac{1}{3}x^3$.

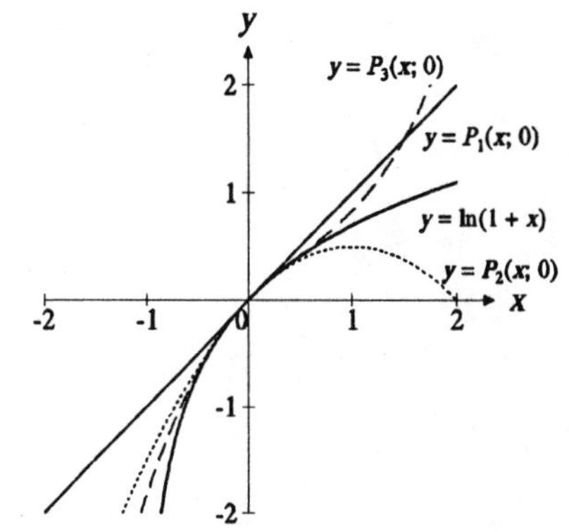

5 $f(x) = e^x$, $f'(x) = f''(x) = f'''(x) = f^{(4)}(x) = e^x$ so $a_0 = f(0) = 1$, $a_1 = f'(0) = 1$, $a_2 = \dfrac{f''(0)}{2} = \dfrac{1}{2}$, $a_3 = \dfrac{f'''(0)}{3!} = \dfrac{1}{6}$, $a_4 = \dfrac{f^{(4)}(0)}{4!} = \dfrac{1}{24}$, and $P_1(x; 0) = 1 + x$, $P_2(x; 0) = 1 + x + \dfrac{1}{2}x^2$, $P_3(x; 0) = 1 + x + \dfrac{1}{2}x^2 + \dfrac{1}{6}x^3$, $P_4(x; 0) = 1 + x + \dfrac{1}{2}x^2 + \dfrac{1}{6}x^3 + \dfrac{1}{24}x^4$.

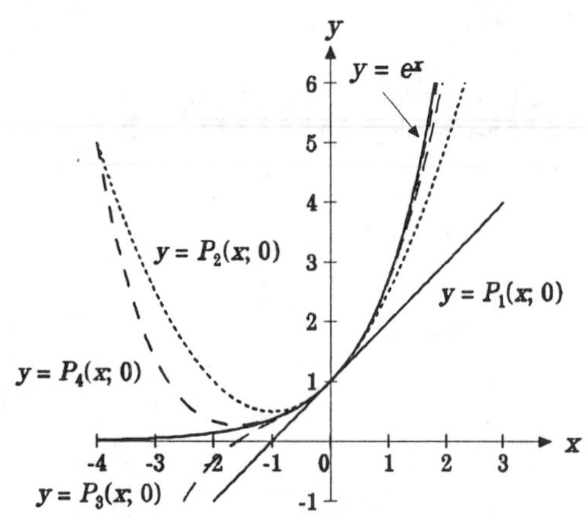

7 $f(x) = \tan^{-1} x, f'(x) = \dfrac{1}{1 + x^2}$,

$f''(x) = \dfrac{-2x}{(1 + x^2)^2}, f'''(x) = \dfrac{2(3x^2 - 1)}{(1 + x^2)^3}$, so $a_0 =$

$f(0) = 0,\ a_1 = f'(0) = 1,\ a_2 = \dfrac{f''(0)}{2} = 0,$

$a_3 = \dfrac{f'''(0)}{3!} = -\dfrac{1}{3}$, and $P_1(x; 0) = x,\ P_2(x; 0) =$

$x,\ P_3(x; 0) = x - \dfrac{1}{3}x^3.$

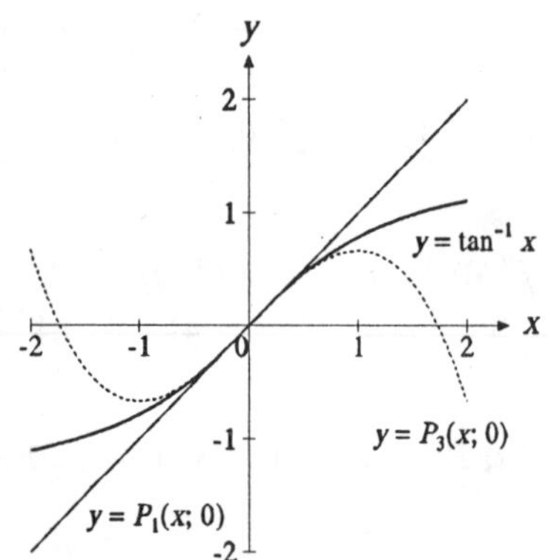

9 $f(x) = \cos x, f'(x) = -\sin x, f''(x) = -\cos x,$

$f'''(x) = \sin x, f^{(4)}(x) = \cos x$ so $a_0 = f(0) = 1,$

$a_1 = f'(0) = 0,\ a_2 = \dfrac{f''(0)}{2} = -\dfrac{1}{2},$

$a_3 = \dfrac{f'''(0)}{3!} = 0,\ a_4 = \dfrac{f^{(4)}(0)}{4} = \dfrac{1}{24}$, and $P_2(x; 0)$

$= 1 - \dfrac{1}{2}x^2,\ P_4(x; 0) = 1 - \dfrac{1}{2}x^2 + \dfrac{1}{24}x^4.$

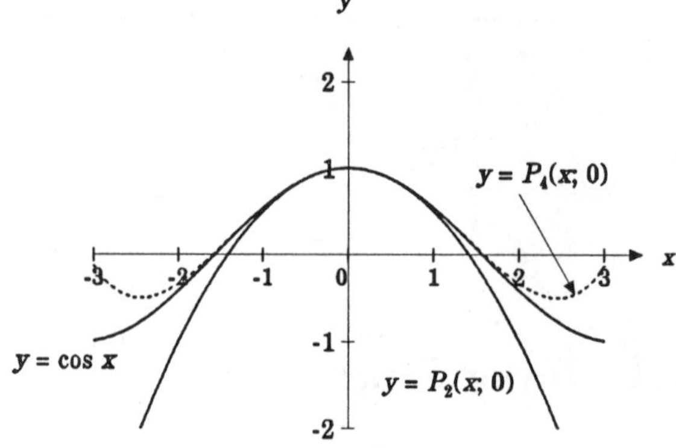

11 $f(x) = \cos x, f'(x) = -\sin x, f''(x) = -\cos x,$

$f'''(x) = \sin x, f^{(4)}(x) = \cos x$ so $a_0 = f\left(\frac{\pi}{4}\right) = \dfrac{\sqrt{2}}{2},$

$a_1 = f'\left(\frac{\pi}{4}\right) = -\dfrac{\sqrt{2}}{2},\ a_2 = \dfrac{f''(\frac{\pi}{4})}{2!} = -\dfrac{\sqrt{2}}{4},$

$a_3 = \dfrac{f'''(\frac{\pi}{4})}{3!} = \dfrac{\sqrt{2}}{12},\ a_4 = \dfrac{f^{(4)}(\frac{\pi}{4})}{4!} = \dfrac{\sqrt{2}}{48}$, and

$P_4\left(x; \dfrac{\pi}{4}\right) = \dfrac{\sqrt{2}}{2} - \dfrac{\sqrt{2}}{2}\left(x - \dfrac{\pi}{4}\right) - \dfrac{\sqrt{2}}{4}\left(x - \dfrac{\pi}{4}\right)^2 +$

$\dfrac{\sqrt{2}}{12}\left(x - \dfrac{\pi}{4}\right)^3 + \dfrac{\sqrt{2}}{48}\left(x - \dfrac{\pi}{4}\right)^4.$

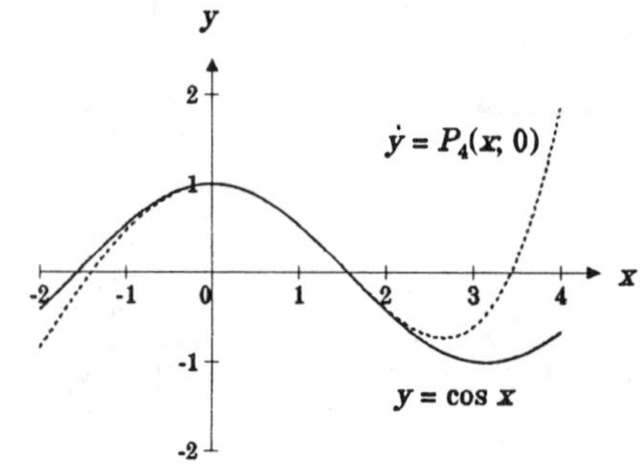

13 $f(x) = \ln(1 + x),\ f'(x) = \dfrac{1}{1 + x},$

$f''(x) = \dfrac{-1}{(1 + x)^2},\ f'''(x) = \dfrac{2}{(1 + x)^3},$

$$f^{(4)}(x) = \frac{-3\cdot 2}{(1+x)^4},\ f^{(5)}(x) = \frac{4\cdot 3\cdot 2}{(1+x)^5}, \ldots,$$

$$f^{(n)}(x) = \frac{(-1)^{n+1}(n-1)!}{(1+x)^n}$$ for $n \geq 1$ so that $a_0 =$

$$f(0) = 0,\ a_1 = f'(0) = 1,\ a_2 = \frac{f''(0)}{2} = -\frac{1}{2},$$

$$a_3 = \frac{f'''(0)}{3!} = \frac{1}{3},\ a_4 = \frac{f^{(4)}(0)}{4!} = -\frac{1}{4},$$

$$a_n = \frac{f^{(n)}(0)}{n!} = (-1)^{n+1}\cdot\frac{1}{n}$$ for $n \geq 1$ and the

Maclaurin series for $\ln(1+x)$ is

$$x - \frac{1}{2}x^2 + \frac{1}{3}x^3 - \frac{1}{4}x^4 + \cdots + (-1)^{n+1}\cdot\frac{1}{n}x^n + \cdots.$$

15 $f(x) = \cos x, f'(x) = -\sin x, f''(x) = -\cos x,$ $f'''(x) = \sin x, f^{(4)}(x) = \cos x, \ldots,$ so $a_0 = f(0) = 1$, $a_1 = f'(0) = 0,$

$$a_2 = \frac{f''(0)}{2!} = \frac{-1}{2!},\ a_3 = \frac{f'''(0)}{3!} = 0,$$

$$a_4 = \frac{f^{(4)}(0)}{4!} = \frac{1}{4!},\ a_5 = \frac{f^{(5)}(0)}{5!} = 0,$$

$$a_6 = \frac{f^{(6)}(0)}{6!} = \frac{-1}{6!},$$ ..., and the Maclaurin series

for $\cos x$ is

$$1 - \frac{x^2}{2!} + \frac{x^4}{4!} - \frac{x^6}{6!} + \cdots + (-1)^n\frac{x^{2n}}{(2n)!} + \cdots.$$

17 $f(x) = e^{-x}, f'(x) = -e^{-x}, f''(x) = e^{-x}, f'''(x) = -e^{-x}, \ldots,$ so $a_0 = f(0) = 1$, $a_1 = f'(0) = -1,$

$$a_2 = \frac{f''(0)}{2!} = \frac{1}{2!},\ a_3 = \frac{f'''(0)}{3!} = \frac{-1}{3!},$$ ..., and

the Maclaurin series for e^{-x} is

$$1 - x + \frac{x^2}{2!} - \frac{x^3}{3!} + \frac{x^4}{4!} - \cdots + (-1)^n\frac{x^n}{n!} + \cdots.$$

19 (a) $f(x) = f'(x) = f''(x) = \cdots = f^{(n)}(x) = e^x$ for $n \geq 0$ so $a_0 = f(0) = 1$, $a_1 = f'(0) = 1,$

$$a_2 = \frac{f''(0)}{2!} = \frac{1}{2!},\ \ldots,\ a_n = \frac{f^{(n)}(0)}{n!} = \frac{1}{n!}$$ for

$n \geq 0$ and $P_{10}(x;0) = 1 + x + \frac{x^2}{2!} + \frac{x^3}{3!} + \frac{x^4}{4!} + \frac{x^5}{5!} + \frac{x^6}{6!} + \frac{x^7}{7!} + \frac{x^8}{8!} + \frac{x^9}{9!} + \frac{x^{10}}{10!}.$

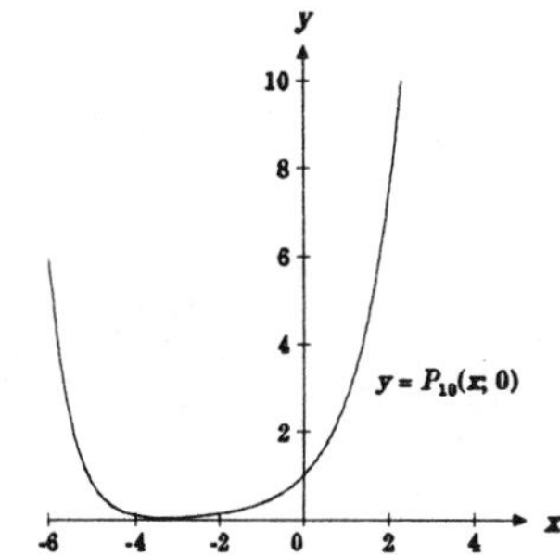

(b)

x	$P_{10}(x;\ 0)$	e^x
1	2.718281801	2.718281828
2	7.388994709	7.389056099
4	54.443104056	54.598150033

21 (a) $f(x) = 2 + 3x - 4x^2, f'(x) = 3 - 8x, f''(x) = -8, f'''(x) = 0$ so $a_0 = f(0) = 2$, $a_1 = f'(0) = 3$, $a_2 = \frac{f''(0)}{2} = -4$, $a_3 = \frac{f'''(0)}{3!} = 0$, and $P_1(x;\ 0) = 2 + 3x$, $P_2(x;\ 0) = 2 + 3x - 4x^2$, $P_3(x;\ 0) = 2 + 3x - 4x^2$.

(b) Since $f^{(n)}(x) = 0$ for $n \geq 3$, the Maclaurin series for $f(x)$ is $2 + 3x - 4x^2 + 0x^3 + 0x^4 + \cdots = 2 + 3x - 4x^2$.

23 (a) $f(x) = (1+x)^3, f'(x) = 3(1+x)^2, f''(x) = 6(1+x), f'''(x) = 6$ so $a_0 = f(0) = 1$, $a_1 = f'(0) = 3$, $a_2 = \frac{f''(0)}{2} = 3$, $a_3 = \frac{f'''(0)}{3!} = 1$

and $P_3(x; 0) = 1 + 3x + 3x^2 + x^3$.

(b) $(1+x)^3 = (1 + 2x + x^2)(1+x)$

$= 1 + 3x + 3x^2 + x^3$

25 $f(x) = \sin x, f'(x) = \cos x, f''(x) = -\sin x, f'''(x) = -\cos x, f^{(4)}(x) = \sin x, f^{(5)}(x) = \cos x, f^{(6)}(x) = -\sin x$, so $a_0 = f\left(\frac{\pi}{6}\right) = \frac{1}{2}$, $a_1 = f'\left(\frac{\pi}{6}\right) = \frac{\sqrt{3}}{2}$,

$a_2 = \frac{f''\left(\frac{\pi}{6}\right)}{2!} = -\frac{1}{4}$, $a_3 = \frac{f'''\left(\frac{\pi}{6}\right)}{3!} = -\frac{\sqrt{3}}{12}$,

$a_4 = \frac{f^{(4)}\left(\frac{\pi}{6}\right)}{4!} = \frac{1}{48}$, $a_5 = \frac{f^{(5)}\left(\frac{\pi}{6}\right)}{5!} = \frac{\sqrt{3}}{240}$,

$a_6 = \frac{f^{(6)}\left(\frac{\pi}{6}\right)}{6!} = -\frac{1}{1440}$, $P_6\left(x; \frac{\pi}{6}\right) =$

$$\frac{1}{2} + \frac{\sqrt{3}}{2}\left(x - \frac{\pi}{6}\right) - \frac{1}{4}\left(x - \frac{\pi}{6}\right)^2 - \frac{\sqrt{3}}{12}\left(x - \frac{\pi}{6}\right)^3 + \frac{1}{48}\left(x - \frac{\pi}{6}\right)^4 + \frac{\sqrt{3}}{240}\left(x - \frac{\pi}{6}\right)^5 - \frac{1}{1440}\left(x - \frac{\pi}{6}\right)^6.$$

27 (a) $f(x) = (1+x)^{1/2}$, $f'(x) = \frac{1}{2}(1+x)^{-1/2}$, $f''(x) = -\frac{1}{4}(1+x)^{-3/2}$, $f'''(x) = \frac{3}{8}(1+x)^{-5/2}$,

$f^{(4)}(x) = -\frac{15}{16}(1+x)^{-7/2}$, so $a_0 = f(0) = 1$,

$a_1 = f'(0) = \frac{1}{2}$, $a_2 = \frac{f''(0)}{2} = -\frac{1}{8}$, $a_3 = \frac{f'''(0)}{3!} = \frac{1}{16}$, $a_4 = \frac{f^{(4)}(0)}{4!} = -\frac{5}{128}$, and

$$P_4(x; 0) = 1 + \frac{1}{2}x - \frac{1}{8}x^2 + \frac{1}{16}x^3 - \frac{5}{128}x^4.$$

(b)

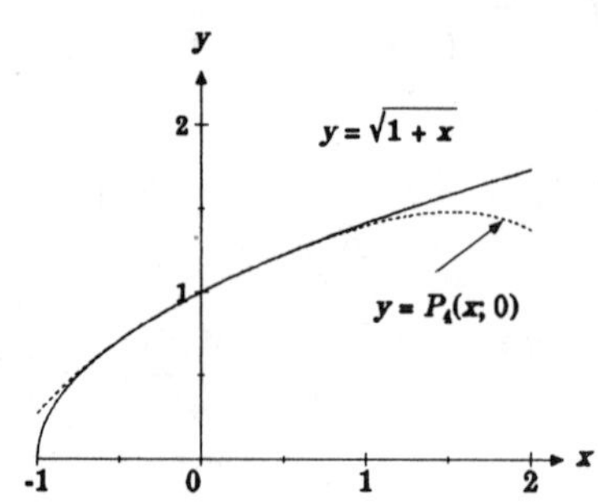

(c)

x	$f(x)$	$P_4(x; 0)$
−1	0.0000	0.2734
−0.5	0.7071	0.7085
−0.1	0.9487	0.9487
0	1.0000	1.0000
0.1	1.0488	1.0488
0.5	1.2247	1.2241
1	1.4142	1.3984

29 No. If there were such a polynomial $p(x) = a_0 + a_1x + a_2x^2 + \cdots + a_nx^n$, then $0 = \frac{d^{n+1}}{dx^{n+1}}(p(x)) = \frac{d^{n+1}}{dx^{n+1}}(\sin x)$. But no derivative of $\sin x$ is the zero function.

31 (a) If m is an even number, then $x^m = (-x)^m$. If $p(x)$ is a polynomial containing only even powers of x, then clearly $p(x) = p(-x)$, that is, $p(x)$ is an even function. Now assume that $p(x)$ is a polynomial that is an even function; we will show that it contains only even powers of x. Let $p(x) = a_0 + a_1x + a_2x^2 + a_3x^3 + \cdots$. Note that $p(-x) = a_0 - a_1x + a_2x^2 - a_3x^3 + \cdots$. Since we assumed that $p(x)$

is even, we must have $a_0 + a_1x + a_2x^2 + a_3x^3 + \cdots = a_0 - a_1x + a_2x^2 - a_3x^3 + \cdots$ for all x. Simplifying this equation, we obtain $2a_1x + 2a_3x^3 + 2a_5x^5 + \cdots = 0$, so $a_1x + a_3x^3 + a_5x^5 + \cdots = 0$ for all x. But from Sec. 2.4 we know that the limit of a polynomial as $x \to \infty$ is infinite unless all of its terms are 0; hence all of the odd-numbered coefficients a_1, a_3, a_5, ..., must be 0. Thus we conclude that $p(x)$ is even if and only if it contains only even powers of x.

(b) If f is an even function, then $P_n(x; 0)$ is an even function:

Fact 1: If g is an even, differentiable function, then $g'(0) = 0$.

Proof: If g is even, then $g(x) = g(-x)$ so that $g'(x) = -g'(-x)$. If $x = 0$, then $g'(0) = -g'(0)$, *i.e.*, $g'(0) = 0$.

Fact 2: If g is an even, differentiable function, then g' is an odd function.

Proof: If g is even, then $g(x) = g(-x)$ so that differentiation gives $g'(x) = g'(-x)\cdot(-1)$, *i.e.*, $g'(x) = -g'(-x)$. Thus, g' is an odd function.

Fact 3: If g is an odd, differentiable function, then g' is an even function.

Proof: If g is odd, then $g(x) = -g(-x)$ so that differentiation gives $g'(x) = -g'(-x)\cdot(-1)$, *i.e.*, $g'(x) = g'(-x)$. Thus, g' is an even function.

Let f be an even function and consider its Taylor polynomial $P_n(x; 0) = f(0) + f'(0)x + \frac{f''(0)}{2!}x^2 + \cdots + \frac{f^{(n)}(0)}{n!}x^n$. Since f is even, it follows from Facts 2 and 3 that $f', f''', f^{(5)}, \ldots$ are odd and $f'', f^{(4)}, f^{(6)}, \ldots$ are even. From Fact 1 it follows that $f'(0) = 0, f'''(0) = 0, f^{(5)}(0) = 0, \ldots$ so that $P_n(x; 0)$ is composed of only even powers of x. By part (a) it follows that $P_n(x; 0)$ is an even function.

33 (a) The function $f(x) = \tan^{-1} x$ is an odd function.

(b) Since $f(x) = \tan^{-1} x$ is an odd function, it follows from Exercise 32 that $P_n(x; 0)$ is odd. Hence, all even powers of x in $P_n(x; 0)$, and therefore the Maclaurin series, itself, will have coefficients of zero.

(c) From Exercise 7, $P_3(x; 0) = x - \frac{1}{3}x^3$.

(d)

x	$P_3(x; 0)$	$\tan^{-1} x$
0	0.0000	0.0000
0.1	0.0997	0.0997
0.2	0.1973	0.1974
0.3	0.2910	0.2915
0.5	0.4583	0.4636
1	0.6667	0.7854
2	−0.6667	1.1071

11.2 The Error in Taylor Series

1 $R_n(x; 0) = \frac{f^{(n+1)}(c_n)}{(n+1)!}x^{n+1}$. For each n, $f^{(n+1)}(x)$ is either $\pm\cos x$ or $\pm\sin x$, so $|f^{(n+1)}(c_n)| \le 1$. Hence, $|R_n(x; 0)| \le \frac{|x|^{n+1}}{(n+1)!}$, which approaches 0 as $n \to \infty$.

3 Assume that $f(x)$ is a polynomial of degree m, so that $f^{(n)}(x) = 0$ for $n \geq m + 1$. Hence,

$$\lim_{n\to\infty} R_n(x; 0) = \lim_{n\to\infty} \frac{f^{(n+1)}(c_n)}{(n+1)!} x^{n+1} =$$

$\lim_{n\to\infty} \frac{0}{(n+1)!} x^{n+1} = 0$, so $f(x)$ is represented by its Maclaurin series.

5 $R_n(x; a) = \frac{f^{(n+1)}(c_n)}{(n+1)!}(x - a)^{n+1} =$

$\frac{e^{c_n}}{(n+1)!}(x - a)^{n+1}$. Since c_n is between a and x, e^{c_n} is between e^a and e^x. Thus, $e^{c_n} \leq M$, where M is the greater of e^a and e^x. So

$|R_n(x; a)| \leq M\frac{|x - a|^{n+1}}{(n+1)!} \to 0$ as $n \to \infty$ and e^x is represented by its Taylor series at $x = a$.

7 $e^{-1} = 1 - 1 + \frac{1}{2!} - \frac{1}{3!} + \frac{1}{4!} - \cdots$ is a decreasing, alternating series, so $R_n <$

$|a_{n+1}| = \frac{1}{(n+1)!}$. If $n \geq 6$, then $R_n < \frac{1}{7!} <$

0.0005. Thus $1 - 1 + \frac{1}{2!} - \frac{1}{3!} + \frac{1}{4!} - \frac{1}{5!} + \frac{1}{6!}$

$= \frac{53}{144} \approx 0.368056$ estimates e^{-1} to at least three decimal places.

9 $\cos 20° = \cos\frac{\pi}{9} =$

$1 - \frac{(\pi/9)^2}{2!} + \frac{(\pi/9)^4}{4!} - \frac{(\pi/9)^6}{6!} + \cdots$ is a decreasing, alternating series, so the error in using n terms is at most the absolute value of the $(n + 1)$st term,

$\frac{(\pi/9)^{2n}}{(2n)!}$. Picking $n = 3$ makes the error less than

$\frac{(\pi/9)^6}{6!} < 0.0005$. Thus $1 - \frac{(\pi/9)^2}{2!} + \frac{(\pi/9)^4}{4!} \approx$

0.9397 estimates cos 20° to at least three decimal places.

11 (a) Let $x_1 = 25$ and $x_2 = 23$ so that $dx = -2$.

The differential $df = f'(25)\cdot dx = \frac{1}{2\sqrt{25}}\cdot(-2)$

$= -0.2$ and $\Delta f = f(23) - f(25)$ with $df \approx \Delta f$, that is, $\sqrt{23} - \sqrt{25} \approx -0.2$, so $\sqrt{23} \approx 4.8$.

(b) $f(x) = \sqrt{x}$, $f'(x) = \frac{1}{2}x^{-1/2}$, $f''(x) = -\frac{1}{4}x^{-3/2}$,

so $a_0 = f(25) = 5$, $a_1 = f'(25) = 0.1$,

$a_2 = \frac{f''(25)}{2!} = -0.001$, and $P_1(x; 25) =$

$5 + 0.1(x - 25)$. Thus, $P_1(23; 25) = 4.8$.

(c) The error in (b) is $|\sqrt{23} - 4.8| \approx 0.00417$.

(d) See part (b). $P_2(x; 25) = 5 + 0.1(x - 25) - 0.001(x - 25)^2$, so $P_2(23; 25) = 4.796$.

(e) The error in (d) is $|\sqrt{23} - 4.796| \approx$ 0.000168.

(f) By Theorem 1, the error is $|R_2(23; 25)| =$

$$\left|\frac{f'''(c_n)}{3!}(-2)^3\right| = \left|\frac{\frac{3}{8}(c_n)^{-5/2}}{3!}(-2)^3\right| = \frac{1}{2}\cdot\frac{1}{c_n^{5/2}}$$

$< \frac{1}{2}\cdot\frac{1}{23^{5/2}} \approx 0.000197$, since $23 < c_n < 25$. As expected, the error calculated in (e) is less than the bound given by Theorem 1.

13 (a) $e^x = 1 + x + \frac{x^2}{2!} + \frac{x^3}{3!} + \cdots + \frac{x^n}{n!} + \cdots$, so if x is in $[0, 2]$, then all terms in the Maclaurin series are positive. Hence, $e^x \geq 1 + x + \frac{x^2}{2!} + \cdots + \frac{x^n}{n!}$ or

$x + \frac{x^2}{2!} + \frac{x^3}{3!} + \cdots + \frac{x^n}{n!} \leq e^x - 1$. In addition, for x in $[0, 2]$, $R_n(x; 0) =$

$\frac{f^{(n+1)}(c_n)}{(n+1)!} \cdot x^{n+1} = \frac{e^{c_n} \cdot x^{n+1}}{(n+1)!} \leq \frac{e^2 \cdot x^{n+1}}{(n+1)!}$

since $0 \leq c_n \leq x \leq 2$. Because $e^x =$

$1 + x + \frac{x^2}{2!} + \cdots + \frac{x^n}{n!} + R_n(x; 0) \leq$

$1 + x + \frac{x^2}{2!} + \cdots + \frac{x^n}{n!} + \frac{e^2 \cdot x^{n+1}}{(n+1)!}$, we have

$e^x - 1 \leq x + \frac{x^2}{2!} + + \cdots + \frac{x^n}{n!} + \frac{e^2 x^{n+1}}{(n+1)!}$.

(b) From part (a) it follows that

$1 + \frac{x}{2!} + \frac{x^2}{3!} + \cdots + \frac{x^{n-1}}{n!} \leq \frac{e^x - 1}{x} \leq$

$1 + \frac{x}{2!} + \frac{x^2}{3!} + \cdots + \frac{x^{n-1}}{n!} + \frac{e^2 x^n}{(n+1)!}$ for

$0 \leq x \leq 2$. We have $\int_0^1 \frac{e^2 x^n}{(n+1)!}\, dx =$

$\frac{e^2}{(n+1)!} \frac{x^{n+1}}{(n+1)}\Big|_0^1 = \frac{e^2}{(n+1)!\,(n+1)}$; if $n = 6$

then this quantity is less than 0.0005. Hence

$\int_0^1 \left(1 + \frac{x}{2!} + \frac{x^2}{3!} + \frac{x^3}{4!} + \frac{x^4}{5!} + \frac{x^5}{6!}\right) dx =$

$\left(x + \frac{x^2}{2!2} + \frac{x^3}{3!3} + \frac{x^4}{4!4} + \frac{x^5}{5!5} + \frac{x^6}{6!6}\right)\Big|_0^1 =$

$\frac{14{,}233}{10{,}800} \approx 1.3179$ estimates $\int_0^1 \frac{e^x - 1}{x}\, dx$

to three decimal places.

15 (a) For $x > 1$, $\int_b^\infty e^{-5x^2}\, dx < \int_b^\infty e^{-5x}\, dx =$

$-\frac{1}{5}e^{-5x}\Big|_b^\infty = \frac{1}{5e^{5b}} < 0.0005$ if $b = 1.2$.

(b) $\int_0^{1.2} e^{-5x^2}\, dx =$

$\int_0^{1.2} \left(1 - 5x^2 + \frac{5^2 x^4}{2!} - \frac{5^3 x^6}{3!} + \cdots\right) dx =$

$\left(x - \frac{5x^3}{3} + \frac{5^2 x^5}{5 \cdot 2!} - \frac{5^3 x^7}{7 \cdot 3!} + \cdots \pm \frac{5^n x^{2n+1}}{(2n+1) \cdot n!} + \cdots\right)$

$= 1.2 - \frac{5(1.2)^3}{3} + \frac{5^2(1.2)^5}{5 \cdot 2!} - \frac{5^3(1.2)^7}{7 \cdot 3!} + \cdots$

is a decreasing alternating series (after the 7th term). For $n = 22$, $\frac{5^n(1.2)^{2n+1}}{(2n+1) \cdot n!} \approx 0.000172$

< 0.0005. Hence $1.2 - \frac{5(1.2)^3}{3} + \frac{5^2(1.2)^5}{5 \cdot 2!}$

$- \frac{5^3(1.2)^7}{7 \cdot 3!} + \cdots - \frac{5^{21}(1.2)^{43}}{43 \cdot 21!} \approx 0.3961$

estimates $\int_0^{1.2} e^{-5x^2}\, dx$ with error less than 0.0005.

(c) From parts (a) and (b) we have $\int_0^\infty e^{-5x^2}\, dx =$

$\int_0^{1.2} e^{-5x^2}\, dx + \int_{1.2}^\infty e^{-5x^2}\, dx \approx 0.3961 +$

$\int_{1.2}^{\infty} e^{-5x^2}\,dx \approx 0.3961$ with error at most

$0.0005 + 0.0005 = 0.001.$

17 The function $f(x)$ is represented by its Maclaurin series for all x since $|R_n(x; a)| =$

$$\left|\frac{f^{(n+1)}(c_n)(x - a)^{n+1}}{(n + 1)!}\right| \leq \frac{(n + 1)|x - a|^{n+1}}{(n + 1)!} =$$

$\frac{|x - a|^{n+1}}{n!}$, which approaches 0 as $n \to \infty$.

19 (a) $\int_0^1 \frac{1}{1 + x^2}\,dx = \tan^{-1}x\Big|_0^1 =$

$\tan^{-1}1 - \tan^{-1}0 = \frac{\pi}{4} \approx 0.7853982$

(b) $h = 1/6$ so Simpson's estimate for

$\int_0^1 \frac{1}{1+x^2}\,dx$ is $S_6 = \frac{1/6}{3}\left[f(0) + 4f\left(\frac{1}{6}\right) + 2f\left(\frac{1}{3}\right) + 4f\left(\frac{1}{2}\right) + 2f\left(\frac{2}{3}\right) + 4f\left(\frac{5}{6}\right) + f(1)\right]$

$\approx 0.7853979.$

(c) $h = 1/6$ so the trapezoidal estimate for

$\int_0^1 \frac{1}{1 + x^2}\,dx$ is $T_6 = \frac{1/6}{2}\left[f(0) + 2f\left(\frac{1}{6}\right) + 2f\left(\frac{1}{3}\right) + 2f\left(\frac{1}{2}\right) + 2f\left(\frac{2}{3}\right) + 2f\left(\frac{5}{6}\right) + f(1)\right] \approx$

0.7842408.

(d) $\frac{1}{1 + x^2}$

$= 1 - x^2 + x^4 - x^6 + x^8 - x^{10} + \cdots$, so

$\int_0^1 \frac{1}{1 + x^2}\,dx$

$\approx \int_0^1 [1 - x^2 + x^4 - x^6 + x^8 - x^{10}]\,dx$

$= \left[x - \frac{x^3}{3} + \frac{x^5}{5} - \frac{x^7}{7} + \frac{x^9}{9} - \frac{x^{11}}{11}\right]\Bigg|_0^1$

$\approx 0.7440115.$

21

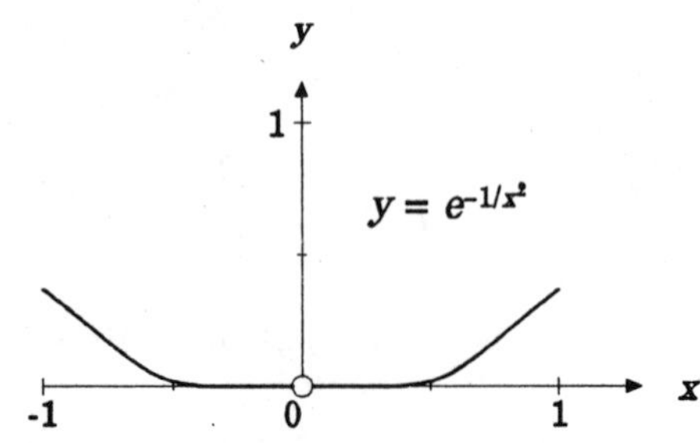

23 If $Q(0) \neq 0$, then $\lim_{x\to 0} \frac{P(x)\cdot e^{-1/x^2}}{Q(x)} = \frac{P(0)\cdot 0}{Q(0)} = 0.$

If $Q(0) = 0$, then factor out of $Q(x)$ the smallest power of x, say x^m, so that $Q(x) = x^m \cdot R(x)$, where $R(x)$ is a polynomial satisfying $R(0) \neq 0$. Thus, by Exercise 22, $\lim_{x\to 0} \frac{P(x)e^{-1/x^2}}{Q(x)} = \lim_{x\to 0} \frac{e^{-1/x^2}}{x^m}\cdot\frac{P(x)}{R(x)} =$

$0 \cdot \frac{P(0)}{R(0)} = 0.$

25 (a) Let $f(x) = e^{-1/x^2}$ for $x \neq 0$ and $f(0) = 0$; then $g(x) = e^{-1/x^2} + e^x = f(x) + e^x$. By Exercise 24, $f^{(n)}(0) = 0$ for all n. Thus, $g(0) = f(0) + e^0 = 1$, $g'(0) = f'(0) + e^0 = 1$, $g''(0) = f''(0) + e^0 = 1, \cdots, g^{(n)}(0) = f^{(n)}(0) + e^0 = 1, \cdots$ and the Maclaurin series for g is $1 + x + \frac{x^2}{2!} + \frac{x^3}{3!} + \frac{x^4}{4!} + \cdots$.

(b) The Maclaurin series for g converges to $g(x)$ for $x = 0$. For all other values of x the series converges to e^x.

27 (a) Since $f(x) = P_n(x; 0) + R_n(x; 0)$ is a

decreasing, alternating series, the error $|f(x) - P_n(x; 0)| = |R_n(x; 0)| \le |a_{n+1}| = \left|\frac{f^{(n+1)}(x)}{(n+1)!} x^{n+1}\right|$.

(b) By Lagrange's form of the remainder, $|f(x) - P_n(x; 0)| = |R_n(x; 0)| = \left|\frac{f^{(n+1)}(c)}{(n+1)!} x^{n+1}\right|$, where c is between 0 and x.

11.3 Why the Error in Taylor Series is Controlled by a Derivative

1 Take the case $b > a$ and $n = 2$. Assume that $R(a) = 0$, $R'(a) = 0$, $R''(a) = 0$, and $m \le R'''(x) \le M$ for x in $[a, b]$, where m and M are constants. We shall show that $\frac{m(b - a)^3}{3!} \le R(b)$. Since $m \le R'''(x)$, by Property 1 and Property 2, $\int_a^t m\, dx \le \int_a^t R'''(x)\, dx$ so that $mx|_a^t \le R''(x)|_a^t$, that is, $m(t - a) \le R''(t) - R''(a) = R''(t)$, or $m(x - a) \le R''(x)$. Similarly, $\int_a^t m(x - a)\, dx \le \int_a^t R''(x)\, dx$ so that $\frac{m}{2}(x - a)^2\Big|_a^t \le R'(x)|_a^t$, that is, $\frac{m}{2}(t - a)^2 \le R'(t) - R'(a) = R'(t)$, or $\frac{m}{2}(x - a)^2 \le R'(x)$. Similarly, $\int_a^t \frac{m}{2}(x - a)^2\, dx \le \int_a^t R'(x)\, dx$ so that $\frac{m}{3!}(x - a)^3\Big|_a^t \le R(x)|_a^t$, that is, $\frac{m}{3!}(t - a)^3 \le R(t) - R(a) = R(t)$. In particular, if $t = b$ then $\frac{m}{3!}(b - a)^3 \le R(b)$.

3 Assume that $R(a) = 0$, $R'(a) = 0$, $R''(a) = 0$, $R'''(a) = 0$, and $m \le R^{(4)}(x) \le M$ for x in $[a, b]$, where m and M are constants. We must show that $\frac{m}{4!}(b - a)^4 \le R(b) \le \frac{M}{4!}(b - a)^4$. Since $m \le R^{(4)}(x) \le M$, $\int_a^t m\, dx \le \int_a^t R^{(4)}(x)\, dx \le \int_a^t M\, dx$, so $mx|_a^t \le R'''(x)|_a^t \le Mx|_a^t$; that is, $m(t - a) \le R'''(t) - R'''(a) = R'''(t) \le M(t - a)$, or $m(x - a) \le R'''(x) \le M(x - a)$. Similarly, $\int_a^t m(x - a)\, dx \le \int_a^t R'''(x)\, dx \le \int_a^t M(x - a)\, dx$ so that $\frac{m}{2}(x - a)^2\Big|_a^t \le R''(x)|_a^t \le \frac{M}{2}(x - a)^2\Big|_a^t$, that is, $\frac{m}{2}(t - a)^2 \le R''(t) - R''(a) = R''(t) \le \frac{M}{2}(t - a)^2$, or $\frac{m}{2}(x - a)^2 \le R''(x) \le \frac{M}{2}(x - a)^2$. Similarly, $\int_a^t \frac{m}{2}(x - a)^2\, dx \le \int_a^t R''(x)\, dx \le \int_a^t \frac{M}{2}(x - a)^2\, dx$, so that $\frac{m}{3!}(x - a)^3\Big|_a^t \le R'(x)|_a^t \le \frac{M}{3!}(x - a)^3\Big|_a^t$, that is, $\frac{m}{3!}(t - a)^3 \le R'(t) - R'(a) = R'(t) \le \frac{M}{3!}(t - a)^3$, or

$\frac{m}{3!}(x - a)^3 \le R'(x) \le \frac{M}{3!}(x - a)^3$. Similarly,

$\int_a^t \frac{m}{3!}(x - a)^3\,dx \le \int_a^t R'(x)\,dx \le$

$\int_a^t \frac{M}{3!}(x - a)^3\,dx$, so that $\frac{m}{4!}(x - a)^4\Big|_a^t \le R(x)\,\Big|_a^t$

$\le \frac{M}{4!}(x - a)^4\Big|_a^t$, that is, $\frac{m}{4!}(t - a)^4 \le$

$R(t) - R(a) = R(t) \le \frac{M}{4!}(t - a)^4$. In particular, if

$t = b$, then $\frac{m}{4!}(b - a)^4 \le R(b) \le \frac{M}{4!}(b - a)^4$.

5 Assume $x \ge a$. Since $f'''(x) \le M$, we have

$\int_a^t f'''(x)\,dx \le \int_a^t M\,dx$. Thus $f''(x)\,\Big|_a^t \le$

$M(t - a)$ and $f''(t) - f''(a) = f''(t) - 3 \le$

$M(t - a)$, or $f''(x) - 3 \le M(x - a)$. Hence

$\int_a^t [f''(x) - 3]\,dx \le \int_a^t M(x - a)\,dx$ and

$(f'(x) - 3x)\,\Big|_a^t \le \frac{M}{2}(x - a)^2\Big|_a^t$, so

$(f'(t) - 3t) - (f(a) - 3a) \le \frac{M}{2}(t - a)^2$.

Simplifying and using x in place of t, we have

$f'(x) - 3x - 1 + 3a \le \frac{M}{2}(x - a)^2$. This implies

that $\int_a^t [f'(x) - 3x - 1 + 3a]\,dx \le$

$\int_a^t \frac{M}{2}(x - a)^2\,dx$, so $\left[f(x) - \frac{3}{2}x^2 - x + 3ax\right]\Big|_a^t$

$\le \frac{M}{3!}(x - a)^3\Big|_a^t$; hence $\left(f(t) - \frac{3}{2}t^2 - t + 3at\right) -$

$\left(f(a) - \frac{3}{2}a^2 - a + 3a^2\right) \le \frac{M}{3!}(t - a)^3$,

$f(x) - \frac{3}{2}x^2 - x + 3ax - 2 + a - \frac{3}{2}a^2 \le$

$\frac{M}{3!}(x - a)^3$, and $f(x) \le \frac{3}{2}x^2 + (1 - 3a)x +$

$(2 - a + \frac{3}{2}a^2) + \frac{M}{3!}(x - a)^3$. For $x \le a$, this

inequality is reversed.

7 (a) The total error E in the left-point method is

$\int_a^b f(x)\,dx - \sum_{k=0}^{n-1} f(a + kh)h =$

$\sum_{k=0}^{n-1}\left(\int_{a+kh}^{a+(k+1)h} f(x)\,dx - f(a + kh)h\right)$. Letting c

denote one of the left endpoints $a + kh$, we

wish to estimate $\int_c^{c+h} f(x)\,dx - f(c)h$.

Adding these estimates for all n values of c

will give an estimate of E.

(b) $E(0) = \int_c^c f(x)\,dx - f(c)\cdot 0 = 0 - 0 = 0$.

$E'(t) = f(c + t) - f(c)$, so $E'(0) = 0$.

$E''(t) = f'(c + t) - 0 = f'(c + t)$.

(c) For $0 \le t \le h$, we have $m_1 \le E''(t) \le M_1$.

Integrating gives $m_1 t \le \int_0^t E''(x)\,dx \le M_1 t$.

But $\int_0^t E''(x)\,dx = E'(t) - E'(0) = E'(t)$, so

$m_1 t \le E'(t) \le M_1 t$. Integrating again gives

$\frac{1}{2}m_1 h^2 \le \int_0^h E'(t)\,dt \le \frac{1}{2}M_1 h^2$. But

$\int_0^h E'(t)\,dt = E(h) - E(0) = E(h)$, so

$\frac{1}{2}m_1h^2 < E(h) < \frac{1}{2}M_1h^2.$

(d) E is a sum of n quantities of the form $E(h)$. Since each of these quantities is between $\frac{1}{2}m_1h^2$ and $\frac{1}{2}M_1h^2$, E is between $n\cdot\frac{1}{2}m_1h^2$ $= \frac{1}{2}m_1(b-a)h$ and $n\cdot\frac{1}{2}M_1h^2 =$ $\frac{1}{2}M_1(b-a)h.$

9 (a) By the discussion preceding Eq. (2) of Sec. 5.4, the trapezoidal estimate is $\sum_{k=0}^{n-1}\frac{h}{2}[f(a+kh)+f(a+(k+1)h)]$. The actual integral can be written as $\int_a^b f(x)\,dx =$ $\sum_{k=0}^{n-1}\int_{a+kh}^{a+(k+1)h} f(x)\,dx$, so the total error is

$\int_a^b f(x)\,dx -$

$\sum_{k=0}^{n-1}\frac{h}{2}[f(a+kh)+f(a+(k+1)h)]$

$$= \sum_{k=0}^{n-1}\left(\int_{a+kh}^{a+(k+1)h} f(x)\,dx - \frac{h}{2}[f(a+kh)+f(a+(k+1)h)]\right).$$

Letting c denote one of the left endpoints, $a + kh$, we wish to estimate $\int_c^{c+h} f(x)\,dx -$ $\frac{h}{2}[f(c)+f(c+h)]$. Adding these estimates for all n values of c will give an estimate of E.

(b) $E(0) = \int_c^c f(x)\,dx - \frac{0}{2}[f(c)+f(c)] = 0 - 0$ $= 0.$

$E'(t) = f(c+t) - \frac{t}{2}f'(c+t) - \frac{1}{2}[f(c)+f(c+t)]$

$= \frac{1}{2}[f(c+t) - f(c)] - \frac{t}{2}f'(c+t)$, so

$E'(0) = 0.$

$E''(t) = \frac{1}{2}f'(c+t) - \frac{t}{2}f''(c+t) - \frac{1}{2}f'(c+t)$

$= -\frac{t}{2}f''(c+t).$

(c) For $0 \le t \le h$, we have $-\frac{t}{2}m_2 \ge E''(t) \ge$ $-\frac{t}{2}M_2$. Integrating and using the fact that $E'(0) = 0$ gives $-\frac{m_2t^2}{4} \ge E'(t) \ge -\frac{M_2t^2}{4}$.

Integrating again and using $E(0) = 0$ gives

$-\frac{m_2h^3}{12} \ge E(h) \ge -\frac{M_2h^3}{12}.$

(d) E is a sum of n quantities of the form $E(h)$. Since each is between $-\frac{m_2h^3}{12}$ and $-\frac{M_2h^3}{12}$, E is between $n\left(-\frac{m_2h^3}{12}\right) = -\frac{1}{12}m_2(b-a)h^2$ and $n\left(-\frac{M_2h^3}{12}\right) = -\frac{1}{12}M_2(b-a)h^2.$

11.4 Power Series and Radius of Convergence

1 $\lim_{n\to\infty}\left|\frac{a_{n+1}}{a_n}\right| = \lim_{n\to\infty}\left|\frac{x^{n+1}/(n+1)^2}{x^n/n^2}\right| =$

$\lim_{n\to\infty}|x|\left(\frac{n}{n+1}\right)^2 = |x|$; convergence requires that $|x| \le 1$; that is, $-1 \le x \le 1$. The series necessarily converges for $-1 < x < 1$; we need check only the endpoints. At $x = 1$ we have $\sum_{n=1}^{\infty}\frac{1}{n^2}$, which is a convergent p series; at $x = -1$, we get $\sum_{n=1}^{\infty}\frac{(-1)^n}{n^2}$, which is absolutely convergent. The radius of convergence is 1, and the region of convergence is graphed below.

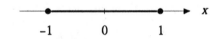

3 $\sum_{n=0}^{\infty}\frac{x^n}{3^n} = \sum_{n=0}^{\infty}\left(\frac{x}{3}\right)^n$ is a geometric series. It converges if and only if $|x/3| < 1$; that is, if and only if $-3 < x < 3$.

5 Use the limit-comparison test: Since

$$\lim_{n\to\infty}\frac{\frac{2n^2+1}{n^2-5}x^n}{x^n} = \lim_{n\to\infty}\frac{2n^2+1}{n^2-5} = 2,$$ the series converges if and only if $\sum_{n=0}^{\infty} x^n$ does. But this is a geometric series; it converges if and only if $|x| < 1$.

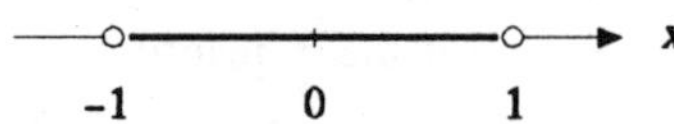

7 $\lim_{n\to\infty}\left|\frac{a_{n+1}}{a_n}\right| = \lim_{n\to\infty}\left|\frac{x^{n+1}/(2n+2)!}{x^n/(2n)!}\right| =$

$\lim_{n\to\infty}\frac{|x|}{(2n+1)(2n+2)} = 0$; the series converges for all x; that is, the radius of convergence is infinite.

9 $\lim_{n\to\infty}\left|\frac{a_{n+1}}{a_n}\right| = \lim_{n\to\infty}\left|\frac{x^{n+1}/(2n+3)!}{x^n/(2n+1)!}\right| =$

$\lim_{n\to\infty}\frac{|x|}{(2n+2)(2n+3)} = 0$; the series converges for all x; that is, the radius of convergence is infinite.

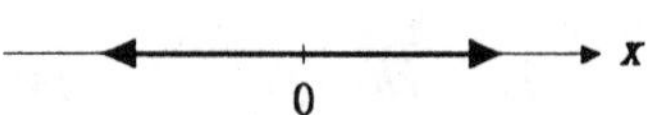

11 $\sum_{n=1}^{\infty}\frac{(-1)^{n+1}x^n}{n}$; the fact that the series is alternating does not enter into the absolute-ratio test, so we will drop the $(-1)^{(n+1)}$: $\lim_{n\to\infty}\left|\frac{a_{n+1}}{a_n}\right| =$

$\lim_{n\to\infty}\left|\frac{x^{n+1}/(n+1)}{x^n/n}\right| = \lim_{n\to\infty}\frac{n}{n+1}|x| = |x|$. For convergence we need $|x| \le 1$; that is, $-1 \le x \le 1$. At $x = 1$, we have the alternating harmonic series, which converges conditionally; at $x = -1$, we have the negative of the harmonic series, which must diverge. The radius of convergence is 1.

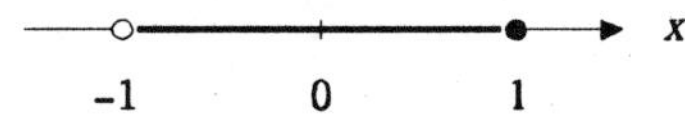

13 Since $\sum_{n=0}^{\infty} a_n x^n$ converges when $x = 9$ and diverges when $x = -12$, its radius of convergence is at least 9 but no greater than 12. Therefore, if $|x| < 9$, the series converges absolutely; if $|x| > 12$, it diverges; and if $9 < |x| < 12$, convergence or divergence cannot be determined from the given information.

(a) At $x = 7$ the series converges absolutely.

(b) At $x = -7$ it converges absolutely.

(c) $x = 9$ could be an endpoint of the interval of convergence if the radius of convergence is 9, but the given information is insufficient to determine absolute convergence.

(d) At $x = -9$ convergence cannot be determined.

(e) At $x = 10$ divergence cannot be determined.

(f) At $x = -15$ it diverges.

(g) It diverges at $x = 15$.

15 Yes. If $x < 0$, let $c = 1 + |x|$. Since $c > 0$, $\sum_{n=0}^{\infty} a_n c^n$ converges. Since $|x| < |c|$, Theorem 1 implies that $\sum_{n=0}^{\infty} a_n x^n$ also converges.

17 $\sum_{n=0}^{\infty} \frac{(x-2)^n}{n!}$; $\lim_{n\to\infty} \left|\frac{a_{n+1}}{a_n}\right| = \lim_{n\to\infty} \left|\frac{(x-2)^{n+1}/(n+1)!}{(x-2)^n/n!}\right|$

$= \lim_{n\to\infty} \frac{|x-2|}{n+1} = 0$, so the radius of convergence is infinite; that is, the series converges for all values of x.

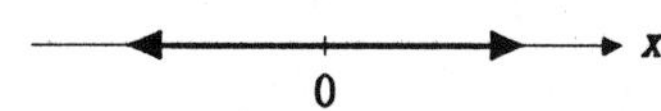

19 $\sum_{n=0}^{\infty} \frac{(x-1)^n}{n+3}$; $\lim_{n\to\infty} \left|\frac{a_{n+1}}{a_n}\right| = \lim_{n\to\infty} \left|\frac{(x-1)^{n+1}/(n+4)}{(x-1)^n/(n+3)}\right|$

$= \lim_{n\to\infty} \frac{n+3}{n+4}|x-1| = |x-1|$ so for convergence we need $|x-1| \le 1$; that is, $-1 \le x - 1 \le 1$ or $0 \le x \le 2$. At $x = 2$, we have $\sum_{n=0}^{\infty} \frac{1}{n+3}$, which diverges because it is a tail of the harmonic series, while at $x = 0$ we have an alternating series that is conditionally convergent. The radius of convergence is 1.

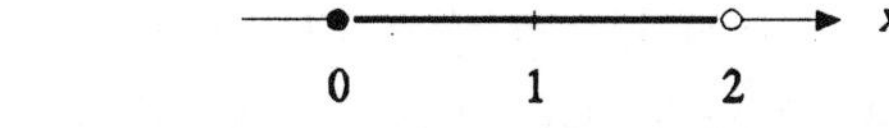

21 $\sum_{n=1}^{\infty} \frac{n(x-2)^n}{2n+3}$; $\lim_{n\to\infty} \left|\frac{a_{n+1}}{a_n}\right| =$

$\lim_{n\to\infty} \left|\frac{(n+1)(x-2)^{n+1}/(2n+5)}{n(x-2)^n/(2n+3)}\right| =$

$\lim_{n\to\infty} \frac{n+1}{n} \cdot \frac{2n+3}{2n+5}|x-2| = |x-2|$, so for convergence we need $|x-2| \le 1$; that is, $-1 \le x - 2 \le 1$ or $1 \le x \le 3$. At $x = 1$ or $x = 3$, the absolute value of the nth term is $\frac{n}{2n+3}$, which does not approach 0 as $n \to \infty$. By the nth-term test for divergence, therefore, the series diverges at

$x = 1$ or 3. The radius of convergence is 1.

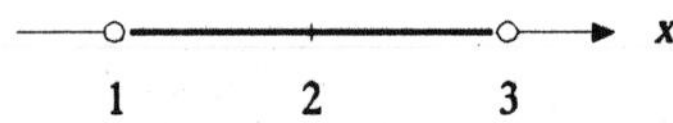

23 $\sum_{n=0}^{\infty} \frac{(x+3)^n}{5^n} = \sum_{n=0}^{\infty} \left(\frac{x+3}{5}\right)^n$ is a geometric series. It converges if and only if $\left|\frac{x+3}{5}\right| < 1$; that is, if and only if $-8 < x < 2$.

x
-8 2

25 $\sum_{n=1}^{\infty} \frac{(x-5)^n}{n^2}$; $\lim_{n\to\infty} \left|\frac{a_{n+1}}{a_n}\right| = \lim_{n\to\infty} \left|\frac{(x-5)^{n+1}}{(n+1)^2} \cdot \frac{n^2}{(x-5)^n}\right|$

$= \lim_{n\to\infty} \left(\frac{n}{n+1}\right)^2 |x-5| = |x-5|$; convergence is guaranteed for $|x-5| < 1$, that is, for $-1 < x - 5 < 1$, or $4 < x < 6$. Convergence could also occur at the endpoints of this interval, but nowhere else. If $x = 4$, then $\sum_{n=1}^{\infty} \frac{(-1)^n}{n^2}$ converges by the alternating series test. If $x = 6$, then $\sum_{n=1}^{\infty} \frac{1}{n^2}$ is a convergent p-series.

x
4 5 6

27 $\sum_{n=0}^{\infty} n!(x-1)^n$; $\lim_{n\to\infty} \left|\frac{a_{n+1}}{a_n}\right| = \lim_{n\to\infty} \left|\frac{(n+1)!(x-1)^{n+1}}{n!(x-1)^n}\right|$

$= \lim_{n\to\infty} (n+1)|x-1| = 0$ if $x = 1$. If $x \neq 1$, the series diverges since $\lim_{n\to\infty} (n+1)|x-1| = \infty$.

29 From equation (4) with $r = 1/2$, the first five terms of the Maclaurin series are

$$1 + \frac{1}{2}x - \frac{1}{8}x^2 + \frac{1}{16}x^3 - \frac{5}{128}x^4 + \cdots.$$

31 From equation (4) with $r = -3$, the first five terms of the Maclaurin series are

$$1 - 3x + 6x^2 - 10x^3 + 15x^4 - \cdots.$$

33 (a) If $\sum_{n=0}^{\infty} a_n x^n$ diverges at $x = 3$, then its radius of convergence cannot be greater than 3. In addition to $x = 3$, therefore, the series also diverges for $|x| > 3$; that is, it diverges for $x \geq 3$ and $x < -3$.

(b) If $\sum_{n=0}^{\infty} a_n(x+5)^n$ diverges at $x = -2$, then its radius of convergence can be no greater than the distance between the points -5 and -2; that is, $|-5-(-2)| = 3$. Thus, in addition to $x = -2$, the series diverges for $|x+5| > 3$; that is, for $-3 > x + 5$ or $3 < x + 5$. Thus the series diverges for $x < -8$ and $x \geq -2$.

35 Use the absolute-ratio test:

$$\left|\frac{x^{2(n+1)+1}/(2(n+1)+1)!}{x^{2n+1}/(2n+1)!}\right| = \left|\frac{x^{2n+3}(2n+1)!}{x^{2n+1}(2n+3)!}\right|$$

$$= \frac{|x|^2}{(2n+2)(2n+3)}.$$ Since

$\lim_{n\to\infty} \dfrac{|x|^2}{(2n+2)(2n+3)} = 0$, the limiting ratio is less than 1. Hence the series converges for all x. The radius of convergence is infinite.

37 (a) Since $(1+x)^{1/2} =$

$1 + \frac{1}{2}x - \frac{1}{8}x^2 + \frac{1}{16}x^3 - \cdots$, $(1+x^3)^{1/2} =$

$1 + \frac{1}{2}x^3 - \frac{1}{8}x^6 + \frac{1}{16}x^9 - \cdots$, so that

$\int_0^1 \sqrt{1+x^3}\,dx$

$$\approx \int_0^1 \left[1 + \frac{1}{2}x^3 - \frac{1}{8}x^6 + \frac{1}{16}x^9\right] dx$$

$$= \left[x + \frac{1}{8}x^4 - \frac{1}{56}x^7 + \frac{1}{160}x^{10}\right]\Bigg|_0^1 \approx 1.1134.$$

(b) If $f(x) = (1+x^3)^{1/2}$, then $f^{(4)}(x) =$

$\dfrac{3}{8}x^2\dfrac{\left(\frac{3}{2}x^6 + 84x^3 - 120\right)}{(1+x^3)^{7/2}}$ so that the maximum value M_4 of $f^{(4)}(x)$ on $[0, 1]$ satisfies $|M_4| \leq$

$\dfrac{3}{8}(1)^2\dfrac{|0 - 120|}{(1+0)^{7/2}} = 45$. The bound on the error is therefore $\left|\dfrac{(b-a)M_4h^4}{180}\right| \leq \dfrac{45}{180}\cdot\dfrac{1}{n^4}$,

which is less than 0.0005 if $n = 6$. Thus, Simpson's estimate with six subdivisions, S_6

$= \frac{1}{18}\left[f(0) + 4f\left(\frac{1}{6}\right) + 2f\left(\frac{1}{3}\right) + 4f\left(\frac{1}{2}\right) + 2f\left(\frac{2}{3}\right) + 4f\left(\frac{5}{6}\right) + f(1)\right] \approx 1.1114$, which

estimates $\int_0^1 \sqrt{1+x^3}\,dx$ to three places.

39 (a) $e^x = \sum_{n=0}^{\infty} \dfrac{x^n}{n!}$; $\lim_{n\to\infty}\left|\dfrac{a_{n+1}}{a_n}\right| = \lim_{n\to\infty}\left|\dfrac{x^{n+1}}{(n+1)!}\cdot\dfrac{n!}{x^n}\right|$

$= \lim_{n\to\infty} \dfrac{|x|}{n+1} = 0 < 1$ for all values of x.

The radius of convergence is $R = \infty$.

(b) $\sin x = \sum_{n=0}^{\infty} \dfrac{x^{2n+1}}{(2n+1)!}(-1)^n$; $\lim_{n\to\infty}\left|\dfrac{a_{n+1}}{a_n}\right|$

$= \lim_{n\to\infty}\left|\dfrac{x^{2n+3}}{(2n+3)!}\cdot\dfrac{(2n+1)!}{x^{2n+1}}\right|$

$= \lim_{n\to\infty} \dfrac{|x|^2}{(2n+3)(2n+2)} = 0 < 1$ for all values of x. The radius of convergence is $R = \infty$.

(c) $\cos x = \sum_{n=0}^{\infty} \dfrac{x^{2n}}{(2n)!}(-1)^n$; $\lim_{n\to\infty}\left|\dfrac{a_{n+1}}{a_n}\right|$

$= \lim_{n\to\infty}\left|\dfrac{x^{2n+2}}{(2n+2)!}\cdot\dfrac{(2n)!}{x^{2n}}\right|$

$= \lim_{n\to\infty} \dfrac{|x|^2}{(2n+2)(2n+1)} = 0 < 1$ for all values of x. The radius of convergence is $R = \infty$.

(d) $\ln(1+x) = \sum_{n=1}^{\infty} (-1)^{n+1}\dfrac{x^n}{n}$; $\lim_{n\to\infty}\left|\dfrac{a_{n+1}}{a_n}\right| =$

$\lim_{n\to\infty}\left|\dfrac{x^{n+1}}{n+1}\cdot\dfrac{n}{x^n}\right| = \lim_{n\to\infty}\dfrac{n}{n+1}\cdot|x| = |x|$, so the radius of convergence is $R = 1$.

(e) $\arctan x = \sum_{n=0}^{\infty} (-1)^n\cdot\dfrac{x^{2n+1}}{2n+1}$; $\lim_{n\to\infty}\left|\dfrac{a_{n+1}}{a_n}\right| =$

$$\lim_{n\to\infty}\left|\frac{x^{2n+3}}{2n+3}\cdot\frac{2n+1}{x^{2n+1}}\right| = \lim_{n\to\infty}\frac{n+\frac{1}{2}}{n+\frac{3}{2}}\cdot|x|^2 =$$

$|x^2|$, so the radius of convergence is $R = 1$.

41 Using the general binomial theorem—Eq. (4)—with r replaced by $\frac{\gamma}{1-\gamma}$, we have

$$(1+x)^{\frac{\gamma}{1-\gamma}} = 1 + \frac{\gamma}{1-\gamma}x + \frac{\gamma(2\gamma-1)}{2(1-\gamma)^2}x^2 + \cdots,$$

so that $\left(1+\left(\frac{\gamma-1}{2}M^2\right)\right)^{\frac{\gamma}{1-\gamma}} =$

$$1 + \frac{\gamma}{1-\gamma}\cdot\frac{\gamma-1}{2}M^2 + \frac{\gamma(2\gamma-1)}{2(1-\gamma)^2}\cdot\frac{(\gamma-1)^2}{4}M^4 + \cdots$$

$$= 1 - \frac{\gamma}{2}M^2 + \frac{\gamma(2\gamma-1)}{8}M^4 + \cdots$$

$= 1 - \frac{1}{2}\cdot\frac{v^2}{RT} + \frac{\gamma(2\gamma-1)}{8}M^4 + \cdots$, where $v^2 = M^2\gamma Rt$.

11.5 Manipulating Power Series

1 $\sin x = \sum_{n=0}^{\infty}(-1)^n\frac{x^{2n+1}}{(2n+1)!}$, so $\cos x = (\sin x)' =$

$$\sum_{n=0}^{\infty}(-1)^n\frac{(2n+1)x^{2n}}{(2n+1)!} = \sum_{n=0}^{\infty}(-1)^n\frac{x^{2n}}{(2n)!}.$$

3 (a) $1 - t^2 + t^4 - t^6 + \cdots$ is a geometric series with initial term 1 and ratio $-t^2$. It converges absolutely if $|-t^2| < 1$. We are given, however, that $|t| < 1$, from which it follows that $t^2 < 1$. Therefore the series converges and its sum is $\frac{1}{1-(-t^2)} = \frac{1}{1+t^2}$.

(b) Integrate the series in (a) term by term from 0 to x: $\int_0^x \frac{dt}{1+t^2} =$

$\int_0^x (1 - t^2 + t^4 - t^6 + \cdots)\,dt$, $\tan^{-1}t\big|_0^x =$

$\left(t - \frac{t^3}{3} + \frac{t^5}{5} - \frac{t^7}{7} + \cdots\right)\Big|_0^x$, so $\tan^{-1}x =$

$x - \frac{x^3}{3} + \frac{x^5}{5} - \frac{x^7}{7} + \cdots$, as was to be shown.

(c) The general term is $\frac{(-1)^n x^{2n+1}}{2n+1}$ for $n \geq 0$.

(d) $\tan^{-1}x = x - \frac{x^3}{3} + \frac{x^5}{5} - \frac{x^7}{7} + \cdots$ is a decreasing, alternating series for $x = 1/2$.

Now $|R_n| < |a_{n+1}| = \left|\frac{(-1)^{n+1}\left(\frac{1}{2}\right)^{2n+3}}{2n+3}\right| =$

$\frac{1}{2n+3}\cdot\left(\frac{1}{2}\right)^{2n+3}$; if $n = 3$, this error is at most $\frac{1}{4608} < 0.0005$. So

$$\frac{1}{2} - \frac{(1/2)^3}{3} + \frac{(1/2)^5}{5} - \frac{(1/2)^7}{7} \approx 0.4635$$

estimates $\tan^{-1}\frac{1}{2}$ to three decimal places.

5 (a) $\tan x = \frac{\sin x}{\cos x}$, so we divide the series for $\sin x$ by the series for $\cos x$.

$$\begin{array}{r|l} & x + \frac{x^3}{3} + \frac{2}{15}x^5 + \cdots \\ \hline 1 - \frac{x^2}{2} + \frac{x^4}{24} - \cdots & x - \frac{x^3}{6} + \frac{x^5}{120} - \cdots \\ & x - \frac{x^3}{2} + \frac{x^5}{24} - \cdots \\ \hline & \frac{x^3}{3} - \frac{x^5}{30} + \cdots \\ & \frac{x^3}{3} - \frac{x^5}{6} + \cdots \\ \hline & \frac{2}{15}x^5 + \cdots \end{array}$$

Thus, the series for tan x is

$$x + \frac{x^3}{3} + \frac{2}{15}x^5 + \cdots.$$

(b) For $f(x) = \tan x$, we have $f'(x) = \sec^2 x$, $f''(x) = 2\sec^2 x \tan x$, and $f'''(x) = 2\sec^4 x + 4\sec^2 x \cdot \tan^2 x$, so that $a_0 = f(0) = 0$,

$a_1 = f'(0) = 1$, $a_2 = \dfrac{f''(0)}{2!} = 0$, and $a_3 =$

$\dfrac{f'''(0)}{3!} = \dfrac{1}{3}$. The Maclaurin series is

$$0 + 1\cdot x + 0\cdot x^2 + \frac{1}{3}x^3 + \cdots = x + \frac{1}{3}x^3 + \cdots.$$

7 e^x is represented by $1 + x + \dfrac{x^2}{2!} + \dfrac{x^3}{3!} + \cdots$ and

sin x is represented by $x - \dfrac{x^3}{3!} + \dfrac{x^5}{5!} - \cdots$, so

$e^x \sin x$ is represented by

$$\left(1 + x + \frac{x^2}{2!} + \frac{x^3}{3!} + \cdots\right) \times \left(x - \frac{x^3}{3!} + \frac{x^5}{5!} - \cdots\right)$$

$$= x + x\cdot x - \frac{x^3}{3!} + \frac{x^2}{2!}x + \cdots$$

$$= x + x^2 - \frac{x^3}{6} + \frac{x^3}{2} + \cdots$$

$$= x + x^2 + \frac{x^3}{3} + \cdots.$$

9 $\cos x = 1 - \dfrac{x^2}{2} + \dfrac{x^4}{24} - \cdots$, so $1 - \cos x =$

$\dfrac{x^2}{2} - \dfrac{x^4}{24} + \cdots$; hence $\dfrac{1 - \cos x}{x^2} = \dfrac{1}{2} - \dfrac{x^2}{24} + \cdots$

and thus $\displaystyle\lim_{x\to 0} \frac{1 - \cos x}{x^2} = \frac{1}{2}$.

11 $\sin x = x - \dfrac{x^3}{6} + \cdots$, so $\sin x^3 = x^3 - \dfrac{x^9}{6} + \cdots$

and $\sin^2 x^3 = \left(x^3 - \dfrac{x^9}{6} + \cdots\right)^2 = x^6 - \dfrac{2x^{12}}{6} + \cdots$

$= x^6 - \dfrac{x^{12}}{3} + \cdots$. Now $\cos x = 1 - \dfrac{x^2}{2} + \cdots$, so

$\cos x^2 = 1 - \dfrac{x^4}{2} + \cdots$ and $(1 - \cos x^2)^3 =$

$\left(1 - 1 + \dfrac{x^4}{2} - \cdots\right)^3 = \dfrac{x^{12}}{8} + \cdots$. We can

therefore write $\displaystyle\lim_{x\to 0} \frac{\sin^2 x^3}{(1 - \cos x^2)^3}$

$$= \lim_{x\to 0} \frac{x^6 - \frac{1}{3}x^{12} + \cdots}{\frac{1}{8}x^{12} + \cdots} = \lim_{x\to 0} \frac{8 - \frac{8}{3}x^6 + \cdots}{x^6 + \cdots}$$

$$= \infty.$$

13 $e^x - 1 = \left(1 + x + \dfrac{x^2}{2} + \cdots\right) - 1 =$

$x + \dfrac{x^2}{2} + \cdots$, so $(e^x - 1)^2 = x^2 + x^3 + \cdots$. Also,

$\sin x^2 = x^2 - \frac{(x^2)^3}{3!} + \cdots = x^2 - \frac{x^6}{6} + \cdots$. Hence

$$\lim_{x\to 0}\frac{(e^x - 1)^2}{\sin x^2} = \lim_{x\to 0}\frac{x^2 + x^3 + \cdots}{x^2 - \frac{x^6}{6} + \cdots} = 1.$$

15 (a) $\sqrt{x}\sin x = x^{1/2}\sum_{n=0}^{\infty}(-1)^n\frac{x^{2n+1}}{(2n+1)!} =$

$\sum_{n=0}^{\infty}(-1)^n\frac{x^{2n+3/2}}{(2n+1)!}$, so $\int_0^1\sqrt{x}\sin x\,dx$

$$= \sum_{n=0}^{\infty}\left(\frac{(-1)^n}{(2n+1)!}\int_0^1 x^{2n+3/2}\,dx\right)$$

$$= \sum_{n=0}^{\infty}\frac{(-1)^n}{(2n+1)!\,(2n+5/2)}$$

$$= \sum_{n=0}^{\infty}\frac{2(-1)^n}{(2n+1)!\,(4n+5)}$$

$$= \frac{2}{5} - \frac{1}{27} + \frac{1}{780} - \cdots.$$

(b) The series in (a) is a decreasing, alternating series, so $|R_n| < |a_{n+1}| = \left|\frac{2}{(4n+9)(2n+3)!}\right|$. If $n = 2$, then this error is at most $\frac{1}{42{,}840} < 0.00005$. Thus $\frac{2}{5} - \frac{1}{27} + \frac{1}{780} \approx 0.36425$ estimates $\int_0^1\sqrt{x}\sin x\,dx$ to four decimal places.

17 (a) $f(x) = \sum_{n=0}^{\infty}n^2x^n$; $\lim_{n\to\infty}\left|\frac{a_{n+1}}{a_n}\right| =$

$\lim_{n\to\infty}\left(\frac{n+1}{n}\right)^2|x| = |x|$, so f is defined for $|x| < 1$ or $-1 < x < 1$. If $x = 1$ or $x = -1$, then the series diverges by the nth-term test for divergence.

(b) The coefficient of x^{100} is 100^2 so that $100^2 = \frac{f^{(100)}(0)}{100!}$, that is, $f^{(100)}(0) = 100^2\cdot 100!$.

19 (a) $\cos x = 1 - \frac{x^2}{2!} + \frac{x^4}{4!} - \frac{x^6}{6!} + \cdots$, so $\cos 2x$

$$= 1 - \frac{(2x)^2}{2!} + \frac{(2x)^4}{4!} - \frac{(2x)^6}{6!} + \cdots + (-1)^n\cdot\frac{(2x)^{2n}}{(2n)!} + \cdots.$$

(b) $\frac{\sin^2 x}{x^2} = \frac{1 - \cos 2x}{2x^2} =$

$$\frac{1}{2x^2}\left[\frac{(2x)^2}{2!} - \frac{(2x)^4}{4!} + \frac{(2x)^6}{6!} - \cdots\right] =$$

$$\frac{2}{2!} - \frac{2^3x^2}{4!} + \frac{2^5x^4}{6!} - \frac{2^7x^6}{8!} + \frac{2^9x^8}{10!} - \cdots + (-1)^n\frac{2^{2n+1}x^{2n}}{(2(n+1))!} + \cdots.$$

(c) $\int_0^1\left[\frac{\sin x}{x}\right]^2 dx \approx \int_0^1\left[1 - \frac{1}{3}x^2 + \frac{2}{45}x^4\right]dx$

$$= \frac{202}{225} \approx 0.8978$$

(d) $\int_0^1\left[\frac{\sin x}{x}\right]^2 dx =$

$$\int_0^1\left[1 - \frac{1}{3}x^2 + \frac{2}{45}x^4 - \frac{1}{315}x^6 + \frac{2}{14175}x^8 - \cdots\right]dx$$

$$= \left[x - \frac{1}{9}x^3 + \frac{2}{225}x^5 - \frac{1}{2205}x^7 + \frac{2}{127{,}575}x^9 - \cdots\right]\Bigg|_0^1$$

$= \frac{202}{225} - \frac{1}{2205} + \frac{2}{127{,}575} - \cdots$, which is a decreasing, alternating series. Thus, a bound on the error using $\frac{202}{225}$ to estimate

$\int_0^1 \left[\frac{\sin x}{x}\right]^2 dx$ is $\frac{1}{2205} < 0.00046$.

21 Assume $\sum_{n=0}^{\infty} a_n x^n = f(x) = \sum_{n=0}^{\infty} b_n x^n$ for $|x| < 1$.

Then by Theorem 4, $0 = f(x) - f(x) =$

$\sum_{n=0}^{\infty} a_n x^n - \sum_{n=0}^{\infty} b_n x^n = \sum_{n=0}^{\infty} (a_n - b_n)x_n$ for

$|x| < 1$, that is, $0 = (a_0 - b_0) + (a_1 - b_1)x + (a_2 - b_2)x^2 + \cdots$. If $x = 0$, then $0 = a_0 - b_0$ or $a_0 = b_0$. Differentiating the series gives $0 = (a_1 - b_1) + 2(a_2 - b_2)x + 3(a_3 - b_3)x^2 + \cdots$ and letting $x = 0$ gives $0 = a_1 - b_1$, or $a_1 = b_1$. Differentiating the series again gives $0 = 2(a_2 - b_2) + 3 \cdot 2(a_3 - b_3)x + \cdots$ and letting $x = 0$ gives $0 = 2(a_2 - b_2)$ or $a_2 = b_2$. Continuing this process, it follows that $a_n = b_n$ for all n.

23 If $5 = \frac{1+x}{1-x}$, then $x = 2/3$. By Exercise 22, ln 5

$$= 2\left(\frac{2}{3} + \frac{\left(\frac{2}{3}\right)^3}{3} + \frac{\left(\frac{2}{3}\right)^5}{5} + \frac{\left(\frac{2}{3}\right)^7}{7} + \cdots + \frac{\left(\frac{2}{3}\right)^{2n+1}}{2n+1} + \cdots\right).$$

Note that $2\left(\frac{\left(\frac{2}{3}\right)^{2k+1}}{2k+1} + \frac{\left(\frac{2}{3}\right)^{2k+3}}{2k+3} + \frac{\left(\frac{2}{3}\right)^{2k+5}}{2k+5} + \cdots\right)$

$$< \frac{2}{2k+1}\left(\left(\frac{2}{3}\right)^{2k+1} + \left(\frac{2}{3}\right)^{2k+3} + \left(\frac{2}{3}\right)^{2k+5} + \cdots\right)$$

$$= \frac{2}{2k+1} \cdot \frac{2}{3}\left[\left(\frac{4}{9}\right)^k + \left(\frac{4}{9}\right)^{k+1} + \left(\frac{4}{9}\right)^{k+2} + \cdots\right]$$

$= \frac{4}{3(2k+1)}\left(\frac{4}{9}\right)^k \cdot \frac{1}{1 - \frac{4}{9}} = \frac{12}{5(2k+1)}\left(\frac{4}{9}\right)^k$, which is less than 0.005 when k is at least 5. Thus

$$2\left(\frac{2}{3} + \frac{\left(\frac{2}{3}\right)^3}{3} + \frac{\left(\frac{2}{3}\right)^5}{5} + \frac{\left(\frac{2}{3}\right)^7}{7} + \frac{\left(\frac{2}{3}\right)^9}{9}\right) \approx 1.606$$

estimates ln 5 to two decimal places.

25 (a) $\sum_{n=0}^{\infty} n^2 x^n$ converges for $|x| < 1$. (See Exercise 17.)

(b) $g(x) = \frac{x^2}{1-x} = x^2\left(\frac{1}{1-x}\right)$

$= x^2(1 + x + x^2 + x^3 + \cdots)$

$= x^2 + x^3 + x^4 + \cdots$, so $g'(x) = \frac{2x - x^2}{(1-x)^2}$

$= 2x + 3x^2 + 4x^3 + \cdots$. Hence $xg'(x) = \frac{2x^2 - x^3}{(1-x)^2}$ and so $(xg'(x))' =$

$\frac{x(x^2 - 3x + 4)}{(1-x)^3} = 2^2x + 3^2x^2 + 4^2x^3 + \cdots$.

It follows that $x(xg'(x))' = \frac{x^2(x^2 - 3x + 4)}{(1-x)^3}$

$= 2^2x^2 + 3^2x^3 + 4^2x^4 + \cdots$. Thus $\sum_{n=0}^{\infty} n^2 x^n$

$= 1^2x + 2^2x^2 + 3^2x^3 + 4^2x^4 + \cdots$

$= x + \frac{x^2(x^2 - 3x + 4)}{(1-x)^3} = \frac{x + x^2}{(1-x)^3}$.

(c) $\sum_{n=0}^{\infty} n^2\left(\frac{1}{3}\right)^n = \frac{3}{2}$ by part (b).

27 (We assume that $\hbar$, ω, k, and T are positive, so $0 < x < 1$.) The denominator is a geometric series with sum $\frac{1}{1-x}$. The numerator is

$x(1 + 2x + 3x^2 + \cdots) = \frac{x}{(1-x)^2}$ by Example 1.

Hence $\langle E \rangle = \frac{\hbar\omega x/(1-x)^2}{1/(1-x)} = \frac{\hbar\omega x}{1-x} = \frac{\hbar\omega}{x^{-1}-1}$

$= \frac{\hbar\omega}{e^{\hbar\omega/kT} - 1}$, as claimed.

11.6 Complex Numbers

1 (a) $(2 + 3i) + (5 - 2i) = (2 + 5) + (3 - 2)i$

$= 7 + i$

(b) $(2 + 3i)(2 - 3i) = 4 - 6i + 6i - 9i^2$

$= 4 - 9(-1) = 13$

(c) $\frac{1}{2-i} = \frac{1}{2-i}\cdot\frac{2+i}{2+i} = \frac{2+i}{2^2 - i^2} =$

$\frac{2+i}{4+1} = \frac{2}{5} + \frac{i}{5}$

(d) $\frac{3+2i}{4-i} = \frac{3+2i}{4-i}\cdot\frac{4+i}{4+i} =$

$\frac{12 + 3i + 8i + 2i^2}{4^2 - i^2} = \frac{12 + 11i - 2}{16 - (-1)}$

$= \frac{10 + 11i}{17} = \frac{10}{17} + \frac{11}{17}i$

3 (a)

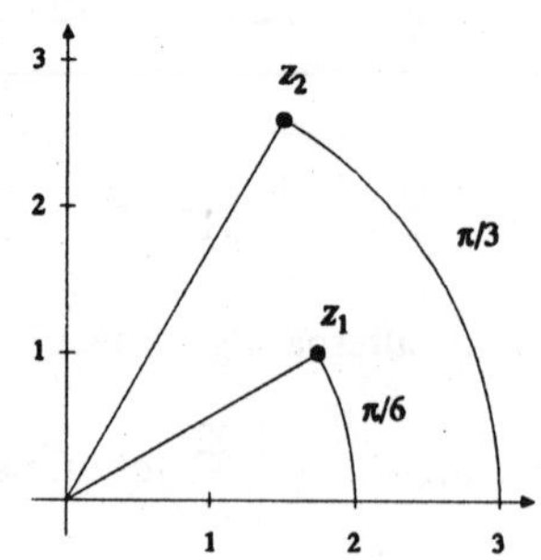

(b) z_1 has magnitude 2 and argument $\pi/6$, which we write in polar coordinates as $z_1 = \left(2, \frac{\pi}{6}\right)$; similarly, $z_2 = \left(3, \frac{\pi}{3}\right)$. Then $z_1z_2 =$

$\left(2\cdot 3, \frac{\pi}{6} + \frac{\pi}{3}\right) = \left(6, \frac{\pi}{2}\right) = 6i.$

(c) $z = r\cos\theta + ri\sin\theta$, so $z_1 =$

$2\cos\frac{\pi}{6} + 2i\sin\frac{\pi}{6} = \sqrt{3} + i$, $z_2 =$

$3\cos\frac{\pi}{3} + 3i\sin\frac{\pi}{3} = \frac{3}{2} + \frac{3\sqrt{3}}{2}i.$

(d) $z_1z_2 = (\sqrt{3} + i)\left(\frac{3}{2} + \frac{3\sqrt{3}}{2}i\right)$

$= \frac{1}{2}[3\sqrt{3} + 3\sqrt{3}\sqrt{3}i + 3i + 3\sqrt{3}i^2] =$

$\frac{1}{2}[3\sqrt{3} + 9i + 3i - 3\sqrt{3}] = \frac{1}{2}[12i] = 6i$

5 In polar form, we have $z = \left(1, \frac{\pi}{3}\right)$, so $z^2 =$

$\left(1^2, \frac{2\pi}{3}\right) = \left(1, \frac{2\pi}{3}\right)$, $z^3 = \left(1^3, \frac{3\pi}{3}\right) = (1, \pi)$,

and $z^4 = \left(1^4, \frac{4\pi}{3}\right) = \left(1, \frac{4\pi}{3}\right)$. The points z, z^2,

z^3, and z^4 are plotted in the graph.

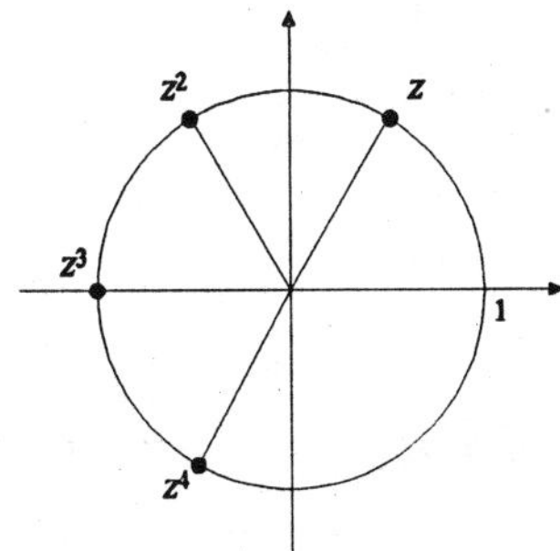

7 z has magnitude 2 and argument $\frac{\pi}{6}$, so (a) z^2 has magnitude $2^2 = 4$ and argument $2\left(\frac{\pi}{6}\right) = \frac{\pi}{3}$, (b) z^3 has magnitude $2^3 = 8$ and argument $3\left(\frac{\pi}{6}\right) = \frac{\pi}{2}$, (c) z^4 has magnitude $2^4 = 16$ and argument $4\left(\frac{\pi}{6}\right) = \frac{2\pi}{3}$, and (d) z^n has magnitude 2^n and argument $\frac{n\pi}{6}$.

(e)

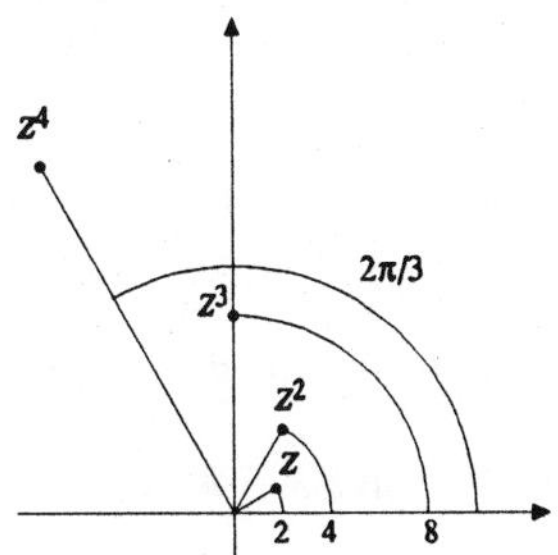

9 The five fifth roots of $32\left(\cos\frac{\pi}{4} + i\sin\frac{\pi}{4}\right)$ all have magnitude $32^{1/5} = 2$. Their arguments are $\frac{\pi/4}{5} = \frac{\pi}{20}$, $\frac{\pi/4 + 2\pi}{5} = \frac{9\pi}{20}$, $\frac{\pi/4 + 4\pi}{5} = \frac{17\pi}{20}$, $\frac{\pi/4 + 6\pi}{5} = \frac{5\pi}{4}$, and $\frac{\pi/4 + 8\pi}{5} = \frac{33\pi}{20}$.

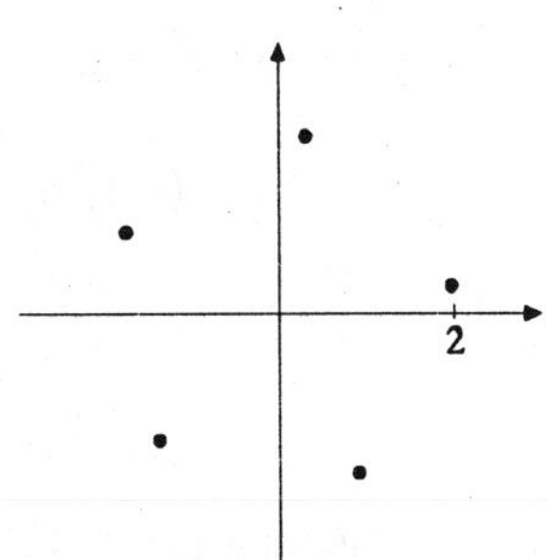

11 zw has magnitude $r \cdot \frac{1}{r} = 1$ and argument $\theta + (-\theta) = 0$, so $zw = 1$.

13 (a) Let $x = 2 + 3i$; then $x^2 - 4x + 13$
$= (2 + 3i)^2 - 4(2 + 3i) + 13$
$= 4 + 12i + 9i^2 - 8 - 12i + 13 = 9 + 9i^2$
$= 9 - 9 = 0$.

(b) By the quadratic formula, the roots of $x^2 - 4x + 13 = 0$ are $\frac{4 \pm \sqrt{16 - 4 \cdot 1 \cdot 13}}{2}$
$= 2 \pm \frac{1}{2}\sqrt{16 - 52} = 2 \pm \frac{1}{2}\sqrt{-36} = 2 \pm \frac{1}{2}(6i) = 2 \pm 3i$.

15 (a) $5 + 5i$ has magnitude $(5^2 + 5^2)^{1/2} = 5\sqrt{2}$ and lies in the first quadrant, so its argument is $\tan^{-1}(y/x) = \tan^{-1}(5/5) = \tan^{-1} 1 = \frac{\pi}{4}$.

(b) $-\frac{1}{2} - \frac{\sqrt{3}}{2}i$ has magnitude $\left(\frac{1}{4} + \frac{3}{4}\right)^{1/2} = 1$ and lies in the third quadrant, so its argument is $\tan^{-1}(y/x) + \pi = \tan^{-1}\sqrt{3} + \pi = \frac{4\pi}{3}$.

(c) $-\frac{\sqrt{2}}{2} + \frac{\sqrt{2}}{2}i$ has magnitude 1 and lies in the second quadrant, so its argument is

$\tan^{-1}(y/x) + \pi = \tan^{-1}(-1) + \pi = -\frac{\pi}{4} + \pi$

$= \frac{3\pi}{4}$.

(d) $3 + 4i$ has magnitude $(3^2 + 4^2)^{1/2} = 5$ and lies in the first quadrant, so its argument is $\tan^{-1}(y/x) = \tan^{-1}(4/3)$.

17 Let $q = (r_1/r_2, \theta_1 - \theta_2)$. Then $qz_2 = (r_1/r_2, \theta_1 - \theta_2)(r_2, \theta_2) = ((r_1/r_2)r_2, (\theta_1 - \theta_2) + \theta_2) = (r_1, \theta_1) = z_1$; that is, $qz_2 = z_1$. Then $q = z_1/z_2$, which shows that (a) the magnitude of z_1/z_2 must be r_1/r_2 and (b) the argument of z_1/z_2 must be $\theta_1 - \theta_2$.

19 (a) $(2 + 3i)(1 + i) = 2 + 3i + 2i + 3i^2 = 2 + 5i - 3 = -1 + 5i$

(b) $\frac{2 + 3i}{1 + i} = \frac{2 + 3i}{1 + i}\cdot\frac{1 - i}{1 - i} =$

$\frac{2 + 3i - 2i - 3i^2}{1^2 - i^2} = \frac{2 + i + 3}{2} = \frac{5 + i}{2}$

(c) $(7 - 3i)(\overline{7 - 3i}) = (7 - 3i)(7 + 3i) = 7^2 - (3i)^2 = 49 + 9 = 58$

(d) $3(\cos 42° + i \sin 42°)\cdot 5(\cos 168° + i \sin 168°)$

$= 3\cdot 5(\cos(42° + 168°) + i \sin(42° + 168°))$

$= 15(\cos 210° + i \sin 210°)$

$= 15\left(-\frac{\sqrt{3}}{2} - \frac{1}{2}i\right)$

(e) $\frac{\sqrt{8}(\cos 147° + i \sin 147°)}{\sqrt{2}(\cos 57° + i \sin 57°)}$

$= 2(\cos(147° - 57°) + i \sin(147° - 57°))$

$= 2(\cos 90° + i \sin 90°) = 2(0 + i) = 2i$

(f) $\frac{1}{3 - i} = \frac{1}{3 - i}\cdot\frac{3 + i}{3 + i} = \frac{3 + i}{3^2 - i^2}$

$= \frac{3 + i}{10}$

(g) $[3(\cos 52° + i \sin 52°)]^{-1}$

$= \frac{1}{3}(\cos(-52°) + i \sin(-52°))$

$= \frac{1}{3}(\cos 52° - i \sin 52°)$

(h) $\left(\cos\frac{\pi}{6} + i \sin\frac{\pi}{6}\right)^{12} = \cos 2\pi + i \sin 2\pi$

$= 1 + i\cdot 0 = 1$

21 The three cube roots of $i = 1\cdot\left(\cos\frac{\pi}{2} + i \sin\frac{\pi}{2}\right)$ all have magnitude $1^{1/3} = 1$. Their arguments are $\frac{\pi/2}{3} = \frac{\pi}{6}$, $\frac{\pi/2 + 2\pi}{3} = \frac{5\pi}{6}$, and $\frac{\pi/2 + 4\pi}{3} = \frac{3\pi}{2}$. Hence the roots are $\cos\frac{\pi}{6} + i \sin\frac{\pi}{6} = \frac{\sqrt{3}}{2} + \frac{1}{2}i$, $\cos\frac{5\pi}{6} + i \sin\frac{5\pi}{6} = -\frac{\sqrt{3}}{2} + \frac{1}{2}i$, and $\cos\frac{3\pi}{2} + i \sin\frac{3\pi}{2} = -i$.

23 $\cos 3\theta + i \sin 3\theta = (\cos\theta + i \sin\theta)^3 =$

$\cos^3\theta + 3\cos^2\theta\, i \sin\theta + 3\cos\theta\, i^2 \sin^2\theta + i^3 \sin^3\theta$

$= (\cos^3\theta - 3\sin^2\theta\cos\theta) + i(3\sin\theta\cos^2\theta - \sin^3\theta)$,

so $\cos 3\theta = \cos^3\theta - 3\sin^2\theta\cos\theta$ and $\sin 3\theta = 3\sin\theta\cos^2\theta - \sin^3\theta$.

25 Let $z_1 = a + bi$ and $z_2 = c + di$.

(a) $z_1z_2 = (ac - bd) + (ad + bc)i$, so $\overline{z_1z_2} = (ac - bd) - (ad + bc)i$; observe that $\overline{z_1} = a - bi$ and $\overline{z_2} = c - di$, so $\overline{z_1}\cdot\overline{z_2} =$

$ac - adi - bci + bdi^2$

$= (ac - bd) - (ad + bc)i = \overline{z_1 z_2}.$

(b) $\overline{z_1 + z_2} = (a + c) - (b + d)i$

$= a - bi + c - di = \bar{z}_1 + \bar{z}_2$

27 Suppose that $\bar{z} = \frac{1}{z}$. Then $|z|^2 = z\bar{z} = z \cdot \frac{1}{z} = 1;$ that is, $|z| = 1$. Conversely, if $|z| = 1$, then $z\bar{z} = |z|^2 = 1$, so $\bar{z} = \frac{1}{z}$. Thus $\bar{z} = \frac{1}{z}$ for points on the unit circle.

29 (a)

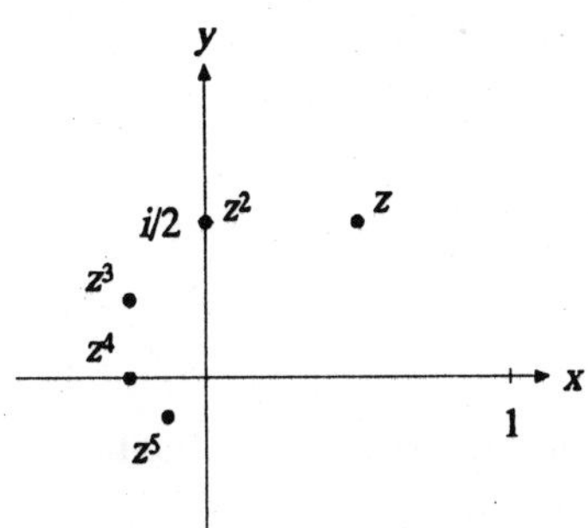

(b) The magnitude of z is $\left(\frac{1}{4} + \frac{1}{4}\right)^{1/2} = \frac{1}{\sqrt{2}}$.

Thus $|z| < 1$, so $|z^n| = |z|^n \to 0$ as $n \to \infty$. Therefore, $z^n \to 0$ as $n \to \infty$.

31 (a) $z^2 = (t + it)^2 = t^2(1 + i)^2$

$= t^2 \cdot 2i = 2t^2 i$

(b) $z^2 = 0 + (2t^2)i$ so $x = 0$, $y = 2t^2$.

(c) The curve is the nonnegative portion of the y axis.

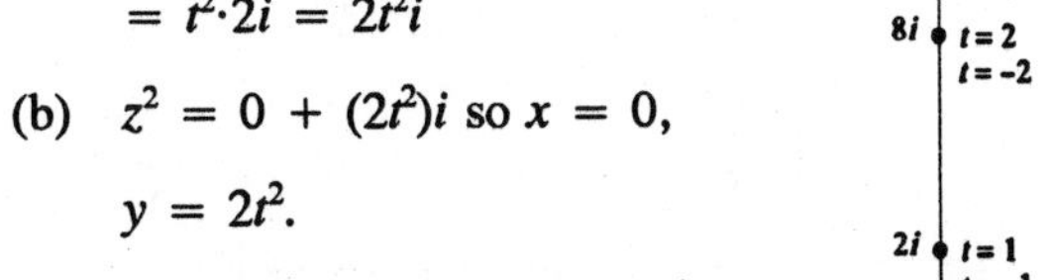

33 (a) $z^2 = (t + i)^2 = (t^2 - 1) + (2t)i$. (See the figure.)

(b) $x = t^2 - 1,\ y = 2t$

(c) From part (b), the curve is the parabola $x = t^2 - 1 = \left(\frac{y}{2}\right)^2 - 1 = \frac{y^2}{4} - 1.$

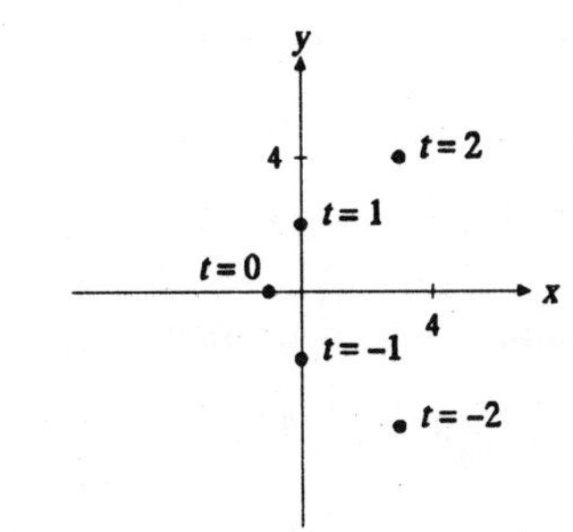

35 $az^2 + bz + c = 0$ implies that $a\left(z^2 + \frac{b}{a}z\right) = -c$, $\left(z^2 + \frac{b}{a}z + \frac{b^2}{4a^2}\right) = -\frac{c}{a} + \frac{b^2}{4a^2}$, $\left(z + \frac{b}{2a}\right)^2 = \frac{b^2 - 4ac}{4a^2}$, and $z = -\frac{b}{2a} \pm \sqrt{\frac{b^2 - 4ac}{4a^2}}$; since $b^2 - 4ac \neq 0$, $\pm\sqrt{\frac{b^2 - 4ac}{4a^2}}$ represents two distinct complex numbers.

37 (a) $x = \frac{3 \pm \sqrt{1}}{2}$, so $x = 2$ or $x = 1$.

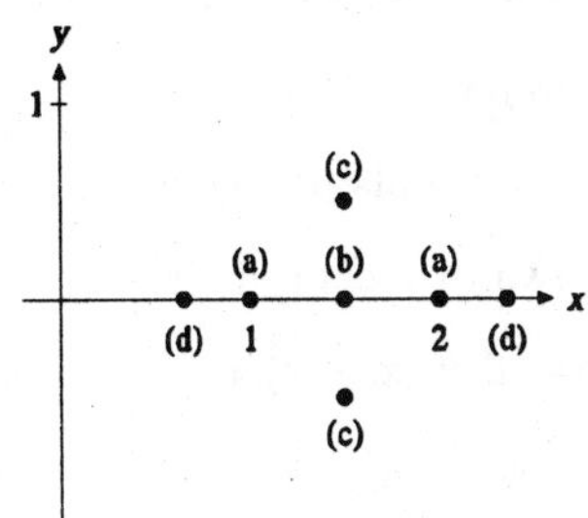

(b) $x = \frac{3 \pm \sqrt{0}}{2} = 1.5$

(c) $x = \frac{3 \pm \sqrt{-1}}{2} = \frac{3}{2} \pm \frac{1}{2}i$

(d) $x = \frac{3 \pm \sqrt{3}}{2} = \frac{3}{2} \pm \frac{\sqrt{3}}{2}$

39 (a) Let $w = u + iv$ and $z = x + iy$ be any two complex numbers. By Exercise 25 we know that $\bar{w} \cdot \bar{z} = \overline{wz}$ and $\bar{w} + \bar{z} = \overline{w + z}$. So if c is any complex number, then $f(\bar{c})$

$$= a_0 + a_1\bar{c} + a_2\bar{c}^2 + a_3\bar{c}^3 + a_4\bar{c}^4$$

$$= a_0 + a_1\bar{c} + a_2\overline{c^2} + a_3\overline{c^3} + a_4\overline{c^4}$$

$$= \bar{a}_0 + \bar{a}_1\bar{c} + \bar{a}_2\overline{c^2} + \bar{a}_3\overline{c^3} + \bar{a}_4\overline{c^4}$$

$$= \overline{a_0} + \overline{a_1c} + \overline{a_2c^2} + \overline{a_3c^3} + \overline{a_4c^4}$$

$$= \overline{a_0 + a_1c + a_2c^2 + a_3c^3 + a_4c^4} = \overline{f(c)}.$$

(Note that $\bar{a}_i = a_i$ because a_i is real.) In particular, if $f(c) = 0$, then $f(\bar{c}) = \overline{f(c)} = \bar{0} = 0$.

(b) Since c is a root of $f(x) = 0$, we have $f(x) = (x - c)g(x)$ for some polynimial g. By (a), $f(\bar{c}) = 0$, so $(\bar{c} - c)g(\bar{c}) = 0$. Since c is not real, $\bar{c} - c \neq 0$, hence $g(\bar{c}) = 0$. Thus $\bar{c}$ is a root of $g(x) = 0$, so $g(x) = (x - \bar{c})h(x)$ for some polynomial h. Therefore $f(x) = (x - c)(x - \bar{c})h(x)$; that is, $(x - c)(x - \bar{c})$ divides $f(x)$.

(c) If all four roots of $f(x) = 0$ are real, let c and d be two of them (possibly equal). It then follows that $(x - c)(x - d)$ divides $f(x)$; but $(x - c)(x - d) = x^2 - (c + d)x + cd$ is a second-degree polynomial with real coefficients, so we are done in this case. On the other hand, if $f(x) = 0$ has a nonreal root c, then $(x - c)(x - \bar{c})$ divides $f(x)$. Note that $(x - c)(x - \bar{c}) = x^2 - (c + \bar{c})x + c\bar{c}$. If $c = a + bi$, then $c + \bar{c} = (a + bi) + (a - bi) = 2a$ and $c\bar{c} = (a + bi)(a - bi) = a^2 + b^2$ are both real, so $(x - c)(x - \bar{c})$ has real coefficients, and again we are done.

11.7 The Relation Between the Exponential and the Trigonometric Functions

1 $z = e^{5\pi i/4} = \cos\frac{5\pi}{4} + i\sin\frac{5\pi}{4} = -\frac{1}{\sqrt{2}} - \frac{1}{\sqrt{2}}i$, so Re z = Im $z = -\frac{1}{\sqrt{2}}$. The number lies in the third quadrant, as shown in the figure.

3 $z = 2e^{\pi i/4} + 3e^{\pi i/6} = 2\cos\frac{\pi}{4} + 2i\sin\frac{\pi}{4} + 3\cos\frac{\pi}{6} + 3i\sin\frac{\pi}{6} = \sqrt{2} + \sqrt{2}i + 3\cdot\frac{\sqrt{3}}{2} + \frac{3}{2}i$, so Re $z = \sqrt{2} + \frac{3\sqrt{3}}{2}$ and Im $z = \sqrt{2} + \frac{3}{2}$.

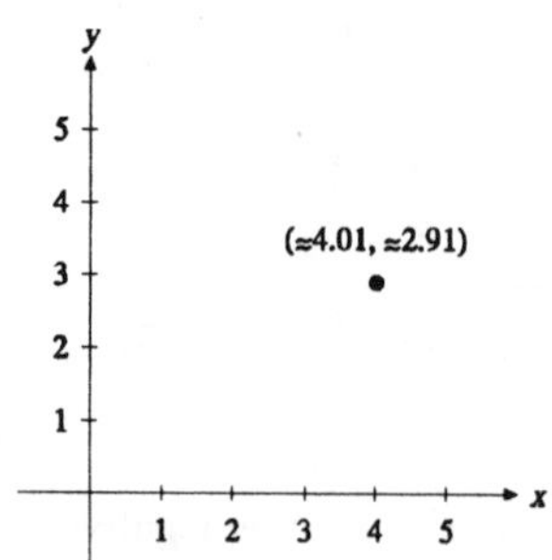

5 $z = e^{\pi i/6} e^{3\pi i/4} = e^{11\pi i/12}$

$= \cos \dfrac{11\pi}{12} + i \sin \dfrac{11\pi}{12},$

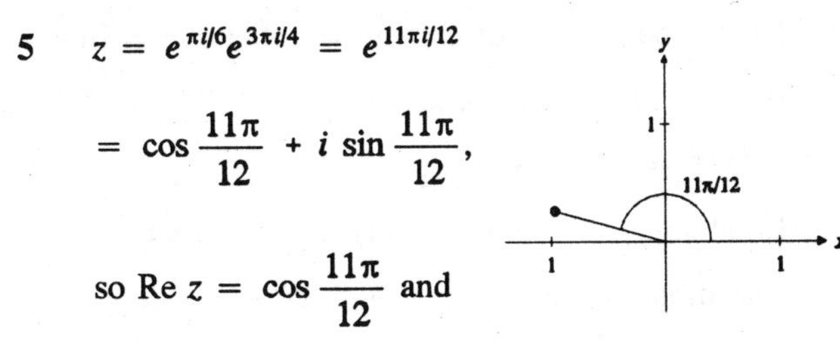

so Re $z = \cos \dfrac{11\pi}{12}$ and

Im $z = \sin \dfrac{11\pi}{12}$.

7 $\dfrac{e^2}{\sqrt{2}} - \dfrac{e^2}{\sqrt{2}} i = e^2\left(\dfrac{1}{\sqrt{2}} - \dfrac{i}{\sqrt{2}}\right) =$

$e^2\left(\cos \dfrac{\pi}{4} - i \sin \dfrac{\pi}{4}\right) = e^2 e^{-\pi i/4}$

9 $5\left(\cos \dfrac{\pi}{6} + i \sin \dfrac{\pi}{6}\right) \cdot 3\left(\cos \dfrac{\pi}{2} + i \sin \dfrac{\pi}{2}\right) =$

$15\left(\cos \dfrac{2\pi}{3} + i \sin \dfrac{2\pi}{3}\right) = 15e^{2\pi i/3}$

11 $e^{(\pi i/4 + 3\pi i)} = e^{\pi i/4} e^{3\pi i}$

$= \left(\cos \dfrac{\pi}{4} + i \sin \dfrac{\pi}{4}\right)(-1)$

$= -\dfrac{1}{\sqrt{2}} - \dfrac{i}{\sqrt{2}}$

13 $e^{2 - (\pi/3)i} = e^2 e^{-\pi i/3}$

$= e^2\left(\cos \dfrac{-\pi}{3} + i \sin \dfrac{-\pi}{3}\right)$

$= e^2\left(\dfrac{1}{2} - \dfrac{\sqrt{3}}{2} i\right)$

15 (a) $|z| = |e^{a+bi}| = |e^a e^{bi}| = |e^a||e^{bi}| = e^a \cdot 1$
$= e^a$

(b) $z = e^a e^{bi} = e^a(\cos b + i \sin b)$, so $\bar{z} =$
$e^a(\cos b - i \sin b) = e^a e^{-bi} = e^{a-bi}$

(c) $z^{-1} = (e^{a+bi})^{-1} = e^{-a-bi}$

(d) Re z = Re$[e^a(\cos b + i \sin b)] = e^a \cos b$

(e) Im z = Im$[e^a(\cos b + i \sin b)] = e^a \sin b$

(f) arg z = arg$[e^a(\cos b + i \sin b)] = b$

17 Let $z = x + iy$. If $e^z = 1$, then $1 = |e^z| = e^x$, so $x = 0$ and $1 = e^z = e^{x+iy} = e^{iy} = \cos y + i \sin y$; hence $\cos y = 1$ and $\sin y = 0$. Now $\sin y = 0$ when y is a multiple of π, and $\cos y = 1$ when y is an even multiple of π. So we have $y = 2\pi n$, where n is any integer; thus $z = yi = 2\pi n i$.

19 (a) $|e^{3+4i}| = |e^3 e^{4i}| = e^3|\cos 4 + i \sin 4| = e^3$.

(b)

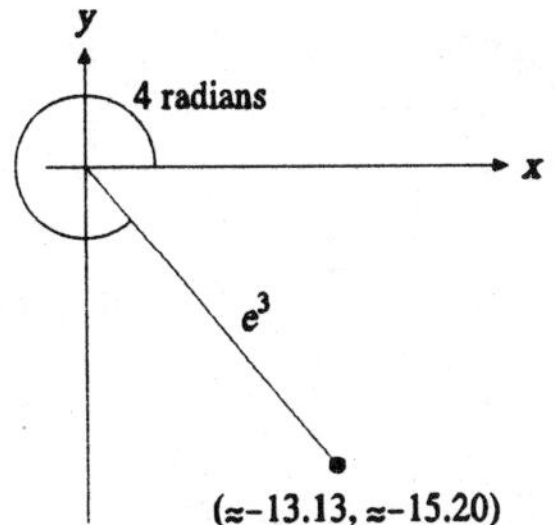

21 We have

$(e^{-i\omega_0 t})^* = (\cos(-\omega_0 t) + i \sin(-\omega_0 t))^* =$

$(\cos(\omega_0 t) - i \sin(\omega_0 t))^* = \cos(\omega_0 t) + i \sin(\omega_0 t)$

$= e^{i\omega_0 t}$. Multiplying by $e^{-i\omega_0 t}$ gives the stated result.

23 Let $w = a + bi$. By Exercise 15, Re$(e^{a+bi}) = e^a \cos b$ and Im$(e^{a+bi}) = e^a \sin b$. We want $e^w = e^{a+bi} = 3 + 4i$, so we must have $e^a \cos b = 3$ and $e^a \sin b = 4$. Now $e^a = |e^w| = |3 + 4i| = \sqrt{3^2 + 4^2} = 5$, so $a = \ln 5$, $\cos b = \dfrac{3}{5}$, and $\sin b = \dfrac{4}{5}$; note that $\tan b = \dfrac{4}{3}$, so $b = \tan^{-1}(4/3)$ is one possibility. We can add multiples of 2π to b without affecting the result; the general solution is $w = a + bi = \ln 5 + (2\pi n + \tan^{-1}(4/3))i$, where n is any integer.

25 Since $\sum_{n=0}^{\infty}\left|\frac{\cos n\theta}{2^n}\right| \le \sum_{n=0}^{\infty}\frac{1}{2^n} = \frac{1}{1-\frac{1}{2}} = 2$, $\sum_{n=0}^{\infty}\frac{\cos n\theta}{2^n}$ is absolutely convergent by the comparison test. To evaluate its sum, let $z = \frac{1}{2}e^{i\theta} = \frac{1}{2}(\cos\theta + i\sin\theta)$; by DeMoivre's law, $z^n = \frac{e^{in\theta}}{2^n} = \frac{1}{2^n}(\cos n\theta + i\sin n\theta)$, so $\sum_{n=0}^{\infty}\frac{\cos n\theta}{2^n}$ is just the real part of $\sum_{n=0}^{\infty} z^n$, which is geometric and sums to $\frac{1}{1-z}$. Note that the summation is valid because $|z| = \frac{1}{2} < 1$. In the derivation that follows, we will also be using the facts that $z\bar{z} = |z|^2 = \frac{1}{4}$ and $z + \bar{z} = 2\operatorname{Re} z = \cos\theta$. We have $\frac{1}{1-z} = \frac{1}{1-z}\cdot\frac{1-\bar{z}}{1-\bar{z}} = \frac{1-\bar{z}}{1-z-\bar{z}+z\bar{z}}$

$= \frac{1-\bar{z}}{1-\cos\theta+\frac{1}{4}} = \frac{4(1-\frac{1}{2}\cos\theta+\frac{1}{2}i\sin\theta)}{4-4\cos\theta+1}$

$= \frac{4-2\cos\theta+2i\sin\theta}{5-4\cos\theta}$. Therefore, $\sum_{n=0}^{\infty}\frac{\cos n\theta}{2^n}$

$= \operatorname{Re}\left(\frac{1}{1-z}\right) = \frac{4-2\cos\theta}{5-4\cos\theta}$.

27 Let the polar form of w be $w = re^{i\theta}$. Let $z = \ln r + i\theta + 2\pi ni$, where n is any integer. Then $e^z = re^{i\theta}e^{2\pi ni} = re^{i\theta} = w$.

29 Combining Theorem 2 of this section with the definition of sinh in Sec. 6.9 gives $-i\sinh(iz) = -i\frac{e^{iz}-e^{-iz}}{2} = \frac{e^{iz}-e^{-iz}}{2i} = \sin z$.

31 (a) $z = \frac{1+i}{\sqrt{2}}$, so $|z| = 1$ and $\arg z = \pi/4$.

Thus $\left|\frac{z^2}{2!}\right| = \frac{1}{2!} = \frac{1}{2}$ and $\arg(z^2/2!) = 2\cdot\frac{\pi}{4} = \frac{\pi}{2}$; $\left|\frac{z^3}{3!}\right| = \frac{1}{3!} = \frac{1}{6}$ and $\arg(z^3/3!) = 3\cdot\frac{\pi}{4} = \frac{3\pi}{4}$; and $\left|\frac{z^4}{4!}\right| = \frac{1}{4!} = \frac{1}{24}$ and $\arg(z^4/4!) = 4\cdot\frac{\pi}{4} = \pi$.

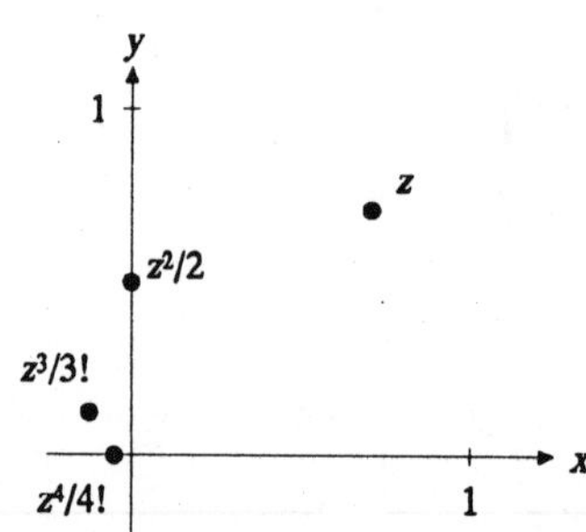

(b) $1 + z + \frac{z^2}{2!} + \frac{z^3}{3!} + \frac{z^4}{4!}$

$= 1 + \frac{1+i}{\sqrt{2}} + \frac{1}{2!}i + \frac{1}{3!}\frac{-1+i}{\sqrt{2}} + \frac{1}{4!}(-1)$

$= \frac{10\sqrt{2}+23}{24} + \frac{7\sqrt{2}+6}{12}i$. (See the graph.)

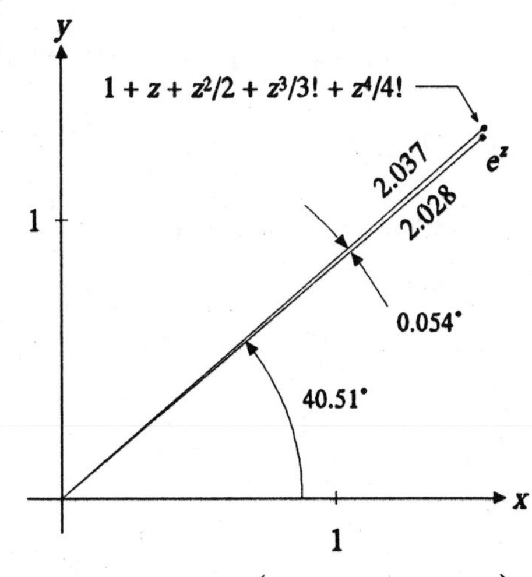

(c) $e^{(1+i)/\sqrt{2}} = e^{1/\sqrt{2}}\left(\cos \frac{1}{\sqrt{2}} + i \sin \frac{1}{\sqrt{2}}\right)$. (See the graph.)

33 (a) $\cos(iz) = \frac{e^{i(iz)} + e^{-i(iz)}}{2} = \frac{e^{-z} + e^{z}}{2}$

$= \cosh z$

(b) $-i \sin(iz) = -i\frac{e^{i(iz)} - e^{-i(iz)}}{2i} = -\frac{e^{-z} - e^{z}}{2}$

$= \frac{e^{z} - e^{-z}}{2} = \sinh z$

11.S Guide Quiz

1 Maclaurin series are centered at $x = 0$, so convergence at $x = 2$ implies that the radius of convergence is at least 2. Therefore the series must also converge for $-2 < x < 2$. The series diverges for $|x| > 2$, that is, for $x < -2$ and $x > 2$.

2 The Taylor series is centered at $x = 2$ and diverges at $x = 5$, so its radius of convergence cannot exceed 3. Thus the series must also diverge for $x > 5$ and $x < -1$.

3 $\frac{1}{1-x} = 1 + x + x^2 + x^3 + \cdots$ converges for $|x| < 1$, but nowhere else. Another example is the series for $\frac{1}{1-x^2}$.

4 Substituting x^2 in the series for $\sin x$ yields $x^2 - \frac{x^6}{3!} + \frac{x^{10}}{5!} - \cdots$, so $\int_0^1 \sin x^2\, dx =$

$\int_0^1 \left(x^2 - \frac{x^6}{3!} + \frac{x^{10}}{5!} - \cdots\right) dx =$

$\left(\frac{x^3}{3} - \frac{x^7}{42} + \frac{x^{11}}{1320} - \cdots\right)\Big|_0^1 =$

$\frac{1}{3} - \frac{1}{42} + \frac{1}{1320} - \cdots$. The result is an alternating series, so the discrepancy between the sum and a partial sum is bounded by the first deleted term. Since $1/1320 \approx 0.00076$, the first two terms suffice to estimate the integral to two decimal places: $\frac{1}{3} - \frac{1}{42} \approx 0.31$.

5 Let m and M be the minimum and maximum values of $f^{(5)}(t)$ for $a \le t \le x$. Let $R(t) = R_4(t; a) = f(t) - P_4(t; a)$, so that $R(a) = R'(a) = R''(a) = R'''(a) = R^{(4)}(a) = 0$ and $R^{(5)}(t) = f^{(5)}(t)$. Thus, $m \le R^{(5)}(u) \le M$, so that $\int_a^t m\, du \le \int_a^t R^{(5)}(u)\, du \le \int_a^t M\, du$. Then $mx\big|_a^t \le R^{(4)}(u)\big|_a^t \le Mx\big|_a^t$, that is, $m(t-a) \le R^{(4)}(t) - R^{(4)}(a) = R^{(4)}(t) \le M(t-a)$, or $m(t-a) \le R^{(4)}(t) \le M(t-a)$. Similarly, integration gives

$m\frac{(t-a)^2}{2} \le R'''(t) \le M\frac{(t-a)^2}{2}$. Repeated integrations result in the following sequence of inequalities:

$$m\frac{(t-a)^3}{3!} \le R''(t) \le M\frac{(t-a)^3}{3!},$$

$$m\frac{(t-a)^4}{4!} \le R'(t) \le M\frac{(t-a)^4}{4!},$$

$$m\frac{(t-a)^5}{5!} \le R(t) \le M\frac{(t-a)^5}{5!}.$$

Hence $R(x) = q\frac{(x-a)^5}{5!}$ for some number q, $m \le q \le M$. Since $f^{(5)}$ is continuous, the intermediate-value theorem implies that $q = f^{(5)}(c)$ for some c in $[a, x]$, so that $R_4(x; a) = R(x) = \frac{f^{(5)}(c)(x-a)^5}{5!}$.

6 (a) By substituting $-x^2$ into the series for $\frac{1}{1-x}$ we obtain $\frac{1}{1+x^2} = 1 - x^2 + x^4 - x^6 + \cdots$, valid for $|x| < R = 1$.

(b) $e^{-x} = 1 - x + \frac{x^2}{2!} - \frac{x^3}{3!} + \frac{x^4}{4!} - \frac{x^5}{5!} + \cdots$, valid for $|x| < R = \infty$.

(c) $\cos x = 1 - \frac{x^2}{2!} + \frac{x^4}{4!} - \frac{x^6}{6!} + \cdots$, valid for $|x| < R = \infty$.

(d) $-\frac{1}{1-x}$ is the derivative of $\ln(1-x)$, so integrate term-by-term the series $-\frac{1}{1-x} = -1 - x - x^2 - x^3 - \cdots$: $\ln(1-x) = -x - \frac{x^2}{2} - \frac{x^3}{3} - \frac{x^4}{4} - \cdots$, valid for $|x| < R = 1$ (and for $x = -1$).

(e) Substitute $2x$ into the series for $\frac{1}{1-x}$ to obtain $\frac{1}{1-2x} = 1 + 2x + 4x^2 + 8x^3 + \cdots$, valid for $|2x| < 1$; that is, $|x| < R = \frac{1}{2}$.

(f) $1 + 3x + 5x^2$ is a polynomial, which is already a Maclaurin series. It has only three terms, so it converges for all values of x and has an infinite radius of convergence.

(g) $\tan^{-1} 2x = 2x - \frac{2^3x^3}{3} + \frac{2^5x^5}{5} - \frac{2^7x^7}{7} + \cdots$ for $|x| \le \frac{1}{2}$. The radius of convergence is 1/2.

7 If $f(x) = a_0 + a_1x + a_2x^2 + a_3x^3 + \cdots$, then $f(0) = a_0$; $f'(x) = a_1 + 2a_2x + 3a_3x^2 + 4a_4x^3 + \cdots$ and $f'(0) = a_1$; $f''(x) = 2a_2 + 3\cdot 2a_3x + 4\cdot 3a_4x^2 + \cdots$ and $f''(0) = 2a_2$ so that $\frac{f''(0)}{2!} = a_2$; $f'''(x) = 3!a_3 + 4\cdot 3\cdot 2a_4x + \cdots$ and $f'''(0) = 3!a_3$ so that $\frac{f'''(0)}{3!} = a_3$. Continuing in this manner, $a_n = \frac{f^{(n)}(0)}{n!}$ for $n = 0, 1, 2, 3, \cdots$.

8 (a) The general term is $\frac{n}{1+n^2}(x-2)^n$. Apply the ratio test to obtain

$$\lim_{n\to\infty}\left|\frac{(n+1)(x-2)^{n+1}/(1+(n+1)^2)}{n(x-2)^n/(1+n^2)}\right| =$$

$$\lim_{n \to \infty} \frac{(n+1)(1+n^2)}{n(n^2+2n+2)} |x-2| = |x-2|,$$

which is less than 1 if and only if $1 < x < 3$. At $x = 3$, the series diverges by limit-comparison with the harmonic series, while at $x = 1$, it is conditionally convergent by the alternating-series test. The series therefore converges for $1 \le x < 3$.

(b) The nth term is $\dfrac{2^n(x+3)^n}{n!}$. Now

$$\lim_{n \to \infty} \left| \frac{2^{n+1}(x+3)^{n+1}/(n+1)!}{2^n(x+3)^n/n!} \right| = \lim_{n \to \infty} \frac{2|x+3|}{n+1} =$$

0, so the series converges for all values of x.

9 (a) The function $f(x) = \sec x$ is even. Since the derivative of an even function is odd and vice versa, it follows that f'', $f^{(4)}$, $f^{(6)}$, ... are even functions and f', $f^{(3)}$, $f^{(5)}$, ... are odd; hence $f'(0) = 0$, $f'''(0) = 0$, $f^{(5)}(0) = 0$, ... Thus a_1, a_3, a_5, ... are all zero in the Maclaurin series $\sum_{n=0}^{\infty} a_n x^n$.

(b)

$$\begin{array}{r|l} & 1 + \frac{x^2}{2} + \frac{5}{24}x^4 + \cdots \\ \hline 1 - \frac{x^2}{2!} + \frac{x^4}{4!} + \cdots & 1 \\ & 1 - \frac{x^2}{2} + \frac{x^4}{24} - \cdots \\ \cline{2-2} & \frac{x^2}{2} - \frac{x^4}{24} + \cdots \\ & \frac{x^2}{2} - \frac{x^4}{4} + \cdots \\ \cline{2-2} & \frac{5}{24}x^4 + \cdots \end{array}$$

Hence $\sec x = \dfrac{1}{\cos x} = 1 + \dfrac{x^2}{2} + \dfrac{5x^4}{24} + \cdots$.

10 (a) $e^{i\theta} = \cos\theta + i\sin\theta$ because the equation is valid when the functions e^x, $\sin x$, and $\cos x$ are replaced with Maclaurin series.

(b) $e^z = e^{x+iy} = e^x e^{iy} = e^x(\cos y + i \sin y) = 2$, so $x = \ln 2$ and $y = 2n\pi$ for some integer n. Hence there are infinitely many solutions, all lying on a vertical line. Some of these solutions are plotted in the accompanying figure.

(c) The solutions to $z^6 = i$ are of the form $z_k = \cos\left(\frac{\pi}{12} + \frac{\pi}{3}k\right) + i \sin\left(\frac{\pi}{12} + \frac{\pi}{3}k\right)$ for $k = 0$, 1, 2, 3, 4, and 5.

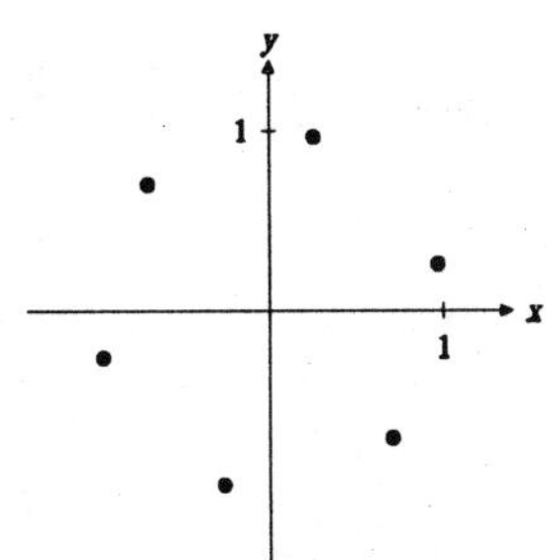

(d) The solutions to $z^2 = 3 + 4i$ are $\pm(2 + i)$.

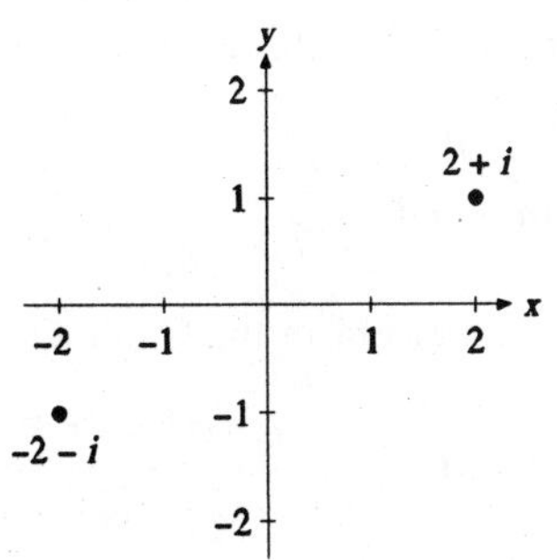

11.S Review Exercises

1 (a) $|R_n| < |a_{n+1}|$ and R_n has the same sign as a_{n+1}.

(b) $R_n = \sum_{k=n+1}^{\infty} ar^k = ar^{n+1}\sum_{k=0}^{\infty} r^k = \frac{ar^{n+1}}{1-r}$

(c) $\int_{n+1}^{\infty} f(x)\,dx < R_n < \int_n^{\infty} f(x)\,dx$

(d) $|R_n| = \left|\sum_{k=n+1}^{\infty} a_k\right| \le \sum_{k=n+1}^{\infty} |a_k| = 27$

(e) $R_n = \sum_{k=n+1}^{\infty} a_k = \sum_{k=n+1}^{\infty} \frac{f^{(k)}(0)}{k!}x^k$

$= \frac{f^{(n+1)}(c)}{(n+1)!}x^{n+1}$, where c is between 0 and x.

3 (a) $\tan^{-1} 3x$

$= 3x - \frac{(3x)^3}{3} + \frac{(3x)^5}{5} - \frac{(3x)^7}{7} + \cdots$, so the

nth nonzero term is $\frac{(-1)^{n+1}(3x)^{2n-1}}{2n-1}$.

(b) $\ln(1+x^2) = x^2 - \frac{x^4}{2} + \frac{x^6}{3} - \frac{x^8}{4} + \cdots$, so

the nth nonzero term is $\frac{(-1)^{n+1}x^{2n}}{n}$.

(c) See Exercise 6(b) in the Guide Quiz. The form of the nth term, if we start with $n = 1$, is $\frac{(-1)^{n-1}}{(n-1)!}x^{n-1}$.

(d) See Exercise 4 in the Guide Quiz. The nth nonzero term is $\frac{(-1)^{n+1}x^{2(2n-1)}}{(2n-1)!}$.

(e) $\cos x^2 = 1 - \frac{x^4}{2!} + \frac{x^8}{4!} - \frac{x^{12}}{6!} + \cdots$ so the nth nonzero term (starting with $n = 1$) is $\frac{(-1)^{n+1}x^{4(n-1)}}{(2(n-1))!}$.

5 (a) $\sin 2x = 2x - \frac{1}{3!}(2x)^3 + \frac{1}{5!}(2x)^5 + \cdots$

(b) $f(x) = \sin 2x$, $f'(x) = 2\cos 2x$, $f''(x) = -4\sin 2x$, $f'''(x) = -8\cos 2x$, $f^{(4)}(x) = 16\sin 2x$, $f^{(5)}(x) = 32\cos 2x$, so $a_0 = f(0) = 0$, $a_1 = f'(0) = 2$, $a_2 = \frac{f''(0)}{2!} = 0$,

$a_3 = \frac{f'''(0)}{3!} = -\frac{4}{3}$, $a_4 = \frac{f^{(4)}(0)}{4!} = 0$,

$a_5 = \frac{f^{(5)}(0)}{5!} = \frac{4}{15}$, and $\sin 2x$

$= 2x - \frac{4}{3}x^3 + \frac{4}{15}x^5 - \cdots$.

(c) $\sin 2x = 2\sin x\cos x =$

$2\left(x - \frac{x^3}{3!} + \frac{x^5}{5!} - \cdots\right)\left(1 - \frac{x^2}{2!} + \frac{x^4}{4!} - \cdots\right) =$

$2\left(x - \frac{x^3}{3!} + \frac{x^5}{5!} - \cdots + \left(-\frac{x^3}{2!}\right) + \frac{x^5}{3!2!} - \cdots + \frac{x^5}{4!} - \cdots\right)$

$= 2\left(x - \frac{2}{3}x^3 + \frac{2}{15}x^5 - \cdots\right)$

$= 2x - \frac{4}{3}x^3 + \frac{4}{15}x^5 - \cdots$.

7 (a) $e^x = 1 + x + \frac{x^2}{2!} + \frac{x^3}{3!} + \cdots$; if $|x| \le 1$,

then for $n \ge 3$, $\left|\frac{x^n}{n!} + \frac{x^{n+1}}{(n+1)!} + \cdots\right| \le$

$$\frac{|x|^n}{n!} + \frac{|x|^{n+1}}{(n+1)!} + \cdots \le$$

$$\frac{1}{n!} + \frac{1}{(n+1)!} + \cdots <$$

$$\frac{1}{n!}\left(1 + \frac{1}{n+1} + \frac{1}{(n+1)^2} + \cdots\right) =$$

$$\frac{1}{n!}\,\frac{1}{1 - \frac{1}{n+1}} = \frac{n+1}{n \cdot n!},$$ which equals 0.01

when $n = 5$. Use five terms.

(b) From part (a), $\frac{n+1}{n \cdot n!} < 0.001$ if $n \ge 7$. Use seven terms.

(c) $e^x = 1 + x + \frac{x^2}{2!} + \frac{x^3}{3!} + \cdots$; if $|x| \le 2$,

then we have $\left|\frac{x^n}{n!} + \frac{x^{n+1}}{(n+1)!} + \cdots\right| \le$

$$\frac{|x|^n}{n!} + \frac{|x|^{n+1}}{(n+1)!} + \cdots \le \frac{2^n}{n!} + \frac{2^{n+1}}{(n+1)!} + \cdots$$

$$\le \frac{2^n}{n!}\left(1 + \frac{2}{n+1} + \left(\frac{2}{n+1}\right)^2 + \cdots\right) =$$

$$\frac{2^n}{n!}\,\frac{1}{1 - \frac{2}{n+1}} = \frac{2^n(n+1)}{(n-1)n!},$$ which is less

than 0.01 if $n \ge 8$. Use eight terms.

(d) From part (c), $\frac{2^n(n+1)}{(n-1)n!} < 0.001$ if $n \ge 10$.

Use 10 terms.

9 For the function $f(x) = \ln(1 + x)$, the Taylor polynomials at $a = 0$ are $P_0(x; 0) = 0$, $P_1(x; 0) = x$, $P_2(x; 0) = x - x^2/2$, and $P_3(x; 0) = x - x^2/2 + x^3/3$. (Since $y = P_0(x; 0)$ coincides with the x axis, it does not appear in the graph.)

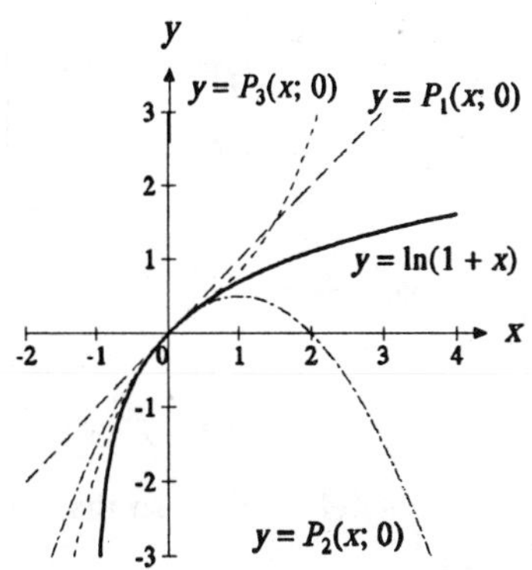

11 For the function $f(x) = \cos x$, the Taylor polynomials at $a = 0$ are $P_0(x; 0) = 1$, $P_1(x; 0) = 1$, $P_2(x; 0) = 1 - x^2/2$, and $P_3(x; 0) = 1 - x^2/2$. (Since $P_1(x; 0)$ coincides with $P_0(x; 0)$ and $P_3(x; 0)$ coincides with $P_2(x; 0)$, they do not appear in the graph.)

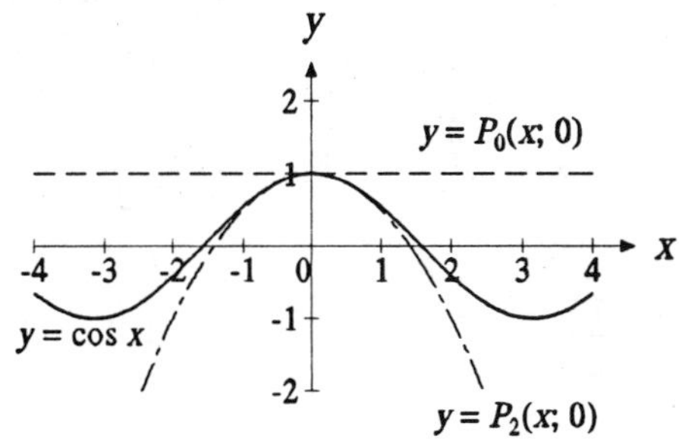

13 For some value of c_n between 0 and x, $|R_n(x; 0)|$

$$= \left|\frac{f^{(n+1)}(c_n)}{(n+1)!}x^{n+1}\right| = \frac{e^{-c_n}}{(n+1)!}|x|^{n+1}$$

$$\le \begin{cases} \dfrac{e^0 \cdot x^{n+1}}{(n+1)!} & \text{if } x > 0 \\ \dfrac{e^{-x} \cdot |x|^{n+1}}{(n+1)!} & \text{if } x \le 0. \end{cases}$$

Note that $\lim_{n\to\infty} \frac{e^0 \cdot x^{n+1}}{(n+1)!} = 0 \cdot 1 = 0$ and

$\lim_{n\to\infty} e^{-x} \cdot \frac{|x|^{n+1}}{(n+1)!} = e^{-x} \cdot 0 = 0$, so that

$\lim_{n\to\infty} R_n(x; 0) = 0.$

15 For $f(x) = \cos x$, $f^{(n)}(x)$ is $\pm \sin x$ or $\pm \cos x$. Therefore, $|R_n(x; 0)| = \left|\frac{f^{(n+1)}(c_n)}{(n+1)!}\cdot x^{n+1}\right| \le \frac{|x|^{n+1}}{(n+1)!} \to 0$ as $n \to \infty$.

17 For $f(x) = \ln(1 + x)$, we have $f^{(n)}(x) = (-1)^{n+1}(n-1)!(1+x)^{-n}$, so that $|R_n(x; 0)| = \left|\frac{f^{(n+1)}(c_n)}{(n+1)!}\cdot x^{n+1}\right| = \left|\frac{(-1)^{n+2}\, n!(1+c_n)^{-(n+1)}}{(n+1)!}\cdot x^{n+1}\right| = \frac{1}{n+1}\cdot\left|\frac{x}{1+c_n}\right|^{n+1}$, where c_n is between 0 and x. If $0 \le c_n \le x < 1$, then $\left|\frac{x}{1+c_n}\right| \le |x| < 1$. If $-\frac{1}{2} < x \le c_n \le 0$, then $|x| < \frac{1}{2}$, $1 + c_n > 1 + \left(-\frac{1}{2}\right) = \frac{1}{2}$, and $\left|\frac{x}{1+c_n}\right| < 1$. In either case, $\frac{1}{n+1}\cdot\left|\frac{x}{1+c_n}\right|^{n+1} \to 0$, so $R_n(x; 0) \to 0$ as $n \to \infty$.

19 $\cos x = 1 - \frac{x^2}{2!} + \frac{x^4}{4!} - \cdots$, so $\cos\frac{1}{3} = 1 - \frac{1}{2}\cdot\frac{1}{9} + \frac{1}{24}\cdot\frac{1}{81} - \cdots$; the series is alternating, so the error will be bounded by the first omitted term. We have $\cos\frac{1}{3} \approx 1 - \frac{1}{18} + \frac{1}{1944} = \frac{1837}{1944} \approx 0.94496$. The first omitted term is $\frac{1}{6!}\left(\frac{1}{3}\right)^6 = \frac{1}{524{,}880} \approx 1.9 \times 10^{-6}$, so the error is no greater than this.

21 $\frac{1}{\sqrt{e}} = e^{-1/2} = 1 - \frac{1}{2} + \frac{1}{2!}\left(\frac{1}{2}\right)^2 - \frac{1}{3!}\left(\frac{1}{2}\right)^3 + \cdots = 1 - \frac{1}{2} + \frac{1}{8} - \cdots$, so $e^{-1/2} \approx \frac{5}{8} = 0.625$. The first omitted term is $\frac{1}{48} \approx 0.0208$, which is a bound on the error.

23 $\sum_{n=1}^{\infty} nx^{n-1}$ is the derivative of the series for $\frac{1}{1-x}$, so its sum is $\frac{1}{(1-x)^2}$; as with $\frac{1}{1-x}$, it converges absolutely for $|x| < 1$ and diverges elsewhere. The radius of convergence is 1.

25 $\sum_{n=0}^{\infty} \frac{x^{2n}}{n!} = e^{x^2}$; the radius of convergence is infinite.

27 $\sum_{n=1}^{\infty} \frac{x^n}{n}$ is the Maclaurin series for $-\ln(1 - x)$. (See Exercise 6(d) of the Guide Quiz.) It converges absolutely for $|x| < R = 1$, and is conditionally convergent at $x = -1$ by the alternating–series test.

29 By the ratio test, $\lim_{n\to\infty}\left|\frac{\dfrac{((n+1)^5+2)(x+1)^{n+1}}{(n+1)^3+1}}{\dfrac{(n^5+2)(x+1)^n}{n^3+1}}\right| = \lim_{n\to\infty}\frac{((n+1)^5+2)(n^3+1)}{(n^5+2)((n+1)^3+1)}|x+1| = |x+1|$,

which is less than 1 if and only if $-1 < x + 1 < 1$; that is, $-2 < x < 0$. Thus the series converges absolutely for x in $(-2, 0)$ and has a radius of convergence of 1. At $x = -2$ or $x = 0$, the series diverges by the nth-term test. (The nth term goes to infinity in absolute value.)

31 (a) $2x + 1 = e^x \approx 1 + x + \frac{1}{2}x^2 + \frac{1}{6}x^3$, so

$\frac{1}{6}x^3 + \frac{1}{2}x^2 - x \approx 0$; hence

$\frac{x}{6}(x^2 + 3x - 6) \approx 0$. Discarding the solution

$x = 0$, we are left with $x \approx \frac{-3 \pm \sqrt{33}}{2}$. The

desired positive solution is therefore $x \approx$

$\frac{\sqrt{33} - 3}{2} \approx 1.372$.

(b) If one approximation is x, the next is

$x - \frac{f(x)}{f'(x)}$, that is, $x - \frac{e^x - 2x - 1}{e^x - 2}$.

Starting with $x_0 = 1$, we find $x_1 \approx 1.39221$, $x_2 \approx 1.27396$, $x_3 \approx 1.25678$, and $x_4 \approx x_5 \approx 1.25643$. The root is about 1.25643.

33 (a) $\int_0^{1/2} \cos x^3 \, dx$

$$= \int_0^{1/2} \left[1 - \frac{(x^3)^2}{2} + \frac{(x^3)^4}{24} - \frac{(x^3)^6}{720} + \cdots\right] dx$$

$$= \int_0^{1/2} \left[1 - \frac{x^6}{2} + \frac{x^{12}}{24} - \frac{x^{18}}{720} + \cdots\right] dx$$

$$= \left(x - \frac{x^7}{14} + \frac{x^{13}}{312} - \frac{x^{19}}{13{,}680} + \cdots\right)\Bigg|_0^{1/2}$$

$$= \frac{1}{2} - \frac{1}{2^7 \cdot 14} + \frac{1}{2^{13} \cdot 312} - \frac{1}{2^{19} \cdot 13{,}680} + \cdots$$

is a decreasing, alternating series. Thus

$\int_0^{1/2} \cos x^3 \, dx = \frac{8{,}935{,}687}{17{,}891{,}328} \approx$

0.49944235554 with error at most $\frac{1}{2^{19} \cdot 13{,}680}$

$\approx 1.394 \times 10^{-10}$.

(b) $\int_0^1 \sin x^2 \, dx$

$$= \int_0^1 \left[x^2 - \frac{(x^2)^3}{6} + \frac{(x^2)^5}{120} - \frac{(x^2)^7}{5040} + \cdots\right] dx$$

$$= \int_0^1 \left[x^2 - \frac{x^6}{6} + \frac{x^{10}}{120} - \frac{x^{14}}{5040} + \cdots\right] dx$$

$$= \left(\frac{x^3}{3} - \frac{x^7}{42} + \frac{x^{11}}{1320} - \frac{x^{15}}{75{,}600} + \cdots\right)\Bigg|_0^1$$

$= \frac{1}{3} - \frac{1}{42} + \frac{1}{1320} - \frac{1}{75{,}600} + \cdots$ is a

decreasing, alternating series. Thus,

$\int_0^1 \sin x^2 \, dx \approx \frac{2867}{9240} \approx 0.3102814$ with

error at most $\frac{1}{75{,}600} \approx 0.0000132$.

(c) $\int_0^{1/2} (1 + x^3)^{1/3} \, dx =$

$$\int_0^{1/2} \left[1 + \frac{1}{3}(x^3) - \frac{1}{9}(x^3)^2 + \frac{5}{81}(x^3)^3 - \cdots\right] dx$$

$$= \int_0^{1/2} \left[1 + \frac{x^3}{3} - \frac{x^6}{9} + \frac{5x^9}{81} - \cdots\right] dx$$

$$= \left(x + \frac{x^4}{12} - \frac{x^7}{63} + \frac{x^{10}}{162} - \cdots\right)\Bigg|_0^{1/2}$$

$$= \frac{1}{2} + \frac{1}{2^4 \cdot 12} - \frac{1}{2^7 \cdot 63} + \frac{1}{2^{10} \cdot 162} - \cdots \text{ is a}$$

decreasing, alternating series. Thus,

$$\int_0^{1/2} (1 + x^3)^{1/3}\, dx \approx \frac{4073}{8064} \approx 0.50508433$$

with error at most $\frac{1}{2^{10} \cdot 162} \approx 0.00000603$.

(d) $\int_1^2 e^{-x^3}\, dx =$

$$\int_1^2 \left[1 + (-x^3) + \frac{(-x^3)^2}{2} + \frac{(-x^3)^3}{6} + \cdots\right] dx$$

$$= \int_1^2 \left[1 - x^3 + \frac{x^6}{2} - \frac{x^9}{6} + \cdots\right] dx$$

$$= \left(x - \frac{x^4}{4} + \frac{x^7}{14} - \frac{x^{10}}{60} + \cdots\right)\Bigg|_1^2$$

$$= \left(2 - 4 + \frac{64}{7} - \cdots\right) - \left(1 - \frac{1}{4} + \frac{1}{14} - \cdots\right)$$

≈ 6.3214. Since $0 < \int_1^2 e^{-x^3}\, dx < 1$, the error is at most 6.3214, which is not very useful.

35 The absolute value of the ratio of consecutive terms is $\dfrac{|x|}{\left(1 + \frac{1}{n}\right)^{2n}}$, which approaches $\dfrac{|x|}{e^2}$ as $n \to \infty$.

Hence $R = e^2$. For $x = \pm e^2$, the ratio equals $\left[\dfrac{e}{(1 + 1/n)^n}\right]^2$ which is greater than 1, so the terms do not approach 0. Hence the series converges if and only if $|x| < e^2$.

37 For $x \neq 0$, the terms do not approach 0. The series converges only for $x = 0$, so $R = 0$.

39 The absolute value of the ratio of successive terms approaches $|x - 3|$ as $n \to \infty$. Hence $R = 1$. For $x = 2$ the series diverges by limit-comparison with the harmonic series. For $x = 4$ it converges by the alternating-series test. Hence it converges if and only if $2 < x \le 4$.

41 Since the absolute value of the ratio of successive terms approaches $|x + 2|$ as $n \to \infty$, $R = 1$. For $x = -3$ and $x = -1$, the nth term does not approach 0. Hence the series converges if and only if $-3 < x < -1$.

43 Since the absolute value of the ratio of successive terms approaches x^2 as $n \to \infty$, $R = 1$. For $x = \pm 1$, the series converges by the alternating-series test. Hence the series converges if and only if $-1 \le x \le 1$.

45 See the proof of Theorem 1 in Sec. 11.4.

47 $\displaystyle\lim_{x \to 0} \frac{\ln(1 + x^2) - \sin^2 x}{\tan x^2}$

$$= \lim_{x \to 0} \frac{\left(x^2 - \frac{x^4}{2} + \cdots\right) - \left(x - \frac{x^3}{6} + \cdots\right)^2}{x^2 + \frac{x^6}{3} + \cdots}$$

$$= \lim_{x \to 0} \frac{\left(x^2 - \frac{x^4}{2} + \cdots\right) - \left(x^2 - \frac{x^4}{3} + \cdots\right)}{x^2\left(1 + \frac{x^4}{3} + \cdots\right)}$$

$$= \lim_{x \to 0} \frac{-\frac{x^4}{6} + \cdots}{x^2(1 + \cdots)} = 0$$

49 $\lim_{x\to 0} \frac{(1-\cos x^2)^5}{(x-\sin x)^{20}} = \lim_{x\to 0} \frac{\left(1 - 1 + \frac{x^4}{2} - \dots\right)^5}{\left(x - x + \frac{x^3}{6} - \dots\right)^{20}}$

$= \lim_{x\to 0} \frac{\frac{x^{20}}{32} - \dots}{\frac{x^{60}}{6^{20}} - \dots} = \infty$

51 Recall that the coefficient of x^{33} in the Maclaurin expansion of $f(x)$ is equal to $\frac{1}{33!}f^{(33)}(0)$. In the given example the coefficient is 2^{33}, so $f^{(33)}(0) = 33!\cdot 2^{33}$.

53 Let $f(x) = \sum_{n=0}^{\infty} a_n x^n$. We are given that the series for $f(3)$ converges, so the radius of convergence is at least 3. By Theorem 1 of Sec. 11.5,

$f'(x) = \sum_{n=1}^{\infty} na_n x^{n-1}$ also converges for $|x| < 3$.

In particular, $2f'(2) = \sum_{n=0}^{\infty} na_n 2^n$ is convergent.

55 (a) $e^{-x}\sin x =$

$\left(1 - x + \frac{x^2}{2} - \frac{x^3}{6} + \frac{x^4}{24} - \dots\right)\left(x - \frac{x^3}{6} + \frac{x^5}{120} - \dots\right)$

$= x - \frac{x^3}{6} + \frac{x^5}{120} - \dots - x^2 + \frac{x^4}{6} - \dots +$

$\frac{x^3}{2} - \frac{x^5}{12} + \dots - \frac{x^4}{6} + \dots + \frac{x^5}{24} - \dots$

$= x - x^2 + \frac{1}{3}x^3 - \frac{1}{30}x^5 + \dots$

(b)

$$\begin{array}{r|l} & x - x^2 + \frac{1}{3}x^3 \quad - \quad \frac{1}{30}x^5 + \dots \\ \hline 1+x+\frac{x^2}{2}+\frac{x^3}{6}+\frac{x^4}{24}+\dots & x - \frac{x^3}{6} + \frac{x^5}{120} - \dots \\ & x + x^2 + \frac{x^3}{2} + \frac{x^4}{6} + \frac{x^5}{24} + \dots \\ \hline & -x^2 - \frac{2}{3}x^3 - \frac{x^4}{6} - \frac{x^5}{30} + \dots \\ & -x^2 - x^3 - \frac{x^4}{2} - \frac{x^5}{6} - \dots \\ \hline & \frac{x^3}{3} + \frac{x^4}{3} + \frac{2}{15}x^5 + \dots \\ & \frac{x^3}{3} + \frac{x^4}{3} + \frac{x^5}{6} + \dots \\ \hline & -\frac{1}{30}x^5 - \dots \end{array}$$

Hence $e^{-x}\sin x = \frac{\sin x}{e^x}$

$= x - x^2 + \frac{x^3}{3} - \frac{x^5}{30} + \dots.$

57 $\int_0^{1/2} x \tan x \, dx$

$= \int_0^{1/2} x\left(x + \frac{1}{3}x^3 + \frac{2}{15}x^5 + \dots\right) dx$

$= \int_0^{1/2} \left[x^2 + \frac{1}{3}x^4 + \frac{2}{15}x^6 + \dots\right] dx$

$= \left(\frac{x^3}{3} + \frac{x^5}{15} + \frac{2}{105}x^7 + \dots\right)\Big|_0^{1/2}$

$= \frac{1}{2^3\cdot 3} + \frac{1}{2^5\cdot 15} + \frac{1}{2^6\cdot 105} + \dots$

$\approx \frac{1}{24} + \frac{1}{480} + \frac{1}{6720} = \frac{59}{1344} \approx 0.438988.$

59 (a) Applying Theorem 2 of Section 11.3, we get

$R_1(x; a) = f(x) - P_1(x; a) = \frac{f''(c_1)}{2!}(x-a)^2$,

where c_1 is between a and x, that is,

$f(x) - \{f(a) + f'(a)(x-a)\}$

$= \frac{f''(c_1)}{2}(x-a)^2$, or $\frac{f(x) - f(a)}{x - a}$

$= f'(a) + \frac{f''(c_1)}{2}(x-a)$. If $\Delta x = x - a$,

then $\frac{f(a + \Delta x) - f(a)}{\Delta x} = f'(a) + \frac{f''(c_1)}{2}\cdot\Delta x$,

where c_1 is between a and $a + \Delta x$.

(b) Applying Theorem 2 of Section 11.3, we get

$R_2(x; a) = f(x) - P_2(x; a) = \frac{f'''(c_2)}{3!}(x-a)^3$,

where c_2 is between a and x, that is,

$f(x) - \{f(a) + f'(a)(x-a) + \frac{f''(a)}{2!}(x-a)^2\}$

$= \frac{f'''(c_2)}{3!}(x-a)^3$. In particular, if we

substitute $a + \Delta x$ and $a - \Delta x$ for x, we have

$f(a + \Delta x) - f(a) - f'(a)\cdot\Delta x - \frac{f''(a)}{2!}(\Delta x)^2$

$= \frac{f'''(\zeta_1)}{3!}(\Delta x)^3$ and $f(a - \Delta x) - f(a) +$

$f'(a)\cdot\Delta x - \frac{f''(a)}{2!}(\Delta x)^2 = -\frac{f'''(\zeta_2)}{3!}(\Delta x)^3$,

where ζ_1 is between a and $a + \Delta x$ and ζ_2 is between a and $a - \Delta x$. Subtracting the two equations yields

$f(a + \Delta x) - f(a - \Delta x) - 2f'(a)\cdot\Delta x$

$= \frac{f'''(\zeta_1) + f'''(\zeta_2)}{3!}\cdot(\Delta x)^3$, so that

$\frac{f(a + \Delta x) - f(a - \Delta x)}{2\Delta x}$

$= f'(a) + \frac{f'''(\zeta_1) + f'''(\zeta_2)}{2}\cdot\frac{(\Delta x)^2}{3!}$. But f''' is

continuous with $\frac{f'''(\zeta_1) + f'''(\zeta_2)}{2}$ between the

values of $f'''(\zeta_1)$ and $f'''(\zeta_2)$. By the intermediate-value theorem there is a number c_2 between $a + \Delta x$ and $a - \Delta x$ satisfying

$\frac{f'''(\zeta_1) + f'''(\zeta_2)}{2} = f'''(c_2)$. Thus,

$\frac{f(a + \Delta x) - f(a - \Delta x)}{2\cdot\Delta x}$

$= f'(a) + \frac{f'''(c_2)}{3!}\cdot(\Delta x)^2$, where c_2 is between

$a + \Delta x$ and $a - \Delta x$.

61 (a) $E(x) = 1 + x + \frac{x^2}{2!} + \frac{x^3}{3!} + \cdots$, so $E(0)$

$= 1 + 0 + 0 + 0 + \cdots = 1$.

(b) $E'(x) = 0 + 1 + \frac{2x}{2!} + \frac{3x^2}{3!} + \cdots =$

$1 + x + \frac{x^2}{2!} + \cdots = E(x)$

(c) $(E(x)E(-x))' = E(x)(E(-x))' + E'(x)E(-x)$

$= E(x)E'(-x)(-1) + E(x)E(-x)$

$= -E(x)E(-x) + E(x)E(-x) = 0$, so

$E(x)E(-x)$ is a constant; by (a), $E(0)E(-0)$

$= E(0)E(0) = 1$, so $E(x)E(-x) = 1$ for all values of x.